Dinesh K. Benbi, PhD
Rolf Nieder, PhD
Editors

Handbook of Processes and Modeling in the Soil-Plant System

Pre-publication REVIEWS, COMMENTARIES, EVALUATIONS . . .

"This book provides a comprehensive overview of our understanding of soils and soil-plant interactions and integrates this with clear and up-to-date reviews of approaches to modeling. Modeling has rapidly become an essential tool for researchers, policymakers, and extension workers, but for each of these groups, the use of models presupposes a knowledge of the relevant subject area. This book helps bridge the gap between understanding the processes in soils and the approaches that are currently used to model them, and as such will be attractive to a wide range of audiences.

The most impressive feature of this book is its breadth of coverage. As you would expect in a book of this sort, it includes sections describing physical, chemical, and biological processes taking place in soils, but also contains chapters on soil formation, radioactivity, agrochemicals, crop growth, and a very thoughtful chapter to end by Dr. Thomas Addiscott on the limitations of modeling approaches. The behavior of soil organic matter is considered in particular detail—and justifiably so—given its importance in influencing other characteristics of the soil environment. Other topical issues such as the exchange of greenhouse gases, soil acidification, and soil erosion are well covered.

This book will be hugely valuable to research scientists working in many disciplines. It will also provide an excellent source of material for students studying at advanced undergraduate and postgraduate levels in soil and environmental sciences. The range and depth of material covered is unique and will form an invaluable reference work for many years to come."

R. M. Rees, PhD
Senior Soil Scientist,
Scottish Agricultural College

More pre-publication
REVIEWS, COMMENTARIES, EVALUATIONS . . .

"Understanding the processes in soil-plant systems and modeling these interactions is of emerging significance, not only for the improvement of sustainable food production, but also for soil and environmental protection issues, the efficient use of limited water resources, ecosystem management, and natural conservation. This book gives an excellent overview of the important topics related to processes in soil-plant systems and their modeling. It describes basic processes as well as the dynamics of plant nutrients, organic substances, agrochemicals and pollutants in soils, soil water dynamics, solute transport, soil erosion, and last but not least, crop growth.

From my view as a soil ecologist, the chapters on dynamic processes and their modeling in soil-plant systems are of particular interest. However, the great value and benefit of this book rises from the overall view of the aspects related to soil-plant systems. Therefore, the book is very commendable for scientists as well as undergraduate and graduate students."

Juergen Boettcher, PhD
Professor for Soil Ecology,
Institute of Soil Science,
University of Hannover, Germany

"With nearly forty contributors, the editors have been able to present a thoroughly comprehensive treatment of dynamic processes in the soil-plant system. This book starts with a consideration of the physical, chemical, and biological processes that play an interdependent role in soils. Attention has been given to the role of organic matter, the importance of water and air dynamics, the various soil degradation processes, and the behavior of radionuclides. There is also a chapter on processes of soil formation. In addition to significant discussions about the dynamics of major and secondary plant nutrients, the vital importance of trace elements is demonstrated. The substantial progress in understanding the possibilities and limitations of modeling is reflected in many chapters. Extensive, up-to-date bibliographic lists include references to literature in languages other than English. This handbook is a valuable reference text that should be an essential library purchase for universities and institutions where soil and its wider implications for food security are important subjects of study."

Hans van Baren
Book Review Editor,
Bulletin of the International Union of Soil Sciences

"This comprehensive book provides the reader with tools to mathematically describe and predict trends for the soil-plant system. Each chapter is written by one or more leading authorities in their field, and as contributors are from several continents, this book exposes the reader to a greater diversity of literature than is usual. Many of the publications cited by authors are very recent and will provide students and others entering new fields of soil-plant research with concise, state-of-the-art reviews."

Bob Gilkes, PhD
Professor of Soil Science,
University of Western Australia

NOTES FOR PROFESSIONAL LIBRARIANS AND LIBRARY USERS

This is an original book title published by Food Products Press® an imprint of The Haworth Press, Inc. Unless otherwise noted in specific chapters with attribution, materials in this book have not been previously published elsewhere in any format or language.

CONSERVATION AND PRESERVATION NOTES

All books published by The Haworth Press, Inc. and its imprints are printed on certified pH neutral, acid free book grade paper. This paper meets the minimum requirements of American National Standard for Information Sciences-Permanence of Paper for Printed Material, ANSI Z39.48-1984.

Handbook of Processes and Modeling in the Soil-Plant System

FOOD PRODUCTS PRESS®

Crop Science
Amarjit S. Basra, PhD
Senior Editor

Mineral Nutrition of Crops: Fundamental Mechanisms and Implications by Zdenko Rengel

Conservation Tillage in U.S. Agriculture: Environmental, Economic, and Policy Issues by Noel D. Uri

Cotton Fibers: Developmental Biology, Quality Improvement, and Textile Processing edited by Amarjit S. Basra

Heterosis and Hybrid Seed Production in Agronomic Crops edited by Amarjit S. Basra

Intensive Cropping: Efficient Use of Water, Nutrients, and Tillage by S. S. Prihar, P. R. Gajri, D. K. Benbi, and V. K. Arora

Physiological Bases for Maize Improvement edited by María E. Otegui and Gustavo A. Slafer

Plant Growth Regulators in Agriculture and Horticulture: Their Role and Commercial Uses edited by Amarjit S. Basra

Crop Responses and Adaptations to Temperature Stress edited by Amarjit S. Basra

Plant Viruses As Molecular Pathogens by Jawaid A. Khan and Jeanne Dijkstra

In Vitro Plant Breeding by Acram Taji, Prakash P. Kumar, and Prakash Lakshmanan

Crop Improvement: Challenges in the Twenty-First Century edited by Manjit S. Kang

Barley Science: Recent Advances from Molecular Biology to Agronomy of Yield and Quality edited by Gustavo A. Slafer, José Luis Molina-Cano, Roxana Savin, José Luis Araus, and Ignacio Romagosa

Tillage for Sustainable Cropping by P. R. Gajri, V. K. Arora, and S. S. Prihar

Bacterial Disease Resistance in Plants: Molecular Biology and Biotechnological Applications by P. Vidhyasekaran

Handbook of Formulas and Software for Plant Geneticists and Breeders edited by Manjit S. Kang

Postharvest Oxidative Stress in Horticultural Crops edited by D. M. Hodges

Encyclopedic Dictionary of Plant Breeding and Related Subjects by Rolf H. G. Schlegel

Handbook of Processes and Modeling in the Soil-Plant System edited by D. K. Benbi and R. Nieder

The Lowland Maya Area: Three Millennia at the Human-Wildland Interface edited by A. Gómez-Pompa, M. F. Allen, S. Fedick, and J. J. Jiménez-Osornio

Biodiversity and Pest Management in Agroecosystems, Second Edition by Miguel A. Altieri and Clara I. Nicholls

Plant-Derived Antimycotics: Current Trends and Future Prospects edited by Mahendra Rai and Donatella Mares

Concise Encyclopedia of Temperate Tree Fruit edited by Tara Auxt Baugher and Suman Singha

Handbook of Processes and Modeling in the Soil-Plant System

Dinesh K. Benbi, PhD
Rolf Nieder, PhD
Editors

Food Products Press®
The Haworth Reference Press
Imprints of The Haworth Press, Inc.
New York • London • Oxford

Published by

Food Products Press® and The Haworth Reference Press, imprints of The Haworth Press, Inc., 10 Alice Street, Binghamton, NY 13904-1580.

Cover design by Jennifer M. Gaska.

Library of Congress Cataloging-in-Publication Data

Handbook of processes and modeling in the soil-plant system / D. K. Benbi, R. Nieder, editors.
p. cm.
Includes bibliographical references and index.
ISBN 1-56022-914-4 (hard : alk. paper) — ISBN 1-56022-915-2 (soft)
1. Plant-soil relationships—Mathematical models. 2. Soil chemistry—Mathematical models. I. Benbi, D. K. II. Nieder, R.

S596.7 .H36 2003
631.4'01'5118—dc21

2002072066

CONTENTS

ABOUT THE EDITORS

D. K. Benbi, PhD, is a soil chemist at Punjab Agricultural University in Ludhiana, India. Dr. Benbi is an Alexander von Humboldt Fellow (Germany) and a former Guest Researcher at the University of Western Australia. He has authored or co-authored over 40 research publications, including three research bulletins and a special volume of symposium proceedings. Dr. Benbi's research interests focus mainly on efficient nutrient management, nutrient dynamics in the soil-plant system, long-term nutrient budgeting, moisture-fertilizer interactions, soil-test crop response correlation, and simulation modeling. He is a co-author of *Intensive Cropping: Efficient Use of Water, Nutrients, and Tillage* (Haworth).

R. Nieder, PhD, is an experienced lecturer and investigator in soil and environmental sciences at the Technical University of Braunschweig, Germany. Dr. Nieder is the author of more than 30 articles in reviewed journals and three technical bulletins. His research interests include nutrient and soil organic matter dynamics (identification and description of the processes) of agricultural and forest ecosystems. He is an expert in the areas of over-fertilization, nutrient accumulation and mobilization, and water/ecosystem pollution. Dr. Nieder has provided practical recommendations for agricultural and forest management, for agricultural service departments, and for legislators. Part of his scientific work has been incorporated into the German soil protection regulations of 1999. Dr. Nieder is a member of the German Soil Science Society and the International Union of Soil Science.

CONTRIBUTORS

Thomas M. Addiscott, PhD, is Professor and Soil Scientist at IACR-Rothamsted, Harpenden, Hertfordshire, United Kingdom.

Luigi Badalucco, PhD, is Professor and Soil Biochemist in the Dipartimento di Ingegneria e Tecnologie Agro-Forestali, Università di Palermo, Palermo, Italy.

Bnayahu Bar-Yosef, PhD, is Soil Chemist and Plant Nutritionist at the Institute of Soil, Water and Environmental Sciences, The Agricultural Research Organization, Bet Dagan, Israel.

Nanthi S. Bolan, PhD, is Associate Professor of Soil and Earth Sciences, Massey University, Palmerston North, New Zealand.

Edwin M. Bridges, PhD, is Professor and Soil Scientist, Hempton, Frankenham, Norfolk, United Kingdom.

Grant E. Cardon, PhD, is Associate Professor of Soil and Crop Sciences in the Department of Soil and Crop Science, Colorado State University, Fort Collins, Colorado.

Pellegrino Conte, PhD, is Soil Scientist in the Dipartimento di Scienze Chimico-Agrarie, Università di Napoli "Federico II," Portici, Italy.

Annunziata Cozzolino, PhD, is Soil Chemist in the Dipartimento di Scienze Chimico-Agrarie, Università di Napoli "Federico II," Portici, Italy.

Robert Edis, PhD, is Soil Scientist at the Institute of Land and Food Resources, The University of Melbourne, Parkville, Victoria, Australia.

Sabine Ehlken, PhD, is Plant Physiologist in the Department of Physics/FB 1, University of Bremen, Bremen, Germany.

Martin H. Gerzabek, PhD, is Professor of Environmental Toxicology and Isotopic Methods in the Department of Environmental Research, Austrian Research Centers Seibersdorf, Austria; Institute for Soil Research, University of Agricultural Sciences, Vienna, Austria.

Duncan J. Greenwood, PhD, is Professor Emeritus and Horticulture Researcher at Horticulture Research International, Wellesbourne, Warwick, United Kingdom.

Ronald J. Hanks, PhD, is Professor Emeritus and Soil Physicist, Department of Plants, Soils, and Biometeorology, Utah State University, Logan, Utah.

Harry F. Hodges, PhD, is Professor Emeritus, Agronomist and Crop Physiologist, Department of Plant and Soil Sciences, Mississippi State University, Mississippi.

Gerrit Hoogenboom, PhD, is Professor of Agrometeorology, Crop Modeling and Systems Analysis in the Department of Biological and Agricultural Engineering, The University of Georgia, Griffin, Georgia.

Klaus Isermann, PhD, is Agricultural Scientist in the Büro für nachhaltige Landwirtschaft und Agrikultur, Hanhofen, Germany.

Per-Erik Jansson, PhD, is Professor of Land and Water Resources in the Division of Land and Water Resources, KTH, Royal Institute of Technology, Stockholm, Sweden.

Tatiana V. Karpinets, PhD, is Senior Scientist, Modeling of Agricultural Systems, All Russian Scientific Research Institute of Agriculture and Protection of Soil, Kursk, Russia.

Rami Keren, PhD, is Professor and Soil Physical Chemist at the Institute of Soil, Water and Environmental Sciences, The Agricultural Research Organization, Bet Dagan, Israel.

Gerald Kirchner, PhD, is Nuclear and Soil Physicist in the Department of Physics/FB 1, University of Bremen, Bremen, Germany.

Ronald G. McLaren, PhD, is Professor of Environmental Soil Science, Centre for Soil and Environmental Quality, Soil, Plant and Ecological Sciences Division, Lincoln University, Canterbury, New Zealand.

Siddhartha S. Mukhopadhyay, PhD, is Pedologist in the Department of Soils, Punjab Agricultural University, Ludhiana, India.

Paolo Nannipieri, PhD, is Professor and Soil Biochemist in the Dipartimento della Scienza del Suolo e Nutrizione della Pianta, Università degli Studi di Firenze, Firenze, Italy.

Sven Ingvar Nilsson, PhD, is Professor of Soil Chemistry and Pedology in the Department of Soil Sciences, Swedish University of Agricultural Sciences, Uppsala, Sweden.

Alessandro Piccolo, PhD, is Professor of Agricultural and Environmental Chemistry in the Dipartimento di Scienze Chimico-Agrarie, Università di Napoli "Federico II," Portici, Italy.

Kambham Raja Reddy, PhD, is Professor of Plant and Soil Sciences, Department of Plant and Soil Sciences, Mississippi State University, Mississippi.

Jörg Richter, PhD, is Professor Emeritus and Soil Scientist, Loc Cappuccini, Radicofani (Siena), Italy.

Surinder Saggar, PhD, is Senior Scientist, Landcare Research, Palmerston North, New Zealand.

Edward L. Skidmore, PhD, is Professor, Soil Scientist, and Soil Physicist in the USDA-ARS Wind Erosion Research Unit, Kansas State University, Manhattan, Kansas.

Keith A. Smith, PhD, is Professor of Biosphere-Atmosphere Research, Institute of Ecology and Resource Management, University of Edinburgh, United Kingdom.

Riccardo Spaccini, PhD, Agricultural and Environmental Chemistry, Dipartimento di Scienze Chimico-Agrarie, Università di Napoli "Federico II," Portici, Italy.

Friederike Strebl, PhD, is Radioecologist in the Department of Environmental Research, Austrian Research Centers Seibersdorf, Seibersdorf, Austria.

Thilo Streck, PhD, is Professor of Biogeophysics in the Department of Soil Science, University of Hohenheim, Stuttgart, Germany.

Kim H. Tan, PhD, is Professor Emeritus, Soil Chemist, and Pedologist, Greensboro, Georgia.

Simon J. van Donk, PhD, is Agricultural Engineer at the USDA-ARS Wind Erosion Research Unit, Manhattan, Kansas.

Robert E. White, PhD, is Professor and Soil Scientist in the Department of Resource Management and Horticulture, Institute of Land and Food Resources, The University of Melbourne, Parkville, Victoria, Australia.

Preface

A number of interrelated processes occur in soils. Some affect soil fertility, productivity, quality, and functioning while others influence soil formation, degradation, and environmental quality. In view of the complexity of the soil-plant system, simulation models are being increasingly used to understand, integrate, and forecast interaction among various processes, predict environmental impact of agricultural practices, land use changes, land degradation, and to provide farmers with management options. Further models with greater levels of detail, as modeling packages, are expected in the near future.

Various modeling concepts have been developed over the years with important applications in agroecosystems. Generally, the authors of different models document the hypotheses and the mathematical development of their models, but the description of the underlying processes and their interaction with other processes is rarely discussed. For responsible use and calibration or development of a model, a thorough understanding of the interdependent and dynamic processes in the ecosystem is essential for a better conceptualization of models.

This book fills the gap by presenting a basic description of the processes in the soil-plant system followed by the modeling approaches adopted to simulate the process with an emphasis on conceptual clarity and applicability. Almost all of the known processes occurring in the soil-plant-atmosphere system are dealt with in a single book for the first time.

The first three chapters highlight the physical, chemical, and biological processes occurring in the soil-plant system. Chapter 1 deals with heat, gas or air, and water flow within a soil-plant-atmosphere system and presents examples in typical ecological models that incorporate physical processes. Chapter 2 enunciates soil chemical properties that are important in the dynamics of soil water, organic, and inorganic constituents. Statistical models are presented for integrating the various interactive processes and provide quantitative means for interpretation of soil reactions. Chapter 3 treats soil as a biological system and focuses on its biological properties, and the processes carried out by soil organisms and their kinetics.

Because of the strong influence of organic matter on soil quality and soil functioning and its role in nutrient supply, water storage, emission of greenhouse gases, and sustainable agriculture, it has a central role in water and matter dynamics in the soil-plant-atmosphere system. Therefore, two chapters (Chapter 4 and Chapter 13) are devoted to different aspects of organic matter in soils. Chapter 4 deals with soil humus and presents recent findings on the conformational structure of humic material and the profound implications that these may have on our understanding of soil organic matter functions and reactivity. Chapter 13 assesses soil carbon pools, fractions and transformations of organic matter, anthropogenic effects on soil organic matter, influence of climate change on soil organic matter, and modeling of soil organic matter dynamics.

Soils are formed through the interaction of a number of factors, such as climate, organisms, relief, parent material, and time. These factors set the parameters within which the processes of soil formation must act. Chapter 5 discusses the processes of surface soil and lower soil horizon formation and the approaches to model these processes.

One of the rarely discussed topics—radioactivity in soil-plant systems—is presented in Chapter 6 with a focus on the deposition and behavior of radionuclides along with the modeling approaches in the soil-plant system. Chapters 7 through 9 enunciate soil degradation processes, soil acidification (Chapter 7), alkalinization (Chapter 8), and erosion (Chapter 9), and their modeling.

Chapters 10 through 19 elaborate on water and matter dynamics in the soil-plant-atmosphere system. The process of water flow discussed in Chapter 1 is further supplemented in Chapter 10 with particular emphasis on modeling of component processes of soil-water dynamics. Chapter 11 ("Solute-Water Interactions") discusses solute transport modeling concepts in the order of process complexity, from piston displacement through stream-tube flow to convection (or advection)-dispersion approaches.

There is an increasing concern about the rising concentration of carbon dioxide and other greenhouse gases in the atmosphere that are forcing global climate change and causing the depletion of ozone in the stratosphere. Chapter 12 ("Soil-Atmosphere Interactions") examines the current knowledge regarding soil-atmosphere gaseous exchange, and the extent to which models are able to describe the phenomena and are used for predictive purposes.

Chapters 14 through 16 describe nitrogen, phosphorus, and potassium dynamics in soils with an emphasis on their forms, transformations, loss mechanisms, and modeling of the relevant processes. Chapter 17 develops an understanding of the nature of the secondary nutrients (sulphur [S], calcium [Ca], and magnesium [Mg]), and the factors influencing their cycling in soils, plants, and animals. A brief description of the inputs and reaction in soils, and plant and animal requirements of S, Ca, and Mg in the agroecosystems is presented as a background to discussion on modeling of these nutrients. In addition to the major and secondary nutrients, several essential micronutrients are required in relatively small amounts for the healthy growth and reproduction of plants and animals. Conversely, excessive concentrations of some of these micronutrients, such as copper (Cu) and zinc (Zn), may prove to be toxic to living organisms. In addition, other nonessential elements, such as lead (Pb), cadmium (Cd), mercury (Hg), and arsenic (As) can also have toxic effects on plants, livestock, and human health. Chapter 18 focuses on these elements by taking into account their forms, solubility, mobility, and bioavailability in soils. Modeling aspects of trace element speciation, distribution between solution and solid phases, and leaching from soil are also discussed. The fate of pesticides with particular reference to modeling of their sorption, degradation, transport, volatilization, and plant uptake is the theme of Chapter 19. Chapters 20 and 21 focus on modeling of crop growth and development and the role of plant growth regulators. Finally, Chapter 22 presents philosophical aspects of modeling and analyzes the potential and limitations of simulation models under different situations.

This book would not have been possible without the active cooperation and dedicated efforts of the contributing authors. We are indeed grateful to them for their diligence and for pooling their special knowledge. On behalf of the chapter authors, we thank John T. Ammons (Chapter 16) and Anne Austin (Chapter 17). We thank Drs. Amarjit S. Basra (University of California, Davis), Hans P. Blume (Kiel, Germany), Gerhard Brümmer (Bonn, Germany), Kurt Bunzl (Neuherberg, Germany), Peter Christie (Belfast, United Kingdom), Hannes Flühler (Schlieren, Switzerland), Fritz Führ (Jülich, Germany), Martin Kaupenjohann (Stuttgart, Germany), Karl Stahr (Stuttgart, Germany), and Hans van Baren (Wageningen, Netherlands) for their help during different stages of preparation of the book. We express our gratitude to a number of reviewers who donated their time and gave valuable comments on various chapters. We thank Hans P. Dauck for help in the preparation of some illustrations. The senior editor is thankful to Alexander von Humboldt Stiftung, Bonn, Germany, for a research fellowship during the course of which the preparation of this book was undertaken.

Introduction

Increasing population pressure, industrialization, intensification of agriculture, and injudicious use of inputs has led to deterioration of soil health and environmental quality. Some of these effects are apparent in the form of land degradation through soil erosion, soil acidification and alkalinization, nutrient and organic matter depletion, groundwater pollution, and emission of greenhouse gases. Soil-plant systems, and thus the food chain, are at further risk from radionuclide contamination due to nuclear weapons tests and emissions from nuclear power plants. Obviously, soils are neither infinite suppliers of nutrients to plants nor indestructible dump sites for surpluses of nutrients, heavy metals, biocide residues, and radionuclides. For the sustainability of the ecosystem it is essential that resources such as air, water, and soil are given prime consideration. Soils are a major component of the ecosystem. A fundamental knowledge about soil structures and processes at the microscopic level is essential to understand and regulate the behavior of the soil-plant-atmosphere continuum. Therefore, it is important to understand the processes in the soil-plant system at a greater level of resolution than before.

The processes occurring in the soil-plant system may be grouped into physical, chemical, and biological processes. The rate, extent, and nature of these processes, besides influencing plant nutrition, determine soil, water, and environmental quality. Geologists, hydrologists, soil physicists, chemists, microbiologists, ecologists, and agronomists study the processes of their respective fields of interest in isolation. However, most of the processes occurring in the soil-plant system are interrelated. The pathway of a process is modified by other processes, properties of soil itself, and environmental conditions. Therefore, a more holistic approach is required to understand and predict interactions among various processes and their integrated impact on soil, plants, and the environment. The systems approach, which relies on the use of models, provides a framework for systematic analysis and synthesis of soil-plant systems at different levels of spatial aggregation, namely field, farm, regional, national, and even global levels.

Models are essentially hypotheses. They are useful not because they reproduce reality, but because they simplify reality and enable the most important processes to be identified, studied, and simulated, and enable outcomes to be predicted in advance (Addiscott, 1993). When used appropriately, a model allows extrapolation from a limited set of data so that the amount of repetitive and time-consuming experimentation can be reduced. In addition, a model can help to identify knowledge gaps. Another major use of models is to assist farmers and growers in designing effective crop, soil, and management strategies for efficient input use. The purposes of models are varied, from the illustrative and instructive, to the predictive. Models are designed to describe how various components of systems interact with one another, and the approach used to construct that description is generally the basis for categorizing models. Models may be empirical (black box approach), mechanistic, or functional.

Empirical models are based on push-button philosophy and describe *what* happens and not *how* it happens. Prerequisites for using empirical models include data sets that have been collected with well-designed sampling procedures. Many of the growth and yield relations with the input of nutrients or water fall in this category. Such models could be used for summarizing the data or for interpolating within the data range from which these are developed. This approach precludes generalization beyond the data sets and the specific conditions for which the model is parameterized.

A mechanistic model attempts to describe in the most fundamental way the possible mechanisms of the underlying processes and their interactions. In this approach, the black box is replaced by a series of white boxes to describe the how and why of the cause-effect relationship. It is not possible to define an absolute level that can be called mechanistic; mechanistic conjectures underlying one conceptual framework may be empirical observations in another. The mechanistic models have been classified as deterministic and stochastic. Deterministic models present a single outcome or unique solution whereas stochastic models give the probability of an outcome. Deterministic models presume that a system or process operates such that the occurrence of a given set of events leads to a uniquely definable outcome. Stochastic models presuppose the outcome to be uncertain and are structured to account for this uncertainty. Practically all the natural systems have intrinsic uncertainties, but these are ignored in the formulation of a deterministic model.

The major advantage of mechanistic models is that they can be transferred to another set of conditions, provided appropriate parameter values are defined. This makes them ideal for scenario building. Mechanistic models need to maintain a balance between complexity and simplicity, with the goal of re-creating realistic behavior without unnecessary details. This has given birth to functional (or summary) models. Functional models treat processes in a more simplified manner, and often rely heavily on broad, empirical, quantitative relationships to give a description. The major advantage of functional models is that they do not rely on many parameters, but they require local calibration. A functional model is more likely to simplify the process than a mechanistic one, but this usually means that its parameters are easier to obtain. Despite the simplification, functional models often give simulations that are at least as good as those of mechanistic models. Typically, functional models are used as management tools whereas mechanistic models are used in research. The choice of a model and the level of complexity needed depend on the objectives, the information, and the time available for model building and evaluation.

With the advent of computers in the last three decades, a multitude of models have been developed for simulating processes in the soil-plant-atmosphere system. Models are being used as tools for policy analysis and as decision support systems. Models describing soil and plant systems have become increasingly valuable for assimilating knowledge gained from experimentation. In the recent past, modeling exercises have been held to compare different models, on common data sets, with regard to their capability to simulate a particular aspect of the soil-plant system such as organic matter turnover (Smith et al., 1997), nitrogen dynamics, or soil-water dynamics (de Willigen, 1991; Diekkrüger et al., 1995). Although such studies provide useful information about the relative abilities of different models to simulate a given data set, they tell very little about the weaknesses and strengths of different models with regard to conceptualization of the underlying processes. However, proper development, evaluation, calibration, validation, and appropriate use of a model must be based on a thorough and comprehensive understanding of the dynamic processes. The use of a model without adequate understanding of the underlying processes is not only unscientific but may also lead to erroneous conclusions.

This book provides a thorough description of almost all of the known processes occurring in the soil-plant system and the state of the art in modeling these processes. Different chapters describe the concepts used to model various processes and their applications for forecasting the influence of different processes on the ecosystem. The book will benefit students, teachers, researchers, planners, and extension workers who are interested in having a comprehensive knowledge of the processes occurring in the soil-plant system and want to exploit modeling as a tool for the accomplishment of their relevant objectives.

REFERENCES

Addiscott, T.M. (1993). Simulation modeling and soil behaviour. *Geoderma 60:* 15-40.

de Willigen, P. (1991). Nitrogen turnover in the soil-crop system: Comparison of fourteen simulation models. *Fertilizer Research 27:* 141-149.

Diekkrüger, B., D. Söndgerath, K.C. Kersebaum, and C.W. McVoy (1995). Validity of agroecosystem models—A comparison of results of different models applied to the same data set. *Ecological Modelling 81:* 3-29.

Smith, P., D.S. Powlson, J.U. Smith, and E.T. Elliott (Eds.) (1997). Evaluation and comparison of soil organic matter models using datasets from seven long-term experiments. *Geoderma 81:* 1-225.

Chapter 1

Physical Processes

Per-Erik Jansson

Models of physical processes have been part of classical physics since the very beginning of modern science. Especially mechanics, as developed by Galileo and later by Newton, are important starting points for the understanding of how physical processes are understood in an ecological context. Mathematical models of environment-related problems have a long history, but many people refer to Darcy (1856) as the first attempt to understand the physics behind the flow of water. During the last 30 years a number of ecological models have been developed in different contexts that all originate from very fundamental assumptions on a few physical principles. Those that developed and those that are using the models are dealing with processes for transfer of heat and mass. Typical structures for such models were simplified systems in which the geometry was reduced to one dimension and included a homogeneous substrate studied under a simple type of influence. Such simple systems were also studied experimentally to understand the mechanism of transfer. It has been possible to simulate the experimental setup through analytical mathematical models.

The broad use of mathematical modeling of complex systems was introduced with the advent of modern, powerful computers and the use of numerical methods rather than exact analytical methods in mathematics. Such models could handle a broad scale of techniques for considering complex systems and introduce various types of boundary conditions and combinations of systems. The early weather forecast models are good examples of such models that are based on physical assumptions and are able to describe the behavior of the real world to an extent that makes them of high practical value. In the late 1960s, models to describe behavior of soil water and heat flow and the transport of water and matter to plants started to develop. All the models that exist today originated from similar assumptions on the physical principles.

The purpose of this chapter is to present the most important physical processes in a soil-vegetation-atmosphere system and to illustrate how these have been considered in typical ecological models that incorporate physical processes. This brief review of how the physical processes are considered in process-oriented models originates to a large extent from a full documentation of the CoupModel by Jansson and Halldin (1979). The model has been extended to include more details and to consider biological processes such as plant growth, carbon dynamics, and nitrogen turnover in an ecosystem. Readers interested in more details on this particular model may read Jansson and Karlberg (2001). The programming approach used for the CoupModel was recently presented by Jansson and Moon (2001). Readers who would like to know more about the background of many physical processes presented here are referred to textbooks on environmental physics (Monteith and Unsworth, 1990), or on soil physics (Jury, Gardner, and Gardner, 1991; Yong and Warkentin, 1975; Hillel, 1980). For a review of modeling aspects refer to Richter (1987).

ENERGY BALANCE AT THE EARTH'S SURFACE

Radiation Principles and Components

The most fundamental law governing climate and all ecological processes is the energy balance that originates from the radiation laws. Energy can be transferred as electromagnetic radiation from any body in the universe. The energy that is transferred as electromagnetic radiation follows a number of well-known principles that have been very useful both for our understanding and for prediction of a number of phenomena. Electromagnetic radiations can be considered as waves and as particles.

The wavelength (λ) of an electromagnetic radiation is given by

$$\lambda = \frac{c}{\nu} \tag{1.1}$$

where ν is the frequency and c is the speed of light. The speed of light, c, can be considered as a constant (approximately 3×10^8 meters per second, ms^{-1}). All waves travel with the same velocity in vacuum. However, when waves travel in a medium the velocity is slower and consequently the wavelength is longer. This process is known as dispersion and is seen when light passes through a prism that splits it into a spectrum of different colors. The range of electromagnetic waves is very wide, varying from gamma rays to electric waves. In most ecological models, there are only two important wavelengths to account for when considering energy flows. They are sometimes called solar and terrestrial radiation. These can be understood from the surface temperature of the bodies from which they are emitted. Both the sun and the earth can be considered as "blackbodies" that emit radiation at the maximum possible intensity for each wavelength. A blackbody not only emits radiation but also completely absorbs all the radiation that is directly incident on it. The blackbody concept includes the assumptions of emissivity and absorptivity. These are defined as ratios between the ability to emit or absorb radiation of a body, in relation to the ability for a perfect blackbody. Natural bodies are not perfect blackbodies but may effectively be considered as greybodies with different emissivities and absorptivities depending on the wavelength.

The intensity *(I)* of radiation emitted from a body can be expressed as:

$$I = \varepsilon \sigma T^4 \tag{1.2}$$

where ε is the emissivity, σ is the Stefan-Boltzman constant (5.67×10^{-8} Joules (second)$^{-1}$ (meter)$^{-2}$ (Kelvin)$^{-4}$, $Js^{-1}m^{-2}K^{-4}$) and T is the surface temperature (K). This law states that the intensity of emitted radiation is proportional to the fourth power of absolute temperature. This means that a relatively small change in temperature will cause a substantial effect on the radiation density from a body such as the earth's surface. The main difference between solar radiation and terrestrial radiation is described by the Stefan-Boltzman law and the Wiens displacement law that relate the wavelength of the radiation to the surface temperature as previously described. The wavelength λ_{max} (m) which has the highest intensity of emission, is expressed as

$$\lambda_{max} = \frac{2.89 \cdot 10^{-3}}{T} \tag{1.3}$$

This means that the radiation emitted by the sun with a surface temperature around 6,000 K has the highest intensity at a wavelength of 0.483 micrometer (μm) whereas the radiation from the earth at its mean temperature corresponds to a wavelength of 10.10 μm.

The particle concept introduced by Planck is useful to describe energy content *(E)* as a stream of particles

$$E = hv \tag{1.4}$$

where h is Planck's constant (6.62×10^{-34} Joules second, Js).

For a number of ecological models there are two important aspects of radiation. One is the total energy balance of the received and the emitted radiation. This is known as the net radiation (R_n) and can be described as

$$R_n = (1-\alpha)R_s - R_{l,o} + R_{l,i} \quad (1.5)$$

where R_s is the shortwave radiation from the sun, $R_{l,o}$ is the long-wave radiation from the earth's surface, $R_{l,i}$ is the long-wave radiation from atmosphere to the earth's surface, and α is the reflectivity of the shortwave radiation, also called albedo. The shortwave radiation can be described as direct radiation from the sun or diffused radiation from the sky. A number of equations have been suggested to estimate the incoming shortwave radiation that includes the influence of solar angle, cloudiness, and atmospheric properties. The outgoing long-wave radiation can, in most cases, be estimated by the Stefan equation (1.2) provided the emissivity and the surface temperature are known. The incoming long-wave radiation is more complicated and it depends to a large extent on the temperature, vapor pressure, and cloudiness of the atmosphere. Similar to the incoming shortwave radiation, a number of empirical equations have been proposed for the incoming long-wave radiation. Attempts have also been made to combine both directions of long-wave radiations to estimate the net long-wave balance.

The other important aspect of radiation in an ecological context is the photosynthetically active part of shortwave radiation. Shortwave radiation consists of a spectrum of wavelengths that are sometimes divided into three regions: the ultraviolet (0.3-0.4 μm), the visible (0.4-0.7 μm), and the infrared (0.7-3 μm). Ultraviolet radiation, which has many harmful effects on biological processes, is to a large extent, absorbed by the atmosphere. Only the visible fraction is useful for plants in photosynthesis. This fraction corresponds roughly to 50 percent of the total energy of the shortwave radiation. Plant physiologists often express this fraction as PAR, photosynthetically active radiation. This fraction can then be understood either as a number of quanta that may be used in the photosynthesis or as an energy flux density.

Energy Balance Equation

A common approach for formulating energy-related processes is to assume conservation of energy based on the first law of thermodynamics. For any terrestrial surface on the earth, the energy conservation based on net radiation can be expressed as

$$R_n = H + L_v E + q_h + P \quad (1.6)$$

where H is the sensible heat flux that corresponds to the temperature of air, L_vE is the latent heat (L_v) flux that corresponds to evaporation (E), q_h is the heat flux to the soil/snow/vegetation, and P is the net heat used for photosynthesis minus the energy released by decomposition or respiration. For most practical purposes the photosynthesis and respiration processes can be ignored in this energy balance since the biological fixation of energy is less than 1 percent of the total net radiation. Other energy balance components are discussed as follows.

Latent and Sensible Heat Flux to Atmosphere

The mechanism for heat transfer in the atmosphere is very complicated and a number of methods and equations have been suggested. The most important mechanism for heat transfer is convection, which can be defined as: mass motion of fluid resulting in transport and mixing of the properties of the fluid. In meteorologically related issues, we normally distinguish between forced convection and free convection. Forced convection is motion induced by mechanical or external forces and by friction of the fluid. An example of forced convection is the motion caused by wind. Free convection is the motion caused by density differences within the fluid and the fact that the fluid is located within a gravity field. Warm air is lighter and tends to rise because of the lower density. The science dealing with behavior of fluids is called fluid mechanics and it involves studying the fluid behavior with different objectives. The atmospheric phenomena, however, are still not fully understood because of the existence of a number of interactions, particularly the irregular structure of the fluid fields. From fluid mechanics we know that the motion may either be laminar or turbulent. A simplified definition is that the laminar flow consists of a fluid in which the velocity gradient is linear and the resistance is approximately the same at all positions within the flow fields. For meteorological conditions laminar flows are rare and they may only exist for short periods of time in the vicinity of different surface elements. Turbulent motion of air controls all the major phenomena. The vertical flux of heat in the atmosphere can be expressed as

$$H = \rho_a C_p K_h \frac{\partial T_p}{\partial z} \tag{1.7}$$

where ρ_a is the air density, C_p is the specific heat of air, K_h is the turbulent exchange coefficient, and $\partial T_p / \partial z$ is the vertical gradient of potential temperature. The potential temperature concept has been introduced to account for the change of temperature that is caused by a change of pressure without any corresponding change in heat (adiabatic change of state). However, close to the soil surface the potential temperature may be substituted by normal temperature as the change in air pressure with height may be small compared to the temperature gradient.

Similarly, the equation for the latent heat flux can be expressed as

$$L_v E = \frac{\rho_a C_p}{\gamma} K_v \frac{\partial e}{\partial z} \tag{1.8}$$

where γ is the psycrometer constant, K_v is the turbulent exchange coefficient for vapor, and $\partial e/\partial z$ is vapor pressure gradient.

However, instead of using the turbulent exchange coefficients, equations (1.7) and (1.8) may be rewritten using a resistance approach. Its main advantage is that we can consider the temperature or vapor pressure difference between two positions instead of using the respective gradients. The fluxes may then be expressed as

$$H = \rho_a C_p \frac{T_s - T_a}{r_{ah}} \tag{1.9}$$

and

$$L_v E = \frac{\rho_a C_p}{\gamma} \frac{e_s - e_a}{r_{av}} \tag{1.10}$$

where T_s and e_s are the temperature and the vapor pressure of the surface, T_a and e_a are the corresponding conditions at a certain position in air. Between the surface and a given position in air, resistance to heat is r_{ah} and the corresponding resistance to vapor pressure is r_{av}.

The equations (1.9) and (1.10) were by Penman (1948) and later Monteith (1965) combined with the energy balance approach. Penman used equation (1.6) after neglecting the energy storage as photosynthesis to derive the well-known Penman equation

$$L_v E = \frac{\Delta(R_n - q_h) + \rho_a c_p \dfrac{(e(T_a) - e_a)}{r_a}}{\Delta + \gamma} \tag{1.11}$$

where Δ is the slope of the saturation vapor pressure versus temperature curve. A common aerodynamic resistance r_a is assumed for both the vapor and heat flux. The most important assumption behind this widely used equation is that vapor pressure gradient and the temperature gradients can be replaced with each other. The vapor pressure at the evaporative surface will always be at saturation. This makes equation (1.11) restricted in its applications to wet surfaces. This type of equation has consequently been extensively used to calculate potential evaporation from an open water surface or from a type of vegetation well supplied with water. Another limitation of the equation is that it requires available net radiation as an input, which is unrealistic since the heat flux to the soil q_h, and also the net radiation itself may be dependent on the heat partitioning of the evaporative surface. Because of these constraints and the shortage of information available from ordinary meteorological networks, numerous forms of the Penman equation have been suggested to estimate potential evaporation (see Burman and Pochop, 1994).

Monteith (1965) also suggested that the equation (1.11) may be useful to estimate actual transpiration from vegetation that is not fully supplied with water by introducing the canopy resistance or the surface resistance r_s:

$$L_v E = \frac{\Delta(R_n - q_h) + \rho_a c_p \dfrac{(e(T_a) - e_a)}{r_a}}{\Delta + \gamma(1 + r_s / r_a)} \tag{1.12}$$

This equation (1.12) is called the Penman-Monteith equation. It is more general with regard to type of vegetation, time steps within a day, development stage of vegetation, and environmental conditions. As in equation (1.11), a similarity between moisture and temperature is implicit in the equation. This means that a certain vapor pressure gradient also corresponds to a given temperature gradient. By rewriting the same equation, the possibility for estimating the sensible heat flux or the surface temperature also exists. However, these applications may be doubtful since the assumption on using the same aerodynamic resistance for both heat and moisture will not be valid for different stability conditions in the atmosphere. A major constraint is the concept of making only one single surface of the combination of surfaces with interface to the atmosphere which is a typical feature of ecosystems. Another limitation is the lack of feedback between the different fluxes at the boundary to the atmosphere, i.e., the net radiation has to be known as an input. These aspects are discussed later in this chapter.

Heat Flow into the Soil/Snow at the Atmosphere Boundary

The heat flux, q_h, in the energy balance equation (1.6) between the atmosphere and the soil-vegetation-snow represents at least one additional equation, which in its simplest form can be written as

$$q_h = -k_h \frac{\partial T}{\partial z} \tag{1.13}$$

where k_h is the thermal conductivity of the soil or the snow. This equation is called the Fourier equation. The heat transfer mechanism in the soil differs substantially compared to air. The soil system is normally a three-phase system of solid, liquid, and gas components. Similar to that in the atmosphere, heat may be transferred by radiation and convection. However, these mechanisms are less important in the soil. Instead, the major mechanism for heat transfer is conduction. This means that heat transfer will take place by interaction between individual molecules and one can study this behavior in the soil without detailed account for mass transfer processes within the soil. Onthe other hand, the soil system represents a substantial storage for heat and the equation for heat flux into the soil is normally not replaced by any resistance form as can be easily done for the atmosphere system. Different options to consider the heat flow into the soil are presented as follows.

HEAT FLOW PROCESSES IN A SOIL PROFILE

The simplest approach to describe heat flow into the soil and within the soil profile is to consider the heat flow independent of any interaction with water flow. Although it is not a perfect approximation, for many applications it may be good enough. This simplified approach is also useful to define the basic thermal concepts that are used to describe heat processes within a soil profile. By using first law of thermodynamics, the principle of energy conservation can be defined by a partial differential equation in one-dimensional form as:

$$\frac{\partial(CT)}{\partial t} = \frac{\partial}{\partial z}(-q_h) \tag{1.14}$$

where the left side represents change in heat storage with time and the right side represents the change with respect to depth of heat flux. The heat capacity, *C*, is defined on a volumetric basis, which means that it has the units Joule per cubic meter Kelvin. This equation may further be expanded to the diffusivity form as:

$$\frac{\partial T}{\partial t} = D\frac{\partial^2 T}{\partial z^2} \tag{1.15}$$

where the thermal diffusivity, *D*, is defined as the ratio between the thermal conductivity K_h and the heat capacity *C*. Mathematically, *D* is identical to diffusion coefficient and has the units square meter per second ($m^2\ s^{-1}$). The reorganization of *C* and K_h outside their respective derivates implies that these do not differ with time or depth. The main advantage of this equation is that it can be easily solved by standard mathematical methods.

GAS/AIR FLOW PROCESSES IN A SOIL PROFILE

Gas flow in the soil is governed by convection and diffusion processes. Their relative contributions have been discussed in a number of publications. Two different driving gradients and two different regulating properties of the soil are of interest to describe the behavior. In most ecological applications the focus is on major gases such as CO_2 or O_2. The mechanism of transport for all gases is similar as long as they are part of the air that in most cases can be considered as an ideal gas. Most of the gases are found both as part of the air as well as dissolved in soil water. A complete transport model has to consider transport in both phases. Other dissolved substances can be described in a similar manner.

Convective airflow rate, q_f is given as:

$$q_f = -\frac{k_i}{\eta_f} \frac{\partial P_a}{\partial z} \tag{1.16}$$

where k_i is the intrinsic conductivity of the matrix, η_f is the dynamic viscosity of the fluid and $\partial P_a / \partial z$ represents the total pressure gradient for fluid. The corresponding mass flow of a gas that constitutes a fraction of gas may be obtained by scaling the equation to the relative amount of that gas.

The corresponding diffusive gas flux rate is given as:

$$q_i = -D_i \frac{\partial C_i}{\partial z} \tag{1.17}$$

where D_i is the diffusion coefficient of the gas and $\partial C_i / \partial z$ is the concentration gradient for a particular gas.

For a soil system, the diffusion coefficient is a function of the diffusion coefficient of the gas within the fluid and the fraction of the soil that is occupied by the fluid:

$$D_i = f_f n D_0 \tag{1.18}$$

where f_f is the volumetric fraction of soil occupied by fluid f, n is the tortuosity factor, and D_0 is the diffusion coefficient in the fluid undisturbed by any matrix. A lot of literature exists with data and theoretical discussion on how to estimate the tortuosity of a soil. Early investigators had suggested that the tortuosity may be around 0.66 (Penman, 1940), but later work indicated that it is dependent on f_f of a soil. It is obvious that diffusive transport is a relatively linear process with respect to volumetric fraction as compared to intrinsic conductivity for a fluid, which is normally highly nonlinear due to different pore size distribution within a soil matrix.

The mass flux rate may be combined with the conservation law to obtain a partial differential equation for a gas or a dissolved constituent within the soil

$$\frac{\partial C_i f_f}{\partial t} = \frac{\partial}{\partial z}(-q_i) + C_i \frac{\partial}{\partial z}(q_f) \quad (1.19)$$

where f_f is the fraction of the fluid, C_i is concentration of the constituent i. The terms on the right side of the equation represent diffusive and convective flow, respectively.

WATER FLOW PROCESSES IN A SOIL PROFILE

Water flow in a soil-plant-atmosphere system is determined by the energy state of water. Water tends to move from high to low energy states according to the second law of thermodynamics, which states that the level of organization will not increase in a system that does not receive external energy. There are many similarities between water and heat flow processes in soil-plant-atmosphere systems. More specific features on soil water dynamics are discussed in Chapter 10.

Assuming an isothermal system, there are two potential fields that influence water flow. The first influence is caused by different positions in the gravitational field and the second is due to different positions in a field of forces that result from the interaction between the liquid water phase and the surrounding solid and gas phases in soil. To describe these different energy states of water the concept of water potential has been widely used. In an unsaturated soil system, water is always attracted by different adsorption processes that tend to be held to the hydrophilic surfaces of soil minerals. In addition, water molecules are more efficiently attracted to each other than to a gas. Based on this, a generalized equation of water flow in a one-dimensional vertical system can be written as:

$$q_w = -k_w\left(\frac{\partial \psi}{\partial z} - 1\right) \quad (1.20)$$

where ψ is the pressure head and k_w is the unsaturated conductivity of the soil. The pressure head is a convenient way of expressing the water potential that is related to the forces that are originating from the interaction with other phases in the soil system and the gravitational water potential gradient represented as unity. The units for pressure head may be in terms of length if the water potential is expressed on weight basis in a gravitational field. The general equation for unsaturated water flow follows from equation (1.20) and law of mass conservation:

$$\frac{\partial \theta}{\partial t} = -\frac{\partial q_w}{\partial z} + s_w \tag{1.21}$$

where the left side represents the change in soil water content θ with time and the right side represents the change in water flow with depth. Compared to the corresponding partial differential equation for heat flow, a source/sink term s_w is added here. This equation is normally referred to as the Richards equation (Richards, 1931). The sink term is added to account for processes such as water uptake by roots or any other loss/gain that may occur because of nonuniform fluxes in directions other than the depth z.

Because of the gravitational field in a vertical direction, equation (1.21) is more difficult to handle from a mathematical point of view than the corresponding equation for heat in a soil profile. Many possible ways exist to rewrite the basic partial differential equation, which has sometimes been demonstrated as useful when trying to optimize numerical methods for solving equation (1.21). Irrespective of which form of this equation is preferred, there are two properties that are necessary to know for solving equation (1.21) for field condition: The water retention curve that relates the pressure head to the water content of a soil and the unsaturated hydraulic conductivity function that expresses how the unsaturated hydraulic conductivity is related to water content.

Modeling of water flow involves problems with regard to validity of basic equation and hydraulic properties of a soil. Problems stem from spatial and temporal scales as well as to the accuracy required. Many authors have demonstrated that there are reasons to believe that it is not appropriate to use equation (1.21):

1. The flow is not uniform within the pore system. Even at a relatively small uniform scale macropores exist that may create at least one additional flow domain that is not in equilibrium with the smaller pore systems. Pore scale models demonstrate that substantial differences in behavior are expected between the different pore domains.
2. The hydraulic properties are not independent of inertia within the systems. This means that the natural dynamic behavior of the system cannot be described from only the current value of the state variables

without accounting for history. Especially, the water retention curve has been demonstrated to have substantial hysteresis involved. This means that the water content is normally higher when we are moving from a saturated to an unsaturated soil, i.e., a desorption process. On the other hand, during a sorption process the water content is lower for a given water potential.

3. The hydraulic properties are extremely nonlinear, especially the unsaturated hydraulic conductivity function. The hydraulic conductivity varies substantially as a result of naturally occurring spatial variability. Since the scale of application is generally larger than the scale of investigation of the hydraulic properties, large uncertainties may arise due to a scaling-up process.
4. Swelling and shrinkage of the system will take place and make the hydraulic properties time dependent and the system boundaries nonstatic.
5. Interaction with other nonisothermal processes makes the behavior uncertain since the additional properties for thermal-induced flow are poorly understood.

One obvious option in such a situation is to introduce a stochastic approach to represent some specific uncertainty or a lumped type of uncertainty representing a number of mechanisms. The other option may be to introduce a new model to account for some of the poorly understood mechanisms as mentioned previously.

Besides these uncertainties with the use of the physically based models for water flow in isothermal conditions, other major uncertainties exist in the definition of boundary conditions. For a one-dimensional approach, the upper and lower boundary conditions are described as follows.

Infiltration

An early attempt to understand infiltration into a soil profile was also made by Green and Ampt (1911). They assumed that water infiltrated from a pond into an unsaturated soil having a uniform moisture distribution prior to start of infiltration and an infiltration rate that was mainly driven by gravitational forces. It was demonstrated that during such conditions, the infiltration rate could be described as:

$$q_{\inf} = K_s \frac{h_0 + L_f}{L_f} \tag{1.22}$$

where h_0 is the height of the water in the pond and L_f is the depth of wetting front. There have been a number of attempts to describe the infiltration process into the soil profile. Normally we distinguish between empirical and physically based approaches. Detailed work on this has been done by Philip (1957). The best understanding of the infiltration process is obtained if the system is simplified from a vertical to a horizontal direction. This allows the extraction of the gravitational component of the water potential and rewriting the partial differential equation for water flow to a diffusivity form. For many field scale problems, simple empirical infiltration equations have been applied to obtain boundary conditions for water flow in the soil.

Drainage

The simplest lower boundary condition to equation (1.21) is to assume either certain water content or water potential at a given depth in an unsaturated soil. However, in many cases this is not practical since the lower boundary also shows dynamic behavior. An alternative approach, therefore, is to measure this lower boundary and use a dynamic lower boundary condition or to try to use a model to predict the behavior at a certain depth in the soil profile. In this approach the lower boundary to the equation (1.21) will be the water potential of the lowest layer that will also control the position of the groundwater table. The mass balance of the lowest layer will be based on both inflow and outflow rate from the saturated soil beneath that layer. Such a position can be allowed to move between different layers in a numerical model in which finite compartments have been identified. Steady state approaches as suggested by Hooghoudt (1940), Ernst (1956), or Youngs (1980) are useful to describe the groundwater level for a given outflow rate.

COUPLED WATER AND HEAT FLOW PROCESSES WITHIN THE SOIL

Although the major heat flow mechanism in soil is conduction, a complete equation has to account for convective sensible and latent heat fluxes as well. The heat flux is then given as:

$$q_h = -k\left(\frac{\partial T}{\partial z}\right) + L_v q_v + C_w q_w \tag{1.23}$$

where the first term on the right side represents conduction as in equation, the second term expresses latent heat of vaporization L_v and the vapor flux rate q_v, and the third term represents heat capacity of water C_w and water flux rate q_w.

The diffusion process mainly regulates the vapor flux rate. The diffusion coefficient is adjusted because of deviations from diffusion in free air by use of a tortuosity parameter n_v. The vapor flow is given by:

$$q_v = -f_a n_v D_0 \frac{\partial C_v}{\partial z} \tag{1.24}$$

where f_a is the fraction of air-filled pores, D_0 is the diffusion coefficient in free air that is given as a function of soil temperature:

$$D_0 = \left(\frac{T + 273.15}{273.15}\right)^{1.75} \tag{1.25}$$

The vapor concentration C_v is computed from vapor pressure

$$C_v = \frac{Me_v}{R(T + 273.15)} \tag{1.26}$$

where M is the molar mass of water, R is the gas constant, T is the soil temperature and the vapor pressure, e_v is given as

$$e_v = e_s e^{\left(\frac{-\psi Mg}{R(T+273.15)}\right)} \tag{1.27}$$

where e_s is the vapor pressure at saturation and ψ is the soil water potential (pressure head). The later expression is based on the basic assumption that the liquid phase is in equilibrium with the gas phase in soil. Because of strong dependence of vapor pressure at saturation on temperature, the vapor flow is regulated by temperature gradient as long as the soil water potential is higher than the wilting point (the water content beyond which plants are no longer able to extract water from soil).

The other major temperature-driven heat and water flow process results from freezing. Freezing causes changes in soil water potential. This may be considered similar to drying which was first described in a numerical model by Harlan (1973). When the temperature is below 0°C, a tendency for liquid flow from warmer to colder sites is expected. Because of the lowered water potential of liquid water in an unsaturated soil, freezing will not occur at 0°C. Thermodynamically, the freezing point is linearly related to the water potential. However, ice will not be formed without crystallization. This means that ice will grow from already-formed ice and many interesting ice phenomena will occur in soil because of the nonuniform structure of ice. The structure that is obtained is mainly related to freezing rate and unsaturated conductivity of liquid water. A rapid freezing rate will cause less redistribution of water especially for a coarse-grained soil wherein a low hydraulic conductivity is encountered. Substantial redistribution of water will be expected for silt soils where capillarity of water will hold high amounts of water in medium-sized pores that could easily be redistributed because of water-potential gradients. However, Lundin (1990) found a clear tendency to overestimate redistribution during freezing if partly frozen systems are considered similar to unfrozen systems. To account for this, an impedance factor is introduced while calculating the hydraulic conductivity of a partially frozen layer, k_{wf}

$$k_{wf} = 10^{-fc_i Q} k_w \tag{1.28}$$

where Q represents thermal quality (ratio between the mass of ice and the total amount of water), fc_i is an impedance parameter, and k_w is the hydraulic conductivity of the layer calculated from unfrozen water content without accounting for occurrence of ice. Recent work has also demonstrated the need for a two-domain approach to describe water flow in a partly frozen system (Stähli, Jansson, and Lundin, 1996). The new approach separates between a low-flow domain, which is the same as used previously when estimating water flow in a partially unfrozen soil and a high-flow domain. The high-flow domain allows rapid flow of infiltrating water provided that air-filled pores were present at the time of infiltration.

For nonfrozen conditions, the liquid water flow is temperature dependent since the dynamic viscosity depends on temperature. This process also causes flow from higher to lower temperature but it is normally of less importance than the vapor flux in an unsaturated soil.

WATER AND HEAT FLOW PROCESSES OF A SOIL-VEGETATION-ATMOSPHERE TRANSFER SYSTEM (SVAT)

A soil-vegetation-atmosphere transfer (SVAT) system is more complex than what has been described for an earth system because of complicated geometry and the number of possible interfaces to the atmosphere. A simple geometry that is commonly used is the so-called single leaf concept and a soil that may be covered with snow. This requires the definition of at least four different surfaces where the energy balance has to be described between the atmosphere and the respective surfaces: a bare soil surface, a snow surface, a dry vegetation surface that corresponds to the leaf of the vegetation, and a vegetation surface that may be wet because of intercepted precipitation. Irrespective of the surface, partitioning of net radiation between the plant canopy and the soil surface below can roughly be described by Beer's law (Impens and Lemeur, 1969):

$$R_{ns} = R_{na} e^{-k_{rn} LAI} \tag{1.29}$$

where R_{na} is the net radiation above the plant canopy, R_{ns} is the net radiation at the soil surface, k_{rn} is an extinction coefficient, and LAI is the leaf area index.

The energy flows and resistances in the soil-plant-atmosphere system are illustrated in Figure 1.1. In reality, coupling and feedback between different processes exists but many of the processes are tricky to understand and quantify. For most SVAT models the upper boundary condition above the vegetation is considered as an independent input to the model. However, recent studies have shown that a high degree of coupling exists and that doubtful results may occur if the properties of the SVAT systems are changed

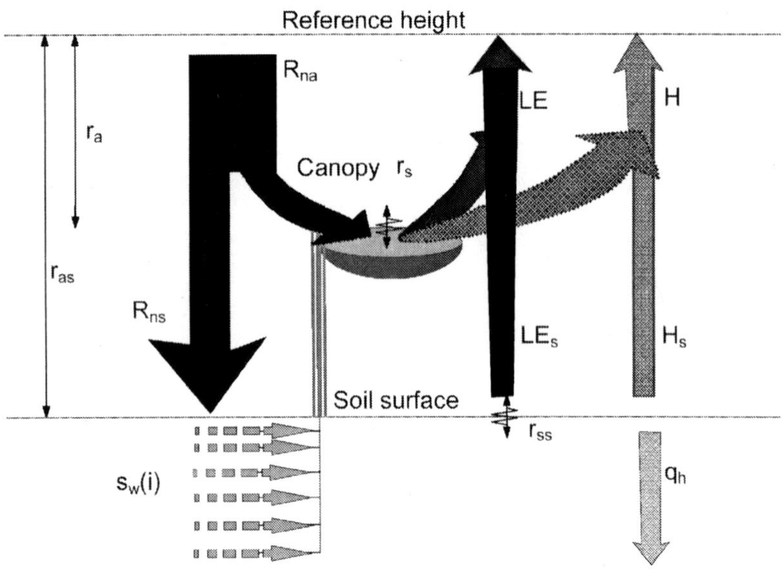

FIGURE 1.1. The energy flows and resistances above a plant canopy and at the snow/soil surface. The symbols correspond to those used in respective equations for describing different fluxes given in text. (*Source:* Jansson and Karlberg, 2001, p. 152.)

without making any changes to the atmospheric input. For example, a change from a wet to a dry area will change the partitioning of the heat flow to the atmosphere and this will also change the boundary condition. Normally, such effects are considered in meteorological models but not in conventional SVAT models. More details on vegetation are presented in Chapter 20.

Bare Soil-Atmosphere Interface

The physically based approach for calculating soil evaporation originates from the idea of solving the equation of heat flux at the soil surface boundary (Alvenäs and Jansson, 1997). According to the law of conservation of energy:

$$R_{ns} = L_v E_s + H_s + q_h \tag{1.30}$$

where R_{ns} is the available net radiation at the soil surface, $L_v E_s$ is the latent heat flux to the air, H_s is the sensible heat flux to the air, and q_h is the heat flux to the soil. The three different heat fluxes are estimated by an iterative procedure in which the soil surface temperature is varied according to a given scheme.

Vapor pressure at the soil surface is given by the surface temperature, T_s, and the water tension of the uppermost layer and an empirical correction factor, e_{corr}, account for a steep gradient in moisture between the uppermost layer and the soil surface:

$$e_{surf} = e_s(T_s) e^{\left(\frac{\frac{-\psi}{100} M g e_{corr}}{R(T_s + 273.15)} \right)} \tag{1.31}$$

where R is the gas constant, M is the molar mass of water, and g is the gravitational constant.

The empirical correction factor depends on a parameter ψ_{eg} and calculated mass balance at the soil surface, δ_{surf}, which is allowed to vary between the parameters s_{def} and s_{excess} given as mm of water:

$$e_{corr} = 10^{(-\delta_{surf} \psi_{eg})} \tag{1.32}$$

$$\delta_{surf}(t) = \max(s_{def}, \min(s_{excess}, \delta_{surf}(t-1) + (I - E_s - q_v)\Delta t) \tag{1.33}$$

where I is the infiltration rate and q_v is the vapor flow from soil surface to the central point of the uppermost soil layer.

The aerodynamic resistance above the soil surface may be calculated as a sum of two components viz function of wind speed and temperature gradient, r_a, which is corrected for atmospheric stability, and a representation of the influence of crop cover, r_{ab}:

$$r_{as} = r_a + r_{ab} \tag{1.34}$$

The aerodynamic resistance, r_a is calculated as:

$$r_a = \frac{\ln^2\left(\frac{z_{ref} - d}{z_o}\right)}{k^2 u} \tag{1.35}$$

where u represents wind speed at a reference height, k is von Karman's "constant," d is the displacement height, and z_o is the roughness length. The roughness length and displacement height can be found in the literature, especially for momentum transfer in the atmosphere. However, these values may differ substantially from the values that are valid for sensible and latent heat flow. The stability correction caused by possible nonneutral conditions may be estimated by different methods. The first method, developed on the results of Pruitt, Morgan, and Lourence (1973), makes a simple correction to the aerodynamic resistance as estimated by equation (1.35). Other methods, suggested by Paulsen (1970) and Beljaars and Holtslag (1991), are physically sound as they allow for independent roughness lengths for momentum and heat. It is not possible to derive the resistance between the soil surface and the crop canopy, r_{ab}, from any consistent micrometeorological theory. A simplified approach suggests that it is proportional to the leaf area index:

$$r_{ab} = r_{alai} LAI \tag{1.36}$$

where r_{alai} is a parameter.

An alternative to using the energy balance is to use the Penman-Monteith equation with appropriate values of both the soil surface resistance and the aerodynamic resistance. This allows for an estimation of surface temperature as pointed out by Monteith (1981) but will not allow for any interaction with the heat flow to the soil. The heat flow has to be estimated from a previous time step or from an independent method if the Penman-Monteith equation is to be used consistently for estimation of both latent and sensible heat flux from a bare soil.

Snow-Atmosphere Interface

The snow surface temperature can be assumed to be equal to the air temperature or it can be estimated by solving the energy balance equation similar to a bare soil. The turbulent fluxes of latent and sensible heat are calculated with the same methods as described for a bare soil. The heat flow into snow differs from the one into the soil due to the different thermal properties. Normally, a snow pack has a low total mass, which means that the heat capacity is very limited. The major heat store is normally in the soil system. A steady state solution may be assumed for the heat flux through the snow pack and to the middle of the topsoil layer. Because of the high porosity and the generally low thermal conductivity of snow, the influence of water vapor flow may be of importance. In case of liquid water in the snow the assumption about a small heat storage capacity is not valid since the latent heat of melting is substantial. However, in such a case, the snow surface temperature is normally close to 0°C and the surface fluxes can be based on that temperature. The vapor pressure is easily estimated for the snow surface assuming thermodynamic equilibrium between the ice or liquid phase and the corresponding saturation vapor pressures.

Vegetation-Atmosphere Interface: Transpiration

Transpiration occurs at a potential rate when neither soil water deficit nor low soil temperature influence the water loss. The potential transpiration may be calculated as described previously by using equation (1.12). The net radiation is reduced to a value as estimated by the Beer's law (1.29).

The canopy/surface resistance for optimal soil moisture conditions is a function of leaf area index *(LAI)*, global radiation (R_{ris}), and vapor pressure deficit $(e_s - e_a)$. A possible model for surface resistance of a nonstressed plant is given by:

$$r_s = \frac{1}{LAI g_l} \tag{1.37}$$

where g_l is the leaf conductance which is given by the Lohammar equation (Lindroth, 1985) as:

$$g_l = \frac{R_{ris}}{R_{ris} + g_{ris}} \frac{g_{\max}}{1 + \frac{(e_s - e_a)}{g_{vpd}}} \tag{1.38}$$

where g_{ris}, $g_{\max}$, and g_{vpd} are parameter values. The Lohammar equation was developed from detailed measurements using only a small fraction of leaves. It has been useful for many different plants but the response due to variations in global radiation and vapor pressure deficit may vary substantially among plants. For a plant canopy, the radiation response has been linear for a wide range (Cienciala, Kucera, and Lindroth, 1998) instead of the nonlinear form described by equation (1.38).

Vegetation-Atmosphere Interface: Intercepted Water

Under similar climatic conditions, evaporation of intercepted water, especially from aerodynamically rough surfaces such as forests, may considerably exceed transpiration. The simplest procedure to calculate this potential evaporation rate when the canopy is wet is to use the Penman-Monteith equation assuming a surface resistance close to zero. In many models, evaporation from a canopy is assumed to be either from the intercepted water or as transpiration through the plant leaves. Studies have shown that there may be an important interaction and competition for available energy at the canopy level. Substantial cooling and wetting of the canopy air will reduce both transpiration and long-lasting evaporation from the interception storage.

SUMMARY

This chapter presented some of the basic theories and assumptions behind major physical processes in ecosystems. The basis is a one-dimensional approach dealing with the energy balance and the corresponding process of radiation and convection. The principles and the basic equations for the single components as well as for coupling between the heat and water processes within a SVAT system gave an overview of how the system can be described to facilitate appropriate ecological models with governing physical environmental conditions.

REFERENCES

Alvenäs, G. and P.-E. Jansson. (1997). Model for evaporation, moisture and temperature of bare soil: Calibration and sensitivity analysis. *Agriculture and Forest Meteorology 88:* 47-56.

Beljaars, A. C. M. and A. A. M. Holtslag. (1991). Flux parameterization over land surfaces for atmospheric models. *Journal of Applied Meteorology 30:* 327-341.

Burman, R. and L.O. Pochop. (1994). *Evaporation, Evapotranspiration and Climatic Data.* Developments in the Atmospheric Science, 22, Amsterdam: Elsevier.

Cienciala, E., J. Kucera, and A. Lindroth. (1998). Water flux in boreal forest during two hydrologically contrasting years; species specific regulation of canopy conductance and transpiration. *Annual Science Forestry 55:* 47-61.

Darcy, H. (1856). *Les fontaines publique de la ville de Dijon.* Paris: Dalmont.

Ernst, L.F. (1956). Calculation of the steady flow of groundwater in vertical cross sections. *Netherlands Journal of Agricultural Science 4:* 126-131.

Green, W.H. and G.A. Ampt. (1911). Studies on soil physics. I. Flow of air and water through soils. *Journal of Agricultural Science 4:* 1-24.

Harlan, R.L. (1973). Analysis of coupled heat-fluid transport in partially frozen soil. *Water Resources Research 9:* 1314-1323.

Hillel, D. (1980). *Fundamentals of Soil Physics*. New York: Academic Press.

Hooghoudt, S.B. (1940). Bijdragen tot de kennis van enige natuurkundige grootheden van de ground No. 7. *Versagen Landbrug Onderzee 42:* 449-541.

Impens, I. and R. Lemeur. (1969). Extinction of net radiation in different crop canopies. *Archive Geophysical BioClimatology, Series B., 17:* 403-412.

Jansson, P.-E. and S. Halldin. (1979). Model for the annual water and energy flow in a layered soil. In *Comparison of Forest and Energy Exchange Models,* ed. S. Halldin, Copenhagen: Society for Ecological Modelling, pp. 145-163.

Jansson, P.-E. and L. Karlberg. (2001). *Coupled Heat and Mass Transfer Model for Soil-Plant-Atmosphere Systems*. Stockholm: Royal Institute of Technology, Department of Civil and Environmental Engineering.

Jansson, P.-E. and D. Moon. (2001). A coupled model of water, heat and mass transfer using object orientation to improve flexibility and functionality. *Environmental Modelling & Software 16:* 37-46.

Jury, W.A., W.R. Gardner, and W.H. Gardner. (1991). *Soil Physics,* Fifth Edition. New York: John Wiley and Sons.

Lindroth, A. (1985). Canopy conductance of coniferous forests related to climate. *Water Resources Research 21:* 297-304.

Lundin, L.-C. (1990). Hydraulic properties in an operational model of frozen soil. *Journal of Hydrology 118:* 289-310.

Monteith, J.L. (1965). Evaporation and environment. In *The State and Movement of Water in Living Organisms,* ed. G.E. Fogg, 19th Symposium of Society of Experimental Biologists, Cambridge: The Company of Biologists, pp. 205-234.

Monteith, J.L. (1981). Evaporation and surface temperature. *Quarterly Journal of the Royal Meteorological Society 107:* 1-27.

Monteith, J.L. and M. Unsworth. (1990). *Principles of Environmental Physics,* Second Edition. London: Arnold.

Paulsen, C.A. (1970). The mathematical representation of wind speed and temperature profiles in the unstable atmospherics surface layer. *Journal of Applied Meteorology 9:* 857-861.

Penman, H.L. (1940). Gas and vapor movements in the soil. I. The diffusion of vapors through porous solids. *Journal of Agriculture Research 30:* 437-462.

Penman, H.L. (1948). Natural evaporation from open water, bare soil and grass. *Proceedings of the Royal Society of London, A, 194:* 120-145.

Philip, J.R. (1957). The theory of infiltration: 1. The infiltration equation and its solution. *Soil Science 83:* 345-357.

Pruitt, W.O., D.L. Morgan, and F.J. Lourence. (1973). Momentum and mass transfer in the surface boundary layer. *Quarterly Journal of Royal Meteorology Society 99:* 370-386.

Richards, L.A. (1931). Capillary conduction of liquids in porous mediums. *Physics 1:* 318-333.

Richter, J. (1987). *The Soil As a Reactor.* Cremlingen: Catena Verlag.

Stähli, M., P.-E. Jansson, and L.-C. Lundin. (1996). Preferential water flow in a frozen soil—A two-domain model approach. *Hydrological Processes 10:* 1305-1316.

Yong, R.N. and B.P. Warkentin. (1975). *Soil Properties and Behaviour.* Developments in Geotechnical Engineering 5. Amsterdam: Elsevier.

Youngs, E.G. (1980). The analysis of groundwater seepage in heterogeneous aquifers. *Hydrological Sciences, Bulletin 25:* 155-165.

Chapter 2

Chemical Processes

Kim H. Tan

This chapter discusses chemical properties in soils that are important in the dynamics of soil water, organics, and inorganics. A number of chemical processes occur in soils. Some affect soil fertility, while others will influence soil formation, degradation, and environmental quality. Almost all of these processes are the result of electrochemical properties of the soil constituents. The electrical charges of the organic and inorganic materials are paramount for the development of water and chemical potentials governing movement of water, ions, and ion uptake by plant roots. They also control the formation of electric double layers and determine interaction, chelation, precipitation, and interparticle attraction phenomena in soils. Adsorption, cation and anion exchange reactions are difficult to realize in the absence of electrical charges.

Statistical models will be used to integrate the various interactive processes providing quantitative means for interpretation of soil reactions and, where possible, for prediction of the environmental impact on agricultural practices. However, an overuse of statistics in modeling, common in many soil chemistry and physical chemistry books, will be avoided. Long derivations of complex formulas using pages of integral and differential equations considered of high scientific value by many authors are neither practical nor informative. Modeling will be presented in this chapter as simply as possible in easy-to-understand language, but without neglecting scientific standards.

ELECTROCHEMICAL PROPERTIES OF SOIL SOLIDS

Solid soil constituents consist of a variety of inorganic and organic compounds. Most of the coarse materials, such as sand, silt, and undecomposed organic material, are chemically inert, though they will affect many soil properties and are important for building up the soil. Since they are coarse in size, they have low specific surface area and do not exhibit colloidal properties. They may participate in a number of soil reactions and exhibit some adsorption capacities, but they are not really chemically active. In contrast, clay and humus, the smallest constituents of soils, exhibit colloidal properties (Tan, 1998, 2000). They have large specific surface areas and display surface chemistries attributed to the presence of electrical charges in their molecules different from the coarse materials. Because of these charges they are chemically very active and are considered the *seat of chemical activity* in soils (Brady, 1990), enabling many soil reactions and processes to occur.

The Types and Nature of Soil Clays

Soil clays are often called *secondary minerals* by virtue of their formation from the weathering of *primary minerals*. The definition of clay is usually based on size (diameter) measurements, which differ considerably from one scientific institution to another. Clay is defined by the United States Department of Agriculture and the International Union of Soil Sciences as inorganic materials < 0.002 mm (< 2 μm) in diameter. This definition differs from that of the Clay Minerals Society and Association Internationale Pour L'etude des Argilles (Guggenheim and Martin, 1995; Hurst and Pickering, 1997), which consider clays to be particles < 74 μm, a size based on the limit of visual resolution. In soil science, a size limit of 74 μm characterizes very fine sand, which is still visible. The size limit of 2 μm for clays is chosen not only because they are too small to be seen with the eye or magnifying glass, but because at this size clays also start to possess colloidal behavior. In pure colloid chemistry, colloidal sizes can be expressed in terms of linear dimensions (μm or nm) or in terms of mass using molecular weights or daltons. The choice depends on the purpose of study and no direct statistical conversion of size into mass units is available, though a few daltons are usually considered equivalent to a diameter of 1 nm. Generally, the size of colloids represents a *continuum,* ranging from 0.001 to 1.0 μm in diameter (Ranville and Schmiermund, 1998), hence, defining upper and lower size limits of colloidal particles is arbitrary. Though a size limit of 0.45 μm is most widely accepted, Ranville and Schmiermund (1998) believe that the range of 2-5 μm would better describe the hydrodynamic behavior of large colloids, which coincides with the usual clay-silt boundary in soil science.

The soil-clay fraction is usually composed of a mixture of different types of clays, and depending upon conditions, one type may dominate in amount over the others. The mixture is usually made up of crystalline and amorphous minerals. The crystalline clays have crystalline structures identifiable by X-ray diffraction and include 1:1, 2:1, and 2:2 layer types of clays. In contrast, the amorphous clays are noncrystalline and include allophane and a wide variety of substances, such as silica gel, sesquioxides, silicates, and phosphates. They are called *amorphous* because they are amorphous to X-ray diffraction analysis, meaning they produce featureless X-ray diffraction patterns. Many scientists object to using the term amorphous because it is always possible that the current method of analysis is inadequate to detect the so-called amorphous clays. Hurst, Schroeder, and Styron (1997) report that poor resolutions in X-ray diffraction analysis could be caused by misorientation of *coherent diffraction domains* (CDDs), and by the extremely small sizes. With well-ordered CDDs, some of these clays have now been detected in very fine microcrystal forms, hence, several soil mineralogists suggest naming them *noncrystalline* or *paracrystalline* instead of amorphous minerals. Lately, the name *short-range-order* (SRO) mineral is becoming popular.

Charged surfaces are the main reasons for the electrochemical properties of clays. Several types of clay surfaces have been recognized, e.g., siloxane, oxyhydroxy, silanol, aluminol, and ferrol surfaces, each exhibiting different types of surface chemistry (Sticher and Bach, 1966; Sposito, 1989; Tan, 1998). The charges on siloxane surfaces are attributed mainly to permanent negative charges developed by isomorphous substitution, whereas those of the other surfaces originate from dissociation of exposed hydroxyl groups. Some refer to these OH groups as functional groups (Zachara and Westall, 1998), but the use of this term is commonly restricted to chemically active groups of organic molecules. The discussion that follows of some of the clay minerals will be based on the division of *permanent-charged* and *variable-charged clays.* Due to space limitations it is not possible to discuss them all, and for more information on the nature, crystal structure, and subdivisions of the clay minerals reference is made to Tan (1998, 2000), and Mackenzie (1975).

Variable-Charged Clays

Clay minerals in this group exhibit *variable* or *pH dependent* electrical charges, formed by dissociation of exposed OH-groups. All of the amorphous clays, e.g., allophane and imogolite, belong to this category. The sesquioxide minerals, both crystalline and amorphous, are also variable in charges, because most of their surfaces are composed of structural OH groups. Goldberg, Lebron, and Suarez (1999) group them separately from the clay minerals under the name *oxides.* This is uncommon in soil science where the term *clay mineral* is used for all the minerals present in the soil's clay fraction, including the iron and aluminum oxide minerals. These minerals, e.g., hematite, goethite, and gibbsite, are more commonly called sesquioxides instead of oxides (Fairbridge and Finkl, 1979; Soil Science Society of America, 1996; Tan, 1998) because of their unique composition composed of 1.5 atoms of oxygen per atom of Al or Fe (Latin sesqui = 1.5 times; Miller and Gardiner, 1998). They are negatively charged in acid soils, but may be electropositively charged in alkaline soils, and hence are amphoteric in reaction. At specific pH values, they can also be neutral (no charge), and the pH at which the minerals exhibit zero charge is called the *zero point of charge* (ZPC). Depending on mineral species, their ZPC may range from 2.0 to 7.0 (Schwertmann and Taylor, 1977). The adsorption capacity of these sesquioxides, ranging from 30 to 300 μmol g^{-1} (Schwertmann and Taylor, 1977), compares favorably with cation exchange capacity (CEC) values of silicate clay minerals. Clay minerals in the oxide family can also be present in soils, e.g., MnO_2 and TiO_2, but they are usually treated separately as miscellaneous oxide minerals because of their minor occurrence in most soils. Of wider occurrence are silica minerals, such as quartz and crystobalite. They are classified as clays when their sizes fall within the definition of soil's clays (< 2 μm). Though their general formula is $n(SiO_2)$, they are not grouped in the oxide family but are classified as *tectosilicates* due to their *framework* crystal structures. These silica minerals are generally inert or chemically inactive and occur extensively in nature as important constituents of the soil's clay fraction where they serve as diluents of the more chemically active clay and humic materials. Their surface charges are usually very small, if not negligible, and the correspondingly small adsorption capacities are more attributable to Si-O broken bonds and Si-OH groups on particle edges.

Permanent-Charged Clays

The crystalline layer silicates with predominant siloxane surfaces are perhaps the best-known clay minerals exhibiting permanent charges. These charges, attributed to isomorphous substitution, are considered not to change with changing soil pH. *Smectite,* formerly called montmorillonite, a 2:1 expanding lattice type of silicate clay, is perhaps the best-known mineral with a high permanent negative charge, arising mainly from isomorphous substitution. Together with a high specific surface area of 700-800 square meters g^{-1} which is exposed on dispersion in water, this negative charge is the reason for smectite to exhibit a high CEC of 70 cmol kg^{-1} or higher. Other examples are vermiculite, illite, and chlorite, with permanent negative charges equivalent to CECs of 100, 30, and 0 cmol kg^{-1}, respectively. On the other hand, kaolinite, a 1:1 lattice type of clay, has only a small permanent charge, but has a high variable charge, causing its CEC to fluctuate with changing pH from 1.0 to 10 cmol(+) kg^{-1}.

The Types and Nature of Soil Organic Matter

A number of organic compounds, especially those making up the humus fraction of soils, are also important contributors of electrical charges in soils. They are soil colloids with high specific surface areas and electrical charges, causing the occurrence of many biochemical reactions of importance for continuation of life in soils. Strictly speaking, both live organic material and dead components are included in the term soil organic matter (Stevenson, 1994), but only the nonliving fraction will be discussed in this chapter. For a more detailed treatise on soil organic matter, reference is made to Tan (1998, 2000). The dead organic fraction is formed by chemical and biological decay of mainly plant materials, and can be divided again into (1) materials in which the anatomy of plant substance is still visible, and (2) completely decomposed materials. The first group is of significance in soil physics, e.g., decreasing bulk density, affecting soil structure, and in protection of surface soils when used as mulch. The term *litter* is often used for this type of organic matter when it lies on the soil surface. The second group is of importance in soil chemistry and soil biochemical processes. Called humus, it is the most chemically active fraction in soils, exceeding even the chemical activity of clays. The term humus is used today to refer to a mixture of decomposed organic compounds and biologically synthesized new compounds, called humic matter. Therefore, humus can be distinguished into a (1) nonhumified and (2) humified fraction (Stevenson, 1994; Tan, 1998, 2000). The nonhumified fraction is composed of a variety of organic substances, e.g., carbohydrates, amino acids, protein, lipids, lignin, nucleic acid, and a variety of organic and inorganic acids. Many of them are colloidal in nature, and may carry electrical charges. The humified fraction, with humic and fulvic acids as the major components, makes up the bulk of humus and, like clay, the humic substances are the building constituents of soils. However, in contrast to clays, almost all of the organic substances are amphoteric in reaction, and their electrical charges are mostly variable or pH dependent in nature. Numerically, their charges are often extremely high, exceeding those of vermiculite and smectite. The pH-dependent charge of humic compounds in terms of cation exchange capacities may vary from a low of a few hundred me (milliequivalents) per 100 g to a high of 1,500 me per 100 g (me per 100 g = cmol kg^{-1}), values that are astronomically high.

The contribution of soil organic matter to the soil's electrical charges has not been fully realized, though it is a well-known fact that soils high in organic matter possess high CECs. Even today many prominent soil scientists consider soil clays as the major contributors of negative charges in soils. Although many of the organic compounds are known to possess electrical charges and affect many chemical and biochemical processes, due to space limitations only two major organic constituents of humus will be discussed very briefly, e.g., amino acids and humic matter. The electrochemical properties of especially humic matter and amino acids are unique. Not only do they contribute in making the soil positively or negatively charged, but they are also responsible for many interactions, complexation, chelation, and detoxification reactions, affecting numerous aspects of importance in the soil ecosystem and in issues on environmental quality. For complete information on the types, nature, and chemical properties of the soil's organic fraction, reference is made to Stevenson (1994) and Tan (1998, 2000).

Amino Acids

These compounds are the fundamental units of protein. Because of the presence of both carboxyl and amino groups in their molecules, amino acids are amphoteric. The carboxyl groups behave as acids, whereas the amino groups behave as bases. Therefore, they can be negatively and positively charged depending on conditions. The pH at which amino acid exhibits equal amounts of negative and positive charges is called the *isoelectric point* (pH_0) or *isoionic point* (pI), and at this point the amino acid is electrically neutral. The isoelectric or isoionic point has a similar meaning as ZPC (pH_0); the only difference is in the symbols pH_0 and pI as indicated. The isoionic point is often used in amino acid reactions because the neutral amino acid ion is considered a *zwitter ion* (German zwitter = double or pair). However, the more general term used for organic reactions in soil science is isoelectric point, whereas ZPC is employed in clay mineralogy. The amino acids obtained by hydrolysis of proteins are α-amino acids (Tan, 1998, 2000) and can be classified into (1) aliphatic, (2) aromatic, and (3) heterocyclic amino acids. In addition to this classification, amino acids can also be distinguished into *neutral, acidic,* and *basic* amino acids. Neutral amino acids contain one NH_2 and one COOH group in their structure, whereas acidic amino acids have two COOH groups or more versus one NH_2 group in their molecules. On the other hand, the basic types of amino acids possess more than one NH_2 group versus one COOH group in their structure (Stevenson, 1994; Tan, 1998).

Humic Matter

The humified organic fraction is known currently as humic compounds or humic matter. In the German and Russian literature it is also called *humus acid* (Scharpenseel, 1966; Orlov, 1985). It is a new product in soils, synthesized during the decomposition of plant and animal residue with or without the assistance of microorganisms. Currently it is known that humic matter is also present in streams, rivers, lakes, oceans, and their sediments, which is classified as *aquatic humic matter.* In addition, humic matter can also be found as geologic deposits, e.g., lignite or leonardite, coal, and oil shale, which are of economic importance as sources for the production of commercial humates used as soil amendments (Lobartini et al., 1992). The chemical reactivity of humic matter can be predicted from its *total acidity value,* which is defined as the sum of carboxyl and phenolic-OH group content. These groups, called *functional groups,* are the sources of negative charges and enable the humic molecule to take part in many chemical reactions, e.g., adsorption, complex formation, chelation, interaction with clays, water and ion bridging or coadsorption. Even neutral humic molecules are capable of reacting with a variety of substances through water-bridging mechanisms.

ELECTRICAL CHARGES OF SOIL CLAYS

As indicated earlier, soil clays ordinarily carry electrical charges, which, depending upon conditions, can be electronegative or electropositive. The most important source for the development of positive charges is a process called *protonation of exposed hydroxyl groups,* which mostly yields variable positive charges. Isomorphous substitution can, under certain conditions, also produce positive charges. Recently, the idea was presented that *chemisorption* of H_2O can yield electrical charges due to "splitting" of water into H^+ and OH^- during adsorption to form a hydroxylated surface, a process called *ionic dissolution* (Singh and Uehara, 1998). However, chemisorption of H_2O is not known to be a form of ionic dissolution in soil chemistry, whereas adsorption of H_2O usually results in *polarization* of the dipolar water molecule (H-O-H). The water molecule in the form of clay-H-O-H will seldom dissociate (split) its *nonadsorbed* proton, but retains it instead, producing surfaces with positive charges.

Isomorphous Substitution and Permanent Negative Charges

This process involves the substitution of atoms in the mineral structure for other atoms without affecting the crystal structure, and takes place in both the silica tetrahedrons and the aluminum octahedrons of silicate clays. Once the electric charge is created, it is not subject to further modification by pH, though in nature exceptions always occur. This is the reason why the charge developed is called *permanent charge* or *constant charge,* and is the main important charge carried by 2:1 clay minerals, such as smectite. A number of scientists object to using the name permanent charge, since they believe that it cannot be used for soil systems containing high amounts of organic matter, intergrade minerals, and allophane. However, the new definition created by the Soil Science Society of America subcommittee on soil chemistry terminology is too long and too wordy and also very confusing or ambiguous. Some suggest naming it *permanent structural charge,* σ_p (Sposito, 1989; Tan, 1998), which can be expressed in the following statistical model:

$$\sigma_p = -(X / M_r) \tag{2.1}$$

where X = layer charge per unit formula, and M_r = relative molecular mass, calculated using the formula weight per unit cell. The symbol σ_p is used here to indicate the permanent charge, instead of σ_0 as used by Sposito (1989), since the suffix p refers more closely to permanent charge than the suffix 0. Using a formula for a unit cell of smectite (montmorillonite) of $(Si_8)(Al_{3.33}Mg_{0.67})O_{20}(OH)_4nH_2O$ and a layer charge per unit formula of 0.50 equivalents (see Mackenzie, 1975; Tan, 1998), the permanent charge (neglecting nH_2O) can be calculated as follows:

$$\begin{aligned} M_r &= (8\times28.1) + (3.33\times27) + (0.67\times24.3) + (20\times16) + (4\times17) = 719 \text{ g} \\ \sigma_p &= -(0.50/719) \text{ equivalents g}^{-1} \\ &= -0.0006954 \text{ equivalents g}^{-1} = -69.5 \text{ me } 100 \text{ g}^{-1} = -69.5 \text{ cmol kg}^{-1} \\ &= -0.695 \text{ mol kg}^{-1} \end{aligned}$$

The permanent charge is negative, hence the use of a negative sign. The unit mol kg^{-1} is the official unit for cation exchange capacity in the *Soil Science Society of America Journal,* though the smaller denomination, cmol per kg, is more commonly used in many other books.

Isomorphous Substitution and Permanent Positive Charges

Isomorphous substitution can also produce electropositive charges when the replacing cation has a larger positive charge than the cation being replaced. However, such a substitution cannot occur in the aluminosilicates, since the Al and Si ions in their crystals possess the highest charges in soils. Perhaps only Mn^{4+} has a higher charge than Al^{3+}, but Mn^{4+} has never been noticed to replace Al^{3+} ions in 1:1 or 2:1 clay minerals. The development of positive charges by isomorphous substitution is commonly known to occur in vermiculite, chlorite, or in the 2:2 lattice type of clays, minerals in the category of magnesium silicates, where Mg^{2+} can be replaced by ions of larger charges, such as Al^{3+} and Fe^{3+}.

Dissociation of Exposed Hydroxyl Groups

Exposed hydroxyl groups are OH groups located on the surfaces of Al-octahedrons. They are present on surface planes of 1:1 types of clays, sesquioxides, and amorphous clays. These OH groups are in contact with the soil solution and tend to dissociate their protons. The release of the H^+ ion leaves one negative charge in the octahedron not neutralized. Such a dissociation reaction is dependent upon pH, and is encouraged by high pH values. The dissociation reaction is usually not that simple, since the proton is strongly bonded in the OH group as an integral part of the Al-octahedron. It needs OH^- ions, available at high pH in the soil solution, to attract and liberate the H^+. Therefore, the magnitude of negative charge increases and decreases accordingly with changing pH values. Because of this, this type of negative charge is called *pH-dependent charge* or *variable charge.* It is the major charge in sesquioxide minerals, amorphous minerals, and 1:1 types of clays, such as kaolinite, though kaolinite also possesses some permanent charges. As indicated previously, minerals with variable charges are often called *variable charge minerals.* Mehlich (1981) prefers using the term pH-dependent charge, but Sposito (1989) retains the term variable charge and adds a quantitative interpretation to it by formulating it in terms of *proton charge,* σ_H:

$$\sigma_H = m_H - m_{OH} \tag{2.2}$$

where m_H = mol L^{-1} of H^+, and m_{OH} = mol L^{-1} of OH^- ions complexed by surface functional groups. The sum of permanent charge σ_p and variable charge σ_H is called the intrinsic charge, σ_i, for which a model can be written as follows:

$$\sigma_i = \sigma_p + \sigma_H \tag{2.3}$$

When $\sigma_H > \sigma_p$, the soil has a variable charge, and such soils are called *variable charge soils.* When, on the other hand, $\sigma_H < \sigma_p$, the soil has a permanent charge and is called a *permanent charge soil.*

Protonation of Exposed Hydroxyl Groups

Protonation is a process involving the addition of H^+ ions to exposed OH groups of clay minerals. The H^+ is weakly adsorbed and does not form H_2O with the OH. Because of this addition of H^+ ions, the OH groups are oversaturated with protons and the clay surface becomes positively charged. Protonation of exposed OH groups occurs only at low pH, since acid conditions are required for supplying the extra protons. At high pH, the hydroxyl groups tend to dissociate their protons which results in development of negative charges, as discussed earlier. This kind of positive charge, called *variable* or *pH-dependent positive charge,* in contrast to permanent positive charges, is important in kaolinite, sesquioxides, and amorphous minerals. Smectite does not have exposed OH groups, and hence exhibits only very small amounts of variable charges. The reactions for dissociation and association of protons can be summarized as follows:

Alkaline soils: $-Al{-}OH + OH^- \leftrightarrows -Al{-}O^- + H_2O$
Acid soils: $-Al{-}OH + H^+ \leftrightarrows -Al{-}OHH^+$

To write these reactions in different forms as suggested by Zachara and Westall (1998):

$SOH = SO^- + H^+$ and
$SOH + H^+ = SOH_2^+$

in which SOH = hydroxylated surfaces of Al or Fe, does not mean an advancement, but amounts only to proliferation of noise. Moreover, the dissociation of H^+, as shown in the first reaction, cannot be realized in the absence of free OH^- ions. Such a dissociation, resulting in the development of a negative charge, can only occur if an OH^- ion is available to attract the H^+ to form H_2O. The affinity of the two ions toward each other, to form H_2O, is so great that it precedes all other reactions. Water is known to be the most stable compound in the environment. It is so stable that only one in 10^7 molecules of water can "split" at any one time into H^+ and OH^- (Pauling, 1964). Therefore, instead of water splitting apart into H^+ and OH^- as contended by Singh and Uehara (1998), the two ions are more likely to combine and form water in the process of the development of surface charges.

Zero Point of Charge or ZPC

The pH at which the mineral has no charge, or has equal amounts of negative and positive charges, is called the zero point of charge (ZPC or pH_0).

As pointed out previously, the variable negative charge carried by clay minerals is high at high pH, but decreases with a decrease in pH. When the pH is continuously decreased, a point will be reached at which the negative charge equals zero. The pH at which this occurs is called ZPC as indicated in the definition. Some authors call it PZC or point zero charge (Theng, 1972), which only boils down to semantics, since the meaning remains the same. Others recognize three types of ZPCs (Sposito, 1989): (1) ZPC, (2) ZPPC, and (3) ZPNC. The ZPC or zero point of charge, is the general ZPC as previously defined. The surface charge at ZPC is then zero, hence $\sigma_p + \sigma_H = 0$. Since the ZPPC is the zero point proton charge, $\sigma_p = 0$, hence $\sigma_H = 0$. The ZPNC (zero point net charge) can be defined as $\sigma_e + \sigma_{os} + \sigma_{is} = 0$, where σ_e = effective surface charge influenced by diffuse double layer ions, σ_{os} = surface charge of outer sphere, and σ_{is} = surface charge of inner sphere. Since outer- and inner-sphere surfaces represent, in essence, the total surface plane of the clay mineral, hence the sum of outer- and inner-sphere charges must equal the total surface charge, defined previously as intrinsic charge, σ_i. Consequently:

$$\sigma_i = \sigma_{os} + \sigma_{is} \tag{2.4}$$

and since:

$$\sigma_i = \sigma_p + \sigma_H;$$

hence

$$\sigma_p + \sigma_H = \sigma_{os} + \sigma_{is} \tag{2.5}$$

This means that at $\sigma_p + \sigma_H = 0$, σ_i also equals zero, or in other words, the sum of outer- and inner-surface charges equals zero. The mineral surface has then no charge at all, and will not be able to attract a diffuse layer of counterions, making the use of σ_e obsolete. Whether this ZPNC as discussed is associated with the ZPNC used by Isbell (1980) and Eswaran and Tavernier (1980) is not known. From this discussion, the question can be raised as to whether such a development in the concept of ZPC amounts to advancing soil chemistry, or whether this makes the problem more complex and confusing than necessary. Nevertheless, the ZPC is considered a specific characteristic of the mineral.

ELECTRICAL CHARGES OF SOIL ORGANIC SUBSTANCES

The Charges in Amino Acids

Depending on soil conditions, amino acids can be negatively and positively charged. The zwitterion, assuming usually a tetrahedral configuration, is schematically illustrated by the following formula:

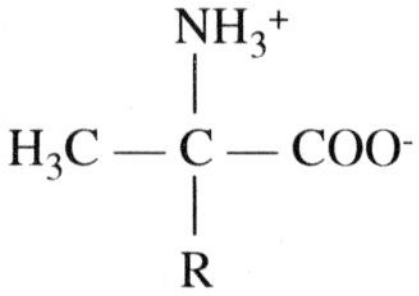

$$H_3C - \underset{\displaystyle R}{\overset{\displaystyle NH_3^+}{\underset{|}{\overset{|}{C}}}} - COO^-$$

in which R represents an H atom, a CH_3 group or an aromatic carbon chain, etc. When amino acids are present in soils with a pH > pH_0, the OH^- ions in the soil solution react with the NH_3^+ group, producing amino acid ions with a negative charge as can be noticed from the following equation:

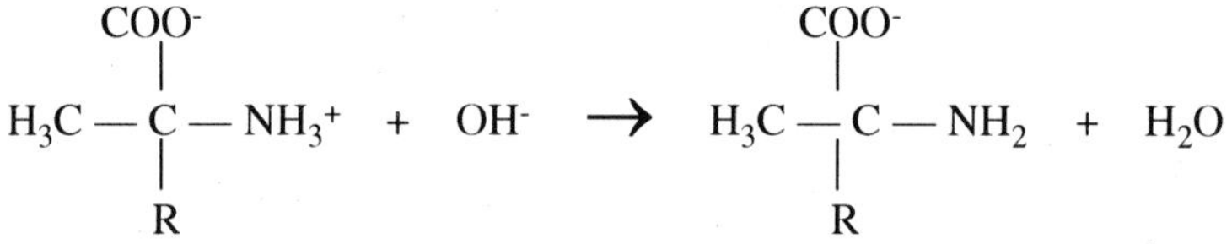

$$H_3C - \underset{\displaystyle R}{\overset{\displaystyle COO^-}{\underset{|}{\overset{|}{C}}}} - NH_3^+ \; + \; OH^- \; \rightarrow \; H_3C - \underset{\displaystyle R}{\overset{\displaystyle COO^-}{\underset{|}{\overset{|}{C}}}} - NH_2 \; + \; H_2O$$

On the other hand, in acidic conditions in which soil pH < pH_0, the large amounts of H^+ ions available in the soil solution will react with the carboxyl group, making the amino acid molecule become positively charged, as illustrated by the reaction:

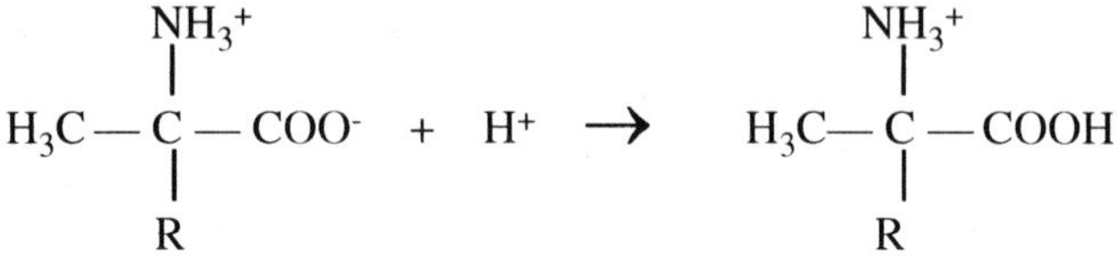

$$\underset{\text{zwitterion}}{H_3C - \underset{\displaystyle R}{\overset{\displaystyle NH_3^+}{\underset{|}{\overset{|}{C}}}} - COO^-} \; + \; H^+ \; \rightarrow \; \underset{\text{cation}}{H_3C - \underset{\displaystyle R}{\overset{\displaystyle NH_3^+}{\underset{|}{\overset{|}{C}}}} - COOH}$$

Some people believe that placing emphasis on the charge contribution of amino acids is misplaced, but next to humic acids, the chelation capacity of amino acids has been reported to be more significant than that of clay minerals (Stevenson, 1994). The fact that metal chelates act as carriers in element transport in soil formation and mineral nutrition of plants makes them even more important. Several of the amino acids have an isoelectric point = 7.0, e.g., alanine, and hence will be positively charged at soil pH = 5-6, common in soils under forest vegetation. This allows an interaction to take place between amino acids and clays, ensuring the accumulation of these organics.

The Charges of Humic Matter

In contrast to clays, surface charges of humic matter are generated by (1) OH groups attached to a carbon chain, in either the form of COOH, -C-OH, or phenolic-OH, and by (2) NH_2 groups. These chemically active groups are called functional groups. In humic matter, the charges are caused by dissociation of carboxyl and phenolic-OH groups. There is debate as to whether, depending on pH, the alcoholic-OH groups will also dissociate as do the phenolic-OHs, since according to basic chemistry the reaction is possible. Protonation of the functional groups in humic matter can only occur with NH_2 groups when present on the surface of the humic molecule. It has not been known to take place with alcoholic- and phenolic-OH groups at pHs normally occurring in soils, whereas the attachment of H^+ to $-COO^-$ and phenolic-O^-, involving covalent bonding, is also different from protonation of NH_2 and -OH groups. Humic substances, composed of humic and fulvic acids, are also amphoteric compounds. Although they can be positively charged, the negative charges are usually of more importance than the positive charges because of the very low ZPC values of humic compounds. This negative charge in humic matter is usually described in terms of *total acidity,* which is defined as the sum of the carboxyl and phenolic-OH groups. Dissociation of COOH in humic matter starts at pH = 3.0, and the humic molecule becomes negatively charged. The humic molecule attains a high negative charge when, especially at pH ≥ 9.0, dissociation of protons from the phenolic-OH groups also takes place. Because this functional group behaves as a *Brønsted acid,* Stevenson (1994) prefers naming it *acidic-OH* rather than phenolic-OH. As can be noticed, the development of negative charges is pH dependent, therefore, this charge is called *pH-dependent* or *variable charge.* In general, the amount of negative charges in humic matter in terms of total acidity is in the range of 500 to 1,500 cmol kg^{-1}, with fulvic acids exhibiting higher total acidities (600 - 1,500 cmol kg^{-1}) than humic acids (500 - 1,000 cmol kg^{-1}). Since total acidity also reflects cation exchange capacity, these values indicate the range of cation exchange capacities of humic matter.

ELECTROCHEMICAL PROPERTIES OF SOIL SOLUTES

Solutes are by definition the dissolved materials in a solution, and water is the solvent in soils. They are important participants in many soil reactions, and without them the reactions would be incomplete. Double layers cannot be formed without the presence of counterions and it is difficult to realize ion adsorption and cation and anion exchange in the absence of charged solutes. These solutes can be inorganic and organic in nature and can be divided into noncharged and charged solutes. For example, sugar dissolved in water is a noncharged solute, whereas amino acids, when truly dissolved (ionic), are charged solutes. Most of the inorganic solutes are electrically charged, e.g., Na^+, Cl^-, Ca^{2+}, and Al^{3+} ions, and their dissolution is called *ionization*. Different concepts are available on ionization, but in basic chemistry it is defined as the interaction between water and a compound yielding ions (Nebergall, Schmidt, and Holtzclaw, 1972). However, this definition is unsatisfactory for explaining that NaCl is already ionic in its solid crystalline form, and that dissolution merely separates the Na^+ from the Cl^- ion. Loss and gain of electrons from the outer shells of atoms are a better reason for atoms to become ions, though these processes are officially called oxidation-reduction reactions (Bartlett, 1998; Tan, 1998). Most of these solutes behave differently than soil colloids. Because charged solutes are ionic, and hence smaller in size than soil colloids, they exhibit ionic behavior and do not have colloidal properties. Their chemical and electrical properties can be described using chemical and electrochemical potentials, properties affecting a number of important chemical processes vital in especially transport and equilibrium reactions.

Chemical Potentials of Soil Solutes

Each chemical species in a reaction mixture is considered to contain a certain amount of (free) energy. This amount of energy per unit amount of ion species, called *chemical potential,* is assigned the symbol μ. It is affected by the pressure *P,* temperature *T,* chemical nature of the ion species, and its mixing ratio with other species, and is usually expressed in the following statistical models:

$$\mu = \mu^o + RT \ln m \text{ (for ideal condition and infinite dilution)} \quad (2.6)$$

$$\mu = \mu^o + RT \ln a \text{ (for nonideal condition)} \quad (2.7)$$

where μ^o = chemical potential of the ion species at standard state, R = gas constant, T = absolute temperature (K), m = moles L^{-1} of ion species, and

a = activity of the ion species in the mixture. For different statistical models and their discussion, reference is made to Tan (1998). Since ln = 2.3 log and R is a known gas constant, at T = 298 K equation (2.6) can be changed into:

$$\mu = \mu^o + 2.3\,RT \log m \quad \text{or} \quad \mu = \mu^o + 1.364 \log m \tag{2.8}$$

A similar change can be enacted with reaction (2.7). The chemical potential as formulated indicates the state of potential energy of the chemical species or component in soils, and its value is independent of external force fields, such as gravity or centrifugal force, hence it remains constant at any distance from the soil surface. It is also constant at equilibrium conditions, characterized by constant concentrations, temperatures, and pressures. Differences may occur at nonequilibrium conditions, owing to differences in concentrations, temperatures, and pressures. These differences in chemical potentials of a species at various locations in soils tend to induce spontaneous movement of the species in the direction of points with lower potentials until an equilibrium is attained. At equilibrium the chemical potentials at the respective locations become similar in value.

Electrochemical Potentials

In contrast to the chemical potentials discussed previously which apply to charged as well as noncharged soil solutes, electrochemical potentials are properties exhibited by only electrically charged solutes or ions, and are formulated by combining the chemical and electrical potentials:

$$E = \mu + zF\psi \tag{2.9}$$

in which E = electrochemical potential, μ = chemical potential, z = valence of the ion, F = Faraday constant, and ψ = electrical potential. By replacing μ by $RT \ln a$, the model for E can be changed into:

$$\mathrm{E} = RT \ln a + zF\psi \quad \text{or} \quad E = 1.34 \log a + zF\psi \tag{2.10}$$

Donnan Equilibrium Law

The electrochemical potential as formulated can be applied to describing ion transport and adsorption in soils. It is particularly useful in predicting ion uptake by plant roots and ion transport from cell to cell in the plant body. Cell compartments in the plant tissue are separated by biological membranes representing barriers for chemical compounds. The porous membranes behave like sieves, favoring penetration of very small particles only.

This kind of passive transport (diffusion and mass flow) of small hydrophilic particles obeys physical and chemical laws. The net ion flux from soil solution into root cells (or from one to another cell in the plant body) will stop as soon as a state of equilibrium is reached. At equilibrium, the system obeys the *Donnan equilibrium law.* With compounds that are not electrically charged, e.g., sucrose, the equilibrium is attained when equal sucrose activities exist on either side of the membrane. However, with ions possessing electrical charges, the electrochemical potential must be equal on both sides of the membrane:

$$E_i = E_o \tag{2.11}$$

$$RT \ln a_i + zF\psi_i = RT \ln a_o + zF\psi_o \tag{2.12}$$

where i = inside the cell, o = in soil solution or outside the cell. Rearranging the equation gives:

$$zF\psi_o - zF\psi_i = RT \ln a_i - RT \ln a_o \tag{2.13}$$

$$\psi_o - \psi_i = \frac{RT}{zF} \ln \frac{a_i}{a_o} \tag{2.14}$$

$\psi_o - \psi_i$ is called the *Donnan* or *membrane potential* (or E).

To maintain electroneutrality on both sides of the membrane, the electrochemical potential of cations also equals the electrochemical potential of anions. Therefore: $E_{cat} = E_{an}$ or

$$(RT/zF) \ln (a_i/a_o)_{cat} = (RT/zF) \ln (a_o/a_i)_{an} \tag{2.15}$$

where *cat* = cations and *an* = anions. Hence:

$$(a_i/a_o)_{cat} = (a_o/a_i)_{an} \text{ or } (a_i)_{cat}(a_i)_{an} = (a_o)_{cat}(a_o)_{an} \tag{2.16}$$

The latter equation reflects the classical *Donnan equilibrium law,* stating that the ion product on either side of the membrane is constant. If KCl or $CaCl_2$ is present in the system, this law states that

$$[(K^+)(Cl^-)]_i = [(K^+)(Cl^-)]_o \text{ or } [(\sqrt{Ca^{2+}})(Cl^-)]_i = [(\sqrt{Ca^{2+}})(Cl^-)]_o \tag{2.17}$$

This model of electrochemical potentials has been developed for equilibrium conditions. However, in living cells, no equilibrium exists between the two sides of the membrane. Metabolism is continuously consuming ions on the inside, and, thereby, is constantly disrupting the equilibrium condition favoring passage of ions from the outside.

Redox Potential

Reduction and oxidation reactions, called redox reactions, occur in almost any soil. It is also known as electrochemistry of soil solutes, but with emphasis on electron transfer resulting in a change of valences. The gain of electrons is defined as reduction, whereas loss of electrons is defined as an oxidation reaction, which can often be presented in a generalized redox reaction as follows:

$$\text{Reduced species} \leftrightarrows \text{Oxidized species} + e^- \tag{2.18}$$

The electrical potential of this reaction is

$$E_h = E^o + \frac{RT}{zF} \ln \frac{\text{oxidized sp}}{\text{reduced sp}} \text{ or } E_h = E^o + \frac{0.059}{z} \log \frac{\text{oxidized sp}}{\text{reduced sp}} \tag{2.19}$$

where E_h = electrode potential measured against a standard hydrogen electrode, E^o = standard potential of the reaction. Since the reaction is an oxidation-reduction reaction, E_h is in this case also the redox potential. Application of the redox model using a classical redox reaction is

$$Fe^{2+} \underset{reduction}{\overset{oxidation}{\rightleftarrows}} Fe^{3+} + e^- \text{ for which the equilibrium constant}$$

$$K_{eq} = Fe^{3+} / Fe^{2+} \tag{2.20}$$

It is common in these general formulas not to include H^+ ions and e^- (Stumm and Morgan, 1981). Using reaction (2.24), the redox potential then assumes the form of a Nernst equation as follows:

$$E_h = E^0 + \frac{RT}{zF} \ln K_{eq} \tag{2.21}$$

Since R = gas constant, T = 298 K, z = 1, F = Faraday constant, and ln = 2.3 log, which are all known constant values, hence *(RT/zF)* ln = 0.059 log, and the redox potential can be changed into the following model:

$$E_h = E^0 + 0.059 \log Fe^{3+}/Fe^{2+} \tag{2.22}$$

ELECTRIC DOUBLE LAYERS

Because of the presence of electrical charges, colloidal materials in soil suspensions can attract ions with opposite charges. Negative clay surfaces will attract cations, whereas positively charged clay surfaces attract anions. Though the following discussion is for simplicity and based only on clay particles, double-layer reactions also occur with organic colloids, e.g., humic matter.

The cations are held on or near the clay surface, and all are free to exchange with other cations. The latter are called exchangeable cations. The negative charge on the clay surface and the swarm of positive counterions in the liquid phase are together called the *electric double layer.* At ZPC no attraction of counterions will take place, hence double layers do not exist. The counterions are attracted to the charged clay surface, but at the same time, are free to distribute themselves by diffusion throughout the solution phase. Attraction and diffusion will eventually come to an equilibrium, and the resulting distribution zone of counterions varies according to the existing theories on double layers. At present, four theories are available, e.g., Helmholtz theory, Gouy-Chapman diffuse double-layer, Stern double-layer, and Yates triple-layer theory (Tan, 1998).

Helmholtz Double Layer

This theory considers the countercharges to be concentrated in a plane parallel to the clay surface. The electric potential, ψ, developed at the solid-liquid interface, is assumed to decrease linearly across the double layer. Considered by many to be the same as an electrokinetic potential, ζ, it is usually formulated as follows:

$$\psi = \frac{4\pi\sigma x}{\varepsilon} \tag{2.23}$$

where σ = surface charge density, x = distance from the clay surface, and ε = dielectric constant of the medium. The value of σ is maximum at the solid surface and decreases with increasing distance from the solid-liquid interface to become zero at the border with the bulk solution, hence $\psi_o = 0$.

Diffuse Double Layer

In contrast to the Helmholtz theory, the diffuse double-layer theory, developed independently by Gouy (1910) and Chapman (1913), differs in the distribution of the counterions in the liquid phase, which are assumed to be dispersed as are the gas molecules in the earth's atmosphere, hence the name *diffuse double-layer* theory. As indicated earlier, the counterions, attracted by the negatively charged clay surface, are at the same time free to diffuse away into the bulk solution to eventually reach a state of equilibrium. Such a system obeys the equilibrium law as discussed previously, stating that the electrochemical potential, *E*, within the double layer *(i)* must equal that at the border or in the bulk solution *(o)*:

$$E_i = E_o$$
$$[RT \ln a + zF\psi]_i = [RT \ln a + zF\,\psi]_o \qquad (2.24)$$

Singh and Uehara (1998) assume that $\psi_i = \psi_o$, which is erroneous. At the border with the bulk solution ψ_o does not equal ψ_i, but $\psi_o = 0$, and equation (2.24) changes into:

$$[RT \ln a + zF\Psi]_i = [RT \ln a]_o$$
$$zF\Psi_i = [RT \ln a]_o - [RT \ln a]_i \quad \text{or} \quad (zF\Psi_i)/(RT) = \ln a_o/a_i \qquad (2.25)$$

Hence:

$$\ln \frac{a_i}{a_o} = -\frac{zF\psi_i}{RT} \rightarrow \frac{a_i}{a_o} = e - zF\psi_i\,/\,RT \quad \text{or} \quad a_i = a_o \exp[(-zF\psi_i)\,/\,(RT)]$$

Since activity $a = \gamma c$, and for dilute concentration $\gamma = 1$, the equation can also be written using concentration units, *c*, in moles L^{-1}:

$$c_i = c_o \exp(-\frac{zF\psi_i}{RT}) \quad \text{or} \quad c_i = c_o \exp(-\frac{zF\psi_i}{kT}) \qquad (2.26)$$

This equation is known as the *Boltzmann equation,* in which k = *Boltzmann constant,* defined as the gas constant *R* per molecule (or *R*/Avogrado number). It is a model showing the concentration of counterions (c_i) to decrease exponentially in the double layer. The statistical derivation resulting in a conclusion that the distribution of counterions obeys the Boltzmann equation brings about the implication that packing of counterions resulting in a flat border, as realized in the Helmholtz double layer, is subject to arguments. Using the law of equilibrium for the system in the Helmholtz double layer, a similar diffuse layer model will be obtained contradicting its flat border-packed layer concept.

The Surface-Charge Density

In a diffuse double-layer model, the surface-charge density, σ, is usually described as follows:

$$\sigma = \sqrt{[2\varepsilon\eta kT)\,/\,\pi]}\,\sinh \frac{ze\psi_i}{2kT} \qquad (2.27)$$

where ε = dielectric constant, η = concentration in number of ions per mL, k = Boltzmann constant, T = absolute temperature (K), z = valence, e = electron charge in esu, and ψ_i = surface potential in Statvolts. The diffuse layer charge, contributed by the counterions, equals σ, but is opposite in sign.

Thickness of the Double Layer

At present, two models are available for describing the diffuse double-layer thickness:

1. Gouy-Chapman model: $1/\chi = \sqrt{(\varepsilon kT)/(8\pi z^2 e^2 \eta)}$ (2.28)

2. Verwey and Overbeek (1948) model: $k = 3\times10^7 z\sqrt{C}$ (2.29)

where the symbols ε, k {except in (2.29)}, T, and z are as defined earlier; e = electronic charge = 4.8×10^{-10} esu, η = concentration in number of ions cm^{-3}, and C = concentration in mol L^{-1}. The value of $1/k$ is used by Verwey and Overbeek as a measure of the thickness of the diffuse double layer, and values of 0.5×10^{-7} to 1×10^{-5} cm have been reported depending on the factors z and C. Since counterions can be attracted only by charged surfaces, this double-layer thickness is apparently stable with permanent-charged minerals, but may fluctuate with variable-charged surfaces and double layers should disappear at ZPC.

Clay particles with thick double layers tend to repel one another and remain in suspension. On the other hand, thin double layers decrease the interparticle distance, making a close approach possible between clay particles. When the interparticle distance decreases to ≤20 Å, it is assumed that *Van der Waals* attraction dominates over the repulsive forces. This results in interparticle attraction causing clay to flocculate, a process that enhances soil-structure formation which is required for a good physical condition in soils.

Stern Double Layer

The diffuse double-layer theory has been developed for application on flat surfaces, but it may apply equally well to rounded or spherical surfaces (Verwey and Overbeek, 1948). However, since the counterions are assumed to be point charges, and, therefore, occupy no spaces, they may reach excessively high concentrations at the solid-liquid interface. Singh and Uehara (1998) indicate that even at surface potentials of 250 mV, excessively high values of counterions were predicted using the models described. Stern made corrections in the double-layer theory by taking into consideration the ionic dimensions and recognized two sublayers in the liquid phase: (1) the Stern layer, where the influence of ionic dimension was the greatest and the counterion distribution was similar to the Helmholtz theory, followed by (2) a diffuse layer with a counterion distribution obeying the Boltzmann equation. The counterion charges, σ_t, then can be distinguished into one for the Stern layer, σ_s, which decreases linearly with distance from the clay surface, and into another one, σ_d, which decreases according to the Boltzmann equation; hence $\sigma_t = \sigma_s + \sigma_d$, in which the subscripts t = total, s = Stern layer, and d = diffuse layer. The counterion charges in the diffuse layer, σ_d, can be calculated using the Gouy-Chapman surface-charge density, σ, formula, but are opposite in sign, whereas σ_s is solved by rearranging the Helmholtz potential: $\psi = (4\pi\sigma\chi)/\varepsilon$ as follows:

$$\sigma = \frac{\varepsilon}{4\pi\chi}\psi \tag{2.30}$$

The counterion charges $\sigma_s = \sigma$, but are opposite in sign.

At issue is why Stern considered dense packing a result of an increase in particle sizes. Basic soil physic rules that a decrease in particle size encourages dense packing. Hence the theory of point charges, though unacceptable by most scientists, is more in line with a development of a densely compacted Helmholtz or Stern layer. Perhaps the statistical models developed from the equilibrium law are insufficient for describing counterion interaction at the solid-liquid interface. In addition, the tendency of creating a *flat* double-layer border with large masses of ions is rather unlikely in systems where kinetic energy is involved with attraction opposed by diffusion dominating the reactions of moving ions. The ions close to the clay surface are more strongly influenced by the surface charge and tend to congregate there as a dense layer of ions with subdued kinetic energy. However, the influence sphere of the surface charge diminishes with increasing distance from the interface and the kinetic energy of the ions will increasingly become the dominating force. Unless this kinetic energy can be triggered off, it is unlikely that the large amounts of counterions can be arranged flat or parallel to the clay surface, as realized by Helmholtz.

Triple Layer

The double-layer theory has been developed for interactions of counterions with silicate clays exhibiting surface properties due to the presence of permanent electrical charges, whereas the triple-layer theory has been created for application with sesquioxides and other types of clays possessing variable charges. The surface properties of such clays are different from those of the silicate clays and their surface charges are usually attributed to adsorption of potential determining ions. This layer of potential determining ions forms the basis for the *triple-layer* model (Yates, Levine, and Healy, 1974), and is considered an integral part of the solid sesquioxidic clay surface (Kleijn and Oster, 1983). It is located, in fact, in the liquid phase, and regarded as the *effective* first layer. The two other layers are a Helmholtz, divided into an *inner* and *outer Helmholtz,* and a diffuse layer. In the absence of adsorption of potential determining ions, Yates, Levine, and Healy (1974) and Kleijn and Oster (1983) believe that a triple layer cannot develop, and in this case the Stern double-layer theory should be applied.

Fused Double Layer

The double-layer theories previously discussed assume that all colloidal particles in suspension are surrounded by individual layers of counterions. This is possible in very dilute suspensions such as those created in laboratory conditions, containing only a very small amount of particles. Such conditions allow the particles to remain in suspension as true individual particles separated from one another by relatively large interparticle distances. In natural conditions, in which puddling of soils causes relatively large amounts of clay to disperse, the clay particles, each exhibiting their electric double layers, are at close distances from one another. The double layers are, in fact, not repelling the clay particles, but two double layers confronting each other are more likely to fuse together to become just one layer. Consequently, two particles at close distance share a counterion layer, thus resulting in the surface of one clay being unable to distinguish whether the counterions belong to its own or to the neighbor's surface. Located or sandwiched between two adjacent particles, the counterions are also unable to distinguish to which surface they belong. This agrees with the concept of cation exchange dictating that Na^+ from one surface can freely exchange for Na^+ from the other surface. The process of sharing can be called *counterion bridging,* similar to *metal* or *water bridging* which is important in the interaction between negatively charged humic matter and clay (Tan, 2000; Stevenson, 1994). If metal bridging is accepted as a process for the interaction between two negatively charged particles, it is conceivable that counterion bridging can also occur between two clay particles.

ADSORPTION AND CATION EXCHANGE REACTIONS

Adsorption is a process of concentrating materials on colloidal surfaces and is one of the reactions attributed to the surface chemistry of soil colloids.

In soils, it is more the type of accumulation of materials at solid-water interfaces, such as with counterions in double layers.

Types of Adsorption

At the present, adsorption can be distinguished into (1) *specific adsorption,* and (2) *nonspecific adsorption.* Specific adsorption, also called chemisorption, is defined as a type of covalent bonding of solutes by the sorbents, whereas nonspecific adsorption is a reaction related to electrostatic bonding (Schwertmann and Taylor, 1977). In addition to such a division, adsorption processes are also recognized as *positive* and *negative.* Positive adsorption refers to accumulation of solutes on the clay surfaces, whereas negative adsorption is a process by which the solutes are repelled from the clay surface and tend to congregate in the bulk solution. The tendency exists to use specific adsorption for complexation of solutes by inner-sphere surfaces of clay minerals, and nonspecific adsorption for complexation of solutes by outer-sphere surfaces of clays (Sposito, 1989; Zachara and Westall, 1998). If a solute does not form a complex with the charged surface, it is adsorbed in the *diffuse-ion swarm.* The formidable statistics accompanying these new developments have convinced many scientists to consider the new theories as important advancements in the science of adsorption chemistry. However, to a larger number of other soil scientists they only result in making the subject very confusing. Questions have been raised about the presence of inner- and outer-sphere surfaces in clay minerals and especially in organic compounds, and what the difference is between a diffuse-ion swarm and ions *complexed* by outer-sphere surfaces. The complexation of ions by outer-sphere surfaces involves an electrostatic bonding mechanism (see Sposito, 1989, p. 131), which is, in essence, nonspecific adsorption, defined previously as attributed to electrostatic attraction. But, so is adsorption of the diffuse-ion swarm, making the problem even more complicated for distinguishing a difference between outer-sphere complexation and electrostatic adsorption of counterions producing the double layers as underscored by the Gouy-Chapman and Stern concepts. Complexation by inner-sphere surfaces adds to the confusion because in the triple-layer theory adsorption in inner spheres is limited to adsorption of potential determining ions only, creating the so-called effective surface, with electrical charges contributed by the adsorbed ions. All the unanswered questions have origins perhaps in regarding complexation similar to adsorption reactions. In basic chemistry, complexation reactions are usually considered to occur with certain cations, and in particular with the transition metals Al, Fe, Mn, Cu, and Zn binding organic compounds. These reactions, yielding the so-called metal-organo complexes, should be viewed as rather different from the regular adsorption of cations in a double-layer region of clay surfaces. The complexed ion usually assumes a central position and must be regarded as an integral part of the solid complex compound (Murmann, 1964; Mellor, 1964). Unless another definition is available, the concept of complexation as outlined in basic chemistry differs from that of adsorption, which emphasizes surface accumulation of counterions.

Adsorption Models

Several methods are available to study adsorption processes in soils; some are very simple, and others are very complex. Since adsorption is an equilibrium reaction, fundamental principles of soil reactions, such as the mass action law or the law of equilibrium, have been applied for interpretation of the process, which is considered as the scientific approach. Apparently, this method has yielded mixed results because of the extreme difficulties obtained when attempts are made to extend it by involving the double-layer concept. In contrast, other methods try to explain adsorption by just accepting the facts obtained without relating them to any basic chemical principle, which are called the empirical methods, e.g., Freundlich and Langmuir equations. Since the latter two are well-established models and closely related to each other, only the Langmuir model will be given here as an example:

$$\frac{x}{m} = \frac{k_1 C}{1 + k_2 C} \tag{2.31}$$

where x = amount adsorbed, m = amount of absorbents, k_1 and k_2 = constants, and C = concentration in equilibrium solution. At low concentrations the value of k_2C becomes so low compared to the factor 1, that it can be neglected, and the equation reverts to the Freundlich model. For a discussion on other classical adsorption models, e.g., BET and Gibbs, reference is made to Tan (1998).

Several scientists regard cation exchange reactions as identical to adsorption reactions, making them more complicated and harder to understand. Impressive names have been used to redistinguish adsorption, such as surface complexation nonelectrostatic model (SC-NEM), surface complexation-electric double-layer (SC-EDL), or the mechanistic approach and the semiempirical approach (Zachara and Westall, 1998). In the new approach, adsorption in inner- and outer-sphere surfaces is recognized as a complex surface reaction, and is defined as formation of a stable molecular unit when an aqueous species reacts with a surface functional group, as illustrated by the reaction models:

$$SOH + M^{m+} \leftrightarrows SOM^{(m-1)} + H^+ \quad \text{(inner sphere)} \tag{2.32}$$

$$SOH + M^{m+} \leftrightarrows SO^- - M^{m+} + H^+ \quad \text{(outer sphere)} \tag{2.33}$$

$$mCX_u + uM^{m+} \leftrightarrows uMX_m + mC^{u+} \tag{2.34}$$

in which SOH = hydroxylated surface sites on Al or Fe oxides, M and C = cations with valences m+ and u+, respectively, and X = fixed-charge sites on layer silicates. These models reflect explicitly cation exchange reactions, sending an improper message of what true adsorption reactions are, as exemplified by a classical adsorption (without exchange) of metals by a magnet or the accumulation of counterions in double layers. In addition, the large positive charge ($SOM^{(m-1)}$) created in reaction (2.32) will more likely repel cations from the outer-sphere surfaces. That these new concepts of adsorption are stretching the basics of soil chemistry a little too far can also be noticed in several other new models as exemplified by an adsorption model, called chemisorption by McBride (1999):

$$S\text{–}OH + M^{n+} = S - O - M - OH^{(n-2)+} + 2H^+ \tag{2.35}$$

The formation of "$-OH^{(n-2)+} + 2H^+$" from merely $S\text{–}OH + M^{n+}$ is beyond comprehension, making the model presented earlier for such a reaction by Zachara and Westall (1998) more acceptable. An even more confusing model of an alleged adsorption reaction of metals by hydroxylated surfaces of Fe oxide minerals is presented by McBride (1999) as follows:

$$[Fe - OH]^{-1/2} + M(H_2O)_6^{n+} \rightarrow [Fe - O - M(H_2O)_5]^{(n+3/2)+} + H_2O^+ \tag{2.36}$$

McBride (1999) claimed that this reaction is different from cation exchange by at least four features, among which are the "release of as many as nH^+ for each M^{n+}," and a "high degree of specificity by particular minerals for particular trace metals" (McBride, 1999, p. B-266). However, these are two factors controlling cation exchange reactions, instead of adsorption processes that seldom obey stoichiometric principles of exchange and selectivity. In addition, the negative charge (–½) on Fe-OH surfaces and H_2O^+ casts doubt on the validity of such a model, because such a statement infers the presence of a permanent negative charge uncommon in sesquioxide minerals, since negative charges are usually the result of dissociation of the exposed OH groups. The reaction is also unbalanced and keeps us wondering about the creation of (n + 3/2) + charge. The conversion of $M(H_2O)_6^{n+}$

into $M(H_2O)_5$ should reduce the n+ charges into (n - 1)+ charges. By assuming M = Al, the following reaction is provided as an example for such a reduction in charges:

$$Al(H_2O)_6^{3+} \leftrightarrows Al(H_2O)_5(OH)^{2+} + H^+ \quad (2.37)$$

Assuming the –½ charge of the Fe–OH is acceptable, McBride's reaction, irrespective of whether it represents an adsorption or exchange model, should then be written properly as:

$$[Fe-OH]^{-1/2} + M(H_2O)_6^{n+} \rightarrow [Fe-O-M(H_2O)_5(OH)]^{(n-2\frac{1}{2})+} + 2H^+ \quad (2.38)$$

Cation Exchange Reaction Models

A variety of models are available for cation exchange, e.g., models based on mass action law. These include the Donnan equilibrium law, thermodynamics, and the diffuse double-layer theory. Due to space limitations, only the *Kerr* and *Gapon* equations, the most frequently used models based on the mass action law, and a thermodynamic model will be given as examples based on the following exchange reaction:

$$2K^+ + \text{Ca-micelle} \leftrightarrows \text{2K-micelle} + Ca^{2+} \quad (2.39)$$

Kerr and Gapon Models

By applying the mass action law, Kerr indicates that the previous reaction can be described by the following equation:

$$\frac{[K^+]^2(Ca^{2+})}{[Ca^{2+}](K^+)^2} = k_{(Kerr)} \quad (2.40)$$

By taking the square root, equation (2.40) is called the Gapon equation:

$$\frac{[K^+](\sqrt{Ca^{2+}})}{[\sqrt{Ca^{2+}}](K^+)} = k_{(Gapon)} \quad (2.41)$$

where [ ] = adsorbed, and ( ) = free ions. The constants $k_{(Kerr)}$ and $k_{(Gapon)}$ are also called the *exchange constant*, k_{ex}, or *selectivity coefficient.* For more details reference is made to Tan (1998). Recently, a more complicated version of Kerr's equation has been presented (Zachara and Westall, 1998):

$$K_{ex} = \{(f_{MXm}X_{MXm})^u\,[C^{u+}]^m\} \,/\, \{(f_{CXu}X_{CXu})^m\,[M^{m+}]^u\} \quad (2.42)$$

where X = mole fraction, [ ] = adsorbed, and ( ) = single ion concentration with f = activity coefficient. Under ideal conditions, Zachara and Westall (1998) consider $f_{MXm} = f_{CXu} = 1$, and the model reverts back to the classical Kerr's equation. This raises the question as to why such a model should be given in the first place, when the Kerr's model is to be used anyway. In addition, the concentration of X is normally in mol L^{-1} when ion activities (fX or γX) are to be used in nonideal conditions, and mole fractions are seldom employed. Nevertheless, when activity units can be applied to the free ion concentrations in equilibrium with the adsorbed fraction, such an approach can be considered a refinement of the Kerr or Gapon equations, which then assume the following models:

$$\frac{[K^+]^2(\gamma_{Ca}Ca^{2+})}{[Ca^{2+}](\gamma_K K^+)^2} = k_{ex(\text{Kerr})} \qquad \frac{[K^+](\sqrt{\gamma_{Ca}Ca^{2+}})}{[\sqrt{Ca^{2+}}](\gamma_K K^+)} = k_{ex(\text{Gapon})} \tag{2.43}$$

where [ ] and ( ) = adsorbed and free ion concentrations, respectively, γ = activity coefficients, and k_{ex} = exchange constant or selectivity coefficient. The ratio:

$$\frac{(\gamma_K K^+)}{(\sqrt{\gamma_{Ca}Ca^{2+}})} \tag{2.44}$$

is used by Beckett (1964) under the name *activity ratio* for the formulation of his quantity/intensity (Q/I) relation in the study of soil potassium status.

Thermodynamic Model

Kinetic or mass action equations of cation exchange reactions have often been confused for thermodynamic models. The only relation that the two models have in common is the application of the law of equilibrium illustrated as follows. For exchange reaction (2.39), the free energy change of the reaction, ΔG_r, can be written as:

$$\Delta G_r = \Delta G_r{}^o + RT \quad \ln \frac{[K^+]^2(Ca^{2+})}{[Ca^{2+}](K^+)^2} \tag{2.45}$$

in which $\Delta G_r{}^o$ = free energy change at standard state

$$\Delta G_r = \Delta G_r{}^o + RT \textit{ in } k_{ex(\text{Kerr})} \text{ or } \Delta G_r = \Delta G_r{}^o + 1.364 \log k_{ex(\text{Kerr})} \tag{2.46}$$

At equilibrium condition $\Delta G_r = 0$, hence:

$$\Delta G_r^o = -1.364 \log k_{ex(\mathrm{Kerr})} \text{ at } 298\mathrm{K} \quad (2.47)$$

SUMMARY

Selected chemical processes have been discussed in relation to electrochemical properties of soil constituents. The electrical charges of both the organic and inorganic materials were emphasized contributing toward a number of soil chemical reactions that deviate from the conventional concept which considers clay only as the seat of soil chemical activity. The importance of variable charges in organic compounds, e.g., humic acids and amino acids, has been addressed to underscore their prominent effect on soil processes. It is an established fact that soils with high humus content always exhibit higher CECs than soils low in humus. This is due to the contribution of the huge charge properties of the organics, which many soil scientists are still questioning. Theories and models on electric charges, double layers, adsorption and cation exchange have been examined, and explanations submitted to distinguish clearly between adsorption, cation exchange, and complex reactions. An attempt has been made to translate into simpler language the complex and very confusing new theories on adsorption and cation exchange without the overuse of statistics. New concepts and revised models have been presented to illustrate the various suggestions and corrections.

REFERENCES

Bartlett, R. J. (1998). Characterizing soil redox behavior. In *Soil Physical Chemistry,* Second Edition, ed. D. L. Sparks. Boca Raton, FL: CRC Press, pp. 371-395.

Beckett, P. H. T. (1964). Studies on soil potassium. I. Confirmation of the ratio law: Measurement of potassium potential. *Journal of Soil Science 15:* 1-8.

Brady, N. C. (1990). *The Nature and Properties of Soils.* Tenth Edition. New York: Macmillan.

Chapman, D. L. (1913). A contribution to the theory of electrocapillarity. *Philosophy Magazine* 25(6): 475-481.

Eswaran, H. and J. Tavernier. (1980). Classification and genesis of oxisols. In *Soils with Variable Charge,* ed. B. K. G. Theng. Lower Hutt: New Zealand Society of Soil Science, pp. 427-442.

Fairbridge, R. W. and C. W. Finkl Jr. (1979). *The Encyclopedia of Soil Science.* Part 1. New York: Van Nostrand Rheinhold.

Goldberg, S., I. Lebron, and D. L. Suarez. (1999). Soil colloidal behavior. In *Handbook of Soil Science,* ed.-in-chief M. S. Sumner. Boca Raton, FL: CRC Press, pp. B-195-B-240.

Gouy, G. (1910). Sur la constitution de la charge électrique à la surface d'un électrolyte. *Annals of Physics (Paris) Ser. 4*(9): 457-468.

Guggenheim, S. and R. T. Martin. (1995). Definition of clay and clay mineral. Joint report of the AIPEA nomenclature and CMS nomenclature committees. *Clays and Clay Minerals 33:* 99-105.

Hurst, V. J. and S. M. Pickering Jr. (1997). Origin and classification of Coastal Plain kaolins, Southeastern USA, and the role of groundwater and microbial action. *Clays and Clay Minerals 45:* 274-285.

Hurst, V. J., P. A. Schroeder, and R. W. Styron. (1997). Review. Accurate quantification of quartz and other phases by powder x-ray diffractometry. *Analytica Chimica Acta 337:* 233-252.

Isbell, R. F. (1980). Genesis and clays of low activity clay alfisols and ultisols. In *Soils with Variable Charge,* ed. B. K. G. Theng. Lower Hutt: New Zealand Society of Soil Science, pp. 397-410.

Kleijn, W. B. and J. D. Oster. (1983). Effects of permanent charge on the electric double layer properties of clays and oxides. *Soil Science Society of America Journal 47:* 821-827.

Lobartini, J. C., K. H. Tan, J. A. Rema, A. R. Gingle, C. Pape, and D. Himmelsbach. (1992). The geochemical nature and agricultural importance of commercial humic matter. *Science of Total Environment 113:* 1-15.

Mackenzie, R. C. (1975). The classification of soil silicates and oxides. In *Soil Components,* Volume 2, Inorganic Compounds, ed. J. E. Gieseking, New York: Springer-Verlag, pp. 1-25.

McBride, M. B. (1999). Chemisorption and precipitation reactions. In *Handbook of Soil Science,* ed.-in-chief M. S. Sumner. Boca Raton, FL: CRC Press, pp. B-265-B-302.

Mehlich, A. (1981). Charge properties in relation to sorption and desorption of selected cations and anions. In *Chemistry in the Soil Environment,* eds. R. H. Dowdy, J. A. Ryan, V. V. Volk, and D. E. Baker. Madison, WI: American Society of Agronomy, Soil Science Society of America, pp. 47-75.

Mellor, D. P. (1964). Historical background and fundamental concepts. In *Chelating Agents and Metal Chelates,* eds. F. P. Dwyer, and D. P. Mellor. New York: Academic Press, pp. 1-50.

Miller, R. W. and D. T. Gardiner. (1998). *Soils In Our Environment.* Upper Saddle River, NJ: Prentice-Hall.

Murmann, R. K. (1964). *Inorganic Complex Compounds.* New York: Holt-Reinhold.

Nebergall, W. H., F. C. Schmidt, and H. F. Holtzclaw Jr. (1972). *College Chemistry.* Lexington, KY: Heath and Company.

Orlov, D. S. (1985). *Humus Acids of Soils.* New Delhi: Amerind Publishing Company.

Pauling, L. (1964). *College Chemistry.* San Francisco, CA: Freeman and Company.

Ranville, J. F. and R. L. Schmiermund. (1998). An overview of environmental colloids. In *Perspectives in Environmental Chemistry,* ed. D. L. Macalady. New York: Oxford University Press, pp. 25-45.

Scharpenseel, W. H. (1966). Aufbau und Bindungsform der Ton-Huminsäure-Komplexe. 2. Hydrothermale Synthese von Ton-Humussäure, sowie anderer organo-mineralischer Komplexe. Untersuchungen am Röntgengerät, I.R. Spectrometer und Elektronenmikroskop. *Zeitschrift Pflanzenernährung, Düngung und Bodenkunde 114:* 3-8.

Schwertmann, U. and R. M. Taylor. (1977). Iron oxides. In *Minerals in Soil Environment,* eds. J. B. Dixon, S. W. Weed, J. A. Kittrick, M. H. Milford, and J. L. White. Madison, WI: Soil Science Society of America, pp. 145-180.

Singh, U. and G. Uehara. (1998). Electrochemistry of the double layer: Principles and application to soils. In *Soil Physical Chemistry,* Second Edition, ed. D. L. Sparks. Boca Raton, FL: CRC Press, pp. 1-46.

Soil Science Society of America. (1996). *Glossary of Soil Science Terms.* Madison, WI: Author.

Sposito, G. (1989). *The Chemistry of Soils.* New York: Oxford University Press.

Stevenson, F. J. (1994). *Humus Chemistry, Genesis, Composition, Reactions.* Second Edition. New York: Wiley and Sons.

Sticher, H. and R. Bach. (1966). Fundamentals in the chemical weathering of silicates. *Soils and Fertilizers 29:* 321-325.

Stumm, W. and J. J. Morgan. (1981). *Aquatic Chemistry. An Introduction Emphasizing Chemical Equilibria in Natural Waters.* Second Edition. New York: Wiley and Sons.

Tan, K. H. (1998). *Principles of Soil Chemistry.* Third Edition. New York: Marcel Dekker.

Tan, K. H. (2000). *Environmental Soil Science.* Second Edition. New York: Marcel Dekker.

Theng, B. K. G. (1972). Formation, properties, and practical application of clay-organic complexes. *Journal Royal Society New Zealand 2:* 437-457.

Verwey, E. J. W. and J. T. G. Overbeek. (1948). *Theory of the Stability of Lyophobic Colloids.* New York: Elsevier.

Yates, D. E. S., S. Levine, and T. W. Healy. (1974). Site-binding model of the electric double layer at the oxide-water interface. *Journal Chemical Society Faraday Transactions I, 70:* 1807-1819.

Zachara, J. M. and J. C. Westall. (1998). Chemical models of ion adsorption in soils. In *Soil Physical Chemistry,* ed. D. L. Sparks. Boca Raton, FL: CRC Press, pp. 47-95.

Chapter 3

Biological Processes

Paolo Nannipieri
Luigi Badalucco

Soil plays a fundamental and irreplaceable role in the biosphere because it governs plant productivity of terrestrial ecosystems, and allows the completion of the biogeochemical cycles. Soil and microorganisms inhabiting soil degrade, sooner or later, all organic compounds including more recalcitrant ones. The aim of this chapter is to discuss soil as a biological system, including its properties, its organisms, the processes carried out by these organisms and the relative kinetics, the main factors affecting these processes, and the approaches to model biological processes. Moreover, we shall try to indicate directions for a better mechanistic understanding of microbial diversity and activity in soil. It is not possible to prepare an exhaustive review as the complexity and vastness of the treated matter exceeds the limits of a single chapter. Therefore, the presentation summarizes the various areas of the topic without a detailed discussion of the underlying mechanisms. Relevant reviews are cited more than the original literature. For further knowledge of mechanisms involved in treated topics, the readers may consult the cited books and reviews.

SOIL AS A BIOLOGICAL SYSTEM

Soil organisms and plants are present everywhere, even in cold (boreal and alpine) or dry (desert) climates. In spite of the fact that soil is generally poor in nutrients and energy sources and microbial life exists in hot spots, it contains more genera and species of microorganisms than other microbial habitats probably as a result of its microbial versatility and adaptation, both physiologically and genetically, to varying environmental conditions (Stotzky, 1997). Torsvik, Sorheim, and Goksoyr (1996) calculated a mean number of about 3.8×10^6 bacteria $clone^{-1}$ g^{-1} dry weight soil by analyzing the reassociation kinetics of complementary strands of DNA extracted from soil.

Organisms inhabiting soils are generally grouped, according to their size and structure, into macrofauna, mesofauna, microfauna, and microbiota. Viruses are the smallest (they can range from 1,000 to 10,000 nm) and simplest biological entities inhabiting soil. They consist of only one type of nucleic acid (RNA or DNA) usually surrounded by protein coats (capsides) (Payne, 1988). Therefore, attractive interactions between viruses and soil colloids are those (dispersion forces, electrostatic interactions, hydrogen bonding, and hydrophobic interactions) of proteins and amino acids with soil colloids (Farrah and Bitton, 1990). Because viruses contain little, if any, metabolic machinery, they are unable to autonomously replicate and must first invade a living host cell. Thus they are grouped into animal, plant, fungal, or bacterial parasites. Viruses bound on clay minerals have been shown to retain their lytic activity (Stotzky, 1986).

Soil microbiota include bacteria, fungi, protozoa, and algae. Bacteria and fungi are generally the most abundant in soil; for example, in a temperate grassland soil the bacterial and fungal biomass amounted to 1-2 and 2-5 t ha^{-1}, respectively (Killham, 1994). Although fewer in number than viruses, and constituting lower biomass than soil fungi, bacteria are nevertheless the most metabolically significant group of organisms in soil. They include actinomycetes and blue-green algae (cyanobacteria). They are assigned to the lower group of kingdom Protista, and are called protokaryotes because they lack a true nucleus. The only chromosome consists of a closed circular loop within the cytoplasm and it is essential for survival and replication of bacteria. In addition to chromosomal DNA, bacteria present small circular DNA molecules, called plasmid DNA, which is characterized by the possibility of being transferred and replicated independently from chromosomal DNA. Many plasmid genes are cryptic because they do not code for any function, whereas other crucial functions such as nitrogen fixation, degradation of recalcitrant organic compounds, and heavy metal or antibiotic resistance are coded by plasmid genes. Nearly all bacteria are surrounded by a cell wall (with the exception of mycoplasma), which contain peptidoglican, a polymer not found in higher organisms (Paul and Clark, 1996). This is a very peculiar polymer containing the enantiomer D-alanine. For this reason, the presence of D-alanine in soil samples may constitute an indicator of the presence of bacteria (Tunlid and White, 1992; Hopkins and O'Dowd, 1997).

Cells of fungi, protozoa, and all other algae except the blue-green algae have a true nucleus surrounded by a membrane; inside the nucleus are the double helixes of DNA. These microorganisms are called eukaryotes.

Bacteria and fungi are highly versatile; they are capable of carrying out almost all known biological reactions. According to Coleman and Crossley (1996), bacteria and fungi are the main decomposers of organic residues entering soil and they differ in their type of growth and exploration of food. Fungi present a central body with long hyphae which can grow and explore and interconnect many different microhabitats. They can decompose organic polymers of organic debris by secreting an array of extracellular enzymes. The released monomers or oligomers are taken up by the hyphae and translocated back through the central body. On the other hand, individual bacterial cells are grouped together in small colonies and can be moved to a different site by water drainage, root growth, or if ingested and then released by soil fauna. Active movement in water films only occurs for bacteria with flagella. Actinomycetes are other prokaryotes resembling gram-positive bacteria in cell diameter and cell wall composition but with a mycelial morphology smaller than that of fungi.

Soil faunas are generally classified according to the body length (Table 3.1) (Coleman and Crossley, 1996). Soil animals can range from microflagellates (1-2 μm) to giant earthworms, whose length can reach several meters. The microfauna (protozoa and small nematodes) usually inhabits water films whereas mesofauna occupies air-filled pores. The macrofauna has the ability to create its own living space by burrowing if the species are endogenic (Capowiez and Belziences, 2001). Indeed, to recolonize a burrow system may be less advantageous for an endogenic species because the use of an abandoned burrow system may only guide the earthworm to already-foraged patches. On the contrary, for anecic earthworm species it seems more advantageous to colonize an abandoned burrow system since it represents a way to save energy and to protect the organism in an unknown environment (being both a shelter and a way to reach the surface). It has been hypothesised that an earthworm can perceive the vicinity of a burrow and the presence of another earthworm by detecting a chemical product, such as "allomone" of the mucus, or the vibrations made by the earthworm occupying the burrow, or the decrease in the soil density near the burrow

TABLE 3.1. Classification of Soil Fauna

Microfauna (up to 0.16 mm)	Protozoa, small Nematodes, small Rotifera, small Tardigrada, small Acari
Mesofauna (from about 0.16 to 10.4 mm)	Insecta, Opiliones, Chelonethi, Diplura, Protura, Collembola, Araneida, Acari, Chilopoda, Diploda, Tardigrada, Isopoda, Mollusca, Rotifera, Enchytraeidae, Nematoda
Macrofauna (longer than 10.4 mm)	Insecta, Chilopoda, Diploda, Insectivora, Isopoda, Mollusca, Enchytraeidae, Lumbricidae

Source: Adapted from Coleman and Crossley, 1996.

wall. The burrowing activity of macrofaunas can have a great influence on gross soil structure (Lee, 1985).

The first characteristic of soil as a biological system is that it is a structured, heterogeneous, and discontinuous system with organisms living in discrete microhabitats (Nannipieri, Ceccanti, and Grego, 1990). The chemical, physical, and biological characteristics of these microhabitats differ both in time and space. Even if the available space is extensive in soil, the "biological space," that is the space occupied by living microorganisms, represents a small proportion, generally lower than 5 percent of the overall available space (Ingham et al., 1985). Indeed, only a few microhabitats present the right set of conditions (available nutrients and energy sources, optimal values of temperature, pH, etc.). According to Hattori (1973) almost 80 to 90 percent of the microorganisms inhabiting soil are located on solid surfaces. Mechanisms by which these microorganisms interact with soil surfaces are not always clear (Stotzky, 1986). Some bacterial cells produce extracellular polysaccharides that interact with clay particles and these clay-polysaccharide complexes can persist even after the microbial death (Chen, 1998; Huang and Bollag, 1998). Electron microscopy techniques with staining procedures have enabled scientists to locate the position of microbial groups, and inorganic and organic colloids in the soil matrix (Forster, 1994; Assmus et al., 1995, 1997).

The domination of the solid phase is another distinctive characteristic of soil with respect to other microbial habitats. This solid phase consists of particles of different sizes, made up of living organisms, inorganic (minerals or amorphous materials), and organic (plant, animal, and microbial residues and humic matter) components, as independent entities and mixed conglomerates. Solid particles are surrounded by aqueous and gaseous phases, the amount and the composition of which fluctuate markedly in time and space. All microorganisms that are aquatic organisms live in water films and their activity depends on the availability of nutrient and energy sources in these water films (Stotzky, 1986, 1997). Since water films generally surround surface-active particles, the distribution of microorganisms in soil is restricted to clay, oxides, hydroxides, or organic particles, but not sandy or silt particles. The status of water molecules changes with the distance from the surface-active particles; it is an ordered film adjacent to the negatively and positively charged surfaces of soil colloids and to their charge-compensating ions. By increasing the distance from these sites, the ordering of water decreases until a distance is reached at which water is no longer held under the attraction of the charges and it is susceptible to gravity (Sposito, 1984). The thickness of the water layer depends on the surface activity of microaggregates (resulting from the cluster of surface-active particles), macroaggregates (resulting from the cluster of microaggregates), or clusters of aggregates and water bridges which can be formed between two different conglomerates. The relative hydrophobicity of some soil colloids can be important in modifying the structure of water films and in affecting the interactions of microorganisms with soil colloids (Stotzky, 1986). Hydrophobic regions, which are the result of the presence in the organic matter of lipids, waxes, and hydrophobic moietes, can interact with hydrophobic regions of the microbial surfaces or may render inorganic particles, such as clay minerals, hydrophobic when complexes between these inorganic and organic components are formed (Stotzky, 1986).

The way in which soil particles are arranged or grouped spatially constitutes soil structure. The link between biological activity and soil structure is strict for several reasons (Ladd et al., 1996). First, both stabilization and degradation of soil structure depends on biological activity. Microaggregates, with diameters of less than 250 μm, are formed by particles held together by different types of bonds, and those held together by polysaccharide-based glues are probably the most important. These polysaccharides can be of plant (mucilages, produced by roots) or microbial (bacterial exopolysaccharides) origin. In response to the formation of soil structure, pores of different sizes are formed and these pores can have different functions (Cambardella and Elliott, 1993; Ladd et al., 1996). Macropores (diameter $> 2 \times 10^{-5}$ m) are responsible for drainage and aeration of soil and are characterized by the presence of roots, and live mesofauna and macrofauna. Mesopores (1×10^{-7}-2×10^{-5} m) contain the available plant water, bacteria, fungi, and root hairs (Ladd et al., 1996). Micropores ($<1 \times 10^{-7}$m) are important for the adsorbed and intercrystalline water.

The adsorption of important biological molecules by surface-active particles is another peculiarity of soil as a biological system. Enzymes can be adsorbed by clay minerals or entrapped by humic molecules and can maintain their activity because they are protected against proteolysis, thermal and pH denaturation (Nannipieri, Sequi, and Fusi, 1996). In this way, stabilized extracellular enzymes can still be active under conditions unfavorable for the activity of soil microorganisms (Nannipieri, Kandeler, and Ruggiero, 2001). Adsorption and binding of DNA on humic molecules, clay and sand particles can protect the adsorbed DNA against degradation by nucleases without inhibiting its transforming ability (Lorenz and Wackernagel, 1987; Khanna and Stotzky, 1992; Pietramellara et al., 1997). Transformation is a mechanism of gene transfer in soil involving a competent bacterial cell, which can take up one of the two strands of extracellular DNA and insert it in the bacterial chromosomal DNA. In such a way the competent cell acquires all or part of the genes associated with the extracellular DNA. The fact that DNA can be adsorbed by soil colloids and protected against microbial degradation has important implications on the use of genetically modified organisms in the terrestrial ecosystems. The recombinant DNA may survive longer than expected if adsorbed by soil colloids or englobated in the genome of soil microbiota.

According to Stotzky (1997), 15 environmental factors affect the ecology, activity, and population dynamics of microorganisms in soil. These factors include: carbon and energy sources, mineral nutrients, growth factors, ionic composition, available water, temperature, pressure, air composition, electromagnetic radiation, pH, oxidation-reduction potential, surfaces, spatial relationships, genetics of the microorganisms, and interaction between microorganisms. It is not possible to discuss the effects of each factor on microbial activities because of space limitations. An example of the complex interrelationships of soil microbial ecology is the effect of the surface-active particles on the availability of organic nutrients in the surface soil layer where plant roots are generally located. The mechanism, by which organic molecules are bound to clays, oxides, hydroxides, and organic particles, determines the tenacity with which these compounds are held and the ability of microbes, usually with the aid of extracellular enzymes, to use these molecules as substrates. It has been hypothesised that surface-active particles, such as clay minerals, concentrate these substrates at the solid-liquid interface; this enrichment can support the growth of microorganisms adsorbed on these surfaces. Indeed, particle-sized fractionation of dispersed soil commonly reveals a concentration of soil organic matter and microbial biomass in fine silt-sized/coarse clay-sized materials (Ladd et al., 1996). Protection of adsorbed microorganisms against predators (Ladd et al., 1996) and protection of organic substrates, such as proteins, peptides, amino acids, polysaccharides, nucleic acids, and nucleotides against microbial degradation (Nannipieri, Sequi, and Fusi, 1996) do not support the hypothesis that adsorbed organic substrates sustain growth of microorganisms adsorbed on active surface particles (Stotzky, 1986).

BIOLOGICAL PROCESSES OCCURRING IN SOIL AND THEIR KINETICS

All known metabolic processes can be carried out by microorganisms inhabiting soil. Organic compounds, even those that are more recalcitrant, are mineralized to CO_2 which returns to the atmosphere, thus closing the cycle that started with the photosynthetic carbon assimilation by plants and the other photosynthetic organisms of terrestrial ecosystems. Synthetic organic pollutants, whose molecular structures differ considerably from natural compounds, can persist for longer periods of time but then are degraded in soil (Dobbins, Marjorie, and Pfaender, 1992). Two different hypotheses have been postulated for explaining the degradation of more dissimilar organic pollutants. The first hypothesis supposes the acquisition of novel degradation pathways by soil microorganisms based on the occurrence of random mutations followed by selection of microorganisms capable of degrading the synthetic compound. This acquisition requires a cluster of "degradative" genes on a single genetic unit, such as a plasmid or a bacteriophage. Indeed, numerous genes involved in the degradation of organic compounds are plasmid-encoded. Bacterial plasmids can be replicated independently of their host chromosome, and they lose whole parts due to the presence of transposable elements and accept insertion sequences, including transposons (Nannipieri et al., 2000). For this reason, and because of the briefness of the replication time (less than one hour under favorable environmental conditions), bacteria are among the most versatile microorganisms of soil. As an example of this versatility, Table 3.2 reports some of the metabolic processes carried out by bacteria with the relative carbon sources, H donors, and H acceptors.

The other hypothesis interpreting the broad metabolic capacity of soil is based on the presence of nonspecific enzymatic systems such as the lig-

TABLE 3.2. Some Metabolic Processes Carried Out by Bacterial Groups with the Relative Carbon Sources, H Donors, and H Acceptors

Bacterial group	**Metabolic process responsible used as energy source**	**Carbon source**	**H donor**	**H acceptor**
Cyanobacteria	Photosynthesis	CO_2	H_2O	$NADP^+$
Aerobic heterotrophs	Respiration	Organic compounds	Organic compounds	Oxygen
Anaerobic heterotrophs	Fermentation	Organic compounds	Organic compounds	NAD^+
Autotrophic nitrifiers	Chemosynthesis	CO_2	NH_3	Oxygen
Autotrophic thiobacilli	Anaerobic respiration	CO_2	H_2S	Oxygen
Denitrifiers	Anaerobic respiration	Organic compounds	Organic compounds	Nitrate
Desulphurizers	Anaerobic respiration	Organic compounds	Organic compounds	Sulphate

Source: Adapted from Killham, 1994; Richards, 1987.

nolytic system of white rot fungi. This system is capable of promoting not only the degradation of the most recalcitrant natural organic polymer (the lignin) but also the polycyclic aromatic hydrocarbons (PAHs) and their derivatives (polychlorinated biphenyl and even dioxin) (Jian and Tso, 1996).

For most recalcitrant compounds, such as pesticides, the degradation pathways in axenic culture under laboratory conditions are largely unknown, as well as the metabolic intermediates and the enzymes involved in most of these pathways (Bollag and Liu, 1990). Various biochemical mechanisms of detoxification can be present when microorganisms degrade organic pollutants, such as halogenated organic compounds, which are added as herbicides, insecticides, fungicides, fumigants, solvents, and other products (Hicks, Stotzky, and Van Voris, 1990; Miller et al., 1997). The microbiological recalcitrance of halogenated compounds is related to the number, type, and position of the halogen substituents and degradation of these compounds depends on the dehalogenation of the molecule (Fetzner and Lingens, 1994). The carbon-halogen bond is cleaved in reactions catalyzed by specific enzymes called dehalogenases. Seven different mechanisms of dehalogenation are known in bacteria (Fetzner and Lingens, 1994):

1. Reductive dehalogenation, which includes essentially two distinct processes: (a) the halogen is replaced by hydrogen; (b) the removal of two halogen substituents from adjacent C atoms with the formation of an additional bond between the C atoms (dihaloelimination or vicinal reduction)
2. Oxygenolytic dehalogenation, where one or two oxygen atoms are incorporated in the molecule in a reaction catalyzed by monoxygenases or dioxygenases
3. Hydrolytic dehalogenation, a reaction catalyzed by halidohydrolases with substitution of the halogen by a hydroxy group derived from a water molecule
4. Thiolytic dehalogenation by which halogen is removed through a reaction catalyzed by a glutathione S-transferase
5. Intramolecular substitution, in which the halogen removal involves the formation of an epoxide
6. Dehydrohalogenation, with elimination of a HCl molecule and formation of a double bond
7. Hydratation catalyzed by a hydratase with addition of a water molecule to a double bond and elimination of the halogen

Microbial consortia in soil are more efficient in degrading recalcitrant compounds than individual microbial species. In a microbial consortium the capabilities of individual species overlap and are integrated in such a way that the consortium as a whole has considerably greater potential than do any of the single species on their own (Zeikus, 1983). Bacteria, yeast, and fungi, which are responsible for the initial transformation of the primary substrates, such as biopolymers or hydrocarbons to monomers or metabolic intermediates, are associated to anaerobic bacteria, which can transform the intermediates to fermentative and low-molecular-weight products (Zeikus, 1983). The initial decomposer can be aerobic or facultative and, under aerobic conditions, its oxygen-scavenging activity creates, inside the consortium, the anaerobic conditions allowing the activity of the anaerobic bacteria. The microbial consortium-degrading cellulose includes cellulolytic fungi degrading the organic polymer to organic acids, which support the activity of noncellulolytic fungi (Trevors and van Elsas, 1997).

In soil, the degradation and polymerization of organic compounds is responsible for the formation of humic molecules, which are characterized by a complex polymeric structure (largely unknown) with aromatic and aliphatic moieties (Stevenson, 1986). These molecules are slowly degraded and thus their "age," estimated by radiocarbon dating, can range from 250 to over 3,000 years (Stevenson, 1986).

Processes such as mineralization of organic carbon or organic nitrogen to CO_2 and NH_4^+, respectively, can be carried out by a variety of microbial species. Other processes such as nitrification can be carried out by a small group of microorganisms, chemolithotrophic nitrifiers (Table 3.2), which use the energy derived in the oxidation of NH_4^+ to NO_2^- and NO_3^- to assimilate the CO_2. Autotrophic nitrifiers can be considered "keystone species" and their disappearance, because of negative impacts by pollutants or bad agricultural management, can compromise the metabolic capacity of soil (Nannipieri et al., 2000). However, biological production of NO_3^-, long believed to be carried out only by the small group of autotrophic nitrifiers, can also occur through the activity of heterotrophic bacteria and fungi, which are capable of oxidizing NH_4^+ and certain organic nitrogen compounds to substituted hydroxylamines, nitroso compounds and, eventually, to NO_3^- (Knowles, 1986). The reactions occurring into heterotrophic nitrifiers are not ATP-coupled and thus they do not provide energy. The finding that the methane monooxygenase (the enzyme catalyzing the oxidation of CH_4 to CH_3OH) of the methanotrophic bacteria can also oxidize NH_4^+ to NH_2OH, the first intermediate of the autotrophic nitrification process, further shows the complex metabolic activity of soil and the enormous capacity of this system to metabolize any compound that reaches soil.

The Michaelis-Menten kinetics have been used to describe the enzyme kinetics in soil according to the following relationship:

$$V = V_{max}[S] / K_m + [S] \tag{3.1}$$

where V_{max} is the maximum reaction rate, K_m is the Michaelis-Menten constant, expressing the affinity of the enzyme toward the substrate, and S is the substrate concentration (Lehninger, Nelson, and Cox, 1993).

This relationship is based on several assumptions and two of these cannot be fulfilled in soil, even with the present enzyme assays involving controlled incubation conditions, such as excess of substrate, optimal pH, and temperature values (Nannipieri and Gianfreda, 1998). Soil is not a physically homogeneous system, as is a solution, and there is more than one enzyme catalyzing the same reaction. Soil extracts have been shown to contain at least two phosphomonoesterases or two proteases active against benzoylargininamide and the K_m and V_{max} values of the two enzymes have been calculated (Nannipieri et al., 1982). Nevertheless, the description of enzyme kinetics of soil slurries or soil columns by the Michaelis-Menten theory has allowed the determination of the K_m values of several enzymes; these values seem to have some ecological and agronomic significance. For example, the K_{m} values of some enzymes decrease with changes in cropping systems (Nannipieri and Gianfreda, 1998). Probably, the calculated values represent weighted values of the various constants of enzymes involved in the measured activities with an unknown factor. Future research should address and measure the contribution of main enzymes catalyzing the same reaction to the overall K_m (Nannipieri and Gianfreda, 1998).

The prevalence of heterogeneous conditions in soil imposes a series of problems, which need to be taken into account for the analytical interpretation of enzyme kinetics (Nannipieri and Gianfreda, 1998). Steric limitations on the penetration of the substrate into the enzymatic active site can affect the K_m of the enzyme and this effect is more important for the high-molecular weight substrates. Intracellular enzymes of microbial cells attached to soil particles as well as extracellular enzymes adsorbed by clay particles or entrapped in humic molecules are subject to a series of microenvironmental effects (Nannipieri and Gianfreda, 1998). For example, if the enzyme is surrounded by a charged support, the concentration of a substrate with an opposite charge will be higher in the microenvironment surrounding the enzyme than that of the soil solution. An opposite behavior will occur with a substrate with the same charge of the microenvironment surrounding the enzyme. The adsorption of the p-nitrophenyl phosphate, the substrate for phosphomonoesterases, by soil colloids was quantified by the Freundlich equation and considered for calculating the true K_m of acid phosphomonoesterase in soil (Cervelli et al., 1973; Trazar-Cepeda and Gil-Sotres, 1988).

The previous examples concern enzyme kinetics involving soluble substrates but substrates are often insoluble in soil. Cellulose, starch, lignin, urea, etc., are present in a solid state and their hydrolysis involves the adsorption of microorganisms or hydrolytic enzymes. According to McLaren and Skujins (1971) the velocity of the enzyme reaction, when the substrate is insoluble, is given by:

$$v = K^*[E_0]n \qquad (3.2)$$

where K^* is a constant related to the distribution coefficient of the enzyme between the solution and the substrate surface, E_0 is initial enzyme concentration, and n is a constant (less than unity). Obviously, the rate of the enzyme reaction depends on the surface of the insoluble substrate.

Burns (1982) proposed an elegant microbial strategy, based on the presence of clay- or humic-enzyme complexes, for the hydrolysis of insoluble organic polymers into soluble and available monomers. The probabilities for the microbial cells to detect these insoluble substrates by direct contact are rather scarce and the release of extracellular enzyme hydrolyzing the insoluble polymer presents several disadvantages. Indeed, the enzyme can be proteolytically degraded in the extracellular soil environment. Even if the enzyme reaches the insoluble substrate, the environmental conditions (mostly pH and temperature) cannot be optimal for the enzyme reaction. If the reaction occurs, an opportunistic cell can use the released soluble reaction products. The presence of the extracellular and immobilized enzymes in soil assures the hydrolysis of these insoluble substrates and decreases the need for microbial cells to synthesize and release the extracellular enzymes.

The mathematical description of the kinetics of enzyme reactions in soil columns has allowed combining the flow of substrates and products through the column with the Michaelis-Menten theory (McLaren, 1978). Several mathematical relationships have been developed by considering the substrate flow through the column, and the pattern of microbial growth along the column profile (McLaren, 1978).

RHIZOSPHERE-MICROBIAL AND SOIL FAUNA-MICROBIAL INTERACTIONS

Soil functioning depends not only on the composition of soil microbiota and chemical and physical properties of soil, but also on microbial interactions. Microorganisms can interact with components of the same group as well as with those of other groups (Trevors and van Elsas, 1997). According to Alexander (1978) microbe-microbe interactions could be categorized as follows:

1. Neutralism, when neither beneficial nor detrimental effects occur between two partners.
2. Symbiosis, when two partners are closely associated and both beneficially profit, generally from a nutritional point of view, from this association. In the case of a nonobligatory symbiosis, the interaction is also called protocooperation.
3. Commensalism, the interaction with beneficial effects for one partner whereas the other one is unaffected. This is often the case for members of microbial consortia described previously.
4. Competition, when one or both the partners suffer from the presence of the other one, generally because both species compete for the same substrate.
5. Amensalism, when one species suppresses the other one by producing antibiotics or toxins: For example, actinomycetes release antibiotics such as streptomycin, erythromycin, neomycin, tetracycline, and amphotericin as secondary metabolites when active growth has stopped.
6. Parasitism and predation involve the direct attack of one species by the other one.

As a result of the discreteness of microhabitats in soil, interactions between microbes probably occur less frequently and more sporadically in soil than in habitats wherein water is continuous (Stotzky, 1997). Even when the pore space is saturated with the aqueous phase or where water bridges are present between adjacent microhabitats, movement of bacteria between microhabitats may be restricted as the surface tension of the ordered water around aggregates may be too great to allow passive movement of cells or even active movement by flagellated cells.

Rhizosphere soil, which surrounds the root, is characterized by a higher number (10 to 20 times) of microorganisms than the bulk soil and this effect is called the rhizosphere effect (Lynch, 1990; Morgan and Whipps, 2001). This effect depends on the flow of organic compounds from roots to rhizosphere soil which increases the concentration of available organic C for the heterotrophic soil microbiota (Toal et al., 2000). The rhizodeposition is grouped into root exudates and root debris and includes amino acids, proteins, monosaccharides, oligosaccharides, polysaccharides, organic acids, fatty acids, growth factors, flavones, nucleotides, and other compounds (Uren, 2001). Root debris include root cap cells and senescent cortical cells. Duffusates, excretions, and secretions are included among root exudates.

Specific compounds released from cells can control the behavior of specific microorganisms. The symbiosis between legumes and rhizobia occurs because there is a specific exchange of molecular signals from the plant root and the specific rhizobial species (Werner, 2001). The sequence of molecular events includes: release of flavonoids from the legumes roots, induction of *nod* genes with production of *nod* factor in the rhizobial cell, chemotaxis of rhizobia, stimulation of flavonoid synthesis in the legume root with phytoalexin and meristem induction, root hair curling, induction of phytoalexin resistance in the rhizobia, degradation of flavonoids and phytoalexin. Exopolysaccharides or lipopolysaccharides present on the external cell surface of the microorganisms permit the anchorment of the rhizobial cell to the plant root surface and permit it to be recognized with respect to other microorganisms, including those that are detrimental, such as pathogens. According to Hungria and Stacey (1997) more than 4,000 flavonoids have been identified in the plant kingdom and some of these function as *nod* gene inducers. Root flavonoids have also been suggested to be molecular signals for the initiation and development of mycorrhizal infection. Mycorrhizal symbiosis is a mutualistic plant-fungus association that can beneficially affect water and nutrient uptake by plants and may protect plants from root diseases and heavy metals pollution (Martin, Perotto, and Bonfante, 2001). Many fungal and plant species can be interested by this symbiosis and the taxonomic position of both plant and fungal partners defines the type of mycorrhiza. Each association is distinguished by specific anatomical and physiological features.

Siderophores, iron-chelating agents, are other important molecules with specific functions in the rhizosphere (Crowley, 2001). They can be secreted by both microorganisms and graminaceous plants (phytosiderophores) in response to iron deficiencies. The competition between plants and microorganisms for iron is very complex and is influenced by many factors such as: the level of siderophore production by the competing organisms, the resistance of siderophores to degradation, the ability of the molecule to solubilize iron by attacking mineral surfaces, and the chelating constants of these complexes and their charge characteristics that can regulate their adsorption by soil colloids (Crowley, 2001).

According to Coleman and Crossley (1996) about 90 percent of the terrestrial net primary production enters the soil system through plant debris, bodies and excretions of animals and, in the case of agricultural fields, with organic residues. The decomposition of these residues releases nutrient elements that can enter into soil food webs or be taken up by plants. This process is mainly the result of microbial activities but it also involves, mainly indirectly and in a minor proportion directly, the fauna (Lavelle, 1997). The effect of microfauna and mesofauna has been quantified in studies in which various elements of soil biota have been suppressed by different biocides (Coleman and Crossley, 1996). The importance of microarthropods varies with litter quality and is less important for the more rapidly decomposing litter. There is an increasing interest in the role of microbiota-fauna interactions in the food chains of soil (Coleman and Crossley, 1996). As soon as soil is wetted by rainfall there is an increase in bacterial biomass followed by a decline due to the increase of protozoa biomass. Protozoa activity is also higher in the rhizosphere where a ready food source is present for the microbial prey. Ingham et al. (1985) found that 70 percent of the fungal and bacterial-feeding nematodes were localized in the rhizosphere and this corresponded to 4 to 5 percent of the overall soil space. Omnivore nematodes can feed on both flagellates and amoebae-type protozoa. The predator-prey interaction between protozoa and bacteria in the rhizosphere can also have an impact on plant nutrition. The use of rhizodeposition by rhizosphere bacteria would result in a marked microbial N immobilization, since the C/N ratio of rhizodeposition is usually high (Badalucco and Kuikman, 2001). Rhizosphere microorganisms use native organic N for this process. The microbial N immobilization around the roots seems to be counteracted by a stimulation of N mineralization promoted by protozoan grazing of bacteria (Badalucco and Kuikman, 2001; Clarholm, 1994). Indeed, when bacteria are ingested by protozoa, release of ammonium occurs and this mineral N can be taken up by the plant root.

MICROBIAL ACTIVITY AND MICROBIAL DIVERSITY

The quantification of microbial diversity in soil has been impeded by the lack of accurate techniques. Microbial number has been determined for a long time by using methods based on the cultivation of microorganisms from soil. Viable counting procedures including both plate counts and the most probable number (MPN) technique enumerate only 1 to 10 percent of the microorganisms inhabiting soil because it is impossible to find nutritional and incubation conditions allowing the cultivation of all microbial species of soil (Torsvik, Goksoyr, and Daae, 1990; Torsvik, Sorheim, and Goksoyr, 1996). The noncultivability of the majority of species present in environmental samples is now well established. Higher counting of bacteria and fungi can be obtained by direct count techniques; however, these techniques do not give any indication on the composition of the respective communities (Bloem et al., 1995). The use of biomarkers, such as ergosterol, has allowed the determination of a subset of microbial biomass, the fungal biomass, but without any indication of the fungal species (Tunlid and White, 1992; Frostegard, Tunlid, and Baath, 1993; Zelles and Alef, 1995). Different subsets of the microbial community of soil can be determined by analyzing phospholipid ester-linked fatty acids (PLFA) (Tunlid and White, 1992). These components of the membranes of all living cells (with the exception of archeobacteria) have been used as biomarkers because they rapidly degrade after cell death, are not present in storage lipids or pollutants, and present as constant proportion of bacterial biomass (Tunlid and White, 1992; Zelles and Alef, 1995; Morgan and Winstanley, 1997). Some of these are specific for a certain group of microorganisms. For example, polysaturated, monosaturated, and branched-chain fatty acids are indicators of fungi, gram-negative and gram-positive bacteria, respectively.

The relationship between microbial diversity and microbial function is largely unknown. Classical concepts of diversity include species richness (the number of species), evenness (the relative contribution that individuals of all species make to the total number of organisms present), and type and relative contribution of the particular species present (Griffith, Ritz, and Wheatly, 1997).

The use of molecular techniques has allowed the detection of unculturable microbial species and the quantification of the bacterial diversity in soil by the reassociation kinetics of complementary deoxyribonucleic acid (DNA) strands (Atlas et al., 1994; Torsvik, Sorheim, and Goksoyr, 1996). However, the use of reassociation kinetics to estimate the range of microbial diversity in soil presents several problems. First, the results of this technique are affected by species evenness, with the most common species playing a major role. Second, it cannot give an absolute value of diversity for eukaryotic microorganisms, such as fungi, because of the presence of repetitive sequences in fungal DNA (Ritz and Griffiths, 1994). A shift in the percent guanine plus cytosine (G+S) can be used to determine changes in microbial community structure (Harris, 1994) but it does not allow determination of diversity parameters (richness, evenness, and composition).

Molecular techniques involve the extraction of DNA from soil and its purification (Trevors and van Elsas, 1997; Ogram, 2000). DNA:DNA probing and amplification by polymerase chain reaction (PCR) of target DNA have been used to detect and amplify, respectively, specific DNA sequences in DNA soil extracts. Both 16S and 23S rRNA genes have been selected as phylogenetical marker molecules because of their universal distribution, structural and functional conservation. Analysis of PCR-amplified DNA fragments extracted from soil by denaturating gradient gel electrophoresis (DGGE) or temperature gradient gel electrophoresis (TGGE) allows a rapid detection of changes in microbial diversity in response to changes in agricultural management, presence of pollutants, or changes in environmental conditions (Heuer and Smalla, 1997). DNA:DNA reassociation kinetics have been used to determine the range of microbial diversity in soil. This approach is based on the assumption that more complex denaturated DNA reassociates at a slower rate than less complex denaturated DNA (Torsvik, Sorheim, and Goksoyr, 1996). Molecular techniques have also been used to better understand the reasons behind the tremendous richness of species present in soil (Tiedje et al., 2001). The bacterial rDNA restriction patterns of surface soils have not shown any dominance of one or a few community members, whereas this occurred in restriction patterns of saturated soils or subsoils. In surface soils this did not occur because the various microbial species inhabiting soil were spatially separated for most of the time due to very low soil moisture (Tiedje et al., 2001). An alternative hypothesis interpreting the high microbial diversity of surface soil is based on the presence of a greater variety and content of organic compounds in surface soil than in deeper soil layers.

Different levels of genetic differences were investigated on soils sampled from different continents in Mediterranean and boreal forest climates to assess if soil microorganisms are present everywhere or are geographically unique (Tiedje et al., 2001). When the comparison was carried out at the level of amplified ribosomal DNA restriction analysis (ARDRA), the microorganisms appeared everywhere, probably because the investigated operon was highly conserved. When the results from replication PCR genomic fingerprintings were compared, it appeared that the genotypes were site-specific and that the genetic distance increased with the corresponding geographical distance of the sampled soils (Tiedje et al., 2001). The fact that microorganisms are endemic means that global microbial diversity is much higher than if microorganisms were present everywhere.

However, the use of molecular techniques presents several problems. First, they require an efficient extraction and purification of DNA, without markedly affecting the DNA size. The direct lysis method is generally more efficient than the indirect method, which is based on the separation of bacterial cells from soil particles followed by the successive lysis of these cells to release the DNA (Ogram, 2000). Indeed, DNA recoveries of over 80 percent have been reported for the former method, which allows small soil samples to be used. The drawbacks of the direct lysis of cells in soil include the higher presence of contaminants, such as humic molecules that inhibit the PCR amplification, and the need of a more exhaustive purification procedure (Table 3.3). Other problems include artifacts produced during the amplification by PCR, the discrepancy between the quantitative composition of rRNA genes within the template DNA and the final amplification product, and the cost and the time required to carry out the procedure (Wintzingerode, Gobel, and Stackebrandt, 1997; Ogram, 2000). In addition, the rDNA-based methods cannot give the total bacterial species richness (the

TABLE 3.3. Problems Occurring During PCR Amplification of Environmental Samples

Problem	Causes
Inhibition of PCR amplification	Presence of contaminants, mainly humic molecules in soil samples
Differential PCR amplification	The following assumptions are required for the same amplification efficiency of all DNA molecules present in the sample: 1. All DNA molecules are equally accessible to primer hybridization and form primer-template hybrids with equal efficiencies. 2. In all templates there is the same extension efficiency of DNA polymerase. 3. All templates are equally affected by the exhaustion of substrates.
Formation of artifacts	1. Formation of chimeras between two different homologous molecules. 2. Formation of mutants that can be deletion ones because of the formation of stable secondary structures or point ones because of misincorporation by DNA polymerases.
Amplification in negative controls	Presence of contaminating DNA
Biased reflection of the microbial diversity	The number of rRNA gene regions (rrn operons) on prokaryotic genomes differs among the various bacterial species

Source: Adapted from Wintzingerode, Gobel, and Stackebrandt, 1997.

number of species) and bacterial evenness (the relative proportion of individual members in different species) because of the very complex community structure of soil microbiota and the limitations of PCR procedure, based on the use of specific primers (Wintzingerode, Gobel, and Stackebrandt, 1997; Ogram, 2000). Some of these limitations can be overcome by using DNA microarrays developed for rapid analysis of microbial communities (Ogram, 2000; Tiedje et al., 2001). Microarrays are not based on PCR amplification. They are constructed by fixing oligonucleotides to a glass or nylon membrane, which then can be hybridized to a sample. The technique is powerful because the DNA microarray can be designed to resolve questions on gene expression, community composition, and genotype analysis (Tiedje et al., 2001). Thus, in the case of community structure analysis, it is possible to use oligonucleotides directed toward various phylogenetic groups of analysis.

Determination of activities of specific genes can be done by measuring specific mRNA concentrations. However, this measurement has not yet developed because mRNA is more susceptible to nuclease degradation of DNA and there is a limited database available on the diversity of bacterial genes contributing to a given function in a community. Therefore, measurements of microbial activities have stopped at the level of the function of the soil sample. The meaning of these measurements has been discussed by Nannipieri, Ceccanti, and Grego (1990). Usually, determinations carried out in laboratory measure microbial rather than biological activity because visible animal and plant debris are removed prior to analysis. Determinations such as CO_2 evolution in the field include the contribution of soil microbiota, active roots, and active fauna. The microbial activity of soil has been determined by measuring activities of entire metabolic processes such as respiration, nitrification and ammonification, the energetic state of cells through the ATP content, the adenylate energy charge, and the thymidine and leucine incorporation rate (Alef and Nannipieri, 1995). The activities of specific reactions such as arginine ammonification or dimethyl sulphoxide reduction, or the activities of a class of enzymes, dehydrogenases, have been taken as indexes of microbial activity. It is hazardous to select a single measurement to represent microbiological activity, which includes an extensive picture of different activities (Nannipieri, Ceccanti, and Grego, 1990). The response of microbial species to different organic compounds, as proposed by the Biolog approach (Garland and Mills, 1991), has limitations as it is culture-dependent and changes in the structure of microbial population can occur during the incubation (Nannipieri, Kandeler, and Ruggiero, 2001; Smalla et al., 1998). These drawbacks have been overcome by considering the complete oxidation to CO_2 of soil treated with 36 different substrates (Degens and Harris, 1997). In this way it is possible to differentiate soils by considering their diverse catabolic response to the additions of the selected organic compounds.

MODELING SOIL BIOLOGICAL PROCESSES

It is impossible to monitor each inorganic and organic compound and to quantify the rates of each abiotic and biotic reaction in a complex system such as soil (Nannipieri, Badalucco, and Landi, 1994). Even if this were possible with the advanced techniques available, it would be very difficult to integrate the huge data and to draw any quantitative conclusions on processes playing an important role in soil quality. Currently, nutrient cycling in soil is quantified by adopting a holistic approach. The system is partitioned into pools, and fluxes among these pools usually represent physical processes such as leaching or volatilization, or abiotic or biotic reactions. Fluxes concerning nutrient transformations include a series of reactions and, generally, only changes in the concentrations of the final products are considered; for example, N mineralization includes a series of biotic and abiotic reactions producing NH_4^+ from N organic compounds which may range from low-molecular weight (amino acids) to high-molecular weight (for example, proteins).

The holistic approach has allowed a better mathematical description of the system with the consequent possibility to simulate processes occurring in soil (Stevenson, 1986). Modeling the turnover of soil organic matter has become very important not only to predict and quantify the release of important nutrients such as N in soil but also to understand and manage the terrestrial C cycle (Paustian, 2001). Soil can act as a sink of atmospheric C-CO_2 which can be sequestered into soil organic matter. The current models describing the mineralization and immobilization of nutrients are based on the following basic principles: (1) microbial biomass is a significant source and sink of nutrients; (2) microbial decomposition and microbial synthesis occur simultaneously in soil; and (3) nutrients in the organic matter are heterogeneous in terms of biological activity because some components cycle very slowly whereas others cycle very rapidly. For this reason, most of the models currently used represent soil organic matter as a multipool system. These pools are characterized by different decomposition rates, governed by first-order kinetics modified by reduction factors, which take into account climatic and main soil properties (Paustian, 2001). Usually these pools are not measurable and the sum of model pools is equal to total organic carbon, a measurable and unambiguous quantity. Terms such as labile, stable, chemically protected organic matter, and physically protected organic matter have been used (see Chapter 13). The terminology is not defined because the physical and chemical significance of the various organic fractions, as well as their functioning, is rather vague. Thus, the very old (high mean age as estimated by carbon dating) organic matter may be protected by chemical interactions with mineral colloids; this organic pool has been considered as chemically protected whereas for others it is a physically protected pool (Nannipieri, Badalucco, and Landi, 1994). It is now believed that the chemical structure of organic molecules, in itself, is insufficient to account for the extreme variation in age and turnover time (Ladd et al., 1996).

A closer linkage between theoretical and analytical representations of organic matter heterogeneity requires a revision of model definitions to coincide with measurable quantities and the application of more functional laboratory fractionation procedures. Further discussion on modeling of soil organic matter dynamics is presented in Chapter 13.

SUMMARY

Soil harbors high population density and enormous microbial diversity under a tremendous range of physical and chemical conditions. It is, therefore, reasonable that many attempts to inoculate beneficial microorganisms have been unsuccessful. Microbial life is not evenly distributed in soil but is concentrated in a few microhabitats with the right set of conditions. The domination of the solid phase is another distinctive characteristic of soil with respect to microbial habitats. Surface-active particles can adsorb important biological molecules such as enzymes and DNA. Enzymes adsorbed by clays or inglobated by humic molecules can maintain their activity because they are protected against proteolysis, thermal and pH denaturation. In this way, stabilized extracellular enzymes can still be active under conditions unfavorable for the activity of soil microorganisms. Adsorption and binding of DNA on humic molecules, clay, and sand particles can protect the adsorbed DNA against degradation by nucleases without inhibiting its transforming ability.

All known metabolic processes can be carried out by microorganisms inhabiting soil. Organic compounds, even the more recalcitrant, are mineralized to CO_2 which returns to the atmosphere, thus closing the cycle started with the photosynthetic carbon assimilation by plants and the other photosynthetic organisms of terrestrial ecosystems. Synthetic organic pollutants, whose molecular structures differ considerably with respect to natural compounds, can persist for longer periods of time but then are degraded in soil.

It is impossible to determine microbial diversity in soil. Microbial number determined by viable counting accounts for only 1 to 10 percent of the microorganisms inhabiting soil. Higher counting of bacteria and fungi can be obtained by direct count techniques, which, however, do not give any indication on the composition of the respective communities. The use of biomarkers such as ergosterol and phospholipid ester-linked fatty acids (PLFAs) has allowed the determination of a subset of microbial biomass. Recently, the use of molecular techniques has allowed the detection of unculturable microbial species and quantified the bacterial diversity in soil by the reassociation kinetics of complementary DNA strands. Microbial genotypes are site-specific and the genetic distance increases with the corresponding geographical distance of the sampled soil. However, molecular techniques are also inaccurate and care is required in interpreting the relative data.

Measurements of microbial activities have stopped at the level of the function of the soil sample because it is not possible to determine the activity of specific genes by measuring specific mRNA concentrations. Activities of entire metabolic processes such as respiration, nitrification, and ammonification, the energetic state of cells through the ATP content, the adenylate energy charge, the thymidine and leucine incorporation rate, the activity of specific reactions such as arginine ammonification or dimethyl sulphoxide reduction or the activity of a class of enzymes, dehydrogenases, have been measured to assess microbial activity. However, it is conceptually wrong to select a single measurement to represent microbiological activity because it includes different metabolic processes and a multitude of different enzyme activities.

Nutrient cycling has been quantified in soil by using the holistic approach that allows for a better mathematical description of the system with the consequent possibility to simulate processes occurring in soil. However, advances in modeling nutrient cycling in soil requires measurement of organic matter heterogeneity.

REFERENCES

Alef, K. and P. Nannipieri, Eds. (1995). *Methods in Applied Soil Microbiology and Biochemistry.* London: Academic Press.

Alexander, M., Ed. (1978). *Soil Microbiology.* New York: John Wiley and Sons.

Assmus, B., P. Hutzler, G. Kirchhof, R. Amman, J.R. Lawrence, and A. Hartmann (1995). In situ detection of *Azospirillum brasiliense* in the rhizosphere of wheat using fluorescently labelled rRNA targeted oligonucleotide probes and scanning confocal laser microbiology. *Applied and Environmental Microbiology 61:*1013-1019.

Assmus, B., M. Schloter, G. Kirchhof, P. Hutzler, and A. Hartmann (1997). Improved in situ tracting of rhizosphere bacteria using dual staining with fluorescence-labeled antibodies and rRNA-targeted oligonucleotides. *Microbial Ecology 33:*32-40.

Atlas, R.M., A. Horowitz, M. Krichevsky, and A.J. Bej (1994). Response of microbial populations to environmental disturbance. *Microbial Ecology 22:*249-256.

Badalucco, L. and P.J. Kuikman (2001). Mineralization and immobilization in the rhizosphere. In *The Rhizosphere: Biochemistry and Organic Substances at the Soil-Plant Interface,* eds. A. Pinton, Z. Varanini, and P. Nannipieri. New York: Marcel Dekker, pp. 159-196.

Bloem, J., P.R. Bolhuis, M.R. Veninga, and J. Wieringa (1995). Microscopic methods for counting bacteria and fungi. In *Methods in Applied Soil Microbiology and Biochemistry,* eds. K. Alef and P. Nannipieri. London: Academic Press, pp. 162-173.

Bollag, J.-M. and S.-Y. Liu (1990). Biological transformation processes of pesticides. In *Pesticides in the Soil Environment: Processes, Impacts and Modeling,* ed. H.H. Cheng. Madison, WI: Soil Science Society of America Book Series, pp. 169-209.

Burns, R.G. (1982). Enzyme activity in soil: Location and a possible role in microbial ecology. *Soil Biology and Biochemistry 14:*423-427.

Cambardella, C.A. and E.T. Elliott (1993). Carbon and nitrogen distribution in aggregates from cultivated and native grassland. *Soil Science Society of America Journal 57:*1071-1076.

Capowiez, Y. and L. Belziences (2001). Dynamic study of the burrowing behaviour of *Aporrectodea nocturna* and *Allobophora chlorotica* interactions between earthworms and spatial avoidance of burrows. *Biology and Fertility of Soils 33:*310-316.

Cervelli, S., P. Nannipieri, B. Ceccanti, and P. Sequi (1973). Michaelis constant of soil acid phosphatase. *Soil Biology and Biochemistry 5:*841-845.

Chen, Y. (1998). Electron microscopy of soil structure and soil components. In *Structure and Surface Reactions of Soil Particles*, eds. P.M. Huang, N. Senesi, and J.-M Bollag. New York: John Wiley and Sons, pp. 155-182.

Clarholm, M. (1994). The microbial loop in soil. In *Beyond the Biomass*, eds. K. Ritz, J. Dighton, and K.E. Giller. Chichester: John Wiley, pp. 221-230.

Coleman, D.C. and D.A. Crossley Jr., Eds. (1996). *Fundamentals of Soil Ecology.* London: Academic Press.

Crowley, D. (2001). Function of siderophores in the plant rhizopshere. In *The Rhizosphere: Biochemistry and Organic Substances at the Soil-Plant Interface,* eds. A. Pinton, Z. Varanini, and P. Nannipieri. New York: Marcel Dekker, pp. 223-261.

Degens, B.P. and J.A. Harris (1997). Development of a physiological approach to measuring the catabolic diversity of soil microbial communities. *Soil Biology & Biochemistry 29:*1309-1320.

Dobbins, D.C., C.M. Marjorie, and F. Pfaender (1992). Subsurface, terrestrial microbial ecology and biodegradation of organic chemicals: A review. *Critical Reviews in Environmental Control 22(1/2):*67-136.

Farrah, R.S. and G. Bitton (1990). Viruses in the soil environment. In *Soil Biochemistry Volume 6*, eds. J.-M. Bollag and G. Stotzky. New York: Marcel Dekker, pp. 529-556.

Fetzner, S. and F. Lingens (1994). Bacterial dehalogenases: Biochemistry, genetics, and biotechnological applications. *Microbiological Reviews 58:*641-685.

Forster, R.C. (1994). Microorganisms and soil aggregates. In *Soil Biota. Management in Sustainable Farming Systems*, eds. C.E. Pankhurst, B.M. Doube, V.V.S.R. Gupta, and P.R. Grace. East Melbourne, Australia: CSIRO, pp. 144-155.

Frostegard, A., A. Tunlid, and E. Baath (1993). Phospholipid fatty acid composition, biomass and activity of microbial communities from two soil types experimentally exposed to different heavy metals. *Applied Environmental Microbiology 59:*3605-3617.

Garland, J.L. and L.A. Mills (1991). Classification and characterization of heterotrophic microbial communities on the basis of patterns of community levels sole-carbon-source utilization. *Applied and Environmental Microbiology 57:*2351-2359.

Griffiths, B.S., K. Ritz, and R.E. Wheatley (1997). Relationship between functional diversity and genetic diversity in complex microbial communities. In *Microbial Communities: Functional versus Structural Approaches,* eds. H. Insam and A. Rangger. Berlin: Springer-Verlag, pp. 1-9.

Harris, D. (1994). Analysis of DNA extracted from microbial communities. In *Beyond the Biomass*, eds. K. Ritz, J. Dighton, and K.E. Giller. Chichester: John Wiley, pp. 111-118.

Hattori, T., Ed. (1973). *Microbial Life in Soil.* New York: Marcel Dekker.

Heuer, H. and K. Smalla (1997). Application of denaturing gradient gel electrophoresis and temperature gradient electrophoresis for studying soil microbial communities. In *Modern Soil Microbiology*, eds. J.D. van Elsas, J.T. Trevors, and M.H. Wellington. New York: Marcel Dekker, pp. 353-373.

Hicks, R.J., G. Stotzky, and P. Van Voris (1990). Review and evaluation of the effects of xenobiotic chemicals on microorganisms in soil. *Advances in Applied Microbiology 35:*195-253.

Hopkins, D. and R.W. O'Dowd (1997). Chirality is a factor in substrate utilization assays. In *Microbial Communities: Functional versus Structural Approaches,* eds. H. Insam and A. Rangger. Berlin: Springer-Verlag, pp. 215-228.

Huang, P.M. and J.-M. Bollag (1998). Minerals-organics-microorganisms interactions in the soil environment. In *Structure and Surface Reactions of Soil Particles,* eds. P.M. Huang, N. Senesi, and J.-M. Bollag. New York: John Wiley and Sons, pp. 3-39.

Hungria, M. and G. Stacey (1997). Molecular signals exchanged between plants and rhizobia: Basic aspects and potential applications in agriculture. *Soil Biology and Biochemistry 29:*819-830.

Ingham, R.E., J.A. Trofymow, E.R. Ingham, and D.C. Coleman (1985). Interactions of bacteria, fungi, and their nematode grazer. Effect of nutrient cycling and plant growth. *Ecological Monography 55:*119-140.

Jian, H. and W.W. Tso (1996). DNA recombination, plasmid dissemination, enzymatic "combustion" and cell immobilization: Their potential effect on environmental biotechnology. In *Environmental Biotechnology:Principles and Application,* eds. M. Moo-Young, W.A. Anderson, and A.M. Chakrabarty. Dordrecht, Netherlands: Kluwer, pp. 28-37.

Khanna, M. and G. Stotzky (1992). Transformation of *Bacillus subtilis* by DNA bound on montmorillonite and effect of Dnase on the transforming ability of bound DNA. *Applied Environmental Microbiology 58:*1930-1939.

Killham, K., Ed. (1994). *Soil Ecology.* Cambridge: Cambridge University Press.

Knowles, R. (1986). Aspects of nitrification and denitrification. In *Perspectives in Microbial Ecology,* eds. F. Megusar and M. Gantar. Ljubljana, Slovenia: Slovene Society of Microbiology, pp. 319-324.

Ladd, J.N., R.C. Forster, P. Nannipieri, and J.M. Oades (1996). Soil structure and biological activity. In *Soil Biochemistry Volume 9,* eds. G. Stotzky and J.-M. Bollag. New York: Marcel Dekker, pp. 23-78.

Lavelle, P. (1997). Faunal activities and soil processes: Adaptative strategies that determine ecosystem function. *Advances in Ecological Research 27:*93-132.

Lee, K.E., Ed. (1985). *Earthworms: Their Ecology and Relationships with Soils and Land Use.* Sydney, Australia: Academic Press.

Lehninger, A.L., D.L. Nelson, and M.M. Cox, Eds. (1993). *Principles of Biochemistry.* New York: Worth Publishers.

Lorenz, M.G. and W. Wackernagel (1987). Adsorption of DNA to sand and variable degradation rates of adsorbed DNA. *Applied and Environmental Microbiology 53:*2948-2952.

Lynch, J.M., Ed. (1990). *The Rhizosphere.* London: Academic Press.

Martin, F.M., S. Perotto, and P. Bonfante (2001). Mycorrhizal fungi: A fungal community at the interface between soil and roots. In *The Rhizosphere: Biochemistry and Organic Substances at the Soil-Plant Interface,* eds. A. Pinton, Z. Varanini, and P. Nannipieri. New York: Marcel Dekker, pp. 263-296.

McLaren, A.D. (1978). Kinetics and consecutive reactions of soil enzymes. In *Soil Enzymes,* ed. R.G. Burns. London: Academic Press, pp. 97-116.

McLaren, A.D. and J.J. Skujins (1971). In *Soil Biochemistry Volume 2,* eds. A.D. McLaren and J.J. Skujins. New York: Marcel Dekker, pp. 1-12.

Miller, L.G., T.L. Connell, J.R. Guidetti, and R.S. Oremland (1997). Bacterial oxidation of methyl bromide in fumigated agricultural soils. *Applied and Environmental Microbiology 63:*4346-4354.

Morgan, J.A.W. and J.M. Whipps (2001). Methodological approaches to the study of rhizosphere carbon flow and microbial population dynamics. In *The Rhizosphere: Biochemistry and Organic Substances at the Soil-Plant Interface,* eds. A. Pinton, Z. Varanini, and P. Nannipieri. New York: Marcel Dekker, pp. 373-409.

Morgan, J.A. and C. Winstanley (1997). Microbial biomarkers. In *Modern Soil Microbiology,* eds J.D. van Elsas, J.T. Trevors, and M.H. Wellington. New York: Marcel Dekker, pp. 331-352.

Nannipieri, P., L. Badalucco, and L. Landi (1994). Holistic approaches to the study of populations, nutrient pools and fluxes: Limits and future research needs. In *Beyond the Biomass,* eds. K. Ritz, J. Dighton, and K.E. Giller. Chichester: John Wiley, pp. 231-238.

Nannipieri, P., B. Ceccanti, S. Cervelli, and C. Conti (1982). Hydrolases extracted from soil: Kinetic parameters of several enzymes catalysing the same reaction. *Soil Biology and Biochemistry 14:*429-432.

Nannipieri, P., B. Ceccanti, and S. Grego (1990). Ecological significance of the biological activity in soil. In *Soil Biochemistry Volume 6,* eds. J.-M. Bollag and G. Stotzky. New York: Marcel Dekker, pp. 293-355.

Nannipieri, P., L. Falchini, L. Landi, and G. Pietramellara (2000). Management of soil microflora. In *Biological Resource Management: Connecting Science and Policy,* eds. E. Balazs, E. Galante, J.M. Lynch, J.P. Schepers, J.-P. Toutant, D. Werner, and P.A.Th.J. Werry. Berlin: Springer-Verlag, pp. 237-255.

Nannipieri, P. and L. Gianfreda (1998). Kinetics of enzyme reactions in soil environment. In *Structure and Surface Reactions of Soil Particles,* eds. P.M. Huang, N. Senesi, and J. Buffle. New York: John Wiley and Sons, pp. 450-479.

Nannipieri, P., E. Kandeler, and P. Ruggiero (2001). Enzyme activities as research tool for microbiological and biochemical processes in soil. In *Enzymes in the Environment: Activity, Ecology, and Application,* eds. R.G. Burns and R. Dick. New York: Marcel Dekker, pp. 1-33.

Nannipieri, P., P. Sequi, and P. Fusi (1996). Humus and enzyme activity. In *Humic Substances in Terrestrial Ecosystems,* ed. A. Piccolo. Amsterdam, Netherlands: Elsevier, pp. 293-328.

Ogram, A. (2000). Soil molecular microbial ecology at age 20: Methodological challenges for the future. *Soil Biology and Biochemistry 32:*1469-1504.

Paul, E.A. and F.E. Clark (1996). *Soil Microbiology and Biochemistry.* San Diego: Academic Press.

Paustian, K. (2001). Modelling soil organic matter dynamics—Global changes. In *Sustainable Management of Soil Organic Matter,* eds. R.M. Rees, B.C. Ball, C.D. Campbell, and C.A. Watson. Wallingford, UK: CABI International, pp. 43-53.

Payne, C.C. (1988). Viruses. In *Microorganisms in Action: Concepts and Applications in Microbial Ecology,* eds. J.M. Lynch and J.E. Hobbie. Boston: Blackwell Scientific Publications, pp. 33-43.

Pietramellara, G., L. Dal Canto, C. Vettori, E. Gallori, and P. Nannipieri (1997). Effects of air-drying and wetting cycles on the transforming ability of DNA bound on clay minerals. *Soil Biology and Biochemistry 29:*55-61.

Richards, B.N. (1987). *The Microbiology of Terrestrial Ecosystems.* Harlow, UK: Longman Group UK Limited.

Ritz, K. and B.S. Griffiths (1994). Potential application of a community hybridization technique for assessing changes in the population structure of soil microbial community. *Soil Biology and Biochemistry 26:*963-971.

Smalla, K., U. Wachtendorf, H. Heur, W.-T. Liu, and L. Forney (1998). Analysis of BIOLOG GN substrate utilization patterns by microbial communities. *Applied and Environmental Microbiology 64:*1220-1225.

Sposito, G., Ed. (1984). *The Surface Chemistry of Soils.* New York: Oxford University Press.

Stevenson, F.J., Ed. (1986). *Cycles of Soil: Carbon, Nitrogen, Phosphorus, Sulfur, Micronutrients.* New York: John Wiley and Sons.

Stotzky, G. (1986). Influence of soil mineral colloids on metabolic processes, growth, adhesion, and ecology of microbes and viruses. In *Interactions of Soil Minerals with Natural Organics and Microbes,* eds. P.M. Huang and M. Schnitzer. Madison, WI: Soil Science Society of America, pp. 305-428.

Stotzky, G. (1997). Soil as an environment for microbial life. In *Modern Soil Microbiology,* eds. J.D. van Elsas, J.T. Trevors, and E.M.H. Wellington. New York: Marcel Dekker, pp. 1-372.

Tiedje, J.M., J.C. Cho, A. Murray, D. Treves, B. Xia, and J. Zhou (2001). Soil teeming with life: New frontiers for soil science. In *Sustainable Management of Soil Organic Matter,* eds. R.M. Rees, B.C. Ball, C.D. Campbell, and C.A. Watson. Wallingford, UK: CABI International, pp. 393-412.

Toal, M.E., C. Yeomans, K. Killham, and A.A. Meharg (2000). A review of rhizosphere carbon flow modelling. *Plant and Soil 222:*263-281.

Torsvik, V., L.J. Goksoyr, and F.L. Daae (1990). High diversity in DNA of soil bacteria. *Applied and Environmental Microbiology 56:*782-787.

Torsvik, V., R. Sorheim, and J. Goksoyr (1996). Total bacterial diversity in soil and sediment communities—A review. *Journal of Industrial Microbiology 17:*170-178.

Trazar-Cepeda, M.C. and G.F. Gil-Sotres (1988). Kinetics of acid phosphatase activity in various soils of Galicia (NW Spain). *Soil Biology and Biochemistry 20:*275-280

Trevors, J.T. and J.D. van Elsas (1997). Microbial interactions in soil. In *Modern Soil Microbiology,* eds. J.D. van Elsas, J.T. Trevors, and M.H. Wellington. New York: Marcel Dekker, pp. 215-243.

Tunlid, A. and D.C. White (1992). Biochemical analysis of biomass, community structure, nutritional status, and metabolic activity of microbial communities in soil. In *Soil Biochemistry Volume* 7, eds. G. Stotzky and J.-M. Bollag. New York: Marcel Dekker, pp. 229-262.

Uren, N.C. (2001). Types, amounts, and possible functions of compounds released into the rhizopshere by soil-grown plants In *The Rhizosphere: Biochemistry and Organic Substances at the Soil-Plant Interface,* eds. A. Pinton, Z. Varanini, and P. Nannipieri. New York: Marcel Dekker, pp. 19-40.

Werner, D. (2001). Organic signals between plants and microorganisms. In *The Rhizosphere: Biochemistry and Organic Substances at the Soil-Plant Interface,* eds. A. Pinton, Z. Varanini, and P. Nannipieri. New York: Marcel Dekker, pp. 197-222.

Wintzingerode, F.V., U.B. Gobel, and E. Stackebrandt (1997). Determination of microbial diversity in environmental samples: Pitfalls of PCR-based rRNA analysis. *Federation of European Microbial Societies (FEMS) Microbiology Reviews 21:*213-229.

Zeikus, J.G. (1983). Metabolic communities between biodegradative populations in nature. In *Microbes in Their Natural Environments,* eds. J.H. Slater, R. Whittenburry, and J.W.T. Wimpenny. Cambridge, UK: Cambridge University Press, pp. 415-432.

Zelles, L. and K. Alef (1995). Biomarkers. In *Methods in Applied Soil Microbiology and Biochemistry,* eds. K. Alef and P. Nannipieri. London: Academic Press, pp. 422-439.

Chapter 4

The Conformational Structure of Humic Substances

Alessandro Piccolo
Pellegrino Conte
Annunziata Cozzolino
Riccardo Spaccini

Humic substances (HS) are natural organic substances that are ubiquitous in water, soil, and sediments. Because of the beneficial effects that HS have on the physical, chemical, and biological properties of soil, their role in the soil environment is significantly greater than that attributed to their contributions to sustaining plant growth. The HS are recognized for their role in controlling both the fate of environmental pollutants and the biogeochemistry of organic carbon in the global ecosystem (Piccolo, 1996). Despite their prominent importance, a better knowledge of the basic chemical nature and reactivity of HS has been elusive up to now because of their large chemical heterogeneity and geographical variability. Because it is a mixture that originates randomly from the decay of plant tissues or microbial metabolism-catabolism or both, the chemistry of humus is not only of utmost complexity but also a function of the different general properties of the ecosystem in which it is formed: vegetation, climate, topography, etc. The tremendous task of advancing the knowledge of humic chemistry and its consequences to other environmental domains still lies ahead of us. It should be obvious, to a world that appreciates the potentials of genetic engineering based on an understanding of DNA structure, that accurate predictions of reactivities and development of related technologies can only be made when there is a basic knowledge of the chemical structure of the reacting molecules.

This chapter reports on recent findings related to the conformational structure of humic material and the profound implications that these may have on our understanding of soil organic matter functions and reactivity.

PARADIGMS IN HUMUS CHEMISTRY

Stevenson (1994) reports that humus includes a broad spectrum of organic constituents. Many have their counterparts in biological tissues, and are distinguished between nonhumic substances and humic substances. The former consist of compounds belonging to the well-known classes of organic chemistry such as amino acids, carbohydrates, lipids, lignin, nucleic acids, etc., while the latter is an unspecified, transformed, dark-colored, heterogeneous, amorphous, and high-molecular-weight material.

The generalized terms humic acids (HAs), fulvic acids (FAs), and humins cover the major fractions still used to describe HS components, but the boundaries between these fractions have not yet been clarified in chemical terms. The HAs are the fraction of HS solubilized in neutral or more often alkaline conditions and precipitates when solution pH is reduced to 1 by acid addition. The FAs are the fraction of humic substances that remain soluble under all pH conditions or the fraction that stays in solution when alkaline soil extracts are adjusted to pH 1. Humin is the fraction of HS that is not soluble in water at any pH value and, because of the difficulty in extraction, has not been, until recently, studied as extensively as HAs and FAs.

Several studies have shown that chemical features such as elemental analysis (C, H, O) and acidic functional groups appear relatively constant for HAs and FAs from many different soils (Kononova, 1961). Though not based on any molecular understanding, such simple correlations have encouraged scientists to consider humic fractions, rather than complex mixtures of nonspecific compounds, as chemical entities having specific properties which can be recognized in different environments. Kononova (1961) introduced the concept that HS are comprised of a system of polymers, based on the observation that elemental composition, optical properties, exchange acidities, electrophoretic properties, and molecular weight characteristics vary consistently with soil classes. Using this concept, the various fractions of HS obtained on the basis of solubility characteristics are imagined to be part of a heterogeneous mixture of molecules, which, in any given soil, range in molecular weight from as low as several hundreds to perhaps over 300,000 daltons (Da) and exhibit a continuum of any given chemical property (Stevenson, 1994).

Despite the existence of data showing that molecular dimensions (as measured by osmometry, viscometry, and diffusion) of some HS were hardly beyond one or two thousand Da (Schnitzer and Khan, 1972), more confidence was placed on the early work of Flaig (1958), who showed, using the ultracentrifuge, that molecular weights were in the range of 30,000-50,000 Da for HAs and about 10,000 Da for FAs. A reason for such bias toward high-molecular-weight structures may be explained by the predominant role of historical hypotheses viewing HS as products of biologically assisted syntheses from compounds derived from degradation of lignin, polyphenols, cellulose, and amino acids. Although there is no direct evidence for the occurrence of such polymer buildup in a natural soil system, many classical laboratory experiments have shown the possibility for either abiotic or biotic condensation of simple molecules into humiclike material (Kononova, 1961). Some of these early laboratory studies have been proposed again in more or less confined conditions in recent times (Haider and Martin, 1967; Flaig, Beutelspacher, and Riets, 1975; Hedges, 1988).

The polymeric view of humic substances includes the general idea of polydispersity by which HS are made of polymers with different molecular weights similar to other natural biological macromolecules such as proteins, polysaccharides, and lignin. The experimental difficulty in isolating chemically homogeneous fractions of HS was approximated to that observed for those biopolymers which are synthesized in the living cell with varying molecular dimensions. This unwarranted similarity supported the polydisperse polymeric understanding of HS and justified the undisputed observation that they are a mixture of compounds. The concept of high-molecular-weight and polydisperse polymers became a paradigmatic part of any descriptive definitions of humic substances proposed thereafter (Schnitzer and Khan, 1972; Malcolm, 1990; Stevenson, 1994).

The assumption, without a sound molecular basis, that HS are polymers has promulgated the use of simple physical-chemical measurements to characterize HS. An example is the E4/E6 index, the ratio of the absorbance of HS at 465 nm to that at 665 nm, introduced by Welte (1955) and reproposed by Kononova (1961) as supposedly indicating a reverse relationship with progressive humification and increased condensation (large content of polycondensated aromatic-ring structures). Despite its popularity and continued application, the E4/E6 ratio has repeatedly been shown not to hold the predicted relationship with molecular weight. At variance with the hypotheses related to this ratio, Campbell et al. (1967) had already indicated that humic material with the lowest mean residence time had the highest E4/E6 ratio, and Handerson and Hepburn (1977), using gel permeation chromatography, showed that humic fractions with large molecular size and low E4/E6 ratio had mainly aliphatic content whereas those with low molecular size had the highest aromatic content. Piccolo (1988) compared E4/E6 ratios of various HS with gel permeation chromatograms and showed that the two methods produced comparable results only after extreme purification of HS. Summers, Cornel, and Roberts (1987) reported on the limitations of E4/E6 ratios and found that the ratio values varied considerably with concentration of ultrafiltered fractions of HS. By comparing HS extracted from soils and soil particles, before and after long-term amendments with organic wastes, Piccolo and Mbagwu (1990) found that the E4/E6 ratio did not account for the increase in molecular size observed in amended samples by gel filtration.

Various reasons exist for why the scientific community thought that HS had a polymeric structure, despite the lack of any sound evidence of their macromolecularity. One reason is the acceptance of Staudinger's (1935) description of macromolecular polymers occurring in living cells. According to Staudinger's view, it seemed convenient to assume that HS are also polymers, notwithstanding the fact that HS are not a product of cellular synthesis as are other biomolecules, but rather products of cell death. The rapid degradation and decomposition in soil of biopolymers liberated from cell lysis is now a well-accepted process from both a biological and a thermodynamic perspective (Jenkinson, 1981; Haider, 1987; Clapp and Hayes, 1999; Spaccini, et al., 2001). Surprisingly, Staudinger's view is currently still critically advocated by defenders of the polymeric nature of HS (Swift, 1999).

A second reason is that a stable polymeric structure accounted for the refractory characteristics of HS in soil, other than the physical protection conferred by inorganic soil particles. The organic carbon in stable humic materials is known to possess long residence times (from 250 up to 3,000 years) in soil (Stevenson, 1994). Moreover, the classical hypothesis of HS formation through a condensation between amino acids and components of degraded lignin spread the assumption that HS had a polymeric structure similar to that of lignin (Waksman, 1936). Lignin is known to be polydisperse in molecular weight with values ranging from <1,000 to several million Da (Goring, 1971) and its resistance to microbial degradation in soil has been repeatedly attributed to the macromolecular structure (Waksman, 1936). Similarly, other classical hypotheses of HS formation, such as the polyphenol theory (Flaig, Beutelspacher, and Riets, 1975) or the melanoidin theory based on the Maillard reaction are based on the same logic that justifies the polymeric paradigm of HS with polymerization processes occurring in laboratory or confined conditions.

Another aspect that has contributed to the polymeric view of HS is their colloidal properties which may be assimilated to those of polyelectrolytes in aqueous media (Flaig, Beutelspacher, and Riets, 1975). Most of the properties of polyelectrolytes, such as processes of flocculation and dispersion, responses to electrolytes, and double-layer behavior were apparently observed also for HS, thereby easily transferring to humic extracts the concept of high molecular weight.

Despite the widely accepted view of the macromolecularity of HS, it has never been unambiguously demonstrated in chemical and physical-chemical terms in soil-extracted humic fractions. The instrumentation for organic and physical chemistry analysis has dramatically improved in the present world of sophisticated biophysical technology. Therefore, the polymeric paradigm of HS should be fully proved beyond any doubt or abandoned for other descriptions.

MODELS OF CONFORMATIONAL STRUCTURE OF HUMIC SUBSTANCES

A determination of the true molecular weight (MW) of a humic fraction is the central problem of HS research. The molecular weight determines the conformational structure (i.e., size and shape) of HS and ultimately their reactivity in soil and the environment. A number of reviews have repeatedly pointed out that there is no accordance among the values obtained by the different methods applied to measure molecular weight of HS (Wershaw and Aiken, 1985; Swift, 1989; Clapp, Emerson, and Olness, 1989; Stevenson, 1994). Often, the differences are not slight but of several orders of magnitude. This undeniable fact has had (with some exception) little impact on those who view HS as polymers. Without questioning the polymeric paradigm, the great difference in values has been attributed to either the variability of HS or the intrinsic limitation of methods when applied to polydisperse systems. However, very few attempts have been made to seriously reduce such polydispersity by classical chemical methods.

Much of the confusion on the molecular weight of HS is a result of the use of sedimentation velocity and diffusion methods performed with an ultracentrifuge, which became fashionable in the years between the 1950s and 1970s with the upsurge of biochemistry research. Despite the clear indication that such semiempirical methods are not suitable for polydisperse systems because of the multiple diffusion coefficients and sedimentation constants of different-sized particles (Wershaw and Aiken, 1985), studies on whole extracts of humic acids were conducted and molecular weight values ranging from 25,000 to more than 200,000 Da were reported (Flaig, 1958; Piret et al., 1960). Interestingly, Flaig and Beutelspacher (1968), also working with polydisperse humic solutions, found that molecular weight varied from 77,000 Da, when 0.2 M NaCl was added to depress repulsive negative charges, to only 2,050 Da in the absence of the background electrolyte. Though the latter is commonly added to polyelectrolytes to reduce interferences from charge repulsion in monodisperse systems, a jump of about two orders of magnitude in MW for the same humic polydisperse system could suggest a molecular association rather than an extreme interference in sedimentation velocity of polyelectrolytes.

In an attempt to reduce the polydispersity of HS, soil-extracted humic acids were extensively fractionated by ultrafiltration and gel permeation and the fractions were subjected to sedimentation-velocity ultracentrifugation (Cameron et al., 1972). The resulting fractions were more homogeneous than the unfractionated solution, but still were not monodisperse. In fact, Cameron et al. reported molecular weights ranging from 2,600 to a surprising 1,360,000 Da. Instead of true polymers, they measured associations of smaller molecules randomly aggregated by different forces during the fractionation procedures. A more successful fractionation into homogeneous fractions may have been achieved if classical solid-liquid or liquid-liquid separations of mixtures of organic compounds had been employed, instead of gel permeation that notoriously separates HS by size exclusion only in specific circumstances (Linqvist, 1967). Moreover, Cameron et al. (1972) overlooked the difficulties in assessing sedimentation coefficients for associations of molecules such as HS. Measurement of sedimentation coefficients of polydisperse materials, which include subunits, invariably leads to erroneous values of molecular weight (Laue and Rodhes, 1990). Other studies using sedimentation-velocity ultracentrifuge studies (Ritchie and Posner, 1982) and even the more mathematically sound equilibrium centrifugation (Posner and Creeth, 1972) also showed polydispersity in HS, and, for this very reason, confirmed the ambiguity of the MW values obtained by ultracentrifuge methods.

More reliability exists in molecular weight values for fulvic acids or organic matter dissolved in waters. There is a general agreement that the molecular weight of these humic molecules is in the range 400-1,500 as determined by a variety of methods (Aiken and Gillam, 1989; Wershaw, 1989; Stevenson, 1994). As shall be seen later, the smaller discrepancy among fulvic acid MW values can be ascribed to their larger hydrophilicity (small, very acidic molecules) that prevents strong molecular associations by hydrophobic forces. However, a process of molecular association in solution can produce a polydisperse system of apparently high molecular weight also for fulvic acids. In fact, fulvic acid molecular weights analyzed by ultrafiltration and gel filtration methods were consistently higher than when measured with osmometry or cryoscopy (Thurman et al., 1982).

If the molecular weight value of HS has been a source of confusion, similar contradictions exist about the shape to be attributed to humic polymeric macromolecules. Ghosh and Schnitzer (1980) reconciled the different views by measuring surface pressures and viscosities of HS at different pHs and neutral salt concentration and adapting the results to relationships (the Flory and Fox and the Staudinger equations) which have been developed for real polymers. They explained the observed behavior of HS (uncharged matter at low pH and polyelectrolytes at high pH) on the basis of the polymeric theory and stated that HS macromolecular configurations are not unique but vary with pH and ionic strength of the medium in which HS are dissolved. They proposed that HS are rigid spherocolloids at high sample concentration and ionic strength and at low pH, whereas they behave as flexible linear polymers at low sample concentration and ionic strength and at high pH. This understanding had two major flaws: it was based on whole humic extracts with full polydispersity and, despite the lack of a direct knowledge of the real molecular structure of HS, data were arbitrarily used in equations specifically derived for polymers. Nevertheless, this reversible coiling model for humic configurations soon became the most widely used to describe HS, though it does not explain all the behavior of HS.

In applications of small-angle X-ray scattering to determine the particle sizes of whole humic acids in solution or of fractions separated by adsorption chromatography on cross-linked dextran gels (see Wershaw, 1989), it was found that HS formed molecular aggregates in solution and their size was a function of pH. It was concluded that the various fractions were chemically different and that the differences in aggregation behavior were a reflection of the interaction of different bonding mechanisms. These findings, coupled with other results showing that humic fractions from different sources have surface-active properties (Hayase and Tsubota, 1983), led to the formulation of a description of HS that was alternative to the random polymeric coil structure. Wershaw (1986) proposed that HS consist of ordered aggregates of amphiphiles composed mainly of relatively unaltered plant polymer segments possessing acidic functionality. In this model, HS are aggregates held together by hydrophobic (π-π and charge-transfer bonds) and H-bonding interactions and the hydrophobic parts of the molecules are in the interior whereas the hydrophilic parts make up the exterior surfaces. Ordered aggregates of humus in soils were depicted to exist as bilayer membranes coating mineral grains and as micelles in solutions. Wershaw's model represented a major breakthrough because it introduced the concept of aggregation of different particle sizes of humic constituents in contrast to the traditional view of humic polydisperse linear polymers. Nevertheless, the issue of HS molecular weight was not yet solved by the micellelike model. The spontaneous aggregation of humic molecules into micellar aggregates was advocated by others (Engebretson and von Wandruszka, 1994) to explain fluorescence quenching of pyrene, but explanation of the results was still based on the polymeric nature of the aggregating humic molecules. However, the classical concept of an ordered micelle is hardly applicable to the heterogeneity of HS. In fact, the critical micelle concentration (CMC) reported in the literature for HS, in the range of 1 to 10 $g.L^{-1}$ (Hayase and

Tsubota, 1983), is much higher than those found for surface-active compounds giving regular micellar structures (Tanford, 1980).

Notwithstanding its limitations, the concept of aggregation of hydrophobic parts of HS could explain: (1) results from light scattering which showed that addition of Cu ions to dilute soil FA increased the amount of light scattered by the solution (Ryan and Weber, 1982); (2) the increased solubility of nonpolar compounds in humic solution because of partition/adsorption in the hydrophobic interior of HS (Carter and Suffet, 1982); and, (3) the further release of humic matter through dialysis bags from an already extensively dialyzed HS when this is treated with an amphiphilic compound such as acetic acid or an electrolyte (De Haan, Jones, and Salonen, 1987; Nardi, Arnoldi, and Dell'Agnola, 1988).

The further explanation of the environmental behavior of HS that was given by a molecular aggregation model in respect to the polymeric paradigm did not prevent recent hypotheses, based on pyrolytic analysis of HS of schematic macromolecular structures for HA of up to about 100,000 Da (Schulten, Leinweber, and Schnitzer, 1998). Despite the many limitations inherent to analytical pyrolysis of HS (Saiz-Jimenez, 1995, 1996), compounds identified by pyrolysis-mass spectrometry techniques as having a molecular mass hardly higher than 500 Da were used in computer molecular models and arbitrarily linked together by covalent bondings which yielded pictured views of large-sized, branched polymers. Such large macromolecules are being presently proposed as models of humic molecules and used to explain the behavior of HS in soil (Schulten and Leinweber, 2000).

THE SUPRAMOLECULAR ASSOCIATIONS OF SELF-ASSEMBLING HUMIC MOLECULES

Both low-pressure and high-pressure size-exclusion chromatography (HPSEC) cannot provide reliable values of molecular weights of HS because no adequate standards are available to simulate the complex arrangement of humic molecules in solution. Nevertheless, several studies have pointed out that HS behavior reflects that of molecular associations rather than of polymers and that some conformational changes can occur during chromatography (De Haan, Jones, and Salonen, 1987; Yonebayashi and Hattori, 1987; Piccolo, Rausa, and Celano, 1990). Size-exclusion chromatography may be then used to compare relative conformational changes of HS when they interact with other solutes.

Piccolo, Nardi, and Concheri (1996a,b) have employed low-pressure size exclusion chromatography to monitor the chromatographic variation of alkaline humic solutions which were brought to pH 2 with addition of different organic acids and then placed on a Biorad P100 Biogel column (nominal molecular range 5-100 KD) under elution with 0.02 M borate ($Na_2B_4O_7$) buffered at pH 9.2. This buffer was chosen because it reduced the HS adsorption on the gel column more than a Tris-(hydroxymethyl) aminomethane (TRIS) or a carbonate/bicarbonate buffer, though it produced exactly the same bimodal distribution as the other buffers. When the control alkaline solution was chromatographed through the gel in the borate buffer, most of the humic material eluted at the void volume (V_0) of the column (> 100 KD). Conversely, when the humic solution was brought to pH 2 with organic acids (all the monocarboxylic and most of the bicarboxylic acids) the total peak absorbance was shifted to elution volumes near the total column volume (V_t) suggesting a nominal MW <25 KD. Mineral acids did not have any apparent influence on the chromatographic performance of the HS. A progressive shift from low to high elution volumes of the band was also observed when the pH of the solution was progressively brought to lower pHs with acetic acid prior to insertion on the column (Figure 4.1, I). The phenomenon was reversible since the band shifted back to low elution volumes when the action of acetic acid was progressively neutralized by addition of 0.5 M KOH (Figure 4.1, II).

Supporters (Swift, 1999) of the traditional polymeric model of HS criticized these results not on the basis of replication of the same experiment but on theoretical and qualitative interpretations of gel-solute interactions. However, the results of Piccolo, Nardi, and Concheri (1996 a,b) could not be attributed to a buffering action of the organic acid toward the alkaline eluent because the amount of the different organic acids varied by two orders of magnitude but the shift to larger elution volumes remained the same for all acids. Nor could an elution delay due to a solid deposition on the gel and subsequent resolubilization by the eluent have been the cause of the shift, since the treated samples remained soluble at low pH and entered the chromatographic elution immediately after deposition on the column. Nevertheless, if this had been the case, a progressive neutralization of the acidic buffering capacity by the alkaline eluent would have caused a smearing out in the column of the polydisperse humic mixture and a diffuse chromatographic band from lower to larger elution volumes, instead of the sharp peak at the total volume (V_t) of the column. Furthermore, the ionic strength effect could not be invoked for the reversible peak shift shown in Figure 4.1 because elution in an ionic strength quencher, such as a borate buffer ten times more concentrated, gave the same chromatographic change upon acetic acid and KOH additions.

Piccolo, Nardi, and Concheri (1996 a,b) considered their findings an expression of the associative nature of humic molecules which aggregate into only apparent high-molecular-size materials. This interpretation was in ac-

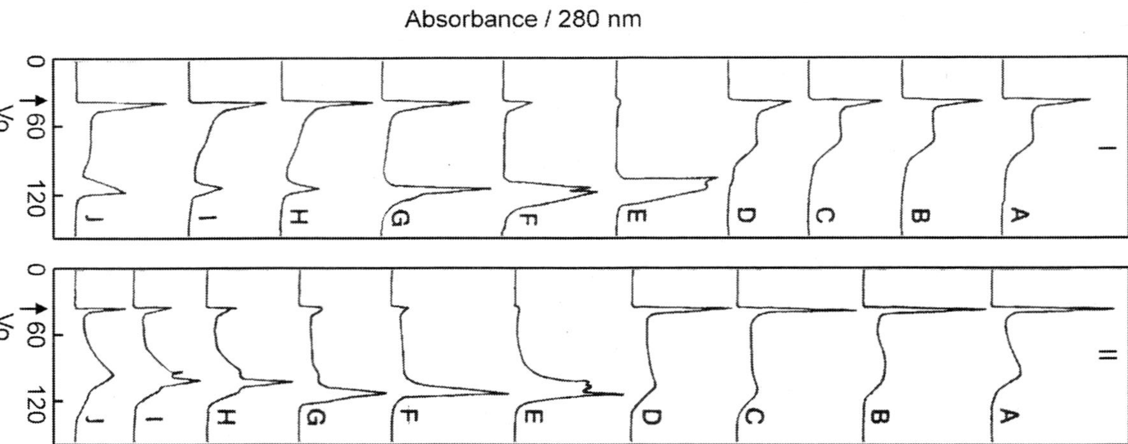

FIGURE 4.1. Low-pressure size exclusion chromatograms of a humic acid treated with acetic acid and eluted with a 0.02 M $Na_4B_2O_7$ solution at pH 9.2 (I) and with a 0.1 M $Na_4B_2O_7$ solution at pH 9.2 (II). Humic acid was treated before elution as follows: (A) dissolved at pH 11.8 (B) titrated with acetic acid to pH 6 (C) to pH 4.5 (D) to pH 3.5 (E) to pH 2 (F) the material brought to pH 2 was further back-titrated with KOH to pH 3.5 (G) back-titrated to pH 4.5 (H) back-titrated to pH 6 (I) back-titrated to pH 8.5 (J) the latter material at pH 8.5 was further roto-evaporated to attempt the elimination of the residual acetic acid. (*Source:* Piccolo, Nardi, and Concheri, 1996a, p. 324.)

cordance with previous research that showed that HS behaved as molecular associations when studied by size-exclusion chromatography (De Haan, Jones, and Salonen, 1987; Yonebayashi and Hattori, 1987; Piccolo, Rausa, and Celano, 1990). Moreover, laboratory observations have indicated that when acetic acid is added to HS that have been already extensively dialyzed, further small-sized components are released during subsequent dialysis (Nardi, Arnoldi, and Dell'Agnola, 1988). These low-molecular-size fractions are a product of a conformational rearrangement and a chemical composition different than the bulk HS. In fact, the separated fractions were found to stimulate specific biological properties in plants, and were more biologically active than the whole humic materials from which they were separated (Piccolo, Nardi, and Concheri, 1992).

Piccolo, Nardi, and Concheri (1996 a,b) have described HS as micellar associations that are stabilized by predominantly hydrophobic forces at pH 7. They reason that the organic acids penetrate into the inner (hydrophobic) core of the micellar structure while neutralizing the HS acidic functions from pH 7 to 2. The association between the organic acids and HS occurs because of the amphiphilic properties of the acids which are able to interact with both the hydrophilic and the hydrophobic domains of humic aggregates. Such interactions are capable of disrupting the weak forces which stabilize the humic conformations and the subsequent chromatographic elution separates the small subunits that form the aggregate and prevents the reaggregation that would have occurred in static conditions.

Despite its widespread use in the recent past, it is generally accepted that low-pressure size-exclusion chromatography has severe limitations in both time requirements and reproducibility in the same gel bed. Hence, Conte and Piccolo (1999b) have compared the capacities of two commercial HPSEC columns to measure accurately and precisely the molecular sizes of different HAs. They found discrepancies in the chromatographic behavior of the same samples in the two columns. They attributed the different UV response between the effluents from these columns to different stabilities of humic associations rather than to differences in column properties. The differences between chromatograms from the two columns can be explained by considering the response of HS and mixture of molecules to electron excitation. Although it is known that humic substances do not strictly follow Beer's law and that their molecular absorptivity varies with molecular size (Stevenson, 1994; Chin, Aiken, and O'Loughlin, 1994), a recognized way to assess molecule mixtures in close association is to vary the resulting molar absorptivity according to the reciprocal orientation of chromophores (Cantor and Schimmel, 1980). This is because the close interaction of the transition dipole moment of an absorbing chromophore and the induced dipoles of neighboring chromophores depend on their reciprocal orientation and may either increase (hyperchromism) or decrease (hypochromism) the molar absorptivity (Cantor and Schimmel, 1980; Freifelder, 1982). The observed decrease in peak absorbance in the effluent from the first column indicates that the molecular absorptivity of the first peak was lower than for the second column. A most probable reason for the decrease in the UV reading (hypochromism) is that a separation of molecules (or chromophores) from the high-molecular-size arrangement may have taken place during elution from the first column, thereby causing an alteration of the reciprocal orientation of the dipole moments among chromophores. This can be attributed to the higher resolution of the first column that is able to disrupt during HPSEC elution of humic materials which are only loosely bound molecular clusters. When gel pores are smaller than the aggregate hydrodynamic radius, small humic molecules are separated from the larger aggregate. The aggregate molar absorptivity is thus decreased, as is the absorbance reading of the eluting fraction at that molecular size.

The findings of Conte and Piccolo (1999b) showed that the behavior of HS in both high- and low-pressure chromatography is consistent with the model for association of small heterogeneous molecules into apparently large high-molecular-size structures. Since the apparent macromolecular structure of HS reflects an assemblage of molecules associated by weak forces, HPSEC chromatograms might provide a measure of the conformational stabilities of humic materials from different sources and an indication of their molecular composition.

An HPSEC experiment was conducted by Conte and Piccolo (1999a) with a TSK 3000SW column to verify the stabilities of various humic solutions. The mobile phase (0.05 *M* $NaNO_3$, pH 7, not absorbing light at 280 nm) was modified by small additions of methanol (4.6×10^{-7} *M* to a pH of 6.97), HCl ($\leq 2.0 \times 10^{-6}$ *M* to pH 5.54), and acetic acid (4.6×10^{-7} *M* to a pH of 5.69) so that the ionic strength of the eluting solution was always kept constant. In fact, all mobile phases had the same ionic strength (I= 0.0504 *M*) because the low ionic changes introduced in solution modified with low amounts of HCl and acetic acid ($\leq 2.0 \times 10^{-6}$ *M*) did not significantly vary the I values. Moreover, sodium humate and fulvate solutions (0.5 $g.L^{-1}$) were previously titrated to pH 7 so that their dissolution in the mobile phase at pH 7 would have prevented any random occurrence of negative charges. This was done to avoid any uncontrolled formation of negative charges on the solute that have been ascribed to change the ionic strength of humic solutions, thereby affecting the volume of sample elution (Swift and Posner, 1971). UV-Vis and refractive index (RI) detectors were used to obtain the chromatograms of the molecular size distribution of the humic materials. The objective was to compare the chromatographic behavior of chromophores (UV at 280 nm) with that of the real humic mass (RI).

Modifications of the mobile phase resulted in progressive decreases in the molecular sizes of humic solutions when going from the control solution to those to which methanol, HCl, or acetic acid were added. The HPSEC chromatograms obtained by means of the UV-Vis and the RI detectors for a HA isolated from a Danish agricultural soil are shown in Figures 4.2 and 4.3, respectively. The decrease in molecular size is shown in the shift toward

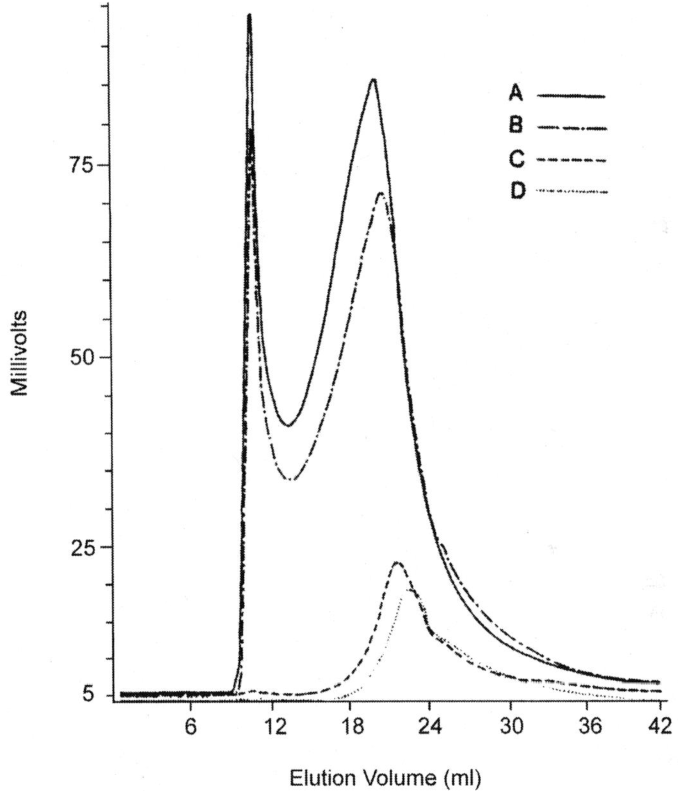

FIGURE 4.2. HPSEC chromatograms of a humic acid from a Danish soil recorded with the UV-Vis detector. A = control mobile phase (0.05 M $NaNO_3$, pH = 7, I = 0.05); B = same as A but 4.6×10^{-7} M in methanol (final pH 6.97); C = same as A but to pH 5.54 with HCl; D = same as A but 4.6×10^{-7} M in acetic acid (final pH 5.69). (*Source:* Reprinted with permission from Conte and Piccolo, 1999a, p. 1686. Copyright 1999 American Chemical Society.)

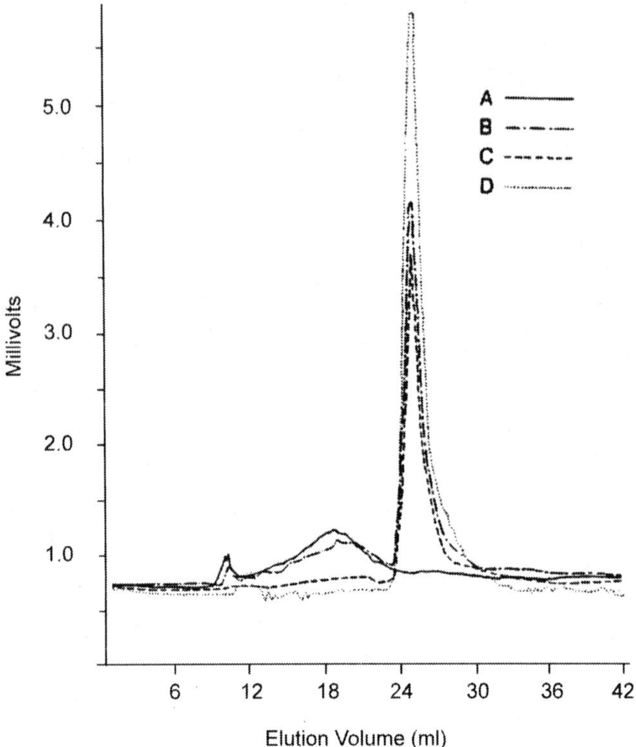

FIGURE 4.3. HPSEC chromatograms of a humic acid from a Danish soil recorded with the RI detector. A = control mobile phase (0.05 M $NaNO_3$, pH = 7, I = 0.05); B = same as A but 4.6×10^{-7} M in methanol (final pH 6.97); C = same as A but to pH 5.54 with HCl; D = same as A but 4.6×10^{-7} M in acetic acid (final pH 5.69). (*Source:* Reprinted with permission from Conte and Piccolo, 1999a, p. 1686. Copyright 1999 American Chemical Society.)

increasingly larger elution volumes for both the UV and RI detectors, and by the concomitant reductions in peak absorbance at the UV-Vis detector. The variation in molecular size distribution in the different mobile phases is reflected in the weight-average molecular weights of the humic material calculated from the chromatograms (Table 4.1).

The observed changes were not attributable to variations in ionic strength because this was kept constant in the different mobile phases, but they could have been attributable to the interactions of the added chemicals with the weakly associated macromolecular structure of the humic matter. In the

TABLE 4.1. Weight-Average Molecular Weight *(M_w)* and Polydispersity (P) of HA from a Danish Soil in Different Mobile Phases As Determined by UV and RI Detectors

	A		B			C			D		
	M_w	P	M_w	P	%	M_w	P	%	M_w	P	%
UV	3500	3.6	34000	3.9	2.8	6500	1.9	81.4	5300	1.4	84.8
RI	51130	5.2	49670	5.3	2.9	3100	1.1	93.4	10720	1.3	79.0

Source: Reprinted with permission from Conte and Piccolo, 1999a, p. 1685.

M_w changes (%) as compared to control mobile phase (A) are reported
A. $NaNO_3$ (pH 7); B. CH_3OH (pH 6.97); C. HCl (pH 5.54); D. AcOH (pH 5.69);
P. Polydispersion: Weight-average molecular weight/Number-average molecular weight *(M_w/M_n)*.

case of the methanol addition, no new ions were introduced, nor were there changes in pH. Hence, the alteration in molecular size distribution of the humic materials can be attributed only to the capacity of the CH_3OH to form both van der Waals bonds with the hydrophobic humic components and hydrogen bonds with the oxygen-containing functional groups of the HA. Thus, a very small amount of methanol in the eluting solution could disrupt the weak forces that temporarily stabilized the humic associations in aggregates. The result can be interpreted as a dispersion of large aggregates into smaller humic molecules, a diffusion of these through smaller gel pores, and an overall decrease in molecular size (Table 4.1). This effect was confirmed by the RI chromatogram (Figure 4.3) which showed a shift of humic mass toward elution volumes typical of lower-molecular-weight materials.

Although the RI detector in the HPSEC studies of HS (Piccolo, Rausa, and Celano, 1990; Rausa, Mazzolari, and Calemma, 1991; von Wandruszka et al., 1999) is an essential means to evaluate mass rather than chromophoric distribution of HS, it may not completely exclude the polymeric structure of HS as proposed by the classical random-coil model. In fact, supporters of that model may argue that the RI results may be due to a dramatic coiling down of large macromolecules rather than dispersion of weakly associated small molecules. Such a hypothetical explanation based on a decrease of the HS hydrodynamic radius had been previously used, but never proved, to account for elution shifts to higher elution volumes (Berden and Berggren, 1990; Chin and Gschwend, 1991).

However, the traditional polymeric model of HS should not explain the experimental observation of the decreased peak intensities (seen in UV chromatograms where the various treatments were applied) compared with those for the control solution (Figure 4.2). In fact, the substantial reduction in peak intensity revealed by the UV detector should be regarded as evidence for the hypochromic effect described previously, as a result of real separation of small molecules for diffusion through pores. Not only were the peaks shifted to higher elution volumes (lower molecular sizes), but the decreases in molecular absorptivities of the humic fractions indicated that the chromophores were drawn apart from each other because of the disrupting effect of methanol on the loosely associated humic structures. If humic samples had been composed of polymeric macromolecules which, despite the constant ionic strength, had coiled down to give the changes observed in the elution profiles, one would have instead expected an increase in molecular absorptivity and UV readings with respect to control solution.

The greater changes produced by decreasing the pH of the control solution (with addition of HCl) to 5.54 were due to a larger disruption (than for methanol) of the humic molecular associations (Figures 4.2 and 4.3). The additional hydrogen ions in this mobile phase protonated the humic carboxylic functionalities (which were in their dissociated forms at pH 7 with the control mobile phase). A number of negative charges were then neutralized and hydrogen bonds were concomitantly formed between the complementary functionalities of the humic molecules. In that way the conformational stability that existed in the control solution was disrupted. Due to the parallel observation that a reduction in peak absorbance (hypochromism) was seen in the UV chromatograms, and that a shift of humic mass to high elution volumes was visible in RI chromatograms, this change could not simply be due to a volume decrease of the random coil, as was previously suggested (Swift and Posner, 1971; Berden and Berggren, 1990; Chin and Gschwend, 1991; von Wandruszka et al., 1999). Again, a more plausible explanation is that the heterogeneous humic conformation collapsed into molecular associations of smaller dimensions, but of greater thermodynamic stabilities than for the control solution. The chemical rationale for this behavior lies in the energy gained in hydrogen bond formation (ranging from 10 to 20 kJ mol^{-1}) compared with van der Waals bonding (Schwarzenbach, Gschwend, and Imboden, 1993). Humic molecules, protonated by HCl addition, abandoned the loose conformation assumed at the pH 7 of the control solution and formed relatively strong intermolecular hydrogen bonds. The concomitant large decrease in molecular size suggests that the weak association of apparently high molecular size, as observed for the HA in the control solution, must therefore have been due predominantly to weak intermolecular hydrophobic forces such as van der Waals, π-π, and CH-p (Nishio, Hirota, and Umezawa, 1998) bonds which hold small molecules together.

The further decrease in molecular size distribution with acetic acid addition (Figure 4.2 and Table 4.1) can be attributed to the methyl group of the acetic acid. As for HCl addition, acidification of the solution favored hydrogen bond formation. Moreover, the weak acidity of CH_3COOH (pKa 4.8) allowed a small number of undissociated species to exist at pH 5.69 and thus the formation of mixed intermolecular hydrogen bonds with humic molecules. As noted already, such energy-driven rearrangements will outweigh the weak humic associations in the control solution that were stabilized mainly by hydrophobic forces. However, the shift to higher elution volumes and the general reduction of molecular absorptivity in the UV chromatogram (hypochromism), as well as the large shift to the total column volume of the humic mass in the RI chromatogram, would suggest that the methyl group (the apolar end) of acetic acid played an additional role in further disrupting the weakly bound humic associations. The apolar methyl group of acetic acid must have altered the residual hydrophobic forces that still stabilized the humic associations even after the hydrogen bonding rearrangement had taken place.

The work of Conte and Piccolo (1999a) on both humic and fulvic acids provides further direct evidence for the conformational model based on the reversible self-association of small humic molecules rather than on the macropolymeric random-coil concept. Furthermore, using HPSEC it was shown that the molecular size distribution of HS must be interpreted by a combination of two factors: the elution volume and the molar absorptivity of the chromatographic peaks. Earlier research failed to address the combination of these factors because either closely similar HAs were analyzed only by UV detection in less-sensitive, low-pressure size-exclusion systems (Swift and Posner, 1971), or poorly UV-absorbing fulvic acids and/or dissolved organic matter (DOM) samples were used in studies with HPSEC systems without the support of a RI detector (Chin and Gschwend, 1991; Berden and Berggren, 1990; Chin, Aiken, and O'Loughlin, 1994). More recent HPSEC investigations that coupled UV, RI, and MALS (multi-angle light scattering) detectors have also indicated different molecular size distribution for HS according to the detector employed (von Wandruszka et al., 1999). Humic size reductions that were reported by Conte and Piccolo (1999a) on modifications of the mobile phase may be attributed to a disruption of weak humic associations giving rise to separated smaller molecules rather than to the compaction of macromolecular coils. Intermolecular hydrophobic interactions appear to be the predominant binding forces for associations of relatively small humic molecules. These are consequently stabilized by the entropy-driven (the hydrophobic effect) tendency to exclude water molecules from humic aggregates, and thus decrease total molecular energy in solution (Tanford, 1980; Israelachvili, 1994).

Although ionic-exclusion interferences were carefully eliminated in the work of Conte and Piccolo (1999a), permanent adsorption of HS on the column was not found to occur (Conte and Piccolo, 1999b). However, a modification of the mobile phase with additions of methanol, HCl, and acetic acid, although in very small amounts, may have produced pore-sized changes which determine non-size-exclusion chromatography.

To verify that mobile phase modification did not alter the exclusion properties of the HPSEC column, Piccolo, Cozzolino, and Conte (2001) subjected polymeric standards of known molecular weights, such as the negatively charged polystirenesulphonates (PSS) and neutral polysaccharides (PYR), to size-exclusion chromatography in the same mobile phases used for HS by Conte and Piccolo (1999a). Moreover, because of the undisputed polymeric nature of both nonionic PYR and polyelectrolytic PSS, a comparison between the behavior of real covalently bound polymers and that of HS in the same chromatographic conditions would have provided information on the humic conformational structure.

Figure 4.4 shows curves relating elution volumes and log of MW obtained for both PYR (RI detector) and for PSS standards (both UV and RI detectors) when these were dissolved and eluted in the four different mobile phases as done by Conte and Piccolo (1999a). No significant differences were shown from the curves obtained for the uncharged PYR standards in the different mobile phases (Figure 4.4, I). The linear equations were similar and the slight differences were well within the statistical analytical significance (2% of relative standard deviation). The curves obtained for the polyelectrolytic PSS standards were found to be very similar in the different elution phases by either the UV or the RI detector (Figure 4.4, II and III). The relative equations for the respective four mobile phases were also very similar.

Regardless of the charge density of the employed polymers, the lack of differences in their chromatographic behavior indicated that the slight variation in the composition of mobile phases was not sufficient to alter either the conformational stability conferred to these macromolecular polymers by strong covalent bondings or their interactions with the stationary phase. Conversely, the size-exclusion chromatograms of three different humic acids (HA1 from a volcanic soil, HA2 from an oxidized coal, and HA3 from a lignite), revealed by either UV or RI detectors, varied dramatically with the composition of mobile phases in both peak absorbance and elution volumes.

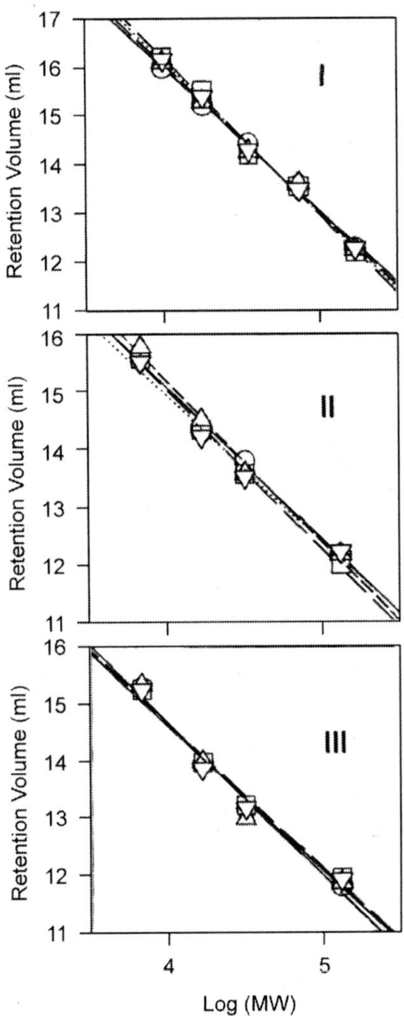

FIGURE 4.4. HPSEC calibration curves by RI detector of polysaccharide (PYR) standards (I) and by both UV (II) and RI (III) detectors of polystyrenesulphonate (PSS) standards of known MW first dissolved and then eluted in different mobile phases. A = (O) control mobile phase (0.05 M $NaNO_3$, pH = 7, I = 0.05); B = (Δ) same as A but 4.6×10^{-7} M in methanol (final pH 6.97); C = (□) same as A but to pH 5.54 with HCl; D = (∇) same as A but 4.6×10^{-7} M in acetic acid (final pH 5.69). (*Source:* Piccolo, Cozzolino, and Conte, 2001, p. 181. © Copyright Lippincott, Williams & Wilkins, 2001.)

The resulting *Mw* and polydispersity values as well as percent reduction in respect to control mobile phase are shown in Table 4.2.

Piccolo, Cozzolino, and Conte (2001) concluded that, while slight modifications in the mobile phase did not affect the column capacity for size exclusion, the substantial difference between the response to HPSEC of the polymeric standards and that of HS in the very same chromatographic conditions proved that humic materials have a different conformational structure. As proposed earlier, this may well be a self-assembling association of relatively small and heterogeneous molecules instead of a coil of polymeric macromolecules.

Piccolo, Cozzolino, and Conte (2001) further studied the decrease of molar absorptivity observed in HS chromatograms with modifications in the mobile phases. To verify that such variations were not limited to the wavelength used to record chromatograms (280 nm), thereby being simply accountable to shifts of peak maxima, UV spectra of HS solutions were recorded over a range of wavelengths. Figure 4.5 shows that the three humic

TABLE 4.2. Weight-Average Molecular Weight (M_w) and Polydispersity (P) of Humic Samples in Different Mobile Phases as Determined by UV and RI Detectors

Sample	A		B			C			D		
	M_w	P	M_w	P	%	M_w	P	%	M_w	P	%
HA1											
UV	23000	2.8	29000	3.5	+26.1	8200	1.9	–64.3	4000	1.2	–82.6
RI	22700	2.9	9200	2.1	–68.8	3400	1.3	–85.2	2250	1.1	–90.2
HA2											
UV	9700	1.7	4200	1.1	–56.7	3600	1.0	–62.9	2900	1.2	–70.1
RI	9500	1.6	2930	1.4	–69.1	2100	1.0	–77.9	2200	1.0	–76.8
HA3											
UV	17000	2.0	7900	1.5	–53.5	9000	1.8	–47.0	3500	1.1	–79.4
RI	16650	3.5	7790	1.7	–53.2	7990	1.5	–52.0	2200	1.0	–86.7

Source: Piccolo, Cozzolino, and Conte, 2001, p. 178. © Copyright Lippincott, Williams & Wilkins, 2001.

M_w changes (%) as compared to control mobile phase (A) are reported.
A. 0.05 M $NaNO_3$ (pH 7); B. As solution A but added with CH_3OH (pH 6.97); C. As solution A but added with HCl (pH 5.54); D. As solution A but added with AcOH (pH 5.69); P. Polydispersity (M_w/M_n).

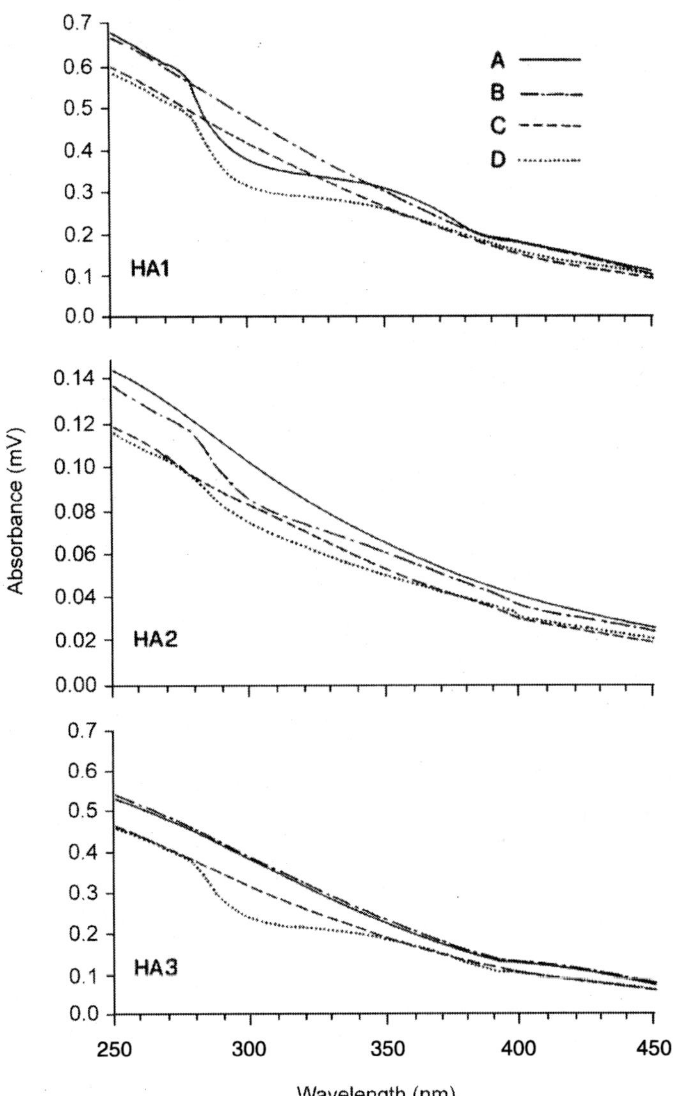

FIGURE 4.5. UV spectra (250-450 nm) of solutions (0.5 g.L^{-1}) of HA1, HA2, HA3 in: A = control solution (0.05 M NaCl, pH = 7, I = 0.05); B = same as A but 4.6×10^{-7} M in methanol (final pH 6.97); C = same as A but to pH 5.54 with HCl; D = same as A but 4.6×10^{-7} M in acetic acid (final pH 5.69). (*Source:* Piccolo, Cozzolino, and Conte, 2001, p. 182. © Copyright Lippincott, Williams & Wilkins, 2001.)

materials produced different absorbance values upon modification of their solutions. Although addition of methanol caused a reduction in molecular absorptivity at all wavelengths in HA2 and only for some wavelengths in HA1, it did not have any effect in HA3. Addition of HCl produced, instead, lower absorbance values than control solution for all HAs, except for the 290-330 nm range in HA1. After adding acetic acid, a significant decrease in absorbance was observed for all humic solutions and over all wavelengths. These results showed that molar absorptivity of the bulk HS varied with composition of the solution and was reduced by adding chemicals which disrupt their weakly stabilized molecular associations.

HPSEC was applied by Piccolo, Conte, and Cozzolino (1999) to confirm previous results obtained by low-pressure gel-permeation chromatography (Piccolo, Nardi, and Concheri, 1996a,b). Humic solutions were titrated to pH 3.5 using either a mineral acid (HCl) or different monocarboxylic (formic, acetic, propionic, and butyrric) acids, and then eluted through a HPSEC Biosep S2000 column (Phenomenex). In this case, a mobile phase of constant composition and pH 7 (0.05 M $NaNO_3$) was used.

All HS used by Piccolo, Cozzolino, and Conte showed a decrease in the UV absorbance of chromatographic peaks when treated with either HCl or monocarboxylic acids. Again, such decreases in peak intensities were attributed to the hypochromic effect arising when the closely associated molecules in the conformational arrangements of the humic materials at pH 7 were separated by the additions of both mineral and monocarboxylic acids. The combined effect of reduction of peak intensity and the shift of peaks to higher elution volumes were taken as evidence of the disrupting effects of the added acids on the original conformations of the HS. As previously explained, molecular separation upon acid treatment was attributed to formation of stronger intermolecular hydrogen bonds which altered the original conformations that were stabilized mainly by weaker hydrophobic interactions. The net effect allowed the separation of smaller molecules during HPSEC elution. Addition of organic acids not only further decreased the peak absorbances of HS, but also enhanced their shifts to higher elution volumes, indicating a more extensive disruption of the original association compared with that obtained for HCl addition.

The work of Piccolo, Conte, and Cozzolino (1999) represents additional evidence that HS do not behave as polymeric random coils. It also indicates that the HPSEC technique can be used to reproducibly decrease the apparently large dimensions of humic associations into fractions of smaller molecular sizes as a result of simple interactions with monocarboxilic acids. It indicated that the extent of size reductions of humic associations depends on the aliphatic chain lengths of the acids and the hydrophobicities of HS. The difference between the effect of HCl observed by HPSEC and that by the low-pressure mode may be attributed to the larger resolution and sensitivity of the HPSEC technique.

Cozzolino, Conte, and Piccolo (2001) studied the effect of organic acids of plant, microbial, or anthropic origin on the molecular size distribution of dissolved HS. They used a Phenomenex Biosep S2000 under elution with a 0.05 *M* $NaN0_3$ solution to evaluate size changes in four different HS upon addition of hydroxy (glycolic and malic), keto (glyoxylic), and sulfonic (benzenesulfonic and methanesulfonic) acids. All HS showed a decrease in peak absorbance when humic matter was dissolved in the HPSEC mobile phase at pH 7 and the pH of the solution was lowered to 3.5 by acid addition before analysis. This effect was generally accompanied by an increase in peak elution volumes. The combination of the two effects led to a comparison of the changes observed for treated samples in relation to the control by measuring the total area of chromatograms (Table 4.3). Results were also explained with a disruption of supramolecular humic associations into smaller-sized but energy-richer conformations brought about by formation of mixed intermolecular hydrogen bonds upon acid treatment.

Table 4.3 shows that the hydroxy-bicarboxylic malic acid was the most effective in disrupting the original humic associations among the carboxylic acids. This was attributed to its greater capacity to form new hydrogen bonds with complementary functions of HS. In fact, malic acid is a bicarboxylic acid (HOOC-CH_2-CHOH-COOH) with two carboxyl groups ($pKa_1 = 3.4$; $pKa_2 = 5.11$) that can be either protonated or partially dissociated at pH 3.5, thereby allowing for a larger number of mixed hydrogen bonds on its oxygen-containing functions than for either the strong hydro-

TABLE 4.3. Variation (%) in Respect to Control of Total Area in HPSEC Chromatograms of Four Different Humic Acids upon Addition of Acids

	Humic substances			
Acid	**HA1**	**HA2**	**HA3**	**HA4**
HCl	–51.5	–14.4	–63.0	–56.6
Glyoxylic	–23.2	–24.0	–20.8	–21.1
Glycolic	–54.6	–46.6	–47.8	–50.6
Malic	–65.3	–69.6	–72.6	–73.1
Methanesulfonic	–73.0	–50.5	–74.6	–54.7
Benzenesulfonic	–36.7	–52.8	–43.2	–25.8

Source: Cozzolino, Conte, and Piccolo, 2001, p. 570.

chloric acid that is only a proton donor or the more weakly acidic monocarboxylic glycolic and glyoxylic acids.

The extent of conformational variation was related not only to pKas of acids but also to the chemical and stereochemical affinity of humic components that may allow penetration of acids into the inner humic domains depending on acid structures. For instance, the strongly acidic methanesulfonic and benzenesulfonic acids showed effects that varied with the humic properties (Table 4.3). Although methanesulfonic acid was more effective in conformational disruption of HS predominantly containing aliphatic and alkyl moieties, the size distribution of an aromatic-rich humic material was equally varied by benzenesulfonic acid because of the probable larger π-π interactions with aromatic humic components (Figure 4.6).

ENZYMATIC POLYMERIZATION OF HUMIC MOLECULES

Understanding humus as a supramolecular association of small molecules means overcoming the limitations imposed by the paradigmatic polymeric model. If HS are seen as weakly bound supramolecular associations, their unstable conformation could then be stabilized in real polymeric structures. This could be achieved by increasing the number of intermolecular covalent bonds via an oxidative coupling reaction catalyzed by oxidative enzymes such as the phenoloxidases. This class of enzymes has been shown to promote, through a free-radical mechanism, oligo- and polymerization of phenols and anilines and hence is believed to contribute to soil detoxification from related organic contaminants (Kim, Fernandes, and Bollag, 1997). However, no evidence of the direct catalytic action of these enzymes on humic molecules has ever been produced. Although covalent binding of contaminants to HS was observed, there was no reason to evaluate a size increase in a humic matter that was assumed to be already polymeric.

Piccolo et al. (2000) attempted to turn a loosely bound humic superstructure into a covalently linked polymer by treating it with horseradish peroxidase (HRP) and hydrogen peroxide (oxidant), a humic material dissolved in 0.1*M* phosphate buffer at pH 7. They used a HPSEC Biosep S2000 column (Phenomenex) to evaluate the changes in molecular size distribution brought about by the oxidative reaction with HRP catalysis. Moreover, addition of acetic acid to the reacted humic mixture to pH 4 before HPSEC injection was used to assess the stability of humic conformation following the polymerization reaction.

Figure 4.7 shows the HPSEC chromatograms obtained with control humic solution before (Figure 4.7, A) and after (Figure 4.7, B) acetic acid addition. The control solution showed only a slight absorption at the void volume (V_0) which is characteristic of high-molecular-sized fractions. Treatment of this so-

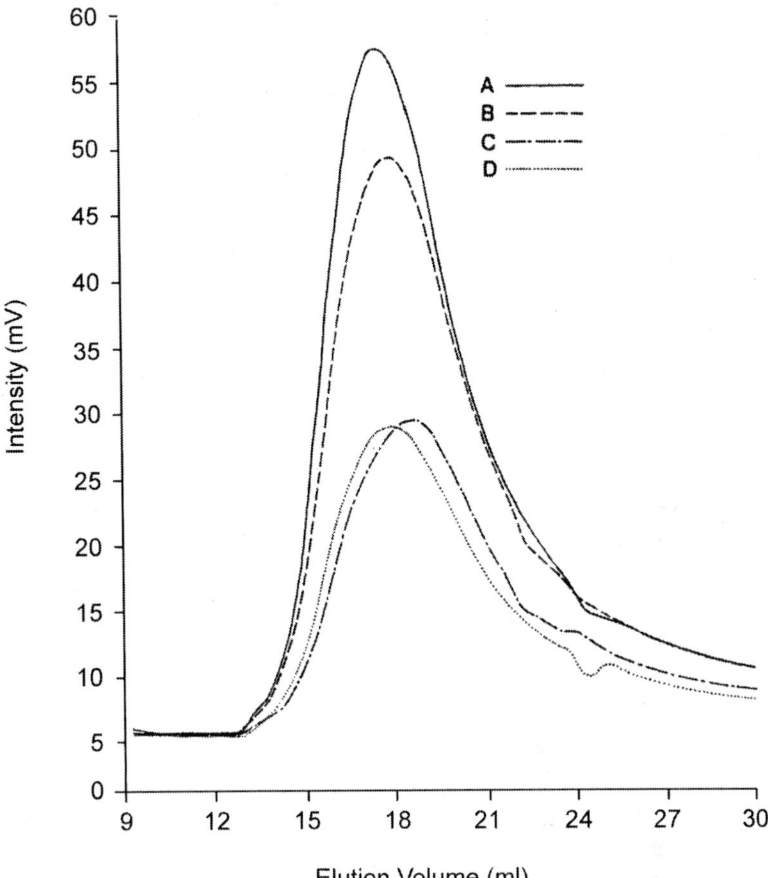

FIGURE 4.6. HPSEC chromatograms of HA2. A = control solution at pH 7; B = same as A but pH lowered to 3.5 with HCl; C = same as A but pH lowered to 3.5 with methanesulfonic acid; D = same as A but pH lowered to 3.5 with benzenesulfonic acid. (*Source:* Cozzolino, Conte, and Piccolo, 2001, p. 569.)

lution with acetic acid to pH 4 decreased the molecular size distribution of the humic material as also observed in other research described previously. Similar behavior was shown by the humic solution when treated with either H_2O_2 (Figure 4.7, C) or HRP (Figure 4.7, E) alone and after acetic acid addition (Figure 4.7, D and F, respectively). The lower intensity of peak absorption (hypochromism) in the latter solutions suggests an influence of both oxidant and enzyme on the relative distance (and dipole orientation) among chromophores. A

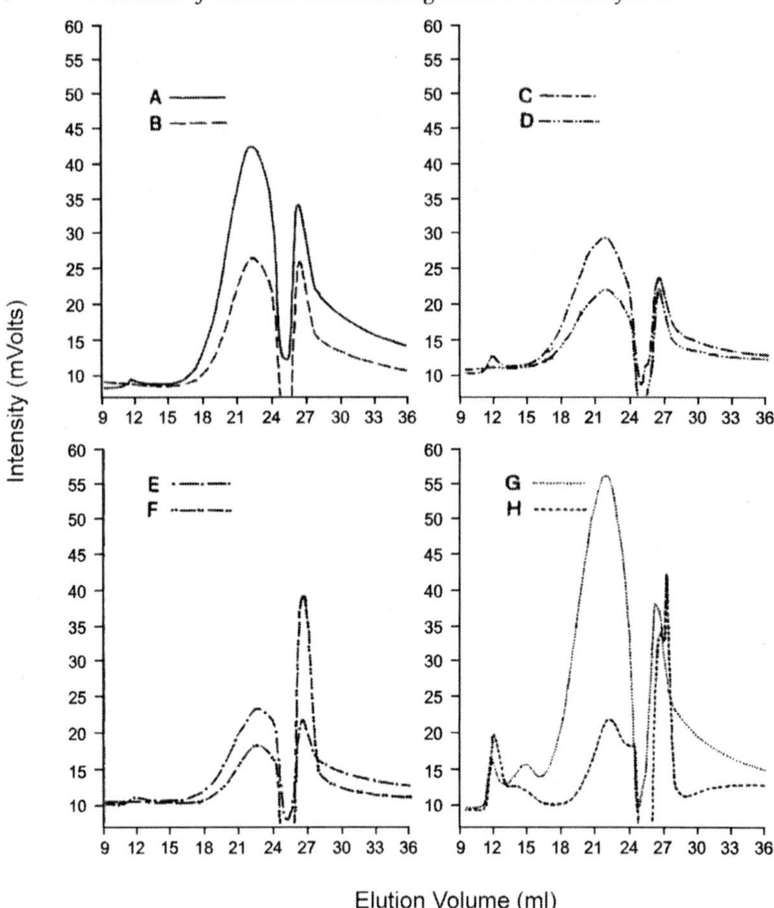

FIGURE 4.7. HPSEC chromatograms of a control humic solution in phosphate buffer at pH 7 (A), of the same solution as in A but added with H_2O_2 (C), of the same solution as in A but added with peroxidase (E), of the same solution as in A but added with both H_2O_2 and peroxidase (G) and of the same solutions but added with acetic acid to pH 4 before injection (B, D, F, H, respectively). (*Source:* Piccolo et al., 2000, p. 392. Copyright 2000 Springer-Verlag.)

degree of disaggregation of the humic supramolecular structure into smaller associations by the presence of peroxidase alone was indicated in the acetic acid-treated sample (Figure 4.7, F) by the concomitant reduction of intensity in the diffused peak and its enhancement in the peak eluted after the solvent hump (around 25 mL).

The chromatogram of the humic solution subjected to the oxidative-coupling reaction with both H_2O_2 and HRP (Figure 4.7, G) was distinctly different from the control chromatograms (Figure 4.7, A, C, E). The peak at V_0 was increased, a new peak appeared at around 14.7 mL, and the large diffused peak was not only more intense than in control solutions but also shifted to lower elution volumes (21.5 mL versus about 22.5 mL). These changes indicated a significant increase in the molecular size of humic material with oxidation catalyzed by HRP.

Treatment of the humic solution with acetic acid confirmed that the size increase was due to a true polymerization of humic molecules via formation of carbon-oxygen or carbon-carbon bonds rather than to a different supramolecular association stabilized by weak forces. In fact, unlike control solutions, the peak at V_0 not only maintained its intensity but it was even increased in the chromatogram of sample treated with acetic acid (Figure 4.7, H). The same behavior was partially shown by the new peak that appeared at 14.7 mL after the polymerization reaction. This suggested that the high-molecular-sized material excluded at these elution volumes was stabilized by stronger forces than in control samples and their macromolecular arrangement could not be disrupted by addition of acetic acid. However, the reduced intensity of the large diffused peak (about 21.5 mL) after acetic acid treatment suggests a hypochromic effect due to chromophores which, being not yet covalently bound in polymeric structures, were separated from their weakly bound associations and eluted at larger elution volumes.

Piccolo et al. (2000) studied infrared spectroscopy using humic samples that underwent an oxidative catalyzed reaction to collect future evidence of the formation of covalent bonds. DRIFT (diffuse reflectance infrared fourier transform) spectra of HRP alone, a control HS, and HS oxidized by HRP catalysis are shown in Figure 4.8 A, B, and C, respectively. In comparison to the control, the DRIFT spectrum of the humic material subjected to oxidative coupling showed a substantial change in the 1500-900 cm^{-1} frequency interval with the appearance of three main bands at 1247, 1097, and 947 cm^{-1} and a decrease in the 1400 and 1227 cm^{-1} bands. The absorptions shown at 1247 and 1097 cm^{-1} were reasonably assigned to bond deformation of aryl and alkyl ethers, respectively, which were formed during free-radical coupling reactions catalyzed by HRP and, hence, confirm the interpretation of HPSEC measurements.

The HPSEC and DRIFT results of Piccolo et al. (2000) suggest that the small heterogeneous molecules present in HS, as in weakly associated superstructures, can be covalently bound into true oligomer or polymers by an oxidative coupling reaction catalyzed by a peroxidase enzyme. The extent of covalent polymerization should be a function of the amount of humic molecules, mainly phenolic or benzencarboxylic acids derived from lignin

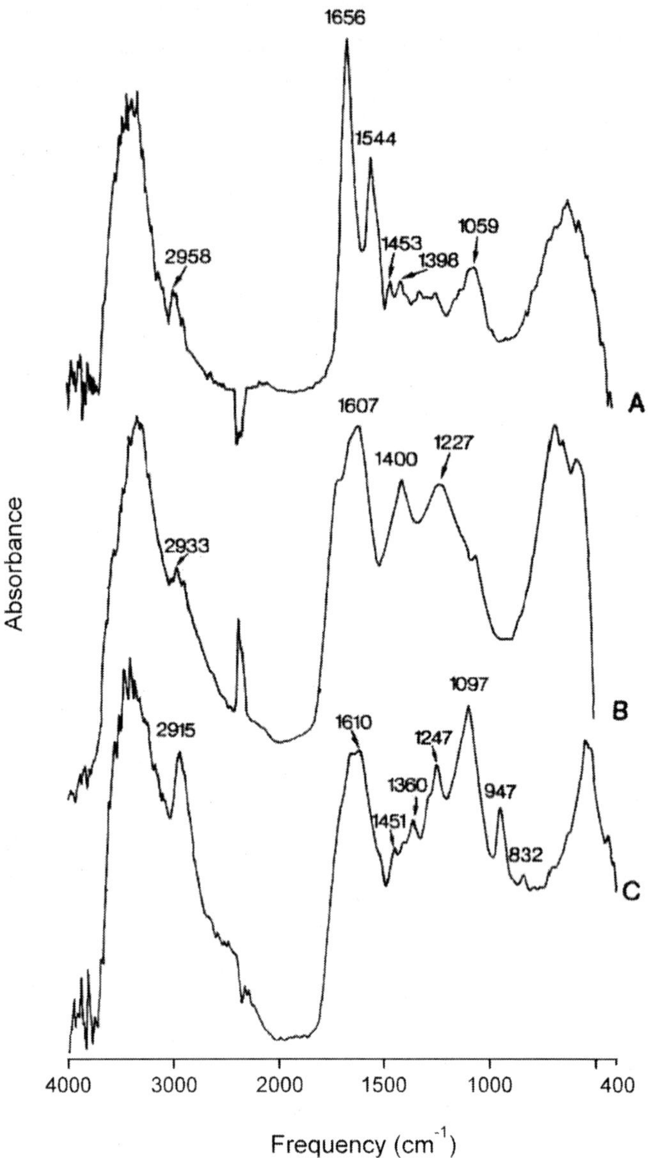

FIGURE 4.8. DRIFT spectra of Horseradish Peroxidase (A), humic acid (B), and humic acid subjected to oxidation catalyzed by Horseradish Peroxidase (C). (*Source:* Piccolo et al., 2000, p. 393. Copyright 2000 Springer-Verlag.)

and microbial biosynthesis which may undergo oxidative coupling reactions. However, other classes of compounds may become confined into the macromolecular conformations of polymerized humus.

Studies showing that humic supramolecular associations can be turned into more stable, covalently linked conformations of truly larger molecular size can be interpreted as additional evidence that HS should not be considered macromolecular polymers as they have been viewed for so long.

COMPUTATIONAL MODELS OF CONFORMATIONAL STRUCTURE OF HUMIC SUBSTANCES

A molecular simulation for the minimization of conformational energy was conducted using HyperchemT 4.0 software (Schulten, Leinweber, and Schnitzer, 1998) to describe the interactions of humic supramolecular associations with an organic acid. Eleven different molecular structures of compounds identified as components of humic substances (Stevenson, 1994) were grouped together in the simulation to form a supramolecular association. The structures represented molecules such as saturated and unsaturated fatty acids, carbohydrates, peptides, lignin derivatives, etc., with molecular weights varying from 116 daltons for a dihydroxybenzene to 504 daltons for a triglucose. The total molecular weight of the eleven molecules was 3,065 daltons.

The geometry of the association was automatically adjusted and its conformational energy was minimized in a vacuum (Figure 4.9, A). Ten molecules of acetic acid were added first to surround the hypothetical supramolecular association (Figure 4.9, B) and then placed within the conformation of the association (Figure 4.9, C). The resulting association energies were calculated by the software to be 114, 91.2, and 84.0 $Kcal.mol^{-1}$, respectivcly. The association of the different molecules also varied its physical appearance with the approach of acetic acid molecules which caused a loosening of intermolecular attractions until some spaces among the molecules were formed. If such a loosened-down molecular association was eluted through a HPSEC column, as in the experiments described previously, some of the molecules, held together only by weak dispersive interactions, would be separated and would come out from the column at larger elution volumes. This computer simulation (Figure 4.9) pictorially shows that the addition of organic acids to humic molecules is capable of reducing the solvation energy and, concomitantly, causing a partial disruption of their association. These results are in line with the experimental HPSEC findings described previously.

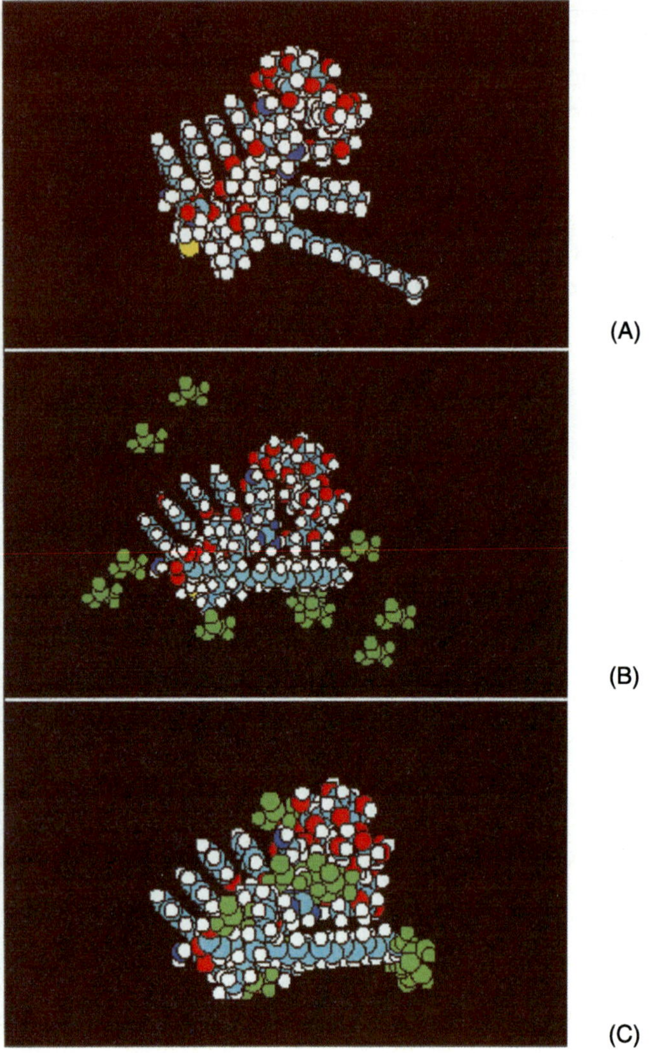

FIGURE 4.9. Computer simulation of the optimum conformational energy *(in vacuo)* for an association of eleven different humic precursors with a total molecular weight of 3,065 daltons. A. upper picture: molecular association with an energy of 114 $Kcal.mol^{-1}$; B. middle picture: molecular association surrounded by ten molecules of acetic acid with an energy of 91.2 $Kcal.mol^{-1}$; C. lower picture: molecular association containing ten molecules of acetic acid with an energy of 84.0 $Kcal.mol^{-1}$.

Conversely, when a covalently linked structure of humic substances, based on the polymeric model of Schulten, Leinweber, and Schnitzer (1998), was placed in the same exercise of molecular simulation, no significant changes in energy content and physical association were noted with the addition of acetic acid. Figure 4.10A shows that a polymeric structure with a molecular weight of 6,326 daltons is not significantly altered (Figure 4.10, B) by the same amount of acetic acid molecules used for the simulation of the weakly bound supramolecular association shown in Figure 4.9. Moreover, the gain in conformational energy was only of 10 $Kcal.mol^{-1}$, passing from 627.40

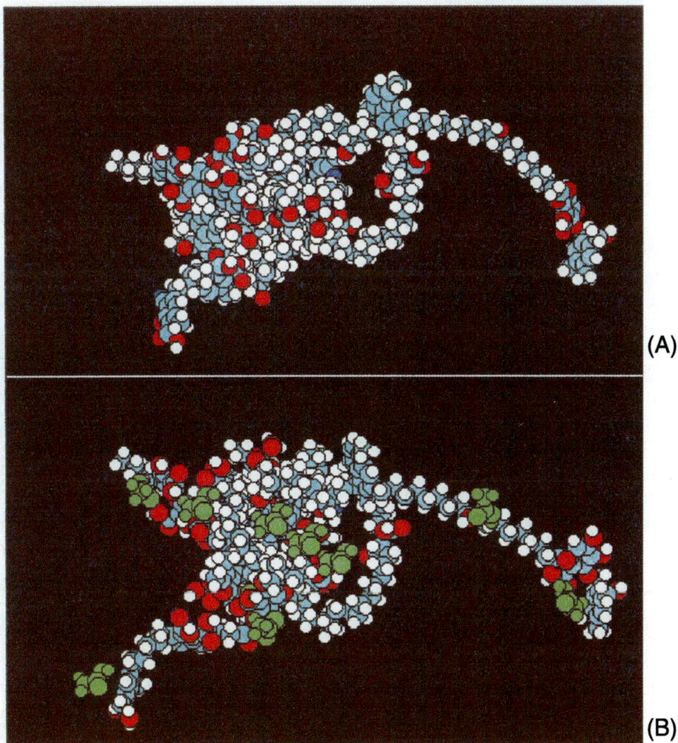

FIGURE 4.10. Computer simulation of the optimum conformational energy *(in vacuo)* of a covalently linked humic polymer (MW = 6,326 daltons) as hypothesized by Schulten, Leinweber, and Schnitzer (1998). A. upper picture: humic polymer having an energy of 627.40 $Kcal.mol^{-1}$; B. lower picture: humic polymer containing ten molecules of acetic acid and having an energy of 617.26 $Kcal.mol^{-1}$.

Kcal.mol^{-1} for the polymer to 617.26 Kcal.mol^{-1} for the same polymer added with acetic acid. Thus, it would be hardly possible, using this hypothetical polymeric model, that the simple addition of acetic acid molecules to such a high molecular weight polymer would provide a conformational rearrangement leading to molecular disruption during HPSEC elution.

CONCEPTS OF SUPRAMOLECULAR ASSOCIATION OF HUMIC SUBSTANCES

The described experiments, which used analytical size-exclusion chromatography, cannot be explained by analytical interferences or the traditional polymeric model of HS. Rather, they can be interpreted with the concept of loosely bound humic supramolecular associations. Using this concept, one can imagine HS as relatively small and heterogeneous molecules of various origins which self-organize in supramolecular conformations. Humic superstructures of relatively small molecules are not associated by covalent bonds but are stabilized only by weak forces such as dispersive hydrophobic interactions (van der Waals, π-π, and CH-π bondings) and hydrogen bonds, the latter being progressively more important at low pHs. Hydrophilic and hydrophobic domains of humic molecules can be contiguous to or contained in each other and, with hydration water, can form apparently large-molecular-sized associations. In humic supramolecular organizations, the intermolecular forces determine the conformational structure of HS and the complexity of the multiple, noncovalent interactions control their environmental reactivity.

The definition given by Lehn (1995) may well be applied to HS: "supramolecular assemblies (are) molecular entities that result from the spontaneous association of a large undefined number of components into a specific phase having more or less well-defined microscopic organization and macroscopic characteristics depending on its nature (such as films, layers, membranes, vesicles, micelles, mesomorphic phases, solid state structures, etc.)" (p. 201).

Using the concept of supramolecular association, the classical definitions of humic and fulvic acids should be reconsidered. Fulvic acids may be regarded as associations of small hydrophilic molecules in which there are enough acidic functional groups to keep the fulvic clusters dispersed in solution at any pH. Humic acids are made by associations of predominantly hydrophobic compounds (polymethylenic chains, fatty acids, steroid compounds) which are stabilized at neutral pH by hydrophobic dispersive forces (van der Waals, π-π, and CH-π bondings). Their conformations grow progressively in size when intermolecular hydrogen bondings are increasingly formed at lower pHs, until they flocculate.

SUMMARY

The clarification of the conformational structure of humic substances represents a major innovation in humus chemistry. The notion that humic substances are not macromolecular polymers, but rather superstructures of only apparent large size and self-assembled by relatively small heterogeneous molecules held together by mainly hydrophobic dispersive (van der Waals, π-π, CH-π) forces, opens up new opportunities to enlarge the knowledge of both their detailed chemistry and their management in the soil and the environment. Chromatographic methods of separation such as HPSEC were found to produce reproducible and more homogeneous fractions of the molecules constituting the humic superstructures. Awareness of the weak forces that cause the self-assembling of humic molecules has allowed the creation of methods based on interactions with chemical species such as amphiphilic organic acids, urea, mono- and polyvalent cations, which can disrupt the apparently large humic associations and obtain fractions that are chemically even simpler and more homogeneous. The novel understanding of humic substances as supramolecular associations has great implications in soil and environmental management. One example is the possibility to turn the loose humic superstructures into real, covalently linked polymers by a catalytic technology that can ensure polymerization of humic molecules in both water and soil environments. Finally, there are a large number of observations indicating that humic molecules released from the large supramolecular associations can influence nutrient uptake by plants and significantly increase crop yields. Combinations of refined chemical analysis of humic molecules with physiological studies on their effect on plants may clarify the mechanism(s) by which soil humic substances increase crop yield.

REFERENCES

Aiken, G.R. and A.H. Gillam (1989). Determination of molecular weights of humic substances by colligative property measurements. In *Humic Substances II. In Search of Structure,* eds. M.H.B. Hayes, P. McCarthy, R.L. Malcolm, and R.S. Swift. New York: John Wiley and Sons, pp. 516-543.

Berden, M. and D. Berggren (1990). Gel filtration chromatography of humic substances in soil solutions using HPLC-determination of the molecular weight distribution. *Journal of Soil Science 41:* 161-172.

Cameron, R.S., B.K. Thornton, R.S. Swift, and A.M. Posner (1972). Molecular weight and shape of humic acid from sedimentation and diffusion measurements on fraction extracts. *Journal of Soil Science 23:* 394-408.

Campbell, C.A., E.A. Paul, D.A. Rennie, and K.J. McCallum (1967). Applicability of the carbon dating method of analysis to soil humus studies. *Soil Science 104:* 217-224.

Cantor, C.R. and P.R. Schimmel (1980). *Biophysical Chemistry. Part II: Techniques for the Study of Biological Structure and Function.* New York: Freeman.

Carter, C.W. and I.H. Suffet (1982). Binding of DDT to dissolved humic material. *Environmental Science and Technology 16:* 735-740.

Chin, Yu-P., G.R. Aiken, and E. O'Loughlin (1994). Molecular weight, polydispersity, and spectroscopic properties of aquatic humic substances. *Environmental Science and Technology 28:* 1853-1858.

Chin, Yu-P. and P.M. Gschwend (1991). The abundance, distribution, and configuration of porewater organic colloids in recent sediments. *Geochimica Cosmochimica Acta 55:* 1309-1317.

Clapp, C.E., W.W. Emerson, and A.E. Olness (1989). Sizes and shapes of humic substances by viscosity measurements. In *Humic Substances II. In Search of Structure,* eds. M.H.B. Hayes, P. McCarthy, R.L. Malcolm, and R.S Swift. New York: John Wiley and Sons, pp. 497-514.

Clapp, C.E. and M.H.B. Hayes (1999). Sizes and shapes of humic substances. *Soil Science 16:* 777-788.

Conte, P. and A. Piccolo (1999a). Conformational arrangement of dissolved humic substances. Influence of solution composition on association of humic molecules, *Environmental Science and Technology 33:* 1682-1690.

Conte, P. and A Piccolo (1999b). High pressure size exclusion chromatography (HPSEC) of humic substances. Molecular sizes, analytical parameters, and columns performance, *Chemosphere 38:* 517-528.

Cozzolino, A., P. Conte, and A. Piccolo (2001). Conformational changes of soil humic substances induced by some hydroxy-, cheto-, and sulphonic acids. *Soil Biology and Biochemistry 33:* 563-571.

De Haan, H., R.I. Jones, and K. Salonen (1987). Does ionic strength affect the configuration of aquatic humic substances, as indicated by gel filtration? *Freshwater Biology 17:* 453-459.

Engebretson, R. and R. von Wandruszka (1994). Micro-organization of dissolved humic acids. *Environmental Science and Technology 28:* 1934-1941.

Flaig, W. (1958). Die Chemie der organischer Stoffe in Boden und deren physiologische Wirkung. *Verhandlungen, II urteil, IV Bodenkunde Gesellschaft 2:* 23-45.

Flaig, W. and H. Beutelspacher (1968). Investigations of humic acids with the analytical ultracentrifuge. In *Isotopes and Radiation in Soil Organic Matter Studies,* ed. International Atomic Energy Agency. Vienna: International Atomic Energy Agency, pp. 23-30.

Flaig, W., H. Beutelspacher, and E. Riets (1975). Chemical composition and physical properties of humic substances. In *Soil Components,* Volume 1, ed. J.E. Gieseking. Berlin: Springer-Verlag, pp. 1-219.

Freifelder, D. (1982). *Physical Biochemistry,* Second Edition. New York: Freeman.

Ghosh, K. and M. Schnitzer (1980). Macromolecular structures of humic substances. *Soil Science 129:* 266-276.

Goring, D.A.I. (1971). Polymer properties of lignin and lignin derivatives. In *Lignins: Occurrence, Formation, Structure and Reactions,* eds. K.V. Sarkanen and C.H. Ludwig. New York: John Wiley and Sons, pp. 695-768.

Haider, K. (1987). The synthesis and degradation of humic substances in soil. In *Transactions 13th Congress of Soil Science,* ed. International Society of Soil Science. Hamburg, Germany: International Society of Soil Science, pp. 644-656.

Haider, K. and J.P. Martin (1967). Synthesis and transformation of phenolic compounds by *Epicoccus nigrum* in relation to humic acid formation. *Soil Science Society of America Proceedings 31:* 766-772.

Handerson, H.A. and A. Hepburn (1977). Fractionation of humic acid by gel permeation chromatography. *Journal of Soil Science 28:* 634-644.

Hayase, K. and H. Tsubota (1983). Sedimentary humic and fulvic acid as surface active substances. *Geochimica Cosmochimica Acta 47:* 947-952.

Hedges, J.I. (1988). Polymerization of humic substances in natural environments. In *Humic Substances and Their Role in the Environment,* eds. F.H. Frimmel and R.F. Christman. Chichester: John Wiley and Sons, pp. 45-58.

Israelachvili, J.N. (1994). *Intermolecular and Surface Forces,* Second Edition. London: Academic Press.

Jenkinson, D.S. (1981). The fate of plant and animal residues in soil. In *The Chemistry of Soil Processes,* eds. D.J. Greenland and M.H.B Hayes. Chichester: John Wiley and Sons, pp. 505-561.

Kim, J.-E., E. Fernandes, and J.-M. Bollag (1997). Enzymatic coupling of the herbicide betazon with humus monomers and characterization of reaction products. *Environmental Science and Technology 31:* 2392-2398.

Kononova, M.M. (1961). *Soil Organic Matter: Its Nature, Its Role in Soil Formation and in Soil Fertility.* New York: Pergamon Press.

Laue, T.M. and D.G. Rodhes (1990). Determination of size, molecular weight and presence of subunits. In *Guide to Protein Purification,* ed. M.P. Deutscher. Methods of Enzymology, Volume 182. San Diego: Academic Press, pp. 566-587.

Lehn, J.-M. (1995). *Supramolecular Chemistry.* Weinheim: VCH.

Linqvist, I. (1967). Adsorption effects in gel filtration of humic acid. *Acta Chimica Scandinavica 21:* 2564-2566.

Malcolm, R.L. (1990). The uniqueness of humic substances in each soil, stream, and marine environments. *Analitica Chimica Acta 232:* 19-30.

Nardi, S., G. Arnoldi, and G. Dell'Agnola (1988). Release of the hormone-like activities from *Allophobora rosea* (sav.) and *Allophobora caliginosa* (sav.) feces. *Canadian Journal Soil Science 68:* 563-567.

Nishio, M., M. Hirota, and Y. Umezawa (1998). *The CH/π Interaction. Evidence, Nature and Consequences.* New York: John Wiley and Sons.

Piccolo, A. (1988). Characteristics of soil humic substances extracted with some organic and inorganic solvents and purified by the HCl-HF treatment. *Soil Science 146:* 418-426.

Piccolo, A. (1996). Humus and soil conservation. In *Humic Substances in Terrestrial Ecosystems,* ed. A. Piccolo. Amsterdam: Elsevier, pp. 225-264.

Piccolo, A., P. Conte, and A. Cozzolino (1999). Effects of mineral and monocarboxylic acids on the molecular association of dissolved humic substances. *European Journal of Soil Science 50:* 687-694.

Piccolo A., A. Cozzolino, and P. Conte (2001). Chromatographic and spectrophotometric properties of dissolved humic substances as compared to macromolecular polymers. *Soil Science 166:* 174-185.

Piccolo, A., A. Cozzolino, P. Conte, and R. Spaccini (2000). Polymerization of humic substances by an enzyme catalyzed oxidative coupling. *Naturwissenschaften 87:* 391-394.

Piccolo, A. and J.S.C. Mbagwu (1990). Effects of different organic waste amendments on soil microaggregates stability and molecular sizes of humic substances. *Plant and Soil 123:* 27-37.

Piccolo, A., S. Nardi, and G. Concheri (1992). Structural characteristics of humic substances as related to nitrate uptake and growth regulation in plant systems. *Soil Biology and Biochemistry 24:* 373-380.

Piccolo, A., S. Nardi, and G. Concheri (1996a). Macromolecular changes of soil humic substances induced by interactions with organic acids. *European Journal of Soil Science 47:* 319-328.

Piccolo, A., S. Nardi, and G. Concheri (1996b). Micelle-like conformation of humic substances as revealed by size-exclusion chromatography. *Chemosphere 33:* 595-660.

Piccolo, A., R. Rausa, and G. Celano (1990). Characteristics of molecular size fractions of humic substances derived from oxidized coal. *Chemosphere 24:* 1381-1387.

Piret, E.L., R.G. White, H.C. Walther, and A.J. Madded (1960). Some physicochemical properties of peat humic acids. *Scientific Proceeding Royal Dublin Society A: 1:* 69-79.

Posner, A.M. and J.M. Creeth (1972). A study of humic acids by equilibrium centrifugation. *Journal of Soil Science 23:* 333-341.

Rausa, R., E. Mazzolari, and V. Calemma (1991). Determination of molecular size distributions of humic acids by high-performance size-exclusion chromatography. *Journal Chromatography 541:* 419-429.

Ritchie, G.S.P. and A.M. Posner (1982). The effect of pH and metal binding on the transport properties of humic acids. *Journal of Soil Science 33:* 233-247.

Ryan, D.K. and J.H. Weber (1982). Fluorescence quenching titration for determination of complexing capacities and stability constants of fulvic acids. *Analytical Chemistry 54:* 986-990.

Saiz-Jimenez, C. (1995). Analytical pyrolysis of humic substances: Pitfalls, limitations and possible solutions. *Environmental Science and Technology 28:* 1773-1780.

Saiz-Jimenez, C. (1996). The chemical structure of humic substances: Recent advances. In *Humic Substances in Terrestrial Ecosystems,* ed. A. Piccolo. Amsterdam: Elsevier, pp. 1-44.

Schnitzer, M. and S.U. Khan (1972). *Humic Substances in the Environment.* New York: Marcel Dekker.

Schulten, H.-R. and P. Leinweber (2000). New insights into organic-mineral particles: Composition, properties, and models of molecular structure. *Biology and Fertility of Soils 30:* 399-432.

Schulten, H.-R., P. Leinweber, and M. Schnitzer (1998). Analytical pyrolysis and computer modeling of humic and soil particles. In *Structure and Surface Reactions of Soil Particles,* eds. P.M. Huang, N. Senesi, and J. Buffle, New York: John Wiley and Sons, pp. 281-324.

Schwarzenbach, R.P., P.M. Gschwend, and D.M. Imboden (1993). *Environmental Organic Chemistry.* New York: John Wiley and Sons.

Spaccini, R., A. Piccolo, G. Haberhauer, M. Stemmer, and M.H. Gerzabek (2001). Decomposition of maize straw in different European soils as revealed by DRIFT spectra of soil particle fractions. *Geoderma 99:* 245-260.

Staudinger, H. (1935). *Die hochmolekularen organischen Verbindungen, Kautschuk und Cellulose.* Berlin: Springer.

Stevenson, F.J. (1994). *Humus Chemistry. Genesis, Composition, Reactions,* Second Edition. New York: John Wiley and Sons.

Summers, R.S., P.K. Cornel, and P.V. Roberts (1987). Molecular size distribution and spectroscopic characterization of humic substances. *Science of the Total Environment 62:* 27-37.

Swift, R.S. (1989). Molecular weight, shape, and size of humic substances by ultracentrifugation. In *Humic Substances II. In Search of Structure,* eds. M.H.B. Hayes, P. McCarthy, R. L. Malcolm, and R.S. Swift. New York: John Wiley and Sons, pp. 449-466.

Swift, R. S. (1999). Macromolecular properties of soil humic substances: Fact, fiction, and opinion. *Soil Science 164:* 790-802.

Swift, R.S. and A.M. Posner (1971). Gel chromatography of humic acid. *Journal of Soil Science 22:* 237-249.

Tanford, C. (1980). *The Hydrophobic Effect: Formation of Micelles and Biological Membranes.* New York: John Wiley and Sons.

Thurman, E.M., R.L. Wershaw, R.L. Malcolm, and D.J. Pinkney (1982). Molecular size and weight measurements of humic substances. *Organic Geochemistry 4:* 27-35.

von Wandruszka, R., M. Schimpf, M. Hill, and R. Engebretson (1999). Characterization of humic acid size fractions by SEC and MALS. *Organic Geochemistry 30:* 229-235.

Waksman, S.A. (1936). *Humus.* Baltimore: Williams and Wilkins.

Welte, E. (1955). New results of humus research. *Angewandte Chemie 67:* 153-155.

Wershaw, R.L. (1986). A new model for humic materials and their interactions with hydrophobic chemicals in soil-water and sediment-water systems. *Journal of Contamination Hydrology 1:* 29-45.

Wershaw, R.L. (1989). Size and shape of humic substances by scattering techniques. In *Humic Substances II. In Search of Structure,* eds. M.H.B. Hayes, P. McCarthy, R. L. Malcolm, and R.S. Swift. New York: John Wiley and Sons, pp. 545-559.

Wershaw, R.L. and Aiken, G.R. (1985). Molecular size and weight measurements of humic substances. In *Humic Substances II. In Search of Structure,* eds.

M.H.B. Hayes, P. McCarthy, R.L. Malcolm, and R.S. Swift. New York: John Wiley and Sons, pp. 477-492.

Yonebayashi, K. and T. Hattori (1987). Surface active properties of soil humic acids. *Science of the Total Environment 62:* 55-64.

Chapter 5

Processes of Soil Formation

Edwin M. Bridges
Siddhartha S. Mukhopadhyay

Unlike plants and animals, soils do not have a genetic basis that controls their development. Instead, they are formed through the interaction of a number of factors, such as climate, organisms, relief, parent material, and time, as proposed originally by Dokuchaev at the end of the nineteenth century and later synthesized into a semiquantitative relationship by Jenny (1941). The importance of the factors of soil formation is that they set the parameters within which the processes of soil formation must act. Although individual processes may be identified, it must be appreciated that the soil forming processes are not mutually exclusive because, for example, podzolization and gleying can take place simultaneously as can gleying and salinization. Current thinking regards the soil forming factors as bundles of chemical, physical, and biological processes. The individual components of the bundles operate at different strengths in different parts of the world (Bridges, 1997).

The soil has been described as the "excited" surface of the earth implying that the factors responsible for soil formation have imparted a vitality into the weathered residue of geological processes and given it an existence of its own. Some authorities have gone as far as to claim that the soil is a living entity, as it can respire by taking in oxygen, release carbon dioxide, and perform other features of an organism. Equally, soil may be regarded as essential for all terrestrial ecosystems and may also be thought of as an ecosystem in its own right in which producers, the plants, provide food (energy) for the many soil-living organisms. These organisms are the primary consumers which in turn provide nourishment for predatory species, the secondary consumers, by creating a food web in which plant material, bacteria, fungi, springtails, mites, earthworms, moles, etc., are all interdependent. All this activity of life in the soil breaks down organic matter, recycles plant nutrients back into the system, and creates humus, the semipermanent organic constituent of soil. Soil fauna and flora are also responsible for the creation of certain types of soil structure as well as homogenization of profiles by mixing.

Thus, soil may be described as: *an integral part of the earth's ecosystems, situated at the interface between the earth's surface and bedrock. It is subdivided into successive horizontal layers with specific physical, chemical, and biological characteristics* (Council of Europe, 1990). These horizontal layers, or horizons, are revealed in a section from the soil surface to the underlying geological parent material; this section is the soil profile. The concept of a soil profile has a long history and is a natural development from the layman's terms topsoil and subsoil. Pedologists have elaborated upon the idea of topsoil, recognizing ochric, umbric, mollic, and fimic A horizons and the histic H horizon of peaty soils. The full definitions of these and other diagnostic horizons are given in the revised legend of the *Soil Map of the World* (FAO-Unesco, 1988). Subsoil horizons are less easily changed by human activity and so have been favored by pedologists for the recognition and classification of soils. A larger number of these horizons have been recognized, including the albic E, cambic B, argic B, spodic B, ferralic B, calcic and gypsic horizons, the development of which is discussed in subsequent paragraphs. Produced by the soil-forming processes, these horizons are fundamental to an understanding of current pedology. The 1988 revised FAO-Unesco classification system is used throughout this book, and this chapter explores the role of the soil-forming processes in the development of these diagnostic horizons, thus providing a direct link between soil characteristics, classification, and the corresponding processes.

FORMATION OF PARENT MATERIAL

Because both sedimentary and igneous rocks are not stable at the surface of the earth, a process of weathering occurs which breaks down the rocks (Chesworth, 1973a; Paton, 1978; Furrer and Sticher, 1999; Wiechmann, 2000). The unconsolidated debris from rock weathering lying upon the earth's surface is called the regolith and it is in the surface of the regolith that soil formation takes place. Until plants colonize it and organic material is added, the regolith is not soil in a pedological sense, only a potential parent material for soil, although engineers refer to it as soil.

Conventionally, weathering is subdivided mainly into physical and chemical processes, but in practice these act together and cannot always be separated (Ollier, 1984). Taking physical weathering first, there are several different physical processes involved in rock disintegration. These include:

Rock shattering: Variations in temperature can result in rock shattering. Surface minerals with different colors and coefficients of expansion cause many small stresses between the constituent crystals, leading to granular disintegration. As rock is a poor conductor of heat, changes of temperature cause expansion and contraction of the surface layer of rock, thus causing strains with the interior. These strains manifest themselves in the detachment of layers of rock parallel to the surface, a process called sheeting or spalling. Layers of rock may also separate parallel to the surface as erosion reduces the burden of overlying rocks.

Crystal growth: An extremely common physical process in cold regions that results in rock disintegration is the growth of ice crystals within the fabric of the rock. Crystals of ice increase in volume by about 9 percent when water freezes in cavities of a rock, prizing the rock apart. With a permeable rock, water may be drawn in and added to existing ice masses, forcing the rock apart as the ice mass grows. A similar process can occur with the growth of salt crystals in arid areas. Salts contained in groundwater are drawn into the rock where they crystallize and proceed to grow, forcing the grains of the rock apart.

Wetting and drying cycles: Simple wetting and drying cycles are reputed to cause physical weathering in certain rock types, but their effectiveness is probably related more to the swelling and contraction of minerals and the chemical role of water.

Biological effects: Growth of plant roots can result in wedging apart of rocks already broken by other physical processes.

Although the processes discussed previously are responsible for physical rock disintegration, the chemical processes of weathering are much more effective in breaking down rock to form a soil parent material. Chemical processes of weathering include:

Solution: Virtually all minerals are to some extent soluble, although the solubility of individual minerals varies greatly. Rainwater percolating through rock and rock debris dissolves any soluble minerals, e.g., common salt or gypsum, and carries them out of the system in solution. The amount of solution depends upon the solubility of the mineral concerned and the amount of water passing through the soil and over the mineral's surface.

Hydration: This process is the addition of water to a mineral that does not dissolve, but usually involves considerable change of volume, e.g., biotite, leading to disruption of the rock structure and allowing further alteration to occur by other chemical processes.

Hydrolysis: This process involves an ionic reaction between minerals and dissociated H^+ and OH^- ions of water; these reactions involve base exchange and also underlie the basis of plant nutrition. Removal of bases from the site of weathering allows further chemical breakdown to occur, and plants, together with acid deposition, provide a source of H^+ ions to effect the exchange.

Carbonation: This is defined as the reaction of carbonate or bicarbonate ions with minerals and as such is a step in the weathering process. The soil atmosphere is rich in CO_2, leading to carbonic acid in the soil moisture that enhances the solution of carbonates and facilitates the base exchange process.

Oxidation: This process occurs when a mineral compound reacts with oxygen to form an oxide. It is possible that a purely chemical reaction occurs, but in many cases, biological agents derive energy from the oxidation reaction. Iron and manganese minerals particularly participate in oxidation reactions within the regolith.

Reduction: This process usually occurs in sites where oxygen is limited, such as in soils saturated with water. In anaerobic conditions, many compounds become more soluble and therefore more mobile within the regolith. On encountering aerobic conditions, reactions can reverse which leads to chemical deposition in processes that also include the role of bacteria.

These weathering reactions can take place separately from the processes of soil formation, but weathering is not confined to one cycle of rock breakdown. It continues as a background to the other processes of soil formation, so that a wide range of minerals is present in fresh parent materials and a decreasing suite of minerals persists as the intensity is increased or the length of weathering is prolonged as seen in Ferralsols, Acrisols, and Lixisols of the intertropical regions. As weathering processes attack minerals, they undergo progressive surface alteration through ionic substitution and leaching conforming to rates of change determined by the laws of chemical thermodynamics. Despite the complexities of the system, models can be developed that enable the evaluation of weathering as a process in soil formation (Kirkby, 1985; Ross, 1989; Sumner, 2000; Minasny and McBratney, 2001; Heuvelink and Webster, 2001).

Plant material provides the second main ingredient of soils (Scheffer and Schachtschabel, 1998a). Under saturated and anaerobic conditions, it may be preserved as peat accumulating as a deposit up to several meters deep where it forms the parent material for Histosols. Under aerobic conditions, plant debris lying upon mineral soils is decomposed as a sequence of fungi and fauna utilize it to satisfy their energy requirements. In the process, humus is formed and plant nutrients are released for recycling.

Chemical studies comparing C:N ratios, the proportions of fulvic to humic acid in the different humic forms, and the use of different chemical extraction techniques have pointed to the role of organic substances in Bh horizon (spodic horizon) formation in podzols. Techniques such as nuclear magnetic resonance and fractionation of organic carbon and nitrogen are beginning to reveal some of the detail of organic breakdown in grassland and woodland soils (Jörgensen and Meyer, 1990; Hopkins, Chudek, and Shiel, 1993; Beyer et al., 1993; Zech and Kögel-Knabner, 1994). Breakdown of organic matter releases organic molecules that are capable of chelating with metals, e.g., iron, and increases their mobility within the soil profile. The process of respiration of plant and animal life within the soil uses oxygen and produces carbon dioxide, which, when dissolved in the soil solution with substances from organic breakdown and root exudates, produces an aggressive solution for further weathering. Animal life in the regolith is responsible for loosening it, thus allowing access for water and gaseous exchange to take place.

All aspects of weathering are significant when a lake is drained, or the sea is impoldered, as a new parent material is revealed in which normal terrestrial soil development can take place. Experience has shown that a number of important changes, called ripening, take place rapidly in this new material as soil formation is initiated. Subaqueous materials are normally rich in organic matter, and as water is lost from the previously saturated material, the volume is reduced which results in shrinkage, cracking, and settling with the development of an initial coarse prismatic structure. Salts are leached, and depending upon the relationship of calcium to sodium salts, the developing soil may be dispersed into a slurry as rain falls upon the new land surface. Formerly anaerobic, the materials exposed to the atmosphere change as iron and manganous minerals are oxidized, resulting in the development of mottles and coatings of these minerals on developing soil surfaces. The biological population changes from aquatic to a terrestrial flora and fauna (Pons and Zonneveld, 1965).

Soils of urban areas are influenced to a great extent by human activity, a process described as metapedogenesis (Yaalon and Yaron, 1966). Although normal soils exist in parks and extensive open spaces, elsewhere they are disturbed as soil is moved from one place to another in leveling operations, with mixed, buried, and inverted profiles. Building materials have been added over the centuries as well as artifacts and wastes. Following the industrial revolution, the range of materials became greater and included many toxic substances. Some of these materials have become parent materials for new soil development, leading to specialized ecological niches for rare plants (Bullock and Gregory, 1991; Hiller and Meuser, 1998).

SURFACE SOIL-FORMING PROCESSES

The processes acting at the earth's surface in the formation of soil horizons are those involved in the breakdown and addition of organic matter to the soil as A, O, or H horizons (Scheffer and Schachtschabel, 1998b). Plants take simple, naturally occurring substances from the soil and atmosphere and with the energy of sunlight convert these into plant tissue by photosynthesis. On dying, much of this material falls to the soil surface, where it is added to the soil system together with plant roots. It is all gradually decomposed in an incompletely understood process that involves many steps with different organisms and complex organic chemical reactions, to provide (with the surface litter) the organic component of soils. The great significance of the organic matter component is that only when it is intimately combined with a parent material does the regolith become a soil.

Organic Additions to the Soil System

Plant litter falling to the soil surface may accumulate to form a litter layer (L), and in many soils development of this layer builds up during the period of the year when deciduous trees shed their leaves or grasses die back. Subsequently, the plant debris is used by soil-living organisms for their nutrition. In the process, it is broken down, converted into humus, and incorporated by earthworms into the soil as mull. The plant material does not persist from year to year, especially in cultivated soils, but the incorporated humus adds greatly to the stability and fertility of the soil. In undisturbed natural and seminatural conditions, especially in acidic soils, the plant litter may accumulate and not be completely broken down in an annual cycle. In such conditions, there may be a second layer of plant debris, of increasing decomposition with depth, referred to as a fermentation layer (F). The combination of the litter and fermentation layers is known as the moder form of organic matter. Finally, in strongly acidic soils, the breakdown of organic matter is restricted mainly to fungal activity, and an absence of earthworms means that the completely decomposed organic matter is not incorporated into the mineral soil. Instead it forms a third surface layer of black amorphous organic material; this is the humus or H layer. A combination of the litter, fermentation, and humus layers makes the mor form, or raw humus (Kubiena, 1953; Scheffer and Schachtschabel, 1998a,b). The symbols L, F, and H were introduced by Hesselman (1926). Further development of these terms was done by Heiberg and Chandler (1941) and by Kubiena (1953). However, these terms for humus forms are not universally accepted or used. The letter O is used by FAO-Unesco (1988) to signify surface organic material that is not permanently saturated.

In places where the decay of plant material is inhibited, as by anaerobic conditions or strong acidity, plant debris may be preserved rather than decomposed. In the saturated conditions of bogs, for example, both conditions occur and unless drained by human intervention, organic material will accumulate as peat. Peat occurs extensively in arctic areas of North America and Eurasia and on upland areas in cool, humid environments where heavy rainfall and continual saturation leads to blanket bog development. Alternatively, accumulation of peat can occur in saturated conditions associated with valleys in lowland situations. These fen peat deposits may be either acid or slightly alkaline depending upon the surrounding geology and supply of calcium-rich drainage waters. When drained, these peats form excellent arable and horticultural soils, but as a result of drainage and cultivation, most of the peat becomes oxidized so that eventually only a dark-colored mineral soil remains. Limited areas of peat occur in tropical coastal areas where exploitation by agriculture has led in many cases to the development of acid sulphate conditions as pyrites oxidize in soils and organic deposits influenced by accumulation in salt or brackish water.

In the FAO guidelines for soil profile description (FAO, 1990), organic horizons are referred to as either O or H horizons. The O horizons are those that develop upon mineral soils, previously described as mull, moder, mor, or as raw humus. These forms of soil organic matter are only saturated for a few days at a time during the year, and contain variably decomposed material with more than 20 percent organic carbon. The H horizon is formed by an organic accumulation that is saturated for prolonged periods or is permanently saturated unless artificially drained (Figure 5.1). The H horizon should have a thickness of more than 20 cm but less than 40 cm and contain 18 percent or more organic carbon if the mineral fraction contains more than 60 percent of clay; lesser amounts of organic carbon are permitted at lower clay contents. The H horizon may be between 40 and 60 cm thick if it consists mainly of sphagnum, or has a bulk density when moist of 0.1 mg m^{-3}.

The form of organic matter previously referred to as mull is characteristic of A horizons in many soils. An A horizon is a mineral horizon formed or forming at or adjacent to the surface that has an accumulation of completely humified organic matter intimately associated with the mineral fraction. Its morphology is acquired mainly by earthworm activity and it lacks the properties of E and B horizons. Incorporated organic matter in A horizons is in the form of fine particles or coatings on the soil minerals. Distribution of the organic matter throughout the A horizon is by biological activity and not through translocation. This mixing gives A horizons a darker color than underlying mineral horizons. According to the prevailing environmental conditions it is possible to identify mollic, umbric, and ochric A horizons.

Both mollic and umbric A horizons have a well-structured, dark-colored appearance with a moderate to high organic matter content. The mollic A is base-rich, and typically occurs in Rendzic Leptosols (Figure 5.2), Chernozems (Figure 5.3), and Kastanozems, and the umbric A horizon (Figure 5.4) is base-poor and occurs in some Leptosols, Fluvisols, Gleysols, and Andosols.

Both horizons have a robust structure that has a chroma of less than 3.5 when moist, a value darker than 3.5 when moist, and 5.5 when dry, measured on the Munsell Soil Color Charts. A mollic A horizon may rest directly on hard rock, a petrocalcic horizon or a duripan, but its thickness must be more than 10 cm. The content of P_2O_5 soluble in 1 percent citric acid must be less than 250 mg kg^{-1} soil to distinguish it from a fimic horizon. The umbric A horizon has the same criteria as the mollic A horizon but differs by having a base saturation (by NH_4OAc) of less than 50 percent. The ochric A horizon is too light in color, has too high a chroma, too little organic carbon, or is too thin to be mollic or umbric, or it may be hard and massive when dry.

Human influences are most strongly felt in the surface horizons of soils and there is strong evidence that the impact of human activities is increasing (Bridges, 1978; Bridges and De Bakker, 1998). Through disturbance and long-continued additions of manures, a fimic A horizon (Figure 5.5) is produced. This is a human-made surface layer 50 cm or more thick commonly containing artifacts such as bits of brick and pottery throughout its depth. The fimic A horizon may meet the criteria for a mollic or an umbric A horizon but is distinguished from these by an acid-extractable P_2O_5 content which is in excess of 250 mg kg^{-1} of soil. The best-known examples are the plaggen soils of the Netherlands and northwest Germany (Pape, 1970; Mückenhausen, 1977).

SOIL-FORMING PROCESSES IN LOWER SOIL HORIZONS

Pedologists have paid greater attention to horizons lower in the soil profile because these are less likely to be affected by changes of land use and management; hence their characteristics are more permanent and of greater use for classification. It follows that the processes involved in the formation of lower horizons are of importance in any study of soil-plant relationships. The processes usually identified as responsible for lower soil horizons are leaching, eluviation, podzolization, gleying, ferralitization, calcification, salinization, alkalization, solodization, rubefication, and pedoturbation. These processes are not always mutually exclusive as has been explained earlier in this chapter.

Leaching

The process of leaching involves the removal of soluble constituents from soils. When rainfall exceeds evapotranspiration, soluble salts in the soil are taken into solution and removed from the profile. However, the process of leaching is also involved in reactions with ions held on the exchange sites of the clay-humus complex. After removal of sodium and calcium salts by solution from a parent material, cations of Na^+, Ca^{++}, Mg^{++}, and K^+ attached to the clay-humus complex begin to be exchanged for H^+ ions, or in extremely acid soils Al^{+++} ions, thus intensifying the acidity of the soil. Cation exchange at colloidal surfaces has been represented by the diffuse double-layer model comprising the Stern layer in which attraction to the colloid surface by cations declines linearly as far as the Helmholtz plane, beyond which in the Gouy layer, cationic attraction declines exponentially. Leaching, or water-transport models, have been employed extensively to explain the aqueous movement of constituents through soils using Darcy's Law for mass flow and Fick's Law for diffusion. Analogues have been used from chromatographic or electrical resistance models and developed into a coherent theory of miscible displacement based upon equations for mass flow and diffusion (Ross, 1989). Anions are attracted to positively charged colloidal surfaces either at broken edges of the clay lattice or where unsatisfied Al^{+++} groups are exposed or on the surfaces of Fe, Mn, or Al oxides or hydroxide films. This process is particularly important for phosphate adsorption, retention, and availability which has been modeled using the Langmiur and Freundlich equations (Sibbesen, 1981; Barrow, 1983). Retention of cationic forms of metals within soils has important implications for land contamination (Barrow, 1986).

The process of leaching, together with swelling and contraction during wetting and drying, or heating and cooling cycles leads to the development of a cambic B horizon. The cambic B horizon (Figure 5.6) is formed by removal of carbonates, alteration of color to a stronger chroma or redder hue, and development of a different structure from the parent material below. The cambic B horizon has a texture that is a sandy loam or finer and contains at least 8 percent clay, and any remaining rock structure must occupy less than half the volume of the horizon. There should be evidence that carbonates have been removed but the cation exchange capacity must be more than 16 cmol+ kg^{-1} clay.

Eluviation

In the past, the term eluviation simply implied the movement of material out of the surface soil horizon, but in recent years the term has been confined specifically to the movement of clay-sized material from the surface horizons to the B horizons of soils.

The French term lessivage and the German tonverlagerung are synonymous with this process which is the mobilization of clay material and its redeposition lower in the profile (Scheffer and Schachtschabel, 1998b). In

Luvisols, Nitisols, Alisols, Lixisols, and Acrisols the argic B horizon (Figure 5.7) is normally related to an overlying eluvial horizon from which the clay has been depleted. This eluvial horizon has a lower clay content and weaker structure but in many cases is not sufficiently well-developed to qualify fully as an albic E horizon.

Clay movement appears to begin in loamy soils after the profile has been acidified by the leaching processes, but in extremely acid conditions, Al concentrations increase and the clay becomes less mobile. Clay is dispersed when rain falls upon a dry soil and it is carried in suspension down into the lower part of the profile. As the water stagnates and gradually drains, the clay is redeposited as an oriented thin layer (argillan or clay cutan) on the faces of soil structures and as linings of pores to give the clay skins that are typical of the process. Over a period of time, the clay content of the lower horizon is increased, and in soils that undergo pedoturbation, the clay becomes mixed with material already present. The clay skins may be observed in the field using a hand lens, or using a microscope to study thin sections in the laboratory. Conditions that favor the movement of clay are acidification of soil, a climate with dry periods, and a subsoil that gradually absorbs the water and allows the deposition of the oriented clay and maintains the stability of the mobile soil colloids (Dixit, Gombeer, and D'Hoore, 1975). In certain layered parent materials and clay-rich subsoils, there is evidence that the clay increase is not the result of eluviation but clay formation in situ by weathering of silicate minerals.

The result of eluviation (or in situ clay formation) is an argic B horizon which is a subsoil horizon that has a distinctly higher clay content than the overlying horizon. This textural differentiation may be due to an illuvial accumulation of clay, destruction of clay in the surface horizon, or a selective erosion of the surface. To be diagnostic, an argic B horizon has to have a texture that is sandy loam or finer with stipulated ratio of clay content with the overlying horizon.

Podzolization

In soils of regions where precipitation exceeds evaporation, and also where deep quartzitic sands occur from arctic climates to the humid tropical regions, the process of podzolization operates (Blume et al., 1997). A very characteristic soil profile results (Figure 5.8) with clearly defined horizons with distinctive physical and chemical properties to match (De Coninck, 1980; Buurman, 1984; Wiechmann, 2000). Podzolization has almost always been preceded by strong leaching which has decalcified the parent material and acidified the soil to the extent that only acidophyllous plants such as heath or coniferous forest will grow satisfactorily. These plants produce an extremely acid litter which breaks down slowly. The faunal population is restricted, and organic breakdown mostly takes place through fungal activity. As a result, the mor type of humus develops with its litter, fermentation, and humus horizons. It tends to lie upon the soil surface as conditions are too acid for earthworms, so incorporation is minimal. Moreover, decomposition products in the canopy drip, and from the decomposing plant debris in the mor humus, are capable of chelating iron and aluminum sesquioxides and carrying them downward in solution from the soil surface (Farmer, Russel, and Berrow, 1980; Anderson et al., 1982; Farmer et al., 1985; Wiechmann, 1978). Acid weathering breaks down clay minerals, commonly mica, chlorite or kaolinite, and aluminum is released. Sand grains are left with a greyish white appearance as they are stripped clean of iron hydroxides by downward-migrating solutions. Thus the surface horizons of the podzol become depleted of clay and iron. Several processes are involved in the precipitation of iron, aluminum, and organic matter in the B horizon including bacteria that intercept and break down the organic linkages, the weakening of chelation bonds as they age, effects of soil wetting and drying, or the result of a pH rise in lower horizons.

The results of the podzolization process produce the albic E horizon and the spodic B horizon. For classification purposes, the presence of a spodic B horizon is mandatory for podzols, but an albic E horizon does not always have to be present.

The spodic B horizon normally lies below an A or an albic E horizon and must lie below a depth of 12.5 cm. (This definition rules out superficial iron pans that may be seen within the A horizons of many soils in humid climates.) (Stahr, 1973; Bridges, 1974).

Oxidation-Reduction Processes

In virtually all regions of the world, some soils are saturated by water for all of the year, and many are seasonally wet. When soils are saturated, the soil-forming processes that operate are different from those in freely drained conditions. The presence of water in the pores drives out the air and when this persists for long periods, bacteria consume the remaining oxygen and the soil becomes anaerobic. The presence of anaerobic conditions and organic-matter-breakdown products in the soil solution increases the solubility and mobility of many minerals in their reduced form, consequently the internal chemistry and morphological appearance of the soil is changed. The resulting morphology is that soils have stagnic or gleyic properties and reductimorphic (redox) features are produced (Schlichting and Schwertmann, 1973).

Soils in which redox processes (E_h) operate are characterized by the segregation of iron. Changes in pH accompany changes in redox so that the soils are capable of denitrification of nitrates and generating phytotoxic substances. Reduction conditions are identified by the value of rH in the equation $rH = Eh\ (mV)/29 + 2pH < 19$ or the appearance of a dark blue color on a freshly broken surface of a field-wet soil sample after spraying it with a solution of 1 percent potassium ferric cyanide, $K_3\ Fe(III)(CN)_6$, or the appearance of a red color on a freshly broken surface of a field-wet soil sample after spraying it with a 0.2 percent dipyridyl solution in 10 percent acetic acid (FAO-Unesco, 1988).

Gleyic conditions: Soils in which significant oxidation-reduction processes occur lie toward the foot of slopes and in low-lying situations in the landscape where a high groundwater table exists (endosaturation); such soils have been called Gleysols or groundwater gley soils. The morphological results of gleying are apparent in the transformation of iron minerals from ferric to the reduced (ferrous) form with a grey rather than a reddish brown color. In permanently waterlogged soils, iron minerals may be removed in solution from the soil altogether, leaving the natural grey or greyish blue colors of the quartz and clay minerals. Because the water table has a seasonally upward and downward movement, a mottled zone frequently overlies the completely gleyed lower horizons.

Stagnic conditions: In plateau situations with little surface relief and slow internal drainage, soils with reductimorphic features may also occur. Such soils are more common in regions with high or seasonally high rainfall (episaturation) and have been referred to as pseudogley soils or surface-water gley soils. Temporary waterlogging occurs as a result of rainfall, but the soil returns to an oxidizing condition as the water drains away and iron minerals are re-precipitated as iron or iron-manganese mottles of high chroma in a matrix having a chroma of 2 or less. Lower horizons normally become less mottled and may even be freely drained. Slow drainage associated with a dense B horizon, or an ironpan, can produce a gleyed albic E horizon as in Planosols or Podzols.

Ferrolysis is a form of gleying with alternating aerobic and anaerobic conditions leading to the development of Planosols with an albic E horizon overlying a dense clayey argic B horizon (Brinkman, 1979). Under anaerobic conditions during the wet season, free iron is reduced to the ferrous state and leached to give the bleached appearance of the albic horizon. Hydrogen ions displace aluminum, magnesium, and other cations from the clay lattice and clay minerals are disrupted and the soil is acidified. When the soil dries, oxidation causes iron hydroxides to change back from ferrous to ferric, liberating hydrogen ions and increasing acidity. These hydrogen ions then begin another cycle of clay mineral disruption by displacing cations and attacking the lattice structure of clay minerals. The result is a soil with an abrupt textural change between eluvial and illuvial horizons, typically found on plateau locations in warm climates.

Ferralitization

In soils of the humid intertropical regions where there has been a period of long and intense weathering before and during the process of soil formation, the ferralic B horizon (Figure 5.9) is developed. In this horizon the soil has become depleted of most minerals except for quartz sand and low activity kaolinitic clays. By definition, a ferralic B horizon must be at least 30 cm thick and have a texture that is a sandy loam or finer. The cation exchange capacity is equal to or less than 16 $cmol^+ kg^{-1}$ clay, or an effective cation exchange capacity equal to or less than 12 $cmol^+ kg^{-1}$ clay (sum of NH_4OAc exchangeable bases plus 1M KCl-exchangeable acidity). In this horizon weathering has reduced the content of weatherable minerals to less than 10 percent in the 50-200 mm fraction, there is less than 10 percent water-dispersable clay, and less than 5 percent of the soil possesses residual rock structure.

In the lower horizons of highly weathered soils, mottled, iron-rich, humus-poor clays referred to as plinthite occur (Figure 5.10). The processes involved in their formation have been given various names including laterization, desilicification, ferralitization, ferritization, and allitization (Nye, 1954; Van Wambeke, 1992). In the most advanced cases, only residual iron and aluminium oxides in the form of laterite or bauxite remain. Irreversible hardening can occur where the plinthite is exposed and becomes petroplinthite and is shown as a petroferric phase on the *Soil Map of the World.*

The main differences between soil formation in the tropical regions and elsewhere in the world are in the speed and effectiveness of reactions at the higher temperatures and the length of time for soil formation. Thus the soils are strongly leached of any soluble material and the base cations attached to the exchange complex of the clays are greatly reduced so that the soil becomes dominated by H^+ and Al^{+++} ions. Destruction of high-activity clay minerals occurs but there is some recombination of their constituents to produce low-activity kaolinitic clays.

Calcification and Gypsum Accumulation

These processes are involved in the accumulation of calcium carbonate and/or calcium sulphate usually in the B or C horizon of soils, but occasionally in the A horizon (Alaily, 1996a,b; Scheffer and Schachtschabel, 1998b). Accumulation of calcium carbonate is characteristic of soils in semiarid areas and continental interior situations (Figure 5.11). The rainfall must be sufficient to wet the soil to a depth of between 0.5 m and 1.5 m, but insufficient to leach the soil. Calcareous dust dissolved by the rainwater is carried down the soil profile, but as the movement of the moisture downward loses momentum, and eventually stops and drying begins, calcium carbonate is precipitated. Calcium carbonate may also be precipitated from lime-rich water drawn upward by capillary attraction from the groundwater. Successive precipitation episodes lead to the development of a calcic horizon that contains at least 15 percent calcium carbonate equivalent and which is at least 15 cm thick. Calcareous accumulations in soils take several forms including soft powdery lime, hard nodules, or layers of rocklike appearance. The soft powdery lime form is precipitated in place from the soil solution, normally with a relationship with the soil fabric. It may interrupt the soil fabric to form weak, roughly spheroidal aggregates, or it may form a white crystalline covering to soil structures. Filaments of calcium carbonate (pseudomycelia) which form and re-form according to changing moisture conditions are not included in the definition of soft powdery lime. When in the form of nodules, calcium carbonate should occupy 5 percent or more of the volume of the soil horizon to be diagnostic for a calcic horizon. Accumulation over a long period of time can result in the development of a petrocalcic horizon, a continuously cemented calcareous accumulation, usually in the lower part of the soil profile but it may appear at the surface if the upper part of the soil profile has been eroded. Mainly formed by coalescence of nodular material, a thin laminar form also may be observed at the upper surface of the petrocalcic horizon where carbonates have been dissolved and re-precipitated. Accumulation of carbonates may also occur in calcic Gleysols under humid climatic conditions in central Europe on loess or moraines of the last major Pleistocene glaciation.

Calcium sulphate is more soluble than calcium carbonate and so pedological enrichment of soil horizons with gypsum is associated with areas with a drier climate than those where carbonate accumulation is common. However, there is considerable geographical overlap, in which case the gypsum-enriched, gypsic horizon normally appears below the calcic horizon. Calcium sulphate in soils normally originates from capillary rise of gypsum-rich groundwater. As the water rises in the soil, it reaches the point where evaporation limits the upward movement and gypsum is precipitated. Gypsum accumulates in the soil uniformly as nests of crystals or as blade-like crystals to form a gypsic horizon. If the soil material is stony, gypsum precipitation may take the form of pendants on the underside of rock fragments.

FIGURE 5.1. Terric Histosol, Wales, United Kingdom. Highly decomposed organic materials with only small amounts of visible plant fibers; very poor drainage. (*Photo:* E.M. Bridges; *Source:* Bridges, 1997.)

FIGURE 5.2. Rendzic Leptosol, Romania. A mollic A horizon overlies limestone with calcium carbonate equivalent of more than 40 percent. (*Photo:* E.M. Bridges; *Source:* Bridges, 1997.)

FIGURE 5.3. Haplic Chernozem, Ukraine. A deep (1 m) mollic A horizon is present, and krotovinas are visible in the C horizon. (*Photo:* D. Gunary; *Source:* Bridges, 1997.)

FIGURE 5.4. Umbric A horizon: dark-colored, deep, base-poor A horizon. (*Photo:* International Soil Reference and Information Centre; *Source:* Bridges, 1997.)

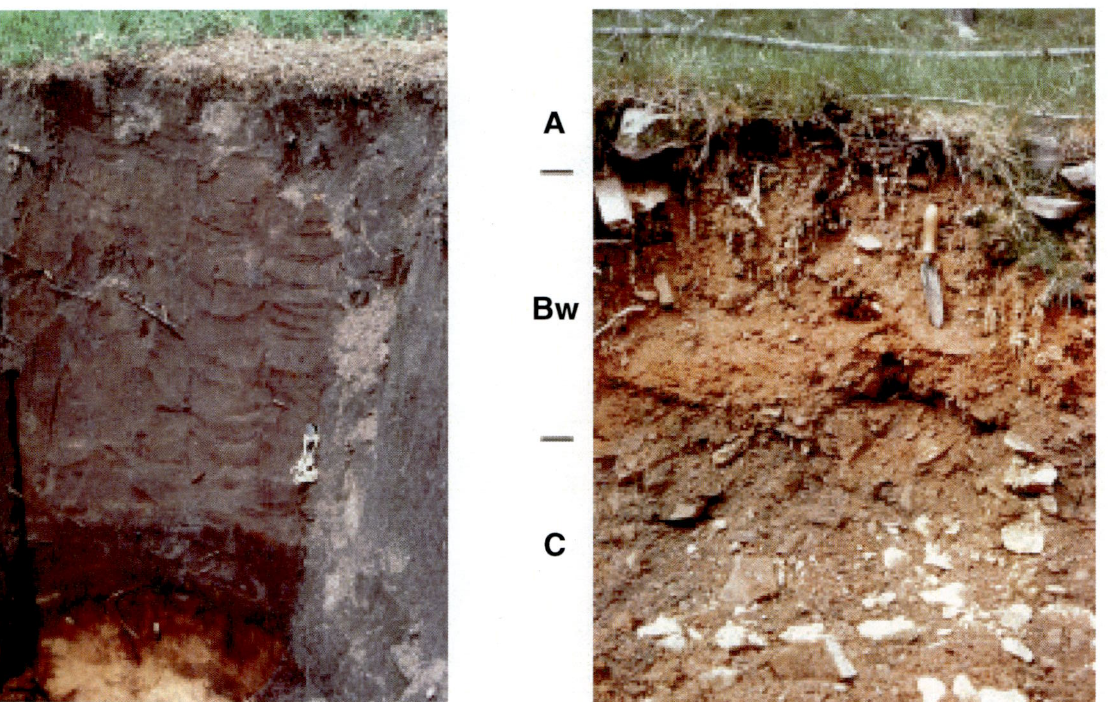

FIGURE 5.5. Fimic A horizon: human-made surface horizon, 50 cm or more thick. (*Photo:* E.M. Bridges; *Source:* Bridges, 1997.)

FIGURE 5.6. Cambic B horizon: color and structure are different from the parent material. (*Photo:* E.M. Bridges; *Source:* Bridges, 1997.)

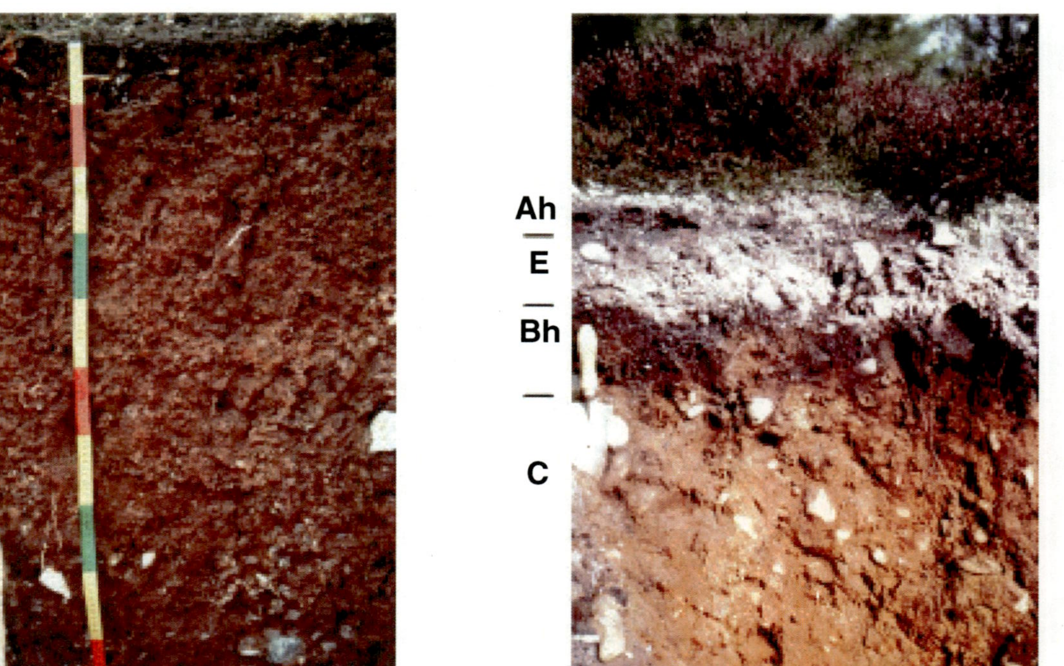

FIGURE 5.7. Acrisol, Jinxian Province, China. The profile has an argic B horizon with low CEC, but the B horizon has less than 50 percent saturation, so it is a haplic Acrisol. (*Photo:* International Soil Reference and Information Centre; *Source:* Bridges, 1997.)

FIGURE 5.8. Haplic Podzol, England, United Kingdom. Beneath a heath vegetation a well-developed E horizon and spodic B horizon can be seen. There is no gleying and no placic horizon. (*Photo:* E.M. Bridges; *Source:* Bridges, 1997.)

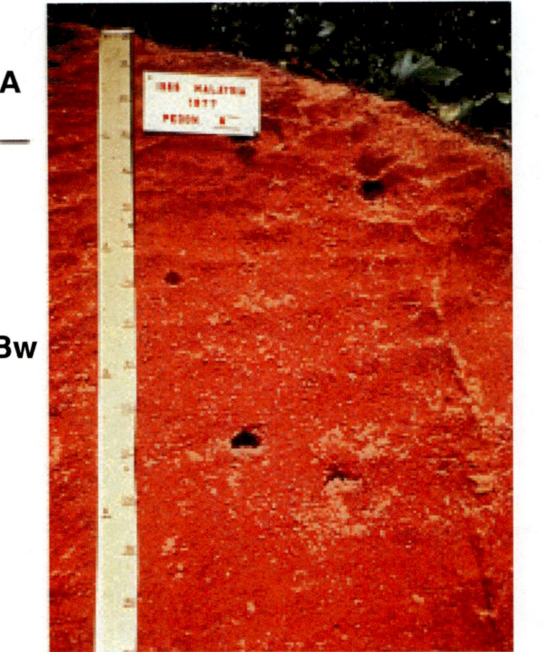

FIGURE 5.9. Rhodic Ferralsol, Malaysia. The presence of a ferralic B horizon and the bright red colors (redder than 5 YR) place it in the Rhodic Soil Unit. (*Photo:* International Soil Reference and Information Centre; *Source:* Bridges, 1997.)

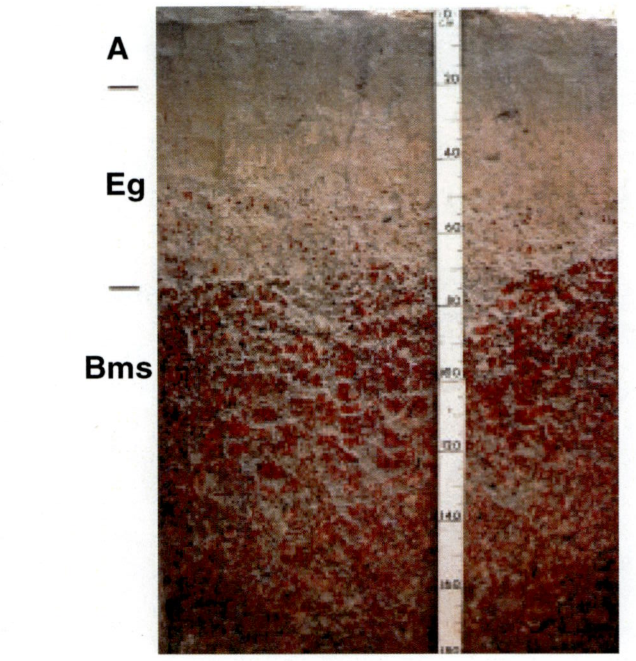

FIGURE 5.10. Dystric Plinthosol, Ghana. Hardened nodules are present at a depth of 80 cm, comprising more than 25 percent by volume; low saturation with bases. (*Photo:* E.M. Bridges; *Source:* Bridges, 1997.)

FIGURE 5.11. Petrocalcic horizon: massive, cemented horizon, indurated with calcium carbonate. (*Photo:* International Soil Reference and Information Centre; *Source:* Bridges, 1997.)

FIGURE 5.12. Vertic properties: parallelepiped blocky structures with polished "slickensides" surfaces. (*Photo:* International Soil Reference and Information Centre; *Source:* Bridges, 1997.)

In areas where ephemeral gypsum lakes occur, e.g., the Tunisian Chott lakes, gypsiferous dust is blown from the dried lake surface onto surrounding soils and incorporated by occasional rains. In receiving gypsum from an aerial source and from capillary rise, thick petrogypsic horizons may be developed which commonly contain more than 60 percent calcium sulphate. The surface horizon of highly gypsiferous soils acquires a characteristic crust that gives it the feeling of walking upon a thick carpet (Van Alphen and De los Rios Romero, 1971; Porta, 1998).

Salinization and Alkalization

Salts that are more soluble than calcium sulphate occur in the soils of arid and semiarid regions of the world (Buol, 1965). The process whereby they accumulate is called salinization; it is both a natural phenomenon and a human-induced form of soil degradation. The use of irrigation has greatly expanded the area of agriculture in dry climates, but accumulation of salts in the soils of irrigated areas through poor farming practices continues to increase the area of salinized soils. Salinization is an economically important process of soil formation, but techniques for reclamation by leaching the salts are simple and well known. Another closely related process is alkalization in which sodium ions come to dominate the exchange positions of the clays, resulting in a structural deterioration of soils that is extremely difficult to rectify. In both cases, growth of crops is sharply restricted and plant growth is limited to a few salt-tolerant plants (Szabolcs, 1989; Scheffer and Schachtschabel, 1998b).

The features of salinization are readily observable as salts effloresce upon the soil surface (the white alkali soils of an earlier American soil classification), and can be seen within the fabric of the soil which usually retains its structural stability. Chemically, saline soils are recognized as having salic properties in which the electrical conductivity of the saturation extract is more than 15 dS m^{-1} at 25° C at some time of the year, within 30 cm of the surface or of more than 4 dS m^{-1} within 30 cm of the surface if the pH (H_2O, 1;1) exceeds 8.5. The electrical conductivity is usually greatest in the soil surface horizon and decreases with depth.

Soils in which the process of alkalization has occurred may be recognized as having sodic properties and possessing a natric B horizon. In these soils (the black alkali soils of the former U.S. classification) humic material becomes dispersed to give the dark color and electrical conductivity is variable throughout the profile. The term sodic properties refers to a soil material with a saturation in the exchange complex of 15 percent or more of exchangeable sodium. The process of alkalization results in a natric B horizon. This has similar properties to the argic B horizon but additionally has exchangeable sodium of more than 15 percent and generally a columnar structure, or a blocky structure with tongues of an eluvial horizon extending downward in which there are uncoated silt or sand grains.

Solodization

When soils rich in calcium ions are leached, either through a fall in the regional water table or through irrigation, the soluble salts in the soil profile are gradually removed and the soil eventually reverts to its regional soil type. If for some reason, calcium-poor, sodium-rich soils are leached of their cations, a dramatic loss of structure and biological productivity occurs. Carbonate and bicarbonate salts are produced by a reaction between carbon dioxide given off by plant roots and organic decomposition. The calcium ions being taken from the exchange complex leave the sodium ions to dominate and to come into equilibrium in the soil solution, raising the pH to above pH 8 or 9 accompanied by loss of structural stability.

Further washing out of salts occurs and the dispersed clay and organic matter begins to move down the profile giving it a banded appearance with lighter and darker horizons. Once the salts have been removed, the soil becomes acidic with a compact structure. A sequence of solonetz–solodized solonetz–solod has been identified in southeastern Europe and Australia where these processes have been observed (Joffe, 1949).

Melanization, Rubefication, and Ferrugination

These terms are used to describe processes involved in color changes of soil horizons. In part, they have been described as results of other processes such as incorporation of organic matter which darkens A horizons, a process described as melanization. Continued weathering of soils changes colors in the lower horizons of Cambisols and Ferralsols in the development of the cambic B horizon and ferralic B horizon. In these horizons, weathering releases iron from minerals which is re-precipitated on the surface of peds or as discrete particles throughout the soil to give a red, or brownish-red coloration. This has been described as rubefication in the Mediterranean area or ferrugination in soils of the intertropical regions. The term leucinization has been suggested for the development of the paler color caused by removal of iron from the A and E horizons.

Pedoturbation

This term is used to describe processes in the soil that tend toward homogenization. The other soil-forming processes discussed previously result in the formation of features of specific horizons that give the soil an isotropic character. The presence of animal life in the soil tends to mix soil material, as for example in a Cambisol compared with a Podzol, where a lack of earthworms in the latter permits the development of clear horizons. The larger soil-living mammals, e.g., rabbits, badgers, foxes, and moles also actively mix soil materials through their burrowing activities.

When soils are repeatedly wetted and dried, the clay minerals in them expand and contract leading to the development of soil structures. Some clays, though, have a large capacity for expansion which can cause heaving within the soil. They crack widely and deeply, allowing topsoil particles to fall into lower soil layers, so that over time the whole soil is turned over and the surface is thrown into a complex of humps and hollows. Soils with this type of microrelief, together with the presence of slickensides on structure faces where peds have rubbed against one another, were previously called gilgai (an Australian term), but now are called Vertisols (Wilding and Puentes, 1988).

The term vertic properties is used for clayey soils which at some time show one or more of the following: cracks, slickensides, or intersecting, wedge-shaped or parallelipiped structural aggregates that are not in a combination or are not sufficiently expressed (Figure 5.12).

Physical disturbance of soil horizons can be observed in the formation of soils in arctic regions where periglacial activity produces such features as "festoons" or "involutions" in layers or horizons of different textures (Tarnocai, Smith, and Fox, 1993). In these regions, as the ice accumulates, it distorts any layers or horizons present. On slopes, segregation of stones into stripes occurs and on level ground polygonal soil structures develop. Ice in vertical cracks can force the soil apart to form an ice wedge which subsequently may be replaced with soil material to give an ice-wedge pseudomorph. At the foot of slopes or where groundwater wells up, the surface may burst open to form frozen, conical hill features known as pingos. Once the ice in these features melts, a water or peat-filled hollow, surrounded by a circular gravel ridge, remains. Many of these features can still be recognized in the lower horizons of soils in formerly periglacial areas.

MODELING SOIL-FORMING PROCESSES

The unraveling of nature and the demand for effective land management have brought forth a need for modeling of soil formation. Sequential soil occurrence and its spatial distribution have offered various paradigms. These paradigms generally have had a verbal or flow chart or at best a quasi-mechanical framework. Recent advances in computer-based geostatistics and general statistics have quickly changed the scenario. Efforts have also been made to explain soil formation in terms of thermodynamics, especially using the concepts of nonlinear dynamic systems (NDS) theory. Modeling efforts in describing soil formation have now entered into an exciting phase of resolving inherent contradictions, linking state factors with soil processes at a scale ranging from landscape to global, and quantifying unresolved parameters.

In the early twentieth century, pedologists attempted to follow the example of natural sciences in formulating soil formation in terms of simple and exact relationships. However, lately, simple conceptual models of soil formation have given way to more complex ones. Hoosbeek and Bryant (1992) grouped these models on the basis of degree of computation, complexity of the structure, and organizational hierarchy. In this section, the same grouping is followed with certain modifications.

Qualitative Functional Models

Jenny (1941) postulated soil (S) in the functional form as:

$$S = f(\text{cl, o, r, p, t, } \ldots) \tag{5.1}$$

where cl represents climate, o represents organisms, r represents topography, p represents parent material, t represents time, and the dots represent local factors. This equation was useful for climo-, topo- and chronosequences, but was not unambiguous enough to explain either S or p and experimentation was difficult in the absence of a well-defined initial state of p. Jenny continued with his attempts and they were presented later on (Jenny, 1980). One of the first attempts was to redefine S as a state of soils and p as the key property (an aggregate of physical, chemical, mineralogical, and organic makeup at an initial state). He presented r as surface configuration or topography and the initial water table, insofar as it influences the moisture thresholds. In another attempt, he restricted S to land ecosystems and therefore the right side of the equation (5.1) was related to system properties (equation 5.2)

$$\underset{\text{(system properties)}}{\text{l, v, a, s}} = \underset{\text{(state factors)}}{f(\text{cl, o, r, p, t, } \ldots)} \tag{5.2}$$

where, for soil system properties of the ecosystem l, v is vegetation property, a is animal property, s is any other soil property, and the dots are local factors. In this equation, l remains a vague term.

Although it is extremely difficult to keep all but one factor constant in an open system such as soils, a large number of pedological studies have used Jenny's equation to explain soil formation in a fairly consistent manner

(Conacher and Dalrymple, 1977; Thompson, Bell, and Butter, 1977). The Jenny model had depicted association of pedon formation with landscape (Stevens and Walker, 1970; Frolking, 1989), but Yaalon (1975) pointed out that the model remained site-specific, failing to depict either landscape or ecosystem unit. Certain studies (Sehgal, Sys, and Bhumbla, 1968; Sehgal, 1970; and Dahlgren et al., 1997) have used this model to explain soil occurrence over a climosequence. The researchers had to select a vast area over which climate change could be convincing. But in the process, selection of sites was unavoidably subjective, possibly open to bias and preconceived notions. These studies ignore soil variability in the physiographic units on the one hand and mapping criteria on the other. Apart from these studies, the assumption of climate as a consistent variable is unfounded in literature. For example, northwest India has experienced four climatic epochs in the past ca 10,000 years, and during the last epoch (ca 3,500 years back) rainfall was twice what it receives today (Pant, Kumar, and Kumar, 1996). Another serious shortcoming of Jenny's models (Jenny, 1941; and his subsequent works summarized into Jenny, 1980) is the interdependency of the state factors. Several authors (Stephens, 1947; Chesworth, 1973b; Runge, 1973; Yaalon, 1975) have expressed doubts that the mystery of soil formation can be solved by using these models.

Nevertheless, state factor models have acted as a leverage for the development of models based on soil geomorphic relationships and soil occurrence over hill slopes. As the technology enabling observation of the earth from space (remote sensing) becomes precise, the models could reduce the cost of soil surveys. Soil geomorphic models hypothesize that the formation of soil column and evolution of earth surface features are controlled by the same factors (Daniels, Gamble, and Cady, 1971; Carson and Kirkby, 1972; Ruhe, 1975; Young and Hammer, 2000). Anjos et al. (1998) grouped the landscape evolution models into three categories based on varying conditions of climate. For humid climates, the Davisian system (Davis, 1899 as cited by Anjos et al., 1998) suggests that after the rapid uplift of a landmass and stream incision, denudation processes begin. This results in the leveling of the landscape, and slope gradient decreases steadily toward the valley sides. For arid and semiarid regions, Kings' model (as cited by Anjos et al., 1998) explains the succession of high-level erosion surfaces separated from each other by steep scarps. The denudation of landmass yields basal pediments which ultimately give rise to pediplains. The central theme in these models is the weathering of rocks and formation of soil parent material rather than soil formation from parent material or soil development. Anjos et al. (1998) reported relevance of Penck's model of slope replàcement in the tropical wet climate (average annual rainfall of 1,250 mm) and confirmed that pedogenic intensity was strongly dependent on characteristics of the geomorphic surfaces. The intensity factors that had influenced soil formation in their investigation were solution loss, degree of weathering, depth of solum, illuviation process, and cation accumulation.

Factorial functional models are predictive in nature and predict soil classes and their spatial arrangement or soil properties and their trends from landscape features (Hewitt, 1993) that can be assessed through remote sensing techniques. Many of their features are associated with surface nature of the landscape and no convincing relationship has been evolved to establish that geomorphic features alone can be used to predict profile characteristics or catena (Jacob and Nordt, 1991).

Qualitative Functional Anthropogenic Model

Although Jenny (1941, 1980) had taken an overview of all organisms, large-scale modification by human beings was underplayed in his models. These changes are fast and often drastic and noticeable within a few decades. The changes may occur due to changes in land use and because of a shift of cropping system. The former has been the second largest contributor to the global climate change. Bidwell and Hole (1965) listed several processes that could be an outcome of anthropogenic activity on soils. Yaalon and Yaron (1966) developed a framework of a process-response model with the natural soil as a parent material and termed the process metapedogenesis. Yaalon and Yaron (1966) considered that the new soil formed is at a steady-state condition and metapedogenetical processes are mostly reversible. Their proposed mathematical relationship could be rewritten as:

$$S_2 = f(S_1, m_1, m_2, m_3, \ldots m_n) \tag{5.3}$$

where S_2 is the new soil produced from the initial state of the soils (S_1) by the action of a number of (n) metapedogenetical factors (m_1, m_2, m_3, ...m_n) acting singly or in combination. These factors could be grouped under topographical, hydrological, chemical, and cultivation and cropping factors at landscape scale. However, it is debatable as to whether the assumption of reversible metapedogenic changes is compatible with soil development from S_1 to S_2 state. Yaalon and Yaron (1966) also could not establish a relationship between "factors" of soil changes with "processes" (or soil-formation processes).

Thermodynamics-Based Models

In a macroscopic dynamic system such as soils, thermodynamics has always been considered a useful tool to explain soil formation. Runge (1973) attempted to develop a predictive energy model assuming soils to be a closed system. His energy considerations included organic matter production (o, renewing vector), amount of water available for leaching or hydrological factors (w, developing vectors) and time, t. The proposed function (equation 5.4) is:

$$S = f(o, w, t) \tag{5.4}$$

The Runge (1973) model, for all practical purposes, became an extension of Jenny's (1941) model but gave an insight into the genesis of B-horizon and soil formation on a large scale. The groundbreaking attempt, however, came when Philips (1998) attempted to use nonlinear dynamic systems (NDS) theory to explain soil formation. In the NDS theory, a system is perceived as a set of interlinked parts that function together as a complex whole with no linear relationship between impetus and response. NDS theory takes a holistic view of the system, which is made up of many chaotic and complex functions but emphasizes the identification of critical components of the system and establishment of functional links between them.

Philips (1998) used functional models of Jenny (1941), but instead of using factorial concepts, relied on the evolutionary model of soil formation earlier proposed by Johnson and Wastson-Stegner (1987). The latter model (equation 5.5) claimed to include both soil progression (P) and regressive pedogenesis (R), two oppositely acting forces developing into a complex, open system which is self-organizing in varying degrees, scales, and variable rates of energy and mass fluxes under changing environmental conditions.

$$S = f(P, R) \tag{5.5}$$

In the Johnson and Wastson-Stegner (1987) model the P and R are not specified, and their relationships either with soil processes or factors of soil formation are not established. Philips (1998) argued in favor of a unifying framework to overcome these limitations and suggested that NDS theory was best suited for it.

Semiquantitative and Quantitative Models

Huggett (1998) favored an evolutionary approach to pedogenesis, incorporating both horizonation and haplodization processes, and suggested the use of chronosequences as a tool for probing the rate and direction of soil formation both in space and relative time. The nonconstancy of environmental conditions and treating pedospheres as dissipative structures are essential characteristics of the model. These features advocate treatment of soil formation as a nonlinear dynamic system.

Scalenghe, Zanini, and Nielsen (2000) used a single multivariate model to describe complex pedogenic evolution in postincisive chronosequence in Italy. They argued that in the evolution of Bt or Bx horizons, from Bw horizon, a factor comprised of particle-size distribution, cation-exchange capacity, and crystallinity ratio (CR) of iron oxides could be an index for chronosequence. The CR was expressed as:

$$CR = (Fe_d - Fe_o)/Fe_t \tag{5.6}$$

where, Fe_d is citrate-bicarbonate-dithionite (CBD) extractable iron, Fe_o is oxalate-extractable iron, and Fe_t is total amount of iron.

Minasny and McBratney (1999) developed a mechanistic model for soil formation and landscape evolution in catena scale for a time period greater than 10 years. The weathering rate, soil formation, and erosional episodes yielded soil thickness which was a function of landscape. The rate of weathering or lowering of bedrock surface ($\partial e/\partial t$) was represented as an exponential decline with thickness of soil, h (equation 5.7):

$$\partial e/\partial t - P_o \exp(-bh) \tag{5.7}$$

where, P_o (length per time, LT^{-1}) is the potential weathering rate of bedrock at $H = 0$ and b (L^{-1}) is an empirical constant. They observed that the relief of the soil surface decreased with time and soil accumulation was greater for the valley than for the hill. The soil thickness was a power series function of time.

One of the recent key developments in the area of soil formation is the large-scale use of parameters other than those conventionally used in soil survey. Some are mineral weathering rates and state, soil processes (both hydrological and chemical), and the use of statistical and geographical information system (GIS) tools, both temporally and spatially (Hoosbeek and Bryant, 1992; McBratney et al., 2000). Important statistical concepts that are finding a place in modeling soil formation are kriging, fuzzy logic, and fractals (De Bruin and Stein, 1998; Crawford, Pachepsky, and Rawls, 1999; McBratney et al., 2000).

Grunwald et al. (2000) used simulation modeling for four different locations in Wisconsin at pedon, catena, catchment, and soil-region scales. He used soil data, including texture, cone index, and depth of soil layers, and portrayed 3-D soil landscape models using virtual reality modeling language (VRML). The use of 3-D portrayals and 2-D map layering using GIS software may allow pedologists now to view soil formation over different scales vividly.

Quantifying soil formation and using mathematical modeling for soil formation remain elusive to pedologists. Future research may be directed toward identifying independent state factors, variables of soil processes, resolving contradictions between parallel hypotheses (such as catena and polypedon), and establishing the relationship of surface soil characteristics (which can be observed through remote sensing techniques) with subsurface and subsoil characteristics. The current need is for a holistic view of soil formation (Bridges and Catizzone, 1996). Pedologists have to go beyond traditional soil survey data so that soil-forming factors and soil processes are integrated into temporal and spatial dimensions over varying scales, and develop meaningful pedotransfer functions (Bouma, 1989).

SUMMARY

This chapter began with a brief introduction to the processes that precede soil formation: the physical and chemical weathering of geological materials and the complex chemistry involved in the breakdown of organic debris. Soil-forming processes were discussed under two headings: surface processes mainly concerned with the role of organic matter in the development of A horizons, and processes that operate lower in the soil profile to produce the range of B horizons recognized as being important for soil identification, classification, and management. The aim of the chapter was to bring together the different strands of soil genesis, classification, and management. It concluded with a review of the role of modeling in explaining the mechanisms and results of the soil-forming processes.

REFERENCES

Alaily, F. (1996a). Carbonate, sulfate, chloride, phosphate. Chapter 2.1.1.3. In *Handbuch der Bodenkunde,* eds. H.P. Blume, P. Felix-Henningsen, W.R. Fischer, H.G. Frede, R. Horn, and K. Stahr. Landsberg: Ecomed, pp. 1-72.

Alaily, F. (1996b). Carbonate, gips und lösliche salze. Chapter 2.1.4.3. In *Handbuch der Bodenkunde,* eds. H.P. Blume, P. Felix-Henningsen, W.R. Fischer, H.G. Frede, R. Horn, and K. Stahr. Landsberg: Ecomed, pp. 1-8.

Anderson, H.A., M.L. Berrow, V.C. Farmer, A. Hepburn, J.D. Russell, and A.D. Walker (1982). A reassessment of podzol formation processes. *Journal of Soil Science 33:*125-136.

Anjos, L.H., M.R. Fernandes, M.G. Pereira, and D.P. Franzmeier (1998). Landscape and pedogenesis of an Oxisol–InceptIsol–Ultisol sequence in southeastern Brazil. *Soil Science Society of America Journal 62:*1651-1658.

Barrow, N.J. (1983). A mechanistic model for describing the sorption and desorption of phosphate by soil. *Journal of Soil Science 34:*733-750.

Barrow, N.J. (1986). Testing a mechanistic model. 2. The effects of time and temperature on the reaction of zinc with soil. *Journal of Soil Science 37:*277-286.

Beyer, L., H.R. Schulten, R. Frund, R. Hempfling, and U. Irmler (1993). Formation and properties of organic matter in forest soils, as revealed by its biological activity, wet chemical analysis, CPMAS 13C-NMR spectroscopy and pyrolysis-field ionization mass spectrometry. *Soil Biology and Biochemistry 25:*587-596.

Bidwell, O.W. and F.D. Hole (1965). Man as a factor of soil formation. *Soil Science 99:*65-72.

Blume, H.P., L. Beyer, M. Bölter, H. Erlenkeuser, E. Kalk, S. Kneesch, U. Pfisterer, and D. Schnieder (1997). Pedogenetic zonation in the Southern circumpolat region. *Advances in Geoecology 30:*69-90.

Bouma, J. (1989). Using soil survey data for quantitative land evaluation. *Advances in Soil Science 9:*177-213.

Bridges, E.M. (1974). Soil formation in a maritime environment: The modifying effect of gleying. *Transactions of the 10th International Congress of Soil Science VI (ii)* 171-178. Moscow: Academia Nauk.

Bridges, E.M. (1978). Interaction of soil and mankind. *Journal of Soil Science 29:*125-139.

Bridges, E.M. (1997). *World Soils.* Third Edition, Cambridge: Cambridge University Press.

Bridges, E.M. and M. Catizzone (1996). Soil science in a holistic framework: Discussion of an improved integrated approach. *Geoderma 71:*275-287.

Bridges, E.M. and H. De Bakker (1998). Soil as an artifact: Human impact upon the soil resource. *Land 1:*197-215.

Brinkman, R. (1979). Ferrolysis, a soil-forming process in hydromorphic conditions. *Agriculture Research Report 887.* Wageningen: Pudoc.

Bullock, P. and P. Gregory (1991). *Soils in the Urban Environment.* London: Blackwell.

Buol, S.W. (1965). Present soil-forming factors and processes in arid and semi-arid regions. *Soil Science 99:*45-49.

Buurman, P. (1984). *Podzols.* New York: Van Nostrand Rheinhold.

Carson, M.A. and M.J. Kirkby (1972). *Hillslope Form and Processes.* London: Cambridge University Press.

Chesworth, W. (1973a). The parent rock effect in genesis of soil. *Geoderma 10:*215-225.

Chesworth, W. (1973b). The residual system of chemical weathering: A model for the chemical breakdown of silicate rocks at the surface of the earth. *Journal of Soil Science 24:*69-81.

Conacher, A.J. and J.B. Dalrymple (1977). The nine unit landsurface model: An approach to pedogeomorphic research. *Geoderma 18:*1-154.

Council of Europe (1990). *European Conservation Strategy.* Recommendations for the 6th European Ministerial Conference on the Environment. Strasburg: Council of Europe.

Crawford, J.W., A.Y. Pachepsky, and W.J. Rawls (1999). Integrating processes in soils using fractal models (editorial). *Geoderma 88:*103-107.

Dahlgren, R.A., J.L. Boettinger, G.L. Huntington, and R.G. Amundson (1997). Soil development along an elevational transect in the western Sierra Nevada, California. *Geoderma 78:*207-236.

Daniels, R.B., E.E. Gamble, and J.G. Cady (1971). The relation between geomorphology and soil morphology and genesis. *Advances in Agronomy 23:*51-88.

De Bruin, S. and A. Stein (1998). Soil-landscape modeling using fuzzy c-means clustering of attribute data derived from a Digital Elevation Model (DEM). *Geoderma 83:*17-33.

De Coninck, F. (1980). Major mechanisms in the formation of spodic horizons. *Geoderma 24:*101-123.

Dixit, S.P., R. Gombeer, and J. D'Hoore (1975). The electrophoretic mobility of natural clays and their potential mobility within the pedon. *Geoderma 13:*325-330.

FAO (Food and Agriculture Organisation) (1990). *Guidelines for Soil Profile Description.* Third Edition. Rome: Author.

FAO (Food and Agriculture Organisation)-United Nations Education, Scientific and Cultural Organisation (1988). *Soil Map of the World.* Revised Legend. Rome: Author.

Farmer, V.C., W.J. McHardy, L. Robertson, A. Walker, and M.J. Wilson (1985). Micromorphology and sub-microscopy of allophane and imogolite in a podzol Bs horizon: Evidence for translocation and origin. *Journal of Soil Science 36:*87-95.

Farmer, V.C., J.D. Russell, and M.L. Berrow (1980). Imogolite and proto-imogolite allophane in spodic horizons: Evidence for a mobile aluminium silicate complex in podzol formation. *Journal of Soil Science 31:*673-684.

Frolking, T.A. (1989). Forest soil uniformity along toposequences in the loess-mantled driftless area of Wisconsin. *Soil Science Society of America Journal 53:* 1168-1172.

Furrer, G. and H. Sticher (1999). Chemische Verwitterungsprozesse Chapter 2.1.3.2. In *Handbuch der Bodenkunde,* eds. H.P. Blume, P. Felix-Henningsen, W.R. Fischer, H.G. Frede, R. Horn, and K. Stahr. Landsberg: Ecomed, pp. 1-16.

Grunwald, S., P. Barak, K. McSweeney, and B. Lowerd (2000). Soil landscape models at different scales portrayed in virtual reality modeling language. *Soil Science 165:*598-615.

Heiberg, S.O. and R.F. Chandler (1941). A revised nomenclature of forest humus layers for the northwestern United States. *Soil Science 52:*87-100.

Hesselman, H. (1926). Studier over Barrskogens Humustacke. *Statens Skogsforsoksanstalt 22:*169-552 (German Summary 508-552).

Heuvelink, G.B.M. and Webster, R. (2001). Modelling soil variation: Past, present and future. *Geoderma 103:*249-272.

Hewitt, A.E. (1993). Predictive modeling in soil survey. *Soils and Fertilizers 56:*305-314.

Hiller, D. and H. Meuser (1998). *Urbane Böden.* Berlin: Springer.

Hoosbeek, M.R. and R.B. Bryant (1992). Towards the quantitative modeling of pedogenesis—A review. *Geoderma 55:*183-210.

Hopkins, D.W., J.A. Chudek, and R.S. Shiel (1993). Chemical characterization and decomposition of organic matter from two contrasting grassland soil profiles. *Journal of Soil Science 44:*147-157.

Huggett, R.J. (1998). Soil chronosequences, soil development and soil evolution: A critical review. *Catena 32:*155-172.

Jacob, J.S. and L.C. Nordt (1991). Soil and landscape evolution: A paradigm for pedology. *Soil Science Society of America Journal 55:*1194.

Jenny, H. (1941). *Factors of Soil Formation.* New York: McGraw-Hill.

Jenny, H. (1980). *The Soil Resource—Origin and Behaviour.* New York: Springer.

Joffe, J.S. (1949). *Pedology.* New Brunswick, NJ: Pedology Publications.

Johnson, D.L. and D. Wastson-Stegner (1987). Evolution model of pedogenesis. *Soil Science 143:*349-366.

Jörgensen, R.G. and B. Meyer (1990). Nutrient changes in decomposing beech leaf litter assessed using a solution flux approach. *Journal of Soil Science 41:*279-293.

Kirkby, M.J. (1985). A basis for soil profile modelling in a geomorphic context. *Journal of Soil Science 36:*97-121.

Kubiena, W.L. (1953). *The Soils of Europe.* London: Thomas Murby & Co.

McBratney, A.B., I.O.A. Odeh, T.F.A. Bishop, M.S. Dunbar, and T.M. Shatar (2000). An overview of pedometric techniques for use in soil survey. *Geoderma 97:*293-327.

Minasny, B. and A.B. McBratney (1999). A rudimentary mechanistic model for soil production and landscape development. *Geoderma 90:*3-21.

Minasny, B. and A.B. McBratney (2001). A rudimentary mechanistic model for soil formation and landscape development. *Geoderma 103:*161-179.

Mückenhausen, E. (1977). *Entstehung, Eigenschaften und Systematik der Böden der Bundesrepublik Deutschland.* 2. Auflag, Frankfurt am Main: DLG Verlag.

Nye, P. H. (1954). Some soil forming processes in the humid tropics. *Journal of Soil Science 5:*7-21.

Ollier, C.D. (1984). *Weathering.* Edinburgh: Oliver and Boyd.

Pant, G.B., K.R. Kumar, and K.K. Kumar (1996). Climate change and variability over South Asia. In *Climate Variability and Agriculture,* eds. Y.P. Abrol, S. Gadgil, and G.B. Pant. New Delhi, India: Narosa, pp.53-68.

Pape, J.C. (1970). Plaggensoils in the Netherlands. *Geoderma 4:*229-255.

Paton, T.R. (1978). *The Formation of Soil Material.* London: Allen and Unwin.

Philips, J.D. (1998). On the relations between complex systems and the factorial model of soil formation (with discussion). *Geoderma 86:*1-42.

Pons, L.J. and I.S. Zonneveld (1965). *Soil Ripening and Soil Classification, Initial Soil Formation of Alluvial Deposits with a Classification of the Resulting Soils.* Publication 13. Wageningen: International Land Reclamation Institute.

Porta, J. (1998). Methodologies for the analysis and characterisation of gypsum in soils: A review. *Geoderma 87:*31-46.

Ross, S.M. (1989). *Soil Processes: a Systematic Approach.* London: Routledge.

Ruhe, R.V. (1975). *Geomorphology.* Boston: Houghton Mifflin.

Runge, E.C.A. (1973). Soil development sequences and energy models. *Soil Science 115:*183-193.

Scalenghe, R., E. Zanini, and D.K. Nielsen (2000). Modeling soil development in a post-incisive chronosequence. *Soil Science 165:*455-462.

Scheffer, F. and P. Schachtschabel (1998a). Organische substanz und bodenorganismen. In *Lehrbuch der Bodenkunde,* eds. P. Schachtschabel, P. Blume, G. Brümmer, K.H. Hartge, U. Schwertmann, K. Auerswald, L. Beyer, W.R. Fischer, I. Kögel-Knabner, M. Renger, et al. Stuttgart: Ferdinand Enke, pp. 45-86.

Scheffer, F. and P. Schachtschabel (1998b). Bodenentwicklung, bodensystematik und bodenverbreitung. In *Lehrbuch der Bodenkunde,* eds. P. Schachtschabel, P. Blume, G. Brümmer, K.H. Hartge, U. Schwertmann, K. Auerswald, L. Beyer, W.R. Fischer, I. Kögel-Knabner, M. Renger, et al. Stuttgart: Ferdinand Enke, pp. 373-470.

Schlichting, E. and U. Schwertmann (1973). *Pseudogley and Gley.* Weinheim: Verlag Chemie.

Sehgal, J.L. (1970). The soils of Punjab (India) I. Geographic conditions. *Pedologie* (Ghent) *20:*178-203.

Sehgal, J.L., C. Sys, and D.R. Bhumbla (1968). A climatic soil sequence from Thar Desert to the Himalayan Mountains in Punjab (India). *Pedologie* (Ghent) *18:* 351-373.

Sibbesen, E. (1981). Some new equations to describe phosphate sorption by soils. *Journal of Soil Science 32:*67-74.

Stahr, K. (1973). Die Stellung der Böden mit Fe-Bändchen in der Bodengesellschaft der Scharzwaldberge. Arbeiten der Geologie und Paläontologie, *Universität Stuttgart 69:*85-183.

Stephens, C.G. (1947). Functional synthesis in pedogenesis. *Transactions of the Royal Society of Amsterdam 71:*168-181.

Stevens, P.R. and T.W. Walker (1970). The chronosequence concept and soil formation. *Quarterly Review of Biology 45:*333-350.

Sumner, M.E. (2000). *Handbook of Soil Science.* Boca Raton: CRC Press.

Szabolcs, I. (1989). *Salt-Affected Soils.* Boca Raton: CRC Press.

Tarnocai, C., C.A.S. Smith, and C.A. Fox (1993). *International Tour of Permafrost-Affected Soils: The Yukon and Northwest Territories of Canada.* Ottawa: Agriculture Canada.

Thompson, J.A., J.C. Bell, and C.A. Butter (1977). Quantitative soil landscape modeling for estimating the area extent of hydromorphic soils. *Soil Science Society of America Journal 61:*971-980.

Van Alphen, J.G. and F. De los Rios Romero (1971). *Gypsiferous Soils.* Wageningen: International Land Reclamation Institute.

Van Wambeke, A. (1992). *Soils of the Tropics: Properties and Appraisal.* New York: McGraw-Hill.

Wiechmann, H. (1978). *Stoffverlagerung in Podsolen.* Stuttgart: Ulmer.

Wiechmann, H. (2000). Gesteine als Ausgangsmaterial der Bodenkunde. Chapter 2.1.2. In *Handbuch der Bodenkunde,* eds. H.P. Blume, P. Felix-Henningsen, W.R. Fischer, H.G. Frede, R. Horn, and K. Stahr. Landsberg: Ecomed, pp. 1-43.

Wilding, L.P. and R. Puentes, eds. (1988). *Vertisols: Their Distribution, Properties, Classification and Management.* Technical Monograph 18. Texas A & M University: Soil Management Support Services.

Yaalon, D.H. (1975). Conceptual models in pedogenesis. Can soil forming functions be solved? *Geoderma 14:*189-205.

Yaalon, D.H. and B. Yaron (1966). Framework for manmade soil changes—An outline of metapedogenesis. *Soil Science 102:*272-277.

Young, F.J. and R.D. Hammer (2000). Soil-landform relationships on a loess-mantled upland landscape in Missouri. *Soil Science Society of America Journal 64:*1443-1454.

Zech, W. and I. Kögel-Knabner (1994). Patterns and regulations of organic matter transformation in soils: Litter decomposition and humification. In *Flux Control in Biological Systems,* ed. E.D. Schulze. San Diego: Academic Press, pp. 303-335.

Chapter 6

Radioactivity in the Soil-Plant System

Martin H. Gerzabek
Friederike Strebl
Sabine Ehlken
Gerald Kirchner

The spontaneous discharge of radiation from unstable atomic nuclei is termed radioactivity. It usually occurs in the form of alpha (α: rays of helium nuclei particles, high energy–low penetration ability) or beta particle radiation (β: electron or positron particles), together with gamma radiation (γ: high energy electromagnetic radiation of short wavelength with high penetration depth in biological systems).The units of activity are Bequerel [s^{-1}], which counts the spontaneous disintegrations of a given radionuclide per second.

About 1,300 different radionuclides (RN) exist that partly originate from natural sources and are partly anthropogenically produced. Natural radionuclides can be generated by activation of stable isotopes via cosmic radiation (e.g., ^{3}H (tritium, 12.3 years physical half-life–$t_{1/2phys}$), ^{7}Be (beryllium, 53.3 days), ^{14}C (carbon, 5,730 y), ^{35}S (sulphur, 87.5 d)), or they originated from the creation of our universe (primordial; ^{40}K (potassium, $t_{1/2phys}$: 1.3 × 10^{9} y); uranium and thorium isotopes and daughter nuclides, e.g., ^{226}Ra (radium, 1,600 y), ^{222}Rn (radon, 3.8 d), ^{210}Pb (22.3 y). Radionuclides from natural decay chains, despite short physical half-lives, can gain high radiological importance because they are continuously produced and thus remain in the environment at a constant level. Typical ranges of some relevant nuclides in Austrian topsoils are listed in Table 6.1.

Artificial radionuclides have entered the human environment as a consequence of atmospheric nuclear weapon tests (global fallout of fission products in the 1950s and 1960s, mainly in Europe of ^{137}Cs (cesium, $t_{1/2phys}$: 30.17 y), ^{90}Sr (strontium, 28.5 y), ^{89}Sr (50.5 d), ^{3}H, ^{54}Mn (manganese, 312 d), ^{65}Zn (zinc, 244 d), ^{85}Kr (krypton, 10.8 y), ^{95}Zr (zirconium, 64 d), $^{103/106}Ru$ (ruthenium, 39.4 d/368 d), ^{129}I (iodine, 1.6 × 10^{7} y), ^{144}Ce (cerium, 284.8 d)). At present, most important sources are routine releases from nuclear power (NPP) and reprocessing plants (^{3}H, ^{14}C, ^{60}Co (cobalt, $t_{1/2phys}$: 5.27 years), ^{54}Mn, ^{89}Sr, ^{90}Sr, ^{95}Zr, ^{85}Kr, ^{238}Pu (plutonium, 87.7 y), $^{239/240}Pu$ (24 y/6,600 y), ^{129}I, ^{95}Nb (niobium, 35 d), $^{103/106}Ru$, ^{110m}Ag (silver, 249.9 d), ^{125}Sb (antimony, 2.77 y), $^{134/137}Cs$ (2.06 y/30.17 y), ^{140}Ba (barium, 12.8 d), ^{144}Ce). Exceedingly high emissions of these isotopes can occur in the aftermath of NPP accidents such as the one that occurred during 1986 in Chernobyl. This accident increased $^{134/137}Cs$ inventories of many European countries considerably. Some typical values of ^{137}Cs and ^{90}Sr soil inventories of different countries are reported in Table 6.1. Other, mainly short-lived isotopes are used in medicine as tracers for diagnostics or as agents for

TABLE 6.1. Typical Ranges of Radionuclide Content in Austrian Soils and Comparison to Other Regions

	Source	Average topsoil (0-20 cm) content (Bq kg^{-1})	Soil inventory (kBq m^{-2})
^{14}C	Natural	330	
^{40}K	Natural	1-1000	
^{210}Pb	Natural	5-50	
^{226}Ra	Natural	5-200	
^{238}U	Natural	5-150	
^{234}U	Natural	9-120	
^{90}Sr	Bomb tests		Austria: range: 1-4
	Nuclear accident		ChNPP 30 km zone: 18-1800
			Kyshtym region: 2000-4000
^{103}Ru	Nuclear accident		Austria: 3.7-74
^{137}Cs	Bomb tests		Austria: 23.4
	Nuclear accident		Denmark: 3.7
			ChNPP 30 km zone: 1500-5000
$^{239/240}Pu$	Bomb tests		Austria: 1-15 × 10^{-2}
	Nuclear accident		ChNPP 30 km zone: > 3.7

Source: Bunzl (1997); Aarkrog et al. (1992); Ageets (1999); United Nations Scientific Committee on the Effects of Atomic Radiation (UNSCEAR) (1988).

Note: All artificial radionuclides corrected for decay to 1986-05-01; ChNPP: Chernobyl Nuclear Power Plant.

radiotherapy (^{99m}Tc ($t_{1/2phys}$: 6 h), ^{131}I, ^{51}Cr (27.7 d), ^{54}Mn, ^{60}Co). Additional risks of radionuclide contamination of the environment may arise from radioactive waste depositories.

DEPOSITION OF RADIONUCLIDES

In most cases, radionuclides enter ecosystems and food chains via deposition. In the early phase after the deposition of a radionuclide, direct contamination of crop and fodder plants is a key concern, particularly if the fallout event takes place during the vegetation period. Root uptake of radionuclides and transfer to shoot is the main source of radionuclide contamination of crops in later vegetation periods. Table 6.2 shows the relative ^{137}Cs concentrations in different agricultural products in 1986 (Chernobyl fallout) and thereafter. The substantial reduction is due to the loss of direct contamination of crops and fodder plants after 1986. Milk and apples showed a less-pronounced decrease from the first to the second year due to remobilization of radiocesium in perennial fodder plants (and the subsequent transfer to milk) or from tree wood to fruit.

On the process level, wet and dry deposition are distinguished. Wet deposition is highly effective and, if rainfall occurs during a fallout event, it contributes to most of the radionuclide deposition on plant and soil surfaces. The wet deposition is described by the "washout factor," which is defined as the fraction of activity deposited per second from the radioactive plume. It depends on the rain intensity. The contamination of plants depends on the growth stage and the prevalent leaf area available for interception. As a default value, a water retention capacity of agricultural crops of approximately 0.2-0.3 mm has been used (Pröhl, 1990). Small rain events of $\leq$1 mm might

TABLE 6.2. Relative Decrease of ^{137}Cs Activity Concentrations in Different Foodstuffs in Austria Following the Chernobyl Fallout

	Percent of the average value			
Year	**Milk**	**Cereals**	**White cabbage**	**Apple**
1986	100.0	100.0	100.0	100.0
1987	9.5	5.1	5.9	9.6
1990	2.1	3.3	2.5	1.5
1993	0.7	2.0	1.3	0.5

Source: Adapted from Mück, 1995.

be retained completely by plants in advanced growth stages with larger leaf area indices (= leaf area per unit ground area). Larger rain events (>10 mm) will be retained only to a small extent, with decreasing importance of the leaf area index (Hoffman et al., 1992).

Dry deposition is defined as the transfer of gases and airborne particles from the surface near air onto soil and plant surfaces. It is often described by the deposition velocity (V_g, [m s^{-1}]), which is defined as the ratio of radioactivity deposited per second (Bq m^{-2} s^{-1}) and the activity concentration in the surface near air (Bq m^{-3}). The deposition velocity depends on the size distribution of the radioactive particles and their chemical form, the meteorological condition during deposition, and the leaf area index of intercepting crops. As a default value 0.005 m s^{-1} is suggested as a deposition velocity for particle sizes of a few micrometers. This value can vary considerably. The amount of deposited radionuclides is calculated by multiplying the deposition velocity by the time integral of the activity concentration in surface near air (Bq s m^{-3}).

Deposition of radioactive particles onto plant surfaces is also possible even after the actual deposition has occurred due to resuspension of contaminated soil particles and their subsequent deposition. Determination of the soil-to-plant transfer of radionuclides can be complicated by soil adhesion onto plant surfaces. This is especially crucial for radionuclides with low plant availability. A brief review on soil adhesion and methods of its determination can be found in Li, Gerzabek, and Mück (1994).

After deposition onto leaf surfaces, radionuclides are partly washed off by later rainfall events, lost by leaf fall or harvest, and partly absorbed into the leaf and translocated to other plant parts. These processes are highly element-specific.

BEHAVIOR OF RADIONUCLIDES IN SOILS

On a long-term scale, soil is a major source of radionuclides entering the food chain or contamination of groundwater through leaching of radionuclides. An additional input pathway to humans can be through inhalation of resuspended contaminated soil particles. There are no general differences between radioactive and stable isotopes of a given chemical element with respect to their behavior in the environment. Soil characteristics having a major impact on radionuclide mobility are: (1) the composition of the soil solution (pH, concentration of inorganic ions, redox potential, concentration of organic substances), (2) physical and chemical soil properties (species/characteristics and contents of clay minerals, oxides and organic matter, surface, and charges of particles), (3) microorganisms and fungi, and

(4) temperature (Bunzl, 1997). Because most of the radionuclides in soil are present in cationic form, therefore, low pH, low clay content, and resulting low cation exchange capacity lead to an increased radionuclide mobility within the soil profile and higher plant uptake. Changes in soil organic matter contents may yield different effects, depending on the ability of a radionuclide to form organic complexes (Frissel, Noordijk, and Van Bergeijk, 1990).

The influence of several soil characteristics on the behavior of some radionuclides in soils is summarized in Table 6.3. Furthermore, the downward movement of radionuclides in soil profile is determined by soil water content. The radionuclides can migrate both in soluble form and as bound to soil particles (colloids). The latter process is especially important for poorly water-soluble radionuclides. According to Bunzl (1987) typical time spans for the migration of different radionuclides to a soil depth of one meter are: 150-4,500 years for Cs, 40-200 years for Sr, and 0.5-20 years for Tc. The migration rates can vary considerably between different soil horizons (Strebl et al., 1996). Bunzl (1997) derived ecological residence half-times (see Modeling Radioactivity in Soil-Plant Systems) for ^{137}Cs (originating from global fallout) in a podzol soil under spruce forest. He found that the

TABLE 6.3. Influence of Soil Characteristics on the Behavior of Different Radionuclides in Soils

	Cs	**Sr, Ra**	**J**	**U, Pu**	**Ru**
Similar ions	K, Rb, NH_4	Ca, Ba		Np	Tc, Mo, Nb, Zr, Y, Co, Fe, Mn, Cr,
Chemical form in soil	Cs^+	Sr^{2+}	J_2, CH_3J, $J^-$$JO_3^-$	PuO_2^{2+}, $Pu(NO_3)^{3+}$	$RuNO^{3+}$, Ru^{3+}, RuO_4^-
	changes in RN mobility and plant availability if . . .				
pH-value decreases	⇑	⇑		⇑	⇑
Clay content decreases	⇑	⇑		⇑	⇑
Sand content decreases	⇓	⇓		⇓	⇓
Humus content low	⇔	⇓	⇑	⇓	⇔
CEC decreases	⇑	⇑			⇑
K_d value (l kg^{-1})	10^2-10^3	10^1-10^2	10^{-1}-10^2	10^2-10^3	10^1-10^3

Source: Compiled from Bunzl (1987); Coughtrey, Jackson, and Thorne (1983); IAEA (1994); Mortvedt (1994).

values varied between approximately two years per centimeter in the litter horizons to 33 years per centimeter in the first two centimeters of the first mineral horizon.

Experimental results have shown that the fixation of radionuclides increases in the following order: $^{95}Tc < {}^{131}I < {}^{85}Sr < {}^{103}Ru < {}^{109}Cd \sim {}^{65}Zn < {}^{57}Co < {}^{141}Ce < {}^{137}Cs$ (Bunzl and Schimmack, 1988).

In the presence of illitic clay minerals *cesium* is strongly fixed in soil and shows a very low vertical mobility or plant availability. This occurs especially in the so-called frayed-edge-sites (FES) located at the edges of micaceous clay minerals which selectively and strongly sorb alkali cations such as potassium or cesium (Cremers et al., 1988). To quantify the effect of FES, the radiocesium interception potential—calculated as the product of Cs to K selectivity coefficient in the FES and their abundance—can be used (Sweek et al., 1990). In addition to clay content, increasing pH values decrease radiocesium mobility. After the Chernobyl-derived contamination, it soon became evident that radionuclide behavior in seminatural environments differs significantly from that in agricultural areas. Especially, the radiocesium migration in the soil profile was slower than that expected based on earlier experience with agricultural soils. Organic soil horizons play an important role for Cs retention in seminatural soils under highland or alpine meadows and forests. Although plant availability of Cs is high, which can be related to a decrease of Cs fixation in clay minerals in the presence of humic substances and to high ammonium concentrations in the soil solution that remove Cs from the exchange sites (Cheshire et al., 2000; Sanchez et al., 1999), migration of Cs in these soils is slow. This may partly be explained by the high microbial/fungal biomass in organic soils, which is able to store Cs (Johanson and Nikolova, 1996). Moreover, the vegetation of such nutrient-limited ecosystems tends to store and recycle deposited radionuclides (as nutrient analogues) in the rooting zone very efficiently.

Strontium is chemically similar to calcium and thus shows much higher mobility than cesium. Gerzabek et al. (1992) reported that a fraction between 18-62 percent of the total ^{90}Sr in Austrian soils is in exchangeable form. The fraction of mobile radiostrontium (and other elements) can be influenced by the presence of hot particles from nuclear power plant accidents or nuclear weapons testing (Gastberger, Steinhäusler, Gerzabek, Hubmer, et al., 2000). In such particles, radionuclides are strongly fixed over a long time. Sorption increases at higher pH values along with the formation of Sr-humic complexes (Juo and Barber, 1970). Calcium competes with strontium for exchange sites. Because of low fixation strength in soils (see K_d value in Table 6.3), migration rates of up to several cm per year are observed in agricultural soils (Haunold, Horak, and Gerzabek, 1987). Recently it has been reported that, similar to Cs, strontium is efficiently retained in alpine soils and is prone to cycling within the vegetation cover (Gastberger, Steinhäusler, Gerzabek, Lettner, et al., 2000).

Some *iodine* isotopes are very short-lived (^{131}I: 8.02 days). Only ^{129}I has some relevance with regard to soil-plant transfer. In soil it occurs in close association with soil organic matter (Gerzabek et al., 1999). Microbial biomass can fix iodine very effectively (Dertinger et al., 1986). As an anion, iodine can be sorbed to Fe and Al hydroxides and clay minerals (Whitehead, 1984). For Austrian soils a positive correlation between clay content and iodine content in soil was reported (Gerzabek et al., 1999).

In soil, *ruthenium* can be found as Ru^{3+} and ruthenate (RuO_4^-). Its mobility is slightly higher than that of cesium. The typical K_d values for ruthenium are between those of Cs and Sr (Bunzl, 1987).

Plutonium has an extremely high radiotoxicity. In soil it can be found in five different oxidation states (III, IV, V, VI, VII). It can occur as complex-bound or as PuO_2. Significant amounts of plutionium seem to be bound to humic substances (Bunzl, 1997). Its K_d values are similar to those of cesium, so are the migration rates in soil, which were observed by Shaw and Wang (1996) in the close vicinity of the Chernobyl NPP.

For *cobalt* at $pH < 5$ the free Co^{2+} is abundant in the soil solution. At higher pH values, soluble organic complexes are formed, and at $pH > 6$, more than 90 percent of Co in the soil solution is complex-bound (Scheffer and Schachtschabel, 1989). Cobalt adsorption occurs on Fe and Mn oxides; occlusions in sesquioxides are possible. Lysimeter experiments show that cobalt mobility in various Austrian soils was significantly higher than radiocesium mobility (Gerzabek et al., 1996).

Radium, a divalent cation, behaves similar to calcium in soil and plants. As compared to Cs, mobility of Ra in soils is higher and it decreases with increasing pH, clay content, and concentration of Mn and Fe hydroxides. Plant availability and plant uptake of radium is distinctly influenced by the essential plant nutrient Ca (Gerzabek, Strebl, and Temmel, 1998).

SOIL-TO-PLANT TRANSFER OF RADIONUCLIDES

Even if released due to major accidents at nuclear installations, radionuclides are present in only trace quantities in soil-plant system. For example, activity concentrations of ^{90}Sr and ^{137}Cs in soil solution are in the order of 1 kBq l^{-1} in the areas heavily contaminated by the Chernobyl fallout. This corresponds to concentrations of ca. 2×10^{-12} M l^{-1}, which is lower by about nine orders of magnitude than concentrations of major cations in soil solution. As a consequence, plant uptake of many radionuclides (including those that are of highest radiological importance) may not primarily depend on their absolute concentration in the soil-plant system, but on the concentration ratio to other micro- and macronutrients. For radioactive cesium and strontium, these competitive effects form the basis of countermeasures aimed at reducing plant uptake and transfer via food chains of these nuclides following a nuclear accident (Howard and Desmet, 1993).

In most soils, concentrations of stable elements and their radioactive isotopes in solution are determined by cation exchange reactions with the soil matrix. Although fair knowledge exists on the chemistry of trace radionuclides in soils, evidence is accumulating that processes in the rhizosphere may considerably alter their availability for plant uptake. Because of different microenvironments in the rhizosphere, the adsorption of radiocesium on clay minerals may increase (Guivarch, Hinsinger, and Staunton, 1999), or remobilization of clay-fixed radiocesium may occur (Thiry, 1997). Only limited data are available on the accumulation of radionuclides by bacteria and soil fungi, which often show elevated concentrations in the rhizosphere compared to the bulk soil. It has been reported that in organic and forest soils a significant fraction of radiocesium may be stored and potentially immobilized by soil fungi (Dighton, Clint, and Poskitt, 1991; Brückmann and Wolters, 1994; Nikolova, Johanson, and Clegg, 2000). However, current knowledge on the effects of roots and rhizosphere organisms on bioavailabilities of radionuclides appears to be incomplete.

Transport of solutes including radionuclides to plant roots is accomplished by convective mass flow and diffusion. Extensive information on the implications of these processes for major nutrients and trace metals has been gained through experimentation and simulation (Tinker and Nye, 2000), but information on the dynamics of radionuclides in the rhizosphere and its implications for root uptake is still missing.

Soil moisture is known to influence both the concentration and the mobility of solutes. A lower soil water content may decrease (Pendleton and Uhler, 1960) or increase (Tikhomirov, 1988) radionuclide uptake by roots. As pointed out by Shalhevet (1973) these differing responses reflect whether soil drying mainly restricts the mobility or increases the concentration of the radionuclide in solution. Recent long-term field studies (Ehlken and Kirchner, 1996) of radiocesium and radiostrontium uptake by meadow vegetation growing on various soil types indicate that, under natural environmental and climatic conditions, periods of low soil water contents are associated with elevated plant-root-uptake rates of both of the radionuclides.

Radionuclides as well as other minerals are taken up by plant roots in ionic form. After passing the plasma membrane of the epidermal or cortical cells, the ions move into the cytoplasma through the plasmadesmata that connect adjoining cells. Having thus crossed the endodermis, ions are released from the xylem parenchyma cells into the apoplasm of the xylem vessel again by moving through a plasma membrane. Novel molecular and electrophysiological techniques reveal that ion transport through plasma membranes of root cortical and xylem parenchyma cells is facilitated by ion pumps (Michelet and Boutry, 1995), carriers (Tanner and Caspari, 1996), and ion channels (White, 1997). From solution culture experiments it is well known that these uptake and translocation mechanisms are subject to competitive and inhibitory interactions for alkaline (Shaw and Bell, 1991; Smolders et al., 1997) and earth-alkaline (Epstein and Leggett, 1954) elements, but also for other metals with essential ions of similar ionic radii (Kawasaki and Moritsugu, 1987). For the alkaline elements, Maathuis and Sanders (1996) pointed out that selectivity sequences observed with intact roots correspond to selectivities measured at plasma membranes. This is valid both at low ion concentrations when carrier transport dominates and at high ion concentrations when movement across the membrane mainly occurs through potassium channels. Moreover, in plasma membranes of plant root cells a second type of channel (termed calcium channel) has been identified, which mediates, although with lower selectivities, the translocation of earth-alkaline elements and other divalent cations including Mn^{2+}, Co^{2+}, Ni^{2+}, Cu^{2+}, Zn^{2+}, and Cd^{2+} (White, 1998). Interestingly, Sr:Ca permeability ratios observed for root cell calcium channels (White, 1998) are in good agreement with data on the discrimination of these elements in root uptake ("observed ratios"), which had traditionally been used to characterize the plant uptake of radiostrontium (Russell and Newbould, 1966).

Translocation describes the distribution of radionuclides subsequent to foliar or root absorption. The xylem is the principal water-conducting tissue of vascular plants: dissolved substances are transported together with the transpiration stream from soil to the roots and further upward. The other long-distance transport system is the phloem. Its principal function is the conduction of assimilates and food from their sites of production (leaves) to the storage organs (fruits, tubers, etc.). These flows can therefore be directed both upward and downward within the plant (Strasburger, 1998). Elements such as K, Cs, and P are readily transported in both xylem and phloem. Mobility in xylem is mainly dependent on ion valence ($K^+ > Ca^{2+} >$ polyvalent cations) and resulting adsorption to the cell wall (Marschner, 1986). Cobalt is translocated in phloem as a negatively charged complex. Due to physiological reasons the alkaline earth metals (Be, Mg, Ca, Ba, Sr, Ra), lead, plutonium, lanthanides, and actinides are discriminated from phloem. They are therefore dependent on xylem mobility and mainly remain at the absorption sites within plants or are accumulated in vegetative plant parts and mostly excluded from edible plant parts (fruits, tubers) (Thiessen et al., 1999).

MODELING RADIOACTIVITY IN SOIL-PLANT SYSTEMS

Radionuclide Transport in Soil

The basic processes controlling mobility of radionuclides (and other trace elements) in soil include convective transport by flowing water, dispersion caused by spatial variations of convection velocities, diffusive movement within the fluid, and physico-chemical interaction with the soil matrix. In addition to these abiotic processes, soil fauna may contribute to the transport of radionuclides in soils (Müller-Lemans and Van Dorp, 1996), but their action under general conditions results in a dispersion-like translocation (Boudreau, 1986) and hence will not be considered separately.

Since the early 1960s scientific interest has focused on predicting the environmental fate of isotopes ^{90}Sr, ^{137}Cs, and $^{238-240}$Pu with long half-lives. As discussed previously, these radioisotopes strongly sorb to the soil matrix, showing K_d values usually well above 100. It can be easily calculated that transport velocities in the unsaturated zone for those radionuclides will be about 1 cm y^{-1} or lower, indicating that a major fraction of the originally deposited activity may remain within the rooting zone for decades. Because of this slow migration velocity, models that simulate radionuclide movement in soils usually do not take into account soil moisture changes in the unsaturated zone, but (often implicitly) assume a constant mean water content. As a second simplification, usually one-dimensional models are applied based on the assumption that deposition rates are spatially uniform, which may be justified for weapons-testing fallout and, at least on a local scale, also for the Chernobyl fallout. With these simplifications, two approaches have become most popular for modeling the migration of radionuclides in soils—a serial compartmental approach and the convection-dispersion equation.

Compartmental models are often considered as a black-box approach (Coughtrey, 1988) and are applied without the need for knowledge on the site-specific hydrological, physico-chemical, and biological processes that the radionuclides are subjected to. Consequently, models of this type have frequently been employed to describe radionuclide migration in soils (e.g., Frissel and Pennders, 1983). Usually, the soil is split into a series of horizontal layers (compartments) which are connected by downward transport rates of the radionuclide under study. It is assumed that (1) the radionuclides are homogeneously mixed in each compartment, (2) transfer rates are time-invariant, (3) sorption of the radionuclide is instantaneous and characterized by a linear isotherm, and (4) fluxes between compartments are donor-controlled. The radionuclide dynamics are described by a system of linear first-order differential equations with constant coefficients. If deposition of the radionuclides from the atmosphere can be approximated by pulselike inputs into the surface compartment, then analytical solutions of this sytem of differential equations are available (Boone et al., 1985; Schuller, Ellies, and Kirchner, 1997). At the same time, numerical methods of solution that can be used with arbitrary deposition histories and initial radionuclide concentrations in the compartments have also been developed (Bunzl et al., 1994). Site-specific values of the transfer rates, from which migration velocities and residence times can easily be calculated, are usually estimated by tuning the model to measured concentrations in soil profile layers. Hence, the thickness of the compartments into which the soil is divided is usually arbitrarily set to the dimensions of the soil layers taken in the field for parameter estimation.

The multicompartmental models are simple chromatographic models that can be interpreted as discrete analogues of the dispersion-convection equation (Naumann and Buffham, 1983). Hence, the compartmental model chosen should not be considered as a black-box approach, but must reflect the basic transport characteristics of the soil studied. As a consequence, the serial compartmental model is applicable only if the transport of the radionuclides is dominated by convection (Kirchner, 1998a). The number of compartments is then a discrete analogue of the dispersion characteristics of the soil under study and hence should not be chosen arbitrarily (Kirchner, 1998a). If the nuclides' translocation in soil is dominated by dispersion-like processes—as is indicated by concentration profiles declining exponentially with depth (e.g., Isaksson and Erlandsson, 1998; Hölgye and Maly, 2000)—a compartmental model with backflow should be chosen; this type of model adequately represents transport driven by vertical concentration differences (Kirchner, 1998a).

A number of studies applying the convection-dispersion equation have been reported in the literature (e.g., Mahara and Miyahara, 1984; Smith, Hilton, and Comans, 1995; Schuller, Ellies, and Kirchner, 1997; Bossew and Strebl, 2001). The deposition history is commonly approximated by pulselike input functions, and effective values of the dispersion coefficient and convection velocity are obtained by fitting the analytical solution of the model equation to measured depth distributions of the radionuclides. By far, the most extensive study has been performed by Bossew (1997), who analyzed more than 1,300 depth distributions taken at 477 sites in Austria. Generally, all nuclides analyzed had a limited mobility in soil with radiocesium having smaller values than ^{106}Ru, ^{125}Sb, ^{110m}Ag, and ^{144}Ce. Bossew (1997) also showed that the parameter values resulting from the fitting procedure could be grouped with regard to soil types and that parameter variability was considerably reduced within groups. This indicates that the fitted parameter values are related to the physics of radionuclide transport in the soils studied and hence can be used for predictive modeling. In general, radionuclide mobilities increased in the sequence humic alpine soils > Podzols > Cambisols > Chernozems.

For various radionuclides and soils a small fraction of the deposited activities was found at greater depths than predicted by the convection-dispersion model (Mahara and Miyahara, 1984; Smith, Hilton, and Comans, 1995; Schuller, Ellies, and Kirchner, 1997). This effect has also been observed with other solutes (e.g., Warrick, Biggar, and Nielsen, 1971; Khan and Jury, 1990; Rudolph et al., 1996) and is generally interpreted to be caused by stochastic spatial variations of the hydraulic properties (Dagan, 1989). A simple stochastic approach is the transfer function model suggested by Jury (1982) and Simmons (1982) and extensively discussed by Jury and Roth (1990). If transport is stationary and linear, all information of the system dynamics is inherent in its impulse-response function, which then can be taken as the probability density function of travel times to a fixed depth or, alternatively, of travel distances in soil during a fixed time period. The transfer function models for solute movement are discussed in detail in Chapter 11 ("Solute-Water Interactions").

Hydraulic conductivities of soils have repeatedly been found to show spatial variations that are distributed approximately lognormally (Nielsen, Biggar, and Erh, 1973; Jury et al., 1987). In soils with such spatial properties, solute transport, which is assumed to occur by convection via individual isolated streamlines, is characterized by lognormal travel time probability density functions. This model, as proposed by Simmons (1982), may be characterized as stochastic-convective transport. It shows dispersion properties that are fundamentally different from those of the familiar convective-dispersive model (Jury and Sposito, 1985): the variance of travel times (which is a convenient measure of a solute's dispersion) grows linearly with transport distance for convective-dispersive transport, but with the square of the distance for the stochastic-convective model.

Recently, Kirchner (1998b) extended the stochastic-convective model to take also into account spatial variability of radionuclide sorption. After fitting both models to the depth distribution of weapons fallout ^{90}Sr in a Podzol, the stochastic-convective model adequately predicted dispersion of the Chernobyl-derived radiostrontium in this soil, but the convective-dispersive model did not. The implications of these two different dispersion models for the translocation of radionuclides are illustrated by Figure 6.1.

Depth distributions, predicted by the two models for weapons fallout of ^{90}Sr 100 years after deposition has started, differ considerably, clearly reflecting the higher dispersion of the stochastic-convective transport process. Although further testing of the potential of the stochastic-convective model for simulating the transport of radionuclides in soils is desirable, Figure 6.1 indicates that for predictive purposes (e.g., of expected arrival of a long-lived radionuclide at a groundwater horizon) its application may be preferable to the convective-dispersive model.

Only a few studies have analyzed the transport of radionuclides in the unsaturated zone using mechanistic models that simulate the impact of soil moisture changes on transport caused by hydrological and biological processes (Hormann and Kirchner, 1998; Brechignac et al., 1999). Butler et al. (1999) studied the upward migration of radionuclides from contaminated groundwater into the rooting zone—a scenario which is important for long-term safety assessments of underground nuclear waste depositories. Employing models of differing complexity, lysimeter experiments covering a

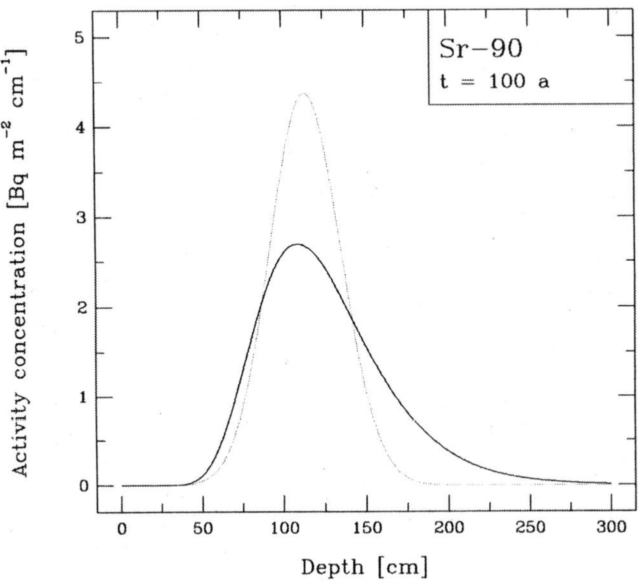

FIGURE 6.1. Predictions of the depth distribution of weapons fallout ^{90}Sr at the time of 100 y after deposition started using the stochastic-convective model (solid line) and the convection-dispersion equation (dotted line); parameters of both models were derived from the depth distribution observed in 1992.

five-year period were simulated. Generally, predictions of physically based models were closer to observations, but showed discrepancies indicating that some processes important for upward translocation of radionuclides are not adequately conceptualized in current mechanistic models.

Radionuclide Soil-to-Plant Transfer

The soil-to-plant transfer is one of the most important pathways leading to human ingestion of radionuclides. Thus, since the 1950s and 1960s many efforts have been made to predict and quantify radionuclide root uptake and to implement suitable models in radioecological/radiological models. Because root uptake of radionuclides and the subsequent translocation to edible plant parts is influenced by many factors (see previous section), its modeling is a compromise between availability of input parameters and scientifically based mechanistic approaches. In the past, most efforts were directed to determining simple radionuclide concentration ratios between crops and soils; more recently, radionuclide fluxes have gained importance in soil-to-plant transfer modeling. A range of different approaches is presented as follows.

The Transfer Factor Concept

The transfer factor concept is the simplest model for the prediction of crop contamination with radionuclides (RN). The transfer factor (TF) is defined as:

$$TF = (\text{RN concentration in plant } [\text{Bq kg}^{-1}]) / (\text{RN concentration in soil } [\text{Bq kg}^{-1}]) \quad (6.1)$$

The radionuclide concentration in soil is always determined on a dry weight basis. With regard to the investigated plant material, most researchers follow the definition provided in the *Handbook of Parameter Values for the Prediction of Radionuclide Transfer in Temperate Environments* (International Atomic Energy Agency [IAEA], 1994), in which radionuclide concentrations are always related to plant dry matter. In earlier literature many TFs were calculated based on plant fresh weight and had to be recalculated into dry-matter-based values using average dry matter content of plants (IAEA, 1994). The transfer factor is based on a fixed "rooting zone": in arable soils the radionuclide concentration in soil is determined in a mixed sample from 0 to 20 cm depth, in pastures the TFs are based on a rooting layer of 0 to 10 cm. Another assumption is the independence of the TF from the absolute radionuclide concentration in soil. This assumption does not appear to hold true under practical conditions (Bunzl et al., 2000). Other simplifications in the transfer factor concept include (1) the artificial definition of the rooting zone, (2) the lack of discrimination between radionuclide pools of different availability in soil, (3) the fact that the TF does not really describe the process of root uptake but merely provides a concentration ratio including, for example, mass loading of plant surfaces with contaminated soil particles, and (4) the omission of plant physiological parameters. The influence of these and additional simplifications discussed previously is reflected by the huge variability of transfer factor values obtained under field and experimental conditions, which for many radionuclides exceeds three orders of magnitude (Coughtrey, Jackson, and Thorne, 1983; IAEA, 1994). Table 6.4 provides literature data on soil-to-plant transfer factors for agricultural crops. Another comprehensive data set published recently can be

TABLE 6.4. Expected Values of Soil-to-Plant Transfer Factors Based on Crop and Soil Dry Weight Basis for Selected Crops and Elements

Element	**Soil type**	**Cereal grain**	**Grass**	**Pea, bean (pods)**	**Potato tubers**	**Green vege-tables**
Co	not specif.	0.0037	0.054	0.03	0.06	0.2
Zn	not specif.	1.2	0.99	0.71	35	—
Sr	clay	0.12	1.1	1.3	0.15	2.7
	sand	0.21	1.7	2.2	0.26	3.0
	peat	0.020	0.34	—	—	—
Tc	not specif.	0.73	76	4.3	0.24	—
Cs	clay	0.01	0.11	0.017	0.07	0.18
	sand	0.026	0.24	0.094	0.17	0.46
	peat	0.083	0.53	—	0.27	—
Ra	not specif.	0.0012	0.08	0.007	0.0011	0.049
Th	not specif.	0.000034	0.011	0.00012	0.000056	0.0018
U	not specif.	0.0013	0.023	—	0.011	0.0083
Np	not specif.	0.0027	0.069	0.018	0.0067	0.037
Pu	not specif.	0.0000086	0.00034	0.000061	0.00015	0.000073

Source: International Atomic Energy Agency (1994), modified.

not specif. = not specified
— = no data available or mean based on very few observations

found in Gerzabek, Strebl, and Temmel (1998). A very careful evaluation of the available Cs and Sr transfer data was published by Nisbet, Woodman, and Haylock (1999), who investigated the effects of aging, pH, organic matter, and exchangeable potassium (or calcium) on Cs and Sr soil-to-plant transfer. The growing awareness of the limitations of the simple empirical transfer factor concept has stimulated the interest in models which better predict plant uptake after soil contamination by radionuclides.

Frissel et al. (1987) attempted to improve the simple TF model for radiocesium, the best-investigated radionuclide with respect to soil-to-plant transfer, by including soil characteristics, aging, and irrigation. The following equation (6.2) including five exponential correction terms was suggested:

$$TF_{act} = TF_{stand} * \exp(-0.64(pH-6)) * \exp(0.11(OM-4)) * \exp(-0.017(depth-20)) * \exp(-0.082(period-2)) * \exp(0.008(irr)) \quad (6.2)$$

where TF*act* represents the actual Cs transfer factor, TF*stand* represents the standard Cs transfer factor, OM represents the soil organic matter [%], depth represents the contaminated horizon thickness [cm], period represents the time elapsed after fallout [y], and irr represents the irrigation [mm]. Based on this formula, the standard TF was calculated using the large TF data bank of the International Union of Radioecologists (IUR) for a soil pH of 6, OM content of 4 percent, a 20-cm-thick contaminated horizon, two years after contamination and without irrigation.

The IUR Radflux database (Mitchell and Donelly, 2000) was designed to produce a readily available, comprehensive compilation of rate constants for the parameterization of dynamic multicompartmental models represented mathematically as sets of first-order linear differential equations. These rate constants (unit [s^{-1}]), representing net transfer between a donor and a receiving compartment, mostly soil to plant, have been taken directly or were calculated from literature data by using soil/plant concentration ratio and available information on exposure time and biomass production. Existing transfer factor databases collated by the UIR and others were incorporated as well after recalculation of values to a consistent format. A calculation example is given as follows:

$$\text{flux soil-plant } [s^{-1}] = -\text{LN}\,(1-(\text{lost fraction [dimensionless]}))\,/\,(\text{exposure time [s]})$$

$$\text{where lost fraction} = \text{inventory in plant } [Bq\ m^{-2}]\,/\,\text{total (soil + plant) inventory } (Bq\ m^{-2}) \quad (6.3)$$

Refined and Alternative Modeling Approaches

For the radiologically important nuclear fission products ^{137}Cs and ^{90}Sr, various approaches have been suggested (e.g., Roca et al., 1997; Absalom et al., 1999; Kirchner and Ehlken, 1999). The common denominator is that they focus on including the influence of soil/soil solution physicochemical interactions on the concentrations of the radionuclides present in solution in bulk soil, which are then assumed to be available to plant roots. Uptake and translocation are not modeled mechanistically, but represented by fairly general empirical relationships. By comparing model predictions with experimental data, the model developers were able to demonstrate that, although they address only some of the relevant processes in soil-plant systems, these recent models constitute a major improvement compared to the traditional transfer factor concept.

Van Dorp, Eleveld, and Frissel (1979) proposed a method for estimating radionuclide concentration in edible plant parts based on the solubility of a given radionuclide in soil water, its ability to enter across root membranes, and its upward movement within the plant with the transpiration stream. This approach was chosen to better represent the closely linked processes of root uptake and translocation into edible plant parts, which are often incorporated in a single soil-plant transfer factor. Measured values of K_d (Bq absorbed per g soil / Bq dissolved per ml solution) are used together with a plant-specific fraction of absorbed radionuclide that is transfered to the edible plant part (F, dimensionless), root selectivity coefficient (S, dimensionless), total production (PT, dry matter; [g cm^{-2} y^{-1}]), P*ep* (production of edible plant part dry matter; [g cm^{-2} y^{-1}]), transpiration coefficient (TC; [ml g^{-1}] dry weight), total amount of nuclide in soil surface (C*t*; [Bq cm^{-2}]), rooting depth of soil profile (L; [cm]), soil water content (Θ; [ml cm^{-3}]), and bulk density of soil (ρ; [g cm^{-3}]) to finally estimate the radionuclide concentration in dry matter of edible plant parts (C*ep*):

$$Cep = F * S * (PT / Pep) * TC * Ct / (L(\Theta + \rho * K_d)) \qquad (6.4)$$

Radionuclide Translocation

A parameter widely applied in assessment models is the translocation factor [m^2 kg^{-1}]. It is defined as the ratio of the activity concentration in the edible plant part [Bq kg^{-1}] to the total activity retained by the plant canopy per unit ground area [Bq m^{-2}]. Some models relate the total activity in the edible plant parts at harvest [Bq m^{-2}] to the total activity initially retained by the plant canopy [Bq m^{-2}]. In this case, translocation is defined in terms of a dimensionless parameter (Thiessen et al., 1999). For leafy vegetables and fodder hay the translocation factor is per definition 1.0 (whole plant consumable). An extended summary on shoot/fruit translocation factors of fruit crops is presented by Watkins and Maul (1995). Generic values often used in assessment models are 0.1, representing a conservative value for most radionuclides. A collection of recommended translocation factors into agricultural crops is presented in IAEA (1994).

Although useful for application in assessment models, the translocation factor is unable to describe the complicated plant-internal transport processes, which are highly variable and depend on developmental stage of plants at the time of deposition, seasonal effects, plant species, and environmental conditions. System dynamic modeling approaches that dynamically describe radionuclide translocation within plants as a consequence of nutrient uptake for biomass production are presented by Acosta et al. (1994; for forest ecosystems) and Brambilla (2001; for tomato plants). Although the first is based on generic values, the latter is parameterized by experimental data for ^{85}Sr, ^{65}Zn, and ^{134}Cs.

Food-Chain Models

From a radiation protection viewpoint the contamination of foodstuffs for human consumption is a crucial issue. Complex assessment models have been developed to predict human exposure via different pathways, e.g., the transfer of radionuclides to food products (PATHWAY, Whicker and Kirchner, 1987; ECOSYS-87, Müller and Pröhl, 1993; FARMLAND, Brown and Simmonds, 1995). Such models address a range of important atmosphere-soil-plant interaction processes (for example deposition, interception, weathering, resuspension, root uptake, fixation and leaching in soils) and subsequent transfer to animal products. In most applications the radionuclide contamination of consumed plant material is calculated by a type of transfer factor. A GIS-based food-chain model with a process-oriented approach—calculating food-crop contamination from a combination of soil characteristics and deposition levels—was developed under the EU fifth framework program (SAVE-IT, Crout et al., 1999; Wright et al., 1998).

Forest Ecosystems

In forest ecosystems nutrient cycling is very important due to the longevity of trees and the nutrient-limited soil conditions. Soil-plant transfer is complicated by mycorrhizal symbioses and species diversity in understorey vegetation. An intercomparison of radioecological forest models is presented by Riesen et al. (1999). Most of these models are based on aggregated transfer factors (T_{agg}-value; [m^2 kg^{-1}]), which relate the contamination of plant material (Bq kg^{-1}) to the total soil inventory (or the inventory in a defined soil depth only) for a given radionuclide [Bq m^{-2}]. One example is the forest model RIFE II (Belli et al., 2000), where radionuclide contamination of trees is modeled dynamically (compartment approach) via rate constants, but the ^{137}Cs concentration in understorey species is calculated by aggregated transfer factors and a consideration of the rooting/mycelium depth of plants/fungi. In combination with the dynamically modeled depth distribution within soil layers (by a linear compartment model), changes of ^{137}Cs contamination over time are also estimated for herbaceous plants and mushrooms.

FUTURE OUTLOOK

Models that describe the transport of radionuclides in the rooting zone are often based on simplifying assumptions. This is generally justified due to the very low mobility of most radionuclides in soils. A major limitation of current models is that they fail to consider spatial inhomogeneity of water flow and radionuclide transport. There is an urgent need to study—both experimentally and by modeling—the impact of preferential flow (e.g., by macropores and by soil cracks) on the transport of radionuclides in soils.

Recently, considerable progress has been made in understanding many of the processes that control the soil–soil solution interaction of radiocesium and radiostrontium and of the mechanisms involved in uptake and translocation of ions by plant roots. For both radionuclides, the physicochemical mechanisms acting at soil/solution interfaces have been successfully integrated in models for predicting plant accumulation. It would be highly desirable to extend these models so that they simulate radionuclide behavior in the rhizosphere, but the present understanding of these processes is too incomplete to allow quantification. We are optimistic that incorporating the growing insight in plasma membrane transport mechanisms and upcoming methods of molecular modeling of pollutant–soil matrix interactions will soon improve our capabilities to understand and predict the behavior of radionuclides in soil-plant systems.

SUMMARY

This chapter provided an overview of the behavior of radiologically relevant radioactive elements in soil-plant systems. In addition to some basic information about the processes of radionuclide deposition, fixation, vertical migration in soils, and their uptake into vegetation, several modeling approaches and typical ranges of parameters were introduced. Linear compartment models and dispersion-convection equations for the description of vertical movement of radionuclides within the soil profile were discussed with respect to their application in radioecological models. The soil-plant transfer concept for the description of radionuclide uptake into plants and food chains as well as advanced modeling approaches were outlined together with a review of available data sources and radioecological and assessment model applications.

REFERENCES

Aarkrog, A., L. Botter-Jensen, Chen Quing Jiang, H. Dahlgaard, H. Hansen, E. Holm, B. Lauridsen, S.P. Nielsen, M. Strandberg, and J. Sogaard-Hansen (1992). *Environmental Radioactivity in Denmark in 1990 and 1991*. Roskilde, Denmark: Riso National Laboratory.

Absalom, J.P., S.D. Young, N.M.J. Crout, A.F. Nisbet, R.F.M. Woodman, E. Smolders, and A.G. Gillett (1999). Predicting soil to plant transfer of radiocesium using soil characteristics. *Environmental Science & Technology 33:* 1218-1223.

Acosta, F.J., J.M. Barandica, F. Lopez, J.M. Serrano, F. Diaz-Pineda, A. Baeza, M. Rufo, and A. Sterling (1994). Hierarchical model of the movement of nutrients and artificial radionuclides in the soil-plant system. In *Nuclear Techniques in Soil-Plant Systems*. Proceedings of an International Conference October, 1994 eds. International Atomic Energy Agency. Vienna, Austria: IAEA.

Ageets, V.Yu. (1999). *Belarus and Chernobyl: The second decade*. Gomel, Belarus: Scientific-Research Institute of Radioecology of the Ministry of Emergencies, Repubic of Belarus.

Belli, M., K. Bunzl, B. Delvaux, M. Gerzabek, B. Rafferty, T. Riesen, G. Shaw, and E. Wirth (2000). SEMINAT—Long-term dynamics of radionuclides in seminatural enviroments—Derivation of parameters and modeling. Final report. *Serie Stato dell' Ambiente 10/2000.* Roma, Italy: Agenzia Nationale per la Protezione dell' Ambiente.

Boone, F.W., M.V. Kantelo, P.G. Mayer, and J.M. Palms (1985). Residence half-times of ^{129}I in undisturbed surface soils based on measured soil concentration profiles. *Health Physics 48:* 401-413.

Bossew, P. (1997). *Migration of Radionuclides in Undisturbed Soil (Final Report)*. Vienna, Austria: Austrian Ministry of Science and Transport (in German).

Bossew, P. and F. Strebl (2001). Radioactive contamination of tropical rainforest soils in southern Costa Rica. *Journal of Environmental Radioactivity 53:* 199-213.

Boudreau, P. (1986). Mathematics of tracer mixing in sediments: I. Spatially-dependent, diffusive mixing. *American Journal of Science 286:* 161-198.

Brambilla, M. (2001). *"Ventomod": A Dynamic Conceptual Model in Order to Assess the Radionuclide Contamination of Processing Tomato Plants.* Seibersdorf, Austria: Austrian Research Center Seibersdorf. Report UL-0005.

Brechignac, F., R. Vallejo, T. Sauras, J. Casadesus, Y. Thiry, N. Waegeneers, S. Forsberg, G. Shaw, C. Madoz-Escande, M.A. Gonze (1999). *Soil-Radionuclides Processes of Interaction and Modelling of their Impact on Contamination of Plant Food Products.* Fontenay-aux-Roses, France: Institut de Protection et de Surete Nucleaire, Report DPRE/SERLAB/99-017.

Brown, J. and J.R. Simmonds (1995). *FARMLAND: A Dynamic Model for the Transfer of Radionuclides Through Terrestrial Foodchains.* London, UK: National Radiation Protection Board. NRPB-Report No. R273. ISBN 0-85951-380-7.

Brückmann, A. and V. Wolters (1994). Microbial immobilization and recycling of ^{137}Cs in the organic layers of forest ecosystems: Relationship to environmental conditions, humification and invertebrate activity. *Science of the Total Environment 157:* 249-256.

Bunzl, K. (1987). Das Verhalten von Radionukliden im Boden. *Deutsche Tierärztliche Wochenschrift 94:* 357-359.

Bunzl, K. (1997). Radionuklide. In *Handbuch der Bodenkunde,* eds. H.-P. Blume, P. Felix-Henningsen, W.R. Fischer, H.-G. Frede, R. Horn, and K. Stahr. Landberg/Lech, Germany: Ecomed. pp. 1-18.

Bunzl, K., B.P. Albers, W. Schimmack, M. Belli, L. Ciuffo, and S. Menegon (2000). Examination of a relationship between Cs-137 concentrations in soils and plants from alpine pastures. *Journal of Environmental Radioactivity 48:* 145-158.

Bunzl, K., H. Förster, W. Kracke, and W. Schimmack (1994). Residence times of $^{239+240}$Pu, ^{238}Pu, ^{241}Am and ^{137}Cs in the upper horizons of an undisturbed grassland soil. *Journal of Environmental Radioactivity 22:* 11-27.

Bunzl, K. and W. Schimmack (1988). Distribution coefficients of radionuclides in the soil: Analysis of the field variability. *Radiochimica Acta 44/45:* 355-360.

Butler, A.P., J. Chen, A. Aguero, O. Edlund, M. Elert, G. Kirchner, W. Raskop, and M. Sheppard (1999). Performance assessment studies of models for water flow and radionuclide transport in vegetated soils using lysimeter data. *Journal of Environmental Radioactivity 42:* 271-288.

Cheshire, M.V., C. Dumat, A.R. Fraser, S. Hillier, and S. Staunton (2000). The interaction of soil organic matter with soil clay minerals. *European Journal of Soil Science 51:* 497-509.

Coughtrey, P.J. (1988). Models for radionuclide transport in soils. *Soil Use and Management 4:* 84-90.

Coughtrey, P.J., D. Jackson, and M.C. Thorne (1983). *Radionuclide Distribution and Transport in Terrestrial and Aquatic Ecosystems; A Critical Review of Data.* Volume 1. Rotterdam, Netherlands: A.A. Balkema.

Cremers, A., A. Elsen, P. De Preter, and A. Maes (1988). Quantitative analysis of radiocesium retention in soils. *Nature 335:* 247-249.

Crout, N.M.J., A.G. Gillett, J.P. Absalom, and S. Young (1999). *SAVE-IT (Spatial and Dynamic Prediction of Radiocaesium Transfer to Food Products)* CD-ROM and outline documentation. Brussels, Belgium: European Commission. DG XII. Available online: <http://www.nottingham.ac.uk/environmental-modelling/SAVE-IToverview.html>.

Dagan, G. (1989). *Flow and Transport in Porous Formations.* New York: Springer-Verlag.

Dertinger, H., A. Müller, K. Nagel, A. Riedl, and S. Strack (1986). *Fixierung von radioaktivem Jod im Boden.* Karlsruhe, Germany: Kernforschungszentrum Karlsruhe KFK-Report 3916 (in German).

Dighton, J., G.M. Clint, and J. Poskitt (1991). Uptake and accumulation of ^{137}Cs by upland grassland soil fungi: A potential pool of Cs immobilization. *Mycological Research 95:* 1052-1056.

Ehlken, S. and G. Kirchner (1996). Seasonal variations in soil-to-grass transfer of fallout strontium and cesium and of potassium in north German soils. *Journal of Environmental Radioactivity 33:* 147-181.

Epstein, E. and J.E. Leggett (1954). The absorption of alkaline earth cations by barley roots: Kinetics and mechanisms. *American Journal of Botany 41:* 785-791.

Frissel, M., H. Noordijk, and K.E. Van Bergeijk (1990). The impact of extreme environmental conditions, as occuring in natural ecosystems, on the soil-to-plant transfer of radionuclides. In *Transfer of Radionuclides in Natural and Semi-Natural Environments,* eds. G. Desmet, P. Nasimbeni, and M. Belli. London, UK: Elsevier. pp. 40-47.

Frissel, M.J. and R. Pennders (1983). Models for the accumulation and migration of ^{90}Sr, ^{137}Cs, ^{239,240}Pu and ^{241}Am in the upper layer of soils. In *Ecological Aspects of Radionuclide Release,* eds. P.J. Coughtrey, J.N.B. Bell, and T.M. Roberts. Oxford, UK: Blackwell Scientific. pp. 63-72.

Frissel, M.J., J.R. Soutjesdijk, A.C. Koolwijk, and H.W. Köster (1987). The Cs-137 contamination of soils in the Netherlands and its consequences for the contamination of crop products. *Netherlands Journal of Agricultural Science 35:* 339-346.

Gastberger, M., F. Steinhäusler, M.H. Gerzabek, A. Hubmer, and H. Lettner (2000). ^{90}Sr and ^{137}Cs in environmental samples from Dolon near the Semipalatinsk nuclear test site. *Health Physics 79:* 257-265.

Gastberger, M., F. Steinhäusler, M.H. Gerzabek, H. Lettner, and A. Hubmer (2000). Soil-to-plant transfer of fallout caesium and strontium in Austrian lowland and Alpine pastures. *Journal of Environmental Radioactivity 49:* 217-233.

Gerzabek, M.H., C. Artner, O. Horak, and K. Mück (1992). *Results of Field Studies on ^{90}Sr and Stable Sr Soil-to-Plant Transfer.* Seibersdorf, Austria: Austrian Research Center Seibersdorf Report 4617. 25 pp.

Gerzabek, M.H., K. Mück, F. Steger, and S.M. Algader (1996). Die Auswaschung von ^{60}Co, ^{137}Cs und ^{226}Ra im Lysimeterversuch. *Die Bodenkultur 47:* 71-80.

Gerzabek, M.H., Y. Muramatsu, F. Strebl, and S. Yoshida (1999). Iodine and bromine contents of some Austrian soils and relations to soil characteristics. *Journal of Plant Nutrition and Soil Science 162:* 415-419.

Gerzabek, M.H., F. Strebl, and B. Temmel (1998). Plant uptake of radionuclides in lysimeter experiments. *Environmental Pollution 99:* 93-103.

Guivarch, A., P. Hinsinger, and S. Staunton (1999). Root uptake and distribution of radiocesium from contaminated soils and the enhancement of Cs adsorption in the rhizosphere. *Plant and Soil 211:* 131-138.

Haunold, E., O. Horak, and M.H. Gerzabek (1987). Umweltradioaktivität und ihre Auswirkung auf die Landwirtschaft. I. Das Verhalten von Radionukliden in Boden und Pflanze. *Die Bodenkultur 38:* 95-118.

Hoffman, F.O., K.M. Thiessen, M.L. Frank, and B.G. Blaylock (1992). Quantification of the interception and initial retention of radioactive contaminants deposited on pasture grass by simulated rain. *Atmospheric Environment 26 A:* 3313-3321.

Hölgye, Z. and M. Maly (2000). Sources, vertical distribution, and migration rates of $^{239,240}Pu$, ^{238}Pu, and ^{137}Cs in grassland soil in three localities of central Bohemia. *Journal of Environmental Radioactivity 47:* 135-147.

Hormann, V. and G. Kirchner (1998). Modeling the transport of radionuclides in unsaturated soils. In *Radioaktivität in Mensch und Umwelt*, ed. Fachverband für Strahlenschutz. Köln, Germany: TÜV-Verlag, Volume II. pp. 838-843 (in German).

Howard, B.J. and G. Desmet (Eds.) (1993). Relative effectiveness of agricultural countermeasure techniques. *Science of the Total Environment (Special Issue) 137:* 1-315.

IAEA (International Atomic Energy Agency) (1994). *Handbook of Parameter Values for the Prediction of Radionuclide Transfer in Temperate Environments.* Vienna, Austria: IAEA Technical Report Series 364.

Isaksson, M. and B. Erlandsson (1998). Investigation of the distribution of ^{137}Cs from fallout in the soils of the city of Lund and the province of Skane in Sweden. *Journal of Environmental Radioactivity 38:* 105-131.

Johanson, K.J. and I. Nikolova (1996). The role of fungi in the transfer of ^{137}Cs in the forest ecosystem. *Mitteilungen der Österreichischen Bodenkundlichen Gesellschaft 53:* 259-266.

Juo, A.S.R. and S.A. Barber (1970). The retention of strontium by soils as influenced by pH, organic matter and saturation cations. *Soil Science 99:* 143-148.

Jury, W.A. (1982). Simulation of solute transport using a transfer function model. *Water Resources Research 18:* 363-368.

Jury, W.A. and K. Roth (1990). *Transfer Functions and Solute Movement Through Soils.* Basel, Switzerland: Birkhäuser Verlag.

Jury, W.A., D. Russo, G. Sposito, and H. Elabd (1987). The spatial variability of water and solute transport properties in unsaturated soil. I. Analysis of property variation and spatial structure with statistical models. *Hilgardia 55:* 1-32.

Jury, W.A. and G. Sposito (1985). Field calibration and validation of solute transport models for the unsaturated zone. *Soil Science Society of America Journal 49:* 1331-1341.

Kawasaki, T. and M. Moritsugu (1987). Effect of calcium on the absorption and translocation of heavy metals in excised barley roots: Multi-compartment transport box experiment. *Plant and Soil 100:* 21-34.

Khan, A.U.-H. and W.A. Jury (1990). A laboratory study of the dispersion scale effect in column outflow experiments. *Journal of Contaminant Hydrology 5:* 119-131.

Kirchner, G. (1998a). Applicability of compartmental models for simulating the transport in soils. *Journal of Environmental Radioactivity 38:* 339-352.

Kirchner, G. (1998b). Modeling the migration of fallout radionuclides in soil using a transfer function model. *Health Physics 74:* 78-85.

Kirchner, G. and S. Ehlken (1999). Soil-to-plant transfer of strontium: Another test of the observed ratio model using data from field experiments. In *Soil-Plant-Relationships,* ed. M.H. Gerzabek. Proceedings of XXIXth Annual Meeting of the European Society for Nuclear Application and International Union of Radioecologists (ESNA/IUR) Austrian Research Center Seibersdorf Report L-209. pp. 194-200.

Li, J.G., M.H. Gerzabek, and K. Mück (1994). An experimental study on mass loading of soil particles on plant surfaces. *Die Bodenkultur 45:* 15-24.

Maathuis, F.J.M. and D. Sanders (1996). Mechanisms of potassium absorption by higher plant roots. *Physiologia Plantarum 96:* 158-168.

Mahara, Y. and S. Miyahara (1984). Residual plutonium migration in soil of Nagasaki. *Journal of Geophysical Research 89:* 7931-7936.

Marschner, H. (1986). *Mineral Nutrition of Higher Plants.* San Diego: Academic Press Inc.

Michelet, B. and M. Boutry (1995). The plasma membrane H^+-ATPase. *Plant Physiology 108:* 1-6.

Mitchell, N.G. and C.E. Donelly (2000). The flux database concerned action. Online publication of the International Union of Radioecologists. <http://www.iur-uir.org/Publications/iur_scientific_ publications.pdf>.

Mortvedt, J.J. (1994). Plant and soil relationships of uranium and thorium decay series radionuclides—A review. *Journal of Environmental Quality 23:* 643-650.

Mück, K. (1995). *Langzeitfolgedosis nach großräumiger Kontamination.* Seibersdorf, Austria: Austrian Research Center Seibersdorf, Report 3605 (in German).

Müller, H. and G. Pröhl (1993). ECOSYS-87: A dynamic model for assessing radiological consequences of nuclear accidents. *Health Physics 64:* 232-252.

Müller-Lemans, H. and F. Van Dorp (1996). Bioturbation as a mechanism for radionuclide transport in soil: Relevance of earthworms. *Journal of Environmental Radioactivity 31:* 7-20.

Naumann, E.B. and B.A. Buffham (1983). *Mixing in Continuous Flow Systems.* New York: John Wiley and Sons.

Nielsen, D.R., J.W. Biggar, and K.T. Erh (1973). Spatial variability of field measured soil-water properties. *Hilgardia 42:* 215-259.

Nikolova, I., K.J. Johanson, and S. Clegg (2000). The accumulation of ^{137}Cs in the biological compartment of forest soils. *Journal of Environmental Radioactivity 47:* 319-326.

Nisbet, A.F., R.F.M. Woodman, and R.G.E. Haylock (1999). *Recommended Soil-to-Plant Transfer Factors for Radiocaesium and Radiostrontium for Use in Arable Systems.* Chilton, UK: National Radiological Protection Board, Report NRPB-R304.

Pendleton, R.C. and R.L. Uhler (1960). Accumulation of ^{137}Cs by plants grown in a simulated pond, wet meadow and irrigated field environments. *Nature 185:* 707-708.

Pröhl, G. (1990). *Die Modellierung der Radionuklidausbreitung in Nahrungsketten nach Deposition von Strontium-90, Cäsium-137 und Jod-131 auf landwirtschaftlich genutzten Flächen.* Neuherberg, Germany: GSF-Forschungszentrum für Umwelt und Gesundheit, Report 29/90.

Riesen, T.K., R. Avila, L. Moberg, and L.M. Hubbard (1999). A review of forest models developed after the Chernobyl NPP accident. In *Contaminated Forests: Recent Developments in Risk Identification and Future Perspectives,* eds. I. Linkov and W.R. Schell. Amsterdam, Netherlands: Kluwer Academic. NATO Science Series. 2. Environmental security; volume 58, pp. 151-160.

Roca, M.C., V.R. Vallejo, M. Roig, J. Tent, M. Vidal, and G. Rauret (1997). Prediction of cesium-134 and strontium-85 crop uptake based on soil properties. *Journal of Environmental Quality 26:* 1354-1362.

Rudolph, D.L., R.G. Kachanoski, M.A. Celia, D.R. LeBlanc, and J.H. Steven (1996). Infiltration and solute transport experiments in unsaturated sand and gravel, Cape Cod, Massachusetts: Experimental design and overview of results. *Water Resources Research 32:* 519-532.

Russell, R.S. and P. Newbould (1966). Entry of strontium-90 into plants from the soil. In *Radioactivity and Human Diet,* ed. R.S. Russell. Oxford, UK: Pergamon Press. pp. 215-245.

Sanchez, A.L., S.M. Wright, E. Smolders, C. Naylor, P.A. Stevens, V.H. Kennedy, B.A. Dodd, D.L. Singleton, and C.L. Barnett (1999). High plant uptake of radiocesium from organic soils due to Cs mobility and low soil K content. *Environmental Science and Technology 33:* 2752-2757.

Scheffer, F. and P. Schachtschabel (1989). *Lehrbuch der Bodenkunde.* Twelfth Edition. Stuttgart, Germany: Ferdinand Enke Verlag.

Schuller, P., E. Ellies, and G. Kirchner (1997). Vertical migration of fallout ^{137}Cs in agricultural soils from southern Chile. *Science of the Total Environment 193:* 197-205.

Shalhevet, J. (1973). Effect of mineral type and soil moisture content on plant uptake of ^{137}Cs. *Radiation Botany 13:* 165-171.

Shaw, G. and J.N.B. Bell (1991). Competitive effects of potassium and ammonium on caesium uptake kinetics in wheat. *Journal of Environmental Radioactivity 13:* 283-296.

Shaw, G. and X. Wang (1996). Caesium and plutonium migration in forest soils of the Chernobyl 30km zone. *Mitteilungen der Österreichischen Bodenkundlichen Gesellschaft 53:* 27-34.

Simmons, C.S. (1982). A stochastic-convective transport representation of dispersion in one-dimensional porous media systems. *Water Resources Research 18:* 1193-1214.

Smith, J.T., J. Hilton, and R.N.J. Comans (1995). Application of two simple models to the transport of ^{137}Cs in an upland organic catchment. *Science of the Total Environment 168:* 57-61.

Smolders, E., L. Sweeck, R. Merckx, and A. Cremers (1997). Cationic interactions in radiocaesium uptake from solution by spinach. *Journal of Environmental Radioactivity 34:* 161-170.

Strasburger, E. (1998). *Lehrbuch der Botanik* [Textbook Botany]. Thirty-fourth Edition. Stuttgart, Germany: Gustav Fischer Publishers.

Strebl, F., M.H. Gerzabek, V. Karg, and F. Tataruch (1996). ^{137}Cs-migration in soils and its transfer to roe deer in an Austrian forest stand. *Science of the Total Environment 181:* 237-247.

Sweek, L., J. Wauters, E. Valcke, and A. Cremers (1990). The specific interception potential of soils for radiocesium. In *Transfer of Radionuclides in Natural and Semi-Natural Environments,* eds. G. Desmet, P. Nassimbeni, and M. Belli. London: Elsevier. pp. 249-266.

Tanner, W. and T. Caspari (1996). Membrane transport carriers. *Annual Review of Plant Physiology and Plant Molecular Biology 47:* 595-626.

Thiessen, K.M., M.C. Thorne, P.R. Maul, G. Pröhl, H.S. Wheather (1999). Modeling radionuclide distribution and transport in the environment. *Environmental Pollution 100:* 151-177.

Thiry, Y. (1997). *Etude du Cycle du Radiocesium en Ecosysteme Forestier: Distribution et Facteurs de Mobilité.* Thesis, Louvain-la-Neuve, Belgium: Université Catholique de Louvain.

Tikhomirov, FA. (1988). Long lived radionuclides in soil plant systems. In *Radionuclides in the Food Chain,* ed. M.W. Carter. Berlin, Germany: Springer Publishers. pp. 136-144.

Tinker, P.B. and P. Nye (2000). *Solute Movement in the Rhizosphere.* Second Edition. Oxford, UK: Oxford University Press.

UNSCEAR (United Nations Committee on the Effects of Atomic Radiation) (1988). *Sources, Effects, and Risks of Ionizing Radiation.* New York: United Nations.

Van Dorp, F., R. Eleveld, and M.J. Frissel (1979). A new approach for soil-plant transfer calculations. In *International Symposium on Biological Implications of Radionuclides Released from Nuclear Industries.* Vienna, Austria: International Atomic Energy Agency, March 26-30, 1979. Report IAEA-SM-237/134.

Warrick, A.W., J.W. Biggar, and D.R. Nielsen (1971). Simultaneous solute and water transfer for an unsaturated soil. *Water Resources Research 7:* 1216-1225.

Watkins, B.M. and P.R. Maul (1995). *Transfer of Deposited Radioactivity from Foliage to Edible Crop Parts.* Report of the work undertaken by Intera Information Technologies under Ministry of Agriculture, Fisheries and Food project reference 1B058. United Kingdom.

Whicker, W.F. and T.B. Kirchner (1987). PATHWAY: A dynamic food-chain model to predict radionuclide ingestion after fallout deposition. *Health Physics 52:* 717-737.

White, P.J. (1997). Cation channels in the plasma membrane of rye roots. *Journal of Experimental Botany 48:* 499-514.

White, P.J. (1998). Calcium channels in the plasma membrane of root cells. *Annals of Botany 81:* 173-183.

Whitehead, D.C. (1984). The distribution and transformations of iodine in the environment. *Environment International 10:* 321-339.

Wright, S.M., B.J. Howard, C.L. Barnett, P. Stevens, J.P. Absalom (1998). Development of an approach to estimating mid- to long-term critical loads for radiocaesium contamination of cow milk in western Europe. *Science of the Total Environment 221:* 75-87.

Chapter 7

Soil Acidification

Sven Ingvar Nilsson

Soil acidification is an inevitable consequence of soil development, and proceeds at a high rate in areas with a humid climate. The rate of soil acidification may be considerably increased by anthropogenic sulphur (S) and nitrogen (N) emissions, since soil acidification is intimately connected with the cycling and transformations of carbon (C), nitrogen, and sulphur species. The strong acids, nitric acid (HNO_3) and sulphuric acid (H_2SO_4), are formed in the atmosphere through oxidation of emitted SO_x and NO_x compounds. Protons (H^+) are deposited to vegetation, soils, and surface waters together with SO_4^{2-} and NO_3^- ions. However, HNO_3 can also be formed in soil through nitrification processes. H_2SO_4 formation is of potentially great importance in soils containing sulphide minerals. Due to various human activities, the soil acidification rate has increased, particularly since World War II. Up to now, soil acidification has mainly been considered a problem of the industrialized countries of Europe and North America, but the problem of high S and N emissions is rapidly spreading to other parts of the world such as southeast Asia (Kuylenstierna et al., 1995). During the 1990s, declines in S deposition have occurred in Europe and North America, and researchers are trying to evaluate the potential for recovery of both soils and freshwaters in these areas.

BUFFERING PROCESSES

To understand the concepts of soil acidity and soil acidification, both the quantity and the intensity variables must be considered. Intensity variables are the concentrations (or activities) of H^+ and aluminum (Al) species in the soil solution, or the ratio between the sum of exchangeable base cations, $\Sigma(Ca^{2+} + Mg^{2+} + K^+ + Na^+)$ and the cation exchange capacity (CEC) expressed as base saturation ($mol_c/mol_c \times 100$). Soil solution pH is the most commonly used intensity variable. A relevant quantity variable is the acid neutralizing capacity of either the soil solution (ANCaq), or the whole soil

(ANCs) at a predefined reference pH. ANCaq is also called the alkalinity. Temporal declines in ANCs are largely equivalent to mineral weathering (Van Breemen, Mulder, and Driscoll, 1983).

ANCaq is expressed either as the equivalent (charge) sum of hydrolyzable cations and anions in $mmol_c\ L^{-1}$ (Equation 7.1), or as the difference between nonhydrolyzable cations and anions (Equation 7.2). An ANCaq value >1 at a reference pH ≥ 5, implies that proton acceptors in the soil solution (carbonate species, hydroxyl ions, and organic anions) dominate over proton donors (i.e., H^+ and hydrolyzable Al species). At lower pH values, Al species also function as proton acceptors.

$$ANCaq = [HCO_3^-] + 2[CO_3^-] + [OH^-] + [RCOO^-] - [H^+] - 3[Al^{3+}] - 2[Al(OH)^{2+}] - [Al(OH)_2^+] \quad (7.1)$$

$$ANCaq = 2[Ca^{2+}] + 2[Mg^{2+}] + [K^+] + [Na^+] + [NH_4^+] - [Cl^-] - [NO_3^-] - 2[SO_4^{2-}] \quad (7.2)$$

The acid neutralizing capacity of the whole soil (ANCs) is usually expressed as the difference between total cations and total anions in oxide form (with [HCl] as an exception). The unit is, for instance, $mmol_c\ m^{-2}$ per unit soil depth.

$$ANCs = 6[Al_2O_3] + 2[CaO] + 2[MgO] + 2[K_2O] + 2[Na_2O] + 4[MnO_2] + 2[MnO] + 6[Fe_2O_3] + 2[FeO] - 2[SO_3] - 2[P_2O_5] - [HCl] \quad (7.3)$$

Silica (SiO_2) is not included in the ANC balance as it occurs as an uncharged dissolved chemical species $Si(OH)_4^0$ after weathering. According to the reactions (Equations 7.4 –7.10), Al and Fe contribute to ANC only at low pH values. [MnO] and [FeO] are only important under reducing conditions. The relation between pH and ANC can be described as a geological titration curve. This curve can be divided into a number of buffer ranges, extending from $CaCO_3$ buffering at high pH, via cation exchange buffering at intermediate pH values, to Al and Fe buffering at low pH values. In very acid soils with pH < 4, such as the acid sulphate soils (e.g., Thionic Fluvisols), with sulphide oxidation as a dominating acidifying process (see Equation 7.10), Fe buffering is particularly important.

The soil buffer intensity (β) with respect to changes in pH can be defined as $\beta = \Delta ANC/\Delta pH$. The inverse value of this expression, i.e., $1/\beta$, indicates the acidification sensitivity in terms of decline in pH at a given decline in ANC. As changes in soluble Al are not explicitly accounted for, the expression is of limited value in most acid soils.

In the buffer reactions that follow, (s) stands for solid compounds (minerals) and ≡ stands for positive or negative surface charge on soil particles. The pH ranges are defined according to Ulrich (1991), and indicate that a certain process is of dominating importance within a particular pH range. Differences in kinetics, e.g., between weathering of silicate minerals (slow kinetics) and cation exchange (rapid kinetics), will determine whether the process is important from a short-term (cation exchange) or a long-term (silicate weathering) perspective.

Weathering of carbonate minerals. 8.6 > pH > 6.2

$$CaCO_3\,(s) + 2H^+ <=> Ca^{2+} + CO_2 + H_2O \quad (7.4)$$

Weathering of silicate minerals. pH > 5

$$CaAl_2Si_2O_8\,(s) + 2H^+ + 6H_2O => 2Al(OH)_3\,(s) + 2H_4SiO_4 + Ca^{2+} \quad (7.5)$$

Cation exchange. 5 > pH > 4.2

$$Ca^{2+}\equiv + 2H^+ \Leftrightarrow Ca^{2+} + 2H^+\equiv \quad (7.6)$$

Weathering of Al oxides and hydroxides. 4.2 > pH

$$Al(OH)_3\,(s) + 3H^+ \Leftrightarrow Al^{3+} + 3H_2O \quad (7.7)$$

Weathering of Fe oxides and hydroxides. 3.2 > pH

$$Fe(OH)_3\,(s) + 3H^+ \Leftrightarrow Fe^{3+} + 3H_2O \quad (7.8)$$

In strongly weathered and acid soils, SO_4^{2-} adsorption on oxide surfaces will significantly affect the proton balance, as protons will be adsorbed together with SO_4^{2-} ions. The H^+/SO_4^{2-} adsorption ratio is 2 (Equation 7.9) or slightly less than 2 (Gustafsson, 1995). The Al and Fe oxides have a zero point of charge > 7 (Sposito, 1989). Therefore, SO_4^{2-} adsorption is important within a broad pH range from strongly acid to weakly acid soils.

$$2H^+ + SO_4^{2-} + 2(\equiv FeOH) \Leftrightarrow (Fe\equiv)_2SO_4 + 2H_2O \quad (7.9)$$

Weathering reactions where nonhydrolyzable anions are formed imply the formation of strong acid. This is exemplified by pyrite weathering, which is the key process in the formation of acid sulphate soils.

$$FeS_2\,(s) + 15/4O_2\,(g) + 7/2H_2O <=> 2SO_4^{2-} + 4H^+ + Fe(OH)_3\,(s) \quad (7.10)$$

Declines in ANCs imply net cation losses from the soil by leaching or plant uptake. Due to the weathering processes during soil development, the soil mineral composition shifts toward more stable minerals, i.e., minerals with successively slower weathering rates dominate as the soil matures.

Transformations of fresh organic matter to stable humic compounds and transformations of primary minerals to secondary clay minerals or to Al and Fe oxides, affect both the cation exchange capacity (CEC) and anion exchange capacity (AEC), and thereby the relative importance of cation and anion exchange (adsorption) as pH buffering processes. Humic compounds will generally enhance the CEC, whereas at the most advanced weathering stages that are found in Alisols, Acrisols, and Ferralsols, the CEC of the clay minerals (mainly kaolinite) is low, and the AEC of the oxides in some Ferralsols may be in the same order of magnitude as CEC or even greater. Table 7.1 presents a simplified scheme to illustrate the relative importance of the different buffer mechanisms.

The ion adsorption properties of organic matter and Al and Fe oxides are strongly pH dependent. The strong positive relationship between CEC and pH is often ignored in soil acidification investigations. Matschonat and Vogt (1997) showed that a pH increase of not more than 0.2 to 0.5 units could increase the CEC of mor humus layers collected from Podzols in Germany and Sweden with approximately 5 $cmol_c\ kg^{-1}$, which in their study corresponded to a relative increase in CEC of 15 to 39 percent. Khanna, Raison, and Falkiner (1986) obtained similar results in their study of organic matter-rich soils in Australia. Unfortunately, the pH-CEC relationship is usually ignored in soil acidification models, despite the fact that it may significantly affect estimates of changes in both exchangeable acidity and base saturation. Due to electrostatic interactions, the CEC of organic matter will also respond to a change in the ionic strength of the soil solution. An increase in ionic strength will increase the CEC and vice versa. In their study, Matschonat and Vogt (1997) could also demonstrate the importance of this phenomenon. They pointed out that in some soils a reduction in acid deposition could cause both an increase in pH and a decrease in ionic strength. The effects of pH and ionic strength on CEC could therefore partly counteract each other.

TABLE 7.1. Dominating Buffer Mechanisms in Young Soils versus Old Soils

	Young soils	**Old soils**
Carbonate weathering	X	
Silicate weathering	X	
Cation exchange	X	X
Al and Fe oxide weathering	X	X
SO_4^{2-} adsorption	X	X

ACID-GENERATING PROCESSES CONNECTED WITH C, N, AND S CYCLING

Carbon

Soil acidification related to C cycling is mainly caused by proton dissociation of carbonic acid ($H_2CO_3^*$) and water soluble organic acids followed by leaching of HCO_3^- and organic anions ($RCOO^-$) and balanced by cations other than H^+.

$$H_2CO_3^* \Leftrightarrow H^+ + HCO_3^- \quad (7.11)$$

$$RCOOH \Leftrightarrow H^+ + RCOO^- \quad (7.12)$$

Nitrogen

The soil acidifying effect of nitrogen transformations is obtained by calculating the input-output balances of NH_4^+ and NO_3^-. Inputs could either be through atmospheric deposition or as N fertilizers (see Chapter 14).

$$NH_4^+ \text{(input)} - NH_4^+ \text{(output)} > 0 \quad (7.13)$$

$$NO_3^- \text{(output)} - NO_3^- \text{(input)} > 0 \quad (7.14)$$

Both equations 7.13 and 7.14 yield a net soil acidification. The NH_4^+ inequality implies that NH_4^+ is retained in the soil and taken up by plants or microorganisms (microbial immobilization). In either case, 1 mol H^+ will be released per mol NH_4^+ taken up. The NO_3^- inequality implies a net NO_3^- formation in the soil followed by NO_3^- leaching and an ensuing average release of 1 mol H^+ per mol of NO_3^- leached.

Sulphur

Sulphur inputs to most soils are mainly from atmospheric deposition. However, sulphide weathering or inputs as S fertilizers are also important. The net acidifying effect of the atmospheric deposition of SO_2 gas may be described as equation 7.15.

$$SO_2 + 0.5\ O_2 + H_2O \Leftrightarrow SO_4^{2-} + 2H^+ \quad (7.15)$$

The SO_2 oxidation takes place either in the vegetation canopy or in the soil. The acidifying effect of the SO_4^{2-} ion input-output balance is analogous to that of NO_3^-.

$$SO_4^{2-}\text{(output)} - SO_4^{2-}\text{(input)} > 0 \tag{7.16}$$

The inequality implies a net release of SO_4^{2-} and H^+ into the soil solution either due to sulphide weathering, net S mineralization, or desorption of previously adsorbed SO_4^{2-}. Net S immobilization, SO_4^{2-} adsorption, or SO_4^{2-} uptake by the vegetation will shift the inequality from > 0 to < 0 and H^+ will consequently be neutralized.

PROTON BUDGETS

Proton budgets for whole ecosystems have been used to evaluate the relative importance of natural versus anthropogenic acidification (Van Breemen, Mulder, and Driscoll, 1983; Binkley and Richter, 1987). Such budgets have repeatedly demonstrated that atmospheric deposition of S and N may give significant contributions (30 to 70 percent) to the total proton load of forest soils in high deposition areas. In a rather unique, 30-year study in a stand of loblolly pine *(Pinus taeda)* growing on an Alisol in South Carolina, atmospheric deposition accounted for 38 percent of the proton load, and the remaining 62 percent was due to excess base cation accumulation in tree biomass (30 percent) and forest floor (14 percent), HCO_3^- leaching (15 percent), and SO_4^{2-} desorption (3 percent). Repeated soil samplings showed a close relationship between net depletion of exchangeable base cations in the mineral soil and proton load (Markewitz et al., 1998).

Forest soils in pristine areas with excessively high rainfall may become strongly acidified mainly because of carbonic acid dissociation and HCO_3^- leaching. This was demonstrated in an investigation of Cambisols and Podzols in New Zealand, which were glaciated 10,000 years ago, and are geologically similar to high elevation soils of similar age in Germany. The pristine New Zealand soils had just as low base saturation values in their deeper soil layers as the German soils, which had been affected by acid deposition for many years (Matzner and Davis, 1996).

For intensively managed arable soils, the acidifying effect of management practices such as N fertilizer applications and of base cation export in harvested plant biomass are usually more important than the atmospheric deposition. N fertilization and base cation export together with carbonic acid dissociation may account for more than 90 percent of the total proton load (Bergström and Gustafson, 1985). Management practices may also be important in some intensively managed forests, for instance where whole-tree harvesting is practiced instead of stem harvesting. Generally, root uptake of base cations is equivalent to the release of 1 mol H^+ into the soil for each mol_c of base cations taken up, whereas the release of base cations from plant litter will generate a corresponding amount of OH^-. Consequently, the base cation export in harvested plant biomass represents a permanent loss of ANCs from the soil, which may be several times greater at whole-tree harvesting compared to stem harvesting (Nilsson, Miller, and Miller, 1982; Johnson, Binkley, and Conklin, 1995).

ALUMINUM CHEMISTRY

Aluminum in Soil Solution

High concentrations of labile and potentially toxic Al is probably one of the most important consequences of soil and water acidification. Most arable crops (Sumner, Fey, and Noble, 1991) and biotic communities in freshwaters (Muniz, 1991) are sensitive to high Al concentration. Several forest species appear to be considerably Al tolerant, however (De Wit, 2000).

Monomeric chemical species of Al such as Al^{3+} and $Al(OH)^{2+}$ are considered to be the most harmful Al forms, whereas the organic complexes have no toxic effects (Sumner, Fey, and Noble, 1991). Proper evaluations of Al toxicity can only be made following a chemical speciation of the total Al measured. In addition to H_2O and OH^- ions, F^-, SO_4^{2-}, and anions of organic acids are important ligands in forming water-soluble Al complexes. In the following equations, H_2O as a ligand has been omitted. L stands for organic ligands in general:

$$Al^{3+} + H_2O <=> Al(OH)^{2+} + H^+ \quad (7.17)$$

$$Al^{3+} + F^- <=> AlF^{2+} \quad (7.18)$$

$$Al^{3+} + SO_4^{2-} <=> Al\ SO_4^+ \quad (7.19)$$

$$Al^{3+} + L^- <=> AlL^{2+} \quad (7.20)$$

The quantification of organic Al complexes is a rather difficult part of Al speciation, because of the great diversity of organic acids involved and the insufficient knowledge concerning the equilibrium constants. Some progress has been made during recent years (Tipping, 1994; Tipping et al., 1995; Lofts et al., 2001) and it has been possible to simulate the equilibria of both dissolved and solid organic and inorganic Al complexes in forest soils by using the Windermere Humic Aqueous Model for soils and sediments (WHAM-S).

In regard to dissolved organic Al complexes, there are well-established chemical methods, based on the separation of inorganic (positively charged) and organic (negatively charged) Al species, which treat soil solutions with strong, acid-type cation exchange resins (Driscoll, 1984; McAvoy et al., 1992). These and similar chemical methods should probably still be preferred to the modeling approach. Inorganic Al complexes, on the other hand, the equilibrium constants of which are well known, can be determined with a reasonable accuracy based on measurements of pH, total inorganic Al, and the total concentrations of F^- and SO_4^{2-}. Equilibrium model packages such as MINTEQA2 (Allison, Brown, and Novo-Gradac, 1991) can be used to iteratively determine the distribution of the inorganic species.

Aluminum Equilibria with Solid Phases

In several widely used soil acidification models such as the Birkenes model (Christophersen, Seip, and Wright, 1982) and MAGIC (Cosby et al., 1985) it is generally assumed that the relationship between Al^{3+} and pH in the soil solution at equilibrium can be modeled as dissolution of a mineral phase, which is commonly represented as gibbsite (e.g., Equation 7.7). However, several researchers who have tried to apply an $Al(OH)_3$ equilibrium as the only Al buffering process, have often found that the soil solution was undersaturated with respect to an $Al(OH)_3$ phase, and that the H^+/Al^{3+} ratio was < 3 (Walker, Cronan, and Bloom, 1990).

The soil acidification models mentioned do not include the weathering of other minerals. In the PROFILE model (Sverdrup, Warfvinge, and Rosén, 1992; Warfvinge and Sverdrup, 1992; Sverdrup and Warfvinge, 1993), the dissolution rates of different primary minerals are modeled as a function of pH, aluminum, and base cations in the soil solution and the partial pressure of CO_2. The dissolution reactions are assumed to occur in a congruent manner. This treatment of mineral weathering processes is not ideal, but represents a significant step forward in soil acidification modeling. However, even in the PROFILE model it is still assumed that the Al^{3+} concentration in the soil solution is solely determined by an $Al(OH)_3$ phase, whose equilibrium constant is allowed to vary by several orders of magnitude between different horizons in the same soil.

A growing body of evidence based on investigations in forest soils of temperate and boreal regions suggests that the solid labile pool of Al could mainly consist of organic complexes, at least in the upper part of the soil profile, which is usually rich in humified organic matter (Walker, Cronan, and Bloom, 1990; Berggren and Mulder, 1995). The organic complexation equilibrium between the solid phase and the soil solution can be written as (Equation 7.21):

$$RAl^{(3-X)} + xH^+ \Leftrightarrow RH_x + Al^{3+}, \qquad (7.21)$$

where $RAl^{(3-X)}$ stands for the organically complexed Al in the solid phase. The equilibrium equation will be (Equation 7.22)

$$K_{H\text{-}Al} = (Al^{3+})\,[RH_x]\,/(H^+)^x[RAl^{(3-x)}] \qquad (7.22)$$

By replacing the ratio of protonated and Al-binding sites, i.e., $[RH_x]/[RAl^{(3-x)}]$ with the log ratio of organically bound Al and organic C, i.e., log (Al_{org}/C), the description of the Al^{3+} solubility could be simplified (for sufficiently small values of $[RAl^{(3-x)}]$) to

$$\log (Al^{3+}) = \log K_{H\text{-}Al} + \log (Al_{org}/C) - x\text{pH} \qquad (7.23)$$

where x is a regression coefficient. An empirical way to quantify the organic Al pool is to measure the pyrophosphate extractable aluminum (Al_p) and organic carbon (C_p) pools, under the assumption that C_p represents the Al-binding part of the organic matter (Simonsson, 1999). Simonsson (1999) arrived at the following equation (7.24) showing the relationship between the ratio Al_p/C_p and log (Al^{3+}) + x pH, based on experimental data from a large number of B_h and B_s horizons in Swedish Podzols:

$$\log (Al^{3+}) + 2\text{pH} = 8.17 + 3.56 \log (Al_p/C_p) \qquad (7.24)$$

The expression was valid for $Al_p/C_p < 0.1$. At greater values of Al_p/C_p there were strong indications of a formation of solid $Al(OH)_3$, showing that the binding capacity of carboxylic and phenolic sites on the organic matter surfaces was exhausted (Kaiser and Zech, 1996). For $Al_p/C_p \geq 0.1$, the left-hand side of the equation shifted from log (Al^{3+}) + 2pH, to log (Al^{3+}) + 3pH, and became independent of the Al_p/C_p ratio. Consequently, a solid $Al(OH)_3$ phase seemed to be formed at high Al_p/C_p ratios. The equilibrium constant for the dissolution of this $Al(OH)_3$ phase was close to $10^{9.40}$.

What is the consequence of organic Al complexation as an Al buffering process? For pedagogical purposes, let us assume an Al_p/C_p ratio of 0.08 and pH of 4.3. Then we could estimate the Al^{3+} activity in two different ways, either by using the "Simonsson equation" (Equation 7.24) or by assuming a gibbsite equilibrium according to equation 7.7 with log K = 9.35, corresponding to microcrystalline gibbsite according to Driscoll and Schecher (1988), i.e., a value close to the empirically found value mentioned previously. This yields:

$(Al^{3+}) = 4.6 \cdot 10^{-5}$ – "Simonsson equation"
$(Al^{3+}) = 3.0 \cdot 10^{-4}$ – microcrystalline gibbsite

Compared to microcrystalline gibbsite, organic matter complexation involves a lower Al^{3+} activity at a given pH. The choice of a correct Al release mechanism is of crucial importance for predictions of Al solubility in soil and water acidification models. In addition to $Al(OH)_3(s)$ imogolite, $(OH)_3Al_2O_3SiOH(s)$ could be an important solid mineral phase in Podzols, particularly in the carbon-poor lower part (Bs) of the B horizon (Gustafsson, Lumsdon, and Simonsson, 1998):

$$(OH)_3Al_2O_3SiOH\ (s) + 3H_2O <=> 2Al(OH)_3(s) + H_4SiO_4 \qquad (7.25)$$

A kaolinite equilibrium (Equation 7.26) may be of importance in old weathered soils such as the Alisols (Cronan et al., 1990).

$$0.5Al_2Si_2O_5(OH)_4\ (s) + 3H^+ <=> Al^{3+} + H_4SiO_4 + 0.5H_2O \qquad (7.26)$$

Knowledge of the vertical differentiation of Al solubility control mechanisms in the soil profile is of great importance for accurate predictions of Al speciation and concentration in runoff water. During periods with a high groundwater level (for instance after snow melt) a large part of the runoff water will be in contact with the upper, humus-rich soil horizons. During such periods the Al concentration is likely to be determined by organic Al complexation. At base flow conditions, however, mineral phase equilibria with, e.g., $Al(OH)_3$, or imogolite could be more important. Although so far, Al equilibria with soil organic matter have only been shown to be important in coarse-textured Podzols, there is no reason why organic Al complexation should not be important within a much wider range of soils with sufficient amounts of humified organic matter in their mineral horizons.

SOIL ACIDIFICATION SENSITIVITY

Numerous attempts have been made to make broad evaluations of soils with respect to their acidification sensitivity in terms of declines in pH or base saturation at a given proton load. Commonly measured soil characteristics related to pH buffering, such as CEC, organic matter content, and clay content have been used. Soil and bedrock geology, climate, and land cover have also been considered (McFee, 1980; Kuylenstierna et al., 1995). According to Petersen (1980), soil groupings such as Arenosols, Fluvisols, Cambisols, Acrisols, Alisols, Podzols, and Ferralsols should include potentially sensitive soil types. Kuylenstierna et al. (1995) made an evaluation of soil acidification sensitivity in tropical regions, and integrated soil type (ranging from low- to high-buffering capacity), soil moisture regime (ranging from very humid/humid to semiarid/arid), and land cover/use (ranging from sensitive to insensitive) into a common index. They found that the most sensitive areas were characterized by old acidic soils such as Ferralsols, Alisols, and Acrisols and high annual rainfall. Particularly in southern Asia, the sensitive areas overlapped with areas of high population density and high emissions of acidifying S and N compounds.

ACIDIFICATION OF ARABLE SOILS

Arable soils in the industrialized countries in Europe and North America are rather young and fertile, but older soils are also found in these parts of the world. The arable soils have become acidified mainly because of intensive management that has caused extensive NO_3^- and base cation leaching, and base cation removal in harvested plant biomass. Arable soils are usually limed on a regular basis to compensate for the acidification connected with N turnover and cropping. The liming effect of the commonly used calcium carbonate (Equation 7.4) advances rather slowly downward in the soil profile, however. Therefore, applications of gypsum ($CaSO_4 \cdot 2H_2O$) which dissolves at a faster rate as compared to $CaCO_3$, have been used alone or in combination with $CaCO_3$ to mitigate severe subsoil acidity in old, strongly weathered soils (Sumner, 1993). The neutralizing reactions in the subsoil related to gypsum are SO_4^{2-} adsorption (Equation 7.9) followed by aluminum precipitation (Equation 7.27).

$$Al^{3+} + 3H_2O <=> Al(OH)_3 + 3H^+ \qquad (7.27)$$

The net result is decline in both exchangeable and soluble Al.

In low-input agricultural systems in the third world tropics, green manure is used both as an organic fertilizer and as a cheap and practical alternative to liming. Various humic compounds formed during the organic matter decomposition may complex Al and make it practically harmless. Decomposition of the green manure also involves a breakdown of low-molecular organic acid anions and release of base cations, leading to an increase in pH, a decline in soluble and exchangeable Al, as well as an increase in ANCs (Hairiah, Adawiyah, and Widyaningsih, 1996; Haynes and Mokolobate, 2001).

In acid tropical soils there is often a close relationship between low pH values, high concentrations of labile Al, and limited P availability, and it is sometimes difficult to make a clear distinction between Al toxicity and P deficiency (Hairiah, 1992). Phosphate ions form strong surface complexes with Al and Fe oxides, and precipitated Al and Fe phosphates are sparingly soluble. Haynes and Mokolobate (2001) point out that green manure amendments may improve P availability by initially increasing the pH as a result of decomposition. This yields a less positive charge on the oxide surfaces and less Al available for Al phosphate precipitation. Soluble humic compounds and low-molecular weight organic acids formed during the decomposition may directly or indirectly compete with phosphate ions for adsorption sites on oxide surfaces. Al complexation by these organic compounds will also contribute to an increased P availability. Further, the added green manure contains phosphorus, which will eventually be released through P mineralization.

ACIDIFICATION OF FOREST SOILS

Soil Acidification and Forest Damage

During the 1980s an intensive debate started in Europe, the United States, and Canada concerning effects of Al on trees growing on acid soils and subjected to high atmospheric loads of acidifying S and N compounds (Ulrich and Pankrath, 1983; Ulrich, 1991; Sverdrup, Warfvinge, and Nihlgård, 1994). Temporal declines in the amounts of exchangeable base cations and accompanying declines in base saturation and pH were documented (Hallbäcken and Tamm, 1986; Berdén et al., 1987; Johnson et al., 1991; Falkengren-Grerup and Tyler, 1992). Proton budgets strongly indicated that the atmospheric deposition accounted for an appreciable share of the total acid load in forest ecosystems (Ulrich, Mayer, and Khanna, 1979; Van Breemen, Mulder, and Driscoll, 1983; Nilsson, 1985). Visual symptoms of forest damage as well as indications of a declining tree nutrient status (foremost Mg deficiency) were reported. These symptoms were claimed to be a new phenomenon and became known as "neuartige Waldschäden" (a new type of forest decline). The Ca/Al ratio in the soil solution or in the exchangeable phase was claimed to be a useful index. The Ca/Al value < 1 was considered as a threshold value, below which tree (root) damage was likely to occur. Based on experiences from agriculture, liming was recommended as a remedy (Ulrich, 1991; Sverdrup, Warfvinge, and Nihlgård, 1994).

Numerous new field experiments with forest liming were initiated, and these experiments together with older field experiments were investigated with respect to liming effects on soil chemical properties, tree vitality, and tree growth (Hüttl, 1987, 1988; Staaf, Persson, and Bertills, 1996). Even if liming significantly altered the soil chemistry with respect to increases in base saturation and declines in labile Al, evaluations of forest liming experiments in both Germany (Hüttl and Schneider, 1998) and Sweden (Persson et al., 1996) did not demonstrate any generally positive effects of liming per se, on either tree vitality or tree growth. Exceptions were those sites where specific base cation deficiencies, particularly Mg deficiency, were documented. At these sites, amendments with either alkalizing or neutral Mg salts (such as $MgSO_4$) were shown to improve tree nutrient status and tree growth (Evers, 1984; Kaupenjohann and Zech, 1989). Mg deficiency is still a problem, and has recently been shown in Norway spruce stands growing on acid soils in the Czech republic (Krám et al., 1995, 1997). However, the role of Al as a toxic element in forest ecosystems has probably been overrated (De Wit, 2000).

Trends in S Deposition and SO_4^{2-} Storage and Leaching

Eriksson, Karltun, and Lundmark (1992) made a regional study of Swedish forest soils and found a clear relationship between an increasing north-south gradient in S deposition and actual amounts of adsorbed SO_4^{2-} in the soils. There was also a relationship between the amounts of adsorbed SO_4^{2-} and temporal decline in base cation storage. During the 1990s, S deposition has declined considerably over most of Europe mainly because of mutual agreements among the European countries to cut down on emissions and closures of coal-fired power plants in Eastern Europe (Fölster, 2001). Decreasing S loads are expected to eventually improve the soil acidity situation. Fölster (2001) documented a recent decline in SO_4^{2-} concentration in soil solutions collected from three Swedish forest Podzols during a 9- to 11-year period. The relative change in SO_4^{2-} concentration was somewhat smaller (amounting to 3 to 4 percent per year of the median SO_4^{2-} concentration) than in the atmospheric deposition, and was accompanied by declines in Ca^{2+} and Mg^{2+} concentrations. More substantial declines in soil solution SO_4^{2-} have been observed at forest sites in Germany (Marschner, Gensior, and Fischer, 1998; Alewell et al., 2000).

Attempts have been made to use simulation models to predict changes in soil SO_4^{2-} fluxes resulting from declines in S deposition (Giesler et al., 1996). However, there are difficulties in tracking changes in SO_4^{2-} storage and leaching over time. Manderscheid, Jungnickel, and Alewell (2000) investigated two forested catchments in Germany and found a substantial spatial variability in the parameters of Langmuir SO_4^{2-} adsorption isotherms, and also in correlations between isotherm parameters and adsorption-determining soil properties. An important conclusion was that a high spatial variability would have severe effects on model predictions of soil SO_4^{2-} fluxes at varying S inputs. The same research group quantified the storage of adsorbed SO_4^{2-} in weathered bedrock (to a maximum depth of 10 m) and found that this SO_4^{2-} pool was in equilibrium with SO_4^{2-} in the surrounding groundwater. The weathered bedrock SO_4^{2-} pool was considerably larger than the SO_4^{2-} pool in the soil profile to a depth of 50 cm. The researchers concluded that it might take several decades before declines in groundwater SO_4^{2-} could be expected as a result of decreasing atmospheric S deposition (Manderscheid et al., 2000).

The Significance of Organic S Turnover

Though much work has been devoted to investigations of S mineralization both in forest soils (David, Vance, and Krzyszowska, 1995; Valeur, 2000) and arable soils (Eriksen, Murphy, and Schnug, 1998), the effect of S mineralization on SO_4^{2-} fluxes in soil is not fully understood. Generally, one distinguishes between "biological" S mineralization of organic compounds with reduced S in C-S bonds, and "biochemical" mineralization of ester sulphates. Some claim that ester sulphates could be temporary sinks for deposited SO_4^{2-} ions, but this hypothesis has seldom been experimentally verified (David, Vance, and Krzyszowska, 1995). One exception is a study by Prietzel et al. (2001), who documented a substantial increase of ester sulphates in one Podzol and one Cambisol after repeated $(NH_4)_2SO_4$ applications during a seven-year period, corresponding to totally 510 kg S ha^{-1}. A net increase in total organic S storage has been observed at sites exposed to dry deposition of SO_2 (Ohtonen, Markkola, and Torvela, 1989), and in a regional study of European forest sites that had been exposed to different levels of ambient S deposition (Erkenberg, Prietzel, and Rehfuess, 1996). Experiments with the addition of diluted H_2SO_4 or sulphate salts have usually not resulted in an increase in total organic S (David, Fasth, and Vance, 1991; Mitchell et al., 1998). Again, the study by Prietzel et al. (2001) is an exception.

Ratios between stable S and O isotopes have been used to evaluate the quantitative significance of S mineralization in the sulphur cycle. In a study by Alewell et al. (1999) in the Hubbard Brook area in New Hampshire, budget calculations and stable S isotope measurements indicated that S mineralization could be a major contributor to the SO_4^{2-} found in stream water. This was also shown in an experiment in a small forested catchment on the Swedish West coast, which had been covered with a roof and treated with "clean" rainwater with a low SO_4^{2-} concentration. Initially, the SO_4^{2-} flux in runoff water was mainly affected by desorption processes (Giesler et al., 1996), but subsequent isotope studies indicated an increasing importance of S mineralization (Torssander and Mörth, 1998).

Trends in N Deposition and N Leaching

N deposition has not declined to the same extent as S deposition, and N saturation (N leaching close to N inputs) has been approached in some forest ecosystems in Europe and North America. Efforts have been made to find diagnostic indices that could be used for predicting the likelihood of extensive NO_3^- leaching. Empirical relationships between leaching and deposition have been used (Dise and Wright, 1995; Nilsson, Berggren, and Westling, 1998), and deposition thresholds have been reported, above which NO_3^- leaching is likely to increase. A drawback is that rather different thresholds can be obtained for different data sets. The C/N ratio of the forest floor (O horizon) has been used as an index and has proved to be more useful for prediction purposes. The rationale of this index is that a low C/N ratio indicates a potentially high net N mineralization, and thereby also a high net nitrification and NO_3^- leaching rate (Dise, Matzner, and Forsius, 1998; Gundersen, Callesen, and de Vries, 1998). Combining variables into one single index has also proved to be useful for separating sites with different N status. Andersson, Berggren, and Nilsson (2002) used the N flux density, i.e., annual N deposition (or N fertilization) plus annual net N mineralization in O horizons of several N fertilized and unfertilized stands of Norway spruce in Sweden. Nohrstedt et al. (1996) compared the nitrogen status of three nearby Norway spruce sites on the Swedish West coast with widely different NO_3^- leaching rates. According to Ring (2001), the differences among the three sites with respect to NO_3^- leaching could be qualitatively explained by using the leaching indices of either Dise, Matzner, and Forsius (1998) or Andersson, Berggren, and Nilsson (2002).

Recently, Asner et al. (2001) published a biogeochemical model which indicated that N saturation may develop much faster in moist tropical forests than in temperate forests. This is because the tropical forest ecosystems usually have an excess of N already at low atmospheric loads of N. The productivity of these systems is limited by other factors such as the phosphorus availability (Vitousek and Sanford, 1986). Model simulations showed that an increase in the atmospheric N load initiated an extensive NO_3^- leaching accompanied by an equally extensive leaching of Ca^{2+}, probably due to H^+/Ca^{2+} exchange, resulting from HNO_3 formation in the soil (Asner et al., 2001). Increased calcium losses are potentially serious in the humid tropics. Nykvist (2000) showed that old, strongly weathered Acrisols on Borneo and elsewhere in the humid tropics had, not unexpectedly, a low storage of exchangeable and even total Ca^{2+}, indicating that commercial forestry would not be sustainable unless Ca^{2+} was added to these soils.

FUTURE PERSPECTIVES

For future experimental and modeling work concerning soil acidification and recovery processes, the following items need consideration.

Spatial Heterogeneity

The studies of, e.g., Manderscheid, Jungnickel, and Alewell (2000) make it seem worthwhile to include geostatistical aspects in the modeling work. This should be relevant for soil chemical properties and processes as well as for input and output fluxes of individual elements.

Inclusion and Description of Soil Processes

Considerable progress has been made concerning the mechanistic understanding and inclusion of mineral weathering processes in simulation models (Sverdrup and Warfvinge, 1993). However, because aluminum is one of the key elements in soil acidification, an accurate description of aluminum-proton interactions is crucial. Therefore, complexation and decomplexation of organic Al should be properly included in all models that predict the Al concentration in soil solution (Berggren and Mulder, 1995; Simonsson, 1999). Also, the pH and ionic strength dependency of CEC may be of critical importance particularly in models that are intended to predict soil recovery after a decrease in acid deposition (Matschonat and Vogt, 1997).

Soil acidification is closely connected with the N and S cycles, and the generation and fluxes of NO_3^- and SO_4^{2-} in the soil. Some progress has been made concerning the understanding of both nitrification (De Boer and Kowalchuk, 2001) and S mineralization processes (Eriksen, Murphy, and Schnug, 1998; Valeur, 2000). Still, there are few mechanistic descriptions available that could be readily included in soil acidification models. Nitrification in acid soils shows a wide spatial variability, which is not explained by the substrate (NH_4^+) level. Concerning S mineralization, there has probably been too much emphasis on studies of microcosms in the laboratory rather than experimental work conducted under field conditions.

SUMMARY

In this chapter, the usefulness of proton budgets for quantifying acidifying and neutralizing processes was illustrated for arable soils and forest soils. Buffering mechanisms, particularly aluminum buffering, were reviewed and the importance of equilibria between soluble Al^{3+} and solid organic Al complexes was emphasized.

Chemical and biological effects in temperate and tropical soils of acidity mitigation practices such as liming and application of green manure were reviewed. Sulphur and N emissions are increasing in densely populated tropical areas with sensitive soils, while S deposition has decreased during the 1990s in industrialized parts of Europe and North America. Recent investigations of recovery in European forest soils indicate that the expected desorption of previously adsorbed SO_4^{2-} and increase in exchangeable base cation storage are rather slow processes with a wide spatial variation. Moreover, the significance of organic S mineralization as a SO_4^{2-} source is rather ambiguous. The N saturation concept and causal relationships among N deposition, nitrification, and NO_3^- leaching were discussed, and the usefulness of qualitative indices for predicting NO_3^- leaching was evaluated. To improve soil acidification models, key processes such as SO_4^{2-} adsorption/desorption and organic S mineralization and nitrification have to be better characterized with respect to the key factors behind their temporal and spatial variability.

REFERENCES

Alewell, C., B. Manderscheid, P. Gerstberger, and E. Matzner (2000). Effects of reduced atmospheric deposition on soil solution chemistry and elemental contents of spruce needles in NE-Bavaria, Germany. *Journal of Plant Nutrition and Soil Science 163:*509-516.

Alewell, C., M.J. Mitchell, G.E. Likens, and H.R. Krouse (1999). Sources of stream sulfate at the Hubbard Brook Experimental Forest: Long-term analyses using stable isotopes. *Biogeochemistry 44:*281-299.

Allison, J.D., D.S. Brown, and K.J. Novo-Gradac (1991). *MINTEQA2/ PRODEFA2. A geochemical assessment model for environmental systems.* EPA /600/3-91/021. Athens, Georgia: U.S. EPA.

Andersson, P., D. Berggren, and S.I. Nilsson (2002). Experimental studies of nitrogen mineralisation and nitrification and model simulation of nitrogen saturation. *Forest Ecology and Management 157:*39-53.

Asner, G.P., A.R. Townsend, W.J. Riley, P.A. Matson, J.C. Neff, and C.C. Cleveland (2001). Physical and biogeochemical controls over terrestrial ecosystem responses to nitrogen deposition. *Biogeochemistry 54:*1-39.

Berdén, M., S.I. Nilsson, K. Rosén, and G. Tyler (1987). *Soil acidification—Extent, causes and consequences.* Report 3292. Solna: National Swedish Environment Protection Board.

Berggren, D. and J. Mulder (1995). The role of organic matter in controlling aluminum solubility in acidic mineral soil horizons. *Geochimica et Cosmochimica Acta 59:*4167-4180.

Bergström, L. and A. Gustafson (1985). Hydrogen ion budgets of four small runoff basins in Sweden. *Ambio 14:*346-348.

Binkley, D. and D. Richter (1987). Nutrient cycles and H^+ budgets of forest ecosystems. *Advances in Ecological Research 16:*1-51.

Christophersen, N., H.M. Seip, and R.F. Wright (1982). A model for streamwater chemistry at Birkenes, Norway. *Water Resources Research 18:*977-996.

Cosby, B.J., G.M. Hornberger, J.N. Galloway, and R.F. Wright (1985). Modeling the effects of acid deposition: Assessment of a lumped parameter model of soil water and streamwater chemistry. *Water Resources Research 21:*51-63.

Cronan, C.S., C.T. Driscoll, R.M. Newton, J.M. Kelly, C.L. Schofield, R.J. Bartlett, and R. April (1990). A comparative analysis of aluminum biogeochemistry in a northeastern and a southeastern watershed. *Water Resources Research 26:*1413-1430.

David, M.B., W.J. Fasth, and G.F. Vance (1991). Forest soil response to acid and salt additions of sulfate: I. Sulfur constituents and net retention. *Soil Science 151:* 136-145.

David, M.B., G.F. Vance, and A.J. Krzyszowska (1995). Carbon controls on Spodosol, nitrogen, sulfur and phosphorus cycling. In *Carbon Forms and Functions in Forest Soils,* eds. W.W. McFee and J.M. Kelly. Madison, Wisconsin: Soil Science Society of America, Incorporated, pp. 329-353.

De Boer, W. and G.A. Kowalchuk (2001). Nitrification in acid soils: Micro-organisms and mechanisms. *Soil Biology and Biochemistry 33:*853-866.

De Wit, H.A. (2000). "Solubility controls and phyto-toxicity of aluminum in a mature Norway spruce forest" (doctoral thesis, Agricultural University of Norway).

Dise, N.B., E. Matzner, and M. Forsius (1998). Evaluation of organic horizon C:N ratio as an indicator of nitrate leaching in conifer forests across Europe. *Environmental Pollution 102:*453-456.

Dise, N.B. and R.F. Wright (1995) Nitrogen leaching from European forests in relation to nitrogen deposition. *Forest Ecology and Management 71:*153-161.

Driscoll, C.T. (1984). A procedure for the fractionation of aqueous aluminum in dilute acid waters. *International Journal of Environmental Analytical Chemistry 16:*267-283.

Driscoll, C.T. and W.D. Schecher (1988). Aluminum in the environment. In *Metal Ions in Biological Systems,* Volume 24: Aluminum and Its Role in Biology, eds. H. Sigel and A. Sigel. New York: Marcel Dekker, pp. 59-122.

Eriksen, J., M.D. Murphy, and E. Schnug (1998). The soil sulphur cycle. In *Sulphur in Agroecosystems,* ed. E. Schnug. Kluwer, Dordrecht: Academic Publishers, pp. 39-73.

Eriksson, E., E. Karltun, and J.-E. Lundmark (1992). Acidification of forest soils in Sweden. *Ambio 21:*150-154.

Erkenberg, A., J. Prietzel, and K.E. Rehfuess (1996). Schwefelausstattung ausgewählter europäischer Waldböden in Abhängigkeit vom atmogenen S-Eintrag. *Zeitschrift für Pflanzenernährung und Bodenkunde 159:*101-109.

Evers, F.H. (1984). Welche Erfahrungen liegen bei Kalium- und Magnesiumdüngungsversuchen auf verschiedenen Standorten in Baden-Württemberg vor? *Allgemeine ForstZeitung 39:*767-769.

Falkengren-Grerup, U. and G. Tyler (1992). Changes since 1950 of mineral pools in the upper C-horizon of Swedish deciduous forests. *Water, Air and Soil Pollution 64:*495-501.

Fölster, J. (2001). *Catchment Hydrochemical Processes Controlling Acidity and Nitrogen in Forest Stream Water.* Silvestria 190. Uppsala: Acta Universitatis Agriculturae Sueciae.

Giesler, R., F. Moldan, U. Lundström, and H. Hultberg (1996). Reversing acidification in a forested catchment in southwestern Sweden: Effects on soil solution chemistry. *Journal of Environmental Quality 25:*110-119.

Gundersen, P., I. Callesen, and W. de Vries (1998). Nitrate leaching in forest ecosystems is related to forest floor C/N ratios. *Environmental Pollution 102:*403-407.

Gustafsson, J.P. (1995). Modeling pH-dependent sulfate adsorption in the Bs horizons of podzolized soils. *Journal of Environmental Quality 24:*882-888.

Gustafsson, J.P., D.G. Lumsdon, and M. Simonsson (1998). Aluminum solubility characteristics of spodic B horizons containing imogolite-type minerals. *Clay Minerals 33:*77-86.

Hairiah, K. (1992). "Aluminum tolerance of Mucuna, a tropical leguminous cover crop" (PhD thesis, Groningen: University of Groningen, The Netherlands).

Hairiah, K., R. Adawiyah, and J. Widyaningsih (1996). Amelioration of aluminum toxicity with organic matter: Selection of organic matter based on its total cation concentration. *Agrivita 19:*158-164.

Hallbäcken, L. and C.O. Tamm (1986). Changes in soil acidity from 1927 to 1982-1984 in a forest area of southwest Sweden. *Scandinavian Journal of Forest Research 1:*219-232.

Haynes, R.J. and M.S. Mokolobate (2001). Amelioration of Al toxicity and P deficiency in acid soils by additions of organic residues: A critical review of the phenomenon and the mechanisms involved. *Nutrient Cycling in Agroecosystems 59:*47-63.

Hüttl, R. (1987). Neuartige Waldschäden, Ernährungsstörungen und Düngung. *Allgemeine Forst-Zeitung 39:*289-299.

Hüttl, R. (1988). "New type" forest declines and restabilization/revitalization strategies. *Water, Air and Soil Pollution 41:*95-111.

Hüttl, R.F. and B.U. Schneider (1998). Forest ecosystem degradation and rehabilitation. *Ecological Engineering 10:*19-31.

Johnson, D.W., D. Binkley, and P. Conklin (1995). Simulated effects of atmospheric deposition, harvesting and species change on nutrient cycling in a loblolly pine forest. *Forest Ecology and Management 76:*29-45.

Johnson, D.W., M.S. Cresser, S.I. Nilsson, J. Turner, B. Ulrich, D. Binkley, and D.W. Cole (1991). Soil changes in forest ecosystems: Evidence for and probable causes. *Proceedings of the Royal Society of Edinburgh 97 B:*81-116.

Kaiser, K. and W. Zech (1996). Defects in estimation of aluminum in humus complexes of podzolic soils by pyrophosphate extraction. *Soil Science 161:*452-458.

Kaupenjohann, M. and W. Zech (1989). Waldschäden und Düngung. *Allgemeine ForstZeitschrift 44:*1002-1008.

Khanna, P.K., R.J. Raison, and R.A. Falkiner (1986). Exchange characteristics of some acid organic-rich forest soils. *Australian Journal of Soil Research 24:*67-80.

Krám, P., J. Hruška, C.T. Driscoll, and C.E. Johnson (1995). Biogeochemistry of aluminum in a forest catchment in the Czech republic impacted by atmospheric inputs of strong acids. *Water, Air and Soil Pollution 85:*1831-1836.

Krám, P., J. Hruška, B.S. Wenner, C.T. Driscoll, and C.E. Johnson (1997). The biogeochemistry of basic cations in two forest catchments with contrasting lithology in the Czech Republik. *Biogeochemistry 37:*173-202.

Kuylenstierna, J.C.I., H. Cambridge, S. Cinderby, and M.J. Chadwick (1995). Terrestrial ecosystem sensitivity to acidic deposition in developing countries. *Water, Air and Soil Pollution 85:*2319-2324.

Lofts, S., C. Woof, E. Tipping, N. Clarke, and J. Mulder (2001). Modelling pH buffering and aluminum solubility in European forest soils. *European Journal of Soil Science 52:*189-204.

Manderscheid, B., C. Jungnickel, and C. Alewell (2000). Spatial variability of sulfate isotherms in forest soils at different scales and its implications for the modeling of soil sulfate fluxes. *Soil Science 165:*848-857.

Manderscheid, B., T. Schweisser, G. Lischeid, C. Alewell, and E. Matzner (2000). Sulfate pools in the weathered substrata of a forested catchment. *Soil Science Society of America Journal 64:*1078-1082.

Markewitz, D., D.D. Richter, H.L. Allen, and J.B. Urrego (1998). Three decades of observed soil acidification in the Calhoun Experimental Forest: Has acid rain made a difference? *Soil Science Society of America Journal 62:*1428-1439.

Marschner, B., A. Gensior, and U. Fischer (1998). Response of soil solution chemistry to recent declines in atmospheric deposition in two forest ecosystems in Berlin, Germany. *Geoderma 83:*83-101.

Matschonat, G. and R. Vogt (1997). Effects of changes in pH, ionic strength, and sulphate concentration on the CEC of temperate acid forest soils. *European Journal of Soil Science 48:*163-171.

Matzner, E. and M. Davis (1996). Chemical soil conditions in pristine Nothofagus forests of New Zealand as compared to German forests. *Plant and Soil 186:*285-291.

McAvoy, D.C., R.C. Santore, J.D. Shosa, and C.T. Driscoll (1992). Comparison between pyrocatechol violet and 8-hydroxyquinoline procedures for determining aluminum fractions. *Soil Science Society of America Journal 56:*449-455.

McFee, W.W. (1980). Sensitivity of soil regions to long-term acid precipitation. In *Atmospheric Sulfur Deposition Environmental Impact and Health Effects,* eds. D.S. Shriner, C.R. Richmond, and S.E. Lindberg. Ann Arbor, MI: pp. 495-506.

Mitchell, M.J., R.B. Santore, C.T. Driscoll, and B.R. Dhamala (1998). Forest soil sulfur in the Adirondack mountains: Response to chemical manipulations. *Soil Science Society of America Journal 62:*272-280.

Muniz, I.P. (1991). Freshwater acidification: Its effects on species and communities of freshwater microbes, plants and animals. Proceedings of the Royal Society of Edinburgh Section B. *Biological Sciences 97:*227-254.

Nilsson, S.I. (1985). Why is lake Gårdsjön acid?—An evaluation of processes contributing to soil and water acidification. *Ecological Bulletin 37:*311-318.

Nilsson, S.I., D. Berggren, and O. Westling (1998). Retention of deposited NH_4^+-N and NO_3^--N in coniferous forest ecosystems in southern Sweden. *Scandinavian Journal of Forest Research 13:*393-401.

Nilsson, S.I., H.G. Miller, and J.D. Miller (1982). Forest growth as a possible cause of soil and water acidification: An examination of the concepts. *Oikos 39:*40-49.

Nohrstedt, H.-Ö., U. Sikström, E. Ring, T. Näsholm, P. Högberg, and T. Persson (1996). Nitrate in soil water in three Norway spruce stands in southwest Sweden as related to N-deposition and soil, stand and foliage properties. *Canadian Journal of Forest Research 26:*836-848.

Nykvist, N. (2000). Tropical forests can suffer from a serious deficiency of calcium after logging. *Ambio 29:*310-313.

Ohtonen, R., A.M. Markkola, and H. Torvela (1989). Total sulfur content in the humus layer of urban polluted forest soil. *Water, Air and Soil Pollution 44:*135-141.

Persson, T., F. Andersson, B. Nihlgård, S.I. Nilsson, H. Staaf, and H. Sverdrup (1996). Behöver skogsmarken kalkas? In *Rapport 4559, Skogsmarkskalkning – Resultat och Slutsatser från Naturvårdsverkets Försöksverksamhet,* eds. H. Staaf, T. Persson, and U. Bertills. Stockholm: National Swedish Environment Protection Board, pp. 234-246. (In Swedish, with a summary in English.)

Petersen, L. (1980). Sensitivity of different soils to acid precipitation. In *Effects of Acid Precipitation on Terrestrial Ecosystems,* eds. T.C. Hutchinson and M. Havas. New York: Plenum Press, pp. 573-577.

Prietzel, J., C. Weick, J. Korintenberg, G. Seybold, T. Thumerer, and B. Treml (2001). Effects of repeated $(NH_4)_2SO_4$ application on sulfur pools in soil, soil microbial biomass, and ground vegetation of two watersheds in the Black Forest/Germany. *Plant and Soil 230:*287-305.

Ring, E. (2001). *Nitrogen in Soil Water at Five Nitrogen-Enriched Forest Sites in Sweden.* Agraria 267. Uppsala: Acta Universitatis Agriculturae Sueciae.

Simonsson, M. (1999). *Mechanisms Controlling the Solubility of Aluminum in B Horizons of Podzolised Soils.* Agraria 200. Uppsala: Acta Universitatis Agriculturae Sueciae.

Sposito, G. (1989). *The Chemistry of Soils.* Oxford, UK: Oxford University Press.

Staaf, H., T. Persson, and U. Bertills, Eds. (1996). *Skogsmarkskalkning—Resultat och Slutsatser från Naturvårdsverkets Försöksverksamhet.* Rapport 4559. Stockholm: Naturvårdsverkets Förlag. (In Swedish, with a summary in English.)

Sumner, M.E. (1993). Gypsum and acid soils: The world scene. *Advances in Agronomy 51:*1-32.

Sumner, M.E., M.V. Fey, and A.D. Noble (1991). Nutrient status and toxicity problems in acid soils. In *Soil Acidity,* eds. B. Ulrich and M.E. Sumner. Berlin: Springer-Verlag, pp. 149-182.

Sverdrup, H. and P. Warfvinge (1993). Calculating field weathering rates using a mechanistic geochemical model—PROFILE. *Applied Geochemistry 8:*273-283.

Sverdrup, H., P. Warfvinge, and B. Nihlgård (1994). Assessment of soil acidification effects on tree growth in Sweden. *Water, Air, and Soil Pollution 78:*1-36.

Sverdrup, H., P. Warfvinge, and K. Rosén (1992). A model for the impact of soil solution Ca/Al ratio, soil moisture, and temperature on tree base cation uptake. *Water, Air, and Soil Pollution 61:*365-383.

Tipping, E. (1994). WHAM—a chemical equilibrium model and computer code for waters, sediments and soils incorporating a discrete site/electrostatic model of ion-binding by humic substances. *Computers and Geosciences 20:*973-1023.

Tipping, E., D. Berggren, J. Mulder, and C. Woof (1995). Modelling the solid-solution *distributions of protons, aluminum, base cations and humic substances in acid soils. European Journal of Soil Science 46:*77-94.

Torssander, P. and C.M. Mörth (1998). Sulfur dynamics in the roof experiment at Lake Gårdsjön deduced from sulfur and oxygen isotope ratios in sulfate. In *Experimental Reversal of Rain Effects: Gårdsjön Roof Project,* eds. H. Hultberg and R. Skeffington. Chichester: John Wiley and Sons Ltd, pp. 185-206.

Ulrich, B. (1991). An ecosystem approach to soil acidification. In *Soil Acidity,* eds. B. Ulrich and M.E. Sumner. Berlin: Springer-Verlag, pp. 28-79.

Ulrich, B., R. Mayer, and P.K. Khanna (1979). Deposition von Luftverunreinigungen und ihre Auswirkungen in Waldökosystemen im Solling. *Schriften der Forstlichen Fakultät, Universität Göttingen 58.*

Ulrich, B. and J. Pankrath, Eds. (1983). *Effects of Accumulation of Air Pollutants in Forest Ecosystems.* Dordrecht: D. Reidel Publishing Company.

Valeur, I. (2000). *Sulphur Dynamics in Forest Soils. Effects of Liming.* Silvestria 138. Uppsala: Acta Universitatis Agriculturae Sueciae.

Van Breemen, N., J. Mulder, and C.T. Driscoll (1983). Acidification and alkalinization of soils. *Plant and Soil 75:*283-308.

Vitousek, P.M. and R. Sanford (1986). Nutrient cycling in moist tropical forests. *Annual Review of Ecology and Systematics 17:*137-167.

Walker, W.J., C.S. Cronan, and P.R. Bloom (1990). Aluminum solubility in organic soil horizons from northern and southern forested watersheds. *Soil Science Society of America Journal 54:*369-374.

Warfvinge, P. and H. Sverdrup (1992). Calculating critical loads of acid deposition with PROFILE—A steady-state soil chemistry model. *Water, Air and Soil Pollution 63:*119-143.

Chapter 8

Soil Alkalinization

Rami Keren

Salinity and alkalinity phenomena are generally pronounced in arid or semiarid regions under irrigation and in regions of poor natural drainage. Alkaline soils have poor physical properties due to high sodicity and high pH. Alkali soils contain an excess of exchangeable sodium and have sodium carbonates. Under conditions of excess salts, the soil pH is seldom higher than 8.5 and the clay particles remain flocculated. As long as excess salts are present, the appearance and properties of these soils are generally similar to those of saline soils. As the concentration of the salts in the soil solution is lowered during rainfall, for example, some of the exchangeable sodium hydrolyzes and forms sodium hydroxide. This may change to sodium carbonate upon reaction with carbon dioxide (CO_2) absorbed (from the soil, air, and atmosphere) and the soil may become strongly alkaline (pH above 8.5). Under these conditions, compaction, slow internal drainage, slow water infiltration into the soil, poor aeration, soil erosion, and waterlogging on the lower lands commonly occur because of soil swelling and dispersion.

Considerable progress has been made in managing and controlling salinity and alkalinity in irrigated lands. However, coping with salinity and alkalinity problems has become more complex because of its effect on soil physical properties. Because of natural hydrological and geochemical factors, as well as irrigation-induced activities, soil salinity, alkalinity, and associated drainage and soil erosion continue to plague agriculture. Of the total cultivated land area in the world (1.5×10^7 km^2, Massoud, 1981) about 0.37×10^7 km^2 is saline. The saline and sodic soils exist in over 100 countries (Szabolcs, 1989). The distribution of saline and sodic soils over six continents is given in Table 8.1.

TABLE 8.1. Distribution of Saline and Sodic Soils over Six Continents

Continent	Saline soil area (1,000 ha)	Sodic soil area (1,000 ha)	Sodic soil area to saline soil area ratio
Australia	38.63	199.7	5.17
Europe	7.8	22.9	2.9
North America	8.16	9.56	1.17
South America	69.41	59.57	0.86
Asia	194.92	121.86	0.63
Africa	53.49	26.95	0.50

Source: Szabolcs, 1989; Rengasamy and Olsson, 1991.

ORIGIN AND DISTRIBUTION OF ALKALI SOILS

Sodicity and Alkalinity

Three types of sodic soils are known to occur: (1) alkaline sodic soils (soils affected by Na salts, $NaHCO_3$, Na_2CO_3); the partial pressure of CO_2 and the concentration of CO_3^{2-} and HCO_3^- are the principal factors responsible for their alkaline pH; (2) neutral sodic soils (soils affected by Na neutral salts, NaCl, Na_2SO_4); these soils are characterized by low concentrations of CO_3^{2-} and HCO_3^- and a predominance of Cl^- and SO_4^{2-} ions; and (3) acid sodic soils—these soils are less common, but generally occur in regions with rainfall >550 mm per year) and are characterized by leaching and low concentrations of Ca and Mg on the exchange complex (Rengasamy and Olsson, 1991). In nature, strongly alkaline soils invariably have high sodicity levels. On the other hand, a sodic soil with high sodium adsorption ratio (SAR) does not necessarily mean a high pH (Kelley, 1948; Beek and Breeman, 1973).

Alkalinity (Alk) is defined as the acid-neutralizing capacity of an aqueous carbonate system (Stumm and Morgan, 1970). In most natural waters, alkalinity equals the total concentration of cations minus the total concentration of anions other than carbonates:

$$Alk = [Na^+] + [K^+] + 2[Ca^{2+} + Mg^{2+}] - [Cl^-] - 2[SO_4^{2-}] \tag{8.1}$$

More conveniently, alkalinity as expressed in equation (8.1) can be represented as

$$Alk = [HCO_3^-] + 2[CO_3^{2-}] + [OH^-] - [H^+] \quad (8.2)$$

The square brackets denote molar concentrations. Alkalinity in excess over the total amount of divalent cations is often referred to as residual alkalinity (RAlk), or residual sodium carbonate (Eaton, 1950).

$$RAlk = Alk - 2[Ca^{2+}] - 2[Mg^{2+}] \quad (8.3)$$

Residual alkalinity is influenced by changes in the relative proportions of monovalent and divalent cations in soil solution due to cation exchange (equation 8.4) and by dilution of soil solution and soil water evaporation and transpiration,

$$\text{Ca-clay} + 2Na^+ = Na_2\text{-clay} + Ca^{2+} \quad (8.4)$$

Under conditions where Alk is greater than the sum of the divalent cations concentration, as evaporation proceeds, precipitation of calcite, dolomite, and some other minerals would remove Ca^{2+} and Mg^{2+} from solutions, leaving behind monovalent cations. The Na^+ and K^+ ions tend to concentrate and increase residual alkalinity and pH of the soil solution.

In the presence of adsorbed Na, the calcareous soil pH increases due to the presence of Na bicarbonate-carbonate (Nakayama, 1970). Alkalinity and equilibrium, partial pressure of carbon dioxide, and Pco_2 adequately define the pH of an aqueous carbonate system and of alkali soils (Ponnamperuma, 1972; Mashhady and Rowell, 1978). The relation between pH and alkalinity of aqueous carbonate systems in equilibrium with different levels of partial pressure of CO_2 is represented by equation (8.5)

$$pH = 7.82 - \log Pco_2 + \log Alk - 0.5\,(I^{1/2}) \quad (8.5)$$

where I refers to the ionic strength of the solution.

When $Na^+ - Ca^{2+}$ exchange reaction occurs in a calcareous soil, the relation between pH and alkalinity of aqueous carbonate systems under equilibrium can be calculated using equation (8.6) (Thorstenson, Fisher, and Croft, 1979):

$$pH = \log K_c + \log\left[\frac{K_{exv}}{k_2}\right] + \log\left[\frac{Na_{ex}^{\,2}}{Ca_{ex} * a^2 Na^+ * aHCO_3^-}\right] \quad (8.6)$$

where K_c, K_{exv} (v - Vanselow expression) and k_2 are the equilibrium constants for the reactions $CaCO_3 = Ca^{2+} + CO_3^{2-}$; $Ca^{2+} + 2Na_{ex} = 2Na^+ + Ca_{ex}$, where ex is the adsorbed phase, and $H^+ + CO_3^{2-} = HCO_3^-$, respectively, and a is the ion activity in solution.

The relationship between soil pH, Na^+-Ca^{2+} exchange and calcite – solution equilibrium for alkali soils can be expressed by the equation (8.7) (Gupta, Chhabra, and Abrol, 1981):

$$\text{Log ESR} = \text{pH} - 5.236 + 0.5\log Pco_2 + \log(Na^+) + 0.5\ I^{1/2} \quad (8.7)$$

where ESR is the exchangeable sodium ratio [(adsorbed Na) / (cation exchange capacity –adsorbed Na)], Pco_2 is the partial pressure of CO_2 in equilibrium with soil solution, and I is the ionic strength of the soil solution.

Origin of Sodic-Alkali Soils

Sodicity (at various pHs) and salinity development in arid and semiarid regions depends both on inherent factors (primary carbonate minerals, saline rocks, saline water, depth of groundwater) and on human activities (irrigation management and reclamation procedures). The main processes by which soluble salts enter the soil and groundwater include weathering of primary and secondary minerals in soils and application of waters containing salts. In coastal areas, saline water intrusion and submergence of the low-lying lands by seawater cause the salinization of groundwater and soils. The importance of each source depends on the type of soil, the climatic conditions, and the agricultural management.

Mineral Weathering

Salt formation takes place during the process of mineral weathering. Gunn and Richardson (1979) indicated that sufficient amounts of sodium and chloride were present in rocks of both terrestrial and marine origin to provide sources of salts and sodium found in many soils. Many soils from arid and semiarid regions and primary minerals such as olivine, hornblende, oligoclase, and others contribute substantial amounts of salts during weathering (Rhoades, Krueger, and Reed, 1968).

Predictions of soil mineral weathering and soil solution compositions can be made using stability diagrams (Kittrick, 1977; Lindsay, 1979). The solubility of these minerals can be calculated using the appropriate equilibrium constants. Limitations of this approach are (1) the availability of adequate data on the solubility or free energy of formation of various minerals as well as species present in soil solution, and (2) the paucity of information on the kinetics of mineral dissolution. Thermodynamic data are generally more accurate for the primary minerals involved in increasing the salinity of soil solution (Rhoades, Krueger, and Reed, 1968) than for the secondary minerals (Kittrick, 1977). The major ions in the dissolved mineral salts are Na^+, Ca^{2+}, Mg^{2+}, K^+, Cl^-, SO_4^{2-}, HCO_3^-, CO_3^{2-}, and NO_3^-.

Equation (8.1) implies that an increase in alkalinity must be accompanied by an equivalent increase in total cations or carbonate ions due to weathering of the mineral carbonates under the influence of CO_2:

$$CaCO_3 + H_2O = Ca^{2+} + 2HCO_3^- \tag{8.8}$$

$$\text{Na-clay} + H_2O + CO_2 = \text{H-clay} + Na^+ + HCO_3^- \tag{8.9}$$

Similarly, reduction of sulphates and nitrates accompanied by oxidation of organic matter also increases alkalinity. Alkalinity may also change as a result of precipitation of alkali earth carbonates and soil water evaporation or dilution of soil solution (Beek and Breeman, 1973).

Atmosphere

Atmospheric contribution to the salt and sodium load of arid and semiarid lands, which is from 10 to 25 percent of the total yearly contribution from weathering (Bresler, McNeal, and Carter, 1982), is often overlooked. It is an important source in highly weathered landscapes that have poor drainage. The composition of atmospheric salt deposition varies with distance from the salt source. On the coasts salt is predominantly NaCl. As demonstrated by a number of researchers and summarized by Isbell, Reeve, and Hutton (1983), annual deposition of chloride decreases markedly away from the oceans and is low at a distance greater than 200 km inland.

Another source for NaCl is aeolian transport of salts from eroded soils and inland areas, which can be very significant on a regional scale (Bowler, 1976). Beattie (1970) demonstrated that the mineralogy of some of these deposits suggested that considerable amounts of salts were transported with mineral grains. Although current aeolian deposition is considerably less than at peak periods in the past, Middleton (1984) and McTainsh, Lynch, and Burgess (1990) have demonstrated that dust movement and deposition may still be significant factors in arid and semiarid environments and it is logical to assume that soluble salts may be similarly affected.

Secondary Salinization and Sodification

Irrigation with poor-quality water, inadequacy of leaching, seepage from canals, together with the presence of a high water table and high evaporation rate, all account for the secondary salinization of irrigated soils.

The development of waterlogged soils is mainly associated with low-lying lands of poor soil physical conditions and internal drainage. These soils are mainly found in the flood and deltaic plains of rivers and valleys (such as the San Joaquin Valley, California, and Western Jezreel Valley, Israel). The accumulation of salts in soils depends on the salinity of the applied water, the salinity of the native soil, and the rate at which salts are leached out of the root zone. If a restricting layer exists close to the soil surface, waterlogging occurs and the accumulation of salt associated with waterlogged soils develops within a comparatively short time, such as decades. If no restricting layer exists and the vadose zone has a large capacity, irrigation may be practiced over a long time, such as centuries, before problems with surface drainage arise.

Another source of saline water is drainage effluent from irrigated areas. Drainage water is used in many countries for irrigation. The salinity level varies, but often the electrolyte concentrations are relatively high. Under arid or semiarid conditions and in regions of poor natural drainage there is a real hazard of salts accumulation in soils.

Secondary alkalinization takes place mainly under the following conditions: (1) accumulation of salts from low-quality irrigation water due to poor drainage and water evaporation, and (2) rising groundwater that transports salts from the deeper soil layers to the soil surface, and deposits due to water evaporation. In both cases, the SAR of the soil solution increases by the factor of $d^{1/2}$ (where d is the concentration factor). Since the exchangeable sodium percentage (ESP) of the soil is related to the SAR of the soil solution, therefore, the final ESP will be greater than the ESP expected from the SAR of the irrigation water.

ION EXCHANGE

The general expression for a thermodynamic equilibrium constant of a given process in an open system at constant temperature and pressure is derived from the equations for chemical equilibrium and chemical potential (Prigogine and Defay, 1954)

$$\Sigma_i v_i \mu_i = 0 \tag{8.10}$$

$$\mu_i = \mu_i^0 + RT \ln a_i \tag{8.11}$$

where v_i is the stoichiometric coefficient of component i, μ_i and μ_i^0 are the chemical potential and standard chemical potential, respectively, R is the gas constant, T is the absolute temperature, and a is the activity, which is a product of activity coefficient and molar concentration of the ion.

Combining equations (8.10) and (8.11) yields

$$\Sigma_i v_i \mu_i^0 + \Sigma_i v_i RT \ln a_i = 0 \tag{8.12}$$

$$\Sigma_i v_i \mu_i^0 = \Sigma_i v_i RT \ln a_i = RT \ln K_e \tag{8.13}$$

where K_e is the thermodynamic equilibrium constant.

The equilibrium constant can be calculated either from known values of standard chemical potentials of formation, or from values of the activities of the components. For ion exchangers, the standard chemical potentials of formation are unknown, so that the equilibrium constant can be calculated only from the values of the activities of the components.

For the exchange reaction

$$2NaX + Ca^{2+} = CaX_2 + 2Na^+ \tag{8.14}$$

the thermodynamic exchange equilibrium constant is

$$K_e = \frac{(a_{Na+})^2 a_{CaX_2}}{(a_{NaX})^2 a_{Ca^{2+}}} \tag{8.15}$$

where X is one equivalent of the negative electrostatic charge of the clay. In equation (8.15), only the activity of the cations in the solution phase is easily computed in terms of molar ion concentration, m, and molar activity coefficient, γ.

Gapon (1933) reduced equation (8.14) to

$$XCa_{1/2} + Na^+ = XNa + 1/2Ca^{2+} \tag{8.16}$$

And the equilibrium constant, K_G (mmol $l^{-1})^{-1/2}$ is

$$K_G = \frac{X_{Na} a_{Ca^{2+}}^{1/2}}{X_{Ca_{1/2}} a_{Na^+}} \tag{8.17}$$

In this example, Na^+ and Ca^{2+} exchange has been shown as these are the dominant cations in irrigation water.

The activity of adsorbed cations cannot be measured or calculated precisely. Thus, it assumes that the activity coefficient for the adsorbed cation is unity. Nevertheless, the equation often holds well over narrow concentration ranges.

Rearranging equation (8.17) yields:

$$\frac{XNa}{XCa_{1/2}} = K_G \frac{a_{Na+}}{a^{1/2}_{Ca^{2+}}} \tag{8.18}$$

The second term on the right side of equation (8.18) is defined as SAR ($0 \leq \text{SAR} \leq \infty$), and the activity is given in mmol l^{-1}. Usually, Ca^{2+} and Mg^{2+} concentrations are lumped together in the SAR expression (SAR = (Na^+) / (Ca^{2+} + Mg^{2+}) ($mmol_c\ l^{-1}$) $^{-1/2}$). There is no theoretical basis for combining the concentration of these two cations, except for the fact that ion valency is a considerably more sensitive parameter for predicting ion exchange relations than that of ion size. Garrels and Christ (1965) have shown, however, the Debye-Huckel distance of closest approach for magnesium to be approximately one-third larger than for calcium. In general, water and soil solution contain two to five times more Ca^{2+} than Mg^{2+}.

Sposito and Mattigod (1977) distinguished between "true" SAR (SAR_t, which includes free ion activities) and "practical" SAR (SAR_p), uncorrected for ion pairs or complexes, or for activity coefficients, (equation 8.19). The "practical" SAR is generally less than the "true" SAR because of the greater effect of ion pairs, complexes, and activity coefficient corrections on Ca^{2+} and Mg^{2+} than on Na^+. The regression equation between the two parameters is

$$(\text{SAR})_t = 0.08 + 1.115\ (\text{SAR})_p \tag{8.19}$$

Rearranging equation (8.18) and introducing the term exchangeable sodium percentage (ESP)

$$\text{ESP} = \frac{100 * X_{Na}}{X_{Na} + X_{Ca_{1/2}}} \qquad 0 \leq \text{ESP} \leq 100 \tag{8.20}$$

yield:

$$\text{ESP} = \frac{100K_G * \text{SAR}}{1 + K_G * \text{SAR}} \tag{8.21}$$

When K_G * SAR >>1, then 1 can be neglected and ESP is approaching 100. When K_G * SAR <<1, then K_G * SAR can be neglected in the denominator and ESP is approaching 100 K_G * SAR.

Therefore, the Na^+ concentration in irrigation water alone is not sufficient for estimating the potential sodicity hazards. From equation (8.21) it appears that the effective diagnostic criteria for appraising the sodicity status are the SAR of the irrigation water and the adsorption coefficient, K_G, for Na-Ca exchange in a given soil (assuming that Ca and Na are the dominant cations in soil). For example, the relationship between ESP and SAR of a soil at equilibrium at various reaction coefficients is given in Figure 8.1. These calculations indicate that K_G for Na-Ca exchange is an important criterion for evaluating water quality for irrigation for a given soil. It is important to note that for soils having $K_G > 0.01$ (mmol$_c$ l^{-1})$^{-1/2}$, there is only one SAR value that is equal to ESP. Positive and negative deviation is observed when SAR is below or above this value, respectively. However, for many montmorillonitic soils, having a K_G value around 0.015 (mmol l^{-1})$^{-1/2}$, the deviation of ESP from SAR of the soil solution is small up to SAR 30.

The ESP of a soil can be estimated from the SAR of the water (at equilibrium) also from the following equation (Richards, 1954):

$$\text{ESP} = \frac{1.475 * \text{SAR}}{1 + 0.0147 * \text{SAR}} \tag{8.22}$$

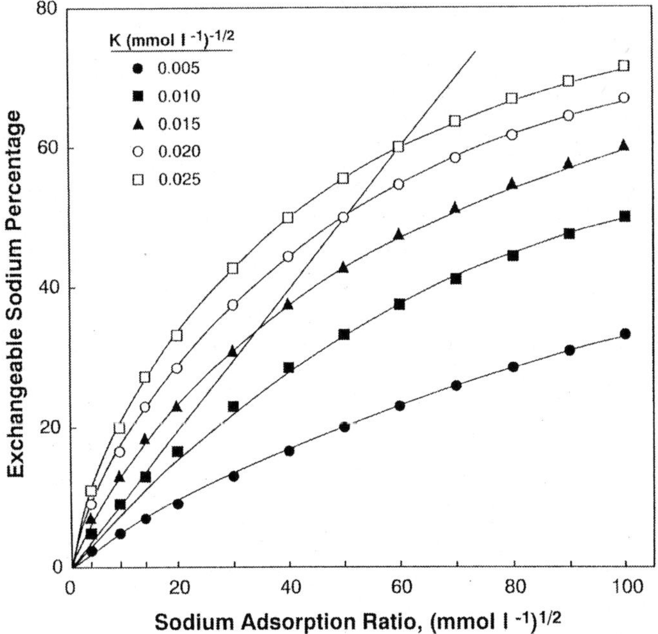

FIGURE 8.1. Relationship between exchangeable sodium percentage (ESP) and sodium adsorption ratio (SAR) for various adsorption coefficients.

Since the SAR is usually determined on an irrigation water or saturation extract of a soil (by adding distilled water), there is an inherent problem of relating these values to the condition that exists in the field. By neglecting salt precipitation or dissolution, mineral weathering, and salts uptake by plants, salts concentration in soil solution increases due to water uptake by roots and water evaporation from bare soil without changing the relative concentration among the ion species in soil solution. The SAR value, however, increases proportionally to the square root of the increase of the total concentration. Thus, if the total electrolyte concentration in solution increases by the factor of d, the SAR will increase by the factor of $(d)^{1/2}$. Since the soil ESP increases with the increase in SAR, it is obvious that the resultant ESP would be greater than the ESP expected from the SAR of the irrigation water. This trend has been observed in lysimeters and under field conditions (Rhoades, 1968).

CLAY SWELLING AND DISPERSION

Smectites are the dominant clay mineral in many semiarid and arid regions and they determine much of the physical properties of soils, such as swelling and dispersion, due to the large specific surface area and the electrostatic charge. These two processes determine, to a large extent, soil structure, hydraulic properties, and soil water retention.

The mechanism of swelling is analogous to the osmotic characteristics of a semipermeable membrane. The ions between the clay platelets in the exchange phase are constrained there by the electrical attraction of the negatively charged clay surfaces. Thus, under a nonequilibrium condition, the gradient for water movement is from the bulk solution to the inner platelet region since the concentration of ions is greater there. The movement of the water into this region creates a hydrostatic pressure that forces the clay to expand and swell. A Na soil with a larger number of ions in the inner platelet region coupled with their weak interaction with the clay is therefore more susceptible to swelling and dispersion than a Ca soil.

Standard Clays

Na and Ca Homoionic Clay Systems

The Van der Waals attraction energy, V_A, between two semi-infinite plates such as 2:1 clay mineral, is given by

$$V_A = -\frac{A}{12\pi}\left[\frac{1}{h^2} + \frac{1}{(h+2t)^2} - \frac{1}{(h+t)^2}\right] \tag{8.23}$$

where A is the Hamaker constant (2.2×10^{-20} Joule, Israelachvili and Adams, 1978), h is the distance between the surfaces of the plates, and t is the thickness of the clay plate. The Van der Waals forces are independent of the charge density on the clay and of the composition and concentration (<1 mol dm^{-3}) of the electrolytes in suspension in the distance range 1 to 15 nm between two adjacent clay plates (Israelachvili and Adams, 1978).

The repulsion energy $V_{R(d)}$ between two clay platelets can be computed from the diffuse double-layer theory (Lyklema, 1982), using equation (8.24)

$$V_{R(d)} = \frac{64nkT}{\chi}\left[\tanh\left(\frac{Ze\psi_s}{4kT}\right)\right]^2 e^{-2\chi d} \tag{8.24}$$

where n is the number of cations per unit volume, k is the Bolzmann constant, T is the temperature, Z is the cation valency, ψ_S, is the Stern layer potential, d is the distance from clay surface to the midway plane between two parallel platelets, e is the electron charge, and χ is the reciprocal Debye length, given by

$$\chi = \left(\frac{2e^2nZ^2}{DkT}\right)^{1/2} \tag{8.25}$$

where D is the dielectric constant of the medium and the other symbols were defined previously.

The repulsion energy depends on charge density of the plate and composition and concentration of the electrolyte in suspension.

Montmorillonite platelets in aqueous suspension may flocculate in three possible modes of particle association (Van Olphen, 1977): (1) association between siloxane planes of two parallel platelets (FF), (2) association between edge surfaces of neighboring particles (EE), and (3) association between an edge surface and a siloxane planar surface (EF). In the presence of electrolyte (e.g., NaCl) at low concentration, the double layers on the Na-montmorillonite particles are well-developed and osmotic repulsion prevents particle association. Thus, a stable suspension of individual platelets is obtained (Van Olphen, 1977). As the concentration of NaCl in the suspension increases, double layers at both the planar and the edge surfaces are compressed, and at the critical flocculation concentration (CFC), both EF and EE association can occur (Heller and Keren, 2001). As the electrolyte concentration increases further, the FF mode of particle association occurs, and oriented aggregates are formed (Keren, Shainberg, and Klein, 1988).

The CFC values for Na-montmorillonite range from 7 to 24 me l^{-1} NaCI (Arora and Coleman, 1979; Oster, Shainberg, and Wood, 1980; Keren, Shainberg, and Klein, 1988; Goldberg and Foster, 1990). The CFC of Na-montmorillonite at pH 9.8, for example (a value that is well above the point of zero charge [PZC] of the edge surfaces of smectites) is much higher (44 me l^{-1}) than that at pH 7 (Keren, Shainberg, and Klein, 1988). The presence of humic substances also causes a marked increase in CFC over the pH range of 4 - 8 (Tarchitzky, Chen, and Banin, 1993). The CFC values for illite range from 55 me l^{-1} NaCl (Oster, Shainberg, and Wood, 1980) to 90 me l^{-1} NaCl (Goldberg and Foster, 1990), depending on pH. Ca-saturated montmorillonite does not expand beyond 2 nm whereas Na-saturated clays expand to complete separation of the platelets in water (Norrish, 1954). Swelling beyond 4 nm leads to dispersion of the clay. Aggregates of Na-montmorillonite swell to a homogeneous gel when they are immersed in solution, but aggregates of K- and NH_4-montmorillonite contain many small, nonswollen sections, suggesting that much of the swelling takes place between nonswelling aggregates (Rowell, 1963). Ca-saturated aggregates take up equal amounts of solution in electrolyte solution and in distilled water, indicating that a stable, nonswelling system exists (Norrish, 1954; Rowell, 1963).

Sodium-saturated kaolinite remains flocculated in a salt-free solution at pH <7 because the positively electrostatic charged edge surfaces interact with the negatively charged planar surfaces (Schofield and Samson, 1954). Under alkaline conditions, however, dispersion takes place.

Mixed Na/Ca Clay Systems

Clay swelling is affected only slightly by an increase in ESP at the ESP range <15. Shainberg, Bresler, and Klausner (1971) demonstrated that clay swelling is limited at ESP <15 because of a demixing phenomenon in which Ca ions concentrate on the internal surfaces of the tactoid while the Na ions are adsorbed on external surfaces (Mering and Glaeser, 1954). It seems that the energy of hydration of the adsorbed ions might be the driving force for this phenomenon (Mering and Glaeser, 1954). Montmorillonite with an ESP >50 swells almost as much as pure Na clay. Smectite clays containing mainly adsorbed Na^+ swell in water because the forces of attraction between the clay platelets are weak and the platelets easily expand as water enters the interplanar region. In contrast, Ca^{2+} promotes binding between clay platelets to form larger particles consisting of several platelets (tactoid).

It is of interest to inquire why the soil matrix responds in an adverse way to such a relatively small amount of exchangeable Na. An adsorbed Na percentage of 15 or less (and corresponding SAR) does not represent a very large fraction of the total exchange capacity that is occupied by Na, especially when we have observed that Ca and Mg dominate because of their divalent nature. The explanation of this phenomenon appears related to the manner in which Na and Ca are distributed between the inner and outer surfaces of the clay. As the concentration of Ca is increased it tends to accumulate on inner surfaces in preference to Na which then accumulates on the outer surfaces. This "demixing" phenomenon was observed by Mering and Glaeser (1954). Thus, Na percentage on the outer surfaces will in fact be much greater than 15 percent and the methods for measuring adsorbed Na will not reveal this fact. Higher Na on outer surfaces will more drastically influence swelling and dispersion processes previously discussed since the outer surfaces will respond first to the processes.

Soil Clays

The general effect of sodium on soil is to weaken the structure that soils possess. Thus, soils with high smectite content may swell considerably in the presence of high adsorbed sodium and low salt concentrations.

Flocculation and dispersion behavior of soil clays differs significantly from that of pure clay systems, because they usually occur as mixtures and complexes with other minerals, oxides, and soil organic matter. Goldberg and Foster (1990) and Frenkel, Fey, and Levy (1992) demonstrated that the CFC values of soil clays are two- to tenfold higher than for pure clay systems. Conversely, in the presence of hydrous oxides (McNeal et al., 1968) or sparingly soluble minerals such as $CaCO_3$ (Shainberg, Rhoades, and Prather, 1981; Shainberg et al., 1981), clay dispersivity is much less severe. Hence, extrapolation from pure clay to soil systems is extremely difficult. Goldberg and Foster (1990) demonstrated that the smallest differences in CFC were found between reference and soil illites, indicating that in many dispersive soils which contain illite or weathered mica (Rengasamy et al., 1984; Miller and Baharuddin, 1986; Miller, Frenkel, and Newman, 1990), illite plays an important role in their dispersibility.

The experimental results of McNeal and Coleman (1966) on soil clays compare favorably with the results with montmorillonite clays. They also found that replacing about 15 percent of adsorbed calcium with sodium had little effect on the swelling of the clay. With a further increase in the ESP of the soil, a very sharp increase in macroscopic swelling was found for two montmorillonitic soils.

Data of McNeal, Norvell, and Coleman (1966) demonstrate the effect of ESP and total electrolyte concentration on the swelling of Gila clay soil. The ameliorating effect of electrolytes on the swelling in the presence of increasing ESP was observed. At high ESP, the swelling was suppressed because the interparticle interaction was reduced. However, the swelling clearly demonstrates the influence of the ESP. The relative hydraulic conductivity, which is the ratio of the hydraulic conductivity of Na-affected soil as compared to the same soil unaffected by Na, decreases as the measured swelling increases.

The ameliorating effect of electrolytes on swelling has considerable practical implications. Even at high ESP values, hydraulic conductivity can be maintained if the salt concentration is adequate to suppress particle interaction. Thus, sodic soils are usually initially reclaimed under high saline conditions to maintain water movement to the drainage system.

Generally, a soil containing nonexpanding clays such as illite (mica), kaolinite, and sesquioxides is less sensitive to the ESP value. In these soils, a total clay analysis could be misleading for interpreting the sodic hazard unless a mineralogical identification is also included in the analysis.

Swelling values for montmorillonitic soils were evaluated using a simplified domain model for characterizing the exchangeable cation distribution on Na-Ca montmorillonites. Interlayer swelling can be calculated from the expression (McNeal, 1968):

$$\chi = (f_{mont})(3.6\times10^{-4})(\text{ESP*})(d^*) \tag{8.26}$$

where χ is the calculated interlayer swelling of montmorillonitic soil, f_{mont} is the weight fraction of montmorillonite in the soil, ESP* is the adjusted ESP and d^* is the adjusted interlayer spacing.

ALKALI SOIL RECLAMATION

The reclamation of alkali soils generally proceeds by increasing Ca ions on the adsorbed phase at the expense of Na and Mg ions. The replaced Na and Mg are removed either below the root zone or out of the soil profile by leaching water. Thus, reclamation requires a certain flow of water through the soil profile and, to be effective, an appropriate profile hydraulic conductivity (HC) must be achieved. The source of Ca for replacing adsorbed Na may be the soil itself, involving the dissolution of Ca-containing minerals, or external sources such as gypsum, or irrigation water containing Ca ions.

Although chemical amendments may be effective in restoring reduced water penetration caused by either excessive sodium or salt concentration in the applied water below a threshold value, they will not be effective if low permeability is caused by soil texture, compaction, or water-restricting layers. Since reclamation requires a certain flow of water through the soil profile, an appropriate infiltration rate (IR) and HC must be achieved.

Soil Amendments

Common examples of amendments that produce Ca in calcareous soils by enhancing the conversion of $CaCO_3$ to the more soluble $CaSO_4$ include sulphuric acid, sulphur and iron, and aluminum sulfates. Because of their relatively low cost and proven effectiveness, gypsum, sulphur, and sulphuric acid are the most common amendments for reclamation.

Gypsum

Gypsum added to an alkali soil can cause permeability changes by increasing electrolyte concentration and by cation exchange effects (Loveday, 1976; Keren and Shainberg, 1981; Keren et al., 1983). In evaluating sodic soil reclamation, equilibrium chemistry and piston movement of soil solution is usually assumed (Oster and Frenkel, 1980). The soil solution in the layer with gypsum, however, reaches equilibrium (in the presence of excess gypsum) only when the contact time between an elemental volume of solution and gypsum fragment surface is sufficiently long, or when the surface area of the gypsum fragments per volume of water is sufficiently large (Keren and O'Connor, 1982). During irrigation or rainfall, this contact time can be short or the surface area of the gypsum fragments per unit volume of water can be small (low gypsum content in soil or large gypsum fragment size); thus the percolated solution can be undersaturated with respect to gypsum. Thus, the benefit of gypsum for reclaiming sodic soils depends not only on the infiltration characteristics of the soil but also on the gypsum dissolution properties.

Some of the factors that influence the rate of gypsum dissolution are the surface area of the gypsum fragments (Figure 8.2), the soil water velocity during leaching, and the electrolyte composition of the soil solution (Kemper, Olsen, and DeMooy, 1975; Keren and O'Connor, 1982).

The rate of gypsum dissolution is written as:

$$\frac{dC}{dt} = k(Cs - C) \qquad (8.27)$$

where dC/dt is the net rate of dissolution, k is the dissolution coefficient, and C_s and C are the solution concentration at saturation and at time t, respectively.

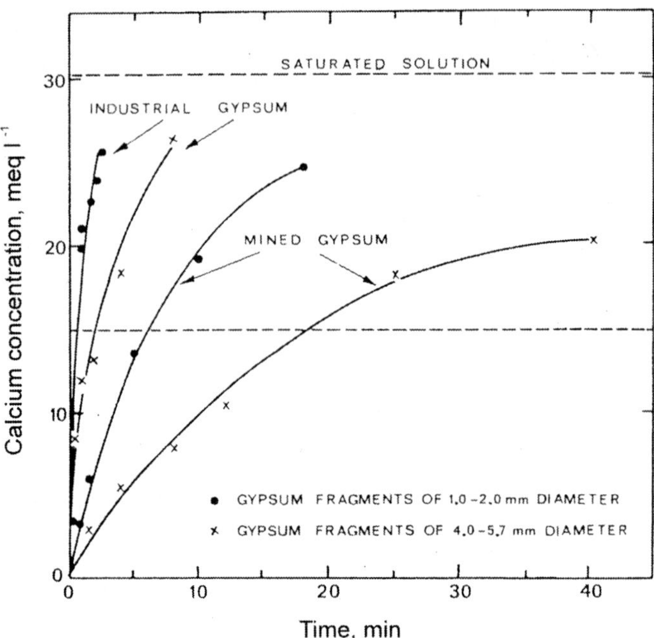

FIGURE 8.2. Calcium concentration in water versus time plots for industrial and mined gypsum samples of two fragment sizes. (*Source:* Keren and Shainberg, 1981, p. 105.)

In accordance with Kemper, Olsen, and DeMooy (1975), Keren and O'Connor (1982) concluded that the rate of gypsum dissolution is strongly influenced by both gypsum content and soil water velocity. Increasing solution velocity increased the dissolution rate coefficient but decreased the contact time between gypsum and the flowing solution; the net effect was a decreasing dissolution rate with increasing soil water velocity.

Integrating equation (8.27) yields

$$\ln\left(1-\frac{C}{C_s}\right)=-kt \tag{8.28}$$

and a plot of ln (1 - C/C_S) versus t should yield a straight line.

The time, t, was calculated from equation (8.28). Since the thickness of the stationery film of solution around the gypsum fragments changes with the soil water velocity, the dissolution rate coefficient for a given surface of gypsum fragment also changes. Thus, under such conditions, the lines obtained from equation (8.28) are not linear. Keren and O'Connor (1982) reported that the left-hand side of equation (8.28) is empirically related linearly to the square root of time (t_c),

$$-\ln\left(1-\frac{C}{C_s}\right)=\alpha t_c^{1/2}+\beta \quad (8.29)$$

where α and β are the slope and intercept, respectively, and t_c is the time for an increment of solution to leave a given soil depth:

$$t_c=\frac{L}{V} \quad (8.30)$$

where L is the length of the layer of soil-gypsum mixture and V is soil water velocity

$$V=\frac{v}{p}$$

where v is the soil water flux and p is the soil porosity.

Combining equations (8.29) and (8.28) gives

$$k=\frac{1}{t_c}\left(\alpha t_C^{1/2}+\beta\right) \quad (8.31)$$

Introducing equation (8.30) in (8.31) yields:

$$k=\left(\frac{V}{L}\right)^{1/2}\left[\alpha+\left(\frac{V}{L}\right)^{1/2}\beta\right] \quad (8.32)$$

The dissolution rate coefficient as a function of water flow velocity can be evaluated for a given surface area of gypsum. Since the permeability of an alkali soil is low, the dissolution rate is relatively high during the initial stage of reclamation. During the reclamation process, however, the permeability increases and the dissolution rate decreases. Thus, the soil regulates gypsum dissolution rate during reclamation. When it is possible to control water flow velocity in soils (e.g., sprinkler irrigation), reduced velocities are to be preferred for increased rate of gypsum dissolution and greater efficiency of cation exchange reaction due to increase of contact time between water and gypsum fragments (Keren and O'Connor, 1982).

The surface area of gypsum fragments has also been invoked to account for differences in infiltration rates of soils resulting from mining and by-products of the phosphorous industry (Keren and Shainberg, 1981). The latter, with a bulk density half of the former, dissolved ten times as fast (Table 8.2). Another factor affecting dissolution rate is the coating of $CaCO_3$ on gypsum fragments in calcareous soils (Keren and Kauschansky, 1981).

Tanji and Deveral (1984) describe representative modeling approaches for the reclamation of sodic soils. The models include many of the known chemical reactions occurring in natural soils as a result of $CaCO_3$ and $CaSO_4$ dissolution, salt leaching, and ion exchange reactions. These models were subdivided into chromatographic and miscible displacement models. The models of Dutt and Terkeltoub (1972) and Tanji et al. (1972) are examples of chromatographic models. Both models assume that local equilibrium occurs between the plates for chemical processes within the solution phase, between the solution and exchanger phases, and between the solution and soil mineral phases. The models compute monovalent-divalent exchange using the Gapon and/or Davis equations. Solubilities of gypsum and $CaCO_3$ are considered in the models. The models of Dutt and Terkeltoub (1972) and Tanji et al. (1972) have been field tested with data from sodic soil reclamation leaching plots. The differences between measured and computed results were found to be no more than the horizontal variations typically found in salt-affected lands. Both models provide powerful tools for quantitative predictions of water and gypsum required to reclaim soil profiles to prescribed levels of salinity and ESP. However, assumptions must be made about the desired level of reclamation: to what level must ESP be reduced, and to what depth of soil is the ESP reduction required? The soil surface, being very susceptible to sodicity, may tolerate only very low ESP (< 3), whereas at depths below the cultivated layer, ESP >10 may be tolerable.

To obtain permanent improvement, however, the cation exchange effect is the important one. The amount of exchangeable sodium to be replaced during

TABLE 8.2. The Dissolution Coefficients of Industrial Gypsum, $K_{(PG)}$ and Mined Gypsum, $K_{(MG)}$, with Different Fragment Sizes

	Fragment size (mm)	
	1.0-2.0	**4.0-5.7**
$K_{(PG)} x 10^4$, s^{-1}	198	58
$K_{(MG)} x 10^4$, s^{-1}	19	6
$K_{(PG)}/K_{(MG)}$	10.4	9.7

Source: Keren and Shainberg, 1981, p. 105.

reclamation depends on the initial ESP (ESP_i), the cation exchange capacity (CEC) ($mol_c\ Mg^{-1}$), soil bulk density (ρb, $Mg\ m^{-3}$), the desired final exchangeable sodium fraction (ESP_f), and the depth of soil to be reclaimed (L, m). Once these parameters have been determined, the amount of exchangeable Na to be replaced per unit area of land (Q_{Na}, $mol_c\ ha^{-1}$) can be calculated from

$$Q_{Na} = 10^4 L\rho_b\,(CEC)(ESP_i - ESP_f) \tag{8.33}$$

The ESP_f value depends on the response of the soil in terms of its physical conditions. The amount of gypsum needed to reclaim a sodic soil (GR, gypsum requirement) (Richards, 1954) in metric tons ha^{-1}, can be calculated from

$$GR = 8.61\times10^{-5}\,Q_{Na} \tag{8.34}$$

The efficiency and rate of exchange, namely, the percentage of applied Ca that exchanges for adsorbed Na, varies with ESP, being much greater at high ESP values (Chaudhry and Warkentin, 1968). Removal of Na at ESP levels below 10 is slow, and part of the applied Ca displaces exchangeable Mg so that the efficiency declines to about 30 percent (Loveday, 1976; Greene and Ford, 1985). Efficiency may also be low (20-40 percent) in fine-textured soils, because of the slowness of exchange of Na inside the structural elements (Manin, Pissarra, and van Hoorn, 1982).

Acids and Sulphur

These amendments are of value for ameliorating soils containing $CaCO_3$ with which the acid interacts to form gypsum (using H_2SO_4) or calcium chloride (using HCl). Sulphur requires an initial phase of microbiological oxidation to produce H_2SO_4.

In some areas, the availability of acid waste products arising from mining and industrial activities is increasing markedly. Use of these waste products as soil amendments may provide a safe means of disposal (Miyamoto, Ryan, and Stroehlein, 1975), and could greatly influence the economics of any amendment proposal.

Being highly corrosive, neither H_2SO_4 nor SO_2 should be added to water that flows through metal or concrete irrigation systems. Miyamoto, Prather, and Stroelhein (1975) report that the concentrated acid, sprinkled directly onto the soil surface, has advantages in better distribution, less destruction of soil aggregates, and more efficient leaching of salts. For elemental sulphur (S), the dust problem has been a serious disadvantage, but this can be overcome by using conventional fluid fertilizer equipment to apply water suspensions containing 55 to 60 percent S (Thorup, 1972).

Other amendments such as sulphur, pyrites, and polysulfides, which must first be oxidized by soil microorganisms, are slower acting than H_2SO_4. Their effectiveness in field experiments has been variable, perhaps being related to the presence or absence of appropriate microbial populations.

Calcium Carbonate

Successful leaching reclamation without amendments has been demonstrated (Jury, Jarrell, and Devitt, 1979). Reclamation under such conditions takes place when the drainage through the soil profile is good, an adequate leaching water is present, and a soil source of Ca exists (e.g., calcium carbonate). The rate at which $CaCO_3$ dissolution in water approaches equilibrium is dependent upon a number of factors which include the surface area-solution volume ratio, the ionic composition of the solution, the ion composition of the adsorbed phase, affinity of the clay minerals to cations, the temperature, and the local partial pressure of CO_2. Amrhein, Jurinak, and Moore (1985) concluded that in soils the kinetics of $CaCO_3$ dissolution is not a simple diffusion controlled or first-order reaction. In these calcareous systems, the transfer of atmospheric CO_2 to solution was a limiting step in the kinetics of dissolution.

$CaCO_3$ particles, isolated from the soils, in water, give ion activity product (IAP) values which are expected for calcite (Suarez and Rhoades, 1982). On the contrary, a supersaturation in respect to calcite was observed in solution extracts from calcareous soils (Levy, 1981). The supersaturation concentration in respect to calcite in soil appears to be due to the presence of silicates in the soil, more soluble than calcite, and is not the result of unstable $CaCO_3$ phases (Suarez and Rhoades, 1982). Plummer, Parkhurst, and Wigley (1979) concluded that there are uncertainties in modeling the kinetics of carbonate chemistry: at low pH the rate depends significantly on the thermodynamic transport constant for H^+ which is not well defined, and reaction site density and controls on pH at the surface interface between the crystal and the bulk solution are also not well understood. Moreover, kinetics of precipitation and dissolution of $CaCO_3$ have not yet been modeled for soil-water systems.

Soil $CaCO_3$ may be dissolved slowly to contribute Ca, especially in the reclamation of saline sodic soils in which its solubility is enhanced (Oster, 1982). However, it has generally been considered of doubtful value to add $CaCO_3$ to nonsaline sodic soils because its dissolution rate is too slow to provide much Ca for exchange, unless an acid or acid former is applied concurrently. Sodic soils containing $CaCO_3$ are common in the arid and semiarid regions of the world. Calcareous soils with moderate ESP levels maintain reasonable physical properties through most of the profile but remain susceptible to dispersion near the surface. This is because the soil solution electrolyte concentration near soil surface may be insufficient to maintain physical structure during raindrop impact. Under such conditions an application of soil amendment on soil surface is necessary to keep infiltration rate sufficiently high.

Successive Dilutions of High-Salt Water

The benefit of an amendment for reclaiming saline-sodic soils depends primarily on the infiltration characteristics of the soil. The use of the successive dilutions of the saline water method may be applied when (1) the soil physical conditions have deteriorated and HC of the soil is so low that the time required for reclamation or the amount of amendment required is excessive, or (2) if a sodic soil is to be leached with a water so low in salinity that water infiltration decreases adversely. The method involves application of successive dilutions of a high-salt water containing divalent cations (Reeve and Bower, 1960). This principle was later successfully tested in experiments, and a monogram was furnished for predicting the reclamation at various stages of leaching (Reeve and Doering, 1966a). In the early phase of reclamation, the high salinity of the water prevents clay dispersion and induces flocculation of the soil colloids. Simultaneously, the Ca ions in the percolating water decrease the sodicity level by exchange reaction. On dilution with high-quality water, the SAR of the irrigation water is reduced by the square root of the dilution factor. To ensure exchange equilibrium to the desired depth in each step of dilution, the total equivalent surface depth of applied water for reclamation should be about nine times the depth of soil to be reclaimed (Reeve and Doering, 1966a,b). This method is particularly effective for soils with the expanding clay type that have extremely low HC. A theoretical analysis for reclaiming sodic soils by the addition of a constant quantity of Ca^{2+} in every step in the successive dilution was reported (Misopolinos, 1985).

Intermittent ponding required less water than continuous ponding to achieve the same degree of leaching (Miller, Biggar, and Nielsen, 1965) and the sprinkling irrigation system has been more efficient than other methods at removing salt from small pores in the soil profile (Nielsen, Biggar, and Luthin, 1966).

The successful application of the successive dilutions method requires a balance between keeping the electrolyte concentration high to reduce reclamation time and keeping the electrolyte concentration low to reduce the amount of amendment required. A practical technique is to apply only two-thirds of the solution depth required for exchange equilibrium with three dilutions of $CaCl_2$, for example, starting with a solution of 250 mol m^{-3} $CaCl_2$

and following with 100 and 50 mol m^{-3}. This is followed by a gypsum application to satisfy the remaining exchange requirement, and completed by leaching with about a 0.3 m depth of water per m depth of soil to be reclaimed. This final step is essential to leach the saline solutions below the root zone.

SUMMARY

Alkaline soils have poor physical properties due to high sodicity and high pH. Under these conditions, slow water infiltration into the soil, soil erosion, slow internal drainage, poor aeration, compaction, and waterlogging on the lower lands commonly occur because of soil swelling and dispersion. The extent of swelling and dispersion of clays depends on the clay mineralogy, the composition of the adsorbed ions, and the salt concentration in solution.

Although salt transport models have provided considerable insight into the chemistry and physics of soil reclamation, suitable mathematical models are not available to describe, on a field scale, the effects of variable rates of water flow in soil among soil pores of differing sizes on the kinetics of reactions involving cation exchange, mineral (e.g., gypsum and calcium carbonate) dissolution, and precipitation and CO_2 dynamics in soil. Therefore, although many of the basic processes are understood, sodic soil reclamation guidelines are mainly based on experimental relations obtained from columns, lysimeters, and field experiments.

Considerable progress has been made in managing and controlling salinity and alkalinity in irrigated lands. However, coping with salinity and alkalinity problems has become much more complex because of effects on soil physical properties and because of increasing environmental constraints. Because of natural hydrological and geochemical factors, as well as irrigation-induced activities, soil salinity, alkalinity, and associated drainage and soil erosion continue to plague agriculture. Control of salinity and alkalinity is therefore essential to establishing sustainable and successful agriculture.

REFERENCES

Amrhein, C., J.J. Jurinak, and W.M. Moore (1985). Kinetics of calcite dissolution as affected by carbon dioxide partial pressure. *Soil Science Society of America Journal 49:*1393-1398.

Arora, H.W. and N.T. Coleman (1979). The influence of electrolytes concentration on flocculation of clay suspensions. *Soil Science 127:*134-138.

Beattie, J.A. (1970). Peculiar features of soils developed in parna deposits in the Eastern Riverina. *Australian Journal of Soil Research 8:*145-156.

Beek, G.G.E.M. and N. Breeman (1973). The alkalinity of alkali soils. *Journal of Soil Science 24:*129-136.

Bowler, J.M. (1976). Aridity in Australia: Age, origins and expressions in aeolian landforms and sediments. *Earth Science Reviews 12:*279-310.

Bresler, E., B.L. McNeal, and D.L. Carter, Eds. (1982). *Saline and Sodic Soils: Principles-Dynamics-Modeling*. Berlin, Germany: Springer-Verlag.

Chaudhry, G.H. and B.P. Warkentin (1968). Studies on exchange of sodium from soils by leaching with calcium sulfate. *Soil Science 105:*190-197.

Dutt, C.R. and R.W. Terkeltoub (1972). Prediction of gypsum and leaching requirements for sodium-affected soil. *Soil Science 114:*93-103.

Eaton, F.M. (1950). Significance of carbonates in irrigation waters. *Soil Science 69:*123-133.

Frenkel, H., M.V. Fey, and G.J. Levy (1992). Organic and inorganic anion effects on reference and soil clay critical flocculation concentration. *Soil Science Society of America Journal 56:*1762-1766.

Gapon, E.N. (1933). Theory of exchange adsorption in soils. *Journal of General Chemistry* (Moscow) *3:*144-152.

Garrels, R.M. and C.L. Christ (1965). *Solutions, Minerals, and Equilibria.* New York: Harper and Row.

Goldberg, S. and H.S. Foster (1990). Flocculation of reference clays and arid-zone soil clays. *Soil Science Society of America Journal 54:*714-718.

Greene, R.S.B. and G.W. Ford (1985). The effect of gypsum on cation exchange in two red duplex soils. *Australian Journal of Soil Research 23:*61-74.

Gunn, R.H. and D.P. Richardson (1979). The nature and possible origins of soluble salts in deeply weathered landscapes of Eastern Australia. *Australian Journal of Soil Research 17:*197-215.

Gupta, R.K., R. Chhabra, and I.P. Abrol (1981). The relationship between pH and exchangeable sodium in a sodic soil. *Soil Science 131:*215-219.

Heller, H. and R. Keren (2001). Rheology of Na-montmorillonite suspension as affected by electrolyte concentration and shear rate. *Clays and Clay Minerals 49:*286-291.

Isbell, R.F., R. Reeve, and J.T. Hutton (1983). Salt and sodicity. In *Soils: An Australian Viewpoint,* eds. Commonwealth Scientific and Industrial Research Organization Publications. Melbourne, Australia: The Academic Press (London), pp. 107-117.

Israelachvili, J.N. and G.E. Adams (1978). Measurements of forces between two mica surfaces in aqueous electrolyte solutions in the range 0-100 nm. *Journal of Chemical Society Faraday Transactions 74:*975-1001.

Jury, W.A., W.M. Jarrell, and D. Devitt (1979). Reclamation of saline-sodic soils by leaching. *Soil Science Society of America Journal 43:*1100-1106.

Kelley, W.P. (1948). *Cation Exchange in Soils.* Baltimore, MD: The Waverly Press.

Kemper, W.D., J. Olsen, and C.J. DeMooy (1975). Dissolution rate of gypsum in flowing water. *Soil Science Society of America Journal 39:*458-463.

Keren, R. and P. Kauschansky (1981). Coating of calcium carbonate on gypsum particle surfaces. *Soil Science Society of America Journal 45:*1242-1244.

Keren, R. and G.A. O'Connor (1982). Gypsum dissolution and sodic soil reclamation as affected by water flow velocity. *Soil Science Society of America Journal 46:*726-732.

Keren, R. and I. Shainberg (1981). The efficiency of industrial and mined gypsum in reclamation of a sodic soil—Rate of dissolution. *Soil Science Society of America Journal 45:*103-107.

Keren, R., I. Shainberg, H. Frenkel, and Y. Kalo (1983). The effect of exchangeable sodium and gypsum on surface runoff from loess soil. *Soil Science Society of America Journal 47:*1001-1004.

Keren, R., I. Shainberg, and E. Klein (1988). Settling and flocculation value of sodium-montmorillonite particles in aqueous media. *Soil Science Society of America Journal 52:*76-80.

Kittrick, J.A. (1977). Mineral equilibria and the soil system. In *Minerals in Soil Environments,* eds. J.B. Dixon and S.B. Weed. Madison, WI: Soil Science Society of America, pp. 1-25.

Levy, R. (1981). Effect of dissolution of alumosilicates and carbonates on ionic activity products of calcium carbonate in soil extracts. *Soil Science Society of America Journal 45:*250-255.

Lindsay, W.L. (1979). *Chemical Equilibria in Soils.* New York: Wiley-Interscience.

Loveday, J. (1976). Relative significance of electrolyte and cation exchange effects when gypsum is applied to a sodis clay soil. *Australian Journal of Soil Research 14:*361-371.

Lyklema, J. (1982). Fundamentals of electrical double layers in colloidal systems. In *Colloidal Dispersions,* ed. J.W. Goodwin. London: Burlington House, Royal Society of Chemistry, Special Pub. No. 43, pp. 47-70.

Manin, M., A. Pissarra, and J.W. van Hoorn (1982). Drainage and desalinization of heavy clay soil in Portugal. *Agricultural Water Management 5:*227-240.

Mashhady and Rowell (1978). Soil alkalinity, equilibria and alkalinity development. *Journal of Soil Science 29:*65-75.

Massoud, F.I. (1981). *Salt Affected Soils at Global Scale and Concepts for Control.* Rome: Food and Agriculture Organization.

McNeal, B.L. (1968). Prediction of the effect of mixed-salt solutions on soil hydraulic conductivity. *Soil Science Society of America Journal 32:*190-193.

McNeal, B.L. and N.T. Coleman (1966). Effect of solution composition on soil hydraulic conductivity. *Soil Science Society of America Journal 30:*308-312.

McNeal, B.L., D.A. Layfield, W.A. Norvell, and J.D. Rhoades (1968). Factors influencing hydraulic conductivity of soils in the presence of mixed-salt solutions. *Soil Science Society of America Journal 32:*187-190.

McNeal, B.L., W.A. Norvell, and N.T. Coleman (1966). Effect of solution composition on the swelling of extracted soil clays. *Soil Science Society of America Journal 30:*313-317.

McTainsh, G., A.W. Lynch, and R.C. Burgess (1990). Wind erosion in Eastern Australia. *Australian Journal of Soil Research 28:*323-339.

Mering, J. and R. Glaeser (1954). On the rule of the valency of exchangeable cations in montmorillonite. *Bulletin de la Societe Francaise de Mineralogie et de Cristallographie 77:*519-530.

Middleton, N.J. (1984). Dust storms in Australia: Frequency, distribution and seasonality. *Search 15:*46-47.

Miller, R.J., J.W. Biggar, and D.R. Nielsen (1965). Chloride displacement in Panoche clay loam in relation to water movement and distribution. *Water Resources Research 1:*63-73.

Miller, W.P. and M.K. Baharuddin (1986). Relationship of soil disperibility to infiltration and erosion of southeastern soils. *Soil Science 142:*235-240.

Miller, W.P., H. Frenkel, and K.D. Newman (1990). Flocculation concentration and sodium/calcium exchange of kaolinitic soil clays. *Soil Science Society of America Journal 54:*346-351.

Misopolinos, N.D. (1985). A new concept for reclaiming sodic soils with high-salt water. *Soil Science 140:*69-74.

Miyamoto, S., R.J. Prather, and J.L. Stroehlein (1975). Sulfuric acid and leaching requirement for reclaiming sodium-affected calcareous soils. *Plant and Soil 3:*573-585.

Miyamoto, S., J. Ryan, and J.L. Stroehlein (1975). Potentially beneficial uses of sulfuric acid in southwestern agriculture. *Journal of Environmental Quality 4:*431-437.

Nakayama, F.S. (1970). Hydrolysis of $CaCO_3$, Na_2CO_3 and $NaHCO_3$ and their combinations in the presence and absence of external CO_2 source. *Soil Science 109:*391-398.

Nielsen, D.R., J.W. Biggar, and J.N. Luthin (1966). Desalinization of soils under controlled unsaturated flow conditions. *Sixth International Congress Communication on Irrigation and Drainage,* New Delhi, India: pp. 19.15-19.24.

Norrish, K. (1954). The swelling of montmorillonite. *Discussion Faraday Society 18:*120-134.

Oster, J.D. (1982). Gypsum usage in irrigated agriculture: A review. *Fertilizer Research 3:*73-89.

Oster, J.D. and H. Frenkel (1980). The chemistry of the reclamation of sodic soils with gypsum and lime. *Soil Science Society of America Journal 44:*41-45.

Oster, J.D., I. Shainberg, and J.D. Wood (1980). Flocculation value and gel structure of Na/Ca-montmorillonite and illite suspension. *Soil Science Society of America Journal 44:*955-959.

Plummer, L.N., D.L. Parkhurst, and T.M.L. Wigley (1979). Critical review of the kinetics of calcite dissolution and precipitation. In *Chemical Modeling in Aqueous Systems,* ed. E.A. Jenne. Washington, DC: American Chemical Society Symposium Series, *93:*537-573.

Ponnamperuma, F.N. (1972). The chemistry of submerged soils. *Advances in Agronomy 24:*29-96.

Prigogine, I. and R. Defay, Eds. (1954). *Chemical Thermodynamics.* London: The Longmans Green and Co.

Reeve, R.C. and C.A. Bower (1960). Use of high-salt waters as a flocculant and source of divalent cations for reclaiming sodic soils. *Soil Science 90:*139-144.

Reeve, R.C. and E.J. Doering (1966a). Field comparison of the high salt-water dilution method and conventional methods for reclaiming sodic soils. *Sixth International Congress Communication on Irrigation and Drainage,* New Delhi, India: pp. 19.1-19.14.

Reeve, R.C. and E.J. Doering (1966b). The high-salt-water dilution method for reclaiming sodic soils. *Soil Science Society of America Journal 30:*498-504.

Rengasamy, P., R.S.B. Greene, G.W. Ford, and A.H. Mehanni (1984). Identification of dispersive behavior and the management of real brown earths. *Australian Journal of Soil Research 22:*413-432.

Rengasamy, P. and K.A. Olsson (1991). Sodicity and soil structure. *Australian Journal of Soil Research 29:*935-952.

Rhoades, J.D. (1968). Mineral weathering correction for estimating the sodium hazard of irrigation water. *Soil Science Society of America Journal 32:*648-651.

Rhoades, J.D., D.B. Krueger, and M.J. Reed (1968). The effect of soil mineral weathering on the sodium hazard of irrigation waters. *Soil Science Society of America Journal 32:*643-647.

Richards, L.A. (1954). *Diagnosis and Improvement of Saline and Alkali Soils.* Washington, DC: U.S Department of Agriculture, Handbook 60.

Rowell, D.L. (1963). Effect of electrolyte concentration on the swelling of oriented aggregates of montmorillonite. *Soil Science 96:*368-374.

Schofield, R.K. and H.R. Samson (1954). Flocculation of kaolinite due to the attraction of oppositely charged crystal faces. *Discussion Faraday Society 18:*138-145.

Shainberg, I., E. Bresler, and Y. Klausner (1971). Studies on Na/Ca montmorillonite systems. I. The swelling pressure. *Soil Science 111:*214-219.

Shainberg, I., J.D. Rhoades, and R.J. Prather (1981). Effect of low electrolyte concentration on clay dispersion and hydraulic conductivity of a sodic soil. *Soil Science Society of America Journal 45:*273-277.

Shainberg, I., J.D. Rhoades, D.L. Suarez, and R.J. Prather (1981). Effect of mineral weathering on clay dispersion and hydraulic conductivity of sodic soils. *Soil Science Society of America Journal 45:*287-291.

Sposito, G. and S.W. Mattigod (1977). On the chemical formation of the sodium adsorption ratio. *Soil Science Society of America Journal 41:*323-329.

Stumm, W. and J.J. Morgan (1970). *Aquatic Chemistry.* New York: Wiley-Interscience.

Suarez, D.L. and J.D. Rhoades (1982). The apparent solubility of calcium carbonate in soils. *Soil Science Society of America Journal 46:*716-722.

Szabolcs, I., Ed. (1989). *Salt-Affected Soils.* Boca Raton, FL: The CRC Press.

Tanji, K.K. and S.J. Deveral (1984). Simulation modeling for reclamation of sodic soils. In *Soil Salinity Under Irrigation-Processes and Management,* eds. I. Shainberg and J. Shalhevet. New York: Springer-Verlag, pp. 238-251.

Tanji, K.K., L.D. Doneen, G.V. Ferry, and R.S. Ayers (1972). Computer simulation analysis on reclamation of salt-affected soil in San Joaquin Valley, California. *Soil Science Society of America Journal 36:*127-133.

Tarchitzky, J., Y. Chen, and A. Banin (1993). Humic substances and pH effects on sodium- and calcium-montmorillonite flocculation and dispersion. *Soil Science Society of America Journal 57:*367-372.

Thorstenson, D.C., D.W. Fisher, and M.G. Croft (1979). The geochemistry of the Fox Hills-Basal Hill creek aquifer in southwestern North Dakota and northwestern South Dakota. *Water Resources Research 15:*1479-1498.

Thorup, J.T. (1972). Soil sulphur application. *Sulphur Institute Journal 8:*16-17.

Van Olphen, H. (1977). *An Introduction to Clay Colloid Chemistry.* Second Edition. New York: John Wiley and Sons.

Chapter 9

Soil Erosion and Conservation

Edward L. Skidmore
Simon J. van Donk

Soil erosion is a worldwide problem. Approximately 90 percent of cropland in the United States is currently losing soil above the sustainable rate. Soil erosion rates in Asia, Africa, and South America are estimated to be twice as high as those in the United States. The Food and Agriculture Organization (FAO) estimates that 140 million ha of high quality soil, mostly in Africa and Asia, will be degraded by 2010 unless better land management practices are adopted (U.S. Global Change Research Information Office, 2001).

Agricultural producers, as well as managers of nonagricultural lands, need to know what to expect in terms of soil erosion by wind and water resulting from alternative management practices. Real-world experiments are too laborious and expensive for erosion assessments and evaluation of alternative management scenarios, so computer models have been developed for this purpose. In this chapter we will focus on basic erosion processes, modeling of these processes, model applications, and erosion control. Wind erosion will be discussed first, followed by water erosion.

EROSION OF SOIL BY WIND

Erosion of soil by wind is a particularly serious problem in many arid and semiarid regions (Figure 9.1). Arid lands comprise about one-third of the world's total land area and are the home of one-sixth of the world's population (Dregne, 1976; Gore, 1979). Areas under agricultural production that are most susceptible to wind erosion include much of North Africa and the Near East, parts of southern and eastern Asia, the Siberian plains, Australia, southern South America, and the semiarid and arid portions of North America (Food and Agriculture Organization, United Nations, 1960).

Extensive soil erosion in the U.S. Great Plains during the last half of the nineteenth century and in the prairie region of western Canada during the

FIGURE 9.1. Severe wind erosion after wildfire in Meade County, Kansas (photograph by Edward Skidmore).

1920s warned of impending disaster. In the 1930s, a prolonged drought culminated in dust storms and soil destruction of disastrous proportions in the prairie regions of both western Canada and the Great Plains (Anderson, 1975; Hurt, 1981; Johnson, 1947; Malin, 1946; Svobida, 1940). More recently, in northern China, drought and overgrazing have caused land degradation with wind erosion resembling the dust bowl days of the 1930s in the United States (Armstrong, 2001). The Sahelian region of West Africa has seen dramatic changes over the past few decades, with decreasing rainfall, vegetation, and wildlife, and increasing wind erosion on the southern fringes of the Sahara desert.

Agricultural lands are adversely impacted by soil tillage that leaves little residue on the soil surface and by cropping systems that leave the soil surface bare for long periods of time, making it more vulnerable to wind erosion. On pastoral rangeland, the composition of pastures subjected to excessive grazing during dry periods deteriorates, the proportion of edible perennial plants decreases, and the proportion of annuals increases. The thinning and death of vegetation during droughts increase the extent of bare ground and surface soil conditions deteriorate, increasing the fraction of erodible aggregates on the soil surface. In rain-fed farming areas, removal of the original vegetation and fallow expose the soil to accelerated erosion.

Erosivity and Erodibility

For erosion to occur, there has to be a force to move the soil. Shearing stress caused by wind is this force in wind erosion and its ability to erode is called erosivity. The extent of erosion also depends on the susceptibility of the soil to erosion, which is called erodibility. As a measure of erosive wind energy Skidmore (1998) defines wind power density as

$$WPD = \rho \left(u_i^{\,2} - u_t^{\,2}\right)^{3/2} \tag{9.1}$$

where *WPD* is wind power density (W m^{-2}), ρ is air density (kg m^{-3}), u_t is threshold wind speed (m s^{-1}), and u_i is measured wind speed (m s^{-1}).

Scientists recognized early that soil erodibility, the susceptibility or ease of detachment and transport by wind, was a primary variable affecting wind erosion. From wind tunnel tests, Chepil (1950) determined relative erodibilities of soils reasonably free from organic residues as a function of apparent specific gravity and proportions of dry soil aggregates of various sizes. Clods larger than 0.84 mm in diameter were immobile in the range of wind speeds used in the tests. In addition to aggregate size distribution, soil erodibility depends on aggregate stability, soil surface wetness, crusting, and amount of loose material on a crust.

Basic Processes

Wind erosion consists of entrainment of loose and abraded particles, followed by their transport and deposition.

Entrainment

The way the first particles are moved has received less attention than the modes of transport. Before 1962, most researchers were satisfied by Bagnold's (1941) description of particles rolling along the surface by direct wind pressure for about 30 cm before starting to bounce off the ground. Bisal and Nielsen (1962) concluded, after observing particles in a shallow pan mounted on the viewing stage of a binocular microscope, that most erodible particles vibrated with increasing intensity as wind speed increased and then left the surface instantly as if ejected.

More recently, particle entrainment has been studied and described in considerable detail (Anderson, Sørensen, and Willetts, 1991; Rasmussen and Rasmussen, 1998). On many agricultural soils, immobile aggregates and crusts are first abraded by incoming sediment before being entrained (Hagen, 1991b; Rice and McEwan, 2001).

Transport

Transport can occur in three modes: surface creep, saltation, and suspension (Figure 9.2). Sand-sized soil particles or aggregates 500-1000 μm in diameter, too large to leave the surface in ordinary erosive winds, are pushed, rolled, and driven by the impacts of spinning particles in saltation. In high winds, the whole surface appears to be creeping slowly forward. The rippling of wind-blown sand has been attributed to unevenness in surface creep flow (Bagnold, 1941). Creep appears nearly passive in the erosion process, but creep-sized aggregates may abrade into the size range of saltation and suspension and, thus, shift modes of transport. Creep aggregates seldom move far from their points of origin (Lyles, 1988).

In saltation, individual particles lift off the surface and follow distinctive trajectories under the influence of air resistance and gravity. Such particles (100-500 μm) rise at fairly steep angles but are too large to be suspended by the flow. They return to the surface where they may abrade themselves or other aggregates on impact, or they may rebound or embed themselves and initiate movement of other particles. Most saltating particles rise no higher than 30 cm (Lyles, 1988).

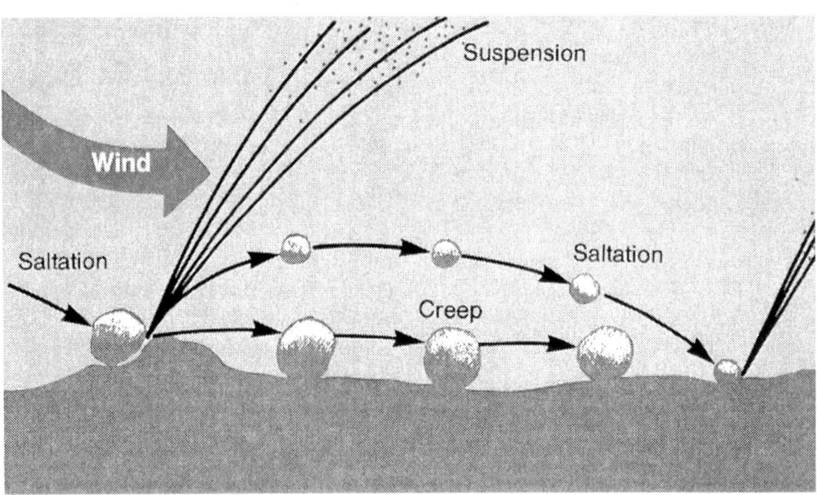

FIGURE 9.2. Airborne sediment transport by means of surface creep, saltation, and suspension (*Source:* USDA, Soil Conservation Service, 1989.)

Suspension refers to the transport of very small soil particles that are generally removed from the local source area. They may be deposited on the neighbor's farm or several states downwind. Suspended particles can range in size from about 2 to 100 μm, with mass median diameter of approximately 50 μm in an eroding field (Chepil, 1957; Gillette and Walker, 1977). Some suspension-sized particles are present in the soil, but most are created by abrasive breakdown during erosion. Because organic matter and some plant nutrients are usually associated with the finer soil fractions, suspension samples are enriched in such constituents compared with the bulk soil source (Lyles, 1988).

The relative amount of sediment in various transport modes varies with downwind distance and field surface. The percentage of sediment in suspension generally becomes greater further downwind. Obviously, there will be a greater percentage in suspension when the source material from the surface is finer.

Deposition

Suspended particles start to settle when wind speed and turbulence slow down. They also move to the surface of the earth by diffusion. Larger and heavier particles settle first. In long-distance transport, particles <20 μm in diameter predominate because the larger particles have significant sedimentation velocities (Gillette, 1977). Suspended material may also be deposited with rainfall. For saltation and creep, deposition occurs when sediment mass flux exceeds transport capacity of the wind for a given surface condition. Standing biomass and soil ridges intercept particles transported in saltation and creep mode. Saltating particles usually are deposited in a fence row, ditch, trap strip, wind break, or on the edge of a vegetated area downwind (Lyles, 1988).

MODELING WIND EROSION

Models can be useful tools to assess erosion and evaluate the effect of alternative management scenarios on erosion. The wind erosion equation (WEQ) proposed by Chepil and Woodruff (1963) and Woodruff and Siddoway (1965) developed as a result of investigations into the mechanics of the wind erosion process, the major factors influencing wind erosion, and the development of wind erosion control methods. The equation expressed in function form is:

$$E = f(I, K, C, L, V) \qquad (9.2)$$

where E is the potential average annual soil loss, I is the soil erodibility index, K is the soil ridge roughness factor, C is the climate factor, L is unsheltered distance across a field, and V is the equivalent vegetative cover.

Solving the functional relationships of the wind erosion equation as presented by Woodruff and Siddoway (1965) required the use of tables and figures. The awkwardness of the manual solution prompted a computer solution (Fisher and Skidmore, 1970; Skidmore, Fisher, and Woodruff, 1970a, b) and development of a slide rule calculator (Skidmore, 1983). After the arrival of personal desktop computers, the model was adapted for use with them (Halsey et al., 1983) and interactive programs (Erickson et al., 1984). Cole, Lyles, and Hagen (1983) adapted the Woodruff and Siddoway (1965) model for simulating daily soil loss by wind erosion as a submodel in EPIC (Williams, Jones, and Dyke, 1984). The latter version was simplified by fitting equations (Skidmore and Williams, 1991) to the figures of Woodruff and Siddoway (1965).

Some limitations in WEQ were recognized (Skidmore, 1976) and various improvements have been incorporated. These include computing erosion by periods (Bondy, Lyles, and Hayes, 1980; Fryrear, 1981; Lyles, 1983; Sporcic and Nelson, 1999); accounting for preponderance, field shape and orientation, and row direction (Skidmore, 1965, 1987; Skidmore, Nossaman, and Woodruff, 1966); improved formulation of the climatic factor and extended climate database (Skidmore and Woodruff, 1968; Skidmore, 1986; Skidmore, Tatarko, and Wagner, 1994); and estimation of small grain equivalents expanded to additional plants and conditions (Lyles and Allison, 1980, 1981; Armbrust and Lyles, 1985; Skidmore and Nelson, 1992).

Despite the many improvements to WEQ, complex interactions between variables are not accounted for in calculation procedures and it is not easily adapted to untested conditions or climates far different from those of the central Great Plains where it was developed. Therefore, the United States Department of Agriculture-Agricultural Research Service (USDA-ARS) appointed a team of scientists to take a leading role in developing the new Wind Erosion Prediction System (WEPS) as a replacement for the "mature" WEQ (Hagen, 1991a; Wagner, 1996; USDA, 1995). Advances in erosion science and the increased power of personal computers have allowed the adoption of more flexible, processed-based erosion prediction technology.

WEPS is a daily time-step computer model that predicts soil erosion via simulation of the physical processes controlling wind erosion. It is intended primarily for soil conservation and environmental planning. WEPS 1.0 is the first implementation of WEPS intended for use by the USDA-Natural Resources Conservation Service (NRCS). It includes a graphical user interface to allow the user to easily select climate stations, specify field dimensions, pick a predominant soil type, and describe wind barriers and management practices applied to an agricultural field. This interface allows the user to quickly assess a site's susceptibility to wind erosion and evaluate the impacts that alternative practices and conditions might have on reducing that susceptibility (Wagner and Tatarko, 2001).

WEPS is modular in design and consists of a main supervisory routine and several submodel components (USDA, 1995). The main routine handles the time steps in the model, reads the input files, controls the individual submodel routines, and generates the output reports. Each of the individual submodel components likewise performs specific duties within the WEPS model.

The erosion submodel decides if erosion can occur based on the current surface roughness, quantity of flat and standing biomass, aggregate size distribution, crust and rock cover, loose erodible material on a crust, and soil wetness. If the surface conditions are susceptible and wind speed is sufficient, erosion and deposition are simulated on a subhourly basis (Hagen, Wagner, and Skidmore, 1999). During erosion, the submodel individually simulates the saltation/creep and suspension components of wind-eroded soil. This approach was used because the saltation/creep component has a defined transport capacity, whereas the suspension component generally continues to increase over the entire length of eroding fields. Individual processes simulated include entrainment of loose material and abrasion of clods and crusts.

The hydrology submodel estimates soil surface wetness, accounts for changes in soil temperature, and maintains a soil-water balance based on daily amounts of snow melt, runoff, infiltration, deep percolation, soil evaporation, and plant transpiration. The soil submodel tracks changes in soil and surface temporal properties in response to various weather processes such as wetting/drying, freezing/drying, freezing/thawing, precipitation amount, and time. The crop submodel simulates the growth of crop plants. It can simulate a variety of crops and plant communities while accounting for water and temperature stresses. It calculates daily biomass production of roots, leaves, stems, and reproductive organs as well as leaf and stem areas.

The decomposition submodel simulates the decrease in crop residue biomass from microbial activity. The decomposition process is modeled as a first order reaction, with temperature and moisture as the driving variables. It maintains separate decomposition pools for residue type (parent material), plant component (stems and roots), location (standing, flat, buried), and residue age. The management submodel simulates the various cultural practices applied to an agricultural field. These include primary and secondary tillage, cultivating, planting/seeding, harvesting, irrigating, and burning and grazing operations. Each individual operation is described as a series of processes that reflect the physical changes in the soil, surface, crop, and residue status.

The modular design of WEPS is intended to allow for easy updating and revising of the science model, or even replacing specific components and/or sections in the future as knowledge and understanding of erosion, climate, plant growth, and other processes improve (Wagner and Tatarko, 2001). Because WEPS did not progress as rapidly as envisioned, it was decided to quick-fix some of the more serious problems associated with WEQ. Hence, a Revised Wind Erosion Equation (RWEQ) (Fryrear et al., 1998; Zobeck et al., 2001) was born. RWEQ is a factor multiplication model with factors similar to those used in WEQ.

Other models emerging on the wind erosion prediction scene include TEAM (Gregory and Darwish, 2001), WEAM (Shao, Raupach, and Leys, 1996), an adaptation of EPIC (Skidmore and Williams, 1991), and models by Berkofsky and McEwan (1994), and Rice et al. (1999).

The Texas Tech Erosion Assessment Methodology (TEAM) model is a mathematical model that simulates the detachment and maximum transport rate of soil and predicts the concentration (loading) of dust into the environment based on a single storm event, modeling the interaction of wind, soil, resistance, and soil roughness factors (Gregory and Darwish, 2001). TEAM also models saltation concentration, particle size distribution, and provides visibility predictions. It has been used primarily to evaluate real-world field and climatic conditions (primarily dust storm prediction and air quality safety analysis).

The overall focus of WEAM (Wind Erosion Assessment Model) is to estimate sand drift and dust entrainment (aerial suspension) of dry, bare soil subject to a given wind condition. The model considers creep, saltation, and suspension, and specifically the mobilization of soil particles due to wind forces (Shao, Raupach, and Leys, 1996). The model is based on empirical equations, is not computer-based, and is primarily a synthesis of recent studies in the wind erosion field involving the physical processes governing sand drift and dust entrainment. The model only predicts wind erosion for a given, static condition; evolution of surface properties due to natural weathering, cultivation/grazing, or abrasion from wind erosion is not provided.

The Erosion Productivity Impact Calculator (EPIC) (Williams, Dyke, and Jones, 1983; Williams, Jones, and Dyke, 1984, 1993) was developed to assess soil erosion and soil productivity. It is a continuous simulation model that is generally used to predict changes in soil productivity and water quality over large temporal domains. The major components in EPIC are weather simulation, hydrology, erosion/sedimentation, nutrient cycling, pesticide fate, plant growth, soil temperature, tillage, economics, and plant environment control. Cole, Lyles, and Hagen (1983) adapted WEQ for simulating daily wind erosion as a submodel in EPIC and later modified it (Skidmore and Williams, 1991). Some recent developments have focused on problems involving water quality and global climate change. Additions to EPIC include the Groundwater Loading Effects of Agricultural Management Systems (GLEAMS) (Leonard, Knisel, and Still, 1987) pesticide fate component, and nitrification and volatilization submodels. These and other developments extend EPIC's capability to deal with a wide variety of agricultural management problems. With these changes it has been renamed as the Environmental Policy Integrated Climate model (Williams et al., 1996).

Berkofsky and McEwan (1994) proposed a partial differential equation for the prediction of rate of change of dust concentration in a thin layer near the ground. Detachment, transport, and deposition were included in the equation. The planetary boundary layer was divided into a surface layer, a transition layer, and an inversion layer. Rice et al. (1999) described a conceptual model for the prediction of wind erosion rates dependent on the distribution of impact energy delivered to the surface by saltating grains, and the distribution of local surface strength.

Model Applications

Scientists and practitioners have used WEQ with various modifications for the past 35 years. NRCS field workers have used the equation extensively to plan wind erosion control practices (Hayes, 1966), classify erodibility of soils (Woodruff and Siddoway, 1965; USDA, 1988), and determine highly erodible land for all farms in the United States in association with the Food Security Act of 1985 (Federal Register, 1992). Hayes (1965) also used WEQ to estimate crop tolerance to wind erosion. The equation is a useful guide to wind erosion control principles as well (Carreker, 1966; Moldenhauer and Duncan, 1969; Woodruff et al., 1972).

Other uses of the equation include: (a) determining spacing for barriers in narrow strip-barrier systems (Hagen, Skidmore, and Dickerson, 1972); (b) estimating fugitive dust emissions from agricultural and subdivision lands (PEDCO-Environmental Specialists, Inc., 1973; Wilson, 1975); (c) predicting horizontal soil fluxes to compare with vertical aerosol fluxes (Gillette, Blifford, and Fenster, 1972); (d) estimating the effects of wind erosion on soil productivity (Lyles, 1974; Williams, Jones, and Dyke, 1984); (e) delineating those croplands in the Great Plains where various amounts of crop residues may be removed without exposing the soil to excessive wind erosion (Skidmore, Kumar, and Larson, 1979); and (f) estimating erosion hazards in a national inventory (USDA, Soil Conservation Service, 1984).

WEPS is intended to be the "tool of choice" for planning soil conservation systems, providing environmental planning and assessment evaluations, and estimating off-site impacts of wind erosion (Wagner, 1996). The NRCS is implementing WEPS 1.0 for use as a conservation and planning tool on agricultural croplands in preparation for the expected 2002 U.S. Farm Bill. However, there are limitations in this initial version that preclude full applicability to a wider range of lands susceptible to wind erosion. Thus, to fulfill additional wind erosion prediction requirements, WEPS needs to be enhanced and extended. Based upon customer and user input, the following needs have been identified: (a) extend WEPS to nonhomogeneous soils and variable topography; (b) incorporate a multispecies plant growth model component into WEPS; (c) incorporate both water (Water Erosion Prediction Project [WEPP]) and wind (WEPS) erosion simulations into a single science model; (d) extend WEPS technology to handle needs for management of range and disturbed lands including military training lands; and (e) extend WEPS technology to handle organic soils.

Lopez (1998) estimated the potential dust flux within an agricultural field in Central Aragon, Spain, using a dust emission model developed by Maticorena and Bergametti (1995). This model is based on the threshold wind shear velocity being a function of aggregate size distribution and roughness length of the soil surface. The observed reduction in soil erodibility with time was probably due to a limited supply of erodible particles at the soil surface. This study underlines the need to consider the temporal variability of the surface conditions in wind erosion research and models.

Lyons et al. (1998) described a system comprised of a physically based wind erosion model driven by data from a high resolution atmospheric model and land surface data from detailed geographical information system (GIS) databases. The model considers the capacity of the wind to entrain and transport particles and the ability of the surface to resist wind erosion through consideration of the effects particle size, frontal area index, topsoil moisture, and surface crusting. The system was applied to an investigation of wind erosion in Australia. It was found to be effective in predicting the timing and location of wind erosion events.

Erosion Control

In general, to control erosion, the capacity of the erosive agent (erosivity) and/or the susceptibility of the surface material (erodibility) must be reduced. Principles for controlling wind erosion include establishing and maintaining sufficient vegetative cover; producing a rough, cloddy surface;

reducing surface wind speed and effective field width with barriers; and stabilizing soil with various materials (Woodruff et al., 1972).

Vegetation—Crops and Crop Residues

Living vegetation or residue from harvested crops protects the soil against wind erosion (Figure 9.3). Standing crop residues provide nonerodible elements that absorb much of the shear stress in the boundary layer. When vegetation and crop residues are sufficiently high and dense to prevent intervening soil-surface drag from exceeding threshold drag, soil will not erode. Rows perpendicular to wind direction control wind erosion more effectively than do rows parallel to wind direction (Englehorn, Zingg, and Woodruff, 1952; Skidmore, Nossaman, and Woodruff, 1966). Flattened stubble, though not as effective as standing, also protects the soil from wind erosion (Chepil, Woodruff, and Zingg, 1955).

Soon after the disastrous "dirty thirties" in the U.S. Great Plains, stubble-mulch systems were demonstrated to be feasible for reducing wind erosion on cultivated land (Duley, 1959). Stubble mulching is a crop residue management system using tillage, generally without soil inversion and usually

FIGURE 9.3. Maintaining crop residue on the surface protects soil from the wind (photograph by Edward Skidmore).

with blades or V-shaped sweeps (McCalla and Army, 1961; Mannering and Fenster, 1983). The goal is to leave a desirable quantity of plant residue on the surface of the soil at all times. Residue is needed for a period of time even after a new crop is planted to protect the soil from erosion and to improve infiltration. The residue used is generally that remaining from a previous crop. Direct seeding into residue and nontillaged areas leaves even more residue on the surface than stubble mulching and, consequently, protects the soil even better against wind erosion.

Early studies quantified specific properties of vegetative covers influencing wind erosion (Chepil, 1944; Chepil, Woodruff, and Zingg, 1955; Siddoway, Chepil, and Armbrust, 1965). These studies led to the relationship presented by Woodruff and Siddoway (1965), showing the influence of an equivalent vegetative cover of small grain and sorghum stubble for various orientations (flat, standing, height). Research efforts have continued to evaluate the protective role of additional crops (Craig and Threlle, 1964; Lyles and Allison, 1981), range grasses (Lyles and Allison, 1980), feedlot manure (Woodruff et al., 1974), and the protective requirements of equivalent residue needed to control wind erosion (Lyles, Schmeidler, and Woodruff, 1973; Skidmore and Siddoway, 1978; Skidmore, Kumar, and Larson, 1979). Using residue for reducing wind erosion is probably the most effective and practical method in many situations, especially with the move toward less tillage that is going on in many parts of the world (Peterson, Westfall, and Cole, 1993; Peterson et al., 1996; Farahani et al., 1998; Anderson et al., 1999).

Tillage

Chepil and Milne (1941), while investigating the influence of surface roughness on drifting dune materials and cultivated soils, found that the initial intensity of drifting was always much less over a ridged than a smooth surface. Ridging cultivated soils reduced the severity of drifting, but ridging highly erosive dune materials was less effective because the ridges disappeared rapidly. The rate of sediment flow varied inversely with surface roughness.

When ridges are mostly gone, vegetative cover is depleted, and the threat of wind erosion continues, a rough, cloddy surface that is resistant to the force of wind can be created on many cohesive soils with appropriate "emergency tillage." Lyles and Tatarko (1982) found that chiseling of growing winter wheat on a silty clay soil greatly increased nonerodible surface aggregates without affecting grain yield. Listers, chisels, cultivators, oneway disks with two or three disks removed at intervals, and pitting machines can be used to bring compact clods to the surface. Emergency tillage is most effective when done at right angles to the prevailing wind direction. Because clods eventually disintegrate (sometimes rapidly), emergency tillage offers, at best, only temporary wind-erosion control (Woodruff, Chepil, and Lynch, 1957; Woodruff et al., 1972).

Barriers

Use of wind barriers is an effective method of reducing field width. Barriers have long been recognized for their value in controlling wind erosion (Bates, 1911). Hagen (1976) and Skidmore and Hagen (1977) developed a model that, when used with local wind data, shows wind barrier effectiveness in reducing wind erosion forces. Barriers will reduce wind forces more than they will wind speed (surface wind shear stress is proportional to wind speed squared). A properly oriented barrier, when winds predominate from a single direction, will decrease wind erosion forces by more than 50 percent from the barrier leeward to 20 times its height. The decrease in force will be greater for shorter distances from the barrier.

Different combinations of trees, shrubs, tall-growing crops, and grasses can reduce wind erosion. Besides the more conventional tree windbreak (Ferber, 1969; Read, 1964; Woodruff et al., 1976), many other barrier systems are used to control wind erosion. They include annual crops, such as small grains, corn, sorghum, Sudan grass, and sunflower (Carreker, 1966; Fryrear, 1963, 1969; Hagen, Skidmore, and Dickerson, 1972; Hoag and Geiszler, 1971), and tall wheatgrass (Aase, Siddoway, and Black, 1976; Black and Siddoway, 1971).

Most barrier systems for controlling wind erosion, however, occupy space that could otherwise be used to produce crops. Perennial barriers grow slowly and are often difficult to establish (Dickerson, Woodruff, and Banbury, 1976; Woodruff et al., 1976). Such barriers also compete with crops for water and plant nutrients (Lyles, Tatarko, and Dickerson, 1983). Thus the net effect for many tree-barrier systems is that their use may not benefit crop production (Frank, Harris, and Willis, 1977; McMartin, Frank, and Heints, 1974; Skidmore, Hagen, and Teare, 1975; Skidmore et al., 1974; Staple and Lehane, 1955). Perhaps the tree-barrier systems could be designed so that the barrier becomes a useful crop, furnishing nuts, fruits, or wood.

Stabilizers

Various soil stabilizers have been evaluated in the search for suitable materials and methods to control wind erosion (Armbrust and Dickerson,

1971; Armbrust and Lyles, 1975; Chepil, 1955; Chepil and Woodruff, 1963; Chepil et al., 1963; Lyles et al., 1969; Lyles, Schrandt, and Schmeidler, 1974). Several tested products successfully controlled wind erosion for a short time but many were more expensive than equally effective wheat straw anchored with a rolling disk packer (Chepil et al., 1963). The following are criteria for surface-soil stabilizers (Armbrust and Lyles, 1975):

1. 100 percent of the soil must be covered.
2. The stabilizer must not adversely affect plant growth or emergence.
3. Erosion must be prevented initially and reduced for the duration of the severe erosion hazard, usually for at least two months each season.
4. The stabilizer should apply easily and without special equipment.
5. Cost must be low enough for profitable use.

Armbrust and Lyles (1975) found five polymers and one resin-in-water emulsion that met all these requirements.

EROSION OF SOIL BY WATER

Erosion of soil by water occurs throughout the world, but especially in the more humid regions. Erosion results in the loss of topsoil that is not easily replaced. Estimates of erosion are essential to issues of land and water management, including sediment transport and storage in lowlands, reservoirs, estuaries, and irrigation and hydropower systems. Another issue is water pollution by contaminants carried with sediment into water bodies.

Erosivity and Erodibility

The potential ability of rainfall to cause erosion is referred to as its erosivity. When raindrops strike bare soil, practically all of the energy is consumed as work done against the soil surface in the disruption of soil aggregates, compaction of the soil surface, and splash of soil particles into the air (Rosewell et al., 2000). Erosivity is related to the kinetic energy of rainfall, which can be related to rainfall amount and intensity. The Universal Soil Loss Equation (USLE) (Wischmeier, 1959) includes a method for calculating storm kinetic energy E and erosivity R. The storm kinetic energy can be computed as the sum of each rain intensity group I_i as:

$$E = (210 + 89 \log I_i) D_i \tag{9.3}$$

where E is kinetic energy (J m^{-2} cm^{-1}), I_i is rainfall intensity (cm h^{-1}) and D_i is rain depth (cm) for intensity group I_i of the rainstorm. The erosivity parameter R, according to USLE, presents for a large group of soils the best linear relation between soil erosion and rainstorm erosivity:

$$R = EI_{30} \tag{9.4}$$

where R is erosivity (J m^{-2} h^{-1}), and I_{30} is the maximum rainfall intensity (cm h^{-1}) for a continuous 30-minute period of rainfall (Morin, 1996). Lal and Elliott (1994) present several ways to estimate and measure rainfall erosivity.

Soil erodibility is a measure of the soil's susceptibility to detachment and transport by the agents of erosion (Lal and Elliot, 1994). Le Bissonnais (1996) defined soil erodibility as the inherent soil property to react to water action in (1) reducing infiltration rate and decreasing soil surface roughness due to aggregate breakdown, i.e., increasing the risk of runoff, and (2) being detached and transported by the resulting runoff. The ease with which the soil matrix yields to a raindrop impact is called detachability of the soil. As the strength of the soil to withstand the erosive force of the impacting drop increases, the detachability of soil decreases (Sharma, 1996).

The response of a soil to erosion processes is complex, and is influenced by soil properties such as texture, structural stability, organic matter content, clay mineralogy, and chemical constituents. Some of these properties, such as organic matter, can be altered over time by land use, management practices, and farming systems. Erosion of the surface layers can expose less-erodible subsoils, which may have different properties than the surface. Consequently, the erodibility of a soil can change with time (Lal and Elliot, 1994). It also varies with soil moisture, temperature, and soil disturbance (Young, Romkens, and McCool, 1990).

Soil texture is important in determining erodibility. Sandy soils have lower runoff rates and are more easily detached, but less easily transported than silt soils. Clay soils are not easily detached, but lower infiltration rates may lead to greater runoff and increased erosion. Silt soils tend to have the greatest erodibilities since particles are easily detached and transported, and consolidation of subsoils, or subsoils with higher clay contents, can lead to greater runoff. The greatest erosion is often associated with high silt loess soils, as found in China's Yellow River watershed (Lal and Elliot, 1994).

Basic Processes

The predominant processes that determine erosion are infiltration, runoff, detachment and transport by raindrops and overland flow (interrill erosion), detachment and transport by concentrated flow (rill erosion), and deposition (Lal and Elliot, 1994).

The process of erosion of soil by water starts with the detachment and transport of soil particles by impact force of raindrops and drag force of overland flow. The dominance of one force over the other determines the controls on the processes of detachment and transport. Raindrop impact provides the primary force needed to initiate detachment of soil particles from the soil mass. Raindrop splash and overland flow transport sediment in a downslope direction. Sediment must be detached from the soil mass or be in a detached state before it can be transported (Sharma, 1996).

The water erosion process can be detachment limited or transport limited (Foster, 1982). The process is transport efficient or detachment limited if all particles generated in an eroding area move across the lower slope boundary. The process is transport limited if all particles detached in an upslope area are not carried across the downslope boundary. Generally, all particles that are detached are not transported out of the eroding area. The process of settling of detached and transported sediments is called deposition. Detachment, transport, and deposition of sediment are three integral processes of soil erosion (Rose, 1985).

Detachment

Soil detachment by raindrop impact is the principal erosion process controlling interrill soil erosion, even though sufficient surface flow must be available for transport of the detached particles (Bradford and Huang, 1996). The detachment capacity of interrill flow is negligible compared to that of raindrop splash (Young and Wiersma, 1973) because of the low shear stresses of the thin sheet flow.

The process of soil detachment by raindrops is best understood by studying the mechanism of soil detachment from the impact of a single drop (Ghadiri and Payne, 1977). Using high-speed photography, Mutchler (1967) and Al-Durrah and Bradford (1982) describe the mechanism of raindrop impact on the soil surface and the resulting soil detachment and sediment splash. When a raindrop impacts a saturated soil surface, a hemispheric cavity is formed on the surface. The vertical compressive stress of the drop is then transformed into lateral shear stress of radial flow of water jetting away from the center of the cavity. At this stage, soil particle detachment is caused by the shear stresses of the radial flow acting on the bottom and sides of the cavity (Sharma, 1996). The amount of soil detachment from the cavity sides will be determined by the magnitude of soil deformation that takes place in the earlier stages of cavity development and by the cohesive forces resisting the shear stresses (Al-Durrah and Bradford, 1982).

Soil detachment by concentrated runoff results in rill erosion (Bryan, 1987). It usually affects only a small proportion of the land surface, but is much more visible than interrill erosion (Rosewell et al., 2000). Rill erosion is erosion in numerous small channels, which can be obliterated by normal tillage. A depth of less than 300 mm may be used as a criterion to distinguish rills from gullies (Houghton and Charman, 1986).

Transport

Commonly known as splash erosion, air splash occurs immediately after raindrop impact initiates soil detachment. The droplets generated after impact radiate outward from the center of the impact while encapsulating solids and carrying them to the landing points. Air splashing of solids declines rapidly as the depth of water covering the soil surface increases, reaching near negligibility at depth of about 2 mm (Moss, 1988). For normal impacts on horizontal surfaces, splash produces only random particle movement. However, air splash can cause net soil transport in one direction, initially, under the influence of slope or wind and, secondly, due to preferential movement of solids from areas of high activity to those of low activity (Sharma, 1996). The apparent absence of thick, extensive deposits attributable to air splash suggests that the mechanism is seldom a major transporting agent (Moss, 1988). However, more recent studies (Erpul, 2001) show that air splash can be of importance in overland flow, especially under inclined (wind-driven) rainfall.

As soon as the process of runoff starts, the overland flow, which is inherently more unidirectional than air splash, begins to transport solids downslope. The solids transported by the flow can be split into suspended and bed loads. Depending on the slope and surface roughness, flow alone can only transport small-sized particles in suspended load. Bed load particles not transported by flow alone remain on the bed until lifted back in the flow by force of raindrops impacting the shallow overland flow (Kinnell, 1988). Sediment may be washed into rills where it may be further transported into gullies, channels, and eventually end up in water bodies.

Gullies are relatively permanent, steep-sided water courses that experience ephemeral flows during rainstorms. Compared with stable river channels which have a relatively smooth, concave-upward, long profile, gullies are characterized by a head cut and various steps or knick points along their course. Gullies have greater depth and smaller width than stable channels, carry larger sediment loads, and display very erratic behavior so that relationships between sediment discharge and runoff are frequently poor. Gullies are almost always associated with accelerated erosion (Morgan, 1995).

Deposition

Sediment settles on the surface when overland flow is obstructed due to surface roughness, plant stalks, and stubble mulches, or when flow turbulence is lowered due to decrease in slope steepness or frequency of rainfall impact (Sharma, 1996). The settling velocity of an aggregate or primary particle is a function of its size, shape, and density. The rate of deposition is related to the velocity of flow and to the concentration and density of a given sediment size (Hairsine and Rose, 1991). Deposition may facilitate the formation of a seal, because of sediment clogging pores, thus reducing infiltration and increasing runoff.

MODELING WATER EROSION

As with wind erosion, computer models have been developed for the simulation of water erosion. The Universal Soil Loss Equation (USLE) (Wischmeier, 1959; Wischmeier and Smith, 1978) predicts average annual soil loss. It is an empirical model based on a large number of experimental data from small plots. USLE was developed to examine the long-term effect of land management options on erosion of soil by water and has been widely used in many parts of the world. The equation:

$$A = RKLSCP \tag{9.5}$$

estimates the average annual soil loss *(A)* from factors that depend on rainfall erosivity *(R)*, soil erodibility *(K)*, topography (*L* and *S*), crop and crop management *(C)*, and erosion control practice *(P)*. Modifications of USLE have resulted in the Revised Universal Soil Loss Equation (RUSLE) (Renard et al., 1991; RUSLE2, Foster et al., 2001) and the Modified Universal Soil Loss Equation (MUSLE) (Williams and Berndt, 1977). Morgan, Morgan, and Finney (1984) developed a model to predict annual soil loss from field-sized areas on hillslopes which, while it retains the simplicity of USLE, encompasses some of the advances in understanding of erosion processes.

The Water Erosion Prediction Project model (WEPP) (Lane and Nearing, 1989) is being developed by an interagency group of scientists in the United States. WEPP is a process-based erosion prediction model meant to replace USLE. It simulates the erosion processes of soil detachment, transport, and deposition, as well as the processes that lead to erosion, including infiltration and runoff. It also simulates plant growth, senescence, and residue decomposition. The effects of tillage processes and soil consolidation are also modeled. Erosion and deposition are calculated based on a predefined value of sediment transport capacity estimated from flow hydraulics, slope, and sediment properties. Net erosion or deposition is estimated by the difference between sediment load and transport capacity (Huang, Darboux, and Zartl, 2001).

WEPP is/will be available in three versions: (1) a hillslope version (Figure 9.4) that predicts soil erosion from a single hillslope of any length. The hillslope can have a complex shape and can include numerous soils and crops along the hillslope; (2) a watershed version that links hillslope elements together with channel and impoundment elements; and (3) a grid or GIS version is envisioned that will link numerous hillslopes to model the erosion and sediment transport processes in large basins. WEPP will model basins that can have a single storm basinwide, and where upland, rather than channel, processes dominate sediment yields.

The European Soil Erosion Model (EUROSEM) (Morgan, Quinton, and Rickson, 1993) simulates erosion, transport, and deposition of sediment over the land surface by interrill and rill processes. It is designed as an event-based model for both individual fields and small catchments. Model outputs include total runoff, total soil loss, the storm hydrograph, and the

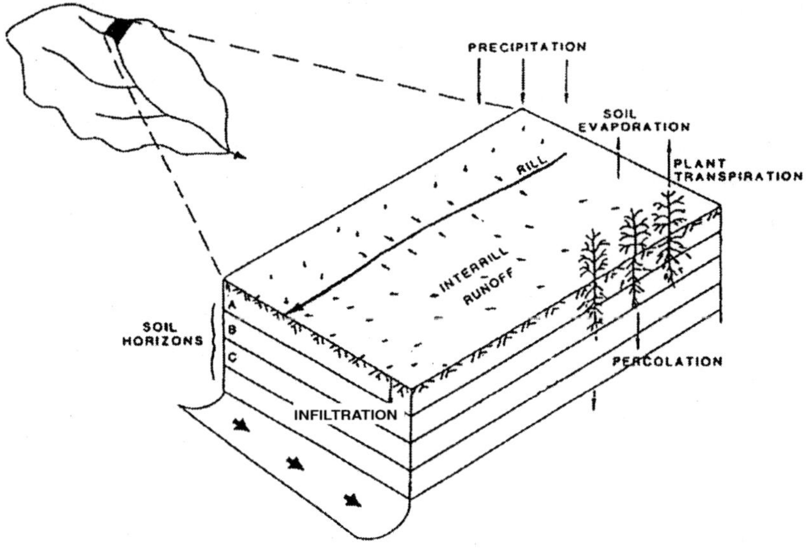

FIGURE 9.4. WEPP hillslope profile within a small watershed, showing hydrological processes (from Savabi and Williams, 1995).

storm sediment graph. EUROSEM simulates the effects of plant cover on rainfall interception, infiltration, flow velocity, and splash erosion (Morgan et al., 1998). It can be used in three different modes (Deinlein and Böhm, 2000). Erosion can be predicted for a (1) single plane or element that represents a small field with reasonably uniform slope, soil, and land cover conditions; (2) consecutive series of multiple planes or cascading elements that represent a heterogeneous slope, with each plane having uniform slope, soil, and land cover characteristics; and (3) small catchment, which is conceptualized as a cascading series of multiple planes and branching channels, with the plane elements contributing their runoff to the channels.

Additional models simulating soil erosion by water include the LImburg Soil Erosion Model (LISEM) (de Roo et al., 1998); the KINematic runoff and EROSion model (KINEROS) (Woolhiser, Smith, and Goodrich, 1990); EROSION-2D/3D (Schmidt, 1996); the Sediment Transport Model (STM-2D/3D) (Biesemans, 2000), which is a modification of EROSION-2D/3D; and the Simulation Model of Overland flow and ERosion Processes (SMODERP) (Dostál, Váška and Vrána, 2000).

Other models simulate complex processes of nutrient cycling or water pollution in addition to soil erosion by water. They include the AGricultural Non-Point Source model (AGNPS) (Young et al., 1994); the Erosion Productivity Impact Calculator (EPIC) (Williams, Dyke, and Jones, 1983; Williams, Jones, and Dyke, 1984, 1990); the Chemicals, Runoff and Erosion from Agricultural Management Systems model (CREAMS) (Knisel, 1980); the Groundwater Loading Effects of Agricultural Management Systems model (GLEAMS) (Knisel, Leonard, and Davis, 1933; Leonard, Knisel, and Still, 1987); and OPUS (Smith, 1992).

Model Applications

Favis-Mortlock and Guerra (2000) studied the influence of global greenhouse-gas emissions on soil erosion in Brazil using WEPP. They concluded, with a considerable band of uncertainty, that erosion rates may rise at the study site. Kincaid and Lehrsch (2001) tested the WEPP hillslope model with data taken under traveling lateral irrigation in southern Idaho. The main parameter affecting infiltration and runoff was hydraulic conductivity. The model was found to predict average runoff and soil loss reasonably well for small slope areas (< 40 m).

Using WEPP, Kalita et al. (2001) conducted a study to predict sediment and runoff from a watershed at the Fort Riley military training area in Kansas. Sediment and runoff data were collected with the help of a cutthroat flume and an automatic sampler installed at the watershed outlet. WEPP model input parameters were collected, measured, and estimated. Sediment and runoff data showed good agreement between measured and predicted results for individual events and excellent agreement for seasonal totals.

EUROSEM was evaluated against data for a watershed in Oklahoma. Model calibration was carried out using data for events from training periods with similar rainfall patterns and soil conditions to that of the test events. Once calibrated, the model was applied to four test events. EUROSEM simulated the total runoff and soil loss from three of these events quite well, but failed to reproduce the hydrographs and sedigraphs (Quinton and Morgan, 1998).

De Roo (2000) conducted a study on flood prevention and soil conservation in the Netherlands. He found LISEM to be a useful tool for planning cost-effective measures to mitigate the effects of runoff and erosion. Stolte et al. (2001) used LISEM to calculate water and sediment discharge in a 1,800 m^2 agricultural watershed on the Loess Plateau of China. Water content of the soil, and water and sediment discharge were measured automatically. Calibration was carried out for one event by adjusting saturated conductivity. Results showed reasonable agreement between measured and calculated water and sediment discharge.

Grunwald and Frede (2000) applied a modified version of AGNPS to three German watersheds with satisfactory results for both hydrology and sediment delivery. SMODERP was used to evaluate plans for the protection of urban areas against surface runoff and sediment from a small agricultural watershed (Dostál, Váška, and Vrána, 2000).

The EROSION-3D model has been successfully applied for estimating yields of sediment and sediment-bound heavy metals of drinking water reservoirs in the Osterzgebirge region of Saxony, Germany (Schmidt and von Werner, 2000). Model simulations showed that sediment yield can be considerably reduced if all agricultural land is managed using conservation tillage. EROSION-3D will be used for simulating erosion and sediment deposition before and after military training periods so that the impacts of training activities on sediment production can be demonstrated (Deinlein and Böhm, 2000). Erosion was simulated in support of land management in the loess belt of Flanders, Belgium, using STM-2D/3D (Biesemans, 2000), implemented in a two-dimensional hillslope and a three-dimensional watershed version.

Erosion Control

To control soil erosion by water, runoff must be eliminated or reduced to a water flow rate that cannot transport detached soil particles. Because water lost as runoff is of no benefit for crop production, controlling runoff is also essential for water conservation purposes. Conservation and minimum tillage, mulches, and cover crops prevent runoff initiation by intercepting raindrops. High infiltration rates are maintained and, therefore, runoff and erosion are minimized. Practices that retain runoff on cropland include contour tillage, furrow dikes, level terraces, and land leveling. These involve some type of soil surface manipulation and retain runoff under small-storm conditions, but water from large storms may overtop and wash out the earthen structures (Unger, 1996).

Under some conditions, it may not be practical or desirable to prevent or retain runoff. However, to control erosion under those conditions, runoff must occur at controlled, nonerosive rates. Practices that result in runoff at controlled rates include land smoothing, strip-cropping, graded furrows, graded terraces, variations of bench terraces, discontinuous parallel terraces, and land imprinting. The objective of these practices is to safely convey excess water from croplands to nearby waterways and streams (Unger, 1996).

An alternative approach to soil conservation is the modification of some soil properties responsible for the susceptibility of soil to erosion. Increasing aggregate stability at the soil surface and preventing clay dispersion are known to control seal formation, increase the infiltration rate, and reduce runoff in cultivated soils. In addition, stable aggregates at the soil surface are less susceptible to detachment by raindrop impact and to transportation by runoff water. Improving aggregate stability and preventing clay dispersion can be done by applying amendments to the soil. Gypsum and synthetic organic polymers are two types of soil amendments that have potential for controlling seal formation, runoff, and water erosion (Levy, 1996).

Organic matter improves the cohesiveness of the soil, increases its water retention capacity, promotes a stable aggregate structure, and reduces soil erosion by water. Organic material may be added as green manures, straw, or as a manure which has already undergone a high degree of fermentation (Morgan, 1995). Risse and Gilley (2001) assembled and summarized information quantifying the effects of manure application on runoff and soil loss resulting from natural precipitation events, and they developed regression equations relating reductions in runoff and soil loss to annual manure application rates. For selected locations on which manure was added annually, runoff was reduced from 1 to 68 percent, and soil loss decreased from 13 to 77 percent.

SUMMARY

Soil erosion by wind and water is a worldwide problem. Approximately 90 percent of cropland in the United States is currently losing soil above the sustainable rate. Erosion rates are even higher in many other countries. Computer models can be useful tools to assess erosion and evaluate the effect of alternative management scenarios on erosion.

The empirical wind erosion equation (WEQ) was developed in the 1960s to identify major factors influencing wind erosion and to develop wind erosion control methods. Advances in wind erosion science and the increased power of personal computers have allowed the development of a processed-based Wind Erosion Prediction System (WEPS) as a replacement for WEQ. WEPS includes submodels for erosion, hydrology, soil, crop, decomposition, and management. It is intended for planning soil conservation systems, providing environmental planning and assessment evaluations, and estimating off-site impacts of wind erosion.

The Universal Soil Loss Equation (USLE) predicts average annual soil erosion by water. It is an empirical model based on a large number of experimental data from small plots. The Water Erosion Prediction Project (WEPP) is a process-based model meant to replace USLE. It describes the processes that lead to erosion, including infiltration and runoff, soil detachment, transport, and deposition, and plant growth, senescence, and residue decomposition. The effects of tillage processes and soil consolidation are also modeled. The European Soil Erosion Model (EUROSEM) simulates erosion, transport, and deposition of sediment over the land surface by interrill and rill processes, and it simulates the effects of plant cover on rainfall interception, infiltration, flow velocity, and splash erosion. Applications of water erosion models have included the prediction of water and sediment discharge in watersheds on agricultural and military training lands, the study of flood prevention and soil conservation, and research on the influence of global greenhouse-gas emissions on soil erosion.

REFERENCES

Aase, J.K., F.H. Siddoway, and A.L. Black (1976). Perennial grass barriers for wind erosion control, snow management, and crop production. Publication No. 78. In *Shelterbelts on the Great Plains—Proceedings of the Symposium*, ed. R.W. Tinus. Denver, CO: Great Plains Agricultural Council, pp. 69-78.

Al-Durrah, M.M. and J.M. Bradford (1982). The mechanism of raindrop splash on soil surfaces. *Soil Science Society of America Journal 46:*1086-1090.

Anderson, C.H. (1975). *A History of Soil Erosion by Wind in the Palliser Triangle of Western Canada.* Historical Series No. 8. Research Branch, Ottawa, Ontario. Canada Department of Agriculture.

Anderson, R.L., R.A. Bowman, D.C. Nielsen, M.F. Vigil, R.M. Aiken, and J.G. Benjamin (1999). Alternative crop rotations for the Central Great Plains. *Journal of Production Agriculture 12:*95-99.

Anderson, R.S., M. Sørensen, and B.B. Willetts (1991). A review of recent progress in our understanding of aeolian sediment transport. *Acta Mechanica* (supplement) *1:*1-19.

Armbrust, D.V. and J.D. Dickerson (1971). Temporary wind erosion control: Cost and effectiveness of 34 commercial materials. *Journal of Soil and Water Conservation 26:*154-157.

Armbrust, D.V. and L. Lyles (1975). Soil stabilizers to control wind erosion. *Soil Science Society of America,* Special Publication No. 7, Soil Conditioners, pp. 77-82.

Armbrust, D.V. and L. Lyles (1985). Equivalent wind erosion protection from selected growing crops. *Agronomy Journal 77*(5):703-707.

Armstrong, R. (2001). *Grapes of Wrath in Inner Mongolia.* Available online: <http://www.usembassychina.org.cn/english/sandt/MongoliaDust-web.htm>.

Bagnold, R.A. (1941). *The Physics of Blown Sand and Desert Dunes.* London: Methuen.

Bates, C.G. (1911). *Windbreaks: Their Influence and Value.* Washington, DC: USDA Forest Service, Bulletin 86.

Berkofsky, L. and I.K. McEwan (1994). The prediction of dust erosion by wind: An interactive model. *Boundary Layer Meteorology 67:*385-406.

Biesemans, J. (2000). "Erosion Modelling as Support for Land Management in the Loess Belt of Flanders." (PhD thesis. Ghent University, Belgium.)

Bisal, F. and K.F. Nielsen (1962). Movement of soil particles in saltation. *Canadian Journal of Soil Science 42:*81-86.

Black, A.L. and F.H. Siddoway (1971). Tall wheatgrass barriers for soil erosion control and water conservation. *Journal of Soil and Water Conservation 26:*107-110.

Bondy, E., L. Lyles, and W.A. Hayes (1980). Computing soil erosion by periods using wind-energy distribution. *Journal of Soil and Water Conservation 35:*173-176.

Bradford, J.M. and C. Huang (1996). Splash and detachment by waterdrops. In *Soil Erosion, Conservation and Rehabilitation,* ed. M. Agassi. New York: Marcel Dekker, pp. 41-60.

Bryan, R. (1987). *Rill Erosion: Processes and Significance.* Catena Supplement 8. Cremlingen, Germany: Catena Verlag.

Carreker, J.R. (1966). Wind erosion in the Southeast. *Journal of Soil and Water Conservation 11:*86-88.

Chepil, W.S. (1944). Utilization of crop residues for wind erosion control. *Science in Agriculture 24:*307-319.

Chepil, W.S. (1950). Properties of soil which influence wind erosion: II. Dry aggregate structure as an index of erodibility. *Soil Science 69:*403-414.

Chepil, W.S. (1955). Effects of asphalt on some phases of soil structure and erodibility by wind. *Soil Science Society of America Proceedings 19:*125-128.

Chepil, W.S. (1957). Sedimentary characteristics of duststorms: III. Composition of suspended dust. *American Journal of Science 255:*206-213.

Chepil, W.S. and R.A. Milne (1941). Wind erosion of soil in relation to roughness of surface. *Soil Science 52:*417-431.

Chepil, W.S. and N.P. Woodruff (1963). The physics of wind erosion and its control. *Advances in Agronomy 15:*211-302.

Chepil, W.S., N.P. Woodruff, F.H. Siddoway, D.W. Fryrear, and D.V. Armbrust (1963). Vegetative and nonvegetative materials to control wind and water erosion. *Soil Science Society of America Proceedings 27:*86-89.

Chepil, W.S., N.P. Woodruff, and A.W. Zingg (1955). *Field Study of Wind Erosion in Western Texas.* Washington, DC: USDA, SCS-TP-125.

Cole, G.W., L. Lyles, and L.J. Hagen (1983). A simulation model of daily wind erosion soil loss. *Transactions of the ASAE 26:*1758-1765.

Craig, D.G. and J.W. Threlle (1964). *Guide for Wind Erosion Control on Cropland in the Great Plains States.* Washington, DC: USDA SCS.

de Roo, A. (2000). Applying the LISEM model for investigating flood prevention and soil conservation scenarios in South-Limburg, the Netherlands. In *Soil Erosion: Application of Physically Based Models,* ed. J. Schmidt. Berlin: Springer-Verlag, pp. 33-41.

de Roo, A., V. Jetten, C. Wesseling, and C. Ritsema (1998). LISEM: A physically-based hydrologic and soil erosion catchment model. In *Modelling Soil Erosion by Water,* eds. J. Boardman and D.T. Favis-Mortlock. NATO-ASI Series I-55. Berlin: Springer-Verlag, pp. 429-440.

Deinlein, R. and A. Böhm (2000). Modeling overland flow and soil erosion for a military training area in southern Germany. In *Soil Erosion: Application of Physically Based Models,* ed. J. Schmidt. Berlin: Springer-Verlag, pp. 163-178.

Dickerson, J.D., N.P. Woodruff, and E.E. Banbury (1976). Techniques for improving survival and growth of trees in semiarid areas. *Journal of Soil and Water Conservation 31:*63-66.

Dostál, T., J. Váška, and K. Vrána (2000). SMODERP—A simulation model of overland flow and erosion processes. In *Soil Erosion: Application of Physically Based Models,* ed. J. Schmidt. Berlin: Springer-Verlag, pp. 135-161.

Dregne, H.E. (1976). *Soils of the Arid Regions.* New York: Elsevier Scientific Publishers Co.

Duley, F.L. (1959). Progress of research on stubble mulching in the Great Plains. *Journal of Soil and Water Conservation 14:*7-11.

Englehorn, C.L., A.W. Zingg, and N.P. Woodruff (1952). The effects of plant residue cover and clod structure on soil losses by wind. *Soil Science Society of America Proceedings 16:*29-33.

Erickson, D.A., P.C. Deutsch, D.L. Anderson, and T.A. Sweeney (1984). Wind driven interactive wind erosion estimator. *Agronomy Abstracts,* 76th Annual Meeting, 247 pp.

Erpul, G. (2001). "Detachment and Sediment Transport from Interrill Areas under Wind-driven Rain." (PhD thesis. Purdue University, West Lafayette, Indiana.)

Farahani, H.J., G.A. Peterson, D.G. Westfall, L.A. Sherrod, and L.R. Ahuja (1998). Soil water storage in dryland cropping intensification. *Soil Science Society of America Journal 62:*984-991.

Favis-Mortlock, D.T. and A.J.T. Guerra (2000). The influence of global greenhouse-gas emissions on future rates of soil erosion: A case study from Brazil using WEPP-CO2. In *Soil Erosion: Application of Physically Based Models,* ed. J. Schmidt. Berlin: Springer-Verlag, pp. 3-31.

Federal Register (1992). 7 CFR Subtitle A (1-1-92 Edition) Part 12 *Highly Erodible Land and Wetland Conservation.* Subpart B Highly Erodible Land Conservation. Washington, DC: National Archives and Research Administration.

Ferber, A.E. (1969). *Windbreaks for Conservation.* Washington, DC: USDA SCS, Agricultural Information Bulletin 339.

Fisher, P.S. and E.L. Skidmore (1970). *WEROS: A Fortran IV Program to Solve the Wind Erosion Equation.* ARS 41-174. Agricultural Research Service, Washington, DC: U.S. Department of Agriculture.

Food and Agriculture Organization, United Nations (1960). *Soil Erosion by Wind and Measures for its Control on Agricultural Lands.* Development Paper No. 71. Rome, Italy: FAO.

Foster, G.R. (1982). Modeling the erosion process. In *Hydrologic Modeling of Small Watersheds,* ed. C.T. Haan. ASAE Monograph No. 5. St. Joseph, Michigan: ASAE, pp. 297-379.

Foster, G.R., D.C. Yoder, G.A. Weesies, and T.J. Toy (2001). The design philosophy behind RUSLE2: Evolution of an empirical model. In *Soil Erosion Research for the 21st Century, an International Symposium and Exhibition,* eds. J.C. Ascough II and D.C. Flanagan. Honolulu, Hawaii, January 3-5, 2001. St. Joseph, MI: ASAE, pp. 95-98.

Frank, A.B., D.C. Harris, and W.O. Willis (1977). Growth and yields of spring wheat as influenced by shelter and soil water. *Agronomy Journal 69:*903-906.

Fryrear, D.W. (1963). Annual crops as wind barriers. *Transactions of the ASAE 6:*340-342, 352.

Fryrear, D.W. (1969). *Reducing Wind Erosion in the Southern Great Plains,* MP-929, September. College Station, TX: Texas Agricultural Experiment Station.

Fryrear, D.W. (1981). Tillage influences monthly wind erodibility of dryland sandy soils. *Proceedings of ASAE Conference on Crop Production with Conservation in the 80s.* December 14-15, Chicago. ASAE Pub. 7-81, pp. 153-163.

Fryrear, D.W., A. Saleh, J.D. Bilbro, H.M. Schomberg, J.E. Stout, and T.M. Zobeck (1998). *Revised Wind Erosion Equation (RWEQ).* Wind Erosion and Water Conservation Research Unit, USDA-ARS, Big Spring, TX: Technology Bulletin No. 1. Southern Plains Area, Cropping Systems Research Laboratory.

Ghadiri, H. and D. Payne (1977). Raindrop impact stress and breakdown of soil crumbs. *Journal of Soil Science 28:*247-258.

Gillette, D.A. (1977). Fine particle emissions due to wind erosion. *Transactions of the ASAE 20:*890-897.

Gillette, D.A., I.H. Blifford Jr., and C.R. Fenster (1972). Measurements of aerosol size distribution and vertical fluxes of aerosols on land subject to wind erosion. *Journal of Meteorology 11:*977-987.

Gillette, D.A. and T.R. Walker (1977). Characteristics of airborne particles produced by wind erosion of sandy soil, High Plains of West Texas. *Soil Science 123:*97-110.

Gore, R. (1979). The desert: An age-old challenge grows. *National Geographic 156:*594-639.

Gregory, J.M. and M.M. Darwish (2001). Test results of TEAM (Texas Tech Erosion Analysis Model). In *Soil Erosion Research for the 21st Century, an International Symposium and Exhibition,* eds. J.C. Ascough II and D.C. Flanagan. Honolulu, Hawaii, January 3-5, 2001. St. Joseph, MI: ASAE, pp. 483-485.

Grunwald, S. and H.G. Frede (2000). Application of modified AGNPS in German watersheds. In *Soil Erosion: Application of Physically Based Models,* ed. J. Schmidt. Berlin: Springer-Verlag, pp. 43-57.

Hagen, L.J. (1976). Windbreak design for optimum wind erosion control. Publication No. 78. In *Shelter-belts on the Great Plains—Proceedings of the Symposium,* ed. R.W. Tinus. Denver, CO: Great Plains Agricultural Council, pp. 31-36.

Hagen, L.J. (1991a). A Wind Erosion Prediction System to meet user needs. *Journal of Soil and Water Conservation 46:*106-111.

Hagen, L.J. (1991b). Wind erosion mechanics: Abrasion of aggregated soil. *Transactions of the ASAE 34:*831-837.

Hagen, L.J., E.L. Skidmore, and J.D. Dickerson (1972). Designing narrow strip barrier systems to control wind erosion. *Journal of Soil and Water Conservation 27:*269-270.

Hagen, L.J., L.E. Wagner, and E.L. Skidmore (1999). Analytical solutions and sensitivity analyses for sediment transport in WEPS. *Transactions of the ASAE 42:*1715-1721.

Hairsine, P.B. and C.W. Rose (1991). Rainfall detachment and deposition: Sediment transport in the absence of flow driven processes. *Soil Science Society of America Journal 56:*234-242.

Halsey, C.F., W.F. Detmer, L.A. Cable, and E.C. Ampe (1983). SOILEROS-A friendly erosion estimation program for the personal computer. *Agronomy Abstracts,* 75th Annual Meeting, 20 pp.

Hayes, W.A. (1965). Wind erosion equation useful in designing northeastern crop protection. *Journal of Soil and Water Conservation 20:*153-155.

Hayes, W.A. (1966). *Guide for Wind Erosion Control in the Northeastern States.* Washington, DC: Soil Conservation Service, U.S. Department of Agriculture.

Hoag, B.K. and C.N. Geiszler (1971). Sunflower rows to protect fallow from wind erosion. *North Dakota Farm Research 28:*7-12.

Houghton, P.D. and P.E.V. Charman (1986). *Glossary of Terms Used in Soil Conservation.* Sydney: Soil Conservation Service of New South Wales.

Huang, C., F. Darboux, and A.S. Zartl (2001). A proposed modification to the WEPP erosion process model concept. In *Soil Erosion Research for the 21st Century, an International Symposium and Exhibition,* eds. J.C. Ascough II and

D.C. Flanagan. Honolulu, Hawaii, January 3-5, 2001. St. Joseph, MI: ASAE, pp. 91-94.

Hurt, R.D. (1981). *The Dust Bowl: An Agricultural and Social History.* Chicago: Nelson Hall.

Johnson, V. (1947). *Heaven's Tableland: The Dust Bowl Story.* New York: Farrar-Straus.

Kalita, P.K., L. Schieferek, S. Bhuyan, P. Woodford, and P. Gipson (2001). Application of WEPP model to military training lands. In *Soil Erosion Research for the 21st Century, an International Symposium and Exhibition,* eds. J.C. Ascough II and D.C. Flanagan. Honolulu, Hawaii, January 3-5, 2001. St. Joseph, MI: ASAE, pp. 119-122.

Kincaid, D.C. and G.A. Lehrsch (2001). The WEPP model for runoff and erosion prediction under center pivot irrigation. In *Soil Erosion Research for the 21st Century, an International Symposium and Exhibition,* eds. J.C. Ascough II and D.C. Flanagan. Honolulu, Hawaii, January 3-5, 2001. St. Joseph, MI: ASAE, pp. 115-118.

Kinnell, P.I.A. (1988). The influence of flow discharge on sediment concentrations in raindrop induced flow transport. *Australian Journal of Soil Research 26:*575-582.

Knisel, W.G. (1980). *CREAMS: A Field Scale Model for Chemicals, Runoff and Erosion for Agricultural Management Systems.* USDA, Conservation Research Report 26. Washington, DC: USDA, Science and Education Administration.

Knisel, W.G., R.A. Leonard, and F.M. Davis (1993). *GLEAMS version 2.1, Part I: Model Documentation.* UGA-CPES-BAED, Pub 5. Tifton, GA: University of Georgia Coastal Plains Experiment Station, Biological and Agricultural Engineering Department.

Lal, R. and W. Elliott (1994). Erodibility and erosivity. In *Soil Erosion Research Methods,* ed. R. Lal. Ankeny, IA: Soil and Water Conservation Society, pp. 181-208.

Lane, L.J. and M.A. Nearing (1989). *USDA-Water Erosion Prediction Project (WEPP): Hillslope Profile Model Documentation,* NSERL Report No. 2. West Lafayette, IN: National Soil Erosion Research Laboratory, USDA-ARS.

Le Bissonnais, Y. (1996). Soil characteristics and aggregate stability. In *Soil Erosion, Conservation and Rehabilitation,* ed. M. Agassi. New York: Marcel Dekker, pp. 41-60.

Leonard, R.A., W.G. Knisel, and D.A. Still (1987). GLEAMS: Groundwater loading effects of agricultural management systems. *Transactions of the ASAE 30:*1403-1418.

Levy, G.J. (1996). Soil stabilizers. In *Soil Erosion, Conservation and Rehabilitation,* ed. M. Agassi. New York: Marcel Dekker, pp. 267-299.

Lopez, M.V. (1998). Wind erosion in agricultural soils: An example of limited supply of particles available for erosion. *Catena 33:*17-28.

Lyles, L. (1974). *Speculation on the Effect of Wind Erosion on Productivity.* Special Report to Task Force on Wind Erosion Damage Estimates. Washington, DC: U.S. Department of Agriculture.

Lyles, L. (1983). Erosive wind energy distributions and climatic factors for the West. *Journal of Soil and Water Conservation 38:*106-109.

Lyles, L. (1988). Basic wind erosion processes. *Agriculture, Ecosystems and Environment 22/23:*91-101.

Lyles, L. and B.E. Allison (1980). Range grasses and their small-grain equivalents for wind erosion control. *Journal of Range Management 33:*143-146.

Lyles, L. and B.E. Allison (1981). Equivalent wind-erosion protection from selected crop residues. *Transactions of the ASAE 24:*405-408.

Lyles, L., D.V. Armbrust, J.D. Dickerson, and N.P. Woodruff (1969). Spray-on adhesives for temporary wind erosion control. *Journal of Soil and Water Conservation 24:*190-193.

Lyles, L., N.F. Schmeidler, and N.P. Woodruff (1973). *Stubble Requirements in Field Strips to Trap Windblown Soil.* Manhattan, KS: Kansas Agricultural Experiment Station, Research Publication 164.

Lyles, L., R.L. Schrandt, and N.F. Schmeidler (1974). Commercial soil stabilizers for temporary wind erosion control. *Transactions of the ASAE 17:*1015-1019.

Lyles, L. and J. Tatarko (1982). Emergency tillage to control wind erosion: Influences on winter wheat yields. *Journal of Soil and Water Conservation 37:*344-347.

Lyles, L., J. Tatarko, and J.D. Dickerson (1983). Windbreak effects on soil water and wheat yield. *American Society of Agricultural Engineers,* Paper No. 83-2074. 1983 Annual Conference.

Lyons, W.F., R.K. Munro, M.S. Wood, and Y. Shao (1998). A broadscale wind erosion model for environmental assessment and management. In *Proceedings of the First International Conference on Ecosystems and Sustainable Development. Advances in Ecological Sciences, Vol. 1,* eds. J.L. Uso and C.A. Brebbia. Southampton, United Kingdom: Computational Mechanics Publications, pp. 275-294.

Malin, J.C. (1946). Dust storms—Part one, two, and three. 1850-1860, 1861-1880, 1881-1900, respectively. *The Kansas Historical Quarterly 14:*129-144, 265-296, 391-413.

Mannering, J.V. and C.R. Fenster (1983). What is conservation tillage? *Journal of Soil and Water Conservation 38:*141-143.

Maticorena, B. and G. Bergamotti (1995). Modeling the atmospheric dust cycle: 1. Design of a soil-derived dust emission scheme. *Journal of Geophysical Research 100:*16415-16430.

McCalla, T.M. and T.J. Army (1961). Stubble mulch farming. *Advances in Agronomy 13:*125-196.

McMartin, W., A.B. Frank, and R.H. Heints (1974). Economics of shelterbelt influence on wheat yields in North Dakota. *Journal of Soil and Water Conservation 29:*87-91.

Moldenhauer, W.C. and E.R. Duncan (1969). *Principles and Methods of Wind Erosion Control in Iowa.* Special Report No. 62. Ames, IA: Iowa State University.

Morgan, R.P.C. (1995). *Soil Erosion and Conservation.* Essex, United Kingdom: Longman.

Morgan, R.P.C., D.D.V. Morgan, and H.J. Finney (1984). A predictive model for the assessment of soil erosion risk. *Journal of Agricultural Engineering Research 30:*245-253.

Morgan, R.P.C., J.N. Quinton, and R.J. Rickson (1993). *EUROSEM: A user guide.* Silsoe, United Kingdom: Silsoe College, Cranfield University.

Morgan, R.P.C., J.N. Quinton, R.E. Smith, G. Govers, J.W.A. Poesen, G. Chisci, and D. Torri (1998). The EUROSEM model. In *Modelling Soil Erosion by Water,* eds. J. Boardman and D.T. Favis-Mortlock. NATO-ASI Series I-55. Berlin: Springer-Verlag, pp. 389-398.

Morin, J. (1996). Rainfall Analysis. In *Soil Erosion, Conservation and Rehabilitation,* ed. M. Agassi. New York: Marcel Dekker, pp. 23-39.

Moss, A.J. (1988). Effects of flow velocity variation on rain-driven transportation and the role of rain impact in the movement of solids. *Australian Journal of Soil Research 26:*443-450.

Mutchler, C.K. (1967). Parameters for describing raindrop splash. *Journal of Soil and Water Conservation 22:*91-94.

PEDCO-Environmental Specialists (1973). *Investigations of Fugitive Dust-sources, Emissions, and Control.* Report prepared under Contract No. 68-02-0044, Task Order No. 9. Cincinnati, OH: US Environment Protection Agency, Office of Air Quality Planning and Standards.

Peterson, G.A., A.J. Schlegel, D.L. Tanaka, and O.R. Jones (1996). Precipitation use efficiency as affected by cropping and tillage systems. *Journal of Production Agriculture 9:*180-186.

Peterson, G.A., D.G. Westfall, and C.V. Cole (1993). Agroecosystem approach to soil and crop management research. *Soil Science Society of America Journal 57:*1354-1360.

Quinton, J.N. and R.P.C. Morgan (1998). EUROSEM: An evaluation with single event data from the C5 watershed, Oklahoma, USA. In *Modelling Soil Erosion by Water,* eds. J. Boardman and D.T. Favis-Mortlock. NATO-ASI Series I-55. Berlin: Springer-Verlag, pp. 65-74.

Rasmussen, K.R. and S. Rasmussen (1998). Initiation of aeolian mass transport in natural winds. *Earth Surface Processes and Landforms 5:*413-422.

Read, R.A. (1964). *Tree Windbreaks for the Central Great Plains,* USDA Agricultural Handbook 250. Washington, DC: USDA.

Renard, K.G., G.R. Foster, G.A. Weesies, and J.P. Porter (1991). RUSLE: Revised Universal Soil Loss Equation. *Journal of Soil and Water Conservation 46:*30-33.

Rice, M.A. and I.K. McEwan (2001). Crust strength: A wind tunnel study of the effect of impact by saltating particles on cohesive soil surfaces. *Earth Surface Processes and Landforms 26:*721-733.

Rice, M.A., I.K. McEwan, C.E. Mullins, and I. Livingstone (1999). A conceptual model of wind erosion of soil surfaces by saltating particles. *Earth Surface Processes and Landforms 24:*383-392.

Risse, L.M. and J.E. Gilley (2001). Modeling the impacts of manure on soil and water losses. In *Soil Erosion Research for the 21st Century, an International Sym-*

posium and Exhibition, eds. J.C. Ascough II and D.C. Flanagan. Honolulu, Hawaii, January 3-5, 2001. St. Joseph, MI: ASAE, pp. 271-274.

Rose, C.W. (1985). Developments in soil erosion and deposition models. *Advances in Soil Science 2:*1-63.

Rosewell, C.J., R.J. Crouch, R.J. Morse, J.F. Leys, R.W. Hicks, and R.J. Stanley (2000). Forms of erosion. In *Soils, Their Properties and Management,* eds. P.E.V. Charman and B.W. Murphy. Sydney: Sydney University Press, pp. 13-38.

Savabi, M.R. and J.R. Williams (1995). *Water Balance and Percolation.* Chapter 5. WEPP Model Documentation. West Lafayette, IN: USDA-ARS National Soil Erosion Research Laboratory, 14 pp.

Schmidt, J. (1996). *Entwicklung und Awendung eines physikalisch begründeten Simulationsmodells für die Erosion geneigter, landwirtschaftlicher Nutzflächen.* Berliner Geographische Abhandlungen 61, Berlin, Institut für Geographische Wissenschaften.

Schmidt, J. and M. von Werner (2000). Modeling the sediment and heavy metal yields of drinking water reservoirs in the Osterzgebirge Region of Saxony (Germany). In *Soil Erosion: Application of Physically Based Models,* ed. J. Schmidt. Berlin: Springer-Verlag, pp. 93-108.

Shao, Y., M.R. Raupach, and J.F. Leys (1996). A model for predicting aeolian sand drift and dust entrainment on scales from paddock to region. *Australian Journal of Soil Research 34:*309-342.

Sharma, P.P. (1996). Interrill erosion. In *Soil Erosion, Conservation and Rehabilitation,* ed. M. Agassi. New York: Marcel Dekker, pp. 23-39.

Siddoway, F.H., W.S. Chepil, and D.V. Armbrust (1965). Effect of kind, amount, and placement of residue on wind erosion control. *Transactions of the ASAE 8:*327-331.

Skidmore, E.L. (1965). Assessing wind erosion forces: Directions and relative magnitudes. *Soil Science Society of America Proceedings 29:*587-590.

Skidmore, E.L. (1976). A wind erosion equation: Development, application, and limitations. In *Atmosphere-Surface Exchange of Particulate and Gaseous Pollutants.* Symposium Proceedings, eds. R.J. Engleman and G.A. Schmel. Springfield, VA: National Technical Information Service, U.S. Department of Commerce, pp. 452-465.

Skidmore, E.L. (1983). Wind erosion calculator: Revision of residue table. *Journal of Soil and Water Conservation 38:*110-112.

Skidmore, E.L. (1986). Wind erosion climatic erosivity. *Climate Change 9:*195-208.

Skidmore, E.L. (1987). Wind erosion direction factors as influenced by field shape and wind preponderance. *Soil Science Society of America Journal 51:*198-202.

Skidmore, E.L. (1998). Wind erosion processes. In *Wind Erosion in Africa and West Asia: Problems and Control Strategies,* eds. M.V.K. Sivakumar, M.A. Zöbisch, S. Koala, and T. Maukonen. Aleppo, Syria: ICARDA, pp. 137-142.

Skidmore, E.L., P.S. Fisher, and N.P. Woodruff (1970a). Computer equation aids wind erosion control. *Journal of Soil and Water Conservation 22:*19-20.

Skidmore, E.L., P.S. Fisher, and N.P. Woodruff (1970b). Wind erosion control: Computer solution and application. *Soil Science Society of America Proceedings 34:*931-935.

Skidmore, E.L. and L.J. Hagen (1977). Reducing wind erosion with barriers. *Transactions of the ASAE 20:*911-915.

Skidmore, E.L., L.J. Hagen, D.G. Naylor, and I.D. Teare (1974). Winter wheat response to barrier-induced microclimate. *Agronomy Journal 66:*501-505.

Skidmore, E.L., L.J. Hagen, and I.D. Teare (1975). Wind barriers most beneficial at intermediate stress. *Crop Science 15:*443-445.

Skidmore, E.L., M. Kumar, and W.E. Larson (1979). Crop residue management for wind erosion control in the Great Plains. *Journal of Soil and Water Conservation 34:*90-96.

Skidmore, E.L. and R.G. Nelson (1992). Small-grain equivalent of mixed vegetation for wind erosion control and prediction. *Agronomy Journal 83:*98-101.

Skidmore, E.L., N.L. Nossaman, and N.P. Woodruff (1966). Wind erosion as influenced by row spacing, row direction, and grain sorghum population. *Soil Science Society of America Proceedings 30:*505-509.

Skidmore, E.L. and F.H. Siddoway (1978). Crop residue requirements to control wind erosion. In *Crop Residue Management Systems,* ed. W. R. Oschwald. Madison, WI: ASA, Special Publication No. 31, pp. 17-33.

Skidmore, E.L., J. Tatarko, and L.E. Wagner (1994). A climate data base for wind erosion prediction. In *Current and Emerging Erosion Prediction Technology: Extended Abstracts from the Symposium.* August 10-11, Norfolk, VA. Ankeny, IA: Soil and Water Conservation Society, pp. 87-89.

Skidmore, E.L. and J.R. Williams (1991). Modified EPIC wind erosion model. In *Modeling Plant and Soil Systems,* eds. R.J. Hanks and J.T. Ritchie. Agronomy Monograph 31. Madison, WI: American Society of Agronomy, pp. 447-459.

Skidmore, E.L. and N.P. Woodruff (1968). *Wind Erosion Forces in the United States and Their Use in Predicting Soil Loss.* Agricultural Handbook 346. Washington, DC: USDA-ARS.

Smith, R.E. (1992). *OPUS: An Integrated Simulation Model for Transport of Nonpoint-source Pollutants at the Field Scale.* Volume 1, Documentation. Washington, DC: USDA-ARS 98.

Sporcic, M. and L. Nelson (1999). *Wind Erosion Equation—Use of Microsoft Excel Spreadsheet.* Technical Notes—Agronomy 4, NRCS. Washington, DC: U.S. Department of Agriculture.

Staple, W.J. and J.H. Lehane (1955). The influence of field shelterbelts on wind velocity, evaporation, soil moisture, and crop yields. *Canadian Journal of Agricultural Science 35:*440-453.

Stolte, J., C.J. Ritsema, E. van den Elsen, and B. Liu (2001). Modeling water flow and sediment processes in a small gully system on the Loess Plateau of China. In *Soil Erosion Research for the 21st Century, an International Symposium and Exhibition,* eds. J.C. Ascough II and D.C. Flanagan. Honolulu, Hawaii, January 3-5, 2001. St. Joseph, MI: ASAE, pp. 326-329.

Svobida, L. (1940). *An Empire of Dust.* Caidwell, ID: Caxton Printers, Ltd.

Unger, P.W. (1996). Common soil and water conservation practices. In *Soil Erosion, Conservation and Rehabilitation,* ed. M. Agassi. New York: Marcel Dekker, pp. 239-266.

U.S. Global Change Research Information Office (2001). Available online: <http://www.gcrio.org/geo/soil.html>.

USDA (1995). *Wind Erosion Prediction System - Technical Description.* Manhattan, KS: USDA-ARS Wind Erosion Research Unit, Kansas State University.

USDA, Soil Conservation Service (1984). *National Resources Inventory.* Washington, DC: Author.

USDA, Soil Conservation Service (1988). *National Agronomy Manual,* 190-V. Washington, DC: Author.

USDA, Soil Conservation Service (1989). *Soil Erosion by Wind.* Agriculture Information Bulletin 555. Washington, DC: Author.

Wagner, L.E. (1996). An overview of the Wind Erosion Prediction System. *Proceedings from the International Conference on Air Pollution from Agricultural Operations,* February 7-9, Kansas City, MO. Ames, IA: MidWest Plan Service, pp. 73-78.

Wagner, L.E. and J. Tatarko (2001). WEPS 1.0 - What it is and what it isn't. In *Soil Erosion Research for the 21st Century, an International Symposium and Exhibition,* eds. J.C. Ascough II and D.C. Flanagan. Honolulu, Hawaii, January 3-5, 2001. St. Joseph, MI: ASAE, pp. 91-94.

Williams, J.R. and H.D. Berndt (1977). Sediment Yield Prediction and Utilization of Rangelands. Documentation and User Guide. USDA-ARS 63, Washington, DC.

Williams, J.R., P.T. Dyke, and C.A. Jones (1983). EPIC: A model for assessing the effects of erosion on soil productivity. In *Analysis of Ecological Systems. State-of-the-art in Ecological Modeling,* eds. W.K. Laurenroth, G.V. Skogerboe, and M. Flug. Amsterdam: Elsevier, pp. 553-572.

Williams, J.R., C.A. Jones, and P.T. Dyke (1984). A modeling approach to determining the relationship between erosion and soil productivity. *Transactions of the ASAE 27:*129-144.

Williams, J.R., C.A. Jones, and P.T. Dyke (1993). The EPIC model. In *EPIC–Erosion/Productivity Impact Calculator: 1. Model documentation*, eds. A.N. Sharpley and J.R. Williams. US Department of Agriculture Technical Bulletin No. 1768. Washington, DC: USDA, pp. 3-92.

Williams, J.R., M. Nearing, A. Nicks, E.L. Skidmore, C. Valentin, K. King, and R. Savabi (1996). Using soil erosion models for global change studies. *Journal of Soil and Water Conservation 51:*381-385.

Wilson, L. (1975). Application of the wind erosion equation in air pollution surveys. *Journal of Soil and Water Conservation 30:*215-219.

Wischmeier, W.H. (1959). A rainfall erosion index for a universal soil loss equation. *Soil Science Society of America Proceedings 23:*246-249.

Wischmeier, W.H. and D.D. Smith (1978). *Predicting Rainfall Erosion Losses: A Guide to Conservation Planning*. USDA, Agriculture handbook 537, Washington, DC: USDA.

Woodruff, N.P., W.S. Chepil, and R.D. Lynch (1957). *Emergency Chiseling to Control Wind Erosion.* Manhattan, KS: Kansas Agricultural Experiment Station, Technical Bulletin 90.

Woodruff, N.P., J.D. Dickerson, E.E. Banbury, A.B. Erhart, and M.C. Lundquist (1976). *Selected Trees and Shrubs Evaluated for Single-row Windbreaks in the Central Great Plains,* USDA-ARS, NC-37. Washington, DC: USDA.

Woodruff, N.P., L. Lyles, J.D. Dickerson, and D.V. Armbrust (1974). Using cattle feedlot manure to control wind erosion. *Journal of Soil and Water Conservation 29:*127-129.

Woodruff, N.P., L. Lyles, F.H. Siddoway, and D.W. Fryrear (1972). *How to Control Wind Erosion.* Agriculture Information Bulletin No. 354. Washington, DC: Agricultural Research Service, U.S. Department of Agriculture.

Woodruff, N.P. and F.H. Siddoway (1965). A wind erosion equation. *Soil Science Society of America Proceedings 29:*602-608.

Woolhiser, D.A., R.E. Smith, and D.C. Goodrich (1990). *KINEROS, a Kinematic Runoff and Erosion Model: Documentation and User Manual.* USDA-ARS 77, Washington, DC: USDA.

Young, R.A., C.A. Onstad, D.D. Bosch, and W.P. Anderson (1994). *Agricultural Non-point Source Pollution Model, Version 4.03: AGNPS User's Guide.* Morris, MN: USDA-ARS North Central Soil Conservation Research Laboratory.

Young, R.A., M.J. Romkens, and D.K. McCool (1990). Temporal variations in soil erodibility. In *Soil Erosion: Experiments and Models,* ed. R.B. Bryan. Catena Supplement 17. Cremlingen-Destedt, Germany: Catena Verlag, pp. 41-53.

Young, R.A. and J.L. Wiersma (1973). The role of rainfall impact in soil detachment and transport. *Water Resources Research 9:*1629-1636.

Zobeck, T.M., S. Van Pelt, J.E. Stout, and T.W. Popham (2001). Validation of the revised wind erosion equation (RWEQ) for single events and discrete periods. In *Soil Erosion Research for the 21st Century, an International Symposium and Exhibition,* eds. J.C. Ascough II and D.C. Flanagan. Honolulu, Hawaii, January 3-5, 2001. St. Joseph, MI: ASAE, pp. 471-474.

Chapter 10

Soil Water Dynamics

Ronald J. Hanks
Grant E. Cardon

A model is a small imitation of the real thing so a discussion of modeling soil water dynamics must be a small imitation of the dynamics of the soil water system. Models help us to partially understand what is happening to the soil that we manage or interact with on a daily basis. If we understand soil water dynamics we can better understand the decisions we must make in the real world. Some examples might be:

When should we irrigate?

How much water is stored in a soil?

How much can we afford to pay for water to irrigate?

The soil water system is very dynamic and responds to many atmospheric, plant, and management factors. Precipitation, *P* (mm), or irrigation, *Irr* (mm), increase the soil water content, *W* (mm mm^{-1}), if the soil conditions promote infiltration. If soil water content is high, or if soil hydraulic conductivity is low, most of the water added may be lost as runoff, *Run* (mm). Drainage, *Dr* (mm), from the soil may occur if soil water content is high. Upward flow, negative *Dr,* of water into the soil may occur if soil water content is low and there is a source of water at some depth. Also, water may be lost from the soil surface as evaporation, *EVAP* (mm), from the soil root zone, or as transpiration, *TRAN* (mm), from plants growing in the soil. These two losses from the soil are often lumped together as evapotranspiration, *ET* (mm). The following "water balance" equation, in one dimension, quantifies this soil water dynamic situation:

$$WCe = WCb + P + Irr - ET - Dr - Run \tag{10.1}$$

where *WCe* is the depth of soil water (mm) in a given depth of soil at the end of a time period and *WCb* is the depth of soil water (mm) in a given depth of soil at the beginning of the time period. The depth of soil must be greater than the total root depth of the plants.

Some of the components of Equation (10.1) are nonsoil factors, such as *Irr* and *P*. Other components, *WCe, WCb, ET, Dr,* and *Run* may be highly dependent on soil factors. *ET,* however, also has limits determined by atmosphere and plant factors. Models consider many of the complications and interactions and thus predict the numerical value of the components over a short or long time period. Until models and computers became available, understanding these many complicated interactions was limited to only one or two factors, with others being ignored.

MODEL CHARACTERISTICS

The model used herein to illustrate the impacts of different conditions is called SOWATET (Soil Water Evapotranspiration) (Hanks, 1991a,b). This is a mechanistic model in one dimension that requires knowledge of soil properties, initial conditions, and boundary conditions to compute a solution. The soil properties needed are the matric head H_m, and hydraulic conductivity *K(W),* versus volume water content *W,* relations as shown in Table 10.1. The equation of flow in one dimension is:

$$\frac{\partial W}{\partial t} = \frac{\partial}{\partial z}\left[K(W)\frac{\partial H}{\partial z}\right] + A(z,t) \tag{10.2}$$

where *t* is time, *z* is depth, *K(W)* is hydraulic conductivity, *H* is soil hydraulic head (sum of matric head, H_m, and gravity head, H_z). *A(z,t),* root water extraction is defined as:

$$A(z,t) = \frac{\left[Hroot + \left(RRES \cdot z\right) - H\left(z,t\right)\right] \cdot RDF(z,t) \cdot K(W)}{\Delta x \cdot \Delta t} \tag{10.3}$$

where *H(z,t)* is H_m at a specific depth and time, *RDF(z,t)* is the proportion of the total active roots in depth increment, *z,* and *x* is the distance between the plant roots and the point in the soil where *H(z,t)* is measured (10 mm assumed). *Hroot* is an effective root water potential at the soil surface and *RRES* is a root resistance term equal to (1 + *Rc*). Rc is a flow coefficient in the plant root system assumed = 0.05. When *RRES* is multiplied by depth, *z,* the product will account for the gravity term and "friction loss" in the root water potential.

TABLE 10.1. Soil Water Properties of Millville Silt Loam

Water Cont. W mm mm^{-1}	Matric Head H_m mm water	Hydraulic Conductivity $K(W)$ mm d^{-1}
0.02	−1.30E+06	2.40E−08
0.04	−6.40E+05	7.20E−08
0.06	−2.90E+05	6.72E−07
0.08	−9.50E+04	5.28E−06
0.10	−4.80E+04	4.56E−05
0.12	−2.70E+04	2.18E−04
0.14	−7.20E+03	1.42E−03
0.16	−3.90E+03	1.08E−02
0.18	−3.00E+03	4.32E−02
0.20	−2.40E+03	1.25E−01
0.22	−2.00E+03	4.56E−01
0.24	−1.70E+03	5.76E−01
0.26	−1.40E+03	1.08E+00
0.28	−1.20E+03	1.87E+00
0.30	−1.00E+03	3.12E+00
0.32	−8.40E+02	5.04E+00
0.34	−6.90E+02	8.16E+00
0.36	−5.60E+02	1.27E+01
0.38	−4.40E+02	1.99E+01
0.40	−3.20E+02	3.12E+01
0.42	−2.20E+02	5.28E+01
0.44	−1.20E+02	1.68E+02
0.46	−3.40E+01	2.88E+02
0.48	0.00E+00	2.88E+02

The major problem with the solution of Equation (10.2) is that both $K(W)$ and H_m are highly dependent on W, as shown in Table 10.1. The value of $K(W)$ at saturation, 299 mm d^{-1} , decreases by several orders of magnitude as W decreases. Also, the value of H_m at saturation, 0.0 mm at saturation, becomes more negative as the soil dries. Numerical methods were used to solve this problem.

Numerical Method

Equation (10.2), approximated numerically (Hanks, 1991b), gives Equation (10.4):

$$\frac{h(i,j+1)-h(i,j)}{\Delta t}C(i,j-1/2)=K(i-1/2,j+1/2)\left[\frac{TM\left[h(i-1,j)-h(i,j)\right]+TT\left[h(i-1,j+1)-h(i,j)\right]+\Delta A}{\Delta A\cdot\Delta C}\right]-K(i+1/2,j+1/2)\left[\frac{TM\left[h(i,j)-h(i+1,j)\right]+TT\left[h(i,j+1)-h(i+1.j+1)\right]+\Delta B}{\Delta B\cdot\Delta C}\right]+A(i,j+1) \tag{10.4}$$

where i is depth, h is the matric head (H_m), $\Delta t=[t(j-1)-t(j)]$, $\Delta A=[z(i)-z(i-1)]$, $\Delta B=[z(i+1)-z(i)]$ and $\Delta C=[z(i+1)-z(i-1)]/2$, $C(ij-1/2)$ is soil water capacity factor or the slope of the W-H_m relation from the data of Table (10.1). This equation allows for depth increments to be unequal. The coefficients TT and TM are used to improve stability of the numerical solution subject to the restriction that $TT + TM = 1$. If $TT = 1$ the solutions tend to be more stable but if $TT = ½$ the solutions are more accurate. Note that the values of $h(j)$ constitute the initial (known) values, and $h(j+1)$ are not known but will be computed.

This model requires an input file containing the information of Table 10.1 as well as the initial conditions (W versus depth at time = 0) and the boundary conditions at the soil surface and at the bottom of the soil. The surface boundary conditions (the potential movement of water in and out of the soil) change with time as weather conditions change. From Table 10.1, and the initial conditions, the initial values of H_m and $K(W)$ can be found for each depth increment.

Many other models have been developed that consider more complicated situations than discussed herein (Kool and van Genuchten, 1991; Richards and Smettem, 1992; Simunek, Huang, and van Genuchten, 1996, 1998).

SIMULATING SOIL-WATER-PLANT-ATMOSPHERE MANAGEMENT DYNAMICS

Example 1. Water Application at a Low Rate

Figure 10.1 shows the computed components of Equation (10.1) over a 20-day period. A uniform *Irr* rate of 240 mm d^{-1} for 0.412 d (10 h) was applied at the beginning of the period. At the beginning, the *W* was assumed to be 0.1 mm mm^{-1} throughout the 0 to 1,000 mm depth. All 100 mm of the water added entered the soil, so *WCe* was 100 mm greater than *WCb*. During the 0.42 d of water addition, all other components of Equation (10.1) were zero. From 0.42 d to 20 d, the potential *ET (POTET)* rate was assumed to be -4.8 mm d^{-1}, so water was removed from the soil. As time went on, cumulative *ET* increased because water was available from soil water storage. As *ET* increased, total *WC* decreased by the same amount. *EVAP* rate was constant for about 2 d then decreased as soil water flow to the soil surface was

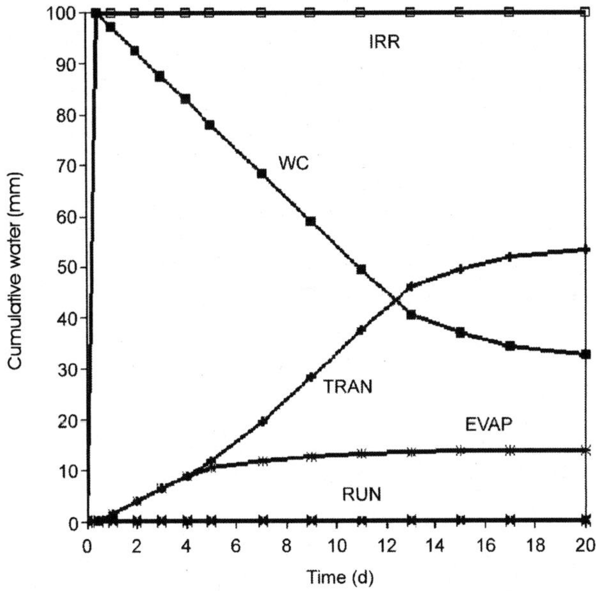

FIGURE 10.1. Water balance components for a 20-day period for Millville silt loam, starting with a uniform water content of 0.1 m m^{-1} and irrigation rate = 240 mm d^{-1} from 0 to 0.412 d and potential evapotranspiration rate = 4.8 mm d^{-1} until 20 d.

limited. However, *TRAN* increased at a constant rate for about 12 d before soil water became limited by water flow to plant roots. This demonstrates that *EVAP* and *TRAN* are quite different processes (Hanks 1991a,b). For this example *P, Dr,* and *Run* were zero for the entire period.

Example 2. Water Application at a High Rate

Example 2 was calculated where *Irr* rate was changed to 1,200 mm d^{-1} for 0.083 d (2 h), instead of 240 mm d^{-1} for 0.412 d. The results (Figure 10.2) show that the amount of water that entered the soil was slightly over 50 mm at the end of irrigation. Since *Irr* was 100 mm, the water that did not go into the soil was lost as runoff *(Run)* and was slightly less than 50 mm. Water was available in the soil to maintain *TRAN* rate constant for about 6 d, or half of the time of Example 1. At 20 d, *TRAN* was also about half that of Example 1. At 20 d *EVAP* was about 10 mm which was more than half that of Example 1 (about 14 mm). Thus the different processes produce different

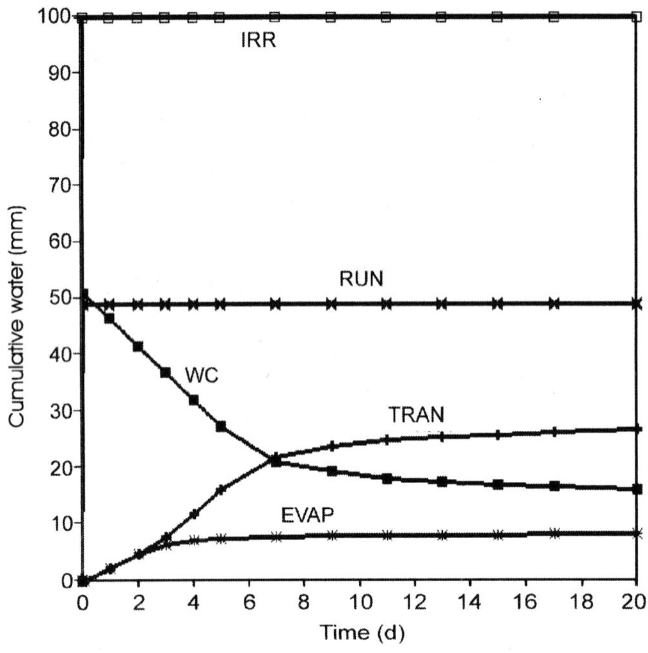

FIGURE 10.2. Same conditions as Figure 10.1 except irrigation rate was 1,200 mm d-1 from 0.0 to 0.083 d.

results. Models help us to understand the processes by allowing us to look at each one separately.

Figure 10.3 shows the infiltration rate as a function of time for Example 2. The rate was constant for a short time so *Run* was 0. Thereafter the infiltration rate decreased with time and *Run* rate increased with time since the application rate was constant.

Redistribution is another process that is impacted by water flow into or out of the soil. Redistribution is a process within the soil in which *W* changes within the soil. The change may be brought about by infiltration or *ET.* Moreover, redistribution may occur when there is no water entering or leaving the soil. Figure 10.4 shows that *W* changed with depth throughout the entire time for Example 1. All of the 100 mm of *Irr* had infiltrated the soil at 0.412 d, and *W* increased from 0.1 to about 0.46 at the surface to about 250 mm depth. At 0.42 d, *Irr* stopped but *W,* in the wetted zone, decreased to about 0.34. Below about 250 mm depth *W* increased. Thus the water stored in the top part of the soil, even after water application stopped, was redistributed into lower layers for many days but at a decreasing rate.

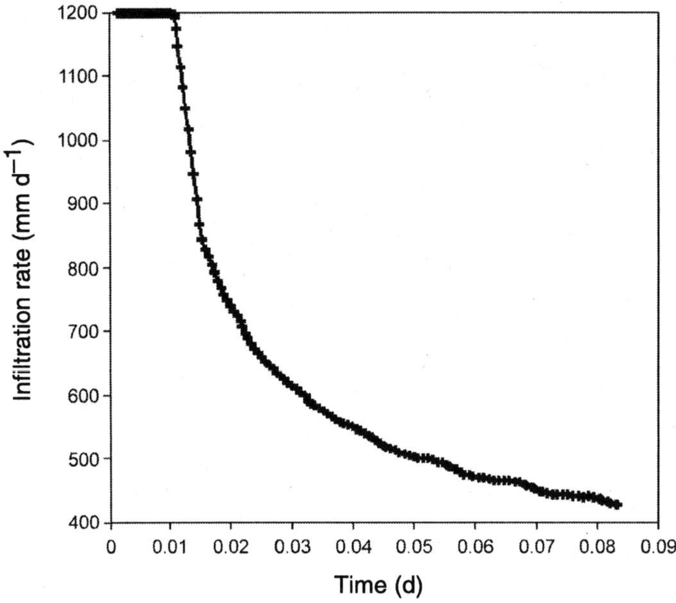

FIGURE 10.3. Infiltration rate as a function of time for the conditions of Figure 10.2 where irrigation was applied 1,200 mm d^{-1} from 0.0 to 0.083 d.

The change in water content shown in Figure 10.4 was not only due to redistribution but was also influenced by water extraction by plant roots, *TRAN*, and evaporation from the soil surface, *EVAP*.

Example 3. Redistribution of Water in the Soil with no ET

A valuable feature of the model is that experiments can be conducted to answer such questions as, "How would redistribution be influenced if there was no loss of water as *ET*?" Figure 10.5 shows redistribution of soil water as a function of time and depth where *ET* was zero. This was accomplished by changing the boundary conditions so that *POTET* was zero. The soil water content profile was the same at 0.412 d as in Figure 10.4, but thereafter the soil water content in the upper wet layers decreased much less with time. After 2 d the water content in the wetted area decreased very little. This illustrates another property of the soil known as *FC* (field capacity). *FC* is de-

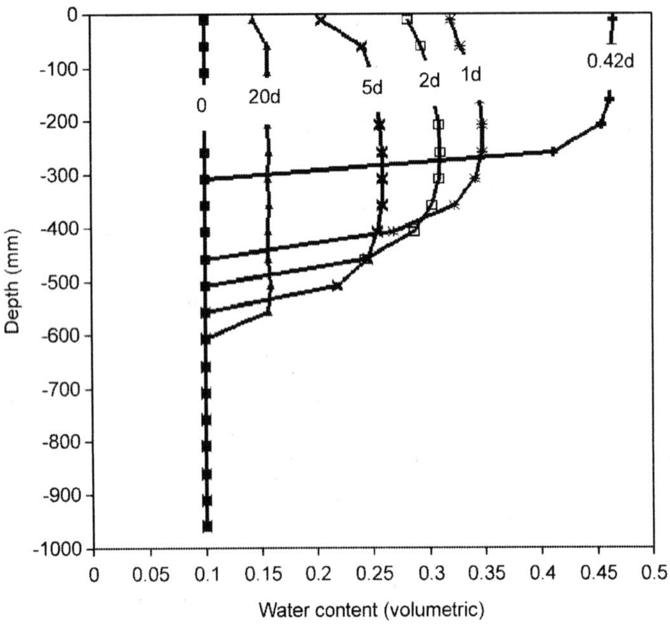

FIGURE 10.4. Soil water content as related to time for the conditions of Figure 10.1 showing redistribution of water as well as water loss to the surface as evaporation and from the root zone as transpiration.

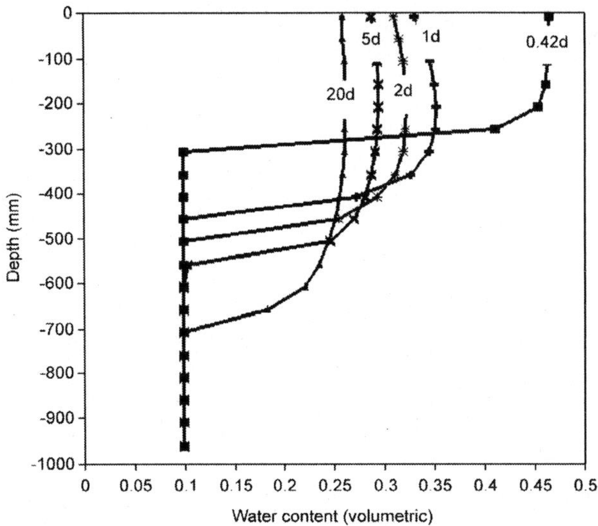

FIGURE 10.5. Soil water content profiles for the conditions of Figure 10.1 with ET of zero.

fined as the water content of the soil after it has been thoroughly wetted and allowed to drain until the drainage rate is small compared to the rate of extraction by plants. This condition is practically attained at about 2 d for most soils. At 2 d the soil water content was nearly constant to a depth of about 450 mm. Comparison of Figure 10.5 (no *ET*) with Figure 10.4 (with *ET*) shows that most of the soil water content change after 2 d in Figure 10.4 must have occurred due to *ET*.

Example 4. Computation of ET and Irr under Sugarcane in Colombia in the Presence of a Water Table

The SOWATET model was modified (Torres and Hanks, 1988) to account for a water table during a wet and dry season in Colombia, South America. There was an impervious layer at depth of 1,900 mm. During the wet season there were large amounts of *P*, compared to *ET*, so there was a buildup of a water table. During the dry season there was little *P* so *Irr* was required. The model was used to tell when to irrigate in conditions which could cause high water contents that might limit plant growth due to lack of aeration. The model called for *Irr* when the soil water content near the top of the root zone dried down to a set value. The simulation required an upper boundary condition, including *P,* and *POTET* for 350 d taken from local data.

The results given in Figure 10.6 show that water uptake from the water table occurred through most of the year to give a total *ET* of about 1,180 mm. The total *Irr* during the same period was about 430 mm. *Irr* was required for the first 50 d but not for 50 to 159 d. The water table depth was about 1,800 mm. After 300 d, the depth to the water table depth decreased due to high rain and lower *ET.*

LIMITATIONS OF MODELS WITH HOMOGENEOUS SOIL WATER PROPERTIES

Soil Structure

Most soils under field conditions have structure changes that occur with time that cause changes in the soil water properties like those shown in Table 10.1. The simulations discussed so far assumed that no structural

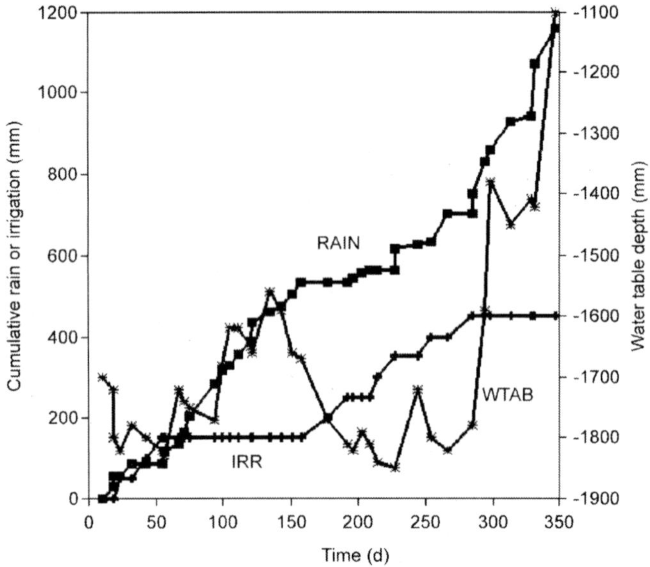

FIGURE 10.6. Simulation of water table depth as influenced by irrigation and precipitation (rain) for the conditions of Cauca Valley, Colombia in 1984 (calculated with the model by Torres and Hanks, 1988).

changes occurred in the soil so there was a single relation between $W\text{-}K(W)$ and $W\text{-}H_m$.

Typically, most agricultural soils will have structural changes from seeding time until harvest. Oliveira (1986) studied the influence that time and various treatments had on infiltration. He found that saturated hydraulic conductivity, one measure of structural change, decreased with time during the season for all treatments. However, the decrease was more for bare soil than it was for strew-mulched soil. Also, runoff rate, another measure of structure change, increased with time during the season for all treatments. Relations similar to those of Table 10.1 were estimated using the Shani et al. (1987) method. This provided relations of $W\text{-}K(W)$ and $W\text{-}H_m$ for each treatment and each time. Model predictions of infiltration and runoff for Millville silt loam (the soil properties shown in Table 10.1) agreed closely to measured data. However, model prediction for another soil, Nibley silty clay loam, were not as good, probably because of extensive soil surface cracking.

Furrow irrigation is one of the oldest methods of irrigation that typically involves soil structural changes with time during the year. It is ironic that it is also one of the most difficult to model. During water application, soil water content changes in three dimensions: (1) between furrows, (2) with depth, and (3) with distance down the furrow. Water is added at the furrow inlet so that it flows down the furrow and eventually to the end of the field. The structural condition of the soil in and near the furrow is quite different than between the furrows and usually changes with time during the season. There is fluid flow of water in the furrows and saturated and unsaturated water flow between and underneath the furrows. Since there is a difference in the amount of time that water is available from the top to the bottom of the field, the amount of water absorbed by the soil is different. If sufficient water is delivered to the furrow inlet to completely wet the root zone over the length of the field, there is usually excess water infiltrated at the inlet end of the field. There may be erosion in the furrow from part of the field, and deposition in other parts of the field. There are changes in soil structure in and near the furrow with time. In some instances, compaction of the furrow before irrigation is done to decrease infiltration so that water flows to the end of the field. Walker and Skogerboe (1987) have developed a two-dimensional model that predicts water flow into the soil, position of the water front in the furrow, and other aspects of furrow irrigation. One of the important parameters is the infiltration from the furrow into the soil. To get good simulations infiltration parameters must be changed with time and treatment. Classical soil parameters that will predict particular situations have not been found. Thus, infiltration parameters are adjusted for the same soil due to structural changes. An empirical infiltration equation that required three parameters was used.

An interesting aspect of furrow irrigation is the effect of surge flow on the efficiency of furrow irrigation in the field. Surge flow is a practice in which water is added to the furrow in surges such as 20 minutes on and 20 minutes off. Figure 10.7 shows a simulation of surge compared to continuous flow for a sandy loam soil near Flowell, Utah, for a nonwheel furrow. The solid lines represent the surge-by-surge advance-recession of the liquid water in the furrows and the dotted line shows the continuous flow treatment. The water in the furrow, for the continuous flow treatment, never reached the bottom of the field. In contrast, the surge treatment reached the bottom of the field after nine cycles. The cumulative infiltration, after 60 minutes, was about twice as much for continuous flow as for surge flow. Moreover, there was water in the furrows for only half of the time for surge flow compared to continuous flow. The average depth of water added to the field was about 100 mm for continuous flow compared to 40 mm for surge flow. Thus the irrigation applied was more evenly distributed over the field for surge flow than for continuous flow.

The same study also compared continuous and surge flow in which the soil had been compacted by tractor wheels when the furrows were being formed. The compaction had a significant effect on infiltration, and thus

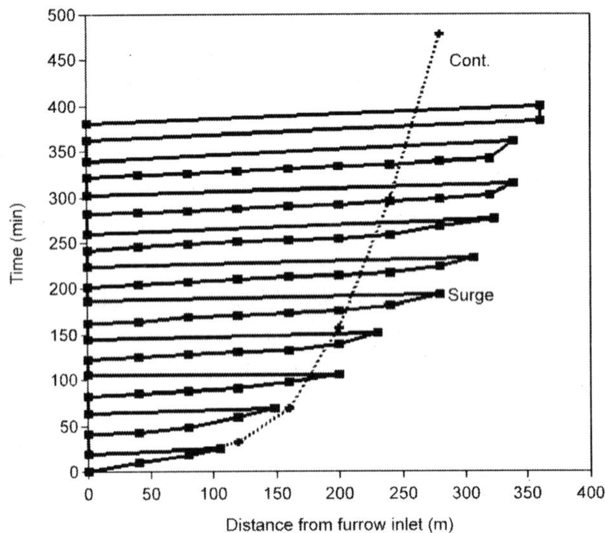

FIGURE 10.7. Comparison of the distance of furrow wetting front with time for continuous (dotted line) and surge flow (solid line) irrigation. Surge flow cycle was 20 min on and 20 min off. Length of field was 360 m. (*Source:* Adapted from Walker and Skogerboe, 1987.)

structure. Cumulative infiltration, after 60 minutes, was 1.24 times greater for the continuous flow, noncompacted soil, compared to the compacted soil.

Spatial Variability

Spatial variability has been shown to be very large for some soil properties, even on soils that have been thought to be fairly uniform (Nielson, Biggar, and Ehr, 1973; Warrick, 1998). Warrick (1998) lists the relative variability of soil properties. Saturated hydraulic conductivity was one of the most variable, bulk density was of low variability, and water content at various matric heads was of medium variability. Because of the high variability of the saturated hydraulic conductivity, there is a question about the value of modeling assuming homogeneous soil properties.

An important role that models can play in evaluating the effect of spatial variability is to make many computer simulations related to the variability of soil properties for important input conditions to see what effect spatial variability has on the intended use of the area. For example, a study of the effect of irrigation amount and frequency and spatial variability is of interest. At a given irrigation frequency, simulations could be run of different irrigation amounts for several sets of $W\text{-}H_m$ and $W\text{-}K(W)$ data, representing spatial variability. The results would show that there is an irrigation amount above which simulated *ET* (and crop growth) would not be influenced by spatial variability. Farmers often irrigate their fields so that the crop looks uniform by overirrigating part of the field to mask any effects caused by spatial variability. If irrigation or rainfall rates are low, all of the water could be absorbed by the soil and plant growth would not suffer if the soil had the ability to store the water in the root zone. However, if irrigation or rainfall rates were high, then spatial variability of saturated hydraulic conductivity would result in differing amounts of water absorbed by the soil and thus differing crop response. Also, it is possible to irrigate excessively when spatial variability is large and get good yields, but considerable drainage below the plant root zone may occur. Models could be used with different properties related to spatial variability to evaluate drainage.

Another approach to minimize the problems of spatial variability is to use models that require less highly variable soils information. Ritchie (1985) and the DSSAT group (Tsuji, Uehara, and Ballas, 1995) have developed a model that has been used extensively. This model depends on soil water-holding characteristics and not water flow characteristics and uses daily time steps. Estimates of upper and lower limits of *W* found under field conditions are needed for each of several layers. The initial *W* is needed as is daily *P* and *Irr* (the same as for the simulations such as Example 1). Daily weather (solar radiation, maximum temperature, minimum temperature, and precipitation) is used to compute *POTET.* This model computes daily *EVAP, TRAN,* growth of roots, water uptake of roots, soil temperature, growth of dry matter and grain. This was used to estimate plant growth as influenced by *TRAN* and *Irr.* The climatic data were from Griffith, Australia, for 1998. The soil was a red-brown clay with very high water-holding capacity, but shallow depth. Figure 10.8 shows the results of this model simulation in which irrigation was changed, by steps, from 0 to about 400 mm. Figure 10.8 shows that yield increased almost linearly as irrigation increased (*P* was about 150 mm) from about 50 mm to 330 mm. Increasing irrigation above 330 mm did not influence yield because water was not limited at that point. Figure 10.9 shows that there is a close relationship between yield and *TRAN.*

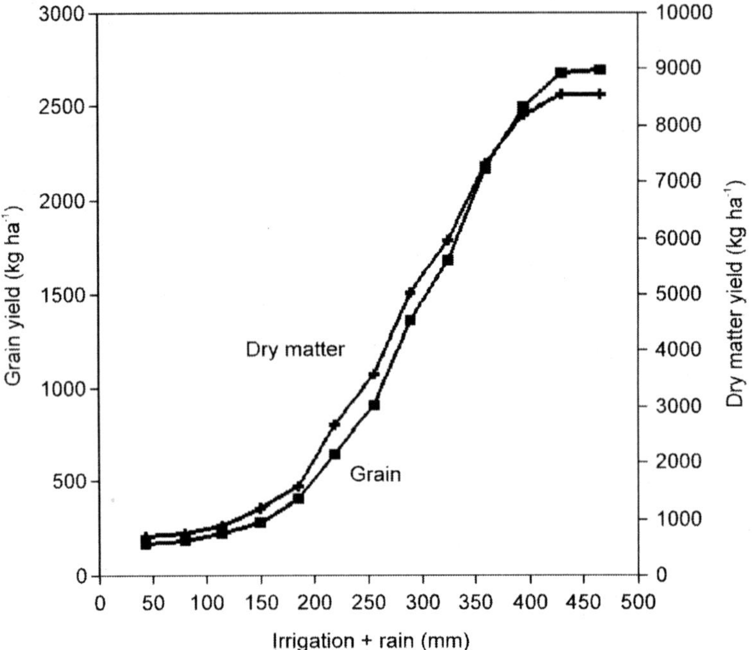

FIGURE 10.8. Simulation of yield of maize (sweet corn) as related to irrigation and rain for several levels of irrigation for Griffith, Australia, 1998, red-brown clay soil.

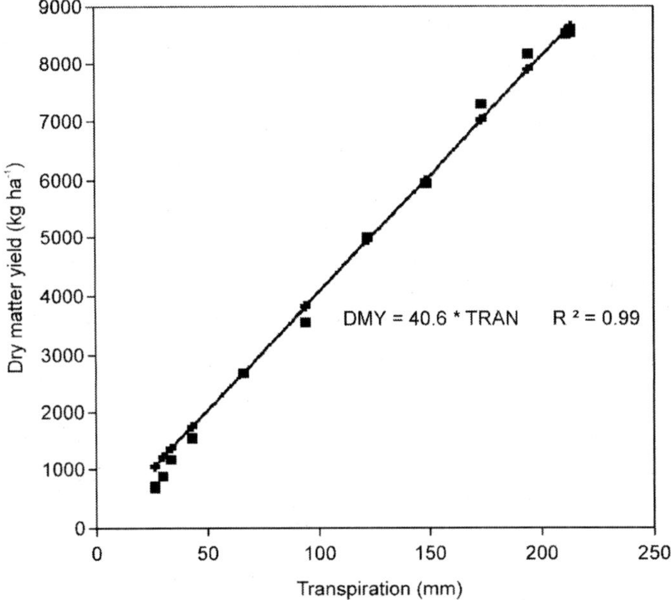

FIGURE 10.9. Simulation of dry matter yield of maize (sweet corn) as related to transpiration for several levels of irrigation at Griffith, Australia, 1998, red-brown clay soil. The solid line shows the regression equation predictions.

Hysteresis

Model prediction of soil water flow is complicated by hysteresis in the W-H_m relation (Topp, 1971). If the soil is wetting up, as for infiltration into a dry soil, the value of H_m, for a given W, is different than if the soil is drying down, as in redistribution or drainage. Common situations in the field, such as infiltration followed by redistribution, involve both wetting and drying, so provisions should be made for transition from a wetting curve to a drying curve. Si and Kachanoski (2000) found that model predictions of soil water storage during drainage were poor if the hydraulic parameters were estimated from infiltration only. They proposed a Haynes' Jump hysteresis model that gave predictions of drainage and soil water storage that were less than measurement errors. They concluded that for practical purposes, the combination of the proposed Haynes' Jump model with the predictions of hysteretic coefficients by either the Parlange model (Parlange, 1976) or the Mualem model (Mualem, 1984) be used to describe hysteresis.

Hopmans, Roy, and Wallender (1991) used a model to evaluate irrigation management as influenced by hysteresis and spatial variability. Kool and van Genuchten (1991) developed a model to consider hysteresis as well as root growth.

The results of model computations, considering hysteresis, for infiltration followed by redistribution, have shown that water content is higher in the redistribution zone than where hysteresis was ignored as done for Examples 1 and 2 (Hillel, 1998). The effect of hysteresis may be compensated for if the W-H_m potential relation and the W-$K(W)$ relation are estimated taking into account values for the field capacity water content (H_m about –2000 mm) as well as the Brooks and Corey (1964) coefficients found during infiltration. These coefficients were determined using the field method of Shani et al. (1987). Without this addition, the model predictions gave a much lower field capacity than measured. The data of Table 10.1 include this modification.

Another method of minimizing hysteresis is to use the model of Ritchie (1985) and the DSSAT group (Tsuji, Uehara, and Ballas, 1995). This model does not require any of the information of Table 10.1, but requires upper and lower limits of *W* measured in the field. Retta and Hanks (1980) use a similar model. These models assume that all rain or irrigation infiltrates into the soil. Drainage is computed if the water added is in excess of the available storage capacity of the soil. If rainfall is excessive so that runoff occurs these models may be in considerable error unless some information is available to estimate the rainfall that infiltrates into the soil.

SUMMARY

Models are very useful for demonstrating factors that influence soil water dynamics. Starting with a dry soil, the addition of water by *Irr* for less than a day allowed *EVAP* and *TRAN* to occur for several days. The dynamics of redistribution were also demonstrated. When infiltration rate was higher, with the same amount of water, *Run* occurred and *EVAP* and *TRAN* were decreased. With the same initial conditions the SOWATET model demonstrated the difference between the loss of water as *EVAP* or *TRAN*. This difference is very difficult to measure under field conditions. The SOWATET model also showed the field capacity phenomenon during redistribution of soil water and the upward flow as a function of time. Demonstration of the influence of upward flow in a field season was computed by a model. Model computations allowed an estimate of the amount of water being used as *EVAP* and *TRAN* and gave information on the amount of *Irr* needed. A two-dimensional model, developed for furrow irrigation, required different estimations of infiltration due to changes in soil structure. Changes in management to increase irrigation efficiency, such as surge, depended on changes in soil water properties. Spatial variability questioned the use of single parameters for field application. Models used with different parameters, related to spatial variability, could help evaluate the effect of spatial variability on other important factors. Many models have been developed to consider hysteresis effects. Simple models depending on soil water storage parameters may be useful if spatial variability is high or hysteresis of importance.

REFERENCES

Brooks, R.H. and A.T. Corey. (1964). *Hydraulic Properties of Porous Media.* Hydrology. Paper No. 3, Ft. Collins, CO: Colorado State University.

Hanks, R. J. (1991a). Infiltration and redistribution. In *Modeling Plant and Soil Systems,* American Society of Agronomy Monograph No. 31, eds. R.J. Hanks and J.T. Ritchie. Madison, WI: American Society of Agronomy. pp. 181-204.

Hanks, R. J. (1991b). Soil evaporation and transpiration. In *Modeling Plant and Soil Systems,* American Society of Agronomy Monograph No. 31, eds. R.J. Hanks and J.T. Ritchie. Madison, WI: American Society of Agronomy. pp. 245-271.

Hillel, D. (1998). *Environmental Soil Physics.* San Diego, CA: Academic Press.

Hopmans, J.W., K.C. Roy, and W.W. Wallender. (1991). Irrigation water management and soil-water hysterisis—A computer modeling study with stochastic soil hydraulic properties. *Transactions ASAE 34:*449-459.

Kool, J.B. and M. Th. van Genuchten. (1991). *HYDRUS: One-Dimensional Variably-Saturated Flow and Transport Model, Including Hysteresis and Root Water Uptake. Version 3.3.* U.S. Salinity Laboratory Research Report No. 124. Riverside, CA: U.S. Salinity Laboratory.

Mualem, Y. (1984). Prediction of the soil boundary wetting curve. *Soil Science 137:*379-389.

Nielson, D.R., J. W. Biggar, and K.T. Ehr. (1973). Spatial variability of field-measured soil-water properties. *Hilgardia 42:*215-259.

Oliveira, Carlos A.S. (1986). "Influence of Diking and Mulch Soil Surface Treatments on Infiltration and Runoff As Affected by Irrigation."(PhD Dissertation. Utah State University.)

Parlange, J.Y. (1976). Capillary hysteresis and the relation between wetting and drying curves. *Water Resources Research 12:*224-228.

Retta, A. and R. J. Hanks. (1980). *Manual for Using Model PLANTGRO.* Research Report 46. Logan, UT: Utah Agricultural Experimental Station.

Richards, B.G. and K.R.J. Smettem. (1992). Modeling water flow in two- and three-dimensional applications. I. General theory for non-swelling and swelling soils. *Transactions ASAE 35:*1497-1504.

Ritchie, J. T. (1985). A user-oriented model of soil water balance in wheat. In *Wheat Growth and Modeling,* NATO-ASI Series, eds. W. Day and R. K. Atkins. New York: Plenum Press. pp. 293-305.

Shani, U., R.J. Hanks, E. Bresler, and C.A.S. Oliviera. (1987). A simple field method for estimating the hydraulic conductivity and matric potential water content relations of soils. *Soil Science Society of America Journal 51:*298-302.

Si, B.C. and R. G. Kachanoski. (2000). Unified solution for infiltration and drainage with hysteresis: Theory and field test. *Soil Science Society of America Journal 64:*30-36.

Simunek, J.K., K. Huang, and M.Th. van Genuchten. (1996). *The SWMS-3D Code for Simulating Water Flow and Solute Transport in Three-Dimensional Variably-Saturated Media. Version 1.* U.S. Salinity Laboratory Research Report No. 139. Riverside, CA: U.S. Salinity Laboratory.

Simunek, J.K., K. Huang, and M.Th. van Genuchten. (1998). *The HYDRUS Code for Simulating the One-Dimensional Movement of Water, Heat, and Multiple Solutes in Variably-Saturated Media. Version 6.0.* U.S. Salinity Laboratory Research Report No. 144. Riverside, CA: U.S. Salinity Laboratory.

Topp, G.C. (1971). Soil water-hysteresis: The domain theory extended to pore interaction conditions. *Soil Science Society of America Proceedings 33:* 219-225.

Torres, J.S. and R.J. Hanks. (1988). Modeling water table contributions to crop evapotranspiration. *Irrigation Science 10:*265-279.

Tsuji, G.Y., G. Uehara, and S. Ballas (Eds.) (1994). *DSSAT version 3.* Honolulu, Hawaii: University of Hawaii.

Walker, W.R. and G.V. Skogerboe. (1987). *Surface Irrigation—Theory and Practice.* Englewood Cliffs, NJ: Prentice-Hall, Inc.

Warrick, A.W. (1998). Spatial variability. In *Environmental Soil Physics,* ed. D. Hillel. San Diego: Academic Press. pp. 655-675.

Chapter 11

Solute-Water Interactions

Robert Edis
Robert E. White

The redistribution of solutes in the soil solution usually represents a loss of material from where it may be beneficial, harmless, or undesirable to a location where it is likely to be deleterious. The dissolved constituents may be nutrients, which when leached result in a significant reduction in soil fertility. Leaching can also lead to excessive quantities of nutrients in groundwater and surface water, degrading the quality of these resources through contamination and eutrophication. Leaching of nitrate (NO_3^-) generated in the soil, together with balancing cations, can accelerate soil acidification. Leaching of biocides, contaminants, and salts also endangers water resources and the well-being of those who use them. A few examples of positive aspects of solute transport in soil include the leaching of excess salts from the root zone, the wash-in of harmful substances from the surface, and the transport of solutes toward pump inlets during clean-up operations. Consequently, the leaching of solutes through soil has important economic, environmental, and social implications. The modeling of solute transport is useful for assessing the degree and rate of resource degradation likely to occur under different management scenarios. The risks associated with land use options, and the likely success of remediation strategies, can also be assessed.

To quantitatively describe the movement of solute, various processes, features, and events that impact solute redistribution must be considered. The processes are physical (water flow, convection, diffusion, volatilization, radioactive decay), chemical (ion-surface interactions, chemical degradation, precipitation-dissolution), microbiological (immobilization, mineralization, transformations, and degradation), and phytological (water extraction, active/passive solute uptake and exclusion, redistribution). The features making up the framework within which the processes operate are themselves subject to change, either through ongoing, gradual processes or through sudden events. These changes may be in response to management (such as clearing, cultivation, surface sealing, or changes in structure by plant roots), or weather (such as wetting and drying, freezing and thawing), and interactions between the two. Many models have been developed that address these concepts at various levels of detail and complexity. The mathematics of these modeling approaches are well described elsewhere, including advanced soil physics texts such as that of Jury, Gardner, and Gardner (1991).

The questions being asked of the models include:

1. How much solute leaches beyond the root zone over a period of time?
2. What will the maximum concentration of solute in the leaching water be?
3. How long will it take for solute to reach the water table?
4. What will be the distribution of solute in the soil profile after a particular time period?

The level of spatial and temporal complexity, both of the system and of the required solution, will be central in deciding the approach adopted. In this chapter, the various concepts employed in describing solute-water interactions and how they have been applied in the field will be discussed. Equations have been kept to a minimum, with references made to appropriate sources for the interested reader. In this review, we consider modeling concepts on the basis of how a model deals with the movement of solute in soil water. Other processes are included as required. The concepts are considered in order of their process complexity, from piston displacement through stream-tube flow to convection (or advection)-dispersion approaches. In all the modeling approaches, there is the underlying concept of the transport volume. The nature of the porosity through which solutes and water move will clearly influence their interaction.

THE TRANSPORT VOLUME

The transport volume is defined as the fluid volume effective in transporting solute during a period of observation. It is usually expressed as the volume fraction, θ_{st}. The occurrence of preferential flow of water and solutes, and the apparent "protection" of solutes inside aggregates from leaching (Addiscott and Cox, 1976), have led to the postulation of a transport volume that is smaller than the total soil porosity (Jury, Sposito, and White, 1986). The transport volume is an operationally defined parameter that can be highly irregular in shape, and may vary in size as the rate of water flow through the soil changes in space and time (White, 1987). Within the transport volume, dissolution/precipitation, adsorption/desorption, diffusion, and biological and chemical transformations can take place.

The transport volume θ_{st} is analogous to the mobile water fraction θ_m used in mechanistic models (White, 1985b,c), which is the fraction of the wetted soil volume through which water containing dissolved solute moves (Coates and Smith, 1964; van Genuchten and Wierenga, 1976). As with θ_{st}, θ_m has been shown in both laboratory and field studies to change with time and to depend on the concentration of the soil solution, initial soil water content, soil water flux, and aggregate size (Nkedi-Kizza et al., 1983; Smettem, 1984; Seyfried and Rao, 1987; Jaynes, Rice, and Bowman, 1988; Clothier, Kirkham, and McLean, 1992; Clothier et al., 1995). Values of θ_{st} have been deduced as the unknown in model calibration experiments (Jarvis et al., 1991); by a curve-fitting procedure in applying mechanistic models (van Genuchten and Wierenga, 1976) or stochastic models (White, Dyson, Haigh, et al., 1986); chosen as an arbitrary fraction of the soil's resident water that is not bypassed by the invading solution (Addiscott, 1977; Corwin, Waggoner, and Rhoades, 1991); estimated from the relationship between mobile water and pressure head (Angulo-Jaramillo et al., 1996) or measured using disc permeametry (Clothier, Kirkham, and McLean, 1992).

Solutes dissolved in the pore water may be unevenly distributed between water that is or is not in the transport volume. Therefore, solute concentrations need to be identified as flux or resident concentrations. A flux concentration is the flux-averaged concentration of solute, such as the concentration of solute in the drainage water. The resident concentration is a volume-averaged concentration, that is, the concentration of solute per unit volume of soil or soil water. Generally, for surface-applied solutes, the flux concentration is higher than the resident concentration, whereas for soil-generated solutes the resident concentration is often higher than the flux concentration (Magesan, White, and Scotter, 1995).

PISTON DISPLACEMENT OF SOLUTE

The simplest way to consider solute movement is with water moving through the soil porosity, uniformly displacing any water previously held there. Because the movement of solute depends on the water flux rather than time, it is often appropriate to consider redistribution as a function of cumulative drainage rather than time. For a nonreactive, conservative solute dissolved in the incoming water, such as irrigation water, the depth z of the solute front (in mm) after I mm of water input is:

$$z = I/\theta \qquad (11.1)$$

where θ is the volumetric water content of the soil during flow (typically taken as the water content at field capacity, corresponding to a matric potential of -10 kPa).

Water moves through soil in response to a potential gradient (Darcy, 1856). The rate at which water moves is a function of the potential gradient, the water-filled porosity, and the size, continuity, and tortuosity of the pores. For a given potential gradient, the flow rate through cylindrical pores varies with the fourth power of the radius. Therefore, a few, large, water-filled pores can strongly dominate the transport of water through the soil, although these large pores will only be full when the soil is at or near saturation. These pores may be old root or worm channels, or be a soil structural development. Near saturation, most of the water flowing through the soil may be restricted to a small volume which is normally empty of water, and, therefore, of solute (Brown, Rose, and Syers, 1995). This phenomenon is called bypass flow, and is a type of preferential flow (White, 1985a). Preferential flow may also occur in response to other factors such as hydrophobicity, which produces "finger flow" through a hydrophobic layer (Ritsema et al., 1998).

The next level of complexity is to consider part of the water-filled porosity as immobile, with solutes either initially distributed through the whole porosity or excluded from the immobile portion. Water and solute travel only through the mobile fraction. The value of θ in equation 11.1 becomes that of the transport volume. A smaller value of θ in equation 11.1 leads to deeper penetration of the solute for the same water input. Alternatively, instead of considering one part of the water-filled porosity mobile, and the other immobile, the two compartments can be allocated different flow rates, and flow partitioned between the two. There will be two arrival times of solute at depth z, one associated with the fast pathway and the other with the slow pathway. At intermediate depths, after a given time, the concentration will be the flow-averaged concentration of the two components. However, the real situation is more complex because water flows through pores of different size, orientation, and tortuosity at a range of velocities. If the pathways are considered as a set of isolated stream tubes, the distribution of vertical velocities can be described mathematically by a probability density function (pdf). Convective transport of solutes in water flowing through stream tubes is the basis of transfer function modeling.

TRANSFER FUNCTION MODELS

Transfer functions are a form of series analysis, and have been popular in engineering fields for many years (Box, Jenkins, and Reinsel, 1994). Transfer functions can be applied to any situation where there is a linear transfer of change from one property to another. That is, when a linear change in one property such as time (the independent variable) occurs, a change is observed in another property (the dependent variable). There is no requirement a priori for the change in one property to affect the other through any particular mechanism. The change in the second property need not be of the same magnitude, type, or direction as the change in the first, or be controlled only by the change in the first. There may be a lag before any change in the second is observed. Some examples include predicting the weather in Melbourne from the weather in Adelaide up to a few days earlier, and the price of Australian wheat from the price of American wheat up to a few weeks earlier (Sniekers and Wong, 1987). A rigorous examination of the theory, principles, and application of series analysis and transfer functions, particularly for business and industrial applications, is given by Box and Jenkins (1976) and Box, Jenkins, and Reinsel (1994).

Transfer function modeling (TFM) is a nonmechanistic approach that has been developed in recent years for application to solute transport problems, and has been described and reviewed in detail by Jury and Roth (1990) and White, Heng, and Edis (1998). TFM has recently been widely used for describing linear processes in soil physics, and particularly solute transport (Jury, 1982; Simmons, 1982). In the case of solute transport, the concentration of solute at the "exit" depth (the dependent property) changes in response to changes in the concentration of solute entering the transport volume (the independent property). The independent driving variable may be either time or cumulative drainage. When a TFM is used for modeling solute transport, the transport phenomenon is considered as stochastic, and the leaching of solute past a particular depth is best described by a pdf of solute travel through the soil, expressed as a function of drainage or time.

A number of stochastic approaches are available, including scaling theories, Monte Carlo simulations, stochastic-continuum models, and stochastic-convective models (Dagan and Bresler, 1979; Jury, 1982; Jury and Roth, 1990; Russo and Dagan, 1991). Jury and Flühler (1992) discussed the appropriateness of these models for simulating solute transport in the unsaturated zone. Any process model, whether it is deterministic or stochastic, conforms to a TFM provided it conserves mass (Sposito et al., 1986). Of the stochastic models, the stochastic-convective stream tube model appears to be most appropriate when solutes are applied over a wide area and observations are made at early times or shallow depths (Butters and Jury, 1989). This model assumes that solute moves by convection at different velocities in individual flow tubes without mixing between adjacent tubes. Once the pdf of solute travel times or path lengths between the input and output surfaces has been defined (the transfer function), the transport of the solute to other depths can be predicted. It also permits the relationship between travel time and travel distance pdfs to be identified.

For many situations, the complex dynamics involved in solute transport may be approximated by simplified models, with averaging and linearization. In a typical agricultural environment, the water flux is highly variable over short periods of time. Chemicals may be surface-applied, soil-generated, or inherited on a spatially uniform scale relative to the scale of interest. Although the local fluxes of water and chemicals are highly variable, they become uniform by averaging over a sufficiently large area. This averaging is performed by the system itself, since the movement of water and chemicals becomes more uniform over time with mixing, which permits the use of linear models of solute transport (Roth and Jury, 1993).

Properties of a TFM

Travel Time and Travel Distance pdfs

The solute transfer function is most readily determined from the response of a defined soil volume to a narrow pulse input of conservative, nonreactive solute at the surface. Consider that the solute pulse is leached with water at a constant flux, and the concentration in the drainage at depth z is monitored relative to cumulative drainage I. The fraction of the applied solute molecules leached below depth z between drainage I and $I + \Delta I$ gives the probability of a molecule reaching depth z in this amount of drainage. For steady flow, the change in the normalized concentration in the drainage with I defines the pdf for solute travel to depth z, known as the drainage-flux pdf $f^f(z,I)$ (because the pdf is expressed in terms of flux concentrations, it is given the superscript f). The normalizing constant is the mass of solute applied to the surface, and provided the cumulative drainage is sufficiently large, the area under the breakthrough curve (BTC) should be unity. The general form of the transfer function equation for such a system, given a flux concentration $C^f(0,I\text{-}I')$ at the input surface, is given by

$$C^f(z,I) = \int_0^I C^f(0, I - I') f^f(z,I) dI' \tag{11.2}$$

where $C^f(z,I)$ = the flux concentration in the drainage at depth z (Jury and Roth, 1990; Jury and Scotter, 1994).

Alternatively, we may consider a soil volume in which a known amount of solute is present at $z = 0$ and time $t = 0$, and monitor the redistribution of this solute through the soil as it is leached with a steady flux of water. The resident solute concentration in the soil is measured at different depths after an amount of drainage I, and the distribution of normalized solute concentration with depth gives the travel distance pdf $f^r(z,I)$. The transfer function equation for this condition, in a semi-infinite system, is given by the equation

$$C^r(z,I) = \int_0^z C^r(z-z',0) f^r(z',I) dz', \tag{11.3}$$

where $C^r(z,I)$ is the normalized resident concentration after drainage I has redistributed the initial resident concentration $C^r(z,0)$. The relationship between the travel distance pdf and the drainage flux pdf, assuming stochastic-convective transport, is

$$f^r(z,I) = \frac{I}{z} f^f(I,z). \tag{11.4}$$

In the flux pdf, I is a variable and z is a parameter, whereas in the travel distance pdf, z is a variable and I is a parameter (Jury and Roth, 1990). Although the soil water content θ and flux q may be assumed constant within a single stream tube, they are likely to vary between stream tubes in a field soil. For this situation, Jury and Scotter (1994) derived the relationship between the travel distance and drainage flux pdfs as

$$f_i^r(z,I) = \frac{LI^2}{E[I,L]z^2} f_b^f(I,z), \tag{11.5}$$

where L is the calibration depth for the drainage flux pdf, and $E[I,L]$ is the first moment of the drainage flux pdf. The subscript b refers to the "boundary value problem," where the pdf is the normalized flux concentration in the drainage in response to a narrow pulse input at the surface, when there is no solute in the soil initially (see equation 11.2). The subscript i refers to the "initial value problem" where the pdf is the normalized resident concentration in the soil at different depths in response to the presence of solute in the soil at $z = 0$ and $I = 0$. The advantage of the relationship in equation 11.5 is that once one pdf has been parameterized by experimentation, the other can be estimated without further measurements.

Depth Scaling

Another property that the stochastic-convective assumption confers on the transfer function is that of depth scaling of solute transport, according to the equation

$$f^f(z, I) = \frac{L}{z} f^f\left(L, \frac{IL}{z}\right) \tag{11.6}$$

For a vertically homogeneous soil, the probability that solute applied to the surface at $z = 0$ and $I = 0$ will appear at depth z after cumulative drainage $\leq I$, is the same as the probability it will reach depth L after drainage $\leq IL/z$ (Jury, 1982).

Parameterizing the pdf

To parameterize a pdf, an appropriate function is fitted to a solute BTC, or to a resident concentration depth profile, using an optimization procedure to obtain the best estimates of the first and second moments of the pdf. The most common procedure is sum of squares optimization, but for some applications other estimation procedures may be more appropriate (Jury and Sposito, 1985). A number of functions has been used to characterize pdfs, the more common of which are discussed as follows.

The Lognormal pdf

In many field and laboratory studies, a lognormal pdf has been found to be most appropriate (Biggar and Nielsen, 1976; Jury, Stolzy, and Shouse, 1982; White, Thomas, and Smith, 1984; White, Dyson, Gerstl, et al., 1986; Dyson and White, 1987; Butters and Jury, 1989). Jury and Roth (1990) described this pdf as a convective lognormal transfer function (CLT) which has the form

$$f(I) = \frac{\exp\left[-(\ln(I) - \mu)^2 / 2\sigma^2\right]}{\sqrt{2\pi}\sigma I}, \tag{11.7}$$

where μ and σ are the mean and standard deviation of the lognormal function. When the log-normal function is parameterized at a depth L, it may be evaluated at a depth z other than the calibration depth by using equation 11.6, which gives

$$f(z, I) = \frac{\exp\left[-(\ln\left(\frac{IL}{z}\right) - \mu_L)^2 / 2\sigma_L^{\ 2}\right]}{\sqrt{2\pi\sigma_L I}} \tag{11.8}$$

where μ_L and σ_L are the mean and standard deviation of the pdf evaluated at the calibration depth L. The pdf can be modified to estimate flux as a function of time or cumulative drainage, or resident concentrations as a function of depth (Jury and Roth, 1990).

The CLT model has two parameters that need to be estimated. They represent statistical properties of the BTC, rather than physical, independently measurable properties. This suggests that construction of a BTC is required for application of the model. However, as will be discussed later, there is potential to estimate these parameters through measured values of θ_{st}.

Exponential pdf

When the transport volume behaves as if it were well-mixed, an exponential function may be appropriate to describe the change in drainage flux concentration of an indigenous solute with increasing I. Such a system occurs in soils subject to frequent water inputs with closely spaced, shallow, subsurface drains. Utermann, Kladivko, and Jury (1990) successfully used an exponential pdf to simulate pesticide transport through the saturated zone of a pipe-drained soil with a shallow water table. Similarly, Scotter, Heng, and White (1991) successfully used an exponential pdf of the form

$$f^f(z, I) = a^{-1} \exp\left(-\frac{I}{a}\right) \tag{11.9}$$

to simulate losses of resident Cl^- to a mole-pipe drainage system in a silt loam soil under pasture. The value of a can be a fitted parameter, or calculated from the transport volume as

$$a = L\theta_{st}, \tag{11.10}$$

where L is the calibration depth (e.g., the drain depth or the bottom of the root zone) (Magesan, Scotter, and White, 1994). Note that equation 11.9 is consistent with the equation developed by White (1987) for estimating NO_3^- concentrations, derived from indigenous soil NO_3^-, that emerged from a drained soil during individual drainage events. It is also consistent with the equation suggested by Raats (1978) for predicting the solute concentration in drainage following a step increase in input concentration to C_o

$$\frac{C^f}{C_o} = 1 - \exp(-r) \tag{11.11}$$

where r is the ratio of drain discharge to soil pore volume.

Since an exponential pdf requires only one parameter, θ_{st}, to be estimated, and this parameter can be obtained from water content data, it has great potential for use in management models. For example, the NO_3^- leaching model nitrate leaching and economic analysis package (NLEAP) (Shaffer, Halvorson, and Pierce, 1991) uses an exponential pdf for predicting N leached for periods of time or cumulative drainage. The governing equation is

$$M = M_o\left(1 - \exp\left(-\frac{kI}{p}\right)\right) \tag{11.12}$$

where M is the amount of N leached, M_o is the amount available for leaching, I is drainage, and p is soil porosity. The leaching coefficient k is used to adjust the porosity to account for immobile water, and the term p/k is thus equivalent to the transport volume $L\theta_{st}$. The NLEAP model has been found to be as successful as the more mechanistic leaching estimation and chemistry model for nitrogen (LEACHN) for predicting amounts of NO_3^- leached (Khakural and Robert, 1993). An almost identical approach is used in the erosion productivity impact calculator (EPIC) leaching model (Williams and Kissel, 1991). Instead of using a leaching coefficient p, the porosity is adjusted by subtracting the volumetric water content at permanent wilting point.

The Burns pdf

Burns (1975) developed a simple functional model for the leaching of a surface-applied solute

$$F = \left\{\frac{I}{(I + \theta\Delta z)}\right\}^{\frac{z}{\Delta z}} \tag{11.13}$$

where F is the fraction of the applied solute leached below depth z in a uniform soil of water content θ by drainage I, and Δz is a notional depth increment. Noting that F was equivalent to the cumulative probability $P(z, I)$ that a solute molecule applied to the surface will appear at depth z after the drainage I, Scotter, White, and Dyson (1993) rewrote this equation in the form of a transfer function

$$P(z, I) = \exp\left(-\frac{z\theta}{I}\right) \tag{11.14}$$

This function has stochastic-convective properties, and by differentiation can be used to obtain flux and resident pdfs for a pulse input of solute, or for a solute that is initially distributed through the soil volume. The resident concentration form of the Burns equation for a solute present at $z = 0$ and $I = 0$ is

$$C^r(z, I) = \frac{M\theta}{I}\exp\left(\frac{-z\theta}{I}\right) \tag{11.15}$$

where $C^r(z,I)$ is the mass of solute per unit volume of soil, and M is the mass of solute per unit area at the surface at $I = 0$ (White, Heng, and Edis, 1998). The flux concentration form for the same initial condition is

$$C^r(z, I) = \frac{Mz\theta}{I^2}\exp\left(\frac{-z\theta}{I}\right) \tag{11.16}$$

Scotter, White, and Dyson (1993) pointed out that the transport volume θ_{st} could be substituted for θ in these equations.

For the initial condition of a solute spread throughout the soil to depth z with a uniform initial resident concentration C^r_o, the resident and flux concentration forms of the Burns equation are, respectively,

$$C^r(z, I) = C^r_o\left[1 - \exp\left(\frac{-z\theta}{I}\right)\right] \tag{11.17}$$

and

$$C^f(z, I) = \frac{C^r_o}{\theta}\left[1 - \left(\frac{z\theta}{I} + 1\right)\exp\left(\frac{-z\theta}{I}\right)\right] \tag{11.18}$$

Scotter, White, and Dyson (1993) also derived equations for the boundary value problem, in which a solute pulse is applied to the soil surface in solution at $I = 0$. Heng and White (1996) used equations 11.16 and 11.17, modified to account for adsorption (see later section), to model the leaching of S fertilizer and soil sulphate in a mole and tile-drained field for three successive years.

Convection-Dispersion Equation

The piston displacement and stream tube approaches rely on dissolved solutes moving at the same rate as water in a stream tube, when not engaged in any interaction with solid surfaces. However, if flow does not occur in isolated tubes, the solutes can mix radially. Then variability in the rate of vertical solute transport, and the spreading of the solute front, can be explained using the concept of hydrodynamic dispersion. Dispersion is the spreading of the solute front partly in response to concentration gradients, but mostly in response to variability in water velocities within and between pores. Solute dispersion resulting from variations in local water velocities is a passive process in that it only occurs when flow is occurring. It has also been called mechanical dispersion (van Genuchten and Wierenga, 1986). The mechanical dispersion coefficient is a function of pore water velocity, and varies with the depth of observation. The combination of diffusion and mechanical dispersion gives the apparent diffusion or diffusion-dispersion coefficient. The convection-dispersion equation (CDE) has been widely used to model solute transport, and is discussed in the literature (van Genuchten and Wierenga, 1986; Wallach, Steenhuis, and Parlange, 1999; Leij and van Genuchten, 2000). Although it is the traditional approach for modeling solute transport, some authors (e.g., Leij and van Genuchten, 2000) recognize that stream flow approaches may be more appropriate in field situations.

For application to petroleum industry problems, Coates and Smith (1964) developed a CDE model for a system in which the water-filled porosity is partitioned into mobile and immobile water, such that one-dimensional (vertical) flow of solute can be written as

$$\theta_m \frac{\partial C_m}{\partial t} + \theta_{im} \frac{\partial C_{im}}{\partial t} = \theta_m D_m \frac{\partial^2 C_m}{\partial z^2} - q_m \frac{\partial C_m}{\partial z} \tag{11.19}$$

where D is the diffusion-dispersion coefficient, z is vertical distance, q is the water flux, and the subscripts m and im refer to mobile and immobile, respectively. Changes in solute concentration in the immobile regions are assumed only to occur through exchange of solute between these regions, according to the equation

$$\theta_{im} \frac{\delta C_{im}}{\delta t} = \alpha(C_m - C_{im}) \tag{11.20}$$

where α is the solute exchange coefficient. Estimation of the parameters D, α, and θ_m ($=\theta_{st}$) in the field can be difficult, and approaches used for field estimation are discussed in a later section.

Sposito et al. (1986) showed that the two-component CDE is a special case of the TFM. The pdf of the CDE is known as a Fickian pdf (Jury and Sposito, 1985) and in terms of I is written as

$$f(I) = \frac{z \exp\left[-(z - vI)^2 / (4DI)\right]}{2\sqrt{\pi D I^3}}, \tag{11.21}$$

where v is the pore water velocity.

The testing of solute transport models in naturally heterogeneous soils is poorly developed. The Fickian pdf was tested in the field by Snow et al. (1994) with limited success, due to the vertical heterogeneity of the soil, which gave values of D dependent upon depth. Correct identification of the governing flow concept requires depth-dependent measurements of solute behavior to be made (Jury and Flühler, 1992). For example, the depth-dependence of solute dispersion is central to discriminating between the CDE and stochastic-convective TFM approaches. This is the result of differences in the fundamental assumptions in the two models regarding mixing between regions of different flow velocity. The CDE assumes that lateral mixing of solute between such regions occurs in a much smaller time than that required for solute to be moved from the inlet to the outlet end of the transport volume (Taylor, 1953). In the CDE, the dispersion coefficient D is constant with depth and time, and the spread of the solute pulse as it is leached through the soil is directly proportional to the depth at which it is observed. In the TFM, however, solute moving in individual stream tubes is assumed to be isolated from that in neighboring tubes, and spreading of the pulse is entirely due to variations in pore water velocity. Based on the TFM assumptions, D should increase linearly with depth, and the spreading of the BTC increases in proportion to z^2.

Patterns of Solute Input

The pattern of solute input affects the initial and boundary conditions for a model. This is particularly the case when an analytical solution is sought. Here we consider three patterns of solute input: narrow pulse, step-change, and square pulse. The patterns of concentration of solute with time at the entry and exit surfaces of the transport volume for the different types of pulse input are shown in Figure 11.1.

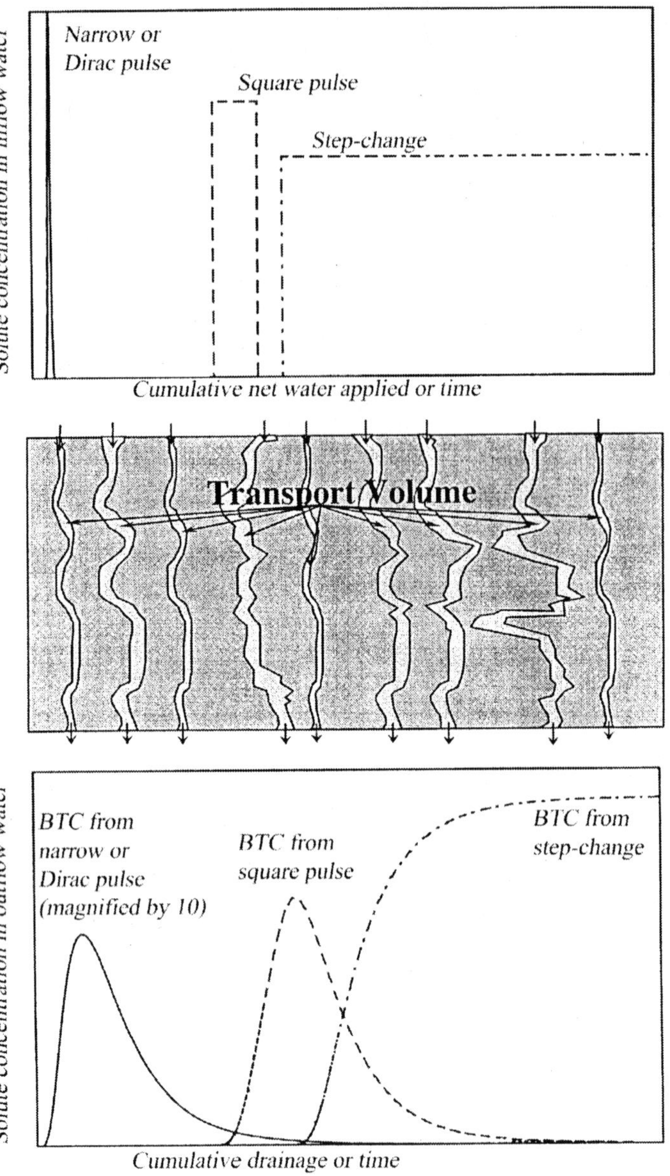

FIGURE 11.1. Types of solute input pulses and typical breakthrough curves after passing through the transport volume of the soil.

Narrow Pulse of Applied Tracer

Consider a quantity of dissolved solute M_0 applied per unit area of soil surface. The amount of water is insignificant compared to the amount of water required to leach a significant proportion of the solute to the depth of interest. This can be treated as a narrow pulse input to the transport volume, and the flux concentration at a particular depth L following cumulative drainage I is calculated from the equation

$$\frac{C^f}{M_o} = f^f(z, I), \tag{11.22}$$

where $f^f(z,I)$ is the drainage flux pdf evaluated at the calibration depth L. The depth concentration profile for the solute, after a given amount of drainage I, can also be obtained (White, Heng, and Edis, 1998). Alternatively, the soil can be sampled at different depths after a given time or drainage to obtain a resident concentration profile (e.g., Vanderborght et al., 1996). This approach is used when there are measurements of resident concentrations at timed intervals obtained by TDR, soil sampling, or suction cup samplers.

Narrow pulse inputs provide the best data for model parameterization using breakthrough curves. Conversely, the outputs from a narrow pulse of a contaminant are relatively sensitive to the parameter estimates.

A Step-Change Input

A step-up change in solute input occurs when the concentration of an applied solute increases from 0 to C_o at $I = 0$ and continues at this concentration for a long time. A step-up change may be associated with the commencement of regular applications of agrochemicals or irrigation with wastewater. This condition has been used to study the movement of solutes and particulates such as bacteria, through large, undisturbed soil cores (White, 1985a). However, a step-up change in concentration of an externally applied solute can also be associated with a step-down change in concentration for an indigenous solute, when that solute is absent from the leaching water. This is a common situation for soil-generated solutes, such as NO_3^- and SO_4^{2-}. The simplest case is when an indigenous solute is uniformly distributed through a defined soil volume at a concentration C_o^r in the transport volume. This is analogous to an initial condition of steady flow through the transport volume of an influent solution containing solute at the same concentration as that in the transport volume. Consider a step change in concentration of the influent solution from C_o to 0 at $I = 0$. Assuming a lognormal drainage flux pdf calibrated at depth L, the solute concentration C^f in the drainage at any depth z is given by

$$C^f(z,I) = \frac{C_o^r}{2}\left\{1 - erf\left[\frac{\ln\left(\frac{IL}{z}\right) - \mu}{\sqrt{2}\sigma}\right]\right\} \tag{11.23}$$

White (1989) and White and Magesan (1991) used this approach to model NO^-_3 leaching from soil cores, as did Heng et al. (1994) and Heng and White (1996) to model the leaching of soil SO^{2-}_4 in the field. However, resident solutes may not be uniformly distributed to the depth of interest, or indeed to the calibration depth of the pdf. For completeness, the expansion of this equation to model leaching from layers of soil is discussed as follows.

A possible situation is that of an indigenous solute uniformly distributed from the soil surface to a depth Z below the surface. The initial conditions are

$$C_o^r = C_o \qquad 0 < z \le Z$$

$$C_o^r = 0 \qquad Z < z \le L$$

where C_o^r is the initial resident concentration of solute in the transport volume to depth Z at $I = 0$ and L is the calibration depth. Consider a step change in concentration of the influent solution from C_o to 0 at $I = 0$. The layer of solute can be treated as a square pulse input with the amount of solute available for leaching per unit area equal to M_o, between the surface and depth Z; that is

$$M_o = C_o^r Z\theta_{st} \tag{11.24}$$

The leading edge (step-up) of the square pulse occurs at Z, and the trailing edge (step-down) occurs at the surface, so that depth scaling of the pdf is required for the leading edge. The flux concentration C^f at depth z following the leaching of the layer of resident solute is therefore described by

$$C_f(z,i) = \frac{C_o^r}{2}\left\{\text{erf}\left[\frac{\ln\left(\frac{IL}{(L-Z)}\right) - \mu}{\sqrt{2}\sigma}\right] - \text{erf}\left[\frac{\ln(I) - \mu}{\sqrt{2}\sigma}\right]\right\} \tag{11.25}$$

A Square Pulse of Applied Solute

A square pulse of applied solute represents the situation in which solute is supplied to the flux volume for a period of time longer than that relevant to a narrow pulse. The integral form of equation 11.8 can be used to predict the flux concentration in the drainage at the calibration depth L, for a step-up change in input concentration from 0 to C_o at the soil surface (Ellsworth et al., 1996). This equation applies for values of I from 0 to ΔI, where ΔI is the size of the square pulse in terms of drainage I; that is, for I ΔI

$$C_f(z, I) = \frac{C_o}{2}\left\{1 + \mathrm{erf}\left[\frac{\ln\left(\frac{IL}{z}\right) - \mu_L}{\sqrt{2\sigma_L}}\right]\right\} \tag{11.26}$$

At the end of the pulse, when there is a step-down change from C_0 to 0 in the influent, the flux concentration at depth L as drainage continues for $I > \Delta I$ is given by

$$C_f(z, I) = \frac{C_o}{2}\left\{\mathrm{erf}\left[\frac{\ln\left(\frac{IL}{z}\right) - \mu_L}{\sqrt{2\sigma_L}}\right] - \mathrm{erf}\left[\frac{\ln\frac{(I - \Delta I)L}{z} - \mu_L}{\sqrt{2\sigma_L}}\right]\right\} \tag{11.27}$$

where μ_L and σ_L are the mean and standard deviation of the lognormal flux pdf (hereafter, the subscript L will be implicit for this pdf). Jury (1982) gave examples of the BTC simulated with equations 11.26 and 11.27 for different square pulse sizes. Similarly, for the same boundary and initial conditions, where the initial solute concentration in the soil is zero, the depth profile of the resident concentration $C^r(z, I)$ for $I \leq \Delta I$ is given by

$$C^r(z, I) = \frac{C_o}{2L}\exp\left(\mu + \frac{\sigma^2}{2}\right)\left\{1 + \mathrm{erf}\left[\frac{\ln\left(\frac{IL}{z}\right) - \mu - \sigma^2}{\sqrt{2\sigma}}\right]\right\} \tag{11.28}$$

and for $I > \Delta I$

$$C^r(z, I) = \frac{C_o}{2L}\exp\left(\mu + \frac{\sigma^2}{2}\right)\left\{\mathrm{erf}\left[\frac{\ln\left(\frac{IL}{z}\right) - \mu - \sigma^2}{\sqrt{2\sigma}}\right] - \mathrm{erf}\left[\frac{\ln\left(\frac{(I - \Delta I)L}{z}\right) - \mu - \sigma^2}{\sqrt{2\sigma}}\right]\right\} \tag{11.29}$$

Step-change and square pulses can also be used to represent the leaching of indigenous solutes. A possible situation is that of an indigenous solute uniformly distributed from the soil surface to a depth Z below the surface. The layer containing solute need not be located at the soil surface, but could be defined between any two depths relative to the surface and the calibration depth. For layers below the surface, each layer can be treated as a separate square pulse, and the output of each is summed at the depth of interest.

In the case of transient flow and depth-dependent water loss due to removal by plant roots, cumulative drainage with depth needs to be estimated using an appropriate model for crop water uptake. Jury and Scotter (1994) give other examples of how the stochastic-convective TFM can be solved for specific initial value and boundary value problems.

SOLUTE REACTIONS DURING TRANSPORT

The discussion so far has assumed that the solutes being transported are nonreactive and conservative. Many chemicals in soil are not conserved and they react with the soil matrix in various ways. The processes receiving most attention in solute transport studies are sorption/desorption at soil surfaces, and microbially mediated reactions such as decomposition and mineralization/immobilization.

Retardation

The rate at which solutes move through the soil may be retarded through reversible reactions such as sorption. A retardation factor R is easily included into both the TFM and CDE models. Sorption slows down the movement of a reactive solute relative to a nonreactive solute. In the stochastic-convective TFM, sorption can be visualized as the lengthening of solute travel times, or an increase in the travel path lengths for drainage. For the simplest case of linear, reversible sorption at equilibrium

$$C_a^r = K_d C_l^r, \tag{11.30}$$

where K_d is the distribution coefficient, C_a^r is the concentration of sorbed solute (mass per unit mass of dry soil), and C_l^r is the solute concentration in the liquid phase. Assuming that a sorbed solute, e.g., SO_4^{2-} behaves identically to a conservative nonreactive solute (e.g., Cl^-) in all processes except adsorption, the pdf of the sorbed solute can be derived from that of the nonreactive, conservative solute using the equation (Jury and Roth, 1990)

$$f_{ad}(z,I)=R^{-1}f\left(z,\frac{I}{R}\right) \tag{11.31}$$

where R is the retardation coefficient defined by

$$R=1+\rho_b K_d/\theta \tag{11.32}$$

Heng and White (1996) used this approach to model SO_4^{2-} leaching in a mole and tile-drained field soil under natural rainfall. For the CDE approach, the parameters D and q are modified to be D/R and q/R in equation 11.19.

Water extraction by plants can also be treated as a process that retards the transport of even nonreactive solutes, particularly in systems where short-lived water input and drainage events are separated by relatively long periods of water extraction after solute application. The plant may take up solute through the roots, or exclude it, but the pores that are emptied of water are also emptied of solute. Effectively, the solute domain in the transport volume is contracted. With subsequent water input and drainage, some water flux occurs in the transport volume with solute, and some occurs through pores that were emptied of solute. Retardation by solute domain contraction is analogous to the delay in the leaching of solutes that are "protected" inside aggregates from bypass flow. Edis, White, and Heng (2000) estimated the value of R in a TFM arising from solute domain contraction on the basis of cumulative evapotranspiration (E_{cum}) and drainage, according to the equation

$$R=\frac{(I+E_{cum})}{I} \tag{11.33}$$

Decomposition of Labile Solutes

Decomposition of labile solutes is a major consideration in solute transport under field conditions. Decomposition may involve chemical, biological, or radioactive processes. Decomposition is assumed to be a first-order process, in that the rate of disappearance of the solute is proportional to the mass of solute M present at any time, that is

$$\frac{\partial M}{\partial T}=-k_c M \tag{11.34}$$

where k_c is a rate coefficient. Integrating equation 11.34 between time $t=0$ and t gives

$$M_t = M_o \exp(-k_c t) \quad (11.35)$$

Consider a narrow pulse of a labile solute injected into the transport volume at $I = 0$. Mass will not be conserved because of decomposition of the solute as it passes through the soil. Thus, the pdf of a conservative solute having the same physical transport characteristics as the labile solute is scaled according to the ratio M_t/M_0, giving the equation:

$$f^f(z, I, k_c) = \exp(-k_c t) f^f(z, I; k_c = 0) \quad (11.36)$$

where $f^f(z, I; k_c = 0)$ is the travel path length pdf of the conservative solute.

Equations 11.31 and 11.36 can be combined to give the travel pdf for a solute that undergoes simultaneous adsorption and decay during transport

$$f_a^f(z, I, k_c) = \frac{\exp(-k_c t)}{R} f^f(z, \frac{I}{R}) \quad (11.37)$$

Utermann, Kladivko, and Jury (1990) and Suter (1997) used this approach to model the leaching of pesticides through soil.

Incorporating Solute Sources and Sinks

Solute input to the transport volume can occur in ways other than through controlled applications, and output can occur through pathways other than leaching. Inputs (sources) include rainfall, fertilizer, dung and urine, and mineralization of organic matter. Outputs (sinks) include plant uptake, microbial immobilization, gaseous losses, and biochemical or radioactive decay. An example of the incorporation of a decay sink in the TFM using a known functional relationship was given in the previous section. If the input or output process is linear with respect to I (zero order), it can be incorporated in the TFM using the principle of superposition as demonstrated by Heng et al. (1994) and Heng and White (1996). For example, in the case of soil SO_4^{2-}

$$C_b = C_m + C_p + C_i + C_s \quad (11.38)$$

where C_b is a net source-sink term, C_m = net mineralization, C_p = plant uptake, C_i = rainfall, and C_s = oxidation of elemental S fertilizer. The combined effect of the sources and sinks is treated as an input to the initial resident concentration in the transport volume, so that the contribution to drainage is given by

$$C^f(z,I) = C_b\left(1 - \int_0^I f^f(z,I)dI\right) \tag{11.39}$$

Where the relationship between the source (or sink) quantity and I changes substantially, as can happen in the case of net mineralization for example, the term C_b must be adjusted at I, I_0, where I_0 is the cumulative drainage at which the change takes effect.

As with models such as LEACHN, which employs a finite difference solution for the CDE, numerical solutions of the TFM are possible. A numerical solution has advantages over the analytical approach when (1) the functions for the input of an applied solute or the source-sink term for a labile solute are complex (i.e., the functions need not be constrained to be linear with time or drainage), (2) the parametric form of the pdf is not easily determined, and (3) the sorption isotherm for a reactive solute is not a simple function. Heng et al. (1994) used a finite difference form of the CLT to model SO_4^{2-} leaching when this solute had been surface-applied (and was assumed not to dissolve immediately), and was also being generated by net mineralization in the soil. The equation used was

$$C^f(m) = \sum_{n=1}^{m} C_{ent}(m-n+1)f(n)\Delta I + C_o^r\left[1 - \sum_{n=1}^{m} f(n)\Delta I\right] \tag{11.40}$$

where C_{ent} is a variable input concentration to the transport volume (including all source-sink terms), $C_0{}^r$ is the initial resident concentration, and m and n are integers, with m defined by

$$m = I/\Delta I \tag{11.41}$$

EXTENSION OF MODELS TO LARGE AREAS

The distribution and flux of solutes in the field vary with time and space. This variability develops through differences in soil, climate, vegetation, animals, management, and their interactions. Heterogeneity occurs at a range of scales, from microscale, time-dependent sorption and precipitation/dissolution reactions, to mesoscale preferential movement of water and chemicals through macropores, to macroscale variability of soils in the landscape, and changes in soil properties with time. Differences between the laboratory and field in the scale of heterogeneity and in characteristic solute mixing times can lead to errors when laboratory results are applied to the field. For example, the lateral dimension of a soil column is normally 10-100 cm. On the other hand, although vertical dimension of the root zone in the field may be of the same order as a soil column, its lateral dimension is of the order of 100-1,000 m.

Soils are not spatially homogeneous. For example, Schulin et al. (1987) observed significant spatial heterogeneity in Br^- concentrations 400 days after the application of a uniform Br^- pulse to a field soil in Switzerland. Similarly, Flury et al. (1994) observed remarkable differences in spatial flow patterns in a dye-tracing experiment. There is often a large change from the scale of intensive measurements to the scale of practical applications. Similarly, there is a need to extrapolate from sites where soil properties and their variability have been intensively studied, to sites where much less is known. Ideally, we seek to define those differences between one site and the next that affect the distribution and flux of solutes, but spatial variability makes this difficult. Also, we seek to develop models for extrapolating from short-term, site-specific observations, to longer periods and larger areas.

To incorporate spatial variability into the modeling of solute-water interactions, spatially referenced models have been developed that make use of geographic information system (GIS) software. The models can be considered as having spatially lumped or spatially distributed parameters. Examples of lumped models are chemicals, runoff, and erosion from agricultural management systems (CREAMS) (Knisel, 1980), groundwater loading effects of agricultural management systems (GLEAMS) (Leonard, Knisel, and Still, 1987), EPIC (Williams, Jones, and Dyke, 1984), LEACHN (Hutson and Wagenet, 1992), chemical movement in layered soils (CMLS) (Nofziger and Hornsby, 1986), NLEAP (Shaffer, Halvorson, and Pierce, 1991) and HYDRUS (Vogel et al., 1995; Simunek, Huang, and van Genuchten, 1998). Spatially distributed models are required to model fluxes of water and solute in more than one dimension, where processes occurring in one dimension impact on what occurs in another. Examples of such models are a real nonpoint source watershed environment response simulation (ANSWERS) (Beasley, Huggins, and Monke, 1980), agricultural nonpoint source pollution model (AGNPS) (Young et al., 1989), simulator for water resources in rural basins (SWRRB) (Arnold et al., 1990), soil and water assessment tool (SWAT) (Arnold, Allen, and Bernhardt, 1993) and TOPOG (Dawes et al., 1997).

Although models are able to simulate water and solutes fluxes in a spatially variable landscape, a stumbling block is the adequacy of the description of variability and the estimation of parameters and input variables. In the case of small-scale solute transport problem, such as a localized chemical spill, accurate parameter and variable values can usually be obtained. But in the case of diffuse solute transport, such as in farm or catchment scale studies, detailed measurement of parameters and input variables is less feasible. Parameter estimation may need to be based on pedotransfer functions (Bouma, Bootlink, and Finke, 1996), which are based on easily measured soil properties, preferably derived from routine soil survey. It follows that models used in such circumstances should have a minimum number of parameters. Not only do the parameter values vary, but the rate of water input also becomes less certain. In such cases, cumulative drainage, estimated from a water balance, may be a suitable independent variable. The transport volume is generally required, and for some models may be the only parameter needed.

The assumptions of stationarity and ergodicity underlie the application of TFM. The stationarity assumption is that the pdf for solute travel does not depend on the time or spatial position at which the transport process is observed. Thus, the moments of the pdf should be applicable over a local area, and the parameter values obtained in one year should be applicable to the same area in other years when seasonal conditions may be different. Stationarity of the transfer function has been observed by Heng et al. (1994) and Magesan, Scotter, and White (1994), where the transport parameters derived from the leaching of Cl^- over one winter season were successfully used to predict Cl^- movement during two other years, and also extended to the leaching of SO_4^{2-}. The ergodicity assumption is that a statistical or ensemble average should equal the time or cumulative drainage average of individual realizations of the stochastic process. This assumption has rarely been tested in solute transport studies. Mallawatantri and Mulla (1996) compared the mean μ and standard deviation σ for Br^- travel, calculated as an average for 40 individual plots, with the average of these parameters for the whole field (57 ha). There was good agreement between the two sets of parameters. However, a true test of ergodicity would be to compare the statistical average of the solute travel moments derived from a number of point measurements in a field, with the independently measured moments of the pdf for solute travel through the whole field.

Simulation models of processes in the unsaturated zone are needed for predicting the effects of soil and land management changes at catchment and regional scales. Although the application of TFMs has mostly been confined to small-scale field studies, its parsimonious parameter requirements, easily obtained input variables, and versatility in incorporating various chemical and biological processes in a simple manner, offer potential for extension to larger scales. To satisfy the need for models for large-scale, complex field systems, new methods of measuring input variables and new or improved approaches for estimating parameters at various scales are needed.

Parameter Estimation in the Field

To date, the key parameters of solute transport models have usually been obtained by calibration from tracer experiments at individual sites. This may involve the simultaneous optimization of all the parameters required. Although this may allow subsequent predictions of solute transport at these sites, the approach is labor intensive and not readily amenable to scaling up or extrapolation to other sites. In the field, BTCs have been constructed from concentrations measured in suction cup samples, and more recently, time domain reflectometry (TDR) has been used to measure in situ bulk electrical conductivity (a surrogate for salt concentration) (Vogeler et al., 1996). The in situ disc permeameter technique of Clothier, Kirkham, and McLean (1992) offers promise for estimating the transport volume and simultaneously obtaining a dynamic adsorption isotherm for a reactive solute (Clothier et al., 1996). However, this kind of intensive and systematic measurement to obtain spatially distributed input data for large-scale models is problematic because the measurements are expensive and time consuming.

When flow is not at a steady state, the mean travel path length (using the drainage flux pdf) is the preferred estimator for calculating the transport volume (Jury, Sposito, and White, 1986; Heng and White, 1996). In the TFM approach, measurements of θ_{st} have been obtained from the drainage flux pdf of solute travel times or path lengths, i.e.,

$$\theta_{st} = \frac{\exp\left(\mu + \sigma^2/2\right)}{L} \tag{11.42}$$

with μ and σ^2 as the mean and variance, respectively, of the lognormal pdf. This suggests that if an independent estimate of θ_{st} and μ and/or σ could be obtained, the pdf could be fully parameterized independently of BTC data.

From a survey of published BTC data (Okom, 1998), the value for σ was found not to vary greatly from soil to soil, to be generally less than 1, and about an order of magnitude smaller than μ. Some examples of σ values are given in Table 11.1. It may be that a value of 0.5 could be allocated to σ in the absence of an actual BTC, to allow the estimation of μ and full parameterization of the convective lognormal TFM (CLT), given an estimation of θ_{st}. There is no apparent correlation between soil texture or scale and the standard deviation σ of the log of solute travel time (or path lengths). The CLT is further desensitized to the value of σ through a codependence of μ and σ. That is, an error in σ is somewhat compensated for by a "correction" error in the resulting estimate of μ.

TABLE 11.1. Literature Values of the Standard Deviation σ for the Lognormal Probability Density Function in the CLT Model of Solute Transport

Soil texture	Experimental system	σ	Source
Silt loam	Field	0.8	Heng, 1991
Clay loam	Field	1.25	Biggar and Nielsen, 1976
Loamy sand	Field	0.57	Jury, Stolzy, and Shouse, 1982
Loamy sand	Field	0.46	Butters and Jury, 1989
Loamy sand	Field	0.4	Butters, 1987
Loamy sand	Field	0.51	El Abd, 1984
Loamy sand to sandy loam	Field	0.37	Mallawatantri and Mulla, 1996
Medium clay	Field lysimeters	0.52	Edis, 1998
Loamy sand	Field scale, depths 0.3-1.8 m	0.28-0.86	Jury and Sposito, 1985
Sand	Field transect	0.37	Vanwesenbeek and Kachanoski, 1991
Layered loam	Plot	0.83	Roth, 1989
Loam	Plot	0.21-0.36	Ellsworth et al., 1996
Fine sandy loam	Plot	0.19-0.28	Costa, Knighton, and Prunty, 1994
Loamy sand	Plots	0.69	Starr et al., 1978
Clay	Plots	0.57	van der Pol, Wierenga, and Nielsen, 1977
Clay loam	Plots	0.47-0.67	Jaynes, Rice, and Bowman, 1988
Stony	Plots	0.21-0.26	Schulin et al., 1987
Loamy sand to sandy loam	Plots	0.44	Mallawatantri and Mulla, 1996
Well-structured clay	Repacked core	0.35-0.55	Edis, 1998
Sandy soil	Intact cores	0.5-0.7	Vanderborght et al., 1996
Humus-rich sandy soil	Intact cores	0.32-0.58	Vanderborght et al., 1997
Silty loam with clay horizon	Intact cores	0.23-0.88	Vanderborght et al., 1997

TABLE 11.1 *(continued)*

Soil texture	Experimental system	σ	Source
Sandy loam with clay horizon	Intact cores	0.19-0.32	Vanderborght et al., 1997
Loamy sand	Intact cores	0.34-0.48	Jury and Sposito, 1985
Well-structured clay	Intact cores	0.38-0.58	Suter, 1997
Sandy soil	Intact cores	0.25-0.32	Suter, 1997
Well-structured clay	Intact cores	0.8	Heng and White, 1996
Silt loam	Intact cores	0.78	White and Magesan, 1991
Loamy sand	Intact cores	0.5-0.7	Vanclooster et al., 1995
Clay loam	Intact cores	0.68	White, 1989

The location of the solute peak in a BTC can be used as an estimate of mean travel time (Zhang, Yang, and Ye, 1996). Estimates of the transport volume are therefore sensitive to tracer-soil interactions that delay the outflow of the tracer. For example, if the peak of an anioinic tracer is delayed through anion adsorption, then the value of θ_{st} will be overestimated, and may exceed the saturated volumetric water content of the soil. θ_{st} will also be overestimated if preferential flow bypasses the soil water containing solute. Calculated transport volumes larger than actual water contents have been observed by Starr et al. (1978), Butters, Jury, and Ernst (1989), Tillman et al. (1991), and Snow et al. (1994).

The Use of Pedotransfer Functions

Pedotransfer functions may allow solute transport parameters to be estimated without additional soil measurements. Simple, routinely measured soil properties such as texture, bulk density, and particle size distribution, and organic matter content could be used to predict soil water properties such as water content-potential relationships (Campbell, 1985; Vereecken et al., 1992). Lee, Horton, and Jaynes (2000) have suggested a method using TDR to estimate both the solute exchange coefficient α and θ_m. Based on measurements of mobile water fraction in a range of soils of different texture, White et al. (1999) constructed a pedotransfer function for the "mobile fraction" θ_m/θ of the form

$$\frac{\theta_m}{\theta} = 1 - 0.0221 clay\left(1 - \frac{bulk\ density}{2.65}\right) \tag{11.43}$$

This function was calibrated using θ_m values measured by the disc permeameter/Br tracer method (Okom, White, and Heng, 2000). The function accounted for 75 percent of the variation in the mobile fraction values in experiments reported in the literature, where θ_m had been measured directly (White et al., 1999). However, this approach is still at a very developmental stage, and there are examples in which the agreement between measured soil hydraulic properties and those predicted from pedotransfer functions has been poor (Espino et al., 1995).

SUMMARY

An understanding of solute-water interactions in soil leading to a quantitative description of behavior is important because of the potential for nutrient loss, remediation and transport of contaminants, and impacts on surface and groundwater quality. These interactions depend on the type of solute, and the rate and pathway of water movement. Solutes range from conservative, nonreactive ions such as Cl^- to biologically labile, nonreactive ions such as NO_3^-, to labile, reactive ions such as NH_4^+ or SO_4^{2-}. The complexity of the pathway depends on the size, shape, and continuity of pores that comprise the pore space. Fundamental to modeling solute-water interactions is the concept of the transport volume—the fractional soil volume that is effective in solute transport during a period of observation. Inputs and outputs of solutes and water occur through the boundary surface of the transport volume, and solute reactions and transformation occur within this volume.

The appropriate choice of model to describe solute-water interactions depends on the objectives and the scale of operation. The simplest model is that of piston displacement, in which the transport volume occupies all the water-filled pore space. Further model development accounts for preferential flow through large pores, as opposed to small pores within aggregates, by partitioning the soil water into "mobile" and "immobile" components. Another approach is based on the concept of water flow at variable velocities through a network of parallel, noninteracting stream tubes. The mathematical description of these concepts ranges from mechanistic-deterministic to stochastic. Of the former, the convection-dispersion equation (CDE) is widely used. In the CDE, pore water velocity and dispersion coefficient can be considered as unique values, or as distribution functions. Terms for reaction of solutes with surfaces and gains and losses in the transport volume can be included. Of the stochastic approaches, the stochastic-convective assumption underlying the lognormal transfer function model (TFM) has been widely tested. The TFM can also incorporate functions for solute reaction and source-sink terms.

Correct parameterization is a major requirement for successful model simulation of solute-water interactions. This becomes more difficult to achieve for large areas because of the time-consuming nature of the measurements, the variability in space and time of parameter values, and input variables such as rainfall. At such scales, pedotransfer functions using readily measured soil properties have promise, but must be more widely tested. Coupling process models with a GIS is appropriate to describe solute-water interactions for large areas, provided the input data and parameters can be spatially referenced.

REFERENCES

Addiscott, T.M. (1977). A simple computer model for leaching in structured soils. *Journal of Soil Science 28:*554-563.

Addiscott, T.M. and D. Cox (1976). Winter leaching of $NO3^-$ from autumn-applied calcium nitrate, ammonium sulphate, urea and sulphur coated urea in bare soil. *Journal of Agricultural Science Cambridge 87:*381-389.

Angulo-Jaramillo, R., J.P. Gaudet, J.L. Thony, and M. Vauclin (1996). Measurement of hydraulic properties and mobile water content of a field soil. *Soil Science Society of America Journal 60:*710-715.

Arnold, J.G., P.M. Allen, and G. Bernhardt (1993). A comprehensive surface-groundwater flow model. *Journal of Hydrology 142:*47-69.

Arnold, J.G., J.R. Williams, A.D. Nicks, and N.B. Sammons (1990). *SWRRB: A Basin Scale Simulation Model for Soil and Water Resources Management.* College Station, TX: Texas A&M University Press.

Beasley, D.B., L.F. Huggins, and E.J. Monke (1980). ANSWERS: A model for watershed planning. *Transactions of the American Society of Agricultural Engineers 23:*938-944.

Biggar, J.W. and D.R. Nielsen (1976). Spatial variability of the leaching characteristic of a field soil. *Water Resources Research 12:*78-84.

Bouma, J., H.W.G. Booltink, and P.A. Finke (1996). Use of soil survey data for modeling solute transport in the vadose zone. *Journal of Environmental Quality 25:*519-526.

Box, G.E.P. and G.M. Jenkins (1976). *Time Series Analysis: Forecasting and Control.* San Francisco: Holden-Day Press.

Box, G.E.P., G.M. Jenkins, and G.C. Reinsel (1994). *Time Series Analysis: Forecasting and Control.* Englewood Cliffs: Prentice-Hall.

Brown, C.D., D.A. Rose, and J.K. Syers (1995). Effects of preferential flow. *British Crop Protection Council Monograph 62:*93-98.

Burns, I.G. (1975). An equation to predict the leaching of surface-applied nitrate. *Journal of Agricultural Science, Cambridge 85:*443-454.

Butters, G.L. (1987). "Field scale transport of bromide through unsaturated soil." (PhD thesis, University of California, Riverside, cited in Jury and Flühler, 1992.)

Butters, G.L. and W.A. Jury (1989). Field scale transport of bromide in an unsaturated soil. 2. Dispersion modelling. *Water Resources Research 25:*1583-1589.

Butters, G.L., W.A. Jury, and F.F. Ernst (1989). Field-scale transport of bromide in an unsaturated soil. 1. Experimental methodology and results. *Water Resources Research 25:*1575-1581.

Campbell, G.S. (1985). *Soil Physics with BASIC: Transport Models for Soil-Plant Systems.* Amsterdam: Elsevier.

Clothier, B.E., L.K. Heng, G.N. Magesan, and I. Vogeler (1995). The measured mobile-water content of an unsaturated soil as a function of hydraulic regime. *Australian Journal of Soil Research 33:*397-414.

Clothier, B.E., M.B. Kirkham, and J.E. McLean (1992). In situ measurement of the effective transport volume for solute moving through soil. *Soil Science Society of America Journal 56:*733-736.

Clothier, B.E., G.N. Magesan, L. Heng, and I. Vogeler (1996). In situ measurement of the solute adsorption isotherm using a disc permeameter. *Water Resources Research 32:*771-778.

Coates, J.H. and B.D. Smith (1964). Dead-end pore volume and dispersion in porous media. *Society Petroleum Engineers Journal 33:*73-84.

Corwin, D.L., B.L. Waggoner, and J.D. Rhoades (1991). A functional model of solute transport that accounts for bypass. *Journal of Environmental Quality 20:* 647-658.

Costa, J.L., R.E. Knighton, and L. Prunty (1994). Model comparison of unsaturated steady-state solute transport in a field plot. *Soil Science Society of America Journal 58:*1277-1287.

Dagan, G. and E. Bresler (1979). Solute dispersion in unsaturated heterogeneous soil at a field scale. I. Theory. *Soil Science Society of America Journal 43:*461-467.

Darcy, H. (1856). *Les fontaines publiques de la ville de Dijon.* Paris: Dalmont.

Dawes, W.R., L. Zhang, T.J. Hatton, P.H. Reece, T.G.H. Beale, and I. Packer (1997). Evaluation of a distributed parameter ecohydrological model (TOPOG-IRM) on a small cropping rotation catchment. *Journal of Hydrology 191:*64-86.

Dyson, J.S. and White, R.E. (1987). A comparison of the convection-dispersion equation and transfer function model for predicting chloride leaching through an undisturbed, structured clay soil. *Journal of Soil Science 38:*157-172.

Edis, R. (1998). "Transfer function modelling of bromide and nitrate leaching through soil with and without plants." (PhD thesis, The University of Melbourne.)

Edis, R.B., R.E. White, and L.K. Heng (2000). Solute transport through a clay soil with and without plants. In *Soil 2000: New Horizons for a New Century,* Australian and New Zealand Second Joint Soils Conference, Volume 2, eds. J. A Adams and A. K. Metherall. Lincoln University, Canterbury: New Zealand Society of Soil Science, pp. 97-98.

El Abd, H. (1984). "Spatial variability of the pesticide distribution coefficient." (PhD thesis, University of California, Riverside, cited in W.A. Jury and H. Flühler, 1992.)

Ellsworth, T.R., P.J. Shouse, T.H. Skaggs, J.A. Jobes, and J. Fargerlund (1996). Solute transport in unsaturated soil: Experimental design, parameter estimation, and model discrimination. *Soil Science Society of America Journal 60:*397-407.

Espino, A., D. Mallants, M. Vanclooster, and J. Feyen (1995). Cautionary notes on the use of pedotransfer functions for estimating soil hydraulic properties. *Agricultural Water Management 29:*235-253.

Flury, M., H. Flühler, W.A. Jury, and J. Leuenberger (1994). Susceptibility of soils to preferential flow of water: A field study. *Water Resources Research 30:*1945-1954.

Heng, L.K. (1991). "Measurement and modelling of chloride and sulphate leaching from a mole-drained soil." (PhD thesis, Massey University, New Zealand.)

Heng, L.K. and R.E. White (1996). A simple analytical transfer function approach to modelling the leaching of reactive solutes through field soil. *European Journal of Soil Science 47:*33-42.

Heng, L.K., R.E. White, D.R. Scotter, and N.S. Bolan (1994). A transfer function approach to modelling the leaching of solutes to subsurface drains. II. Reactive solutes. *Australian Journal of Soil Research 32:*85-94.

Hutson, J.L. and R.J. Wagenet (1992). LEACHN: Leaching estimation and chemistry model—A process-based model of water and solute movement, transformations, plant uptake and chemical reactions in the unsaturated zone. *Research Series 92-3.* Cornell University, Ithaca, NY: Department of Soil, Crop and Atmospheric Science.

Jarvis, N.J., P.E. Jansson, P.E. Dik, and I. Messing (1991). Modelling water and solute transport in macroporous soil. 1. Model description and sensitivity analysis. *Journal of Soil Science 42:*59-70.

Jaynes, D.B., R.C. Rice, and R.S. Bowman (1988). Independent calibration of a mechanistic-stochastic model for field-scale solute transport under flood irrigation. *Soil Science Society of America Journal 52:*1541-1546.

Jury, W.A. (1982). Simulation of solute transport using a transfer function model. *Water Resources Research 18:*363-368.

Jury, W.A. and H. Flühler (1992). Transport of chemicals through soil: Mechanisms, models and field applications. *Advances in Agronomy 47:*141-201.

Jury, W.A., W.R. Gardner, and W.H. Gardner (1991). *Soil Physics* (Fifth Edition). New York: John Wiley and Sons.

Jury, W.A. and K. Roth (1990). *Transfer Functions and Solute Movement Through Soil: Theory and Applications.* Basle, Switzerland: Birkhauser Verlag.

Jury, W.A. and D.R. Scotter (1994). A unified approach to stochastic-convective transport problems. *Soil Science Society of America Journal 58:*1327-1336.

Jury, W.A. and G. Sposito (1985). Field calibration and validation of solute transport models for the unsaturated zone. *Soil Science Society of America Journal 49:*1331-1341.

Jury, W.A., G. Sposito, and R.E. White (1986). A transfer function model of solute transport through soil. 1. Fundamental concepts. *Water Resources Research 22:*243-247.

Jury, W.A., L.H. Stolzy, and P. Shouse (1982). A field test of the transfer function model for predicting solute transport. *Water Resources Research 18:*369-375.

Khakural, B.R. and P.C. Robert (1993). Soil $NO3^-$ leaching potential indices: Using a simulation model as a screening system. *Journal of Environmental Quality 22:*839-845.

Knisel, W.G., Ed. (1980). *CREAMS: A Field Scale Model for Chemicals, Runoff, and Erosion from Agricultural Management Systems.* United States Department of Agriculture, Conservation Research Report No. 26. Washington, DC: USDA.

Lee, J., R. Horton, and D.B. Jaynes (2000). A time domain reflectometry method to measure immobile water content and mass exchange coefficient. *Soil Science Society of America Journal 64:*1911-1917.

Leij, F.J. and M.T. van Genuchten (2000). Solute transport. In *Handbook of Soil Science,* ed. M. E. Sumner. Boca Raton, FL: CRC Press, pp. A183-A227.

Leonard, R.A., W.G. Knisel, and D.A. Still (1987). GLEAMS: Groundwater loading effects of agricultural management systems. *Transactions of the American Society of Agricultural Engineers 30:*1403-1418.

Magesan, G.N., D.R. Scotter, and R.E. White (1994). A transfer function approach to modelling the leaching of solutes to subsurface drains. I. Non-reactive solutes. *Australian Journal of Soil Research 32:*69-83.

Magesan, G.N., R.E. White, and D.R. Scotter (1995). The influence of flow rate on solute concentrations in soil drainage. *Journal of Hydrology 172:*23-30.

Mallawatantri, A.P. and D.J. Mulla (1996). Uncertainties in leaching risk assessments due to field averaged transfer function parameters. *Soil Science Society of America Journal 60:*722-726.

Nkedi-Kizza, P., J.W. Biggar, M. Th. van Genuchten, P.J. Wierenga, H.M. Selim, J.M. Davidson, and D.R. Nielsen (1983). Modeling tritium and chloride-36 transport through an aggregated oxisol. *Water Resources Research 19:*691-700.

Nofziger, D.L. and A.G. Hornsby (1986). A microcomputer-based management tool for chemical movement in soil. *Applied Agricultural Research 1:*50-57.

Okom, A.E.A. (1998). "Estimating soil hydraulic properties and the solute transport volume using surrogate variables." (PhD thesis, The University of Melbourne, Melbourne.)

Okom, A.E.A., R.E. White, and L.K. Heng (2000). Field measured mobile water fraction for soils of contrasting texture. *Australian Journal of Soil Research 38:*1131-1142.

Raats, P.A.C. (1978). Convective transport of solutes by steady flows II. Specific flow problems. *Agricultural Water Management 1:*219-232.

Ritsema, C.J., L.W. Dekker, J.L. Nieber, and T.S. Steenhuis (1998). Modelling and field evidence of finger formation and finger recurrence in a water repellent sandy soil. *Water Resources Research 34:*555-567.

Roth, K. (1989). Stofftransport im wasserungesättigten Untergrund natürlicher heterogener unter Feldbedingungen. (Doctoral thesis, Zurich: Eidgenössisch-Technische Hochschule, ETH.)

Roth, K. and W.A. Jury (1993). Modeling the transport of solutes to groundwater using transfer functions. *Journal of Environmental Quality 22:*487-493.

Russo, D. and G. Dagan (1991). On solute transport in a heterogeneous porous formation under saturated and unsaturated water flows. *Water Resources Research 27:*285-292.

Schulin, R., M. Th. van Genuchten, H. Flühler, and P. Ferlin (1987). An experimental study of solute transport in a stony field soil. *Water Resource Research 23:* 1785-1794.

Scotter, D.R., L.K. Heng, and R.E. White (1991). Two models for the leaching of a non-reactive solute to a mole drain. *Journal of Soil Science 42:*565-576.

Scotter, D.R., R.E. White, and J.S. Dyson (1993). The Burns leaching equation. *Journal of Soil Science 44:*25-33.

Seyfried, M.S. and P.S.C. Rao (1987). Solute transport in undisturbed columns of an aggregated tropical soil: Preferential flow effects. *Soil Science Society of America Journal 51:*1434-1444.

Shaffer, M.J., A.D. Halvorson, and F.J. Pierce (1991). Nitrate leaching and economic analysis package (NLEAP): Model description and application. *In Managing Nitrogen for Groundwater Quality and Farm Profitability,* eds. R.F. Follett, D.R. Keeney, and R.M. Cruse. Madison, WI: Soil Science Society of America, pp. 285-322.

Simmons, C.S. (1982). A stochastic-convective transport representation of dispersion in one-dimensional porous media systems. *Water Resources Research 18:*1193-1214.

Simunek, J., K. Huang, and M. Th. van Genuchten (1998). The HYDRUS code for simulating the one-dimensional movement of water, heat, and multiple solutes in variably saturated media, Version 6.0. *Research Report No.144.* United States Salinity Laboratory, Riverside, CA: USDA-ARS.

Smettem, K.R.J. (1984). Soil-water residence time and solute uptake. 3. Mass transfer under simulated winter rainfall conditions in undisturbed soil cores. *Journal of Hydrology 67:*235-248.

Sniekers, P. and G. Wong (1987). The causality between U.S.A. and Australian wheat prices. *Review of Marketing and Agricultural Economics 55:*37-50.

Snow, V.O., B.E. Clothier, D.R. Scotter, and R.E. White (1994). Solute transport in a layered field soil: Experiments and modelling using the convection-dispersion approach. *Journal of Contaminant Hydrology 16:*339-358.

Sposito, G., R.E. White, P.R. Darrah, and W.A. Jury (1986). A transfer function model of solute transport through soil 3. The convection-dispersion equation. *Water Resources Research 22:*255-262.

Starr, J.L., H.C. DeRoo, C.R. Frink, and J.Y. Parlange (1978). Leaching characteristics of a layered field soil. *Soil Science Society of America Journal 42:*386-391.

Suter, H.C. (1997). "Reactivity and mobility of the organophosphate pesticide phosmet in soils." (PhD thesis, The University of Melbourne, Melbourne.)

Taylor, G.I. (1953). The dispersion of soluble matter flowing through a capillary tube. *Proceedings of the London Mathematical Society 2:*196-212.

Tillman, R.W., D.R. Scotter, B.E. Clothier, and R.E. White (1991). Solute movement during intermittent water flow in a field soil and some implications for irrigation and fertilizer application. *Agricultural Water Management 20:*119-133.

Utermann, J., E.J. Kladivko, and W.A. Jury (1990). Evaluating pesticide migration in tile-drained soils with a transfer function model. *Journal of Environmental Quality 19:*707-714.

van der Pol, R.M.P., P.J. Wierenga, and D.R. Nielsen (1977). Solute movement in a field soil. *Soil Science Society of America Journal 41:*10-13.

van Genuchten, M.Th. and P.J. Wierenga (1976). Mass transfer studies in sorbing porous media. I. Analytical solutions. *Soil Science Society of America Journal 40:*473-480.

van Genuchten, M.Th. and P.J. Wierenga (1986). Solute dispersion coefficients and retardation factors. In *Methods of Soil Analysis, Part 1. Physical and Mineralogical Methods,* Monograph number 9, ed. A. Klute. Madison, WI: American Society of Agronomy, Soil Science Society of America, Agronomy, pp. 1025-1054.

Vanclooster, M., D. Mallants, J. Vanderborght, J. Diels, J. Van Orshoven, and J. Feyen (1995). Monitoring solute transport in a multilayered sandy lysimeter using time domain reflectometry. *Soil Science Society of America Journal 59:*337-344.

Vanderborght, J., C. Gonzalez, M. Vanclooster, D. Mallants, and J. Feyen (1997). Effects of soil type and water flux on solute transport. *Soil Science Society of America Journal 61:*372-389.

Vanderborght, J., M. Vanclooster, D. Mallants, J. Diels, and J. Feyen (1996). Determining convective lognormal solute transport parameters from resident concentration data. *Soil Science Society of America Journal 60:*1306-1317.

Vanwesenbeek, I.J. and R.G. Kachanoski (1991). Partial scale dependence of in situ solute transport. *Soil Science Society of America Journal 55:*3-7.

Vereecken, H., J. Diels, J.V. Orshoven, J. Feyen, and J. Bouma (1992). Functional evaluation of pedotransfer functions for the estimation of soil hydraulic functions. *Soil Science Society of America Journal 56:*1371-1378.

Vogel, T., R. Zhang, M. Th. van Genuchten, and K. Huang (1995). The HYDRUS code for simulating the one-dimensional water flow, solute transport, and heat movement in variably saturated media, Version 5.0. *Research Report No. 140.* United States Salinity Laboratory, Riverside, CA: USDA-ARS.

Vogeler, I., B.E. Clothier, S.R. Green, D.R. Scotter, and R.W. Tillman (1996). Characterizing water and solute movement by Time Domain Reflectometry and disc permeametry. *Soil Science Society of America Journal 60:*5-12.

Wallach, R., T.S. Steenhuis and J.-Y. Parlange (1999). Modeling the movement of water and solute through preferential flow paths. In *The Handbook of Groundwater Engineering,* ed. J.W. Delleur, Boca Raton, FL: CRC Press.

White, R.E. (1985a). The analysis of solute breakthrough curves to predict water redistribution during unsteady flow through undisturbed structured clay soil. *Journal of Hydrology 79:*21-35.

White, R.E. (1985b). The influence of macropores on the transport of dissolved and suspended matter through soil. *Advances in Soil Science 3:*95-120.

White, R.E. (1985c). A model for nitrate leaching in undisturbed structured clay soil during unsteady flow. *Journal of Hydrology 79:*37-51.

White, R.E. (1987). A transfer function model for the prediction of nitrate leaching under field conditions. *Journal of Hydrology 92:*207-222.

White, R.E. (1989). Prediction of nitrate leaching from a structured clay soil using transfer functions derived from externally applied or indigenous solute fluxes. *Journal of Hydrology 107:*31-42.

White, R.E., J.S. Dyson, Z. Gerstl, and B. Yaron (1986). Leaching of herbicides through undisturbed cores of a structured clay soil. *Soil Science Society of America Journal 50:*277-283.

White, R.E., J.S. Dyson, R.A. Haigh, W.A. Jury, and G. Sposito (1986). A transfer function model of solute transport through soil. 2. Illustrative applications. *Water Resources Research 22:*248-254.

White, R.E., L.K. Heng, and R.B. Edis (1998). Transfer function approaches to modeling solute transport in soils. In *Physical Nonequilibrium in Soils. Modeling and Application,* eds. H. M. Selim and L. Ma. Chelsea, MI: Ann Arbor Press, pp. 311-348.

White, R.E. and G.N. Magesan (1991). A stochastic-empirical approach to modelling nitrate leaching. *Soil Use and Management 7:*85-94.

White, R.E., A.E. Okom, R.B. Edis, and L.K. Heng (1999). Estimating a soil's solute transport volume from surrogate variables. *Abstracts.* Salt Lake City: American Society of Agronomy, Crop Science Society of America, Soil Science Society of America, p. 177.

White, R.E., G.W. Thomas, and M.S. Smith (1984). Modelling water flow through undisturbed soil cores using a transfer function model derived from ^{3}HOH and Cl^{-} transport. *Journal of Soil Science 35:*159-168.

Williams, J.R., C.A. Jones, and P.T. Dyke (1984). A modelling approach to determine the relationship between erosion and soil productivity. *Transactions of the American Society of Agricultural Engineers 27:*129-144.

Williams, J.R. and D.E. Kissel (1991). Water percolation: An indicator of nitrogen-leaching potential. In *Managing Nitrogen for Groundwater Quality and Farm Profitability,* eds. R.F. Follett, D.R. Keeney, and R.M. Cruse. Madison, WI: Soil Science Society of America, pp. 5-8.

Young, R.A., C.A. Onstad, D.D. Bosch, and W.P. Anderson (1989). AGNPS: A non-point source pollution model for evaluating agricultural watersheds. *Journal of Soil and Water Conservation 44:*168-173.

Zhang, R., J. Yang, and Z. Ye (1996). Solute transport through the vadose zone: A field study and stochastic analyses. *Soil Science 161:*270-277.

Chapter 12

Soil-Atmosphere Interactions

Keith A. Smith

Rising concentrations of carbon dioxide and certain other gases in the atmosphere that stem from human activity are bringing about major changes to the global environment. The gases cause global warming, deplete the concentration of ozone in the stratosphere that acts as a shield against excessive exposure to ultraviolet radiation (UV) at the earth's surface, and also contribute to acid deposition. The gases that are most important in this context are carbon dioxide (CO_2), methane (CH_4), nitrous and nitric oxides (N_2O and NO), and chlorofluorocarbons (CFCs).

Although most of the anthropogenic CO_2 emissions come from the combustion of fossil fuels (5.5 Gt CO_2-C y^{-1}), there is also a substantial contribution, estimated to be 1.6 ±1.0 Gt C y^{-1}, from land use change, mainly in the tropics (Houghton et al., 1996). This comes from biomass burning and increased mineralization of soil organic carbon following the conversion of forest or native grassland to agriculture. The biosphere also plays a massive role in the global cycling of carbon dioxide. The total quantity of CO_2-C exchanged annually between terrestrial ecosystems (vegetation and soils combined) and the atmosphere is about 60 Gt in each direction, an order of magnitude greater than the emissions from fossil fuels (Figure 12.1). The imbalance between the uptake and emission (1.3 Gt C y^{-1}) is relatively small compared with these fluxes, but (together with the comparable ocean sink) it is responsible for a substantial reduction in the rate of increase in the atmospheric concentration of CO_2 (Houghton et al., 1996). Most of the methane and N_2O emissions to the atmosphere are of biogenic origin, with about a third of the methane and nearly two-thirds of the N_2O originating in soils (Prather et al., 1995). The CFCs are entirely of industrial origin and unaffected by land-atmosphere processes, and so are not considered further here.

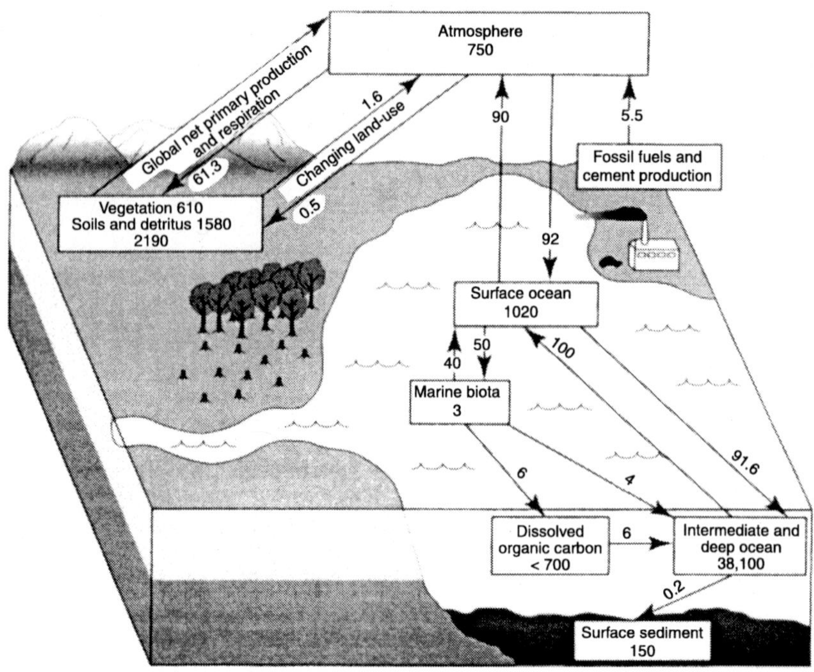

FIGURE 12.1. The global carbon cycle, showing the reservoirs (in Pg C) and the fluxes (in Pg C y^{-1}). (*Source:* Houghton et al., 1996, p. 77.)

Much experimental work has been undertaken in the last few years in an attempt to quantify these emissions and to identify the key factors that govern them. However, fluxes vary greatly between ecosystems and are often subject to great variations, both spatial and temporal, within ecosystems. Consequently, there is an essential role for modeling to scale up observations made on relatively small areas to larger regions and ultimately to a global level, and also to be able to predict the effects of environmental changes and changes in land management on future emissions. This chapter examines current knowledge regarding soil-atmosphere gaseous exchange and the extent to which models are able to describe the key phenomena to be used for predictive purposes. The number of models available is far too great to cover comprehensively, so examples are used to illustrate different types of models and modeling approaches applicable to different scales to give an overview of the state of the art.

CARBON DIOXIDE EXCHANGE

This section is restricted to the fluxes of CO_2 between terrestrial ecosystems and the atmosphere to complement (rather than to duplicate) the coverage of soil organic matter dynamics in Chapter 13.

In all terrestrial ecosystems, except those with bare land resulting from agricultural or forestry operations, soil emissions of CO_2 generally take place within a two-way exchange between the land and the atmosphere. This exchange is usually known as the net ecosystem exchange (NEE) and involves the component exchanges of CO_2 from the plant canopy, its elements, and the ground surface. The NEE is the net sum of the gains of C in photosynthesis by the vegetation, the losses of C from respiration by above-ground plant tissues, and losses by below-ground roots, mycorrhizas, and heterotrophic microorganisms (the "soil respiration"). The most important of these fluxes are the gains by plant photosynthesis and losses through soil respiration (Moncrieff, Jarvis, and Valentini, 2000). Chamber/cuvette techniques, together with infrared CO_2 analysis, are employed to measure soil respiration (Rayment and Jarvis, 1997; Fang and Moncrieff, 1998), and respiration/photosynthesis by leaves, stems, and shoots (Ryan, 1990; Law, Ryan, and Anthoni, 1999). Unless the soil respiration component of NEE can be satisfactorily determined, the overall carbon balance of the ecosystem and the quantification of it as a net source or a net sink for CO_2 cannot be established. Measurements can only relate to very small areas of the soil surface, and so modeling has a key role to play. By being able to simulate soil fluxes and their response to soil and environmental factors, there is then the potential for scaling up emission estimates to the landscape scale and beyond, as well as for predicting possible feedback effects on emissions due to climate change.

Contrasting approaches have been used to model soil CO_2 emissions. For forest soils, a relatively simple regression model relating emissions to temperature, soil water content, and a constant that is dependent on texture and density, was described by Hanson et al. (1993). The model explained between 50 and 74 percent of the variability in the flux data, and performed similarly for data from different topographical locations. Kicklighter et al. (1994) established that there was an exponential relationship between monthly mean CO_2 emissions from temperate forest soils and monthly mean air temperatures:

$$M_{CO2}\ \text{flux} = 27.46 \times \exp\,[\,0.06844(MT_{air}) \qquad (12.1)$$

where M_{CO2} flux is the monthly evolution of CO_2 from soils, in g C m^{-2} month^{-1}, and MT_{air} is the mean monthly air temperature in °C. They then used a georeferenced database of such temperatures to provide estimates of monthly CO_2 fluxes for temperate forests around the world. The relationship between predicted and observed fluxes given by the calibrated monthly flux model is compared with other modeling approaches in Figure 12.2.

The model of Zamolodchikov and Karelin (2001) employed a GIS approach to estimate CO_2 fluxes over the entire Russian tundra zone. Eight

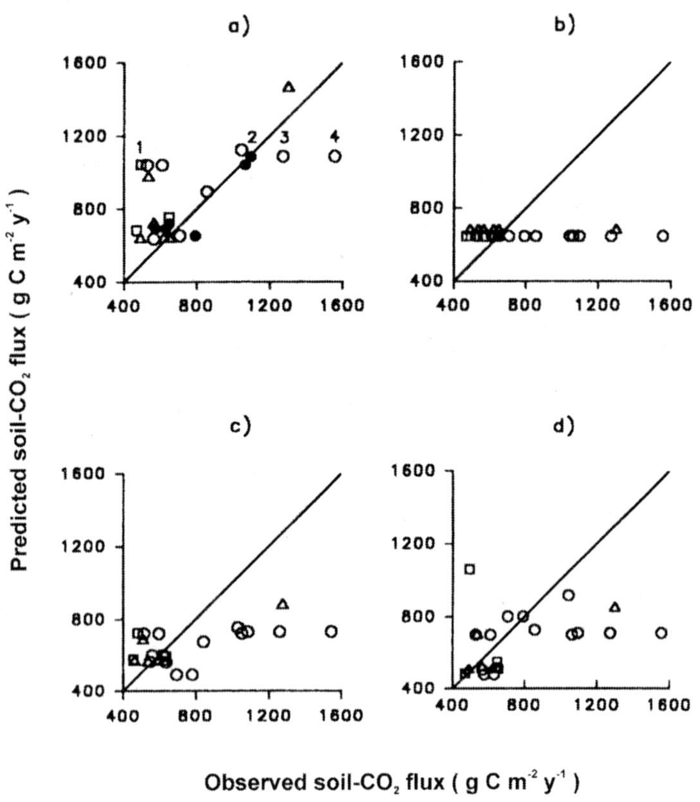

FIGURE 12.2. Observed versus predicted annual soil CO_2 fluxes for 21 mature forest lands across the globe. Predicted fluxes based on (a) calibrated monthly soil CO_2 flux; (b) mean biome fluxes; (c) annual soil CO_2 flux model based on annual mean air temperature; (d) annual soil CO_2 flux model based on annual mean air temperature and annual precipitation. Triangles: coniferous forests; circles: deciduous forests; squares: mixed forests; solid symbols: represent data used to develop daily CO_2 flux model. (*Source:* Kicklighter et al., 1994, p. 1308.)

geographical regions were divided into tundra landscape types (2-13 per region, totaling 69). Six years of field data on CO_2 fluxes and their environmental controls were collected from 423 sample plots between 65° and 74°N and 63° and 172°E, together with data on phytomass and diurnal temperature variations. Using stepwise multiple regression, equations for gross primary production (GPP) and gross ecosystem respiration (GR) were obtained, which explained 67 and 75 percent of the variance, respectively. The relationships between the simulated and real carbon fluxes are shown in Figure 12.3.

A more detailed, process-based model has been described by Fang and Moncrieff (1999). It takes into account the production of CO_2 by the respiration of plant roots and microorganisms and the transport of gases in the soil: CO_2 from soil to atmosphere and O_2 in the opposite direction. Figure 12.4 shows the components of the Fang and Moncrieff model of the production and transport of CO_2 in soil (PATCIS). The major inputs of carbon are leaf

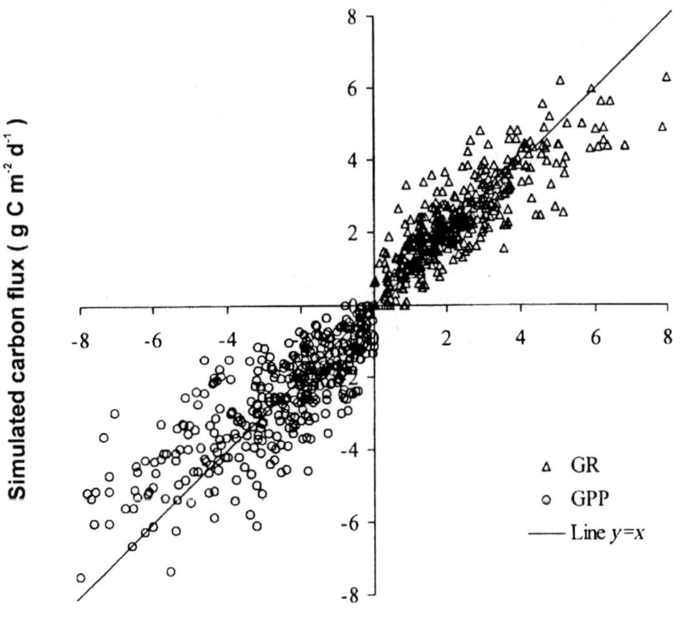

FIGURE 12.3. Modeled versus measured values of gross primary production (GPP) and gross respiration (GR) for Russian tundra sites. The line is the 1:1 relationship. (*Source:* Zamolodchikov and Karelin, 2001, p. 152.)

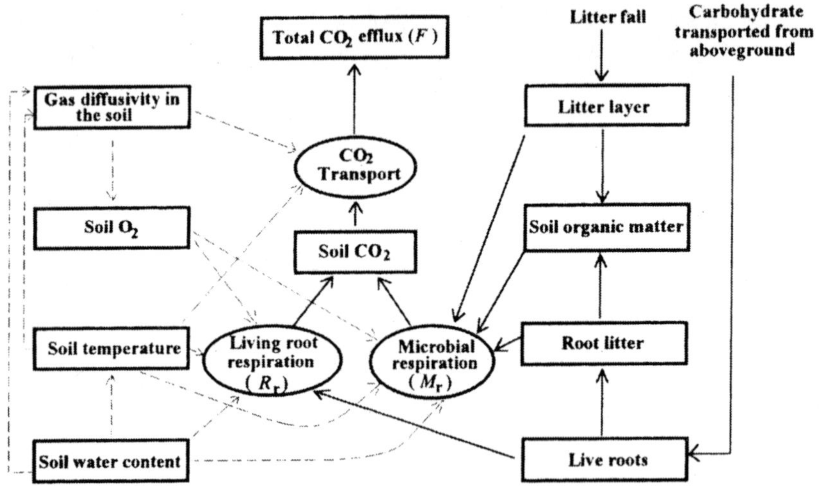

FIGURE 12.4. Diagram of CO_2 efflux model. Solid lines and arrows represent carbon flows; dashed lines and arrows represent influence of environmental factors on respiration or gas transport. Rectangles indicate state variables; ellipses indicate processes. (*Source:* Fang and Moncrieff, 1999, p. 227.)

and root litter, and the only C loss considered is as CO_2 emitted from the surface.

Soil respiration is described by the Arrhenius relationship

$$f(T) = \exp(-E/RT) \tag{12.2}$$

where E is the activation energy for respiration in kJ mol^{-1}, R is the universal gas constant, and T is the absolute temperature in K. E is a variable in this model and decreases with increasing temperature. The effect of soil water content on respiration is given by equation (12.3):

$$df(W)/dW = a\,[f(W)_{max} - f(W)] \tag{12.3}$$

where W is soil water content, and a is a parameter defining the maximum increase in the rate of soil respiration with soil water content. In the equation, $f(W)_{max} = 1$ is the maximum value of $f(W)$ when soil water content does not limit respiration. Integrating equation (12.3) gives

$$f(W) = 1 - \exp(-aW + c) \tag{12.4}$$

where c is an integration constant.

Oxygen concentrations in soil air at different depths are simulated using the method of Campbell (1985) and the corresponding respiration rates are calculated via the Michaelis-Menten equation. Fang and Moncrieff (1999) parameterized and validated the model with data from a slash pine plantation in Florida. The simulated daytime soil CO_2 profiles agreed well with the measured ones, and simulated hourly fluxes also agreed closely with the measured ones: 0.205 and 0.085 mg CO_2 m^{-2} s^{-1}, compared with 0.200 and 0.09 mg m^{-2} s^{-1}; average daily values were even closer. The results are illustrated in Figure 12.5.

SOILS AS A SOURCE/SINK FOR METHANE

Methane contributes 15 percent of the global warming resulting from the increasing concentrations of greenhouse gases in the atmosphere making it the second most important greenhouse gas after CO_2 (Houghton et al.,

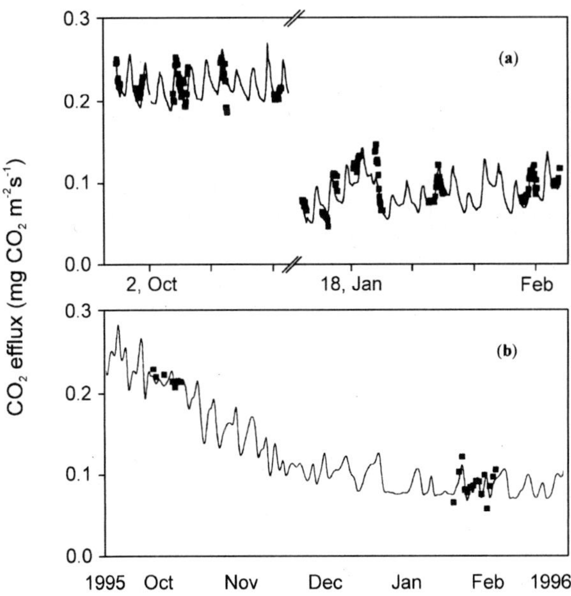

FIGURE 12.5. Comparison between simulated and measured hourly (a) and daily average (b) CO_2 efflux in a slash pine plantation in Florida. Solid lines: simulated fluxes; squares: measured fluxes. (*Source:* Moncrieff and Fang, 1999, p. 241.)

1996). More than a third of all methane emissions come from soils as a result of the microbial breakdown of organic compounds in strictly anaerobic conditions. This process occurs in natural wetlands, in flooded fields used for paddy rice production, and in landfills, as well as in the gut of some species of soil-dwelling termites. The rice and landfill sources make up an important part of anthropogenic methane emissions and have led to the concentration in the atmosphere more than doubling from c. 0.7 (microliters per liter) ppmv in the preindustrial era to just over 1.7 $\mu l\ l^{-1}$ today (Prather et al., 1995).

Emission from Natural Wetlands

The total global emission of methane from wetlands is estimated to be 115 Tg CH_4 y^{-1}, but with an uncertainty range of 55-150 Tg (Prather et al., 1995). The mean value corresponds to just over 20 percent of CH_4 emissions from all sources. There have been relatively few measurements in the tropics and subtropics, compared with the number carried out in boreal regions (Roulet et al., 1992; Whalen and Reeburgh, 1992), and modeling therefore has an important role in assessing the relative contributions from different latitudinal zones.

Rates of emission of methane from wetlands are affected by many factors: soil water status and temperature, soil type, pH, Eh, nutrient inputs, and the presence of adapted vascular plants. These plants have a well-developed system of intracellular air spaces (aerenchyma) in stems, leaves, and roots (Figure 12.6). This allows the transport of oxygen from the atmosphere to the root meristems and also serves as a pathway for the movement of methane from the soil into the atmosphere (Lloyd et al., 1998). Actual emissions to the atmosphere are less than the quantities of methane actually produced in flooded soils because of the occurrence of methane oxidation in overlying aerobic layers. For example, emissions are characteristically higher from areas of a peatland surface where the water table is at or above the soil surface than from raised hummock areas (Daulat and Clymo, 1998).

In two process-based models by Arah and Stephen (1998) and Walter and Heimann (2000), surface irregularities are ignored and only variations with depth are considered and the models are essentially one-dimensional. In the latter model, the three different transport mechanisms by which methane can move from the zone of formation to the atmosphere, i.e., diffusion, plant-mediated transport, and ebullition, are modeled explicitly. Daily values of soil temperature, water table, and net primary productivity (NPP) are used, and at permafrost sites the thaw depth is included. Testing of the model using experimental data from five wetland sites in North and Central

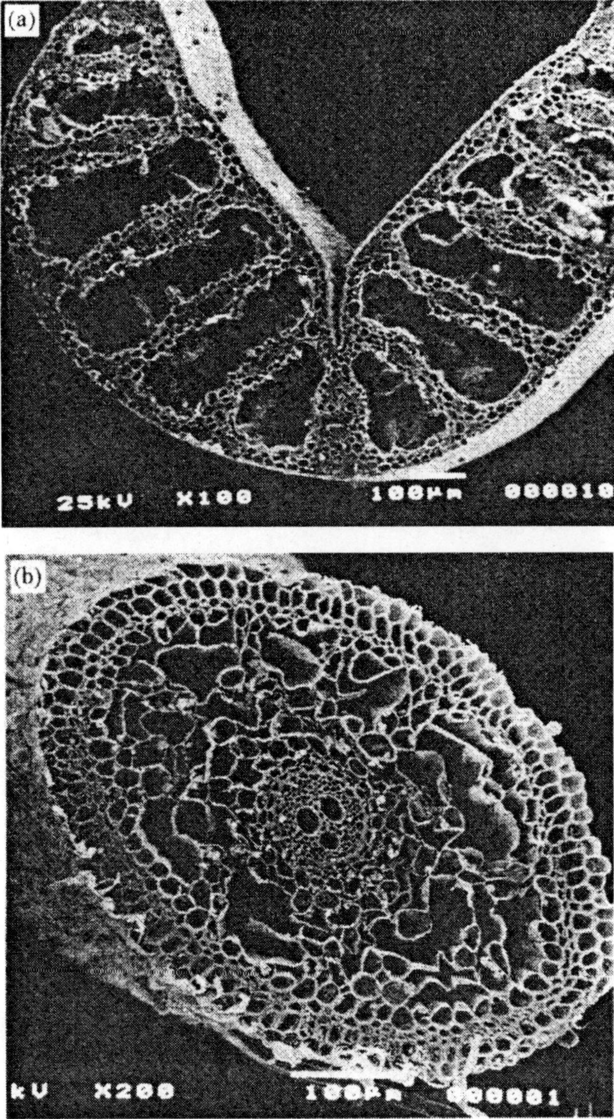

FIGURE 12.6. Scanning electron micrographs of transverse sections through a typical wetland vascular plant, *Eriophorum angustifolium,* showing the presence of large air spaces through which gas diffusion takes place: (a) leaf, (b) root from 40-50 mm depth. Bar = 100 µm. (*Source:* Adapted from Lloyd et al., 1998, p. 3235.)

America and Europe, representing a wide range of environmental conditions, showed a satisfactory fit with experiment in most cases. The comparison between measured and modeled emission values for a peatland site in Minnesota is shown in Figure 12.7.

The WMEM model (wetland methane emission model) of Cao, Marshall, and Gregson (1996) is based on the hypothesis that plant primary production and soil organic matter (SOM) decomposition act to control the supply of substrate needed by methane-producing microorganisms. Net primary production, deposition of litter C into the soil, and SOM decomposition are calculated with the terrestrial ecosystem model (Raich et al., 1991). The wetland environment is assumed to be always saturated. The major environmental factors regulating the production of methane are assumed to be water table position and temperature. Methane flux decreases exponentially with decreasing water table height (as shown in Figure 12.8), and an appropriate function for this is included in the model.

A temperature function assumes a Q_{10} of 2.0 for methane production and an optimum temperature of 30°C. Methane production is assumed to occur only during thaw/wet seasons. Oxidation occurs above the water table and in the rhizosphere where oxygen is likely to be available, and in the model the methane oxidation rate (MOR) is calculated from

$$MOR_t = MPR_t \, [0.60 + 0.30 \, GPP_t/GPP_{max}] \quad (12.5)$$

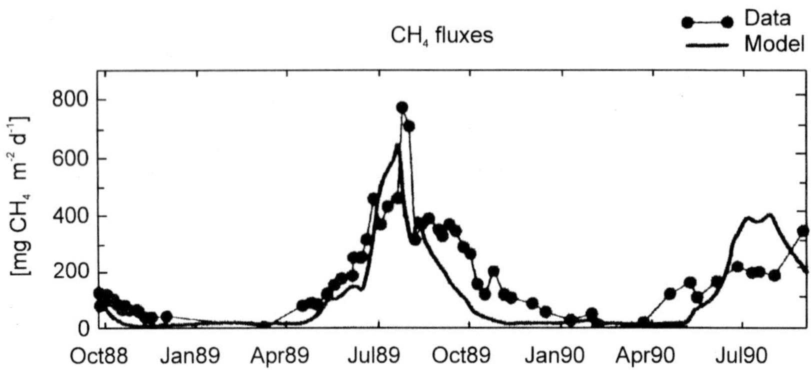

FIGURE 12.7. Comparison between modeled methane emission (thick line) and measured emission (dots), for a natural wetland in Minnesota (data of Dise, 1993). (*Source:* Adapted from Walter and Heimann, 2000, p. 754.)

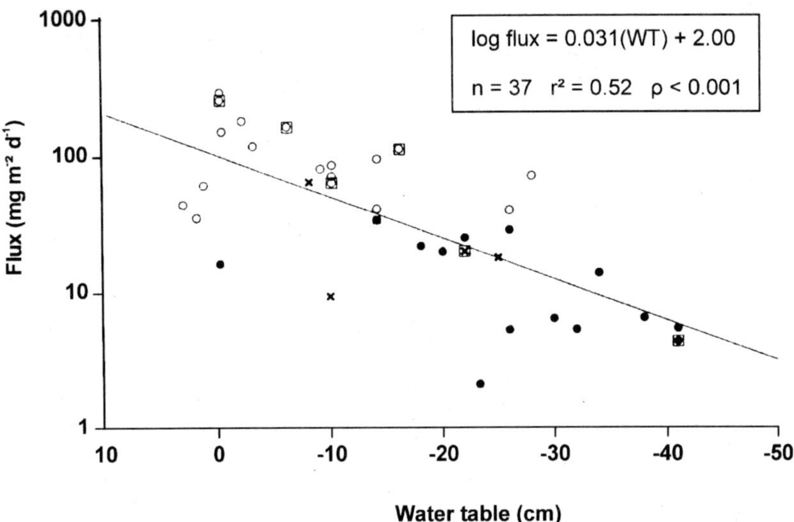

FIGURE 12.8. Relationship between wetland water table and methane flux, using data from six published studies. Open circles: graminoid and pond sites; solid circles: woody sites; crosses: open low shrub fen sites; squares: a mixture of site types (from author's own data). (*Source:* Liblik et al., 1997, p. 490.)

where MPR is the methane production rate and GPP_{max} is the maximum gross primary production occurring during any month. The balance between methane production and oxidation determines the rate of emission into the atmosphere. Data on wetland location, area, and vegetation were derived from Matthews and Fung (1987), and together with climate information were organized in a GIS with a resolution of $0.5° \times 0.5°$. The emission estimates obtained with WMEM by Cao, Marshall, and Gregson (1996) ranged from 3.6 g CH_4 m^{-2} y^{-1} for nonforested bogs (mainly >50°N) to 38.2 g CH_4 m^{-2} y^{-1} for forested swamps (mainly tropical). The model predicted a total emission of 92 Tg CH_4 y^{-1}, of which 56 percent came from tropical wetlands. This prediction is well within the 55-150 Tg range cited by Prather et al. (1995).

Emissions from Rice Paddies

Estimates of the total CH_4 emission from rice paddies amount to 50 ± 20 Tg y^{-1} (Neue, 1997), or about half that from natural wetlands. In rice fields, the most important factors affecting the size of methane emissions are water management, the amount of decomposable organic matter present, and the cultivar of rice being grown (Neue, 1997). Soil temperature, redox potential and pH, and the type and quantity of mineral fertilizer applied also have an effect on the emission rate. Substantial methane emissions from a rice paddy only begin when the soil redox potential has dropped to a very low level (–200 mV), following the reduction of Fe(III) and sulphate. Typically this may take two weeks or so. The quantities of methane which are then released depends on the supply of degradable organic matter. Although some of this substrate material comes from plant exudates from growing roots, the incorporation of straw or green manure into the soil (widely practiced in rice-growing countries) has a greater impact on CH_4 emission. Van der Gon and Neue (1995) found that the emission was about four times greater throughout the growing season from plots fertilized with up to 20 t ha^{-1} of green manure than in urea-fertilized plots. Higher soil temperature speeds up the initiation of CH_4 formation but this may alter the time course of emissions rather than the total over the season.

Nouchi et al. (1994) measured and modeled methane emissions from experimental plots with different fertilizer/straw treatments. Multiple linear regression gave the following equation for the methane flux due to gas bubbles at 1,300 h on sunny summer days:

$$F = 19.34Ra + 8.13T - 197.85 \qquad (12.6)$$

where F is the flux, Ra is the hourly global solar radiation in MJ m^{-2} h^{-1}, and T is the temperature in °C at 1,300 h. Daily flux was calculated from the observed diurnal cycle in emissions. Transport of methane through the plants, via similar aerenchyma channels to those shown in Figure 12.6 for natural wetland plants was assumed to be governed by molecular diffusion:

$$F = (C_s - P_a/H).D \qquad (12.7)$$

where C_s is the concentration in the soil solution, P_a is the concentration in the atmosphere, H is the Henry's Law constant for methane solubility in water, and D is the conductance (the reciprocal of the resistance) due to the biomass of the rice. P_a/H is negligible compared with C_s, so equation (12.7) reduces to

$$F = C_s.D \qquad (12.8)$$

Equation (12.8) was parameterized using the measured values for the conductance and the flux was simulated. The results obtained from the diffusion model and from the linear regression equation are shown in Figure 12.9.

Two other models of methane emission from rice deserve mention here. The semiempirical model of Huang, Sass, and Fisher (1997) and Sass, Fisher, and Huang (2000) predicts emissions on the basis of net rice productivity, cultivar character, soil texture and temperature, and organic matter amendment. It has been validated against measurements from irrigated rice paddy soils in three continents. Matthews, Wassmann, and Arah (2000) have developed a simulation model describing the main processes involved in methane emission from flooded rice fields by incorporating into an existing crop simulation model (Rice Crop Estimation through Resource and Environmental Synthesis [CERES-Rice]) a model describing the steady-state concentrations of methane and oxygen in soils (Arah and Kirk, 2000). It explains the seasonal patterns of methane emissions in an experiment involving mid- and end-season drainage and additions of organic material at the International Rice Research Institute (IRRI) in the Philippines.

Methane from different sources has different values for the ratio of the stable carbon isotopes, $^{13}C/^{12}C$. This can be used, in principle, to identify the origins of the methane emitted in a particular region of the world. This isotopic signal for methane from rice paddies changes during the course of a

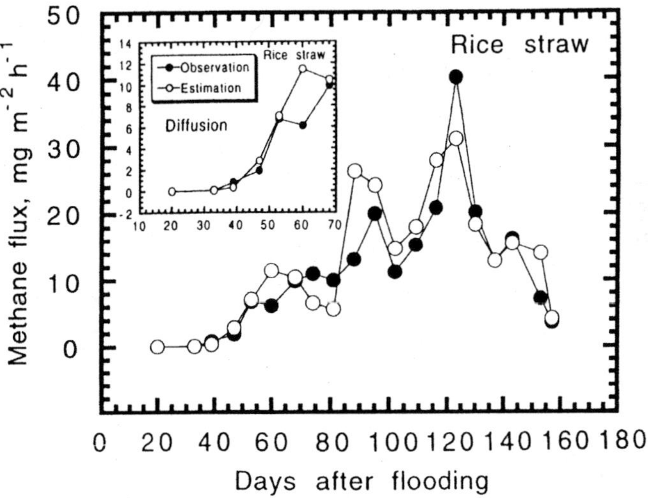

FIGURE 12.9. Actual and simulated methane emissions from a rice paddy plot after incorporation of rice straw. (*Source:* Nouchi et al., 1994, Fig. 6, p. 205. Reproduced with kind permission from Kluwer Academic Publishers.)

growing season, and this variation has been successfully modeled (Marik et al., 2001), using a simple box model adapted from Chanton et al. (1997).

Landfills are estimated to contribute another 40 Tg CH_4 y^{-1} to emissions. Some details of models of the rate of emission, and of the fraction of the produced methane that is oxidized within landfill cover soil may be found in Smith and Bogner (1997).

Oxidation of Atmospheric Methane in Soils

In addition to the oxidation in aerobic soil layers of methane diffusing upward from underlying anaerobic environments, there is also the quite distinct process of oxidation of atmospheric methane. The size of this soil sink for atmospheric methane was estimated by the Intergovernmental Panel on Climate Change (IPCC, 1997) at 30 (range 15-45) Tg CH_4 y^{-1}—6 percent of the estimated sink due to reaction with OH in the atmosphere (Watson et al., 1990). A recent analysis (Smith et al., 2000) of the available sink strengths for different ecosystems has shown consistent median values of between 1 and 2 kg CH_4 ha^{-1} y^{-1}, but with skewed (lognormal) distributions. This leads to a similar global mean estimate to that of Watson et al. (1990), but with a much greater uncertainty range.

The more complex oxidation models simulate microbial dynamics (Grant, 1999) and methane oxidation gradients in the soil (Dörr, Katruff, and Levin, 1993). Ridgwell, Marshall, and Gregson (1999) developed a process-based model in which methane oxidation is controlled by gas diffusivity at high rates of microbial activity and by microbial activity at high diffusivities. The global soil sink calculated by their model is 20-51 Tg CH_4 y^{-1}, with a preferred value of 38 Tg. Dry tropical systems are predicted to account for almost a third of this total.

The model of Potter et al. (1996) is simpler; it assumes that soil gas diffusivity is the major control, and thus it predicts maximum methane oxidation rates at low soil water contents. This is an oversimplification in that there is good evidence that oxidation rates decline as soils become very dry and microbial activity is inhibited (Dobbie and Smith, 1996). Del Grosso, Parton, Mosier, Ojima, Potter, et al. (2000) have developed the Potter et al. model to take account of this constraint. The model is more empirically based than the Ridgwell, Marshall, and Gregson model, using several data sets from the United States and one from the United Kingdom to parameterize it. The well-established decrease in oxidation rate as a result of soil disturbance is also included. The model predicted the variation in oxidation with soil water content in a German coniferous forest, but gave more mixed results with data for arable and formerly arable land in Scotland (Figure 12.10). So far, the model has not been used to estimate the global sink.

OXIDES OF NITROGEN: N_2O AND NO

Nitrous oxide, N_2O, is a greenhouse gas and is also one of the substances that destroys stratospheric ozone (Crutzen, 1976). Ice-core data show that the concentration of N_2O in the atmosphere has been increasing by about 0.25 percent per year, from about 280 ppbv in preindustrial times to 314 ppbv today. Direct emissions of N_2O from agricultural soils are now estimated to be 2.1±1.7 Tg N_2O-N y^{-1}, with a similar amount emanating from

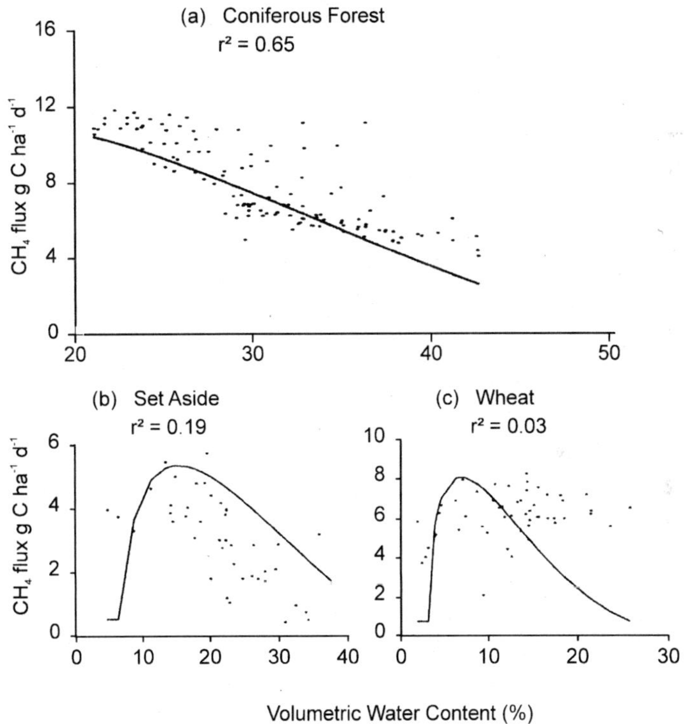

FIGURE 12.10. Measured (dots) and simulated (lines) oxidation rates of atmospheric methane in surface soils. (a) spruce forest, Germany; (b,c) former and current arable land, respectively, in Scotland (including factor for reduction in oxidation due to agricultural disturbance). (*Source:* Adapted from Del Grosso, Parton, Mosier, Ojima, Potter, et al., 2000, pp. 1013, 1014.)

nitrates leached from agricultural soils into surface and groundwaters (Mosier et al., 1998). Atmospheric nitric oxide, NO, a substantial proportion of which comes from soils, is also environmentally important. It reacts with CO and hydrocarbons in the atmosphere to form tropospheric ozone, and is a precursor of acid rain.

Direct emissions of N_2O and NO from soils result primarily from microbially driven nitrification and denitrification processes (Figure 12.11). Nonbiological chemodenitrification occurs in subsoils, but is generally a less important source of emissions (Granli and Bøckman, 1994). The process of denitrification is further described in Chapter 14.

Modeling of N_2O and NO Emissions

Models for N trace gas emissions vary greatly in complexity. One very simple approach is the flux model for N_2O currently used by the Intergovernmental Panel on Climate Change (IPCC, 1997). This is based on a regression analysis by Bouwman (1996) of whole-season studies of N_2O emissions reported in the literature:

$$F = 1 + (1.25 \pm 1.0).N/100 \qquad (12.9)$$

where F is the annual flux in kg N_2O-N ha^{-1}, and N is the quantity of fertilizer N applied in kg ha^{-1} (IPCC, 1997). Thus, agricultural land is regarded as having a "background" emission rate of 1 kg N_2O-N ha^{-1} y^{-1}, due to the

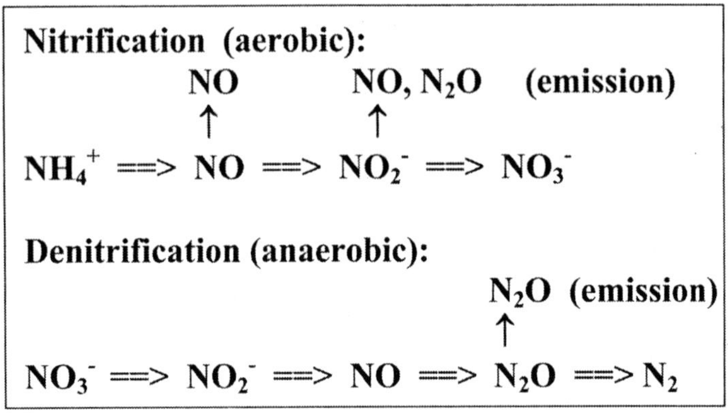

Nitrification (aerobic):

NO ↑ (from NO); NO, N_2O ↑ (from NO_2^-) (emission)

NH_4^+ ==> NO ==> NO_2^- ==> NO_3^-

Denitrification (anaerobic):

N_2O ↑ (from N_2O) (emission)

NO_3^- ==> NO_2^- ==> NO ==> N_2O ==> N_2

FIGURE 12.11. Nitrification and denitrification pathways, showing steps at which N_2O and/or NO emissions occur.

quantity of N cycling which goes on, with or without fertilizer application, and to this is added an emission factor of 1.25±1.0 percent of the N applied. However, new compilations of experimental data show that the emission factor varies even more widely (Dobbie, McTaggart, and Smith, 1999; Smith, Bouwman, and Braatz, 2002).

The most complex models, mechanistically, are those which calculate the soil anaerobic fraction in which denitrification can occur (Smith, 1980; Leffelaar, 1979, 1986; Arah and Smith, 1989; Renault and Stengel, 1994). These diffusion-based models are likely to require too much detailed parameterization to be applicable at sufficiently large scales to be useful for flux prediction purposes. Their role is a different one, i.e., to provide a better understanding of the processes responsible for emissions, and the sensitivity of the fluxes to changes in particular variables. Qualitatively, the trends predicted by these models (see Figure 12.12) are borne out by observation, in particular the very large Q_{10} values for N_2O emission that are sometimes observed (Brumme, 1995; Smith, 1997).

Several "process-oriented" simulation models of intermediate complexity with several features in common have been developed over the last few years. Each simulates (1) soil climate dynamics, (2) plant growth, nutrient uptake, and litter fall, (3) decomposition of soil organic matter, and (4) nitrogen mineralization and transformations. Examples are:

- The version of the CENTURY soil organic carbon cycling model dealing with emissions of nitrogen oxides, CENTURY-NGAS (Parton et al., 1996), now further developed in the denitrification model of Del Grosso, Parton, Mosier, Ojima, Kulmala, et al. (2000);
- The denitrification-decomposition model (DNDC) (Li, Frolking, and Frolking, 1992a,b);
- ExpertN (Engel and Priesack, 1993);
- NASA CASA, the NASA-Ames version of the Carnegie-Ames-Stanford-Approach (CASA) model (Potter, Riley, and Klooster, 1997).

The following comparison of these models is condensed from the review by Frolking et al. (1998).

Inputs from the earlier Century ecosystem model (Parton et al., 1994) that were used to drive the NGAS model include soil water-filled pore space (WFPS), soil temperature, soil CO_2 fluxes, N availability, and nitrification rates. The site-specific parameters include the sand, silt and clay content, the field capacity and wilting point, the saturated hydrologic conductivity, and the soil water potential versus water content curve. The hydrologic conductivity and water potential curves are estimated from the soil texture us-

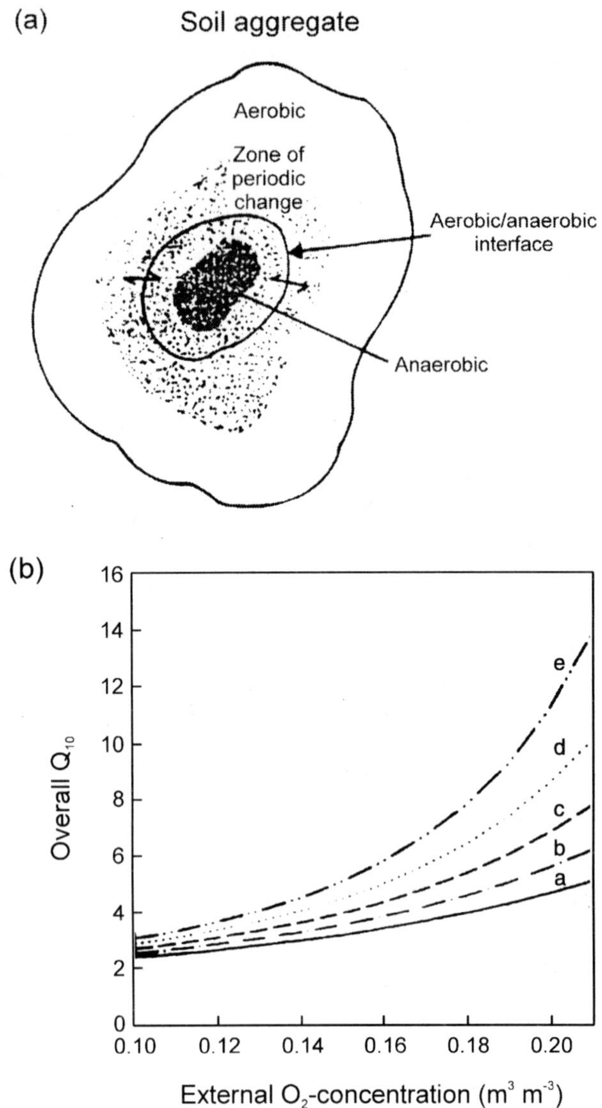

FIGURE 12.12. Diagram of (a) fluctuating anaerobic zone in a soil aggregate (*Source:* Tiedje et al., 1984), (b) predicted overall Q_{10} values for denitrification and N_2O production, using model of Smith (1980); a: D/Q = 1,000; b: D/Q = 1,100; c: D/Q = 1,200; d: D/Q = 1,300; e: D/Q = 1,400. (*Source:* Adapted from Smith, 1997, p. 332.)

ing standard equations, while the wilting point and field capacity are estimated from observed soil water data. The only site-specific curve for the denitrification model is the water-filled pore space curve for total denitrification gas flux (see Parton et al., 1996).

The DNDC model was developed to predict N_2O fluxes produced by nitrification and denitrification, and CO_2 fluxes produced by decomposition and root respiration. The soil climate submodel uses daily meteorological data to predict soil temperature and moisture profiles, soil water flow, and soil water uptake by plants. The crop/vegetation growth submodel simulates the growth of various crops from planting to harvest. The decomposition submodel has four soil carbon pools—litter, labile humus, passive humus, and microbial biomass; each pool has a fixed decomposition rate and C:N ratio. Decomposition rates are influenced by soil texture, soil temperature and moisture, and potentially by nitrogen limitations. The denitrification submodel in DNDC is activated by rain events, flooding (as in irrigated rice agriculture), and soil freezing. For any initiation of denitrification the initial status of the available NO_3^- and soluble carbon pools is provided by the decomposition submodel. The rates for each step in the denitrification reduction are a function of soluble C, soil temperature (or Eh for "frozen" soils), soil pH, N-substrate availability, and denitrifier biomass. As the soil dries following rain, the denitrifying portion of each model layer decreases with soil water content.

ExpertN consists of modular components for soil water flow, soil heat, N transport, and crop growth. For each model component and model unit, several distinct, interchangeable submodels are available and additional, user-defined submodels can be included easily by the supported use of dynamic link libraries. Soil water flow and water-filled pore space can be calculated by numerically solving Richard's equation and using representations of the hydraulic functions for soil water retention and unsaturated conductivity as given by the LEACHM model (Hutson and Wagenet, 1992). Potential evapotranspiration is calculated following Penman (1948); actual evaporation is determined using the current matric potential and conductivity. LEACHM is also used in the calculation of ammonium and nitrate transport and N turnover; the latter calculation uses an approach based on that of Johnsson et al. (1987). The transport of N_2O is simulated by solving the convection-diffusion equation for the gaseous N_2O-N concentration. The reduction of N_2O to N_2 is modeled as a first-order kinetic reaction, with a dependence on soil water content and a rate constant of 2 per d.

NASA CASA simulates seasonal patterns in carbon fixation, nutrient allocation, litter fall, soil nitrogen mineralization, net CO_2 exchange, and soil N_2O and NO production. Other soil trace gas fluxes (i.e., CH_4 and CO uptake) are simulated, using a modified version of Fick's first law, together with a soil water balance model (Potter et al., 1996). First-order equations simulate exchanges of decomposing plant residue (metabolic and structural fractions) at the soil surface. Active (microbial biomass and labile substrates), slow (chemically protected), and passive (physically protected) fractions of the SOM are represented. The soil profile is treated as three layers: surface organic matter, topsoil, and subsoil to rooting depth. These layers can differ in soil texture, moisture holding capacity, and carbon-nitrogen dynamics. Daily water balance in the soil, plus drainage output and soil water-filled pore space (WFPS) in each layer are calculated. Mineralization fluxes from litter, microbial, and soil organic matter pools contribute to a common mineral N pool. Gaseous nitrogen losses are modeled as a "leaky pipe" (Firestone and Davidson, 1989) where emissions are a function of the N mineralization rate (pipe flow rate), and soil water content (size of "holes" through which gas can "leak"). Production and emission of both NO and N_2O occur at intermediate levels of WFPS. At higher moisture levels, N_2O emissions increase exponentially with WFPS, while NO emission declines. Under very wet soil conditions, production of only N_2 occurs. The potential loss of either N_2O-N or NO-N as a percentage of total mineralized nitrogen has a default setting of 2 percent.

The predictive capacities of the four models have been explored through a simulation exercise involving three data sets from contrasting environments: the semiarid prairie land of Colorado, the cool, temperate, and moist conditions of southeast Scotland, and the more continental climate of Germany (Frolking et al., 1998). In most cases the simulated N_2O emissions were within a factor of about 2 of the observed annual emissions, but even when the models produced similar N_2O fluxes they often produced very different estimates of gaseous loss as N_2 and NO. Accurate simulation of soil water content was identified as a key requirement for predicting N_2O fluxes. All the models simulated the general pattern of low background fluxes interspersed by high fluxes following fertilization at the Scottish site, but they were not able to simulate the large pulses of N_2O emitted during winter thaws at the German site. All the models except DNDC simulated the very low fluxes at the dry site in Colorado. Figure 12.13 shows the measured N_2O fluxes and the modeling results for the Colorado and Scottish sites.

Li et al. (1994) used DNDC to model N_2O emissions from all agricultural land in Florida, simulating the emissions from the five major crops, pasture, and fallow land. A high rate of mineralization of organic matter in drained organic soils in this warm climate results in high nitrate concentrations and much enhanced N_2O emission (Duxbury et al., 1982). This effect was successfully captured by the model, which predicted that the organic soils were responsible for 38 percent of all the agricultural emissions, from only 9 per-

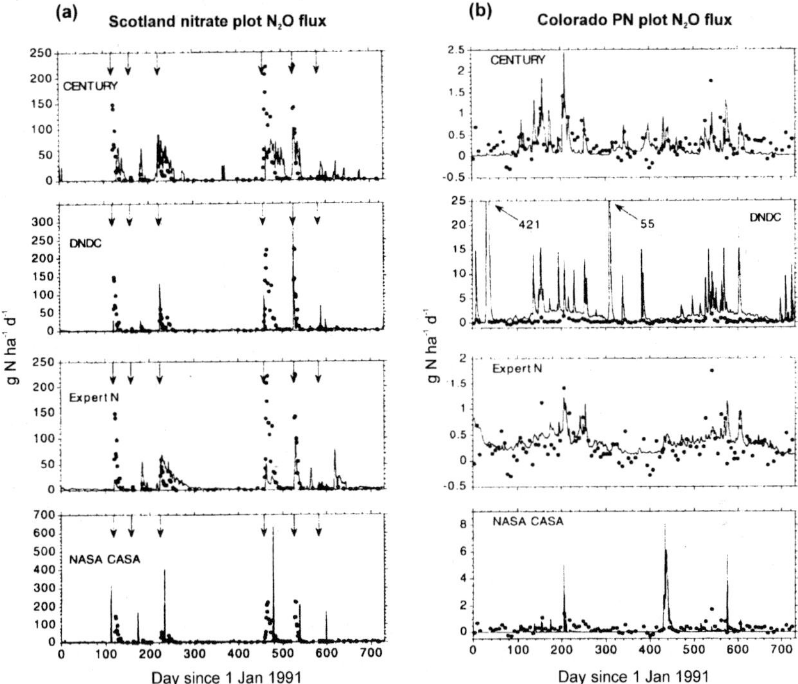

FIGURE 12.13. Measured N_2O fluxes (solid circles) at sites in Scotland (a) and Colorado (b), and simulations (lines) by four different models. (*Source:* Adapted from Frolking et al., 1998, Figs. 4, 5, pp. 88, 90. Reproduced with kind permission from Kluwer Academic Publishers.)

cent of the area. In a further upscaling, Li, Narayanan, and Harriss (1996) applied DNDC to virtually the whole United States, predicting an annual N_2O flux of 0.9-1.2 Tg N. The modeling also predicted that a 2°C rise in mean annual temperature would increase emissions from Iowa maize fields by 33 percent.

N_2O fluxes from N-fertilized grassland in the United Kingdom (and those from cereal-growing sites in the same region) have been successfully predicted by a summary model that relates the flux to soil water-filled pore space, mineral N content, and temperature. Depending on the values of these variables on different dates, a boundary-line method is used to attribute logarithmically varying values to the corresponding N_2O emissions, followed by linear interpolation between the predicted emission values (Figure 12.14; Conen, Dobbie, and Smith, 2000).

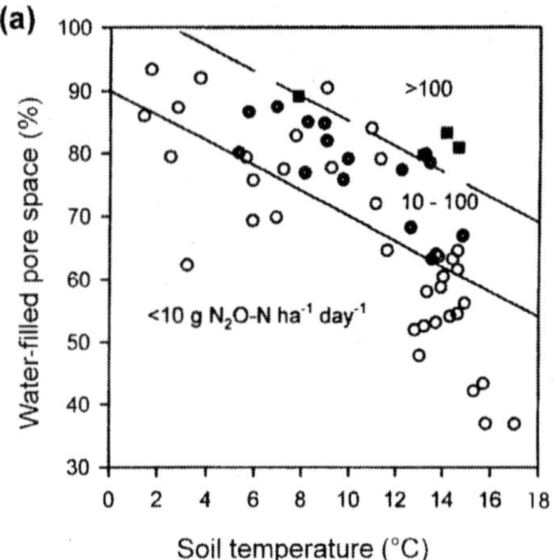

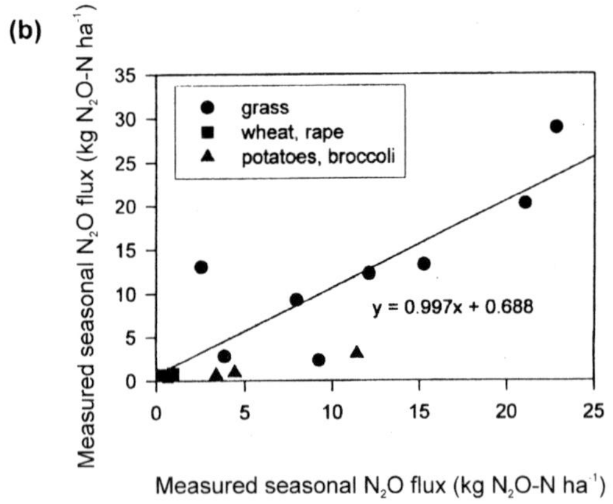

FIGURE 12.14. (a) Defining lower limits to soil WFPS and temperature below which N_2O flux is limited to <10 (solid line) or <100 (dashed line) g N_2O-N/ha/d. Symbols indicate WFPS/temperature combinations at which observed N_2O flux was <10 (open circles), 10-100 (solid circles) and >100 (solid squares) g N_2O-N/ha/d, when $N_{mineral}$ was not limiting. (b) Comparison of predicted seasonal N_2O emissions from different crops and sites in Scotland, based on (a) above, and measured emissions. Points for potato and broccoli crops excluded from regression. (*Source:* Adapted from Conen et al., 2000, pp. 419, 423.)

In addition to the modeling of fluxes from agricultural soils, there has been a considerable development in the modeling of fluxes from forest soils. Li et al. (2000) have coupled the DNDC model to a forest physiology model of photosynthesis, evaporation, and net primary production in forest ecosystems (PnET) (Aber and Federer, 1992), which predicts photosynthesis, respiration, organic carbon production and allocation, and litter production, and also to a model for predicting nitrification-induced NO and N_2O production. In the combined model (PnET-N-DNDC), three submodels predict soil climate, forest growth, and turnover of soil organic matter. Two further submodels calculate nitrification and denitrification. A so-called "anaerobic balloon" concept is employed to calculate the anaerobic status of the soil and divide it into aerobic and anaerobic fractions. Nitrification is only allowed to occur in the aerobic fraction, while denitrification only occurs in the anaerobic fraction. The size of the balloon is defined by the simulated oxygen partial pressure, calculated from oxygen diffusion and consumption in the soil. The mathematical routines are different and simpler than those used by Smith (1980) to calculate the anaerobic fraction; conceptually, however, there are similarities with this earlier model. Although the allocation of processes entirely to one or another volume fraction in PnET-N-DNDC may be a simplification of the true situation, it is a useful device for simulating soil environments in which aerobic and anaerobic microsites coexist, and where nitrification and denitrification can go on simultaneously.

Stange et al. (2000) validated PnET-N-DNDC by comparing its predictions with data sets from seven temperate forest ecosystems in the United States and Europe. Differences between predicted and measured N_2O fluxes were <27 percent, except for one site, while the differences for NO (for three sites) were <13 percent. Results for the NO measurements and predictions at the Höglwald Forest, southwest Germany, are shown in Figure 12.15.

Li et al. (2000) report that the anaerobic balloon concept has also been incorporated into the earlier DNDC model and applied to agroecosystems, with the outcome being improved fits between model prediction and measurement, and good results for emissions from submerged soils such as those in rice paddies. This is a very encouraging development that is likely to see further exploitation in the future.

PAST, CURRENT, AND FUTURE TRENDS

Although experimental measurements of greenhouse gas fluxes between soils and the atmosphere have been made for over 20 years, during the earlier part of this period, associated modeling work was somewhat limited. Although detailed, process-based models of anaerobic processes were published and contributed to the understanding of the mechanisms underlying emissions, they were not used to simulate observed fluxes or to predict

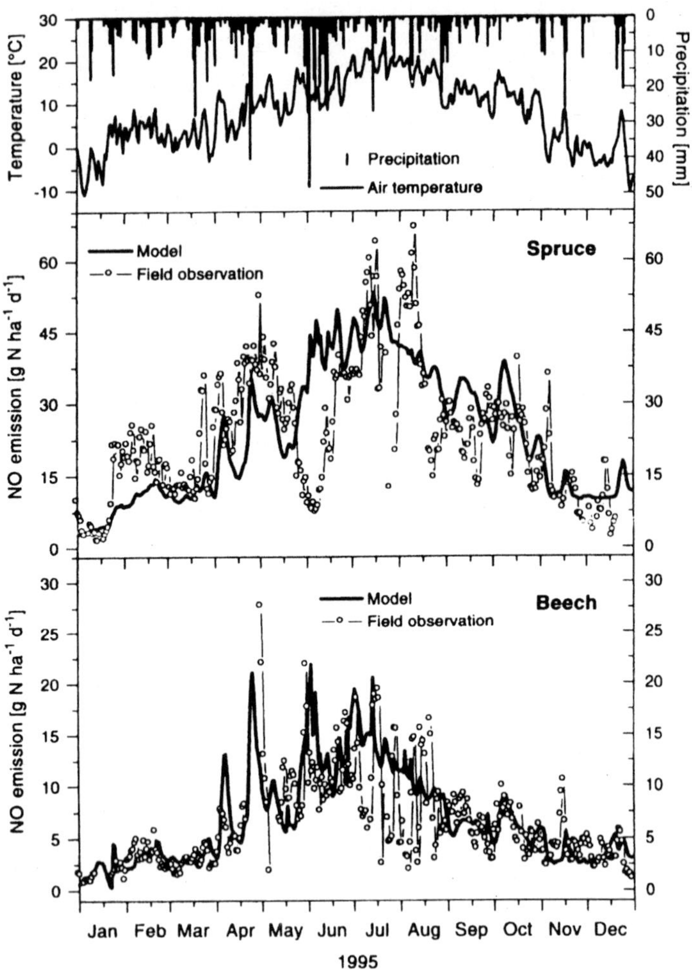

FIGURE 12.15. Precipitation, air temperature, and measured and simulated NO emission rates from soils in the spruce and beech stands of the Höglwald Forest, Germany, in 1995 (data of Gasche and Papen, 1999). (*Source:* Stange et al., 2000, p. 4396.)

fluxes where measurements had not been made. The last few years, in contrast, have been characterized by the appearance of a succession of models of varying degrees of complexity, which have largely succeeded in capturing both the observed temporal dynamics of trace gas fluxes and their magnitude when integrated over time. Some of these models and their applications have been summarized in this chapter.

An increasing effort is being made to use existing models for predictions of fluxes at regional and national scales, and there is still more to be done in model development—for example, to find explanations for those occasions when predictions and observations do not match, and to extend the range of soil types, climates, and ecosystems for which predictions can be made with some confidence. Nevertheless, if the rate of advancement over the next few years is as great as that of the last decade, we should be able to reduce the uncertainties regarding the global budgets for the biogenic greenhouse gases and predict future emissions with a new level of confidence.

SUMMARY

This chapter reviewed developments in the modeling of the exchange between land surfaces and the atmosphere of greenhouse gases including carbon dioxide, methane, and nitrous oxide, and the tropospheric ozone precursor, nitric oxide. Carbon dioxide is emitted from soils by microbial and root respiration as part of the natural C cycle; CH_4 is emitted both from wetland soils and taken up from the atmosphere in aerated topsoils; N_2O and NO are emitted as a result of soil nitrification and denitrification processes. Models of these gaseous exchanges range from simple regression relationships between emissions and variables such as temperature and soil water content, to much more complex, process-based models that take into account such factors as microbial kinetics, gas diffusion coefficients, and the partitioning of substrates between competing processes. The available models are also applicable to a wide range of spatial scales.

The models discussed include three dealing with CO_2 exchange: one example of a regression model, one large-scale regression model involving GIS, and a process-based model. Several process-based models of CH_4 emission from natural wetlands and rice paddies, and semiempirical models of rice emissions were discussed, together with models of differing complexity simulating the oxidation of atmospheric CH_4 in soils. The models of N_2O and NO emission covered a range from a simple regression model relating N_2O flux to the amount of fertilizer N applied to agricultural soils, to mechanistically complex models which calculate the fraction of the soil volume in which anaerobic denitrification can take place. Figures illustrating the degree of agreement achieved between simulated and measured fluxes were included for all four gases.

REFERENCES

Aber, J.D. and C.A Federer (1992). A generalized lumped-parameter model of photosynthesis, evaporation and net primary production in temperate and boreal forest ecosystems. *Oecologia 92:*463-474.

Arah, J.R.M. and G.D. Kirk (2000). Modeling rice-mediated methane emissions. *Nutrient Cycling in Agroecosystems 58:*221-230.

Arah, J.R.M. and K.A. Smith (1989). Steady-state denitrification in aggregated soils: A mathematical model. *Journal of Soil Science 40:*139-149.

Arah, J.R.M. and K.D. Stephen (1998). A model of the processes leading to methane emission from peatland. *Atmospheric Environment 32:*3257-3264.

Bouwman, A.F. (1996). Direct emissions of nitrous oxide from agricultural soils. *Nutrient Cycling in Agroecosystems 46:*53-70.

Brumme, R. (1995). Mechanisms of carbon and nutrient release and retention in beech forest gaps. III. Environmental regulation of soil respiration and nitrous oxide emissions along a microclimatic gradient. *Plant & Soil 168-169:*593-600.

Campbell, G.S. (1985). *Soil Physics with BASIC: Transport Models for Soil-Plant Systems.* New York: Elsevier.

Cao, M., S. Marshall, and K. Gregson (1996). Global carbon exchange and methane emissions from natural wetlands: Application of a process-based model. *Journal of Geophysical Research 101:*14,399-14,414.

Chanton, J.P., G.J. Whiting, N.E. Balir, C.W. Lindau, and P.K. Bollich (1997). Methane emission from rice: Stable isotopes, diurnal variations, and CO_2 exchange. *Global Biogeochemical Cycles 11:*15-27.

Conen, F., K.E. Dobbie, and K.A. Smith (2000). Predicting N_2O emissions from agricultural land through related soil parameters. *Global Change Biology 6:*417-426.

Crutzen, P.J. (1976). The influence of nitrogen oxides on the atmospheric ozone content. *Quarterly Journal of the Royal Meteorological Society 96:*320-325.

Daulat, W.E. and R.S. Clymo (1998). Effects of temperature and water table on the efflux of methane from peatland surface cores. *Atmospheric Environment 32:* 3207-3218.

Del Grosso, S.J., W.J. Parton, A.R. Mosier, D.S. Ojima, A.E. Kulmala, and S. Phongpan (2000). General model for N_2O and N_2 gas emissions from soils due to denitrification. *Global Biogeochemical Cycles 14:*1045-1060.

Del Grosso, S.J., W.J. Parton, A.R. Mosier, D.S. Ojima, C.S. Potter, W. Borken, R. Brumme, K. Butterbach-Bahl, P.M. Crill, K.E. Dobbie, and K.A. Smith. (2000). General CH_4 oxidation model and comparisons of CH_4 oxidation in natural and managed systems. *Global Biogeochemical Cycles 14:*999-1019.

Dise, N.B. (1993). Methane emission from Minnesota peatlands: Spatial and seasonal variability. *Global Biogeochemical Cycles 7:*123-142.

Dobbie, K.E., I.P. McTaggart, and K.A. Smith (1999). Nitrous oxide emissions from intensive agricultural systems: Variations between crops and seasons, key driving variables, and mean emission factors. *Journal of Geophysical Research 104:*26,891-26,899.

Dobbie, K.E. and K.A. Smith (1996). Comparison of CH_4 oxidation rates in woodland, arable and set aside soils. *Soil Biology & Biochemistry 28:*1357-1365.

Dörr, H., L. Katruff, and I. Levin (1993). Soil texture parameterization of the methane uptake in aerated soils. *Chemosphere 26:*697-713.

Duxbury, J.M., D.R. Bouldin, R.E. Terry, and R.L. Tate III (1982). Emissions of nitrous oxide from soils. *Nature 298:*462.

Engel, Th. and E. Priesack (1993). Expert-N, a building block system of nitrogen models as a resource for advice, research, water management and policy. In *Integrated Soil and Sediment Research: A Basis for Proper Protection,* eds. H.J.P. Eijsackers and T. Hamers. Dordrecht: Kluwer. pp. 503-507.

Fang, C. and J.B. Moncrieff (1998). An open-top chamber for measuring soil respiration and the influence of pressure differences on CO_2 efflux measurements. *Functional Ecology 12:*319-325.

Fang, C. and J.B. Moncrieff (1999). A model for soil CO_2 production and transport. 1: Model development. *Agricultural & Forest Meteorology 95:*225-236.

Firestone, M.K. and E.A. Davidson (1989). Microbiological basis of NO and N_2O production and consumption in soil. In *Exchange of Trace Gases Between Terrestrial Ecosystems and the Atmosphere,* eds. M.O. Andreae and D.S. Schimel. New York: John Wiley and Sons. pp. 7-21.

Frolking, S.E., A.R. Mosier, D.S. Ojima, C. Li, W.J. Parton, C.S. Potter, E. Priesack, R. Stenger, C. Haberbosch, P. Dörsch, H. Flessa, and K.A. Smith (1998). Comparison of N_2O emissions from soils at three temperate agricultural sites: Simulations of year-round measurements by four models. *Nutrient Cycling in Agroecosystems 52:*77-105.

Gasche, R. and H. Papen (1999). A three-year continuous record of nitrogen trace gas fluxes from untreated and limed soil of a N-saturated spruce and beech forest ecosystem in Germany, 2, NO and NO_2 fluxes. *Journal of Geophysical Research 104:*18,505-18,520.

Granli, T. and O.C. Bøckman (1994). Nitrous oxide from agriculture. *Norwegian Journal of Agricultural Sciences,* Supplement No. *12:*1-128.

Grant, R.F. (1999). Simulation of methanotrophy in the mathematical model ecosys. *Soil Biology & Biochemistry 31:*287-297.

Hanson, P.J., S.D.Wullschleger, S.A. Bohlman, and D.E. Todd (1993). Seasonal and topographic patterns of forest floor CO_2 efflux from an upland oak forest. *Tree Physiology 13:*1-15.

Houghton, J.T., L.G. Meiro Filho, B.A. Callander, N. Harris, A. Kattenberg, and K. Maskell (Eds.) (1996). *Climate Change 1995—The Science of Climate Change.* Cambridge: Cambridge University Press.

Huang, Y., R.L. Sass, and F.M. Fisher (1997). Methane emission from Texas rice paddy soils. 1. Quantitative multi-year dependence of CH_4 emissions on soil, cultivar, and grain yield. *Global Change Biology 3:*479-489.

Hutson, J.L. and R.J. Wagenet (1992). *LEACHM—Leaching Estimation and Chemistry Model. Version 3.0.* Department of Soil, Crop, and Atmospheric Sciences, Research Service No. 93-3, Ithaca, NY: Cornell University.

Intergovernmental Panel on Climate Change (IPCC) (1997). *Guidelines for National Greenhouse Gas Inventories.* Paris: OECD.

Johnsson, H., L. Bergström, P. Jansson, and K. Paustian (1987). Simulated nitrogen dynamics and losses in a layered agricultural soil. *Agriculture, Ecosystems & Environment 18:*333-356.

Kicklighter, D.W., J.M. Melillo, W.T. Peterjohn, E.B. Rastetter, A.D. McGuire, and P.A. Steudler (1994). Aspects of spatial and temporal aggregation in estimating regional carbon dioxide fluxes from temperate forest soils. *Journal of Geophysical Research 99:*1303-1315.

Law, B.E., M.G. Ryan, and P.M. Anthoni (1999). Seasonal and annual respiration of a ponderosa pine ecosystem. *Global Change Biology 5:*169-182.

Leffelaar, P.A. (1979). Simulation of partial anaerobiosis in a model soil in respect to denitrification. *Soil Science 128:*110-120.

Leffelaar, P.A. (1986). Dynamics of partial anaerobiosis, denitrification, and water in a soil aggregate: Experimental. *Soil Science 142:*352-366.

Li, C., J. Aber, F. Stange, K. Butterbach-Bahl, and H. Papen (2000). A process-oriented model of N_2O and NO emissions from forest soils: 1. Model development. *Journal of Geophysical Research 105:*4369-4384.

Li, C., S. Frolking, and T.A. Frolking (1992a). A model of nitrous oxide evolution from soil driven by rainfall events: 1. Model structure and sensitivity. *Journal of Geophysical Research 97:*9759-9776.

Li, C., S. Frolking, and T.A. Frolking (1992b). A model of nitrous oxide evolution from soil driven by rainfall events: 2. Model applications. *Journal of Geophysical Research 97:*9777-9783.

Li, C., S.E. Frolking, R.C. Harriss, and R.E. Terry (1994). Modeling nitrous oxide emissions from agriculture: A Florida case study. *Chemosphere 28:*1401-1415.

Li, C., V. Narayanan, and R.C. Harriss (1996). Model estimates of nitrous oxide emissions from agricultural lands in the United States. *Global Biogeochemical Cycles 10:*297-306.

Liblik, L.K., T.R. Moore, J.L. Bubier, and S.D. Robinson (1997). Methane emissions from wetlands in the zone of discontinuous permafrost: Fort Simpson, Northwest Territories, Canada. *Global Biogeochemical Cycles 11:*485-494.

Lloyd, D., K.L. Thomas, J. Benstead, K.L. Davies, S.H. Lloyd, J.R.M. Arah, and K.D. Stephen (1998). Methanogenesis and CO_2 exchange in an ombrotrophic peat bog. *Atmospheric Environment 32:*3229-3238.

Marik, T., H. Fischer, F. Conen, and K.A. Smith (2001). Seasonal variations in stable isotope ratios in methane from rice fields. *Global Biogeochemical Cycles* (in press).

Matthews, E. and I. Fung (1987). Methane emission from natural wetlands: Global distribution, area and environmental characteristics of sources. *Global Biogeochemical Cycles 1:*61-86.

Matthews, R.B., R. Wassmann, and J. Arah (2000). Using a crop/soil simulation model and GIS techniques to assess methane emissions from rice fields in Asia. *Nutrient Cycling in Agroecosystems 58:*141-159.

Moncrieff, J.B. and C. Fang (1999). A model for soil CO_2 production and transport. 2: Application to a Florida *Pinus elliotte* plantation. *Agricultural & Forest Meteorology 95:*237-256.

Moncrieff, J.B., P.G. Jarvis, and R. Valentini (2000). Canopy fluxes. In *Methods in Ecosystem Science,* eds. O.E. Sala, R.B. Jackson, H.A. Mooney, and R.A. Howarth. New York: Springer-Verlag. pp. 161-180.

Mosier, A.R., C. Kroeze, C. Nevison, O. Oenema, S. Seitzinger, and O. Van Cleemput (1998). Closing The Global Atmospheric N_2O Budget: Nitrous Oxide Emissions Through The Agricultural Nitrogen Cycle. *Nutrient Cycling In Agroecosystems 52:*225-248.

Neue, H.U. (1997). Fluxes of methane from rice fields and potential for mitigation. *Soil Use & Management 13:*258-267.

Nouchi, I., T. Hosono, K. Aoki, and K. Minami (1994). Seasonal variation in methane flux from rice paddies associated with methane concentration in soil water, rice biomass and temperature, and its modelling. *Plant & Soil 161:*195-208.

Parton, W.J., A.R. Mosier, D.S. Ojima, D.W. Valentine, D.S. Schimel, K. Weier, and A.E. Kulmala (1996). Generalized model for N_2 and N_2O production from nitrification and denitrification. *Global Biogeochemical Cycles 10:*401-412.

Parton, W.J., D.S. Ojima, C.V. Cole, and D.S. Schimel (1994). A general model for soil organic matter dynamics: Sensitivity to litter chemistry, texture, and management. In *Quantitative Modeling of Soil Forming Processes,* Soil Science Society of America Special Publication 39, Madison, WI: Soil Science Society of America. pp. 147-167.

Penman, H.L. (1948). Natural evaporation from open water, soil and grass. *Proceedings of the Royal Society of London, Series A, 193:*372-383.

Potter, C.S., P.A. Matson, P.M. Vitousek, and E.A. Davidson (1996). Process modeling of controls on nitrogen trace gas emissions from soils worldwide. *Journal of Geophysical Research 101:*1361-1377.

Potter, C.S., R.H. Riley, and S.A. Klooster (1997). Simulation modeling of nitrogen trace gas emissions along an age gradient of tropical forest soils. *Ecological Modelling 97:*179-196.

Prather, M., R. Derwent, D. Ehhalt, P. Fraser, E. Sanhueza, and X. Zhou (1995). Other trace gases and atmospheric chemistry. In *Climate Change 1994: Radiative Forcing of Climate Change and an Evaluation of the IPCC IS92 Emission Scenarios,* eds. J.T. Houghton, L.G. Meira Filho, J. Bruce, H. Lee, B.A. Callender, E. Haites, N. Harris, and K. Maskell. Cambridge: Cambridge University Press. pp. 73-126.

Raich, T.M., E.B. Rastetter, J.M. Melillo, D.W. Kicklighter, P.A. Steudler, B.J. Peterson, A.L. Grace, B. Moore III, and C.J. Vorosmarty (1991). Potential net primary productivity in South America: An application of a global model. *Ecological Applications 1:*399-429.

Rayment, M.B. and P.G. Jarvis (1997). An improved open chamber system for measuring soil CO_2 effluxes in the field. *Journal of Geophysical Research 102:* 28,779-28,784.

Renault, P. and P. Stengel (1994). Modeling oxygen diffusion in aggregated soils: I. Anaerobiosis inside the aggregates. *Soil Science Society of America Journal 58:*1017-1023.

Ridgwell, A., S.J. Marshall, and K. Gregson (1999). Consumption of atmospheric methane by soils: A process-based model. *Global Biogeochemical Cycles 13:*59-70.

Roulet, N.T., T. Moore, J. Bubier, and P. Lafleur (1992). Northern fens: Methane flux and climate change. *Tellus 44B:*100-105.

Ryan, M.G. (1990). Growth and maintenance respiration in stems of *Pinus contorta* and *Picea engelmannii. Canadian Journal of Forest Research 20:*48-57.

Sass, R.L., F.M. Fisher, and Y. Huang (2000). A process-based model for methane emissions from flooded rice fields: Experimental basis and assumptions. *Nutrient Cycling in Agroecosystems 58:*249-258.

Smith, K.A. (1980). A model of the extent of anaerobic zones in aggregated soils, and its potential application to estimates of denitrification. *Journal of Soil Science 31:*263-277.

Smith, K.A. (1997). The potential for feedback effects induced by global warming on nitrous oxide emissions by soils. *Global Change Biology 3:*327-338.

Smith, K.A. and J. Bogner (eds.) (1997). *Report of Joint North American-European Workshop on Measurement and Modelling of Methane Fluxes from Landfills,* Argonne National Laboratory, Argonne, Illinois. Cambridge, MA: IGAC.

Smith, K.A., A.F. Bouwman, and B. Braatz (2002). "Nitrous oxide: Direct emissions from agricultural soils." Background paper for IPCC Workshop on Good Practice in Inventory Preparation: Agricultural Sources of Methane and Nitrous Oxide. February 1999. Kanagawa, Japan: IPCC.

Smith, K.A., K.E. Dobbie, B.C. Ball, L.R. Bakken, B.K. Sitaula, S. Hansen, R. Brumme, W. Borken, S. Christensen, A. Priemé, et al. (2000). Oxidation of atmospheric methane in Northern European soils, comparison with other ecosystems, and uncertainties in the global terrestrial sink. *Global Change Biology 6:*791-803.

Stange, F., K. Butterbach-Bahl, H. Papen, S. Zechmeister-Boltenstern, C. Li, and J. Aber (2000). A process-oriented model of N_2O and NO emissions from forest soils. 2. Sensitivity analysis and validation. *Journal of Geophysical Research 105:*4385-4398.

Tiedje, J.M., A.J. Sexstone, T.B. Parkin, N.P. Revsbech, and D.R. Shelton (1984). Anaerobic processes in soil. *Plant & Soil 76:*197-212.

Van der Gon, H.A.C.D. and H.U. Neue (1995). Influence of organic matter incorporation on the methane emission from a wetland rice field. *Global Biogeochemical Cycles 9:*11-22.

Walter, B. and M. Heimann (2000). A process-based, climate-sensitive model to derive methane emissions from natural wetlands: Application to five wetland soils, sensitivity to model parameters, and climate. *Global Biogeochemical Cycles 14:*745-765.

Watson, R.T., H. Rohde, H. Oeschger, and U. Siegenthaler (1990). Greenhouse gases and aerosols. In *Climate Change: The IPCC Assessment,* eds. J.T. Houghton, G.J. Jenkins, and J.J. Ephraums. Cambridge: Cambridge University Press, pp. 1-40.

Whalen, S.C. and W.S. Reeburgh (1992). Inter-annual variations in tundra methane emissions: A 4-year time series at fixed sites. *Global Biogeochemical Cycles 6:*137-159.

Zamolodchikov, D.G. and D.D. Karelin (2001). An empirical model of carbon fluxes in Russian tundra. *Global Change Biology 7:*147-161.

Chapter 13

Soil Organic Matter Dynamics

Rolf Nieder
Dinesh K. Benbi
Klaus Isermann

The significance of soil organic matter (SOM) to soil fertility, crop productivity, and terrestrial cycling of carbon (C), nitrogen (N), and sulphur (S) has long been recognized. At the beginning of permanent agriculture, fields were cropped for two years, followed by a fallow year that served to rebuild soil fertility. As population pressure on land use increased and the fallow was eliminated, SOM declined in cultivated soils. As a consequence, new management practices were introduced to ameliorate soil fertility, and crops such as clover and alfalfa became common rotation crops. Although biological N_2 fixation was not discovered until the late nineteenth century, N_2-fixing legumes had been important crops for many centuries. In many agricultural systems, important means to maintain or increase SOM have been the incorporation of crop residues, animal wastes, and green manures. In the twentieth century, their significance was dramatically altered by the increased use of mineral N fertilizers.

However, organic matter imports desirable physical environments to soils by favorably affecting soil structure expressed through soil porosity, aggregation and bulk density, and soil water storage (Benbi et al., 1998). Organic matter also exerts a significant influence on chemical properties of soils (Duxbury, Smith, and Doran, 1989), nutrient availability (Benbi and Biswas, 1997; Nieder, 2000; Nieder and Richter, 2000), cation exchange capacity (van Dijk, 1966), and retention and mobilization of metals (Stevenson, 1986).

Intensified land use by humans has increased the exchange of C, N, and S between the land and atmosphere. The importance of the C transfer from soils to the atmosphere lies not only in the global carbon cycle and the consequences of global warming, but also in the potential of soils to produce food, fiber, building materials, and fuel. Of particular concern are the changes in land use for tropical regions where the human population is increasing most rapidly. During the nineteenth century until the 1940s, changes in carbon emissions from terrestrial ecosystems were dominated by the expansion of agriculture in the middle and high latitudes. Since the 1950s, emissions from the tropics have increased due to changes in land use. Losses from soils that are already of low fertility are clearly of concern in relation to future productivity.

ASSESSMENT OF SOIL CARBON POOLS

Global Pool Size

The global pool of SOM is estimated to contain about 1,500 Pg (1 Pg = petagram = 10^{15} g = 1 billion tons) of carbon (Eswaran, Van den Berg, and Reich, 1993; Batjes, 1996, 1997; Watson et al., 1995). This compares with estimates of 600 to 700 Pg C in aboveground biomass of vegetation (Melillo et al., 1990; Sombroek, 1990; Schimel et al., 1994), 800 Pg C in atmosphere, and about 40,000 Pg C in the oceans (Schimel et al., 1994). The geologic C pool comprises 5,000 Pg C with 4,000 Pg C as coal, 500 Pg C as gas, and 500 Pg C as oil (Lal, 2000). Most organic carbon in soils is associated with organic matter, although charcoal may be an important constituent in ecosystems subject to frequent fires. Reserves of inorganic carbon (as carbonate) stored in soils have been estimated to be about 720 Pg C (Sombroek, Nachtergaele, and Hebel, 1993).

Global calculations of the soil organic carbon (SOC) pool are complicated by a number of factors, notably: (1) partly high spatial variability in the SOC content of soils, (2) the limited knowledge of the extent of different kinds of soils, (3) unavailability of data on bulk density and coarse fragments, and (4) the confounding effect of vegetation and land use changes. The information is especially incomplete for organic soils (Histosols) of northern latitudes. Estimates of the soil C pool are also constrained by the lack of information on charcoal C in soils. Relative amounts of charcoal C may be substantial in fire-dependent ecosystems (e.g., tropical savannas). The most reliable estimate of global soil distribution, the 1:5 M *Soil Map of the World* (FAO, 1971-1981), has been used to compile a data set of soil carbon pools (Figure 13.1).

Climate and Vegetation Effects

Climate and vegetation are important factors controlling SOM pools. Mean SOC and N contents have been estimated using the Holdridge (1947) classification scheme (Table 13.1). Among forest ecosystems, mean values

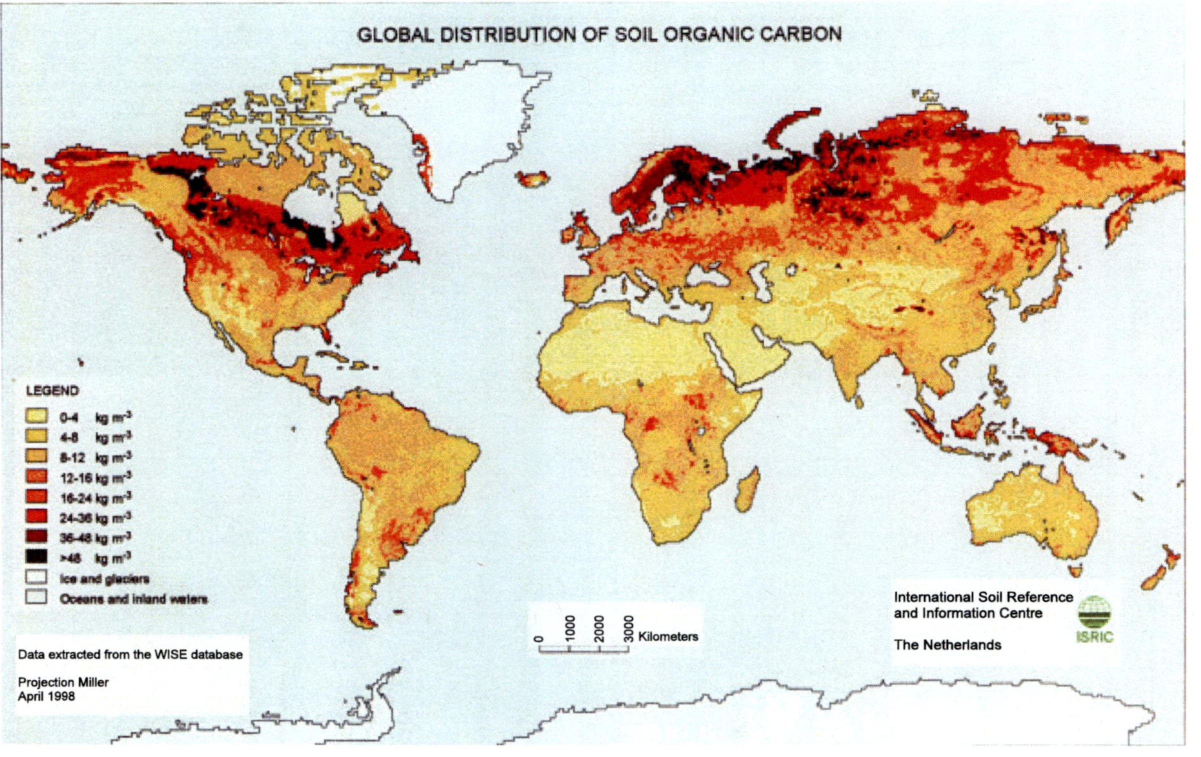

FIGURE 13.1. Soil organic carbon density map derived from WISE database (kg C m^{-2} to a depth of 1.0 m) (*Source:* Batjes, 1997, p. 59, reproduced with permission of ISRIC.)

TABLE 13.1. Estimates of Organic Carbon (C) and Nitrogen (N) Pools for the Depth Intervals 0-30 cm and 0-100 cm (Corrected for Coarse Fragments), Aggregated as per Holdridge Life Zone

	Depth interval (cm)				
	0-30		**0-100**		
Holdridge life zone	**C (Pg)**	**N (Pg)**	**C(Pg)**	**N(Pg)**	**Area[1]**
Tundra	54.0	3.9	158.8	8.3	10.16
Cold Parklands	16.1	1.5	35.7	3.4	2.78
Forest Tundra	74.2	5.3	180.2	11.7	8.72
Boreal Forest	152.1	11.1	345.9	23.4	14.94
Cool Desert	9.6	1.3	19.5	3.0	4.00
Steppe	35.3	4.3	69.1	9.2	7.35
Temperate Forest	63.9	6.2	120.3	11.7	9.90
Hot Desert	36.6	4.9	66.1	11.4	20.80
Chaparral	21.8	2.5	42.5	5.4	5.62
Warm Temperate Forest	15.9	1.6	30.4	3.1	3.20
Trop. Semiarid	27.0	3.3	53.1	7.1	9.53
Trop. Dry Forest	61.1	6.6	118.3	13.7	14.85
Trop. Seasonal Forest	70.8	6.7	133.0	13.5	15.08
Trop. Rain Forest	45.7	4.0	89.0	8.0	8.46
All ecosystems	684.1	63.0	1462.0	133.0	135.39

Source: Adapted from Batjes (1997).

[1]Area in 10^6 km^2, excluding land glaciers

for soil carbon and nitrogen increase from the lowland tropics to the boreal region. Lowland tropical forests are not greatly different from temperate forest soils in terms of soil carbon and nitrogen content. High rates of SOM production in the tropics are accompanied by high rates of decomposition. Although plant production is lower at high elevations, larger SOM accumulations occur in montane tropical forests because decomposition is inhibited (Grubb, 1971). Though less dramatic, a similar pattern is true for mountains of the temperate zone (Hanawald and Whittaker, 1976). Low temperatures retard decomposition in tundra and boreal areas. Soils of these regions (including Histosols) worldwide contain the largest SOM accumulations. Because turnover in the surface layers is markedly slower at high latitudes, it is not surprising that litter on the soil surface increases from one percent of the total detritus in tropical forests to 13 percent in boreal forests (Schlesinger, 1977).

Temperate steppe (prairie) soils, such as those in the United States and the former USSR, contain very large amounts of SOM. After extensive plant and root growth in spring, decomposition of the produced organic material is inhibited due to drying of the soil in summer and autumn months, and subsequent freezing in the winter period. Although a major portion of grassland detritus is found in deep parts of the soil as a result of death of the roots, a large percentage of the detritus in temperate zone forests is rapidly mineralized on the forest floor. This may explain why there is more SOM in the soils of temperate zone grasslands than in those of forests of the same region. The SOM contents of tropical grasslands and dry forests are much lower, partly because frequent fires may limit the amount of plant debris in these ecosystems. Desert soils contain extremely low amounts of soil organic carbon and nitrogen because lack of water limits plant growth and the production of SOM.

According to earlier literature, SOM and its accumulation was of minor importance with respect to soil formation in permafrost soils (Cryosols, according to *World Reference Base for Soil Resources;* FAO, 1998). This may be the reason why Antarctica appears completely white on the map of the global C distribution (Figure 13.1). In contrast, Blume et al. (1997) and Beyer et al. (1998) documented that the SOM contents in mineral topsoils of the ice-free coastal antarctic region was much greater, as expected from the earlier literature. These findings may require completion of the present estimate of the global SOM pool.

Carbon and Nitrogen Contents in Soil Units

The data in Table 13.2 provide mean figures for global assessment of soil carbon and nitrogen pools in the upper 30 and 100 cm of the FAO-UNESCO soil units (FAO, 1971-1981). Mean SOC content in the upper 100 cm of the various soils ranges from 3.1 kg C m^{-2} for sandy Arenosols to 77.6 kg C m^{-2} for Histosols. The large values for the latter are due to the slow decomposition of organic material under water-saturated conditions, particularly when mean soil temperatures are low. Small amounts of organic C and N are encountered in Xerosols (4.8 kg C m^{-2}) and Yermosols (3.0 kg C m^{-2}) from the arid regions where plant growth is limited. Humic soils (not given in Table 13.2) like well-drained humic Nitosols (18.2 kg C m^{-2}) or poorly drained humic Gleysols (29.3 kg C m^{-2}) contain much larger amounts of organic C

Table 13.2. Estimates of Organic Carbon (C) and Nitrogen (N) Pools for the Depth Intervals 0-30 cm and 0-100 cm (C/N for 0-30 cm, 30-50 cm, and 50-100 cm) by FAO-UNESCO Soil Units (in kg m^{-2})

	Depth interval (cm)						
	0-30		0-100		0-30	30-50	50-100
Soil unit	**C**	**N**	**C**	**N**	**C/N**	**C/N**	**C/N**
Acrisols	5.1	0.48	9.4	1.10	13.2	10.1	8.9
Cambisols	5.0	0.58	9.6	1.12	11.5	9.7	9.0
Chernozems	6.0	0.88	12.5	1.70	10.8	10.7	9.4
Podzoluvisols	5.6	0.54	7.3	0.76	13.6	7.4	7.5
Rendzinas	13.3	1.05	—	—	11.2	—	—
Ferralsols	5.7	0.46	10.7	0.97	14.3	12.6	11.8
Gleysols	7.7	0.75	13.1	1.34	12.6	11.2	10.4
Phaeozems	7.7	0.71	14.6	1.51	11.4	10.0	8.9
Lithosols	3.6	0.42	—	—	11.1	—	—
Fluvisols	3.8	0.50	9.3	1.23	11.2	11.3	10.4
Kastanozems	5.4	0.68	9.6	1.78	10.6	8.8	8.6
Luvisols	3.1	0.45	6.5	1.03	11.6	9.9	9.4
Greyzems	10.8	0.96	19.7	1.92	8.9	11.0	8.6
Nitosols	4.1	0.49	8.4	1.00	12.6	9.8	8.6
Histosols	28.3	1.61	77.6	4.01	25.8	29.8	22.3
Podzols	13.6	0.81	24.2	1.39	23.8	21.5	24.5
Arenosols	1.3	0.22	3.1	0.52	14.2	12.6	9.5
Regosols	3.1	0.45	5.0	0.70	13.5	9.6	10.2
Solonetz	3.2	0.45	6.2	1.11	12.2	10.5	8.8
Andosols	11.4	0.91	25.4	1.99	13.3	13.8	14.3
Rankers	15.9	2.18	—	—	17.1	—	—
Vertisols	4.5	0.50	11.1	1.23	13.3	12.5	12.5
Planosols	3.9	0.41	7.7	1.00	11.5	10.3	7.9
Xerosols	2.0	0.33	4.8	0.58	9.9	9.2	7.0
Yermosols	1.3	0.15	3.0	0.37	11.1	10.5	10.9
Solonchaks	1.8	0.27	4.2	0.75	11.7	9.2	8.5

Source: Adapted from Batjes (1996).

than the means for Nitosols (8.4 kg C m^{-2}) and Gleysols (13.1 kg C m^{-2}). The large values for Andosols (25.4 kg C m^{-2}) can be explained by the protection of SOM by allophane (Mizota and Van Reeuwijk, 1989). Generally, the stabilizing effect of inorganic particles on SOM decreases in the sequence allophane > amorphous and poorly crystalline Al-silicates > smectite > illite > kaolinite (van Breemen and Feijtel, 1990).

The mean C/N ratios across the soil units and depth intervals range from 7.0 for Xerosols to 29.8 for Histosols. The general trend of values in Table 13.2 suggests a decrease in C/N ratio with depth, which reflects a greater degree of breakdown and older age of the humus stored in the lower parts of the profile.

ORGANIC MATTER FRACTIONS AND TRANSFORMATIONS

Sources of SOM

The organic carbon found in the soil is primarily plant derived. Each year enormous quantities of organic residues are deposited into terrestrial environments. Williams and Gray (1974) estimate that plant litter alone accounts for between 1.0 and 15.3 t of organic material ha^{-1} y^{-1}. The root biomass in the top 30 cm of soil is estimated to be 440-1,575 g m^{-2}. In addition, animals and their excreta and dead microbial cells provide a significant residue. In order to maintain a steady state condition between assimilation and mineralization a continuous degradation is needed. The enormous flexibility as to the nature and the number of substrates that microorganisms can utilize is responsible for the continuity of nutrient cycling.

Chemical composition of entire plant residues varies widely among plant species. The organic components on a dry weight basis comprise 5 to 30 percent of water-soluble simple sugars, amino acids, and aliphatic acids; 10 to 50 percent cellulose; 10 to 30 percent hemicellulose; 5 to 30 percent lignin; 1 to 8 percent fats, waxes, oils, and resins; and 1 to 20 percent proteins (Reddy, Feijtel, and Patrick, 1986). The rate of organic matter breakdown depends on the relative proportion of each of the fractions. The order of rapidity of decomposition is given as follows: water soluble organics > hemicelluloses > celluloses > lignin (Tenny and Waksman, 1929).

The lower the N content of a substrate or the wider its C/N ratio, the slower is the rate of decomposition. Typical C/N ratios of common organic substrates and of fossil carbon deposits and their degradability and residence times (geological age) are given in Table 13.3. The mean residence times of organic substrates vary from < one day (glucose) to a few hundred years (stable SOM, woody products). Relative to fossil carbons, their resi-

TABLE 13.3. Typical C/N Ratios of Common Organic Substrates, Their Degradability and Residence Times (Geological Age for Fossil Carbon Deposits)

Organic substance	C/N ratio	Degradability	Residence time
Soil organic matter	8-13	Very low	<5-10^3 y
Farmyard manure	15-30	High	<5 y
Hay, grass, litter, etc.	20-70	High	<5 y
Deciduous leaf litter	20-60	High	<5 y
Coniferous forest floor	30-40	Moderate	1-10 y
Straw	60-120	Moderate	<1-10 y
Pine needle litter	80-130	Moderate	<1-10 y
Bark	100-1500	Low	10-10^2 y
Wood	200-1500	Low	10-10^2 y
Chemical compounds:			
Glucose	No N	Very high	<1-10 d
Cellulose	No N	Moderate	<1-10 y
Lignin	No N	Low	10-10^2 y
Fossil organic			Geological age
C pools:			
Peat, lignite			>10^3-10^5 y
Hard coal			>10^8-10^9 y

Source: Compiled from Fog (1988) and Sauerbeck (2001).

dence time is very short, with the consequence that their carbon storage potential is limited both in time and amount. Accordingly, even if carefully preserved, both forests and soils have a finite capacity to sequester carbon, which gets saturated within less than 100 years.

Roots and Rhizodepositions

The actual data on organic matter in roots underestimate the total addition of organic matter by a crop to the soil as the observations exclude root exudates and the loss of parts of the roots during the growing season. Estimates of the proportion of the total assimilated carbon that is released to the soil, e.g., by cereals, range from 16 to 33 percent. Recent research has suggested that root turnover is relatively short for many temperate species, for example ~30 percent of grass and clover roots survive for < three weeks under U.K. field conditions (Watson et al., 2000). Although there are now reliable estimates of root turnover for many tree and agricultural species (Black et al., 1998; Watson et al., 2000), there is still a lack of quantitative data on organic matter inputs from this source.

Roots, root exudates, and other root-borne organic substances are referred to as rhizodeposits, and the process of their release is rhizodeposition. Rhizodeposition is almost always measured using ^{14}C to distinguish recently deposited C from the large, unlabeled SOM background. Rhizosphere processes play an important role in C sequestration in terrestrial ecosystems. Table 13.4 shows that rates of rhizodeposition during a whole growth period vary among crops and range from 1,000 to 2,900 kg C ha^{-1} under cereals and from 800 to 4,400 kg C ha^{-1} under permanent grassland. Typically, root respiration is not distinguished from rhizodeposition and es-

TABLE 13.4. Total Below-Ground Translocated C (Roots, Exudates, and Co_2 Produced by Root Respiration) During the Growth Period of Cereals and Grasses Quantified with Tracer Techniques

Crop	Period	Input (kg C ha^{-1})	Country	References
Spring wheat	VP[1]	1500	Australia	Martin and Puchridge (1982)
Spring barley	127 d	1650	Denmark	Jensen (1993)
Winter barley	VP	2370	"	Jensen (1994)
Winter wheat	153 d	2300	Germany	Johnen and Sauerbeck (1977)
	VP	1000-1600	"	Knof (1985)
Lolium perenne	95 d	500-650	"	Kuzyakov, Kretzschmar, and Stahr (1999)
Winter wheat and spring barley	VP	1460-2250	Netherlands	Swinnen, van Veen, and Merckx (1995)
Lolium perenne	VP	840-1660	"	van Ginkel, Gorrisen, and van Veen (1997)
Winter wheat	VP	1200-2900	UK	Whipps (1990)
Different grasses	VP	2450-4430	New Zealand	Saggar, Hedley, and Mackay (1997)

[1]VP: Whole vegetation period

TABLE 13.5. Gross Chemical Composition of Animal Manures

Constituent	Typical range in percent
Ether-soluble compounds	1.8-2.8
Cold water-soluble compounds	3.2-19.2
Hot water-soluble compounds	2.4-5.7
Hemicellulose	18.5-23.5
Cellulose	18.7-27.5
Lignin	14.2-20.7
Total N	1.1-4.1
Ash	9.1-17.2
C/N	15-30

Source: Adapted from McCalla, Peterson, and Lue-Hing (1977).

timates (of rhizodeposition + root respiration) for a variety of plant species and experimental conditions have averaged 58 percent of the total C imported into the soil (Whipps, 1990). Recovery of C in the soil at harvest was considerably lower, averaging 20 percent of the total C imports (Whipps, 1990).

Animal Wastes

The feces of farm animals consist mostly of undigested food that has escaped bacterial action during passage through the body. This undigested food is mostly cellulose or lignin fibers (Table 13.5), although some modification of the lignin to humic substances occurs. The feces also contain the cells of microorganisms. Nitrogen in manure solids occurs largely in organic forms (undigested proteins and the bodies of microorganisms). The C/N ratio of farmyard manure usually ranges between 15 and 30. Liquid manure may also contain significant amounts of NH_4^+ which forms from urea through hydrolysis.

Animal wastes are more concentrated than the original feed in lignins and minerals. Lipids are present along with humiclike substances. Manures also contain a variety of trace organics such as antibiotics and hormones. The manure applied to cropland varies greatly in nutrient content, depending on animal type, ration fed, amount and type of bedding material, and storage condition.

Sewage Sludge

Sewage sludge obtained from biological treatment of domestic sewage, which is the material available for land disposal from most municipal treatment plants, is a stabilized product. The solid portion of sewage sludge consists of roughly equal parts organic and inorganic material. The latter includes N, P, K, S, Cl, Zn, Cu, Pb, Cd, Hg, Cr, Ni, Mn, B, and others. The heavy metals such as Zn, Cu, Pb, Cd, Hg, Cr, and Ni can be toxic at some concentrations. The organic component is a complex mixture consisting of (1) digested constituents that are resistant to anaerobic decomposition, (2) dead and live microbial cells, and (3) compounds synthesized by microbes during the digestion process. The organic material is rather rich in N, P, and S. The C/N ratio of digested sludge ranges from 7 to 12, but is usually about 10.

Dissolved Organic Matter

Dissolved organic matter (DOM) is a continuum of organic molecules of different sizes and structures that pass through a filter of 0.45 μm pore size. It represents the most reactive and mobile form of organic matter in soils and plays an important role in the biogeochemistry of C, N, and P, the transport of soil pollutants, and in pedogenesis. It is mainly composed of high-molecular weight, complex humic substances. Only small proportions of DOM, mostly low-molecular weight substances such as organic acids, sugars, and amino acids can be identified chemically (Herbert and Bertsch, 1995). The origin and fate of DOM in soil profiles is related to soil biological activity and sorption/desorption processes within the soil matrix. Dissolved organic matter mobilization may contribute to C accumulation in groundwater. The solubility of DOM depends on the chemical composition, soil pH, and ionic composition of the soil solution. The knowledge about the formation and fate of DOM in soils and its response to changing environmental conditions is often inconsistent. Most of the information available is on soils of temperate and cool mixed deciduous forests, and temperate, cool, montane and boreal coniferous forests.

Sources of DOM

Dissolved organic matter originates from plant litter, soil humus, microbial biomass, and root exudates. Although the release of DOM has been researched extensively, it is still not clear from which part DOM originates—recent litter or from stable organic matter of A-horizons. Zsolnay (1996) suggests that humified organic matter is the major source of DOM because of the relatively high proportion of humus in relation to litter in soils. McDowell and Likens (1988) also hypothesize that leaching and microbial decay of humus rather than of recent litter is largely responsible for DOM production in the forest floor. In contrast, Qualls, Haines, and Swank (1991) report that the greatest net increases in the fluxes of DOM occur in the upper part of the forest floor (O_l horizon) in a deciduous forest. Recent litter fall contributes significantly to the temporal variation in the DOM production rate. The highest levels of hydrophilic neutral fraction of DOM (mainly simple sugars and nonhumic-bound polysaccharides) occur after litter fall. According to Michalzik and Matzner (1999), the greatest amounts of DOM occur in the O_l horizon. The lower organic layers (O_f O_h) function as a sink for the organic solutes, leading to a substantial decrease of DOM in the downward flux. Substrate quality significantly influences DOM production. The rhizosphere is often associated with a large carbon flux attributable to root exudation and turnover.

Recent studies of fractionation and structural analysis of DOM in soil solutions have shown that microbial metabolites constitute a significant proportion of DOM (Kalbitz et al., 2000). Fungi are the most important agents in the process of DOM production, probably because of incomplete degradation of organic matter by this group. The carbohydrate fraction of DOM is chemically different from that of plant residues or bulk humus in that DOM carbohydrates have a higher proportion of hexosesugars and deoxysugars than pentose sugars.

Dynamics of DOM in Soils

The dynamics of DOM in soil solution are mainly controlled by the quality of the original organic substrate. Most studies examining the interactions between substrate quality and DOM production have focused on the concentrations of nitrogen, lignin, polyphenols, or some combination of these chemical constituents. Important compartments for DOM mobilization are forest canopies. In forest floors (O-horizons), DOM concentrations often vary between 20 and 60 mg C l^{-1} (McDowell and Likens, 1988; Guggenberger and Zech, 1993). Calculated DOM fluxes from the forest floor into the mineral soil vary between 150 and 370 kg C ha^{-1} y^{-1} (Guggenberger and Zech, 1993). In Podzols, the A-horizons serve as an additional DOM source. In the mineral soil, DOM is usually retained very effectively. Concentrations of DOM in the subsoil vary between 0.5 and 12 mg C l^{-1} (McDowell and Likens, 1988; Guggenberger and Zech, 1993). The retention of DOM in the mineral soil is mainly controlled by the content of Fe/Al oxides/hydroxides. Fe oxides retain DOM more effectively than Al oxides. To a lower extent, clay can also immobilize DOM.

In the humid tropics, DOM fluxes in the canopy floor are higher than in temperate zones due to higher precipitation and higher leaf-area indices. In the mineral soil, DOM inputs and fluxes may be higher in the tropics than in temperate climates. Upland soils rich in oxides (e.g., Ferralsols) retain DOM effectively. Arenosols with thick E-horizons (so-called Giant Podzols) export significant quantities of DOM into rivers thus creating black water streams. Mean DOM concentration of the Rio Negro river (representing mainly Arenosol areas) is 10.8 mg C l^{-1}, whereas mean DOM concentration of the Amazon river (representing Ferralsol areas) is 4.8 mg C l^{-1} (Ertel et al., 1986). In floodplain soils, frequent flooding causes intensive leaching from the soil, leading to DOM concentrations in groundwaters and rivers of 2-30 mg C l^{-1} (Frangi and Lugo, 1985). The estimated watershed export calculated from these concentrations amounts to up to 380 kg C ha^{-1} y^{-1}, which exceeds DOM leaching in temperate climates.

Under natural vegetation types, the soil production of DOM is much lower than under agricultural or artificial forest systems. In arable soils, fertilization with organic compounds has been shown to increase the DOM content by a factor of 2.7 to 3.2, depending on the kind of fertilizer added (Gregorich et al., 1998). Organic farming results in a higher DOM content than conventional farming. In manured soils, soluble C increases sharply after manure application, followed by a gradual decrease. No information is available about the effects of tillage on DOM dynamics in soils, although one would expect DOM release to increase after soil tillage as a result of enhanced mineralization.

Model predictions on DOM dynamics at the field scale are still uncertain because of a lack of data on the relationship between DOM dynamics and environmental factors in the field. The main objective of future studies should be on factors controlling DOM quality and quantity, and on sources and sinks of DOM. The studies also need to concentrate on DOM dynamics in soils of different use and in climate zones other than the temperate areas.

Labile and Stable SOM Fractions

Soil organic matter actively contributes to gaseous exchanges with the atmosphere (see Chapter 12) and to nutrient cycling in the soil-water-plant system. Microbial biomass accounts for about 2 to 5 percent of total C. The mean turnover time for microbial biomass C is 0.13-0.24 y in soils of the humid tropics and 1.4-2.5 y in soils of temperate regions (Houot et al., 1991) which demonstrates the important role of temperature in controlling soil carbon dynamics. Growing plant roots, a source of readily decomposable photosynthates for microbial biomass, have a priming effect on the breakdown of organic residues added to soil (Lynch, 1990).

The mean residence time of SOM is about 22 y (Post, Emanuel, and King, 1992). The turnover time of SOM increases with soil depth, ranging from several years for litter to 15-40 y in the upper 10 cm and over 100 y below a depth of 25 cm (Harrison, Harkness, and Bacon, 1990). Mean residence times for organic carbon of 2,000-5,000 have been reported for soils rich in allophane (Wada and Aomine, 1975), reflecting the importance of mineralogy, active aluminum, and iron hydrous oxides in preserving organic materials in the soil. Similarly, Ferralsols derived from basic or ultrabasic parent materials have consistently higher organic matter contents than those derived from acid materials. Numerous studies have shown that the fractionation of soil gives significantly different SOM values for the sand, silt, and clay fractions, with the fine-textured fractions containing the oldest and highest amounts of SOM (Batjes and Sombroek, 1997).

The composition of the various SOM constituents is mainly controlled by residue inputs and climate gradients, edaphic factors, soil mineral types and contents, and other physical and chemical properties of the soil that affect the activity of decomposer organisms (Theng et al., 1989). The chemical composition and physical state of the residue, the site of placement in soil, and tillage operations are additional factors that affect SOM stability. Animal manures are much more effective than plant residues in contributing to labile SOM pools because they have already been subjected to rapid initial decomposition. The size of the labile SOM pool derived from plant residue additions is smaller and responds more quickly to changes in residue inputs in tropical than in temperate soils (Sauerbeck and Gonzales, 1977). With constant, long-term annual residue inputs of 1,000 kg C ha^{-1}, the labile SOM pool is estimated to contain a maximum of 2,500 kg C ha^{-1}, 250 kg N ha^{-1}, and 25 kg P ha^{-1}, if the C/N and C/P ratios are 10 and 100, respectively.

The Decomposition Process

Organic matter decomposition in soil is influenced by numerous physical, chemical, and biological factors controlling the activity of microorganisms and soil fauna. Soil temperature and moisture are the most important environmental factors. The increase in decomposition rate caused by temperature is usually described by an Arrhenius equation. Optimal moisture conditions in most soils are around 55 to 60 percent water-filled pore space (Doran, Mielke, and Stamatiadis, 1988), with decomposition decreasing as the soil dries. Water contents near or at saturation inhibit decomposition due to reduced diffusion and availability of oxygen. Both soil temperature and moisture are affected by management including cropping intensity, crop type, residue management, irrigation, and tillage.

Because of the complex nature of organic remains, numerous species of microorganisms are involved in the decomposition process. Part of the organic material is completely mineralized, part is incorporated into microbial tissue, and part is converted into humic substances. Native humus is mineralized simultaneously. Thus, although prodigious quantities of organic residues may be returned to the soil, decomposition does not necessarily lead to an increase in SOM content. The latter depends on the difference between the amount of C that enters the soil and the amount of C that leaves the soil through leaching, erosion, or decomposition. Most of the C is lost from the system through decomposition. Results from field experiments obtained for dry matter and C retention after the first year or growing season are summarized in Table 13.6. From 10 to 66 percent of the dry matter and from 20 to 45 percent retention of applied C has been observed, depending on location (climate and soil), positioning of the plant material beneath or on the soil surface, and plant material involved. Considerable differences in C retention are particularly found for soils of colder and warmer regions.

Several stages can be delineated in the decomposition of organic residues (Stevenson, 1986). Earthworms and other soil animals play an important role in reducing the size of fresh plant material. Soil animals are reported to contribute 1 to 5 percent to total heterotrophic respiration in coniferous forest soils, 3 to 13 percent in deciduous forest soils, and 5 to 25 percent in mesic grasslands (Persson, 1989). They generally contribute less to energy turnover in dry grasslands than in mesic ones. The low figures in coniferous forest soils reflect a situation in which fungivores, bacterivores, and carnivores are dominant components of the soil fauna, while litter and root feeders are scarce. In deciduous forests, litter feeders, such as earthworms, are abundant, and in grasslands the abundant occurrence of both litter feeders and root-feeding arthropods may explain the very high figures of respiration (Persson, 1989).

Further transformations are carried out by enzymes produced by microorganisms. The initial phase of microbial attack is characterized by rapid loss of readily decomposable organic substances. Simple sugars, amino acids, most proteins, and certain polysaccharides decompose very quickly. Cellulose, hemicellulose, and lignin have been found to decompose more slowly under anaerobic than aerobic conditions. Decomposition rates have been found to be significantly correlated with the initial lignin content or the initial C/N ratio. Depending on the nature of the soil microflora and quantity of synthesized microbial cells, the amount of substrate C utilized for cell synthesis varies from 10 to 70 percent. Molds and spore-forming bacteria

TABLE 13.6. Dry Matter and Carbon Retained from Plant Material Applied to Field Soils

Plant material positioning	Period	Method	% DM retained	Country	Reference
Winter wheat straw/surface	330 d	Litter bag	45-66	Germany	Becker and Meyer (1973)
Winter wheat and rye straw/buried	300 d	""	30	"	Schröder and Gewehr (1977)
Summer barley straw/buried	300 d	""	10	"	"
Oat straw/surface	1 y	""	40	UK	Harper and Lynch (1981)
Oat straw/buried	1 y	""	20	"	"
			% C retained		
Winter wheat straw/buried	1 y	Litter bag	23-34	Germany	Nieder and Richter (1989)
Winter wheat straw/buried	1 y	^{14}C-recovery	31	"	Sauerbeck and Gonzales (1977)
Maize (corn)/buried	1 y	""	33	Austria	IAEA (1968)
Winter wheat straw/buried	1 y	""	31	UK	Jenkinson (1971)
Wheat straw/buried	1 y	""	35-45	Canada	Shields and Paul (1973)
Ryegrass/buried	1 y	""	20	Nigeria	Jenkinson and Ayanaba (1977)
	2 y	""	14	"	

DM: Dry matter

are especially active in consuming proteins, starches, and cellulose. In subsequent phases, organic intermediates and newly formed biomass tissues are attacked by a wide variety of microorganisms, with production of new biomass. The final stage of decomposition is characterized by gradual decomposition of the more resistant plant parts, such as lignin, in which actinomycetes and fungi play a major role.

The mass change of C in the soil (from residues plus native humus) can be expressed mathematically by a number of equations, usually involving one or more kinetic rate constants of decomposition, represented as k values (e.g., Equation 13.1)

$$C_{ti} = C_0 e^{-k_1 t} + C_{added} e^{-k_2 t}, \qquad (13.1)$$

where C_{ti} is the amount of soil C at time i, C_0 is the amount of soil C at time 0, k_1 is the decomposition rate constant of the total soil C pool before amendment of C added, C_{added} is the amount of C added, and k_2 is the decomposition rate constant of the added C. The decomposition process is often seen as a series of first-order reactions for the various C fractions, each with its own size and decomposition rate (see Modeling Soil Organic Matter Dynamics in this chapter).

FACTORS AND PROCESSES AFFECTING ACCUMULATION AND DEPLETION OF SOIL ORGANIC MATTER

The SOM level depends on the input and output of biomass C into the soil. Sources of C input include the amount of aboveground and belowground biomass returned to the soil, and any other biosolids (e.g., manure, compost, sludge). The output or losses of C out of the soil system includes soil respiration, oxidation, soil erosion, and leaching. There are several factors (climate, soil, vegetation, land use, soil, water and vegetation management) and processes (aggregation, erosion, leaching, acidification, salinization, eluviation, illuviation, mineralization, humification, bioturbation) that affect SOM pool and its depletion. The delicate equilibrium is easily disturbed by anthropogenic perturbations (e.g., land use change, deforestation, burning, agricultural activities). In general, conversion of natural to agricultural systems leads to depletion of the SOM pool. Most soils lose as much as 50 percent of their original pool within 5 to 10 years of conversion from natural to agricultural systems in the tropics and from 40 to 50 years in temperate climates. Depending on the original SOM pool, the new equilibrium is attained after losing 20-50 t C ha^{-1}. The magnitude of loss is more in soils with high rather than low original SOM pool.

Agricultural Management Effects on Soil Organic Matter

Optimizing SOM Contents

From an economic, social, and ecological point of view, it must be the objective of cultivation measures to make use of the soil and its organic matter in a sustainable way. The urgency of this is especially highlighted by the fact that the natural and near-natural soils of the forest, grasslands, swamps, wetlands, volcanoes, etc. have lost about 25 percent of their original SOM of 222 Pg C since their cultivation around the year 1700. The amount and quality of SOM must be optimized. Organic matter quality determines the mineralization and immobilization processes which influence the availability of nutrients (especially of N, P, and S).

All considerations with respect to SOM quality make it necessary to differentiate between at least two fractions, one of them being more or less inert. This means that the inert humus to a high degree does not participate in mineralization, while the other is decomposable (mineralizable) and mainly determined by the conditions of the agricultural cultivation ("nutritional" humus). Any change of the SOM level in the soil concerns mostly its decomposable part. After management changes, up to more than 50 years may be necessary to obtain a new equilibrium, depending on the original SOM level.

In contrast to plant nutrients such as P, K, Ca, and Mg, it has not been common to classify extractable fractions as standard values for optimum SOM contents. But evaluations of long-term field experiments (Körschens, 1984) in eastern Germany (the former GDR) have provided decisive data that may be relevant for practical purposes. For sandy and loamy arable soils, optimum SOM ranges could be derived for different fractions (see Table 13.7).

Optimum contents of carbon and nitrogen in soils are within a relatively narrow range. Above the optimum ranges, environmental pollution occurs; below the ranges, yields are insufficient. Effects of the SOM on the yield may reach 10 percent in sandy soils and up to 5 percent in clayey soils, whereby the percentage is the result of a comparison between an optimal mineral fertilization and an optimum combination of organic and mineral fertilization. In order to uphold the optimum SOM conditions (Table 13.7), it is necessary to add about 2 t reproduction-efficient organic substance ha^{-1} y^{-1}. This corresponds to 2 t dry matter (dry farmyard manure or solid matter from liquid manure), equivalent to 10 t fresh farmyard manure or the yearly amount of excreta produced by 1 gross weight or life weight unit of 500 kg. If land conservation is considered in cultivation processes combined with a reduction of farming intensity, the addition of organic substances can be reduced. At present, there is no comparable knowledge for optimizing SOM contents of grassland and forest soils, but this could be obtained by similar procedures introduced here.

Soil Tillage

The SOM content of mineral soils with constant rotation is in a state of quasi-equilibrium if the plowing frequency and depth are approximately

TABLE 13.7. Optimum SOM Conditions of Sandy and Loamy Arable Soils Derived from Seven Long-Term Field Experiments Under Common Management in Eastern Germany (Average Yearly Temperatures: 6-10°C, Average Precipitation: 400-800 mm y^{-1})

Concentrations/quantities	
Total SOC	0.7% (sandy soil with 3% clay) to 2.5% (loamy soil with 21% clay)
Total SON	0.07% (sandy soil with 3% clay) to 0.25% (loamy soil with 21% clay)
Hot water-soluble C (C_{hwl})	25-30 mg 100 g^{-1} soil
Decomposable SOC ($SOC_{dec.}$)	0.4 (0.2-0.6)% = 16 (8-24) t.ha^{-1} 30 cm^{-1}
Decomposable SON ($SON_{dec.}$)	0.04 (0.02-0.06)% = 1.6 (0.8-2.4) t ha^{-1} 30 cm^{-1}
Mineralized SOC ($SOC_{min.}$)	680 (320-960) kg ha^{-1} a^{-1} 30 cm^{-1} (= 4% of SOC_{dec})
Mineralized SON ($SON_{min.}$)	68 (32-96) kg ha^{-1} a^{-1} 30 cm^{-1} (= 4% of SON_{dec})
Qualities	
SOC:SON	10:1 (8:1-12:1)
SOC:SOS	100:1 (70:1-140:1)
SOC:SOP	150:1 (100:1-200:1)

Source: Compiled from Körschens (1995, 1997); Schulz (1997); Körschens and Schulz (1999); Isermann and Isermann (1999).

SOC = soil organic carbon; SON = soil organic nitrogen; SOS = soil organic sulphur; SOP = soil organic phosphorus

constant (Nieder, 1998). Increase in tillage frequency leads to increased SOM losses through intensive soil aeration (Grant, 1997). Increasing the plowing depth creates a long-term buildup of humus. After dilution of SOM in the A_p due to mixing of underlying C-poor subsoil material, the C-, N-, S-, and P-mineralization potential is decreased. Subsequently, SOM contents present before the onset of deeper plowing are supposed to be reestablished within several decades (Nieder and Richter, 2000). For example, during the last three decades (1970 to 1998), 300-1,100 kg C ha^{-1} y^{-1} and 30-100 kg N ha^{-1} y^{-1} have been accumulated in the deepened plow layers of western Germany (Nieder and Richter, 2000). In most of these arable soils, the buffering capacity for reactive compounds such as C, N, S, and P and of other (organic or inorganic) pollutants has reached its limit because organic matter equilibria have been reestablished.

Due to the recent SOM accumulation, the N mineralization potentials of arable soils have increased significantly. According to optimized N mineralization parameters from long-term incubation experiments, the N mineralization potentials (N_1 plus N_2; N_1: a more available N fraction, N_2: a more resistant N fraction) increased from about 800 kg N ha^{-1} 30 cm^{-1} in the late 1970s to about 1,200 kg N ha^{-1} 35 cm^{-1} in 2000 (Nieder, Dauck, and Benbi, 2001). This means that a significant part of the newly accumulated N has become part of the active pool of SOM. The changes in the N (and C) mineralization potentials are of particular interest for N fertilization strategies as well as for assessing N and C sequestration in arable soils.

Conservation tillage is defined as a system that has at least 30 percent or more crop residues covering the soil at planting (Conservation Technology Information Center, 1996). Conservation tillage induces stratification of SOM and related nutrients and enhances the size and activity of soil microbial biomass in the upper part of the surface soil. The magnitude of these changes depends on soil texture, climate, and cropping system. The new equilibrium in soil properties by conversion from plow till to no till may be attained over a period of 10-20 y (Frede, Beisecker, and Gäth, 1994). The equilibrium status may be reached more quickly in coarse-textured soils of the tropics than in heavy-textured soils of the temperate climates.

Rates of C accumulation in soils under no till or conservation till reported in the literature vary widely, ranging from below 0 to 1.3 t C $ha^{-1}y^{-1}$ in the upper 15 cm (Reicosky et al., 1995). In a number of long-term experiments (up to 20 y under no till) conducted in humid regions, zero tillage increased soil organic C on an average by 3 t ha^{-1} (Paustian et al., 1997). Differences in SOM between conventional and zero-tillage systems in semi-arid regions are generally small because conventional tillage is less intensive and shallower than in humid regions (Unger, 1991). Experiments in the subhumid and humid tropics have demonstrated a high potential for no-till systems to maintain higher SOM levels compared to conventional cultivation. No till may also contribute to a more effective use of increased C inputs from crop residues or mulches in maintainng SOM levels.

Inorganic and Organic Fertilizers

Mineral fertilizers enhance crop growth and C inputs from increased organic residue production. Positive effects on the soil C balance have been observed due to increased mineral N application rates (Table 13.8). In addition to direct effects of fertilization on C inputs, high N availability may also enhance the formation of recalcitrant humic substances in soils (Fog, 1988).

TABLE 13.8. Soil Organic C Responses to Mineral N Fertilizer Application in Long-Term Field Experiments

Rotation	Period	N application (kg ha^{-1}) SOC (% of control)				Country	Reference
Mixed rotation	20 y		*0*	*112-128*		Germany	Körschens and Müller (1996)
			100	107			
Continuous rye	112 y		*0*	*40*		"	Garz (1995)
			100	113			
Mixed rotation	81 y		*0*	*30-50*		"	Welte and Timmermann (1976)
			100	111			
Continuous wheat	135 y		*0*	*144*		UK	Jenkinson, Bradbury, and Coleman (1994)
			100	123			
	31 y		*60*	*120*		Norway	Uhlen (1991)
Cereals			100	102			
Cereal + row crops			100	102			
2 ley + 4 arable			100	108			
4 ley + 2 arable			100	104			
Cereal/root crops	30 y		*0*	*80*		Sweden	Paustian, Parton, and Persson (1992)
straw removed			100	118			
straw added			100	116			
sawdust added			100	115			
Continuous maize	12 y		*0*	*67*	*200*	USA	Barber (1979)
			100	106	107		
Mixed rotation	22 y	*0*	*160*	*320*	*480*	India	Benbi and Biswas (1997)
mineral N		100	112	115	111		
mineral N + FYM		100		162			

Note: SOC levels are given in % of control (or least fertilized) treatment.

Changes in SOM contents due to altered organic fertilizer application occur very slowly. If the changes are in dimensions relevant to practice, it takes more than 10 years until they can be proved. Numerous, long-term field experiments have shown that with realistic and practically feasible amendments, the SOM levels, compared with unfertilized controls, can hardly be increased by more than 30 percent over decades. Assuming a practically achievable increase of about 0.3 percent C in the 690 million hectares of arable mineral soils in the temperate climate zone, this overall C sequestration would be in the order of about 6 Pg C within a period of 50 years (Sauerbeck, 1993). Even though this applies only to the temperate climate zone, this figure may be more realistic than the global 20-30 Pg C estimate of potential C sequestration in soils in the Intergovernmental Panel of Climate Change (IPCC) Report (Cole et al., 1996), since the practical chances to sequester carbon in tropical arable soils are comparatively small (Batjes, 1998; Batjes and Sombroek, 1997).

Fallow and Crop Rotation Effects

In regions with food surpluses (e.g., European Union [EU], United States, Canada), about 25 million ha (15 percent of total cropland) have been taken out of production by government set-aside programs since the 1980s (Paustian et al., 1998). Enrollments in the U.S. Conservation Reserve Program are for ten-year periods after which time the land may be returned to annual crop production. The EU agricultural set-aside programs are for one to five years and can include annual cropping for nonfood production (e.g., oil seed for fuel).

Winter cover crops (if climate permits) decrease erosion and provide additional inputs of carbon thereby increasing SOM contents. Estimates for the U.S. corn belt suggest an additional C accumulation of 4,000 kg C in 100 years from the use of winter cover crops (Lee, Phillips, and Liu, 1993). The use of perennial forage crops can significantly increase SOM contents, due to high root C production, lack of tillage disturbance, and protection from erosion. If arable soils revert to grassland, SOM contents in upper soil horizons could reach levels comparable to their precultivation condition. Increases up to 550 (Jenkinson, 1971) and to 1,000 (Jastrow, 1996) kg C ha^{-1} y^{-1} have been documented for cultivated land planted to grassland. SOM accumulation would continue only until soils reached a new equilibrium value, most of which would be realized over a 50- to 100-year period. Considering the area of cropland with real or potential surpluses (about 640 million ha in Europe, United States, Canada, former Soviet Union, Australia, Argentina) and assuming recovery of the SOM originally lost due to cultivation (25-30 percent), a permanent set aside of 15 percent of this land area (about 96 million ha) might accumulate 1.5-3 Pg C in SOM (Paustian et al., 1998).

Bare fallow is used extensively in semiarid areas of the world (e.g., Canada, United States, Australia, former Soviet Union) to offset rainfall variability and increase soil water storage. During bare fallow periods, mineralization of SOM is generally faster than under a crop, due to higher soil moisture, and there is no input of crop residues. Eliminating or reducing bare fallow through better water management could significantly increase SOM in semiarid croplands and decrease soil erosion.

Cereals generally produce the most residues among annual crops while crops such as grain legumes, dry beans, and root crops produce less. Thus, SOM levels tend to be lower under maize-soybean rotations compared with continuous maize (Paustian, Collins, and Paul, 1997). Changes in SOM for six rotations including maize, sugar beet, navy bean, oats, and lucerne have been directly correlated to amounts of residue returned and the frequency of maize in the rotation (Zielke and Christensen, 1986). The inclusion of perennial forages (i.e., leys) in rotation increases SOM levels relative to the ones with annual crops alone. Experiments in Europe with three or more years of ley within annual crop rotations have shown up to 25 percent more SOM compared to rotations with only annual crops (van Dijk, 1982; Nilsson, 1986). Carbon accumulation by perennials is attributed to the high relative allocation of organic materials belowground, greater transpiration leading to drier soils, the formation of stable aggregates within the network of grass roots, and the absence of soil disturbance by tillage (Paustian et al., 1997).

Lowland rice systems differ markedly from upland crop systems in C and N cycling processes. Under irrigated rice double cropping, SOM levels tend to be stable or to increase (Nambiar, 1994). Organic matter content is generally lower in rice-upland crop rotations such as rice-wheat or rice-maize. When crop residues are not incorporated in these systems, the SOM amounts can decrease to the point of reducing the supply of N through mineralization-immobilization turnover which could lead to low grain yields.

Land Use Changes

Most land use changes (e.g., changes from forest or grassland to arable land, draining of wetland soils for agricultural uses) have been and still are a significant source for the release of former plant and soil carbon into both the atmosphere and the drainage zone. The reason for decreasing SOM contents as a consequence of cultivation are manifold, but altogether boil down to a reduced input of plant biomass into cropland as compared with unmanaged ecosystems and an enhanced decomposition of the existing SOM.

Conversion of Grassland to Arable Land

Cultivation of grassland stimulates soil C loss because it accelerates oxidation of SOM by increased microbial activity, and it exposes soils to more extensive erosion and this accelerates removal of the C-rich surface soil. Simulation of cultivation effects across the Great Plains has shown that the greatest loss in SOM is in the wetter and warmer part (southwest) of the region, which reflects the greater initial amounts of SOM found in this area (Cole et al., 1989). Peterson and Vetter (1971), using total soil N as an index, report that wheat-fallow systems in the Great Plains had organic N losses of 20 to 30 percent (mass basis). Based on changing C/N ratios when soils are cultivated, these N losses would compare to C losses of about 40 percent. Decreasing the tillage intensity results in less SOM losses after cultivation of grassland.

Under wet-temperate conditions, sites that had been under grassland for many years and were converted to arable land have shown a drastic decline in SOM. The decline was even sharper on soils kept bare because there were no inputs of C from crop roots or residues and only minimal inputs from weeds (Jenkinson et al., 1987).

Drainage of Wetlands

The total land area covered by bogs, swamps, marshes, and alluvial areas is estimated to be between 500 and 600 $\times$ 10^6 ha (Bouwman, 1990). The FAO/UNESCO soil map shows an area of 240 $\times$ 10^6 ha of Histosols (peat soils) worldwide.

Histosols As Sink for Carbon. Northern peatlands (of cool and temperate climate) contain about one third (450 Pg; Gorham, 1991) of the global store of soil organic C, with carbon accumulation rates of 20 to 40 g C m^{-2} y^{-1} over the last 5,000 to 10,000 years (Tolonen and Turunen, 1996). A collation of studies of CO_2 exchange in northern peatlands by Frolking et al. (1998) reveals that the summer uptake rates through photosynthesis are small (15-30 g CO_2 m^{-2} d^{-1}) compared with uptake in forests, grassland, and crops (50-200 g CO_2 m^{-2} d^{-1}; Ruimy et al., 1995). Some Histosol areas are also found in warmer (subtropical and tropical) climates. The mean increase in peat layers is approximately 0.5 to 1.0 mm per year. This means an increase of 5 to 10 m^3 of peat per hectare. According to the FAO classification system, Histosols contain more than 30 percent organic matter and are at least 40 cm thick. Accumulation of organic matter in Histosols represents a geological sink for nitrogen and carbon.

Organic Matter Decomposition in Reclaimed Histosols and Gleysols. Besides soil temperature, the position of the water table influences the rate of organic matter decomposition. In nondrained peat profiles, the cool temperatures and anoxic conditions retard the rate at which organic matter decomposes. Emission rates of carbon dioxide from peatland are relatively small (commonly 5-10 g CO_2 m^{-2} d^{-1}) compared with the other systems. Another factor is that plant tissues, e.g., mosses and shrubs of northern peatlands, provide a substrate that decomposes slowly, with exponential decomposition, or k values, between 0.02 and 0.25 y^{-1} (Scanlon and Moore, 2000).

Up to 35 × 10^6 ha of wetlands (including Histosols) have been drained globally (Bouwman, 1990). Drainage of organic soil deposits results in a decrease in surface elevation (subsidence). Drained organic soils are subsiding at the rate of several cm per year (Netherlands: 1.75 cm y^{-1}; Quebec in Canada: 2.07 cm y^{-1}; Everglades in the United States: 3 cm y^{-1}; San Joaquin Delta in the United States: 7.6 cm y^{-1}; Hula Valley in Israel: 10 cm y^{-1}; Terry, 1986). Among the reasons for subsidence are shrinkage due to drying, loss of the bouyant force of groundwater, compaction, wind erosion, burning, and microbial oxidation. Microbial oxidation is the predominant cause of Histosol subsidence. Approximately 73 percent of the loss of surface elevation in Everglades Histosols is caused by microbial oxidation (Terry, 1986). An assumed C release from drained wetlands by oxidation of the organic material of 10 t C ha^{-1} y^{-1} yields a global annual C release of 0.05 to 0.35 Pg C. Drainage of 10^6 ha of Gleysols causes an extra release of 0.01 Pg C y^{-1}. The resulting global release from Histosols and Gleysols ranges between 0.03 and 0.37 Pg C y^{-1}.

Based on an average storage rate of 200 kg C ha^{-1} in wetlands of the cool climate and assuming an area of about 350 × 10^6 ha, the annual accumulation before disturbance has been calculated as 0.06-0.08 Pg C y^{-1} (Armentado and Menges, 1986). The total area drained in the period 1795-1980 was 8.2$10^6$ ha for crops, 5.5 x 10^6 ha for pasture, and 9.4 x 10^6 ha for forests. In the tropics, about 4 percent of the wetland area was reclaimed in the period 1795-1980. The annual shift (loss of sink strength and gain of source strength) in the global C balance is 0.063-0.085 Pg C due to reclamation of Histosols in cool regions. Including tropical Histosols, the global shift would be 0.15-0.184 Pg C y^{-1} (Bouwman, 1990). The potential to increase C levels in soils under cultivation is largely restricted to upland soils. Restoring C sinks in artificially drained wetland soils is unlikely unless they are taken out of agricultural production and reverted to natural wetlands.

Changing of Fire Regime in Fire-Adapted Ecosystems

Fire is a conspicuous feature of many ecosystems. Although estimates of organic materials burned annually vary widely, savanna fires are thought to be one of the largest sources of CO_2 from biomass burning (Hao and Liu, 1995) which would otherwise have been incorporated into the much larger SOM pool. In many parts of the world, human activities have led to changes in fire frequency in savanna regions, some of which probably date from prehistoric times (Bird, 1995). Given the large area covered by savannas globally (1.3-1.9×10^8 ha), and the importance of the SOM pool in modulating changes in the size of the atmospheric CO_2 pool, it is possible that anthropogenic changes in fire frequency may have altered the dynamics of the savanna SOM pool, with a significant consequent impact on the global carbon cycle. There are only a few studies on the effect of fire on soil carbon stocks in savanna ecosystems. These suggest that complete fire suppression will increase SOM contents, while increasing fire frequency in an area previously subjected to a lower fire frequency may lead to a decline in SOM.

Deforestation

Uncontrolled, large-scale conversions of forest to arable or pasture land as in the semiarid and humid tropics are a relatively recent land use change. The loss of C from biomass and soils due to conversion of native ecosystems to agricultural use in the tropics is the second largest (after fossil fuel) source of CO_2 input to the atmosphere (Paustian et al., 1998). Loss of SOM following land clearing is due to the combined effects of increased rates of decomposition with cultivation and reduced amounts of plant residues returning to the soil. It is estimated that the use of already existing arable land and pasture in the tropics currently contributes a net efflux of 90-230 Tg (1 Tg = teragram = 10^{12} g = 10^6 tonnes) C y^{-1} (Lal and Logan, 1995). Because current rates of deforestation are high (Skole et al., 1994), the single most significant mitigation option related to agriculture in the tropics is to reduce the pressure for converting forested land to agriculture. Breaking the cycle of deforestation requires that the productivity and sustainability of existing agricultural lands are improved.

Shifting Cultivation

Shifting cultivation represents a broad group of land use systems in the humid tropics based on clearing of the original forest and a few years of crop production with declining SOM contents and a subsequent recovery fallow period of a few years (i.e., bush fallow) or for longer periods (classical shifting cultivation, based on secondary forest succession). The decrease in SOM as a consequence of cultivation is of the same order as that for temperate regions (e.g., Europe, North America), averaging 25 to 30 percent of the initial SOM contents (Davidson and Ackermann, 1993).

Conversion to Pasture Land

In much of the tropics (particularly the semiarid tropics) pasture land is the predominant land use system. The degree to which improved pasture management is practiced (i.e., introduction of grasses and legumes, soil fertility maintainance) has a major impact on SOM levels. In many regions, poor management has resulted in overgrazing and nutrient deficiencies, leading to soil erosion and SOM losses (Eden et al., 1991). Well-managed pastures may be able to maintain or even increase SOM levels compared with native forest. In comparing more than 25 sets of paired pasture and mature forest sites, Lugo and Brown (1993) found that soil C stocks under pasture varied from 60 to 140 percent compared to those under forests and that on average, SOM under pasture was not significantly different than under mature forest. In Colombia, Fisher et al. (1994) reported very large belowground C increases of 25 to 70 t ha^{-1} within 5 to 10 y after establishing pastures of deep-rooting African grasses.

It appears that the C sequestration potential in moist, tropical pastures can be significant under favorable conditions. Fisher et al. (1995) suggest that improved pastures which have replaced native savannas throughout South America could account for an additional sequestration of 100-500 × 10^6 t C y^{-1} in these tropical soils. Substantial soil C inputs may be attributed to the deep-rootedness of grasses in improved tropical pastures. Fisher et al. (1995) found that the large increase in SOM under improved tropical pastures (up to 70 t C ha^{-1}) was associated with a substantial increase in the C:N ratio, giving ratios in the SOM of 33:1 compared with usual SOM values of ~12:1. It is thus likely that only partial decomposition of roots occurred leading to the increased SOM content.

Introduction of Monoculture Plantations and Agroforests

In the humid tropics, monoculture plantations with perennial crops including industrial trees, rubber, oil palm, coconut, sugarcane, and pineapple are an important land use. If crop residues are returned and supplemented with nutrient inputs, these systems maintain adequate SOM levels. Agroforests and other mixed perennial systems are among the most sustainable types of systems for the humid tropics (Gouyon, de Foresta, and Levang, 1993). Many of their ecological functions, such as SOM storage, are close to those of natural secondary forests of similar age. Agroforests, in which recalcitrant woody tissues and tree leaves with high C:N ratios and high lignin content are added, have been proposed as a means of building SOM stocks and regulating soil nutrient availability.

Reforestation and Afforestation

Reforestation or afforestation of bare land, redundant arable land, or degraded pastures has been proposed as a measure for sequestering atmospheric CO_2 in growing trees and increasing SOM (Watson et al., 1995). In temperate regions, large-scale afforestation of agricultural land is possible only if adequate supplies of food, fiber, and energy can be obtained from the remaining area. This is currently possible in the European Union and the United States through intensive farming systems. In Europe, afforestation of 30 percent of (surplus) arable land (total arable land area: 40.6×10^6 ha) would increase total soil organic C stocks by 3.58 Pg over 100 years (Smith, Powlson, Glendining, et al., 1997). Tree growth additionally sequesters C in wood. Jenkinson (1971) estimated the C in standing woody biomass to be three times that found in soil in natural woodland regeneration experiments. Standing woody biomass would therefore accumulate 10.74 Pg C over 100 years (Smith, Powlson, Glendining, et al., 1997). This is sequestered only temporarily, unless converted to durable bioproducts. However, if management intensity decreases because of environmental concerns or changes in policy (Enquete Commission, 1995), this option may no longer be available.

Land Degradation

Land degradation can occur as a result of both natural and anthropogenic influences (see Chapters 7 and 9). Natural phenomena that lead to disturbed land such as landslides, dune development, glacier retreat, and vulcanism are local or regional and may therefore not contribute to significant changes in global C cycling. The area of ecosystems disturbed by human activities was estimated to occupy up to $2,000 \times 10^6$ ha worldwide by 1983 and the rate of disturbance was 5 to 7×10^6 ha per year (Oldeman, 1994). Land-degrading processes caused mainly by unsuitable land use and management practices include soil erosion and deposition, surface mining, salinization and acidification, and SOM and nutrient depletion. In disturbed systems, biomass and SOM are generally depleted relative to native ecosystems and well-managed agroecosystems. Reducing land degradation and restoring existing degraded land are significant options to conserve SOM. In tropical and subtropical areas of China, about 48×10^6 ha are classified as wasteland, of which 80 percent is considered suitable for forestry and 8 percent and 6 percent could be restored as cropland and pasture, respectively (Li and Zhao, 1998). In India, more than 100×10^6 ha are classified as degraded and greatly depleted in SOM. At present, there is no systematic global inventory of SOM on degraded lands.

Erosion and Deposition Effects

Erosion and deposition affect SOM distribution in two main ways. First, they redistribute considerable amounts of SOM within a toposequence or a field. Second, erosion and deposition drastically alter the mineralization process in landscapes. Whereas erosion and deposition only redistribute SOM, mineralization results in a net loss of C from the soil system to the atmosphere. The decrease of SOM content on eroded sites especially affects soil quality in a negative way. The permanent erodic output of A_p material induces SOM dilution in cultivated soils. The SOM content decreases due to annual plowing at constant depth and mixing with underlying, C-poor subsoil material. Erosion results in decreased primary productivity, which in turn adversely affects SOM storage because of the reduced quantity of organic C returned to the soil as plant residues. The annual SOM supply in the eroded soils are roots, crop residues, and organic manures containing proteins, lipids, polysaccharides, and lignins (Beyer, Blume, and Köbbemann, 1999). This causes a relative enrichment of litter compounds in eroded soils. In contrast, the A_p of colluvic soils (FAO, 1998: Colluvi-Cumulic Anthrosol) receive SOM from the annual supply of the eroded topsoil materials as well. These are rich in humus and lead to a dominance of humic compounds (Beyer, Blume, and Köbbemann, 1999). In addition, SOM in the A_p of colluvic soils is not diluted by tillage because of the colluvic materials underlying the A_p. Colluvic soils usually contain a larger proportion of SOM in labile fractions because this material can be easily transported. If the accumulation of soil material in depositional areas is extensive, the net result of the burial of this active SOM would be increased SOM storage because decomposition is substantially slowed.

Areas of the tropics are particularly prone to water erosion due to high-intensity rainfall, and C displaced by erosion is estimated to be about 1.6 Pg C y^{-1} for the tropics as a whole (Lal, 1995). Annual C transport from tropical soils to oceans may make 0.16 Pg C or about 10 percent of the total C displaced. The use of management practices that reduce soil erosion may be one of the most effective strategies to maintain or even increase the world's carbon storage in soils.

Mine Spoil Reclamation

Open cast, i.e., surface mining, activities result in a drastic disturbance of large land areas, even entire landscapes. A common characteristic of reclamation areas is the lack of vegetation and SOM. Since production of SOM and its decomposition is considered a key component for carbon and nutrient cycling in terrestrial ecosystems, the course of SOM development has received considerable attention in reclamation and restoration research (Rumpel, Kögel-Knabner, and Hüttl, 1999; Wali, 1999).

In naturally revegetated, 45-year-old chronosequences located in the mixed grass prairie of North Dakota, organic C showed a rate of increase of 131 kg C ha^{-1} y^{-1} (Wali, 1999). This accumulation rate was about one half that reported for mine spoils in Saskatchewan (282 kg C ha^{-1} y^{-1}) by Anderson (1977) and for eastern Montana (256 kg C ha^{-1} y^{-1}) by Schafer and Nielsen (1979). In a chronosequence of young sandy mine soils under forest (11- and 17-year-old Scots pine) in Germany, carbon accumulated mainly in the forest floor (O_l and O_f horizon), whereas in the forest soil of an older, 32-year-old site, most of the carbon was accumulated in the upper mineral horizon (A_i horizon). C accumulation measurements showed a cumulative C production of 50.2 t over 32 years (Rumpel, Kögel-Knabner, and Hüttl, 1999) corresponding to a mean rate of increase of 1,570 kg C ha^{-1} y^{-1}.

In soils reclaimed for agriculture (e.g., loess-derived soils), C accumulation rates over 25 y were between 0.49 t C ha^{-1} y^{-1} (crop residues removed) and 1.16 t C ha^{-1} y^{-1} (high rates of manure application). The C accumulation rates decreased with time (Delschen, 1999). It is important to determine whether the young SOM of reclaimed soils will have properties similar to those of older SOM in undisturbed topsoils.

Soil Salinization and Acidification

In saline systems, organic matter decomposition rates may be lower and potential soil carbon storage higher than in freshwater systems. In a high saline environment, C losses from the soil via microbial respiration and leaching may be much less than from systems with lower salinity. Reduction in soil microbial respiration due to increased salinity has been reported for various salts (Malik and Haider, 1977). Saline soils can store significant quantities of C particularly if halophyte crop residues are plowed annually into the soil (Glenn et al., 1993). If the 35×10^6 ha of salt-affected wasteland in India were reclamated, up to 2 Pg soil organic carbon could be sequestered (Gupta and Rao, 1994).

Soil acidification may be caused by fertilization, leaching, fallowing, increase in organic matter and N contents, deforestation, and acid rain (see Chapter 7). Substantial increases in SOM and total soil N contents are believed to cause acidification by increasing the levels of carboxylates as well as leaching during summer when autumn- and winter-active pastures are dormant. In pasture-cropping systems, where little or no N fertilizer is applied, acidification may occur during the N-building phase.

The temperate and boreal zones are characterized by a cool and wet climate where temperature is a factor that limits microbial activity and mineralization in soils. A dry and warm year influences an ecosystem not only by drought, but also by increased mineralization. If the amount of nitrate that is formed by nitrogen mineralization cannot be taken up by plants, the result will be a temporary accumulation of nitric acid. Guggenberger and Beudert (1989) have observed that the DOM concentrations below the forest floor undergo high variations with regard to the seasonal pattern. The highest concentrations of DOM have been found when periods of drought were followed by heavy rains. Besides N mineralization, the subsequent mineralization push causes a pronounced decrease in pH and an increase in DOM.

Long-term acidification of forest soils leads to the liberation of ionic aluminum from the soil minerals into the soil solution. The presence of aluminum in the soil solution causes inhibition of the repolymerization of organic substances in the humus cycle, while the breakdown is not affected. This process leads to a long-term increase in DOM concentration and an accumulation of inorganic nitrogen in the mineral soil (Eichhorn and Hüttermann, 1999). In contrast, a decrease in soil pH on many sites causes an increase in the thickness of the forest floor as the roots do not penetrate the acid mineral soil and organic residues are decomposed slower on the soil surface than in the soil. Long-term experiments (1966-1995) in Germany (Solling Mountains) have shown that the SOM pool in the forest floor on average has increased by 700 kg C ha^{-1} y^{-1} under a 150-year-old beech forest and by 1,400 kg C ha^{-1} y^{-1} under a 110-year-old spruce forest (Meesenburg, Meiwes, and Bartens, 1999). Besides acidity, stand age (Böttcher and Springob, 2001) and accumulation of nitrogen due to atmospheric N depositions (Nieder, Wachter, and Isermann, 2000) are major factors that influence the quantity and the quality of organic layers of forest soils.

SOIL ORGANIC MATTER AND CLIMATIC CHANGE

Climate is a major factor in soil formation and SOM dynamics, and it regulates biodiversity and community changes. Climatic changes may be accompanied by increases in air temperature, changes in seasonal rainfall distribution, increased burnings, and elevated N and S depositions. This makes prediction of changes in SOM dynamics difficult, especially when factors such as biodiversity and community changes are considered. Since the Q_{10} for litter decomposition (1.3-4.0) is greater than the Q_{10} for primary production (1.0-1.5), litter decomposition is projected to proceed more rapidly under scenarios with increased temperature (Kohlmaier, Janececk, and Kindermann, 1990), provided that water and nutrient supply remain adequate. But the effects of temperature and water availability on soil respiration may be smaller than those associated with the CO_2 fertilization effect (Klein Goldewijk et al., 1994).

Impacts of Global Warming on SOM

Global warming and the concomitant change in the water regime are likely to affect all soil processes that will lead to a wide range of soil and plant responses. The quantitative evaluation of the predicted climate change on soil conditions is difficult because of the complex, interactive fluxes of soil-water relations, vegetation, and land use. Changes in SOM will depend on topography, composition of the mineral soil component, and crop or vegetation cover, all of which vary from place to place. Changing climate may create complex climate-soil patterns that have not previously been observed.

Soil temperature affects the rates at which SOM is decomposed, nutrients are released and taken up, and plant metabolic processes proceed. Soil organic matter and C/N ratio diminish in warmer conditions as a result of accelerated microbial activity. Nitrification is accelerated in warm soils if they are well aerated. Denitrification, which occurs in soil layers with restricted aeration, also increases with soil temperature. Unless additional nutrients are supplied to cropped systems, the soil's own reserves may be depleted. It may take decades before these changes lead to new SOM equilibria but it is hard to predict their net effect on SOM.

Effect of Elevated CO_2 Concentration on SOM

The effect of elevated atmospheric CO_2 concentrations on SOM dynamics and nutrient flows depends on the relative sensitivity of photosynthesis, and autotrophic and heterotrophic respiration. The relative growth-enhancing effects of atmospheric CO_2 enrichment is greatest when resource limitations and environmental stresses are most severe (Idso and Idso, 1994). Elevated atmospheric CO_2 concentrations increase the performance of C_3 plants relative to their C_4 competitors, stimulate biological N_2 fixation, and enhance the capacity of plants to access limited resources such as water and nutrients (Sage, 1995).

Under increased atmospheric CO_2 concentration, plant growth may become nutrient-limited in the long term. This problem should not occur in inherently fertile soils, whereas in chemically poor tropical soils (e.g., Acrisols and Ferralsols) it can be addressed by larger applications of fertilizers. There may be a feedback mechanism in which elevated CO_2 concentration causes an increase in substrate release into the rhizosphere by non-mycorrhizal plants, leading to mineral nutrient sequestration by the expanded microflora and a consequent nutritional limitation for plant growth (Diaz et al., 1993).

Combined Effects of Elevated CO_2 and Increased Temperature on SOM

Using the present knowledge, it is hard to predict net effects of global change on SOM. Under elevated atmospheric CO_2 concentrations, the effects of increased temperature on the allocation pattern of C between roots and leaves can be largely explained by the effect of root temperature on the roots' capacity to transport water. Changes in plant communities associated with changes in climate affect litter quality, which may lead to either positive or negative feedbacks to the atmosphere depending on whether lignaceous perennials or nonlignaceous annuals take over. Particularly, grasslands are prone to SOM accumulation under increased CO_2 concentration (Jongen et al., 1995). Ecosystems such as tropical savannas may become considerable sinks for soil carbon if soil erosion does not increase. Because the quality and degradability of plant residues may decrease under increased CO_2 concentration, this would lead to a SOM accumulation (Sombroek, 1995).

Elevated CO_2 concentration may have positive effects on SOM contents, assuming that ecosystems have time to adapt to increased photosynthesis and water-use efficiency (Brinkman and Sombroek, 1996). This only applies if the incidence of extreme weather events does not increase drastically, the amounts of weatherable minerals in soils remain adequate, the soil moisture regime remains favorable for plant growth, and the land resources are properly managed. These prerequisites may not apply in increasingly large sections of the world. Changes in precipitation and wind regimes can lead to changes in the rate of soil erosion by wind and water, and the loss of SOM. A study on the erosion potential for cropland, pastureland, and rangeland in the United States illustrates that average sheet and rill erosion may change by +2 to +16 percent in croplands, –2 to +10 percent in pasturelands, and –5 to +22 percent in rangelands, depending on the climate-change scenarios (Phillipps, White, and Johnson, 1991). If semiarid grasslands become more arid, wind erosion and SOM losses will increase.

A GLOBAL INVENTORY OF SOIL ORGANIC MATTER

The SOM content of virgin as well as of managed soils varies considerably depending on climate, soil type, and site conditions (Table 13.9). Global

TABLE 13.9. Original C Content of Different Soil Groups and Its Historical Decline As a Consequence of Cropping

Soil groups	Cultivated area (× 10^6 ha)	Soil C 0-100 cm (kg m^{-2})	C mass virgin (Pg C)	C mass cultivated (Pg C)	Historic C loss 1700-2000 (Pg C)
Temperate forest soils (Luvisols, Podzoluvisols, Podzols)	308	6.9-14.2	24.4	18.1	6.3
Temperate grassland soils (Chernozems, Phaeozems, Kastanozems, Greyzems)	352	12.5-18.4	49.8	36.9	12.9
Tropical forest soils (Cambisols, Acrisols, Ferralsols, Nitosols, Planosols)	439	7.1-14.5	47.3	35.1	12.2
Tropical grasslands/savanna soils (Acrisols, Vertisols)	161	9.4-17.8	21.4	15.9	5.5
Shallow, saline, sodic, and arid soils (Regosols, Leptosols, Arenosols, Solonchaks, Solonetz)	308	2.6-7.1	17.7	13.1	4.6
Wetlands/paddy soils (Gleysols)	89	11.9	10.6	7.8	2.8
Histosols	39	112	43.6	35.6	8.0
Andosols	31	23.7	7.3	5.4	1.9
Total	1727		222	168	54

Source: Adapted from Paustian et al. (1997).

soil inventories provide an estimate of the overall amounts which are stored in these soils before their cultivation (Post et al., 1990; Sombroek, Nachtergaele, and Hebel, 1993). In the roughly 1.7 × 10^9 ha that have been converted to cropland this quantity was originally about 220 Pg organic C. Assuming an average C loss of 26 percent from the top 30 cm of soils due to their cultivation yields a historical loss of some 54 Pg C, of which 43 Pg C stems from upland soils and 11 Pg C from cultivated wetland soils, i.e., Histosols and Gleysols (Paustian et al., 1998). Similar soil C losses due to past land conversion to agricultural use were estimated by Houghton and Skole (1990). They estimated a net reduction in terrestrial organic C stocks of 275 Pg since the beginning of settled agriculture, of which 42 Pg was attributed to soils.

These global estimates of historic C losses (40-60 Pg) from cultivated soils provide a reference level for C sequestration potential. In most cases, C levels in agricultural soils, even with improved management, are unlikely to exceed their native condition. Thus, the SOM levels of precultivation conditions can be viewed as a practical upper limit of SOM storage potential. There are notable exceptions to this, such as the plaggen soils of northwest Europe and the Terra Preta do Indio soils of the Amazon, where intensive management and high levels of organic matter additions were practiced over long time periods, resulting in greatly enhanced SOM pools (Sombroek, Nachtergaele, and Hebel, 1993).

Potential to Increase SOM Levels in Cultivated Soils

The potential to increase SOM levels in soils under permanent cultivation is largely restricted to upland soils. Restoring SOM sinks in artificially drained wetland soils (Histosols, Gleysols) is unlikely unless they are taken out of agricultural production and reverted to natural wetlands. Using the assumed historical C loss from the world's mineral soils (excluding wetlands and Histosols) of about 54 Pg C (Table 13.9) as a reference, and assuming a practically achievable restoration by about one half, this would correspond to 20-30 Pg C, or an average C accumulation of 0.4-0.6 Pg C y^{-1} over the next 100 years (Paustian et al., 1997, 1998). Including the potential for SOM accumulation in surplus agricultural lands, which are reverted to grasslands and wetlands, increases the annual estimate to 0.2-0.4 Pg C y^{-1} (Paustian et al., 1997).

MODELING SOIL ORGANIC MATTER DYNAMICS

The turnover of organic matter in soils can be measured on both short- and long-term time scales, but it is the long-term trends that determine the sustainability of the system. There are no short-term or one-season models for C directly analogous to those for N (as discussed in Chapter 14), because of the differing roles played by the two elements in plants and the environment. There are, however, a number of models that chart the progress during the first few months of organic matter introduction into the soil (e.g., Knapp, Elliott, and Campbell, 1983; Molina et al., 1983; van Veen, Ladd, and Frissel, 1984; Sallih and Pansu, 1993). These have been validated against data from relatively short-term laboratory or field incubations.

A variety of long-term models have been proposed to describe organic matter turnover in soils. The models have been developed on the presumption that SOM is represented by a changing continuum (Janssen, 1984;

Bosatta and Ågren, 1985) or by a number of compartments distinguished on the basis of accessibility of organic C to microorganisms. Therefore, the SOM models may be categorized as noncompartment and compartment models.

Noncompartment Models

Bosatta and Ågren (1985) have developed a scheme to derive a set of general models depending upon two microbial functions, viz. the microbial efficiency of substrate utilization and the microbial rate of substrate utilization. They view decomposition as a continuum beginning with the input of fresh organic matter and leading to the formation of refractory humic substances. To formalize this idea, they have used a continuously varying quality variable q, spanning from zero, wherein the most refractory materials are placed, to one, where the most easily decomposable materials are placed. The mathematics of this approach is complex and has not yet been applied to long runs of data from agricultural soils (Jenkinson, 1990). The model becomes mathematically equivalent to the others wherever a sufficient range of (usually geometrically distributed) pool reactivities is considered (Arah and Gaunt, 2001).

Janssen (1984) has introduced the concept of "apparent initial age" of the organic material for predicting accumulation and decomposition of "young" soil organic matter. This age is related to the humification coefficient and varies from 1 year for green matter to 14 years for some peats. Ionenko, Batsula, and Golovachev (1986) describe humification in the form of humification and mineralization processes. The humification rate is interpreted as the vector sum of the rates of decomposition and synthesis. Independent testing of the model has not been encountered.

Compartment Models

In the compartmentalization approach, the organic matter is divided into conceptual compartments varying from one (Jenny, 1941; Woodruff, 1949) to several, and sometimes 10 (McGill et al., 1981). The breakdown of organic matter in each compartment is assumed to follow first-order kinetics which may be generalized as:

$$\frac{dX}{dt} = -kX \tag{13.2}$$

where X is soil C or N content at a given time, k is first-order decomposition rate constant (per time), and t is time.

Jenny (1941) presents the simplest formulation, incorporating single compartments (Equation 13.3) to represent the rate of change of organic C or N *(dX/dt)* with time in cultivated soils.

$$\frac{dX}{dt} = A - kX \tag{13.3}$$

where *A* is rate of addition of C or N (mass per time). The solution of the differential equation is

$$X = X_E + (X_0 - X_E)e^{-kt} \tag{13.4}$$

where X_0 is the initial C or N content and X_E the content at equilibrium (when *A* and *kX* are equal and $dX/dt = 0$). This equation has been used to analyze SOM (C or N) dynamics under a variety of rotations and climates. This model is better applied to organic N than to organic C because most plant materials entering the soil have a large C:N ratio. This means that most of the N tends to be retained in the soil while much of the incoming C is lost rapidly as CO_2 (Jenkinson, 1990).

The major problem with the use of this equation is that the decomposition constant *(k)* does not allow for changes in the decomposition rate that would result from changes in the composition of soil organic matter. It has long been recognized that decomposing SOM is intrinsically heterogeneous because a number of compounds of different characteristics and qualities are contained in the original material and are produced during the decay process. If the proportions of the constituents in SOM change over time, then the value of *k* also changes with time. Strategies to accommodate changing values of *k* range from adding more components, to developing time-dependent ways to modify *k* (Russell, 1975), using decay series in which residual portion becomes more stable (Powers, Wallingford, and Murphy, 1975), or decay series in which *k* is an exponential function of time (Mathers and Goss, 1979).

Hénin and Dupuis (1945) suggest a simple model (Equation 13.5) in which the annual input of plant carbon, A_c, decomposes very rapidly, forming a quantity of humus carbon fA_c, identical in nature to that already in the soil. The constant f is described as the isohumic coefficient. It is usually about 0.3 for material from agricultural crops but much greater for peat and similar materials

$$C = fA_c / k + (C_0 - fA_c / k)e^{-kt} \tag{13.5}$$

Here *C* is the quantity of organic carbon at time *t* and C_0 the initial quantity; *k* is the fraction of *C* that decomposes each year. Jenkinson and Johnston

(1977) have found that this equation provides a good fit to changes in organic C occurring over the 10-100 year period and have concluded that this approach gives more realistic values to A_c than the assumption that the whole input of plant carbon joins the carbon already in the soil.

Generally, the decay of plant material in soil is very rapid during the first few months but becomes much slower later. The decay of ^{14}C-labeled ryegrass leaves in soil in the field during 10 years was found (Jenkinson, 1977) to be very well represented by the two-compartment model (Equation 13.6).

$$C = 71e^{-2.83t} + 29e^{-0.087t} \tag{13.6}$$

Jenkinson (1990) concluded from estimates of the equilibrium C content computed through this model that a further, even slower compartment was needed.

In the multicompartment models, small, homogeneous pools with high turnover rates and pools of greater size with slower turnover rates governed by first-order kinetics are distinguished. The decomposition rates are generally modified by climatic and edaphic reduction factors. Most of the multicompartment models that simulate long-term SOM dynamics in soils follow a similar general structure. These models partition organic matter into two main pools viz. recently added organic material such as plant residue or litter and native soil organic matter (SOM). Each of the main pools is further subdivided into different fractions or components. A generalized scheme of organic-matter partitioning into different fractions is presented in Figure 13.2. The nomenclature used, the number of pools considered, and the associated decomposition rate constants vary considerably in different models (Table 13.10).

Plant residue or litter is generally divided into two compartments: (1) metabolic or labile or decomposable plant material, and (2) structural or resistant plant material (Hunt, 1977; Jenkinson et al., 1987; Parton et al., 1987; Hansen et al., 1991; Sallih and Pansu, 1993). In contrast, some models operate with one litter input pool only (Molina et al., 1983; Nicolardot, Molina, and Allard, 1994), while others (van Veen and Paul, 1981; Stroo et al., 1989; Verberne et al., 1990) rely on three input pools: (1) a readily decomposable pool (water solubles, sugars, proteins), (2) a pool of structural cell wall materials (cellulose, hemicellulose), and (3) a pool of resistant lignified residues. The pool consisting of structural or resistant plant material adds directly to one of the SOM pools while the other inputs pass the microbial biomass before entering the SOM module. Bottner, Pansu, and Sallih (1999) consider an additional compartment for living plant material to include the effect of rhizodeposition.

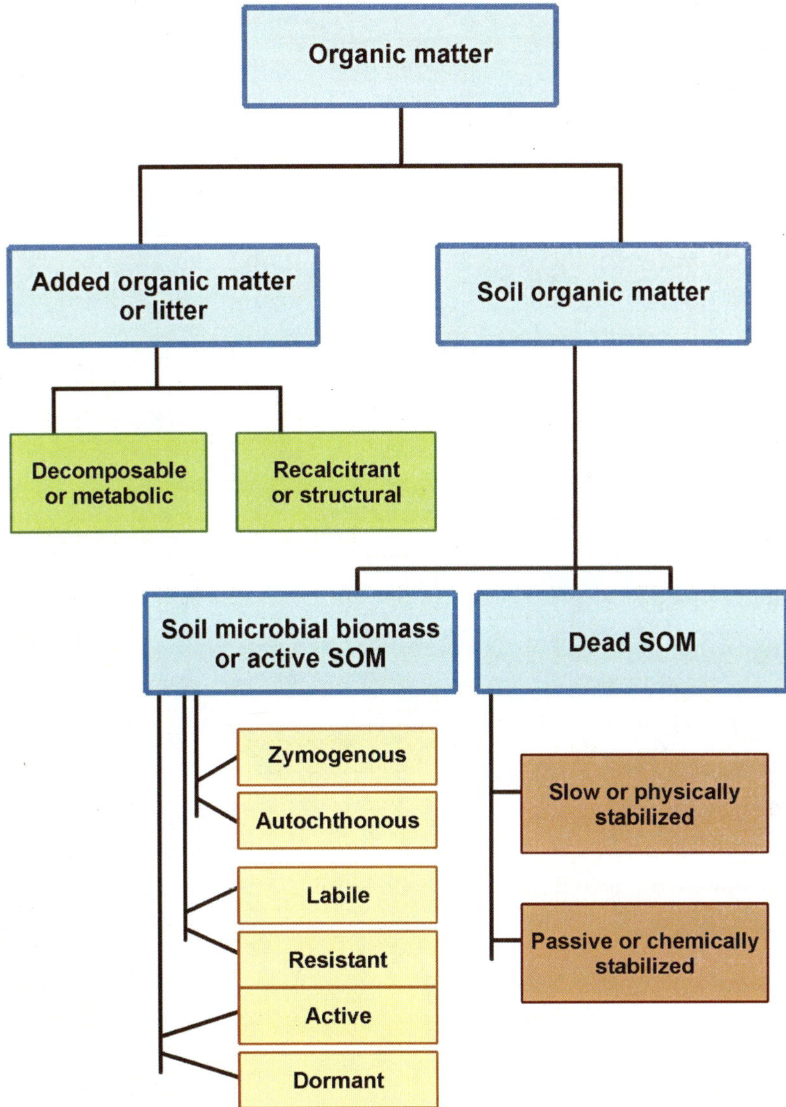

FIGURE 13.2. Schematic representation of organic-matter partitioning into conceptual pools as considered in different SOM models.

TABLE 13.10. Functional Pools of Soil and Added Organic Matter in Different Models

Functional pool	C/N	k (w^{-1})
CENTURY (Parton et al., 1987)		
AOM Metabolic	10-25	0.35 at 35°C
AOM Structural	150	0.094
Active SOM	8	0.14
Slow SOM	11	0.0038
Passive SOM	11	0.00013
Rothamsted model (Jenkinson, 1990)		
DPM	—	0.192 at 9.3°C
RPM	—	0.0058
Microbial biomass	—	0.0127
Humified OM	—	3.85×10^{-4}
Inert OM	—	—
Verberne et al. (1990)		
DPM	6	1.4 at 25°C
SPM	150	0.7
RPM	100	0.14
Microbial biomass	8	
Protected biomass	—	0.035
Nonprotected biomass	—	3.5
Protected active OM	10	0.002
Nonprotected active OM	15	0.07
Stabilized OM	10[a] 20[b]	0.56×10^{-5}
DAISY (Hansen et al., 1991)		
AOM 1	100	0.074 at 15°C
AOM 2	—	0.735
Biomass pool 1	6	0.0105
Biomass pool 2	10	0.105
SOM pool 1	10	2.8×10^{-5}
SOM pool 2	12	1.5×10^{-3}

ANIMO (Rijtema and Kroes, 1991)		
Humus	16	2.7×10^{-4} at 15°C
Fraction 2	12	0.063
Fraction 3	58	0.0044
Fraction 4	76	0.037
Fraction 5	76	0.133
Fraction 6	24	0.075
Fraction 7	24	0.0082
SWATNIT (Vereecken et al., 1991)		
Litter	8	0.007 at 15°C
Manure	10	0.007
Humus	12	7×10^{-5}
NCSOIL (Molina et al., 1983; Nicolardot, Molina, and Allard, 1994)		
Plant residues		
Labile microbial biomass		2.31 at 280°C
Resistant microbial biomass		0.28
Humads		0.042
Stable SOM		7×10^{-5}

OM = organic matter; SOM = soil organic matter; AOM = added organic matter; DPM = decomposable plant material; RPM = resistant plant material; SPM = structural plant material; k = first-order rate coefficient (per week)

The models differ not only in the number of litter fractions considered, but also the manner in which the litter fractions are quantified. In the CENTURY model (Parton et al., 1987), partitioning of litter into metabolic and structural components is based on lignin to nitrogen ratio, such that the fraction of newly formed litter going into the structural components increases with the increasing lignin-to-N ratio of the litter. All lignin is assumed to be part of the structural fraction. The current version of the CENTURY model (Parton et al., 1994) treats surface and root litter separately. In the model of Verberne et al. (1990), the residues are split into decomposable, structural, and resistant fractions based on a fixed C:N ratio of each fraction and the C:N ratio of the entire residue. In the Rothamsted model, litter quality is incorporated as differing proportions of the decomposable and resistant components, which are determined from curve fitting of short-term decomposition data. Typical DPM/RPM ratios used are 1.44 for most agricultural and improved grasslands and 0.25 for deciduous and tropical woodlands (Coleman and Jenkinson, 1996).

Paustian, Ågren, and Bosatta (1997) compared three SOM models viz. CENTURY (Parton et al., 1987), Rothamsted (Jenkinson et al., 1987), and Q model (Ågren and Bosatta, 1987) and found that these yield quite different results with respect to the composition of SOM. They proposed a conceptual framework in which litter quality can be broken down into chemical, physical, and inhibitory factors.

The native soil organic matter pool is further divided into soil microbial biomass (responsible for transformation of added organic matter) and one or more pools of dead soil organic matter. Although the CENTURY model (Parton et al., 1987) includes the microbial biomass together with microbial metabolites and other SOM with a short turnover time (1-5 years) into a pool termed active SOM, the model of Sallih and Pansu (1993) makes a distinction between microbial biomass and other labile compounds. In most of the other models, the soil microbial biomass (SMB) is subdivided into two or more components such as nonprotected (labile, dynamic) and physically protected (resistant or stable) biomass (Molina et al., 1983; van Veen, Ladd, and Frissel, 1984; Verberne et al., 1990; Hansen et al., 1991); cell walls and cytoplasm (Paustian and Schnürer, 1987); or labile cell carbon and assimilated live biomass (Knapp, Elliott, and Campbell, 1983). In contrast, the PHOENIX model (McGill et al., 1981) considers biomass as population of bacteria and fungi dividing each compartment into metabolic and structural components.

It is not only the microbial population heterogeneity and its composition but also the activity state of the microorganisms that may influence the soil organic matter turnover. Soil microorganisms can change from the active to the dormant state and vice versa. It is even possible to estimate the activity state of biomass (van de Werf and Verstraete, 1987; Panikov et al., 1991). Although some (Hunt, 1977; Blagodatsky, Blagodatsky, and Rozanova, 1994) distinguish between active and inactive biomass, others (Jenkinson et al., 1987; Kersebaum and Richter, 1994) partition soil biomass into zymogenous (growing very fast on readily available carbon) and autochthonous (biomass that reacts slowly and consumes the most resistant fraction of carbon). The current version of the Rothamsted model (Jenkinson, 1990) operates with only one biomass pool. Following the original idea of Winogradsky (1949), microbial biomass needs to be split into two pools, a fast-responding, zymogenous pool and a rather indolent autochthonous pool. Furthermore, two metabolic states may be distinguished, an active and a dormant one. According to this line, a Pirt type model (Pirt, 1975) may be developed in different ways: one perhaps according to Kersebaum and Richter (1994), and another according to Yevdokimov, Blagodatsky, and Kudeyarov (1993) to determine which one better describes the measured data.

The dead soil organic matter is further divided into two or more pools based on stabilization mechanism, bioavailability, and biochemical and kinetic parameters (Figure 13.2). Generally, it is divided into slow or physically stabilized pools with turnover times of a few decades and passive or chemically stabilized pools that remain in the soil for hundreds or thousands of years (Jenkinson and Rayner, 1977; Parton et al., 1987). The physically stabilized pools are assumed to consist of compounds protected against biological attack by adsorption to soil colloids or entrapment within soil aggregates whereas the chemically stabilized pools include compounds with a chemical structure resistant to biological attack (Hansen et al., 1991). In contrast, the current Rothamsted model (Jenkinson, 1990) operates with only one dynamic SOM pool (termed humified organic matter), and a small pool of inert SOM (charcoal, etc.) not engaged in biological turnover. Most SOM turnover models postulate a pool of passive, old, inactive or stable SOM that is often quantitatively very significant with as much as half of the SOM being effectively left out of consideration by model dynamics. Christensen (1996) reports that these pools can be considered as "parking lots" for SOM that are left during model calibration and included to improve model performance rather than to satisfy well-defined operational concepts.

Van Veen and Paul (1981) and Verberne et al. (1990) have introduced the concept of physically protected organic matter that has a much lower decomposition rate than nonphysically protected organic matter. As protection against decomposition appears more effective in soils with high clay content, most of the SOM models use clay or soil texture as a surrogate for physical protection and do not mechanistically treat the physical protection of organic matter in soil. The net rate of decomposition of organic matter depends not only on soil texture but also on the degree to which the protective capacity of the soil is already occupied. The rate at which organic matter becomes protected depends on both the amount of free organic matter and the degree to which the protective capacity is filled (Hassink and Whitmore, 1997). The upper limit of C content associated with primary particles <20 μm may be interpreted as the capacity of soil to protect C (Rühlmann, 1999). Hassink and Whitmore (1997) have presented a model that describes physical protection explicitly as a function of the capacity of clay particles and aggregates to hold organic matter. The model closely follows the buildup and decline of organic matter in 10 soils to which grass residues were added each year for a period of 10 y and then left without addition for a further 10 y.

Though the distribution of SOM within conceptual pools is an important consideration in developing a better understanding of SOM dynamics, a major limitation is that most of the conceptual pools do not correspond to experimentally verifiable fractions. Questions exist as to what may be the reasons for widely different assumptions with regard to pool numbers, size, decomposition behavior, and how these have been parameterized.

The Reality of Conceptual Organic Matter Pools

Some have questioned the usefulness of conceptual organic matter pools that cannot be physically or chemically quantified. Exceptions are some of the input pools and the microbial biomass. Attempts have been made to establish linkages between the two either by devising advanced laboratory fractionation procedures to match measurable organic matter fractions with model pool definitions or by revising model pool definitions to coincide with measurable quantities (Christensen, 1996). Elliott, Paustian, and Frey (1996, p. 164) have coined the phrase "modeling the measurable or measuring the modelable" to represent the two approaches. Organic C content of long-term bare fallow soils (Rühlmann, 1999) and radiocarbon dating of soil fractions (Falloon and Smith, 2000) have been used to define stable SOM pool.

Soil organic matter fractions, isolated by physical fractionation procedures, have been related to conceptual pools considered in the existing SOM turnover models (Cambardella and Elliott, 1992; Buyanovsky, Aslam, and Wagner, 1994). Physical fractionation procedures based on differential densities and sizes that separate coarse or light fractions from fine fractions provide a relationship between density or the size of fraction and their ages (Balesdent, Mariotti, and Guillet, 1987; Balesdent, Wagner, and Mariotti, 1988; Martin et al., 1990). Fractions that are 53-2,000 μm may provide an accurate estimate of the slow pool, while those finer than 53 μm may provide an accurate estimate of the passive pool (Cambardella and Elliott, 1992). Alternatively, fractions of various densities (Hassink, 1994) may be recovered by density separation using sequentially heavier liquids, but this approach is time consuming and presents difficulties when soil particles are miscible with the heavy liquid.

The introduction of physically isolated SOM fractions as experimental equivalents to pools in models will lead to a redefinition of pools and the material fluxes between them, including rate modifiers associated with soil structure. This will lead to a redefinition of the structure of current models and the factors that regulate the material fluxes between compartments. Christensen (1996) presents a revised model structure based on soluble and insoluble input fractions and separate aboveground and belowground input routes. It includes two dynamic pools of particulate organic matter (free and intra-aggregate light fraction OM), one pool of inert light fraction OM, and two pools of organo-mineral-associated SOM (silt and clay SOM). Gaunt et al. (2001) presents an alternative approach using analytically defined pools and measurement of ^{13}C- and ^{15}N-stable isotope tracers to derive model parameters. However, introduction of these proposed pools in model structures has not been accomplished yet.

Classifying Models

Different classification schemes have been proposed to compare and organize information about soil organic matter models. Jenkinson (1990) classifies models for the turnover of organic matter in soils into four classes: (1) *single-homogeneous-compartment models* that assume a single compartment and have been applied mainly to organic N (e.g., Jenny 1941; Woodruff, 1949); (2) *two-compartment models* in which incoming plant carbon is split into two compartments, each decomposing by a first-order process, but one much faster than the other (e.g., Jenkinson, 1977); (3) *noncompartmental decay models* which assume that decomposition is a continuum, organic matter moving down a quality scale as it decays, fresh decomposable organic matter having a high quality (set at 1), with the most resistant material present in the system being of zero quality (e.g., Bosatta and Ågren, 1985); and (4) *multicompartmental models* in which material in a compartment is assumed to decay by first-order kinetics so that the rate of decomposition in a particular compartment is deemed to be a feature of organic matter itself. These may be short-term models (Smith, 1979; McGill et al., 1981; van Veen, Ladd, and Frissel, 1984) or longer-term models (Russell, 1975; van Veen and Paul, 1981; Parton et al., 1987; Jenkinson et al., 1987).

Paustian (1994) classifies several multicompartment models of soil biology and biogeochemical processes into *organism-oriented,* and *process-oriented* models. Organism-oriented models focus on flow of energy and matter through food webs, whereas process-oriented models, which are most commonly used, focus on processes controlling energy and matter transformations. Organisms are represented in process models as a form of genetic biomass or active organic matter. Paustian (1994) groups information about models under four headings: (1) nonliving organic matter pools, (2) organism components, (3) rate kinetics, and (4) special attributes.

McGill (1996) has proposed a scheme consisting of a partially catenary sequence of attributes to organize information about ten SOM models. He has grouped information about models under the following eight headings:

1. Model type (dynamic or static)
2. Scale over which a model will function
3. Soil horizons
4. Regulation by soil properties
5. Biotic component
6. Litter/ SOM distinction
7. Litter and SOM compartmentalization
8. Control (process level or biological activity of organisms)

This scheme (McGill, 1996) reveals convergence in the kinetic compartmentalization, growing use of clay content, and inclusion of an inert organic matter component.

Although descriptive model classifications and comparisons are very informative, it is desirable, whenever possible, to supplement visual/graphical methods for model evaluation with quantitative measures of model performance. The use of statistical methods makes it possible to determine common strengths and weaknesses of particular models. Quantitative methods to evaluate and compare SOM models have been presented by Smith, Smith, and Addiscott (1996) and will be further discussed in Chapter 22.

Evaluation, Comparison, and Application of Soil Organic Matter Models

With the increased use of models for regional, national, and global applications, evaluation of models to determine generality (or lack thereof) becomes crucial (Paustian, 2001). The first important step toward the evaluation of SOM models is to test model simulations against real data over the entire domain of their application. Existing, long-term data provide an important avenue for checking whether projections from a particular model are correct or not. Jenkinson et al. (1987) tested the Rothamsted model against data from five long-term field experiments situated on Rothamsted and Woburn experimental farms. There was a reasonably good agreement between model simulations and measured data, but it was in no way perfect. Müller et al. (1998) evaluated a DAISY soil organic matter submodel against data from a one-year field study with incorporation of chopped barley straw, bluegrass, and maize into sandy loam soil. Significant differences between values predicted from the model and measured values of soil respiration, soil mineral N, and soil microbial biomass C and N were observed. Suggestions were made to include the concept of soil microbial residues (SMR), temporarily protected against recycling via the microbial turnover and mineralization, and the changing composition of light particulate matter (LPOM) in the DAISY model.

A major difficulty in validating models of soil organic matter dynamics is the inability to quantify some of the functional pools, particularly the slow pool consisting of as-yet-unknown physical properties and chemical composition. Consequently, the models have to be calibrated by adjusting the rate coefficients and pool sizes (site-specific) to fit the measured data. Parton et al. (1994) determined maximum decay rate of passive SOM by adjusting the values to match observed soil C levels for different soil textures at two sites in the Great Plains. Nicolardot, Molina, and Allard (1994) calibrated NCSOIL with data from a long-term laboratory incubation experiment. To fit the model to experimental data, the decomposition rates had to be reduced by 60 to 70 percent after 35 to 85 days of incubation depending on the soil and the treatment. Generally, the performance of SOM models depends on site-specific calibration. In a NATO Advance Research Workshop, nine long-term, process-oriented, multicompartment SOM models were evaluated and compared using twelve data sets from long-term experiments representing a variety of climate, soil, and land use conditions (Powlson, Smith, and Smith, 1996; Smith, Powlson, Smith, et al., 1997). In terms of overall performance of models across all data sets in simulating SOC dynamics, models fell into two groups with one group (SOMM, ITE, and Verberne) performing significantly less well than the other (RothC, CANDY, DNDC, CENTURY, DAISY, and NCSOIL) because of differences in the level of site-specific calibration used by the two group of models (Smith, Smith, et al., 1997). Therefore, in the future, a major challenge in SOM modeling will be to free simulations from the calibration process and to devise experimental methods that will provide initial values relevant to the dynamic requirements of the model (Molina et al., 1997).

The models for SOM turnover in soils are being used to consider possible mitigation options to sequester carbon in soil or vegetation as a means of decreasing the rate at which atmospheric CO_2 concentration rises (Powlson, 1996; Falloon et al., 2001). SOM models have been applied at the global scale to explore the potential role of soils in C uptake or release during the period over which the burning of fossil fuel (the last 150 years and into the next several centuries) is involved in altering global environmental conditions (Post, King, and Wullschleger, 1996). SOM models have also been used to study the impact of different crop and land management practices on soil C stabilization and nutrient dynamics in different regions of the world (Jenkinson et al., 1987; Paustian, Parton, and Persson, 1992; Parton and Rasmussen, 1994; Metherell et al., 1995), to estimate net primary production of a soil, and to assess the impact of soil texture on SOM dynamics (Parton et al., 1995). A number of concepts used in the different SOM models have been incorporated into nutrient cycling models that are linked to crop plant production models such as erosion-productivity impact calculator (EPIC; Williams, Pulman, and Dyke, 1985); nitrogen-tillage residue management (NTRM; Shaffer et al., 1992); and denitrification and decomposition model (DNDC; Li, Frolking, and Frolking, 1992). Some of the models (e.g., CENTURY, DNDC) have also been used to simulate trace gas fluxes.

The models for SOM turnover in soils have not been put to a rigorous test in comparison to field data. Particular attention needs to be given to the regulation of SOM dynamics by soil properties to increase the universality of models as well as spatial distribution of organic materials and their decomposers. However, since most of the SOM is allocated to model pools that cannot be determined experimentally, there is an urgent need to identify measurable SOM fractions that qualify as candidates for compartments in SOM turnover models (Christensen, 1996). Smith, Smith, et al. (1997) have identified a number of areas that require further research and evaluation including linking of SOM models to other, more complex submodels to examine the effect of global climate change on ecosystems, and determining the sensitivity of models to changing temperature and atmospheric CO_2 concentrations.

SUMMARY

Organic matter in world soils constitutes one of the large global C pools comprising about 1,500 Pg. The SOM level depends on the input and output of biomass C into the soil. Sources of C input include the amount of aboveground and belowground biomass returned to the soil, and any other biosolids (e.g., manure, compost, sludge). The output or losses of C out of the soil system include soil respiration, oxidation, soil erosion, and leaching. There are several factors (climate, soil, vegetation, land use, soil, water and vegetation management) and processes (aggregation, erosion, leaching, acidification, salinization, eluviation, illuviation, mineralization, humification, bioturbation) that affect SOM pool and its depletion. The delicate equilibrium is easily disturbed by anthropogenic perturbations (e.g., land use change, deforestation, burning, agricultural activities). Global warming may additionally increase CO_2 evolution from soils.

The historic loss of soil organic C pool as a consequence of cropping between the year 1700 and today amounts to about 54 Pg. The estimates of SOC loss provide a reference point about the potential of C sequestration through adoption of recommended agricultural practices. In most soils it is possible to resequester 60 to 80 percent of the historic C loss. Principle strategies of C sequestration include reducing losses of C from soils, increasing C concentration in soils, and improving C yields through agricultural intensification. Practices that increase SOM level include afforestation, residue return, conservation tillage, deeper plowing, no or prescribed burning, return to natural ecosystems, restoration of wetlands, rotational and controlled grazing, restoration of degraded soils, and establishing cover crops.

Modeling SOM dynamics is gaining recognition as a key part of efforts better to understand and manage the terrestrial carbon cycle. Most of the models share some basic assumptions which include the representation of SOM as multiple pools with differing inherent decomposition rates, governed by first-order rate constants modified by climatic and edaphic reduction factors. The models were originally conceived to describe processes at the ecosystem or field scale. New approaches continue to be explored, and some of the major recent trends in SOM modeling and its application to environmental problems were discussed in this chapter.

REFERENCES

Ågren, G.I. and E. Bosatta (1987). Theoretical analysis of the long-term dynamics of carbon and nitrogen in soils. *Ecology 68:*1181-1189.

Anderson, D.W. (1977). Early stages of soil formation of glacial till mine spoils in a semiarid climate. *Geoderma 19:*11-19.

Arah, J.R.M. and J.L. Gaunt (2001). Questionable assumptions in current soil organic matter transformation models. In *Sustainable Management of Soil Organic Matter,* eds. R.M. Rees, B.C. Ball, C.D. Campbell, and C.A. Watson. Wallingford, UK: CABI Publishing, pp. 83-90.

Armentado, T.V. and E.S. Menges (1986). Patterns of change in the carbon balance of organic soil-wetlands of the temperate zone. *Ecology 74:*755-774.

Balesdent, J., A. Mariotti, and B. Guillet (1987). Natural ^{13}C abundance as tracer for soil organic matter dynamics studies. *Soil Biology and Biochemistry 19:*25-30.

Balesdent, J., G.H. Wagner, and A. Mariotti (1988). Soil organic matter turnover in long-term field experiments as revealed by the ^{13}C natural abundance in maize field. *Soil Science Society of America Journal 52:*118-124.

Barber, S.A. (1979). Corn residue management and soil organic matter. *Agronomy Journal 71:*625-627.

Batjes, N.H. (1996). Total carbon and nitrogen in the soils of the world. *European Journal of Soil Science 47:*151-163.

Batjes, N.H. (1997). World soil carbon stocks and global change. In *Proceedings of the Workshop (held in Nairobi, Kenya, 4-8 September 1995) "Combating Global Climate Change by Combating Land Degradation,"* eds. V.R. Squires, E.P. Glenn, and A.T. Ayoub. Wageningen, Netherlands: ISRIC (International Soil Reference and Information Centre), pp. 51-78.

Batjes, N.H. (1998). Mitigation of atmospheric CO_2 concentration by increased carbon sequestration in the soil. *Biology and Fertility of Soils 27:*230-235.

Batjes, N.H. and W.G. Sombroek (1997). Possibilities for carbon sequestration in tropical and subtropical soils. *Global Change Biology 3:*161-173.

Becker, K.W. and B. Meyer (1973). Abbau, natürliche Boden-Inkorporation und Ertragswirkungen von Ernte-Rückstandsdecken (Stroh, Rübenblatt) auf Ackerparabraunerden aus Löss. *Göttinger Bodenkundliche Berichte 26:*1-38.

Benbi, D.K. and C.R. Biswas (1997). Nitrogen balance and N recovery after 22 years of maize-wheat-cowpea cropping in a long-term experiment. *Nutrient Cycling in Agroecosystems 47:*107-114.

Benbi, D.K., C.R. Biswas, S.S. Bawa, and K. Kumar (1998). Influence of farmyard manure, inorganic fertilizers and weed control practices on some soil physical properties in a long-term experiment. *Soil Use and Management 14:*52-54.

Beyer, L., H.P. Blume, and C. Köbbemann (1999). Colluvisols under cultivation in Schleswig-Holstein. 3. Soil organic matter transformation after translocation. *Journal of Plant Nutrition and Soil Science 162:*61-69.

Beyer, L., K. Pingpank, M. Bölter, and R.D. Seppelt (1998). Small-distance variation of carbon and nitrogen storage in mineral Antarctic Cryosols near Casey Station (Wilkes Land). *Journal of Plant Nutrition and Soil Science 161:*211-220.

Bird, M.I. (1995). Fire, prehistoric humanity, and the environment. *Interdisciplinary Science Review 20:*141-154.

Black, K.E., C.G. Harbron, M. Franklin, D. Atkinson, and J.E. Hooker (1998). Differences in root longevity in some tree species. *Tree Physiology 18:*259-264.

Blagodatsky, S.A., E.V. Blagodatsky, and L.N. Rozanova (1994). Kinetics and strategy of microbial growth in chernozemic soil as affected by different long-term fertilization. *Microbiology 63:*298-307.

Blume, H.P., L. Beyer, M. Bölter, H. Erlenkeuser, E. Kalk, S. Kneesch, U. Pfisterer, and D. Schneider (1997). Pedogenic zonation in the southern circumpolar region. *Advances in GeoEcology 30:*69-90.

Bosatta, E. and G.I. Ågren (1985). Theoretical analysis of decomposition of heterogeneous substrates. *Soil Biology and Biochemistry 17:*601-610.

Böttcher, J. and G. Springob (2001). A carbon balance model for organic layers of acid forest soils. *Journal of Plant Nutrition and Soil Science 164:*399-405.

Bottner, P., M. Pansu, and Z. Sallih (1999). Modelling the effect of active roots on soil organic matter turnover. *Plant and Soil 216:*15-25.

Bouwman, A.F. (1990). Exchange of greenhouse gases between terrestrial ecosystems and the atmosphere. In *Soils and the Greenhouse Effect,* ed. A.F. Bouwman. Chichester: John Wiley and Sons, pp. 61-129.

Brinkman, R. and W.G. Sombroek (1996). The effects of global change conditions in relation to plant growth and food production. In *Global Climatic Change and Agricultural Production: Direct and Indirect Effects of Changing Hydrological, Soil and Plant Hydrological Processes,* eds. F. Bazzaz and W.G. Sombroek. Chichester: John Wiley and Sons, pp. 48-63.

Buyanovsky, G.A., M. Aslam, and G.H. Wagner (1994). Carbon turnover in soil physical fractions. *Soil Science Society of America Journal 58:*1167-1173.

Cambardella, C.A. and E.T. Elliott (1992). Particulate soil organic matter changes across a grassland cultivation sequence. *Soil Science Society of America Journal 56:*777-783.

Christensen, B.T. (1996). Matching measurable soil organic matter fractions with conceptual pools in simulation models of carbon turnover: Revision of model structure. In *Evaluation of Soil Organic Matter Models Using Existing Long-Term Datasets,* eds. D.S. Powlson, P. Smith, and J.U. Smith. Heidelberg, Germany: Springer, pp. 143-159.

Cole, C.V., C. Cerri, K. Minami, A. Mosier, N. Rosenberg, and D. Sauerbeck (1996). Agricultural options for mitigation of greenhouse gas emissions. In *Climate Change 1995—Impacts, Adaptions and Mitigation of Climate Change: Scientific-Technical Analyses. Contribution of Working Group II to the Second Assessment Report of the Intergovernmental Panel of Climate Change,* eds. R.T. Watson, M.C. Zinyovera, and R.H. Moss. Cambridge: Cambridge University Press, pp. 745-771.

Cole, C.V., J.W.B. Steward, D.S. Ojima, W.J. Parton, and D.S. Schimel (1989). Modeling land use effects of soil organic matter dynamics in the North American Great Plains. In *Ecology of Arable Land,* eds. M. Clarholm and L. Bergström. Dordrecht, Netherlands: Kluwer Academic Publishers, pp. 89-98.

Coleman, K. and D.S. Jenkinson (1996). RothC-26.3—A model for the turnover of carbon in soil. In *Evaluation of Soil Organic Matter Models Using Existing Long-Term Datasets,* eds. D.S. Powlson, P. Smith, and J.U. Smith. Heidelberg, Germany: Springer, pp. 237-245.

Conservation Technology Information Center (1996). *1995 CRM Executive Summary.* West Lafayette, IN: Author.

Davidson, E.A. and I.L. Ackermann (1993). Changes in soil carbon inventories following cultivation of previously untilled soils. *Biogeochemistry 20:*161-164.

Delschen, T. (1999). Impacts of long-term application of organic fertilizers on soil quality parameters in reclaimed loess soils of the Rhineland lignite mining area. *Plant and Soil 213:*43-54.

Diaz, S., J.P. Grimme, J. Harris, and E. McPherson (1993). Evidence of a feedback mechanism limiting plant response to elevated carbon dioxide. *Nature 364:*616-617.

Doran, J.W., L.N. Mielke, and S. Stamatiadis (1988). Microbial activity and N cycling as regulated by soil water-filled pore space. In *Proceedings of the 11th International Soil Tillage Research Organization,* July 11-15, Paper No. 132. Edinburgh, UK: ISTRO, 1988.

Duxbury, J.M., M.S. Smith, and J.W. Doran (1989). Soil organic matter as a source and a sink of plant nutrients. In *Dynamics of Soil Organic Matter in Tropical Ecosystems,* eds. D.C. Coleman, J.M. Oades, and G. Uehara. Honolulu, HI: University of Hawaii Press, pp. 33-67.

Eden, M.J., P.A. Furley, D.F.M. McGregor, W. Milliken, and J.A. Ratter (1991). Effect of forest clearance and burning on soil properties in northern Roraima, Brazil. *Forest Ecology and Management 38:*283-290.

Eichhorn, J. and A. Hüttermann (1999). Mechanisms of humus dynamics and nitrogen mineralization. In *Going Underground—Ecological Studies in Forest Soils,* eds. N Rastin and J. Bauhus. Trivandrum, India: Research Signpost, pp. 239-277.

Elliott, E.T., K. Paustian, and S.D. Frey (1996). Modeling the measurable or measuring the modelable: A hierarchical approach to isolating meaningful soil organic matter fractionations. In *Evaluation of Soil Organic Matter Models Using Existing Long-Term Datasets,* eds. D.S. Powlson, P. Smith, and J.U. Smith. Heidelberg, Germany: Springer, pp. 161-179.

Enquete Commission (1995). *Protecting Our Green Earth. How to Manage Global Warming through Environmentally Sound Farming and Preservation of the World's Forests.* Bonn, Germany: Economia Verlag.

Ertel, J.R., J.I. Hedges, A.H. Devol, R.R. Richey, and M.N. Ribeiro (1986). Dissolved humic substances of the Amazon River System. *Limnological Oceanography 31:*739-754.

Eswaran, H., E. van den Berg, and P. Reich (1993). Organic carbon in soils of the world. *Soil Science Society of America Journal 57:*192-194.

Falloon, P.D. and P. Smith (2000). Modeling refractory soil organic matter. *Biology and Fertility of Soils 30:*388-398.

Falloon, P.D., P. Smith, J. Szabo, L. Pasztor, J.U. Smith, K. Coleman, and S. Marshall (2001). Soil organic matter sustainability and agricultural management—Predictions at the regional level. In *Sustainable Management of Soil Organic Matter,* eds. R.M. Rees, B.C. Ball, C.D. Campbell, and C.A. Watson. Wallingford, UK: CABI Publishing, pp. 54-59.

FAO (Food and Agriculture Organization) (1971-1981). *Soil Map of the World 1:5.000.000.* Volumes 1-10, Rome: FAO-UNESCO, 59 pp.

FAO (Food and Agriculture Organization) (1998). *World Reference Base for Soil Resources.* ISSS-ISRIC-FAO. World Soil Resources Report number 84. Rome, Italy: FAO, 88 pp.

Fisher, M.J., I.M. Rao, M.A. Ayarza, C.E. Lascano, J.L. Sanz, R.J. Thomas, and R.R. Vera (1994). Carbon storage by introduced deep rooted grasses in the South American Savannas. *Nature 371:*236-238.

Fisher, M.J., I.M. Rao, C.E. Lascano, I.J. Sanz, R.J. Thomas, R.R. Vera, and M.A. Ayarza (1995). Pasture soils as carbon sink. *Nature 376:*472-473.

Fog, K. (1988). The effect of added nitrogen on the rate of decomposition of organic matter. *Biological Reviews 63:*433-462.

Frangi, J.L. and A.E. Lugo (1985). Ecosystem dynamics of a subtropical floodplain forest. *Ecological Monography 55:*351-369.

Frede, H.G., R. Beisecker, and S. Gäth (1994). Long-term impacts of tillage on the soil ecosystem. *Journal of Plant Nutrition and Soil Science 157:*197-203.

Frolking, S., J.L. Bubier, T.R. Moore, T. Ball, L.M. Bellisario, A. Bhardwaj, P. Carroll, P.M. Crill, P.M. Lafleur, J.H. McCaughey, et al. (1998). The relationship between ecosystem productivity and photosynthetically active radiation for northern peatlands. *Global Biochemical Cycles 12:*115-126.

Garz, J. (1995). C- und N-Umsetzungen in Dauerversuchen auf Sandlöss-Braunschwarzerde in Halle. In *Strategien zur Regeneration belasteter Agraröko-*

systeme des mitteldeutschen Schwarzerdegebietes, eds. M. Körschens and E.G. Mahn. Stuttgart, Leipzig: Teubner, pp. 463-498.

Gaunt, J.L., S.P. Sohi, H. Yang, N. Mahieu, and J.R.M. Arah (2001). A procedure for isolating soil organic matter fractions suitable for modelling. In *Sustainable Management of Soil Organic Matter,* eds. R.M. Rees, B.C. Ball, C.D. Campbell, and C.A. Watson. Wallingford, UK: CABI Publishing, pp. 83-90.

Glenn, E.P., V.R. Squires, M.W. Olsen, and R.J. Frye (1993). Potential for carbon sequestration in the drylands. *Water, Air and Soil Pollution 70:*341-355.

Gorham, E. (1991). Northern peatlands: Role in the carbon cycle and probable responses to climatic warming. *Ecological Applications 1:*185-192.

Gouyon, A., H. de Foresta, and P. Levang (1993). Does jungle rubber deserve its name? An analysis of rubber agroforestry systems in southeast Sumatra. *Agroforestry Systems 22:*181-206.

Grant, F.R. (1997). Changes in soil organic matter under different tillage and rotations: Mathematical modeling in ecosystems. *Soil Science Society of America Journal 61:*1159-1175.

Gregorich, E.G., P. Rochette, S. McGuire, B.C. Liang, and R. Lessard (1998). Soluble organic carbon and carbon dioxide fluxes in maize fields receiving spring-applied manure. *Journal of Environmental Quality 27:*209-214.

Grubb, P.J. (1971). Interpretation of the "Massenerhebung" effect on tropical mountains. *Nature 229:*44-45.

Guggenberger, G. and G. Beudert (1989). Zur Dynamik des gelösten Kohlenstoffs (DOC) in unterschiedlich immissionsbelasteten Waldstandorten. *Mitteilungen der Deutschen Bodenkundlichen Gesellschaft 59:*367-372.

Guggenberger, G. and W. Zech (1993). Dissolved organic carbon control in acid forest soils of the Fichtelgebirge (Germany) as revealed by distribution patterns and structural composition analyses. *Geoderma 59:*109-129.

Gupta, R.K. and D.L.N. Rao (1994). Potential of wastelands for sequestering carbon by reforestation. *Curriculum Science 66:*378-380.

Hanawald, R.B. and R.H. Whittaker (1976). Altitudinally coordinated patterns of soils and vegetation in the San Jakinto Mountains, California. *Soil Science 121:*114-124.

Hansen, S., H.E. Jensen, N.E. Nielsen, and H. Svendsen (1991). Simulation of nitrogen dynamics and biomass production in winter wheat using the Danish simulation model DAISY. *Fertilizer Research 27:*245-259.

Hao, W.M. and M.H. Liu (1995). Spatial and temporal distribution of tropical biomass burning. *Global Biogeochemical Cycles 8:*495-503.

Harper, S.H.T. and J.M. Lynch (1981). The kinetics of straw decomposition in relation to its potential to produce the phytotoxin acetic acid. *Journal of Soil Science 32:*627-637.

Harrison, A.F., D.D. Harkness, and P.J. Bacon (1990). The use of bomb-^{14}C for studying organic matter and N and P dynamics in a woodland soil.,In *Nutrient Cycling in Terrestrial Ecosystems: Field methods, application and interpretation,* eds. A.F. Harrison, P. Ineson, and O.W. Heal. London, New York: Elsevier Applied Sciences.

Hassink, J. (1994). Active organic matter fractions and microbial biomass as predictors of N mineralization. In *Nitrogen Mineralization in Agricultural Soils,* eds. J.J. Neeteson and J. Hassink. Wageningen, Netherlands: AB-DLO, pp. 1-15.

Hassink, J. and A.P. Whitmore (1997). A model of the physical protection of organic matter in soils. *Soil Science Society of America Journal 61:*131-139.

Hénin, S. and M. Dupuis (1945). Essai de bilan de la matière organique du sol. *Annals of Agronomy 15:*17-29.

Herbert, B.E. and P.M. Bertsch (1995). Characterization of dissolved and colloidal organic matter in soil solution: A review. In *Carbon Forms and Functions in Forest Soils,* eds. J.M. Kelly and W.W. McFee. Madison, WI: SSSA, pp. 63-88.

Holdridge, L.R. (1947). Determination of world plant formations from simple climate data. *Science 105:*367-368.

Houghton, R.A. and D.L. Skole (1990). Carbon. In *The Earth as Transformed by Human Action,* eds. B.L. Turner, W.C. Clark, R.W. Kates, J.F. Richards, J.T. Mathews, and W.B. Meyer. Cambridge, UK: Cambridge University Press, pp. 393-408.

Houot, S., R. Chaussod, C. Hounemenou, E. Barriusos, and S. Bourgeois (1991). Differences induced in the soil organic matter characteristics and microbial activity by various management practices in long-term field experiments. In *Diversity of Environmental Biogeochemistry,* ed. J. Berthelin. Amsterdam, Netherlands: Elsevier, pp. 435-443.

Hunt, H.W. (1977). A simulation model for decomposition in grassland. *Ecology 58:*451-455.

IAEA (International Atomic Energy Agency) (1968). Isotopes and radiation in soil organic matter studies. Vienna, Austria: FAO/IAEA.

Idso, K.E. and S.B. Idso (1994). Plant responses to atmospheric CO_2 enrichment in the face of environmental constraints: A review of the past 10 years' research. *Geoderma 69:*153-203.

Ionenko, V.I., A.A. Batsula, and Y.A. Golovachev (1986). Kinetics of humification. *Soviet Soil Science 18 (2):*33-42.

Isermann, K. and R. Isermann (1999). Bodenkundliche Anforderungen an die fachliche Praxis einer nachhaltigen Landwirtschaft/Landnutzung aus der Sicht ihrer Nährstoffhaushalte. *Mitteilungen der Deutschen Bodenkundlichen Gesellschaft 91:*59-62.

Janssen, B.H. (1984). A simple method for calculating decomposition and accumulation of "young" soil organic carbon. *Plant and Soil 76:*297-304.

Jastrow, J.D. (1996). Soil aggregate formation and the accrual of particulate and mineral-associated organic matter. *Soil Biology and Biochemistry 28:*665-676.

Jenkinson, D.S. (1971). Studies on the decomposition of ^{14}C-labelled organic matter in soil. *Soil Science 111:*64-70.

Jenkinson, D.S. (1977). Studies on the decomposition of plant material in soil. V. *Journal of Soil Science 28:*424-434.

Jenkinson, D.S. (1990). The turnover of organic carbon and nitrogen in soil. In *Philosophical Transactions of Royal Society of London B 329.* London: Royal Society of London, pp. 361-368.

Jenkinson, D.S. and A. Ayanaba (1977). Decomposition of carbon-14 labelled plant material under tropical conditions. *Soil Science Society of America Journal 41:*912-915.

Jenkinson, D.S., N.J. Bradbury, and K. Coleman (1994). How the Rothamsted Classical Experiments have been used to develop and test models of the turnover of carbon and nitrogen in soil. In *Long-term Experiments in Agricultural and Ecological Sciences,* eds. R.A. Leigh and A.E. Johnston. Wallingford, UK: CABI Publishing, pp. 117-138.

Jenkinson, D.S., P.B.S. Hart, J.H. Rayner, and L.C. Parry (1987). Modeling the turnover of organic matter in long-term experiments at Rothamsted. *INTECOL Bulletin 15:*1-8.

Jenkinson, D.S. and A.E. Johnston (1977). Soil organic matter in the Hoosfield continuous barley experiment. *Report, Rothamsted Experimental Station 1976,* ed. Rothamsted Experimental Station. Part 2, pp. 87-101.

Jenkinson, D.S. and J.H. Rayner (1977). The turnover of soil organic matter in some of the Rothamsted classical experiments. *Soil Science 123:*298-305.

Jenny, H. (1941). *Factors of Soil Formation.* New York: McGraw-Hill.

Jensen, B. (1993). Rhizodeposition by $^{14}CO_2$-pulse-labelled spring barley grown in small field plots on sandy loam. *Soil Biology and Biochemistry 25:*1553-1559.

Jensen, B. (1994). Rhizodeposition by field-grown winter barley exposed to $^{14}CO_2$ pulse-labelling. *Applied Soil Ecology 1:*65-74.

Johnen, B.G. and D.R. Sauerbeck (1977). A tracer technique for measuring growth, mass and microbial breakdown of plant roots during vegetation. In *Soil Organisms as Components of Ecosystems,* eds. U. Lohm and T. Persson. Stockholm, Sweden: Ecological Bulletins, pp. 366-373.

Jongen, M., M.B. Jones, T. Hebeisen, H. Blum, and G. Hendrey (1995). The effects of elevated CO_2 concentration on the root growth of *Lolium perenne* and *Trifolium repens* grown in a FACE system. *Global Change Biology 1:*361-371.

Kalbitz, K., S. Solinger, J.H. Park, B. Michalzik, and E. Matzner (2000). Controls on the dynamics of dissolved organic matter in soils: A review. *Soil Science 165:*277-304.

Kersebaum, K.C. and O. Richter (1994). A model approach to simulate C and N transformations through microbial biomass. In *Nitrogen Mineralization in Agricultural Soils,* eds. J.J. Neetson and J. Hassink. Wageningen, Netherlands: AB-DLO, pp. 221-229.

Klein Goldewijk, K., J.G. van Minnen, G.J.J. Kreilemann, M. Vloedbeld, and R. Leemans (1994). Simulating the carbon flux between the terrestrial environment and the atmosphere. *Water, Air and Soil Pollution 76:*199-230.

Knapp, E.B., L.F. Elliott, and G.S. Campbell (1983). Carbon, nitrogen and microbial biomass interrelationships during the decomposition of wheat straw: A mechanistic simulation model. *Soil Biology and Biochemistry 15:*455-461.

Knof, G. (1985). Eine Gerätekombination zur Bestimmung des Kohlenstoffs und seines ^{14}C-Anteils in Wurzeln und anderen pflanzlichen Substanzen. *Archiv für Acker- und Pflanzenbau und Bodenkunde 29:*23-30.

Kohlmaier, G.H., A. Janececk, and J. Kindermann (1990). Positive and negative feedback loops within the vegetation/soil system in the response to a CO_2 greenhouse warming. In *Soils and the Greenhouse Effect,* ed. A.F. Bouwmann. Chichester, UK: John Wiley and Sons, pp. 415-422.

Körschens, M. (1984). *Dauerfeldversuche der DDR—Übersicht, Entwicklung und Ergebnisse von Feldversuchen mit überwiegend mehr als 20 Jahren Versuchsdauer sowie Übersicht über wichtige Dauerversuche der Welt.* Berlin, Germany: Akademie der Landwirtschaftswissenschaften der DDR.

Körschens, M. (1995). Zur Frage optimaler Humusgehalte in Ackerböden. *VDLUFA-Schriftenreihe 40:*157-160.

Körschens, M. (1997). Dependence of soil organic matter (SOM) on location and management, and its influence on yield and soil properties (in German). *Archives of Agronomy and Soil Science 41:*435-463.

Körschens, M. and A. Müller (1996). The static experiment Bad Lauchstädt, Germany. In *Evaluation of Soil Organic Matter Models Using Existing Long-Term Datasets,* eds. D.S. Powlson, P. Smith, and J.U. Smith. Berlin, Germany: Springer, pp. 369-376.

Körschens, M. and E. Schulz (1999). Die organische Bodensubstanz. *UFZ-Bericht 13:*1-46.

Kuzyakov, Y., A. Kretzschmar, and K. Stahr (1999). Contribution of *Lolium perenne* rhizode-position to carbon turnover of pasture soil. *Plant and Soil 213:*127-136.

Lal, R. (1995). Global soil erosion by water and carbon dynamics. In *Soils and Global Change,* eds. R. Lal, J. Kimble, E. Levine, and B.A. Steward. Boca Raton, FL: CRC/Lewis Publishers, pp. 131-142.

Lal, R. (2000). Soil carbon and the accelerated greenhouse effect. In *Poland Agriculture and Water Quality Protection. Proceedings of the conference "Scientific basis to mitigate the nutrient dispersion into the environment,"* ed. A. Sapek. Falenty, Poland: Falenty IMUZ Publisher, pp. 106-118.

Lal, R. and T.J. Logan (1995). Agricultural activities and greenhouse gas emissions from soils of the tropics. In *Soil Management and Greenhouse Effect,* eds. R. Lal, J. Kimble, E. Levine, and B.A. Steward. Boca Raton, FL: CRC/Lewis Publishers, pp. 293-307.

Lee, J.J., D.L. Phillips, and R. Liu (1993). The effects of trends in tillage practices on erosion and carbon content of soils in the US Corn Belt. *Water, Air, and Soil Pollution 70:*389-401.

Li, C., S. Frolking, and T.A. Frolking (1992). A model of nitrous oxide evolution from soil driven by rainfall events: Model structure and sensitivity. *Journal of Geophysical Research (Atmospheres) 97:*9759-9776.

Li, Z. and Q. Zhao (1998). Carbon dioxide fluxes and potential mitigation in agriculture and forestry of tropical and subtropical China. *Climatic Change 40:*119-133.

Lugo, A.E. and S. Brown (1993). Management of tropical soils as sinks or sources of atmospheric carbon. *Plant and Soil 149:*27-41.

Lynch, J.M. (1990). *The Rhizosphere.* Chichester, UK: Wiley Interscience.

Malik, K.A. and K. Haider (1977). Decomposition of carbon-14-labelled plant materials in saline-sodic soils. In *Soil Organic Matter Studies,* ed. International Atomic Energy Agency, IAEA-SM-211/23. Vienna, Austria: IAEA, pp. 215-225.

Martin, A., A. Mariotti, J. Balesdent, P. Lavelle, and R. Vuattoux (1990). Estimate of the organic matter turnover rate in a savanna soil by the ^{13}C natural abundance. *Soil Biology and Biochemistry 22:*517-523.

Martin, J.K. and D.W. Puchridge (1982). Carbon flow through the rhizosphere of wheat crops in South Australia. In *The Cycling of Carbon, Nitrogen, Sulphur and Phosphorus in Terrestrial and Aquatic Ecosystems,* eds. I.E. Galbally and J.R. Freney. Canberra, Australia: Australian Academy of Science, pp. 77-81.

Mathers, A.S. and D.W. Goss (1979). Estimating animal waste application to supply crop nitrogen requirements. *Soil Science Society of America Journal 43:*364-366.

McCalla, T.M., J.R. Peterson, and C. Lue-Hing (1977). Properties of agricultural and municipal wastes. In *Soils for Management of Organic Wastes and Waste Waters,* eds. L.F. Elliott and F.J. Stevenson. Madison, WI: American Society of Agronomy, pp. 9-43.

McDowell, W.H. and G.E. Likens (1988). Origin, composition, and flux of dissolved organic carbon in the Hubbard Brook Valley. *Ecological Monography 58:*177-195.

McGill, W.B. (1996). Review and classification of ten soil organic matter (SOM) models. In *Evaluation of Soil Organic Matter Models Using Existing Long-Term Datasets,* eds. D.S. Powlson, P. Smith, and J.U. Smith. Heidelberg, Germany: Springer, pp. 111-132.

McGill, W.B., H.W. Hunt, R.G. Woodmansee, and J.O. Reuss (1981). PHOENIX, a model of the dynamics of carbon and nitrogen in grassland soils. *Ecology Bulletin (Stockholm) 33:*49-115.

Meesenburg, H., Meiwes, K.J., and H. Bartens (1999). Veraänderung der Elementvorräte im Boden von Buchen- und Fichtenökosystemen im Solling. *Freiburger Forstliche Forschung 7:*109-114.

Melillo, J.M., T.V. Callaghan, F.I. Woodward, E. Salati, and S.K. Sinha (1990). Effects on ecosystems. In *Climate Change: The IPCC Scientific Assessment,* eds. J.T. Houghton, G.J. Jenkins, and J.J. Ephraums. Cambridge, UK: Cambridge University Press, pp. 283-310.

Metherell, A.K., C.A. Cambardella, W.J. Parton, G.A. Peterson, L.A. Harding, and C.V. Cole (1995). Simulation of soil organic matter dynamics in dryland wheat-fallow cropping systems. In *Soil Management and Greenhouse Effect,* eds. R. Lal, J. Kimble, E. Levine, and B.A. Stewart. *Advances in Soil Science.* Boca Raton, FL: CRC Press, pp. 259-270.

Michalzik, B. and E. Matzner (1999). Fluxes and dynamics of dissolved organic nitrogen and carbon in a spruce (*Picea abies* Karst.) forest ecosystem. *European Journal of Soil Science 50:*579-590.

Mizota, C. and L.P. van Reeuwijk (1989). *Clay Mineralogy and Chemistry of Soils Formed under Volcanic Material in Diverse Climatic Regions.* Wageningen, Netherlands: ISRIC.

Molina, J.A.E., C.E. Clapp, M.J. Shaffer, F.W. Chichester, and W.E. Larson (1983). NCSOIL, a model of nitrogen and carbon transformations in soil: Description, calibration and behavior. *Soil Science Society of America Journal 47:*85-91.

Molina, J.A.E., G.J. Crocker, P.R. Grace, J. Klir, M. Körschens, P.R. Poulton, and D.D. Richter (1997). Simulating trends in soil organic carbon in long-term experiments using the NCSOIL and NCSWAP models. *Geoderma 81:*91-107.

Müller, T., J. Magid, L.S. Jensen, H. Svendsen, and N.E. Nielsen (1998). Soil C and N turnover after incorporation of chopped maize, barley straw and bluegrass in the field: Evaluation of the DAISY soil-organic submodel. *Ecological Modelling 111:*1-15.

Nambiar, K.K.M. (1994). *Soil Fertility and Crop Productivity Under Long-Term Fertilizer Use in India.* New Delhi: Indian Council for Agricultural Research.

Nicolardot, B., J.A.E. Molina, and M.R. Allard (1994). C and N fluxes between pools of soil organic matter: Model calibration with long-term incubation data. *Soil Biology and Biochemistry 26:*235-243.

Nieder, R. (1998). Bodenbearbeitung und Nährstoffaustrag. *KTBL Arbeitsbericht 266:*91-116.

Nieder, R. (2000). Nährstoffakkumulation in Ackerkrumen vor dem Hintergrund des Boden-, Klima- und Gewässerschutzes. *Zeitschrift für Kulturtechnik und Landentwicklung 41:*49-56.

Nieder, R., H.P. Dauck, and D.K. Benbi (2001). Mineralization of newly accumulated nitrogen. In *Plant Nutrition—Food Security and Sustainability of Agro-Ecosystems,* eds. W.J. Horst, M.K. Schenk, A. Bückert, N. Claassen, H. Flessa, W.B. Frommer, H. Goldbach, H.W. Olfs, V. Römheld, B. Satelmacher, et al. Dordrecht, Netherlands: Kluwer Academic Publishers, pp. 940-941.

Nieder, R. and J. Richter (1989). Die Bedeutung der Umsetzung von Weizenstroh im Hinblick auf den C- und N-Haushalt von Löss-Ackerböden. *Zeitschrift für Pflanzenernährung und Bodenkunde 152:*415-420.

Nieder, R. and J. Richter (2000). C and N accumulation in arable soils of West Germany and its influence on the environment—Developments 1970 to 1998. *Journal of Plant Nutrition and Soil Science 163:*65-72.

Nieder, R., H. Wachter, and K. Isermann (2000). Erhöhte Stoffausträge bald auch aus Waldböden? *Allgemeine Forstzeitschrift/der Wald 11:*594-599.

Nilsson, L.G. (1986). Data of yield and soil analysis in the long-term soil fertility experiments. *Journal of the Royal Swedish Academy of Agriculture and Forestry 18:*32-70.

Oldeman, L.R. (1994). The global extent of soil degradation. In *Soil Resilience and Sustainable Land Use,* eds. D.J. Greenland and I. Szabolcs. Wallingford, Oxford, UK: CABI Publishers, pp. 99-118.

Panikov, N.S., M.V. Paleyeva, S.N. Dedysh, and A.G. Dorofeyev (1991). Kinetic methods for biomass and activity determination for different groups of soil microorganisms. *Soviet Soil Science 8:*109-120.

Parton, W.J., D.S. Ojima, C.V. Cole, and D.S. Schimel (1994). A general model for soil organic matter dynamics: Sensitivity to litter chemistry, texture and management. In *Quantitative Modeling of Soil Forming Processes,* Special Publication 39, eds. R.B. Bryant and R.W. Arnold. Madison, WI: Soil Science Society of America, pp. 147-167.

Parton, W.J. and P.E. Rasmussen (1994). Long-term effects of crop management in wheat-fallow: II CENTURY model simulations. *Soil Science Society of America Journal 58:*530-536.

Parton, W.J., D.S. Schimel, C.V. Cole, and D.S. Ojima (1987). Analysis of factors controlling soil organic matter levels in Great Plains grasslands. *Soil Science Society of America Journal 51*: 1173-1179.

Parton, W.J., J.M.O. Scurlock, D.S. Ojima, D.S. Schimel, D.O. Hall, and Scope-program Group Members (1995). Impact of climate change on grassland production and soil carbon worldwide. *Global Change Biology 1:*13-22

Paustian, K. (1994). Modeling soil biology and biochemical processes for sustainable agriculture research. In *Soil Biota Management in Sustainable Farming Systems,* eds. Z.E. Pankhurst, B.M. Doube, V.V.S.R. Gupta, and P.R. Grace. Melbourne, Australia: CSIRO Information Services, pp. 182-193.

Paustian, K. (2001). Modelling soil organic matter dynamics—Global challenges. In *Sustainable Management of Soil Organic Matter,* eds. R.M. Rees, B.C. Ball, C.D. Campbell, and C.A. Watson. Wallingford, UK: CABI Publishing, pp. 43-53.

Paustian, K., G.I. Ågren, and E. Bosatta (1997). Modelling litter quality effects on decomposition and soil organic matter dynamics. In *Driven by Nature: Plant Litter Quality and Decomposition,* eds. G. Cadish and K.E. Giller. Wallingford, UK: CABI Publishing, pp. 313-335.

Paustian, K., O. Andren, H.H. Janzen, R. Lal, P. Smith, G. Tian, H. Tiessen, M. Van Noordvijk, and P.L. Woomer (1997). Agricultural soils as a sink to mitigate CO_2 emissions. *Soil Use and Management 13:*230-244.

Paustian, K., C.V. Cole, D. Sauerbeck, and N. Sampson (1998). CO_2 mitigation by agriculture: An overview. *Climate Change 40:*135-162.

Paustian, K., H.P. Collins, and E.A. Paul (1997). Management controls on soil carbon. In *Soil Organic Matter in Temperate Agroecosystems: Long-Term Experiments in North America,* eds. E.A. Paul, K. Paustian, E.T. Elliott, and C.V. Cole. Boca Raton, FL: CRC Press, pp. 15-49.

Paustian, K., W.J. Parton, and J. Persson (1992). Modeling soil organic matter in organic-amended and nitrogen-fertilized long-term plots. *Soil Science Society of America Journal 56:*476-488.

Paustian, K. and J. Schnürer (1987). Fungal growth response to carbon and nitrogen limitation: A theoretical model. *Soil Biology and Biochemistry 19:*613-620.

Persson, T. (1989). The role of animals in C and N mineralisation. In *Ecology of Arable Land,* eds. M. Clarholm and L. Bergström. Dordrecht, Boston, London: Kluwer Academic Publishers, pp. 185-189.

Peterson, G.A. and D.J. Vetter (1971). Soil nitrogen budget of the wheat-fallow area. *Farm Ranch Home 12:*23-25.

Phillipps, D.L., D. White, and C.B. Johnson (1991). *Climate Change and Soil Erosion in the United States.* Corvallis, OR: USEPA.

Pirt, S.J. (1975). *Principles of Microbe and Cell Cultivation.* New York: John Wiley and Sons.

Post, W.M., W.R. Emanuel, and A.W. King (1992). Soil organic matter dynamics and the global carbon cycle. In *World Inventory of Soil Emission Potentials,* WISE Report 3, eds. N.H. Batjes and E.M. Bridges. Wageningen, Netherlands: ISRIC, pp. 107-119.

Post, W.M., A.W. King, and S.D. Wullschleger (1996). Soil organic matter models and global estimates of soil organic carbon. In *Evaluation of Soil Organic Matter Models Using Existing Long-Term Datasets,* eds. D.S. Powlson, P. Smith, and J.U. Smith. Heidelberg, Germany: Springer, pp. 201-222.

Post, W.M., T.H. Peng, W.R. Emanuel, A.W. King, V.H. Dale, and D.L. DeAngelis (1990). The global carbon cycle. *American Science 78:*310-326.

Powers, W.L., G.W. Wallingford, and L.S. Murphy (1975). Formulas for applying organic wastes to land. *Journal of Soil and Water Conservation 30:*286-289.

Powlson, D.S. (1996). Why evaluate soil organic matter models? In *Evaluation of Soil Organic Matter Models Using Existing Long-Term Datasets,* eds. D.S. Powlson, P. Smith, and J.U. Smith. Heidelberg, Germany: Springer, pp. 3-11.

Powlson, D.S., P. Smith, and J.U. Smith (Eds.) (1996). *Evaluation of Soil Organic Matter Models Using Existing Long-Term Datasets.* Heidelberg, Germany: Springer.

Qualls, R.G., B.L. Haines, and W.T. Swank (1991). Fluxes of dissolved organic nutrients and humic substances in a deciduous forest. *Ecology 72:*254-266.

Reddy, K.R., T.C. Feijtel, and W.H. Patrick (1986). Effect of soil redox conditions on microbial oxidation of organic matter. In *The Role of Organic Matter in Modern Agriculture,* eds. Y. Chen and Y. Avnimelech. Netherlands: Martinus Nijhoff Publishers, pp. 117-155.

Reicosky, D.C., W.D. Kemper, G.W. Langdale, C.L. Douglas, and P.E. Rasmussen (1995). Soil organic matter changes resulting from tillage and biomass production. *Journal of Soil Water Conservation 50:*253-261.

Rijtema, P.E. and J.G. Kroes (1991). Some results of nitrogen simulations with model ANIMO. *Fertilizer Research 27:*189-198.

Rühlmann, J. (1999). A new approach to estimating the pool of stable organic matter in soil using data from long-term field experiments. *Plant and Soil 213:*149-160.

Ruimy, A., P.G. Jarvis, D.D. Baldocchi, and B. Saugier (1995). CO_2 fluxes over plant canopies and solar radiation: A review. *Advanced Ecology Research 26:*1-68.

Rumpel, C., I. Kögel-Knabner, and R.F. Hüttl (1999). Organic matter composition and degree of humification in lignite-rich mine soils under a chronosequence of pine. *Plant and Soil 213:*161-168.

Russell, J.S. (1975). A mathematical treatment of the effects of cropping system on soil organic nitrogen in two long-term sequential experiments. *Soil Science 120:* 37-44.

Sage, R. (1995). Was low atmospheric CO_2 during the pleistocene a limiting factor for the origin of agriculture? *Global Change Biology 1:*93-106.

Saggar, S., C. Hedley, and A.D. Mackay (1997). Partitioning and translocation of photosynthetically fixed ^{14}C in grazed hill pastures. *Biology and Fertility of Soils 25:*152-158.

Sallih, Z. and M. Pansu (1993). Modeling of soil carbon forms after organic amendments under controlled conditions. *Soil Biology and Biochemistry 25:*1755-1762.

Sauerbeck, D. (1993). CO_2-emissions from agriculture: Sources and mitigation potentials. *Water, Air, and Soil Pollution 70:*381-388.

Sauerbeck, D. (2001). CO_2 emissions and C sequestration by agriculture—Perspectives and limitations. *Nutrient Cycling in Agroecosystems 60:*253-266.

Sauerbeck, D. and M.A. Gonzales (1977). Field decomposition of carbon-14-labelled plant residues in various soils of the Federal Republic of Germany and Costa Rica. In *Soil Organic Matter Studies. Proceedings of the FAO/IAEA Symposium Volume I.* Vienna, Austria: IAEA, pp. 159-170.

Scanlon, D. and T. Moore (2000). Carbon dioxide production from peatland soil profiles: The influence of temperature, oxic/anoxic conditions and substrate. *Soil Science 165:*153-160.

Schafer, W.M. and G.A. Nielsen (1979). Soil development and plant succession on 1- to 50-year old strip mine spoils in southeastern Montana. In *Ecology and Coal Resource Development* Volume 2, ed. M.K. Wali. New York: Pergamon Press, pp. 541-649.

Schimel, D., I. Enting, M. Heimann, T. Wigley, D. Raynaud, D. Alves, and U. Siegenthaler (1994). CO_2 and the carbon cycle. In *Climate Change 1994,* eds. J.T. Houghton, L.G. Meira Filho, J. Bruce, H. Lee, B.A. Callander, E. Haites, N. Harris, and K. Maskell. Cambridge, UK: Cambridge University Press, pp. 35-71.

Schlesinger, W.H. (1977). Carbon balance in terrestrial detritus. *Annual Review of Ecology and Systematics 8:*51-81.

Schröder, D. and H. Gewehr (1977). Stroh- und Zelluloseabbau in verschiedenen Bodentypen. *Zeitschrift für Pflanzenernährung und Bodenkunde 140:*273-284.

Schulz, E. (1997). Characterization of soil organic matter (SOM) regarding the degree of decomposibility and the importance for transformation processes of nutrients and pollutans (in German). *Archives of Agronomy and Soil Science 41:*465-483.

Shaffer, M.J., K. Rojas, D.G. deCoursey, and C.S. Hebson (1992). *NTRM, a Soil-Crop Simulation Model for Nitrogen, Tillage, and Crop-Residue Management.* USDA-ARS, Conservation Research Report 34-1. Springfield, VA: National Technical Information Service.

Shields, J.A. and E.A. Paul (1973). Decomposition of ^{14}C-labelled plant material under field conditions. *Canadian Journal of Soil Science 53:*297-306.

Skole, D.L., W.H. Chomentovski, W.A. Salas, and A.D. Nobre (1994). Physical and human dimensions of deforestation in Amazonia. *Bioscience 44:*314-322.

Smith, J.U., P. Smith, and T.M. Addiscott (1996). Quantitative methods to evaluate and compare soil organic matter (SOM) models. In *Evaluation of Soil Organic Matter Models Using Existing Long-Term Datasets,* eds. D.S. Powlson, P. Smith, and J.U. Smith. Heidelberg, Germany: Springer, pp. 181-199.

Smith, O.L. (1979). An analytical model of the decomposition of soil organic matter. *Soil Biology and Biochemistry 11:*585-606.

Smith, P., D.S. Powlson, M.J. Glendining, and J.U. Smith (1997). Potential for carbon sequestration in European soils: Preliminary estimates for five scenarios using results from long-term experiments. *Global Change Biology 3:*67-79.

Smith, P., D.S. Powlson, J.U. Smith, and E.T. Elliott (eds.) (1997). Evaluation and comparison of soil organic matter models using datasets from seven long-term experiments. *Geoderma* (Special Issue) *81:*1-225.

Smith, P., J.U. Smith, D.S. Powlson, W.B. McGill, J.R.M. Arah, O.G. Chertov, K. Coleman, U. Franko, S. Frolking, D.S. Jenkinson, et al. (1997). A comparison of the performance of nine soil organic matter models using datasets from seven long-term experiments. *Geoderma* (Special Issue) *81:*153-225.

Sombroek, W.G. (1990). Soils on a warmer earth: Tropical and subtropical regions. In *Soils On a Warmer Earth,* eds. H.W. Scharpenseel, M. Schomaker, and A. Ajoub. Amsterdam, Netherlands: Elsevier, pp. 157-174.

Sombroek, W.G. (1995). Aspects of soil organic matter and nutrient cycling in relation to climate change and agricultural sustainability. In *Nuclear Techniques in Soil-Plant Studies for Sustainable Agriculture and Environmental Preservation.* Proceedings. Vienna, Austria: IAEA, pp. 15-26.

Sombroek, W.G., F.O. Nachtergaele, and A. Hebel (1993). Amounts, dynamics and sequestering of carbon in tropical and subtropical soils. *Ambio 22:*417-426.

Stevenson, F.J. (1986). *Cycles of Soil Carbon, Nitrogen, Phosphorus, Sulfur, Micronutrients.* New York: John Wiley and Sons.

Stevenson, F.J. and E.T. Elliott (1989). Methodologies for assessing the quality and quantity of soil organic matter. In *Dynamics of Soil Organic Matter in Tropical Ecosystems,* eds. D.C. Coleman, J.M. Oades, and G. Uehara. Honolulu, HI: University of Hawaii Press, pp. 78-105.

Stroo, H.F., K.L. Bristow, L.F. Elliott, R.I. Papendick, and G.S. Campbell (1989). Predicting rates of wheat residue decomposition. *Soil Science Society of America Journal 53:*91-99.

Swinnen, J., A. van Veen, and R. Merckx (1995). Carbon fluxes in the rhizosphere of winter wheat and spring barley with conventional vs. integrated farming. *Soil Biology and Biochemistry 27:*811-820.

Tenny, F.G. and S.A. Waksman (1929). Composition of natural organic materials and their decomposition in the soil: IV. The nature and rapidity of decomposition on the various organic complexes in different plant materials under aerobic conditions. *Soil Science 28:*55-84.

Terry, R.E. (1986). Nitrogen transformations in Histosols. In *The Role of Organic Matter in Modern Agriculture,* eds. Y. Chen and Y. Avnimelech. Netherlands: Martinus Nijhoff Publishers, pp. 55-69.

Theng, B.K.G., K.R. Tate, P. Sollins, N. Moris, N. Nadkarni, and R.L. Tate (1989). Constituents of organic matter in temperate and tropical soils. In *Dynamics of Soil Organic Matter in Tropical Ecosystems,* eds. D.C. Coleman, J.M. Oades, G. Uehara, NifTAL project. Honolulu, HI: University of Hawaii Press, pp. 5-32.

Tolonen, K. and J. Turunen (1996). Accumulation rates of carbon in mires in Finland and implications for climate change. *Holocene 6:*171-178.

Uhlen, G. (1991). Long-term effects of fertilizers, manure, straw and crop rotation on total-N and total-C in soil. *Acta Agriculturae Scandinavica 41:*119-127.

Unger, P.W. (1991). Organic matter, nutrient, and pH distribution in no- and conventional-tillage semiarid soils. *Agronomy Journal 83:*186-189.

van Breemen, N. and T.C.J. Feijtel (1990). Soil processes and properties involved in the production of greenhouse gases with special reference to soil taxonomic systems. In *Soils and the Greenhouse Effect,* Ed. A.F. Bouwman. Chichester: John Wiley and Sons, pp. 195-223.

van de Werf, H. and W. Verstraete (1987). Estimation of active soil microbial biomass by mathematical analysis of respiration curves: Development and verification of the model. *Soil Biology and Biochemistry 19:*253-260.

van Dijk, H. (1966). *The Use of Isotopes in Soil Organic Matter Studies.* New York: Pergamon Press.

van Dijk, H. (1982). Survey of Dutch soil organic matter research with regard to humification and degradation rates in arable land. In *Land Use Seminar on Soil Degradation, Wageningen, October 1980,* eds. D. Boels, D.B. Davies, and A.E. Johnston. Rotterdam, Netherlands: Balkeme, pp. 133-143.

van Ginkel, J.A., A. Gorrisen, and J.A. van Veen (1997). Carbon and nitrogen allocation in *Lolium perenne* in response to elevated atmospheric CO_2 with emphasis on soil carbon dynamics. *Plant and Soil 188:*299-308.

van Veen, J.A., J.N. Ladd, and M.J. Frissel (1984). Modelling C and N turnover through the microbial biomass in soil. *Plant and Soil 76:*257-274.

van Veen, J.A. and E.A. Paul (1981). Organic carbon dynamics in grassland soils. I. Background information and computer simulation. *Canadian Journal of Soil Science 61:*185-201.

Verberne, E.L.J., J. Hassink, P. De Willigen, J.J.R. Groot, and J.A. van Veen (1990). Modelling organic matter dynamics in different soils. *Netherlands Journal of Agricultural Science 38:*221-238.

Vereecken, H., M. Vanclooster, M. Swerts, and J. Diels. (1991). Simulating water and nitrogen behavior in soils cropped with winter wheat. *Fertilizer Research 27:*233-243.

Wada, K. and S. Aomine (1975). Soil development during the quaternary. *Soil Science 116:*170-177.

Wali, M.K. (1999). Ecological succession and the rehabilitation of disturbed terrestrial ecosystems. *Plant and Soil 213:*195-220.

Watson, C.A., J.M. Ross, U. Bagnaresi, G.F. Minotta, F. Roffi, D. Atkinson, K.E. Black, and J.E. Hooker (2000). Environment-induced modifications to root longevity in *Lolium perenne* and *Trifolium repens. Annals of Botany 85:*397-401.

Watson, R.T., M.C. Zinyovera, R.H. Moss, and D.J. Dokken (eds.) (1995). *Climate Change 1995. Impacts, Adaptations and Mitigation of Climate Change.* Intergovernmental Panel on Climate Change (IPCC). Cambridge, UK: Cambridge University Press.

Welte, E. and F. Timmermann (1976). Fertilité du sol e bilan de l'azote dans l'essai permanent de fumure "Ewiger Roggenbau" (culture continue de seigle) á Halle/Saale. *Annales Agronomiques 27:*721-742.

Whipps, J.M. (1990). Carbon economy. In *The Rhizosphere,* ed. J.M. Lynch. Chichester: John Wiley and Sons, pp. 59-97.

Williams, J.R. and T.R.G. Gray (1974). Decomposition of litter on the soil surface. In *Biology of Plant Litter Decomposition,* eds. C.H. Dickinson and G.J. Pugh. New York: Academic Press, pp. 611-632.

Williams, J.R., J.W. Pulman, and P.T. Dyke (1985). Assessing the effect of soil erosion on productivity with EPIC. In *Proceedings National Symposium on Erosion and Soil Productivity* (December 1984), New Orleans, LA, ed. ASAE. St Joseph, MI: American Society of Agricultural Engineers, pp. 215-226.

Winogradsky, S. (1949). *Microbiologie du Sol: Problèmes et Méthodes.* Paris: Masson et Compagnie.

Woodruff, C.M. (1949). Estimating the nitrogen delivery of soil from the soil organic matter determination as reflected by Sanbourn field. *Soil Science Society of America Proceedings 14:*208-212.

Yevdokimov, I.V., S.A. Blagodatsky, and V.N. Kudeyarov (1993). Microbiological immobilization, remineralization and plant uptake of fertilizer nitrogen. *Eurasian Soil Science 25:*64-72.

Zielke, R.C. and D.R. Christensen (1986). Organic carbon and nitrogen changes in soil under selected cropping systems. *Soil Science Society of America Journal 50:*363-367.

Zsolnay, A. (1996). Dissolved humus in soil waters. In *Humic Substances in Terrestrial Ecosystems,* ed. A. Piccolo. Amsterdam, Netherlands: Elsevier, pp. 171-223.

Chapter 14

Nitrogen Dynamics

Dinesh K. Benbi
Jörg Richter

Nitrogen (N) was isolated by the British physician Daniel Rutherford in 1772 and recognized as an elemental gas by the French chemist Antoine Laurent Lavoisier about 1776. Gaseous nitrogen makes up the largest portion of the earth's atmosphere. Nitrogen is an essential nutrient and is viewed as a central element because of its role in substances such as proteins and nucleic acids that form living material. It is a part of all the essential constituents of cells: the chlorophyll that is essential for photosynthesis, the nucleic acids, DNA and RNA in which the pattern for plant growth and development are encoded, the proteins which include the enzymes that catalyze all biochemical processes, and the cell walls that hold cells together. It has significance in environmental and health problems. As a result of nitrogen's critical roles and low supply, the management of N resources is an extremely important aspect of crop production.

Nitrogen exists in many different forms in soils, plants, animals, and the atmosphere. Three major forms of N in soil are organic, ammonium (NH_4^+), and nitrate (NO_3^-). Under undisturbed conditions, mineral N ($NH_4^+ + NO_3^-$) usually constitutes between 1 to 2 percent of total soil N. It is, nevertheless, this N that is available for direct uptake by plants. Ammoniacal N occurs in two forms: ammonium ions (NH_4^+) and ammonia gas (NH_3). At high pH the equilibrium between NH_4^+ and NH_3 is in favor of NH_3, especially at high temperatures (see section Ammonia Volatilization in this chapter). NH_4^+ can be held on negatively charged sites of clay minerals and organic compounds. This reduces its mobility in soil. NO_3^- is not held by negatively charged soil and organic matter particles. Therefore, NO_3^- in excess of plant uptake is at risk to loss by leaching. The nitrogen forms, such as ammonia (NH_3), ammonium (NH_4^+), nitrate (NO_3^-), nitrogen gas (N_2), and organic N compounds undergo several transformations in the environment as part of what is called the N cycle. An understanding of N cycle (Figure 14.1) can help to illustrate the importance of N and its fate in the environment.

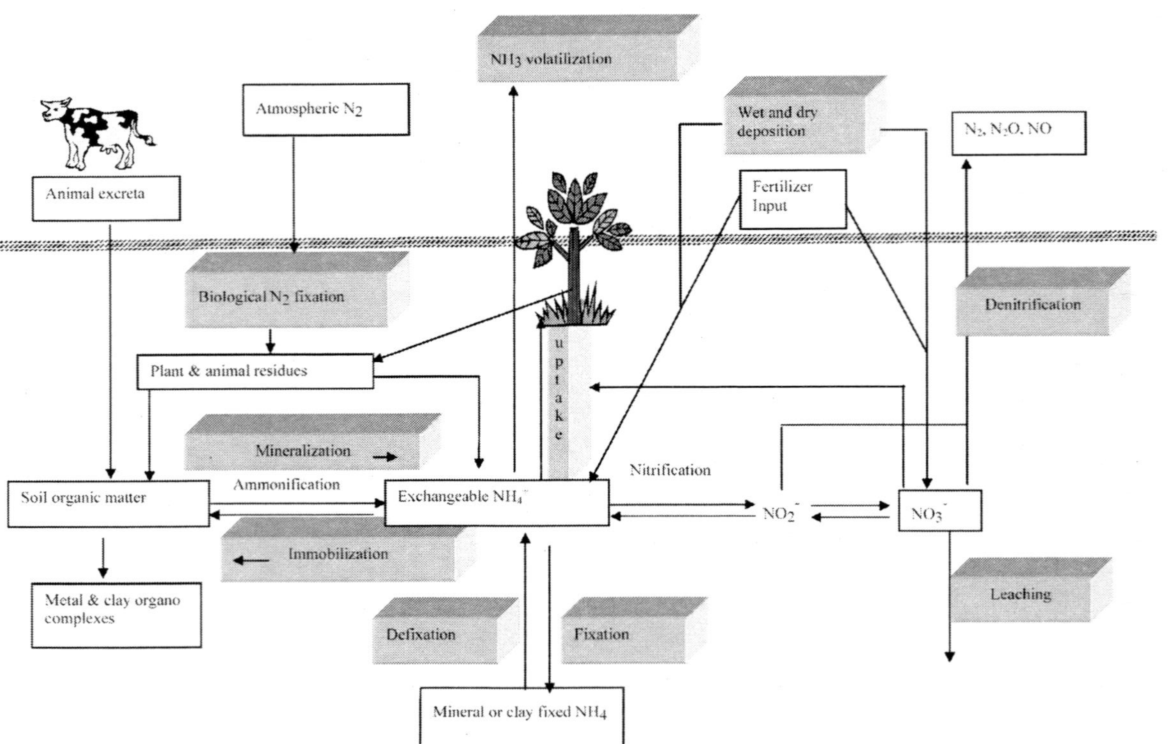

FIGURE 14.1. Nitrogen cycle in the soil-plant-atmosphere system.

Stevenson (1986) and Wilson (1988) have reviewed the N cycle and its component processes. The N cycling processes include atmospheric wet and dry deposition, biological nitrogen fixation by legumes, fertilization (N input processes), erosion and runoff, leaching, ammonia volatilization, denitrification (N loss processes), mineralization and immobilization, nitrification, fixation and release of NH_4^+ (N transformation processes), and plant uptake.

ASSESSMENT OF SOIL N POOLS

Global Pool Sizes

Table 14.1 (adapted from Jenkinson, 1990) gives estimates of the quantities of N in the global N cycle. Atmosphere contains the largest amount of N (3.9×10^9 million tonnes). Total N in an agrosystem can be distributed over the three main pools: soil, plant, and animal. Soils form a major repository of N within terrestrial ecosystems. The N in the soil is mainly in the organic form. Soil organic matter accounts for 1.5×10^5 million tonnes, which is an order of magnitude greater than that in plant biomass. Estimates of soil N

TABLE 14.1. Estimates of the Active Pools in the Global N Cycle

	Million tonnes N
Air	
N_2	3.9×10^9
N_2O	1.4×10^3
Land	
Plants	1.5×10^4
Animals	2.0×10^2
Soil organic matter	1.5×10^5
Sea	
Plants	3.0×10^2
Animals	2.0×10^2
In solution or suspension	1.2×10^6
Dissolved N_2	2.2×10^7

Source: Adapted from Jenkinson, 1990, p. 56.

pools for different climate and vegetation classes are presented in Table 13.1 in Chapter 13.

The major additions of N to the soil occur through the processes of wet and dry deposition and by the action of microorganisms that fix atmospheric N_2. As is apparent from Figure 14.1, soil and atmosphere interact through the movement of NH_3, N_2O, NO, and N_2 gases. There is a bidirectional exchange of NH_3 between the biosphere and the atmosphere and NH_3 flux depends on the difference between the surface concentration and the concentration in the air overlying that surface. Emission occurs if the former exceeds the latter (Asman, Sutton, and Schjørring, 1998). Total global NH_3 emissions from different sources have been estimated to be of the order of 50 Mt N y^{-1} with highest emission densities occurring in India, China, and Western Europe (Schlesinger and Hartley, 1992; Bouwman et al., 1997). Total global NO_x emissions are estimated to be of the order of 40 Mt N y^{-1} (Lee et al., 1997).

Nitrogen Content in Different Soils

Estimates of soil N content in the upper 30 and 100 cm of the FAO-UNESCO soil units are presented in Table 13.2 in Chapter 13. The relationship between soil organic C and N content varies in different soils, yielding widely varying C/N ratios. Generally, the C/N ratios are narrower for Xerosols and wider in Histosols. The C/N ratios for Xerosols and Histosols range from 9.9 to 25.8 in the top 30 cm soil, 9.2 to 29.8 in the 30-50 cm depth, and 7.0 to 22.3 in the 50-100 cm depth, respectively. Mean N content in the upper 100 cm of various soils ranges from 0.52 kg N m^{-2} for sandy Arenosols to 4.01 kg N m^{-2} for Histosols. Small amounts of N occur in Xerosols (0.58 kg N m^{-2}) and Yermosols (0.37 kg N m^{-2}) from the arid regions where plant growth is limited (see Table 13.2).

NITROGEN INPUT PROCESSES

Wet and Dry Deposition

Nitrogen can be added to the soil from the atmosphere through rain, snow, and hail (wet deposition) or dust and aerosols (dry deposition). This input originates mainly from previously emitted NH_3 and NO_x. Deposition occurs in the form of NH_3 and NH_4^+ (collectively termed NH_x) and as NO_x and its reaction products: gaseous nitric acid (HNO_3) and particulate NO_3^-. Dry deposition of NH_3 is most important close to a source and wet deposition of NH_4^+ is most important some distance downwind from the source.

Far from the source, the deposition of NH_4^+ is on an annual average halved approximately every 400 km (Ferm, 1998). In parts of Europe with high NH_3 emissions, such as the Netherlands, Belgium, and Denmark, dry deposition of NH_3 represents the largest contribution to total NH_x deposition. In countries with low NH_3 emission densities only wet deposition of NH_4^+ from remote sources dominates the deposition (Asman, Sutton, and Schjørring, 1998).

The quantities of NO_3^- and NH_4^+ in precipitation over land have increased during the last century. Measurements of precipitation composition at the Rothamsted Experimental Station have shown an increase of NO_3^- and NH_4^+ -N in precipitation from 1 and 3 kg N ha^{-1} y^{-1}, respectively, in 1855 to a maximum of 8 and 10 kg N ha^{-1} y^{-1} in 1980, decreasing to 4 and 5 kg N ha^{-1} y^{-1} in 1995 (Goulding et al., 1998). Goulding (1990) estimates that 50-60 kg N ha^{-1} y^{-1} enters cereal growing systems at Rothamsted. This input is partly offset by gaseous losses giving a net annual input of about 40 kg N ha^{-1} y r^{-1}. Recently, Goulding et al. (1998) computed the total deposition of all N species to winter cereals at Rothamsted to be 43.3 kg N ha^{-1} y^{-1}, 84 percent as oxidized species, 79 percent as dry deposited. Nitrogen deposition to woodlands is likely to be up to 100 kg N ha^{-1} y^{-1} and in some cases even more. The fate of this deposited N has been estimated using computer simulation models for the N cycle of the Broadbalk Continuous Wheat Experiment at Rothamsted, United Kingdom. These show that out of 45 kg N deposited ha^{-1} y^{-1}, around 5 percent is leached, 12 percent is denitrified, 30 percent is immobilized in the soil organic matter, and 53 percent is taken off in the crop (Goulding et al., 1998). Thus, deposited N makes a significant contribution to environmental pollution especially under woodlands, being emitted back to the environment as nitrate and nitrous oxide.

The deposition of N can lead to negative effects on ecosystems, which are determined by the magnitude of deposition and the type of ecosystem. Excessive deposition of NH_x to oligotrophic ecosystems may lead to a shift of plant species to more nitrophilic ones (Bobbink et al., 1992). The major effects of deposited N include soil acidification (leading to nutrient imbalances and mobilizing aluminium and toxic metals), saturated woodland ecosystems, increased nitrous oxide emissions, reduced methane oxidation rates, and altered balance of nitrification and mineralization/immobilization. The negative effects can essentially be reduced by controlling the emissions.

In view of the problems with measurements of long-distance transport of N species and large spatial and temporal variability, it is necessary to have atmospheric transport models with chemical reactions and deposition rates.

Fertilization

Fertilizers are used on farms to maintain optimum soil fertility and to correct known plant nutrient deficiencies. Plants take up N either as NH_4^+ or NO_3^-. The inorganic nitrogenous fertilizers may contain one or the other ion or both. Urea forms of fertilizers contain N as amide which in the soil is rapidly converted into NH_4^+. The common nitrogenous fertilizers are divided into six groups: ammoniacal fertilizers, nitrate fertilizers, ammonium- and nitrate-containing fertilizers, amide fertilizers, nitrogen solutions or liquid fertilizers, and slowly available or slow-release fertilizers. The choice of fertilizer material and amount, time, and method of its application varies according to the differing need and perceptions of the soils and crops, and to factors controlling transformations and utilization of fertilizer N. Urea is the most common form of inorganic fertilizer used in agricultural systems. Urea either from animal urine or applied as fertilizer is rapidly hydrolyzed to ammonium carbonate (Equation 14.1). In most soils, this reaction is catalyzed by the enzyme urease:

$$H_2N\text{-}CO\text{-}NH_2 + 2H_2O \rightarrow (NH_4)_2CO_3 \tag{14.1}$$

All soils seem to contain the enzyme urease, although the activity varies from soil to soil. Plants and plant litter on the soil surface appear to contribute to the urease activity in the soil. Urease activity and thus the rate of urea hydrolysis is strongly related to the organic matter content and the cation exchange capacity of the soil. Urease activity is markedly affected by temperature, pH, and urea concentration and is decreased by air drying and rewetting of the soils. The ammonium ions released on urea hydrolysis are rapidly nitrified to NO_3^- in most soils.

The continued use of ammoniacal fertilizers for protracted periods of time leads to the acidification of soil with attendant yield losses due to aluminium (Al) and manganese toxicities. The acidity arises during the nitrification process and the quantity of acidity produced depends on the particular N carrier used. For further reading on soil acidification refer to Chapter 7.

The general philosophy on fertilizer N usage is to determine or estimate N availability from all other sources, estimate crop N requirement, and apply sufficient fertilizer N to meet any deficit that might exist between these two numbers. As it is not possible to determine N availability from all the sources and their potential losses from the soil-plant system, simulation models are used to predict fertilizer N requirements of crops and optimize nutrient allocation from different sources. Empirical approaches such as nutrient production functions are also used for computing fertilizer requirements of crops under different soil, climate, and management conditions

(Benbi et al., 1993; Benbi and Brar, 1994). For a detailed description of approaches for optimal allocation of nutrients refer to Prihar et al. (2000).

Biological Nitrogen Fixation

Biological nitrogen fixation (BNF) of atmospheric N_2 (dinitrogen) is accomplished by procaryotic organisms living freely (nonsymbiotic fixation) or in association with higher plants (symbiotic fixation). Almost one-quarter of the estimated total fixation is accomplished by the root nodule bacterium, *Rhizobium,* in association with agriculturally important forages and grain legumes. The remainder is fixed by blue-green algae, bacteria, and actinomycetes living either freely in soil or in association with aquatic ferns, grasses, or shrubs.

Equation 14.2 illustrates the overall reaction in the enzymic reduction of atmospheric N_2 to NH_3. The enzyme complex responsible for this reduction is extremely sensitive to oxygen.

$$\underset{\text{(Dinitrogen)}}{N_2} \xrightarrow[2e^-]{2H^+} \underset{\text{(Diamide)}}{NH{=}NH} \xrightarrow[2e^-]{2H^+} \underset{\text{(Hydrazine)}}{H_2N - NH_2} \xrightarrow[2e^-]{2H^+} \underset{\text{(Ammonia)}}{2NH_3} \qquad (14.2)$$

Cobalt and copper are essential trace elements for host legumes and their associated bacteria (Riley and Dilworth, 1985; Seliga, 1993). Their absence may restrict development of free-living *Rhizobia* in rhizoshere, growth and nodulation of host plant, and impair the nodule functions. Considerable energy is required by nodulated legumes to support fixation which in the nitrate-fed plant would be targeted to produce more photosynthetic capacity (Mahon and Child, 1979). A cost of fixation of 6.5 g C g^{-1} N (Mahon, 1979; Ryle, Powell, and Gordon, 1979) translates into a theoretical loss of 15 to 20 kg dry matter for every kg N fixed (LaRue and Patterson, 1981). However, extra cost of N_2 fixation can be safely carried by most field-grown crops with little loss of production (Herridge and Bergersen, 1988).

The role of N_2 fixation in the cycling of N in agricultural systems remains undefined. A wide range of values has been presented on the amount fixed by different legumes. The amounts range from 0-320 kg N ha^{-1} for soybean, 0-125 kg for common bean, 47-120 kg for cowpea, 94-222 kg for groundnut, and 13-69 kg for pigeon pea (Herridge and Bergersen, 1988). Burns and Hardy (1975) calculated average fixation to be approximately 140 kg N ha^{-1} y^{-1}. LaRue and Patterson (1981) proposed an average value of 75 kg N ha^{-1} y^{-1}. On a global scale, estimates of BNF range from 44 to 200 Tg (=10^{12} g) N y^{-1} (Søderlund and Rosswall, 1982); 140 Tg N y^{-1} being the most commonly used estimate.

When the amount of crop N derived from fixation exceeds that harvested with the seed, legume cropping also represents a means of stabilizing, and in some cases improving, the N fertility of the soil. This can result in lowering of fertilizer N use ranging from 50-100 kg N ha^{-1} for the following crop (Herridge and Bergersen, 1988). Fixed N enters the soil-N pool as stable organic N and is resistant to short-term losses via leaching and denitrification. Accurate assessment of N_2 fixation in the field can help in efficient N management. Though it is difficult to estimate the amount of N_2 fixed by a legume crop, advances in methodology in recent years have allowed estimates of seasonal fixation to be made for most of the common species. These measurements can help in the development and calibration of models for predicting BNF.

NITROGEN LOSS PROCESSES

Erosion and Runoff

Soil erosion by wind and/or water leads to the loss of topsoil that contains most of the nutrients essential for plant growth. Erosion loss of N is important both to on-site productivity and off-site environmental impacts. Nutrient loss with sediment increases as slope increases. Erosion losses of total N from cultivated crop systems vary widely from 1 to 100 kg N $ha^{-1}y^{-1}$. Depending on the general fertility level, annual N loss as high as 124 kg ha^{-1} has been reported (Peng, Wang, and Yu, 1996). More than 90 percent of this N loss occurred as organic matter in eroded sediments and organic suspensions (Rose and Dalal, 1988). The nutrients in the sediment mainly move adsorbed to soil particles. Nitrogen can also be lost in runoff water in soluble form, mostly as nitrate. Total N in eroded sediments depends on many factors including the concentration in the source and the mix and extent of erosion processes and their effect on selective sizes. Generally the N concentration in eroded sediments is higher than that of the parent soil, and the enrichment ratio (ER, i.e., nutrient concentration in sediment/concentration in soil) declines as sediment yield increases (Rose and Dalal, 1988). The concentration of soil nutrients depends on the concentration of the sediment in the runoff water, and is 1 to 2 orders of magnitude greater than that in runoff (Liu, Li, and Zhao, 2000). Practices that affect soil erosion will have an impact on N losses in runoff.

Tillage is one of the agricultural practices that has a major influence on N losses in runoff. The losses are generally smaller with no till as compared to conventional or reduced tillage. Chichester and Richardson (1992) found that N losses in runoff were 4 and 8 kg ha^{-1} y^{-1} for no till and conventional till, respectively. Similarly, Sharpley et al. (1991) reported runoff N losses in sorghum watersheds to be 7.3, 1.0, and 0.8 kg N ha^{-1} y^{-1} with conventional, reduced, and no tillage, respectively. For an elaborate discussion on the process of erosion and its modeling see Chapter 9.

Leaching

The movement of water through soil can result in the transport of N down out of the rooting zone of the plant. This process of N loss is called leaching and it usually occurs when nitrogen is in the nitrate (NO_3^-) form since negatively charged nitrate moves freely with the soil water unless the soils have significant anion exchange capacity. Nitrogen leaching can be of serious environmental concern in areas where N can enter groundwater. Nitrates in drinking water are considered harmful at concentrations >10 mg NO_3^--N l^{-1} (milligram $liter^{-1}$). Leaching of N also reduces the efficiency of fertilizer use since N will no longer be available for plant uptake. The rate and extent of NO_3^- loss through leaching depend on climatic, soil, plant, and management factors. Among the climatic factors, rainfall, evaporation, and temperature are the most important, which may affect directly through their effect on the downward flux of water or indirectly through their effect on soil NO_3^- content, and hence on the NO_3^- concentration of the leaching water. Among the soil factors, soil texture and soil structure interact to influence leaching of NO_3^-. Generally, NO_3^- leaches more rapidly from sandy than clay soils. Kolenbrander (1969) reports that the amount of NO_3^- leached decreases from 60 kg N $ha^{-1}y^{-1}$ for soils with <10 percent clay to 10 kg N ha^{-1} y^{-1} for soils with >35 percent clay. The nature of soil determines the time taken for the leached NO_3^- to reach the groundwater. In clayey soils with a deep water table, it may take years for the downward moving NO_3^- to reach groundwater, whereas in sandy soils with a high water table, it may reach the groundwater in a few days or weeks. The effect of texture is greatly modified by soil structure and the microscale distribution of NO_3^- in the soil (White, 1985). If NO_3^- is held within soil aggregates it will be protected from leaching when bypass or preferential flow occurs (Thomas and Phillips, 1979), however, if NO_3^- is held on the outside of aggregates, bypass flow will cause it to leach faster than it would by uniform displacement (Addiscott and Cox, 1976).

There is a linear relationship between leaching and N fertilization (Benbi, 1990). Walther (1989) has surveyed the results of long-term experiments from 64 experimental stations with regard to the amount of N leached under permanent grassland and arable crop rotations. The leaching below grassland was independent of fertilization up to 200 kg N ha^{-1}, above that the leaching increased linearly. In contrast, under arable land, the leaching increased linearly from zero with the amount of fertilization. Kolenbrander (1981) has developed diagrams based on a number of fertilization experiments to show the relationship between the amount of leaching and fertilizer on the basis of groundwater recharge of 300 mm y^{-1}. Nitrate leaching is likely to be higher in agricultural systems based to a greater extent on organic inputs than those in which a larger proportion of the crop's N requirement is met from accurate and well-timed applications of inorganic fertilizers. Powlson et al. (1989) have shown that nitrogen balances for Broadbalk and Hoosfield experiments in the United Kingdom indicate large losses of N where farmyard manure (FYM) has been applied for a long period. Nitrogen balances from Broadbalk winter wheat and Hoosfield spring barley experiments showed that 124 kg NO_3-N ha^{-1} would be leached from the FYM treatment compared to 25 kg ha^{-1} from the inorganic fertilizer treatment. Therefore, agricultural systems based to a greater extent on organic inputs are likely to cause more nitrate pollution.

Measures to reduce the leaching loss of NO_3^- below the root zone include balanced fertilization (Benbi, Biswas, and Kalkat, 1991), manipulation of water applications and rooting depth, appropriate cropping sequence, and the use of slow-release fertilizers and nitrification inhibitors (Prihar et al., 2000).

Modeling Nitrate Movement in Soils

Although a full discussion on solute transport processes and modeling approaches is presented in Chapter 11, only a brief description of the underlying processes and their modeling is presented in this chapter. The movement of NO_3^- through soil is governed by three mechanisms viz. (1) convection or mass flow with the moving water (Equation 14.3), (2) diffusion (Equation 14.4), and (3) hydrodynamic dispersion:

$$J_{S_c} = J_w C \tag{14.3}$$

where J_{S_c} is solute transported by convection (mass per unit area per unit time), J_w is the soil solution or water flux, and C is the solute concentration (mass per solution volume). Equation 14.3 is generally called the piston-flow model and works best in homogenous, structureless soils.

The diffusive flux of solute J_{S_D} in one dimension is described by Fick's law of diffusion, which may be written as (Equation 14.4):

$$J_{S_D} = -D_{SW} \partial C / \partial Z \tag{14.4}$$

where D_{sw} is the binary diffusion coefficient of solute in water. In soil this equation must be modified by including a tortuosity factor to consider the effect of decreased cross-sectional area and increased path length that a solute must move through in the soil. Hydrodynamic dispersion occurs as a consequence of averaging the soil water flux in the convective transport of the solute. The magnitude of this term depends greatly on the space scale over which convection has been averaged (Jury and Nielsen, 1989).

A number of approaches have been used to model the transport of NO_3^- through soil and have been reviewed in detail by Addiscott and Wagenet (1985). The classical physical approach lies in using the Richards equation to model flow of water and the convection-dispersion equation to derive the flow of solute from it. Jury and Nielsen (1989) divide modeling approaches into process models and stochastic models. Process models develop a description of transport based on mass conservation and flux laws, leading to differential equations to predict values of water and solute variables as functions of position and time. These models include description of water and solute flow processes at different levels of complexity (see Chapter 11) and may include convective-dispersive flow or piston flow. In contrast, stochastic models describe the variables as random functions that depend on distribution of values of the soil properties that determine their movement. Stochastic models predict concentration averages and variance, and are used to calculate the probability of having a given value at a given depth or time (Jury and Nielsen, 1989). Two types of stochastic models, e.g., the transfer function model and the Monte Carlo model have been used to model NO_3^- transport in soils. In the transfer function approach, the volume of soil between the soil surface and the outflow surface where solute is monitored is characterized with regard to its solute transport properties by the solute travel time probability density functions (pdf). The transfer function approach has the advantage in that it is general enough to encompass all linear transport models and it does not require the direct measurement of soil water transport properties (Jury and Nielsen, 1989). These models attempt to simulate spatially variable field processes with minimum input data. White (1987) has applied transfer function models to predict nitrate and chloride leaching to mole and tile drains under one ha fertilized field over several seasons. In the Monte Carlo approach, simulations for a specific process model, such as the convection-dispersion equation, are run by drawing values of the parameters randomly from their probable distributions and the field-averaged solute concentrations.

Addiscott and Wagenet (1985) classify solute leaching models into the following categories: (1) models that are deterministic and mechanistic, (2) models that are deterministic and functional, (3) models that are stochastic and mechanistic, and (4) nonmechanistic stochastic (transfer function) models. Models in the first category consider solute transport in soil to be the result of mass flow and diffusion and are based on the convection-dispersion equation. They are usually based on rate parameters. Examples of such models are Nielsen and Biggar (1962), van Genuchten and Wierenga (1976), and Childs and Hanks (1975). Models that are deterministic and functional embrace a variety of methods for simplifying the mechanistic approach such as considering piston flow, ignoring the effect of dispersion and diffusion, and mass conservation of solute and water in soil layers, etc. Examples of these models are Tanji et al. (1972), Burns (1974), and Addiscott (1977). Stochastic models are designed to accommodate variability of hydraulic properties. Models that are stochastic and mechanistic use randomly distributed input values in a mechanistic framework to produce a distribution of output values. The models in this category provide a conceptual framework for the development of further mechanism-based stochastic models. Examples of such models offered are Dagan and Bresler (1979) and Amoozegar-Fard, Nielsen, and Warrick (1982). The nonmechanistic stochastic models are based on the application of transfer functions to hydrological processes.

Nitrate leaching primarily depends on drainage rates and on evapotranspiration, soil texture, and fertilization rates and distribution. Correct simulation of leaching, therefore, depends primarily on the quality of estimates of evapotranspiration (ET) and to a much lesser extent on the quality and method of modeling water movement. There are principally two ways of modeling water movement: the "rate" approach and the "capacity" approach (Addiscott and Wagenet, 1985). A rate model for solute movement combines the description of several transport processes. It first defines the instantaneous rate of change of water content in terms of the product of a hydraulic gradient and a rate parameter, the hydraulic conductivity, and then defines the rate of change of solute concentration in terms of two other rate processes, convection and diffusion. A capacity model defines changes in amounts of solute and water content by using capacity factors such as volumetric water content at field capacity. Rate models are driven by time whereas capacity models are usually driven by the amounts of rainfall, irrigation, or evaporation. Usually deterministic-mechanistic models are based on rate parameters, while deterministic-functional models are based on capacity parameters (Addiscott and Wagenet, 1985). The capacity-type models were found to be slightly superior to the mechanistic ones in simulating water movement and storage (De Willigen, 1991). In contrast, Jabro et al. (1995) did not find any improvement in prediction of NO_3^- leaching by including capacity type approach to the N version of model LEACHM.

Although widely variable approaches have been adopted for modeling nitrate leaching, few data sets are available for testing a range of models and few models have been tested on a range of soil types (Addiscott and Wagenet, 1985). Soil variability appears to be the major problem with modeling and measurement of nitrate leaching. Addiscott (1996) calls these parallel problems. Because of the small-scale heterogeneity that gives rise to mobile and immobile categories of water, both measurements and modeling are easiest in homogenous sandy soils and most difficult in strongly structured clay soil. There are also parallels at plot and field scale in the problems caused to experimenters by lognormal distributions of nitrate concentrations and those caused to modelers by non-linearty in models (Addiscott, 1996).

Ammonia Volatilization

When urea and ammonium forms of fertilizers (such as ammonium nitrate, ammonium sulfate, ammoniated phosphates, and anhydrous ammonia) are applied on moist soil surfaces, they may undergo a series of chemical conversions to ammonia. The ammonia gas then escapes to the atmosphere rather than becoming a plant nutrient. This loss process, termed volatilization, is governed by a number of equilibria that may be represented as (Equation 14.5):

$$NH^+_{4(exch)} \longleftrightarrow NH^+_{4(soil\ solution)} \xleftrightarrow{K} NH_{3(soil\ solution)} \xleftrightarrow{K_H} NH_{3(gas,\ soil)} \xleftrightarrow{K_a} NH_{3(atmosphere)} \quad (14.5)$$

Factors that promote these reactions toward the right will increase ammonia (NH_3) volatilization. The equilibrium between ammonium (NH_4^+) and NH_3 in solution can be represented as (Equation 14.6):

$$NH_4^+ \underset{K}{\longleftrightarrow} NH_3 + H^+ \quad (14.6)$$

The value of equilibrium constant *(K)* depends on temperature. Different empirical relationships have been presented to calculate its value at a given temperature (e.g., Equation 14.7; Emerson et al., 1975; Equation 14.8; Beutier and Renon, 1978):

$$\log_{10} K = -0.09018 - \frac{2729.92}{T} \quad (14.7)$$

$$\ln(K) = -177.95 - \frac{1843.22}{T} + 31.434\ln(T) - 0.0545T \quad (14.8)$$

where *T* is absolute temperature (°K). Both these equations essentially yield the same values of *pK* (i.e., the negative $\log_{10}$ K). The values of *pK* are 10.1

at 0°C, 9.7 at 10°C, 9.4 at 20°C, and 9.1 at 30°C (Bates and Pinching, 1950). The relative proportion of NH_4^+ and NH_3 in solution at a given pH can be derived from the Equation (14.9):

$$pK - pH = \log_{10} \frac{[NH_4^+]}{[NH_3]} \tag{14.9}$$

which may be written as:

$$[NH_3]_{solution} = \frac{[NH_3 + NH_4^+]_{solution}}{1+10^{(pK-pH)}} \tag{14.10}$$

where $[NH_3]_{solution}$ and $[NH_3+NH_4^+]_{solution}$ are aqueous ammoniacal N concentrations in water. Equation (14.10) can be used to calculate NH_3 in solution at different pH values and temperatures. The relationship shows that as the pH increases the relative concentration of ammonium decreases and that of NH_3 increases. Similarly, increasing the temperature also increases the proportion of NH_3 (Figure 14.2).

Ammonia is in equilibrium between liquid and gas phases according to Henry's law which may be expressed as (Equation 14.11):

$$[NH_3]_{solution} = K_H \cdot p(NH_3) \tag{14.11}$$

where $p(NH_3)$ is the partial pressure of ammonia (Pascals, Pa) and $[NH_3]_{solution}$ is the concentration of dissolved ammonia (moles liter^{-1}, mol l^{-1}), and K_H is the Henry's law constant (mol Pa^{-1}l^{-1}) which is a function of temperature. The Henry's law constant is related to the Henry coefficient (K_h) that determines the partition of NH_3 concentration in the liquid and the gaseous phase (Equation 14.12)

$$K_H = \frac{K_h}{RT} \tag{14.12}$$

where R is the ideal gas constant (=8.31 m^3 Pa K^{-1} mol^{-1} or 82.1 cm^3 atm K^{-1} mol^{-1}). The Henry's coefficient is related to temperature (Equation 14.13; Hales and Drewes, 1979) as

$$K_h = 10^{(1477.7/T - 1.69)} \tag{14.13}$$

Combining Equations 14.11 and 14.12 yields:

$$p(NH_3) = \frac{RT[NH_3]_{solution}}{K_h} \tag{14.14}$$

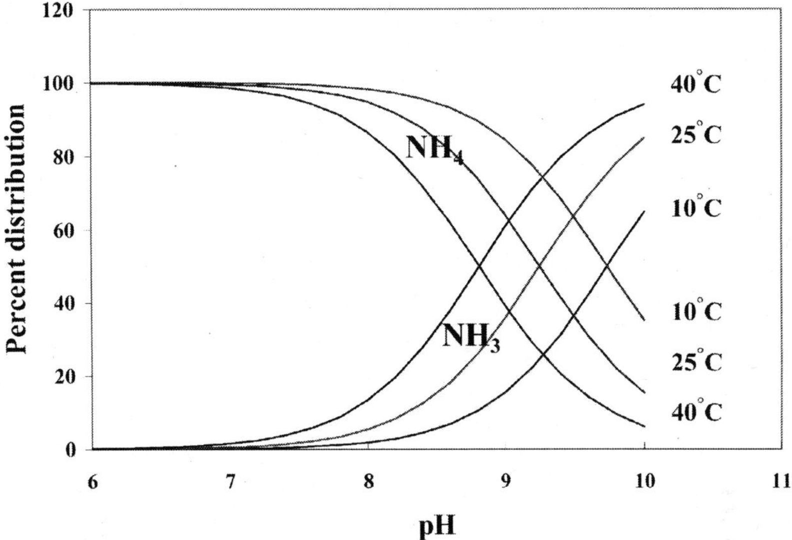

FIGURE 14.2. Distribution between NH_3 and NH_4 at different pH values and temperatures.

The rate at which NH_3 volatilizes from a liquid into the air depends upon the driving force existing between the two, i.e., the partial pressure in the liquid phase as compared to that in the air. Denmead, Freney, and Simpson (1982) used the following aerodynamic formula (Equation 14.15) to calculate the flux density *(F)* of NH_3 away from the surface:

$$F = K_a (p_o - p_z) \tag{14.15}$$

where K_a is a transfer coefficient, p_z is the partial pressure of NH_3 at the reference height *z,* and p_o is the equilibrium NH_3 vapor pressure of the solution. The value of K_a depends on many factors, of which wind speed, temperature, surface roughness, and surface area are the most important.

The loss of NH_3 from soil to atmosphere is influenced by a variety of soil, environmental, and management factors. The important soil factors include pH, buffering capacity, cation exchange capacity (CEC), calcium carbonate ($CaCO_3$) content, soil texture, and organic matter content. The dominant environmental and management factors include wind speed, temperature, soil moisture, and source and method of N application. In submerged soils, losses of NH_3 are directly related to the concentration of aqueous NH_3 in the floodwater, which in turn is a function of total ammoniacal N, the pH, and the temperature of the floodwater (Vlek and Stumpe, 1978; Vlek and Craswell, 1979).

Small amounts of ammonia, ranging from 0.5 to 15 kg N ha^{-1} y^{-1}, are also emitted from plant surfaces (Sutton, Schjørring, and Wyers, 1995). It has long been known that plants can both emit and take up NH_3 from air. The presence of NH_4^+ in leaf tissues results in the existence of an NH_3 "compensation point" concentration (a concentration where NH_3 is neither emitted nor taken up) for substomatal tissues, so that both emission and deposition are possible from stomata. The value of compensation point concentration has been reported to range from as low as 0.5 μg NH_3-N m^{-3} for an N-limited Colorado forest (Langford and Fehsenfeld, 1992) to 23 μg NH_3-N m^{-3} for wheat at late grain filling stage (Morgan and Parton, 1989). The compensation point changes with stage of crop growth, diurnal cycle, and temperature, which complicates evaluation of NH_3 emissions from plants. For each 5°C increase in temperature the compensation point doubles (Asman, Sutton, and Schjørring, 1998). Estimating long-term bidirectional NH_3 fluxes between plant and atmosphere is still uncertain, though it is now possible to apply a single model concept to a range of ecosystem types and satisfactorily infer NH_3 fluxes over diurnal time scales (Sutton, Schjørring, and Wyers, 1995). For detailed information on the subject, readers are referred to Farquhar, Wetselaar, and Weir (1983), Sutton, Pitcaim, and Fowler (1993), Holtan-Hartwig and Bøckman (1994), and ECETOC (1994).

Modeling Ammonia Volatilization

The cumulative NH_3 volatilization as a function of time after urea application follows an exponential relationship reaching an asymptote as time approaches infinity. Different formulations of the exponential function such as Gompertz, logistic, or sigmoidal have been used to describe the shape of the curve (Stevens, Laughlin, and Kilpatrick, 1989; Demeyer, Hofman, and van Cleemput, 1995; Sommer and Ersboll, 1996). Hengnirun et al. (1999) used a first-order kinetic equation (Equation 14.16) to simulate NH_3 volatilization from the soil and the surface applied manure:

$$A_t = A_0 \exp(-K_v t) \tag{14.16}$$

where A_t is the total ammoniacal N remaining in soil at time, t; A_0 is the amount of ammoniacal N in soil at $t = 0$; K_v is the first-order rate constant adjusted to incorporate the effect of temperature, CEC, and air flow rate. Although the empirical models can be used to smoothen data and compare results from different treatments under a given set of soil, environmental, and management conditions, they do not allow for an explanation of the underlying processes and cannot be extrapolated to conditions other than the ones from which they are developed. Attempts have been made to relate NH_3 volatilization losses to one or more of these factors (Whitehead and Raistrick, 1990; Moal et al., 1995; Sommer and Ersboll, 1996; Harmel et al., 1997), but there is difficulty in expressing the individual and interactive effects of a large number of factors. Although soil factors determine the NH_3 volatilization potential, the actual volatilization in the field is of stochastic nature and varies with time and space depending on the environmental and management conditions. The problem is exacerbated by the interdependence of one factor on the other and the likelihood that their individual effects are nonlinear.

A few process-based models have been developed to predict NH_3 losses in the soil-plant system from application of fertilizers to upland conditions (Rachhpal-Singh and Nye, 1986a; Sadeghi et al., 1988), for submerged conditions (Bouwmeester and Vlek, 1981; Moeller and Vlek, 1982; Jayaweera and Mikkelsen, 1990), and for manure and slurry application (van der Molen et al., 1990; Génermont and Cellier, 1997; Huijsmans and de Mol, 1999; Sommer and Olesen, 2000). These models predict NH_3 losses by considering the mechanism of NH_3 volatilization (as discussed in the preceding section) along with situation-specific processes.

The deterministic model developed by Rachhpal-Singh and Nye (1986a) for predicting NH_3 volatilization from urea applied to noncalcareous soils includes simultaneous numerical solution of three continuity or flow equations to describe the diffusive movement of urea (Equation 14.17), ammoniacal N (Equation 14.18), and soil alkalinity (Equation 14.19), involving both solution and gas phase diffusion in the soil. In the course of a time step, the model calculates concentration profiles of urea, bicarbonate, and ammoniacal N as well as the amount of N volatilized as NH_3:

$$\frac{\partial U_T}{\partial t} = D_U \theta f \frac{\partial^2 U}{\partial x^2} - V \tag{14.17}$$

$$\frac{\partial A_T}{\partial t} = D_A \theta f \frac{\partial^2 A}{\partial x^2} + D_g \theta_g f_g \frac{\partial^2 NH_{3(g)}}{\partial x^2} + V \tag{14.18}$$

$$\frac{\partial [S]}{\partial t} = \frac{\partial}{\partial x}\left[\sum \theta f D_B \frac{\partial [B]}{\partial x}\right] + R \tag{14.19}$$

where, U_T, A_T, and S represent concentration of urea, ammoniacal N species, and bases in whole soil, respectively. U, A, and B represent their respective concentrations in soil solution and D_U, D_A, and D_B their diffusion coefficients; θ is the soil moisture content; f the diffusion impedance factor; x the vertical distance from surface; t is time. $NH_{3(g)}$ is the concentration of ammonia in the gaseous phase, and D_g, θ_g, and f_g the gaseous diffusion coefficient, volume fraction, and impedance factor. V is the rate of urea hydrolysis per unit volume of whole soil and R is the rate of formation of soil bases by hydrolysis of urea.

Sensitivity analysis of the model by Rachhpal-Singh and Nye (1986b) showed that the diffusion of bicarbonate ion to the soil surface, to neutralize the acid generated when NH_4^+ is volatilized as NH_3 was the main process controlling the rate of NH_3 volatilization.

Roelcke et al. (1996) modified this model to extend its use to calcareous soils by including pH-buffering action of the soil carbonates in calcareous soils. Comparison of model results with measurements in laboratory column experiments at 18°C showed that the measured NH_3 losses agreed well with the model simulations in the first 10 days following fertilizer applications. The model could not simulate the subsequent losses properly.

The model of Rachhpal-Singh and Nye (1986a) does not allow for convective transport through downward and upward movement of water and nitrification of the ammonium released on hydrolysis of urea. Another problem with the use of this model in the field is that the parameters describing the transfer of NH_3 to the atmosphere were defined for laboratory experiments only and most of the input parameters such as urease activity, transfer coefficient, etc., have to be generated in laboratory experiments. Kirk and Nye (1991a,b) expanded the model to account for the effects of water drainage and evaporation on NH_3 volatilization.

To simplify input, Sadeghi et al. (1988) in their model for ammonia volatilization (AMOVOL), describe urea hydrolysis using a Michaelis-Menten equation modified by the major factors influencing the process. The constants for the chemical equilibria in the model are given as functions of temperature. Similarly, the diffusion coefficients for all the chemical species in water and air are described as a function of temperature. The main limitation of the model is that during a simulation the soil temperature and moisture are kept constant, whereas in the field both these variables change with time and the validity of various temperature functions in the field is not certain.

The models for NH_3 volatilization from flooded soils predict NH_3 loss as a function of floodwater chemistry and atmospheric conditions (Bouwmeester and Vlek, 1981; Moeller and Vlek, 1982; Jayaweera and Mikkelsen, 1990). The model (Equation 14.20) presented by Bouwmeester and Vlek (1981) considers the mass transfer of the air-water interface and is similar to the penetration theories of Higbie and Danckwerts (Danckwerts, 1970). The essential difference in their analysis from that of Higbie and Danckwerts' is that the liquid elements of surface are assumed to have a known time of exposure depending on the wind velocity and the location of the element in the rice paddy:

$$\overline{Q} = \frac{\overline{A}}{\beta t_d}\left(e^{\beta^2 Dt_d} erfc\beta\sqrt{Dt_d} - 1 + \frac{2}{\sqrt{\pi}}\beta\sqrt{Dt_d} \right) \quad (14.20)$$

where $\overline{Q}$ is average volatilization rate per unit area, $\overline{A}$ is the ammoniacal N concentration in the bulk liquid, and $t_d = F/U_d$ is the time during which the water chemistry, wind, and water conditions are assumed to remain steady depending on the fetch, F, and surface drift velocity, U_d; and

$$\beta = \frac{K_a K_H}{D(1+B)} \quad (14.21)$$

where K_a is the bulk transfer coefficient of NH_3 in air, K_H is the Henry's constant, and D is the molecular diffusivities ($=2.5 \times 10^{-5}$ cm^2 sec^{-1}at 25°C for both NH_3 and NH_4^+); and $B = H^+/K$ where H^+ is the hydrogen ion concentration in the system and K_a is the equilibrium constant.

The model (Bouwmeester and Vlek, 1981) described a data set obtained in a laboratory wind-water tunnel, simulating rice paddies. The model was used to study some of the factors that control the rate of NH_3 volatilization such as floodwater chemistry and meteorological conditions. In the development of this model, it was assumed that pH, and therefore the NH_3/NH_4^+ ratio, was constant throughout the diffusion layer. However, Quinn and Otto (1971) contended that a hydrogen ion gradient exists in such a diffusion layer. Therefore, Moeller and Vlek (1982) developed a simulation model which predicts a gradient in the hydrogen ion concentration through the liquid-phase diffusion barrier. At moderate pH the ammonia-diffusion flux in the liquid phase is augmented by ammonium ion diffusion. The existence of a pH gradient in the liquid-diffusion barrier offsets this flux augmentation and may enhance the effect of solution buffer capacity upon differential volatilization rate. The model results agreed with the rate of NH_3 volatilization loss experimentally observed in carbonate-free solutions.

Jayaweera and Mikkelsen (1990) based their model on the two-film theory of mass transfer (Whiteman, 1923). They included depth of floodwater as one of the primary variables in addition to the other four (floodwater NH_4^+-N concentration, pH, temperature, and wind speed) considered by earlier researchers. They derived the following relationship (Equation 14.22) to determine the rate of NH_3 volatilization from a flooded system:

$$\frac{d[NH_4^+]}{dt} = k_a \left\{ \frac{k_d(A-[NH_3]_{aq})}{k_a[H^+]+k_{vN}} \right\} - k_d(A-[NH_3]_{aq}) \tag{14.22}$$

where A is ammoniacal N concentration, $[NH_3]_{aq}$ is aqueous ammonia concentration, $[H^+]$ is hydrogen ion concentration in floodwater at equilibrium, k_d and k_a are dissociation and association rate constants for $NH_4^+/ [NH_3]_{aq}$ equilibrium, and k_{vN} is the first-order volatilization constant for NH_3. The volatilization rate constant is calculated as a ratio of the overall mass transfer coefficient for NH_3 *(K_{oN})* and mean depth of floodwater *(d):*

$$k_{vN} = \frac{K_{oN}}{d} \tag{14.23}$$

To estimate K_{oN}, it is necessary to determine the Henry's law constant for NH_3, and the gas and liquid phase exchange constant for NH_3. Validation of the model against data from wind tunnel and field experiments showed a close match between the predicted and the measured values (Jayaweera, Mikkelsen, and Paw U, 1990).

Génermont and Cellier (1997) have presented a phenomenological model for estimating NH_3 volatilization from slurry applied to bare soil under field conditions using simple meteorological data. The model includes submodels to simulate ammoniacal N transfers and equilibria between ammoniacal N species, as well as heat and water transfer in the soil. The aqueous ammoniacal N transfers between the soil layers are described by the classical convection-diffusion equation (Bear, 1972) and the gaseous NH_3 transfer by the analogous diffusive part of the equation. Ammonia gas flux in the field is derived from the advection model of Itier and Perrier (1976) that could be used to calculate the NH_3 fluxes for field areas from some m^2 to several hectares. In this respect, the model is considered to be superior to that of Van der Molen et al. (1990) that could simulate NH_3 volatilization from slurry application over plots of only several hundred square meters in size. Génermont and Cellier (1997) found that under field conditions, their model simulated fairly well the NH_3 fluxes, including the total NH_3 volatilized, the decrease in daily loss, and their short-term (<1h) variations due to the influence of meteorological conditions on soil surface temperature and atmospheric diffusion. Sensitivity analysis of the model showed that soil pH determined not only the total NH_3 loss but also the pattern of loss. Interestingly, pH was not simulated in the model and so a constant pH (as input) was used over the whole volatilization period. The pH after slurry application has been found to vary with the changing surface water content due to drying (Sommer and Olesen, 2000).

From the foregoing analyses, it is apparent that though most of the models for NH_3 volatilization explicitly incorporate the underlying theory of volatilization, these are difficult to parameterize under field conditions. Although such models are valuable for understanding the processes involved in NH_3 volatilization, they may have limited value particularly because of the influence of a number of interacting environmental and management factors on NH_3 volatilization in the field. Independent tests of these models against the field data have not been encountered so far.

Denitrification

Denitrification is an anaerobic bacterial process by which nitrate is reduced to nitrite (NO_2) and further reduced to nitrous oxide (N_2O) or dinitrogen (N_2) which is lost to atmosphere as a gas (Equation 14.24). The process of denitrification in soil has been reviewed in detail by several researchers, including Firestone (1982), Nieder, Schollmayer, and Richter (1989), and Granli and Bøckman (1994).

$$NO_3^- \rightarrow NO_2^- \rightarrow NO \rightarrow N_2O \rightarrow N_2 \qquad (14.24)$$

There has been some doubt if nitric oxide (NO) is a true intermediate or a by-product (Amundson and Davidson, 1990) in the process. Depending on conditions, intermediate products can accumulate and eventually escape.

Many microorganisms can use NO_3^- as their primary electron acceptor for obtaining energy from organic compounds when low O_2 availability restricts their metabolism (heterotrophic denitrification; Equation 14.25).

$$5(CH_2O) + 4NO_3^- + 4H^+ \rightarrow 5CO_2 + 7H_2O + 2N_2 + energy \qquad (14.25)$$

The majority of soil bacteria seem able to denitrify (Umarov, 1990), but denitrifying bacteria exhibit a variety of incomplete reduction pathways: some bacteria produce only N_2, while others give a mixture of N_2O and N_2, and some only N_2O (Robertson and Kuenen, 1991). There are reports (Shoun et al., 1992) that many fungi are also capable of evolving N_2O under anaerobic conditions.

Some microorganisms can obtain energy by using NO_3 for oxidation of inorganic compounds, e.g., S^{+2}, Fe^{+2} (autotrophic denitrification). This occurs when NO_3^- diffuses into zones rich in FeS, e.g., sediments in shallow waters (Granli and Bøckman, 1994). However, heterotrophic denitrification is the more prevalent of the two processes as a source for N_2O.

Nonbiological denitrification, called chemodenitrification, may also occur wherein NO_2^- can react with organic compounds (e.g., amines) to form N_2, NO_2, and N_2O (Bremner and Nelson, 1968). N_2O can also form in reactions between NO_3^-/NO_2^- and some inorganic compounds (e.g., Fe^{2+}, Cu^{2+}). These reactions may be important for slow denitrification of groundwater (Van Cleemput, Uytterhaegen, and Baert, 1987).

A number of soil, plant, environmental, and management factors interact to influence rate and extent of denitrification in soils. Presence of NO_3^-, the presence of decomposable organic matter, suitable temperature, and absence of oxygen at the microsite where the process is taking place facilitate loss through denitrification. Since denitrification requires anaerobic conditions, there is an inverse relationship between the rate of denitrification and O_2 concentration (e.g., Focht, 1974; Arah et al., 1991). The inverse relationship is more pronounced at high (34.5°C), rather than at low (19.5°C), temperature (Focht and Verstraete, 1977). Reduction of N_2O to N_2 is more prone to inhibition by O_2 than reduction of NO_3^- to N_2O, thus the N_2O/N_2 ratio decreases with increasing O_2 concentration. The O_2 concentration in soil depends on soil water content, diffusion of O_2 into the soil, and consumption of O_2 by soil microorganisms and plant roots (Smith, 1990). Diffusion of O_2 in soil is determined by texture, management (e.g., tillage), and water content. Sufficiently low oxygen diffusion rates that promote rapid denitrification are most likely under waterlogged conditions, as in rice paddies and in pasture systems with compacted soil. The denitrification losses are likely to be greater in fine-textured, aggregated, and frequently wetted soils than in coarse-textured soils (Nieder, Schollmayer, and Richter, 1989).

Generally, denitrification rate increases as soil water content rises largely due to its effect on aeration. Other factors such as temperature, NO_3^- concentration, soil texture, and compaction influence the effect of soil water on denitrification rate. Increased denitrification rate with increased soil water content seems most marked above about 60 percent water-filled pore space. Therefore, the denitrification losses are generally higher from rice paddy fields as compared to upland conditions (Xing and Zhu, 1997). The soil moisture content also affects the relative yield of intermediates (N_2O and NO) with less gas loss and more complete reaction in diffusion-limiting wet conditions (Firestone and Davidson, 1989).

Several studies have shown that denitrification N loss is positively correlated to available C (Stanford, Vander Pol, and Dzienia, 1975; Weier et al., 1993), water-soluble C, total C (Beauchamp, Gale, and Yeomans, 1980), and NO_3 concentration in soil (Weier et al., 1993), provided other factors are favorable. The denitrification rate increases with increasing soil NO_3^- concentrations up to a certain level and then becomes constant which could be described by Michaelis-Menten kinetics. Production of N_2O by nitrification is also enhanced as the soil concentration of the substrate, NH_4^+, increases. Hence application of N fertilizers or manures is usually followed by an increase in N_2O emission which is likely to be greater with urea than nitrate as the fertilizer source (Duxbury, 1990).

Greater rates of denitrification are usually observed with zero tillage compared to plowed soils (Nieder, Schollmayer, and Richter, 1989). The increase is related to increased soil organic matter and higher levels of available C in the topsoil, as well as to greater soil densities and decreased soil aeration (Myrold, 1988). The rate of denitrification is usually low under environmental conditions reported to favor production of N_2O relative to N_2 (e.g., low temperature, low pH, and presence of O_2). Thus N_2O is the favored reaction product under conditions that are marginal for denitrification. N_2O is one of the greenhouse gases that is considered to be forcing a global climate change. For a further discussion on greenhouse gases see Chapter 12.

In flooded soils, denitrification rate is regulated by the NO_3^- supplying capacity of the system (Rao, Jessup, and Reddy, 1984) and time of flooding relative to nitrate production (Freney and Denmead, 1992). On the other hand, flux of NO_3^- is governed by the denitrification rate in the reduced soil layer, flood water depth, and NH_3 concentration in the floodwater and oxidized soil layer. Denitrification in floodwater has been found to increase with increased availability of available C in the underlying anaerobic soil layer (Engler and Patrick, 1974). Water management in lowland rice can influence the extent of N losses due to denitrification. Intermittent dry spells (aerobic conditions) could result in greater total N loss from the soil than would occur under continuous anaerobic conditions (Wijler and Delwiche, 1954). Denitrification losses have been found to be reduced in the presence of rice plants than in systems without plants (Reddy and Patrick, 1980) mainly because of competition for NO_3^- uptake between denitrifying bacteria and rice roots.

Modeling Denitrification

Rates of denitrification vary greatly in both time and space, and this makes it very difficult to measure in field experiments. Considerable research effort is being directed toward improving the methods of measurement. Some methods of indirect measurement are available (Hauck and Weaver, 1986; Nieder, Schollmayer, and Richter, 1989). The most common one is using acetylene as an enzyme inhibitor (the acetylene inhibition method, AIM), but N budget studies are also used. Use of isotope methods (marking the fertilizer N with ^{15}N and measuring the isotope enrichment in the products) permits direct measurement of denitrification rates, but is not as common as AIM due to cost. Because of problems with measurement, there is an important role for modeling to integrate the interactive effects of different soil, plant, environmental, and management factors for predicting denitrification losses on the regional and ultimately on the global level.

A number of different approaches have been used to develop models for predicting denitrification losses in soil. Although a detailed description of different modeling approaches is presented in Chapter 12, only a brief description is given in this chapter. The models for denitrification range in complexity from using first-order kinetics with respect to NO_3 concentration (Hagin et al., 1976) to a mechanistic description of soil anaerobic fraction in which denitrification can occur (Arah and Smith, 1989; Renault and Stengel, 1994). Denitrification losses have been simulated by considering denitrification rate *(dG/dt)* to be a function of NO_3 concentration *(N)*, water extractable organic carbon *(C_w)*, degree of soil water saturation (θ), and temperature (Rolston et al., 1984; Equation 14.26)

$$dG / dt = k_1 \theta f_w f_a C_w N \tag{14.26}$$

where k_1 is the denitrification rate coefficient and f_w and f_a are empirical functions to account for water content and temperature effects, respectively.

Several models of intermediate complexity are also available (McConnaughey and Bouldin, 1985; Li, Frolking, and Frolking, 1992; Parton et al., 1996; Del Grosso et al., 2000; Riley and Matson, 2000). Parton et al. (1996) grouped models for soil denitrification into three distinct classes viz. (1) microbial growth models that include an explicit description of microbial dynamics responsible for nitrification, denitrification, and other nutrient cycling processes (e.g., Hunt, 1977; McGill et al., 1981; Li, Frolking, and Frolking, 1992; Riley and Matson, 2000); (2) simplified process models that represent different N cycling processes as a function of soil water, temperature, and pH controls on microbial activity without representing the microbial dynamics (Bradbury et al., 1993; Parton, Mosier, and Schimel, 1988; Bergstrom and Beauchamp, 1993; Potter et al., 1996); and (3) soil structural models that simulate denitrification by describing soil physical processes such as diffusion of gases and solutes into soil aggregates and explicitly representing the distribution of soil aggregates. Models of this type assume that denitrification primarily occurs in anaerobic soil aggregates and that diffusion of O_2, N_2O, and NO_3 into and out of soil aggregates needs to be represented (Arah and Smith, 1989; Smith, 1990; Arah, 1990). A comparison of the predictive capacity of different models is presented in Chapter 12.

NITROGEN TRANSFORMATION PROCESSES

Nitrogen Mineralization and Immobilization

There is an internal N cycle in soil that has major impacts on N loss processes, maintenance of soil organic matter, recycling of crop residues and manures, and fertilizer N management in agricultural systems. It involves the conversion of organic forms of N to NH_3 or NH_4^+ and NO_3^- by a process called mineralization. The first step in the process, called ammonification, is an enzymatic process, and involves conversion of organic N to NH_3 (Equation 14.27). It is carried out exclusively by heterotrophic microorganisms. The subsequent conversion of NH_3 to NO_3^-, termed nitrification, is mediated primarily through two groups of autotrophic bacteria (*Nitrosomonas* and *Nitrobacter*) according to the following pathways (Equation 14.28 and 14.29):

$$Organic\,N \rightarrow NH_3 \underset{-H^+}{\overset{+H^+}{\rightleftarrows}} NH_4^+ \tag{14.27}$$

$$NH_4 + 1\tfrac{1}{2}O_2 \underset{Nitrosomonas}{\overset{6e-}{\rightarrow}} NO_2^- + 2H^+ + H_2 \tag{14.28}$$

$$NO_{2^-} + \tfrac{1}{2}O_2 \underset{Nitrobacter}{\overset{2e-}{\rightarrow}} NO_3^- \tag{14.29}$$

The process of nitrification is generally fast and any NH_3 produced due to ammonification is rapidly nitrified to NO_3. The main factors that affect nitrification in soil are temperature, moisture, pH, and the substrates NH_4^+, O_2, and CO_2 (Stevenson, 1982). The process may be retarded at low and high pH or at water potentials below 1.5 mega pascals, MPa (Justine and Smith, 1962) or in the presence of high concentrations of heavy metals (Benbi and Richter, 1996). Nitrification is optimal at mesophilic temperatures and slightly alkaline pH. The effect of temperature on nitrification is generally modeled by using the Arrhenius equation (Addiscott, 1983).

Mineralization is always coupled with immobilization which operates in the reverse direction (see Figure 14.1) with the soil microbial biomass (SMB) assimilating inorganic N forms and transforming them into organic N constituents in their cells and tissues during the oxidation of suitable C substrates. However, immobilized N is likely to be available subsequently for mineralization as the microbial population turns over. The continuous transfer of mineralized N into synthesized organic matter and the release of immobilized N back into inorganic forms is known as mineralization-immobilization turnover or MIT (Jansson and Persson, 1982). Immobilization primarily occurs from NH_4^+ pool, however, small organic compounds such as amino acids may also be immobilized at a microsite scale (Drury, Voroney, and Beauchamp, 1991). This is called direct hypothesis.

Total release of NH_4^+ through microbial activity prior to any immobilization back into the organic forms is termed gross mineralization. The difference between gross mineralization and immobilization constitutes net mineralization (or net immobilization). An increase in mineral N level with time indicates net mineralization and a decrease suggests net immobilization. Since gross mineralization is difficult to measure, most often it is the net mineralization that is measured or estimated. Generally during an annual cycle there is net mineralization, although over the shorter term immobilization may occur when available C concentration in the soil increases, for example after addition of organic materials such as crop residues with wider C/N ratios and/or release of C through root exudates. Organic materials having C/N ratios less than 20 lead to net mineralization and those with C/N greater than 30 result in net immobilization by microorganisms. Under conditions favorable for microbial activity, rapid decomposition occurs (see Chapter 13) with concurrent release of C as CO_2, and consumption of mineral N by microorganisms resulting in net immobilization of N. As the C/N ratio of the decomposing material is lowered to about 20, net mineralization occurs.

Long-term accumulation of organic N in soils may also be referred to as immobilization. This occurs due to a buildup of SOM (Nieder, Dauck, and Benbi, 2001) and is not a direct result of SMB activities. It is quite usual for a soil to have net mineralization and to release mineral N, and to accumulate organic N during the year, provided that the N input was greater than removal in harvested crops, plus that accumulated in organic matter. This is particularly the case in undisturbed grassland soils where there is usually long-term accumulation of organic N through this means.

In addition to substrate availability, microbial biomass, and soil texture, temperature and moisture are the most important factors that dictate N mineralization rates in soils. Water potentials between –0.01 to –0.03 MPa and temperature around 35°C are considered optimum for N mineralization from SOM. However, the two have been shown to interact and the effect of moisture on microbial activity is enhanced at higher temperatures (Quemada and Cabrera, 1997).

Modeling Nitrogen Mineralization

Mineralization is one of the most difficult processes to model. Problems stem partly from the fact that both the N source (soil organic N consisting of plant and animal residues) and the microbes that mineralize N are poorly characterized. This situation is exacerbated by the fact that a number of soil and environmental conditions which control the rates and products of mineralization are not well understood (Seyfried and Rao, 1988). Modeling approaches differ according to the level of detail at which these processes are considered, and may be classified into the following two main groups:

1. Simple functional models to predict net N mineralization
2. Mechanistic models for simulating mineralization-immobilization turnover in soils

Simple Functional Approaches. Simple functional models do not take into account the basic processes influencing mineralization. These models predict net N mineralization and do not consider the processes of ammonification and nitrification separately. The parameters for the models are obtained from laboratory incubation studies by fitting N mineralization data to the time of incubation. Therefore, the models are parameterized from the same type of data they are intended to predict. Kinetic equations of varying complexity with different numbers of mineralizable N fractions are fitted to the data.

Stanford and Smith (1972) defined soil N mineralization potential as the quantity of soil organic N susceptible to mineralization at a rate of mineralization *(k)* according to first-order kinetics:

$$\frac{dN}{dt} = -kN \tag{14.30}$$

Where N is concentration of mineralizable substrate (N), and t is time. Integration of this equation between time t_o and t, yields:

$$N_t = N_0 e^{-kt} \tag{14.31}$$

where N_0 is the initial substrate concentration or the potentially mineralizable nitrogen and N_t is the substrate concentration at time t. The equation may be modified by substituting $N_t = (N_0 - N_m)$, where N_m is the N mineralized in time t, as:

$$N_m = N_0(1 - e^{-kt}) \tag{14.32}$$

Using Equation (14.32), Stanford and Smith (1972) found that the estimated values of N_0 at 35°C for 39 soils from the United States ranged from 18 to 305 mg kg^{-1} soil. Values of k ranged from 0.035 to 0.095 week^{-1} with a weighted average of 0.054 week^{-1}. However, widely varying k values have been reported for different soils indicating that soils differ not only in the amount of active organic N but also in their turnover rate by microorganisms.

The first-order single compartment (FOSC, Equation 14.32) kinetic model has been used to describe N mineralization kinetics of soils under different land use, crop, and climatic conditions. However, Lindemann and Cardenas (1984) have found the first-order kinetic model to be inadequate to describe N mineralization kinetics in soils treated with sewage sludge. The fit to the data was improved by adding an exponent *(b)* to the time variable in the equation (Marion, Kummerow, and Miller, 1981):

$$\log(N_0 - N_m) = \log N_0 - kt^b \tag{14.33}$$

A number of studies have demonstrated the inadequacy of the single pool first-order model to describe N mineralization kinetics in soils (Molina, Clapp, and Larson, 1980; Nuske and Richter, 1981; Deans, Molina, and Clapp, 1986; Singh and Benbi, 2000). There are systematic deviations between first-order fitted curves and measured values. The model underestimates mineralization at the beginning and end of the incubation period, and overestimates at intermediate times (Bonde and Rosswall, 1987; Seyfried and Rao, 1988).

Models have been developed (Molina, Clapp, and Larson, 1980; Juma and Paul, 1981; Nuske and Richter, 1981; Van Veen and Frissel, 1981; Deans, Molina, and Clapp, 1986) that describe net N mineralization by dividing the mineralizable soil organic N into different fractions. Each fraction is assumed to mineralize according to first-order kinetics. This may be generalized as:

$$N_m = \sum_{i=1}^{n} N_{0i}(1 - e^{-k_i t}) \tag{14.34}$$

where i represents a specific N fraction, n is the total number of fractions, N_{0i} is the potentially mineralizable N in the i-th fraction, and k_i is the mineralization rate constant for the i-th fraction. In the simplest form of the multi-fraction approach, two main fractions of organic N are assumed to mineralize at different rates (Nuske and Richter, 1981; Nordmeyer and Richter, 1985; Deans, Molina, and Clapp, 1986). One fraction consists of N compounds of easily decomposable plant material (dpm), e.g., residues mainly from the last crop, while the second fraction represents the more resistant or recalcitrant plant material (rpm), which accumulates within the soil. The dpm mineralizes at a faster rate than the rpm.

$$N_m = N_{dpm}(1 - e^{-k_{dpm}t}) + N_{rpm}(1 - e^{k_{rpm}t}) \tag{14.35}$$

where N_{dpm} and N_{rpm} represent potentially mineralizable N in the easily decomposable and resistant organic N fractions, respectively, and k_{dpm} and k_{rpm} are the corresponding rate constants. Various researchers have confirmed that the first-order double compartment (FODC) model (Equation 14.35) is superior to the single compartment (FOSC) model in describing N mineralization (Lindemann and Cardenas, 1984; Cabrera and Kissel, 1988; Diaz-Fierros et al., 1988). For dried soil samples, Richter et al. (1982) and Nordmeyer and Richter (1985) have adopted a three-fraction approach that accounts for dead biomass N, dpm, and rpm.

Beauchamp et al. (1986) have suggested that the apparent presence of a small extra pool of mineralizable N (that mineralizes relatively fast) might be an experimental artifact as a consequence of drying and rewetting the soil rather than due to N mineralization from a separate soil organic N fraction. They have proposed that the "artifact mineralization" could be accounted for by modifying the first-order kinetic model so that a fitted curve through the data points does not pass through the origin but intercepts the Y-axis at N_e.

$$N_m = N_0 - (N_0 - N_e)e^{-kt} \tag{14.36}$$

The model is based on the assumption that the killed microbial biomass N pool *(N_e)* is effectively used up by seven days (usually the time by which the first measurement is made) and N mineralized subsequently originates entirely from the soil organic N pool. Beauchamp et al. (1986) argue that, since most of the N flush has already occurred by that time, there is little basis for selecting a kinetic model. However, it has been shown that drying and rewetting not only generates or enlarges an N pool that mineralizes rapidly according to first-order kinetics, but also increases the size of a more stable slowly mineralizable N pool (Inubushi and Wada, 1987; Cabrera, 1993).

Instead of the two-fraction double-exponential model (=parallel first-order), which describes N mineralization from two independent initial sources, Andrén and Paustian (1987) applied a consecutive first-order model (Equation 14.37) to barley straw decomposition. The model is based on the assumption that the recalcitrant fraction is converted to a labile fraction (e.g., through solubilization of structural polymers) from which N mineralization occurs.

$$N_m = N_{rpm}(1-e^{-k_{rpm}t}) + N_{rpm}\alpha(1-e^{-k_{rpm}t}) - N_{rpm}\alpha(1-e^{k_{dpm}t}) + N_{dpm}(1-e^{-k_{dpm}t}) \quad (14.37)$$

in which,

$$\alpha = \frac{k_{rpm}}{k_{dpm} - k_{rpm}}$$

This model is essentially a reparameterization of the double exponential model and would, therefore, yield the same values for the rate constants. However, the fractions of recalcitrant and labile material present at any given time will differ.

Some researchers have suggested that, with the incubation methods, it may not be possible to quantify the N mineralization potential of soils because the slowly mineralizable N component may in fact follow zero-order kinetics (Bonde and Rosswall, 1987; Bonde, Schnürer, and Rosswall, 1988; Lindemann, Connell, and Urquhart, 1988). They advocate the use of a mixed first- and zero-order (FOZO) kinetic model.

$$N_m = N_1\left(1-e^{-k_1 t}\right) + Kt \quad (14.38)$$

in which N_1 represents the amount of mineralizable N in the easily decomposable pool at the start of incubation and k_1 is the rate constant, and K is the zero-order rate constant. The first-order term of the model is interpreted as accounting for pretreatment effects (e.g., air drying). Therefore, it may be argued that in the absence of air drying, i.e., in field-moist soils, net N mineralization may be described by zero-order only, rather than by combined first- plus zero-order kinetics (Tabatabai and Al-Khafaji, 1980; Addiscott, 1983).

$$N_m = Kt \quad (14.39)$$

The implication of this approach is that there may be any number of N pools and each may be sufficiently large enough not to be significantly depleted by mineralization. Addiscott (1983) has found that N mineralization in Rothamsted soils that are kept moist prior to incubation could be described by the zero-order kinetic model. However, other researchers have documented the applicability of first-order kinetic models to net N mineralization in field-moist samples as well.

In situations where crop residues or organic manures of wider C/N ratios are added to soils, initially, N immobilization may occur followed by net mineralization with the narrowing down of the C/N ratio. The description of such data requires the inclusion of a term for initial lag phase in the model (Bonde and Lindberg, 1988; Diaz-Fierros et al., 1988). Bonde and Lindberg (1988) have modified the FOSC and FOZO models to accommodate the initial lag phase to describe the data of Chae and Tabatabai (1986).

$$N_m = N_{dpm}\left(1-e^{\left(-h_1 t - h_2 t^2/2\right)}\right) + N_{rpm}\left(1-e^{-kt}\right) \qquad (14.40)$$

$$N_m = N_0\left(1-e^{\left(-h_1 t - h_2 t^2/2\right)}\right) + Kt \qquad (14.41)$$

where, h_1 and h_2 are rate constants for the mixed-order model, and other terms are as defined previously. This model is similar to the one presented by Brunner and Focht (1984) for microbial decomposition of C substrates, referred to as three-half-order (3/2 order), as it is suitable for substrate metabolism with (pseudo–second-order) or without (pseudo–first-order) growth.

Empirically determined equations such as polynomials and parabolic functions have also been proposed to describe net N mineralization in soils (Broadbent, 1986; Marion and Black, 1987). Broadbent (1986) has observed that compared to the first-order kinetic model, a parabolic function ($Y=aX^b$) provides a better fit to the data obtained by various investigators in laboratory, greenhouse, and field experiments. Although empirically determined equations have been found to provide a better fit to the data compared to single-pool first-order kinetic models, no physical meaning can be attached to the regression coefficients.

Benbi and Richter (2001) made a critical assessment of these kinetic models based on previously published data from Griffin and Laine (1983), Lindemann and Cardenas (1984), Chae and Tabatabai (1986), and Richter et al. (1989). They concluded that (1) the single-pool first-order kinetic model is inadequate to describe net N mineralization in soils as it provides the worst fit to the data and systematically deviates from the measured cumulative N mineralization; (2) in disturbed soil samples, a minimum of two pools of mineralizable N is considered essential to significantly contribute toward N mineralization, irrespective of whether the soil samples were dried and subsequently rewetted or kept field moist prior to incubation; and (3) for precisely estimating potentially mineralizable N, the incubation needs to be continued until the mineralization rate approaches a small constant value.

Large differences in N mineralization potentials (N_0) and the rate of mineral N production could originate from differences in time of sampling, probably due to seasonal variation in microbial biomass and the amount of undecomposed crop residues in the soil. Microbial biomass has a central role in organic matter transformations. Microorganisms act both as a source (mineralization) and a sink (immobilization) for N during microbial biomass turnover. Therefore, models for N mineralization kinetics should, ideally, include a component for microbial biomass. This would lead to the second-order rather than the first-order kinetics as the N mineralization rate would depend on the concentration of both the substrate as well as the microbial biomass.

The amount of N mineralized during laboratory incubation experiments and the parameter estimates for different models depend on methods used in pretreating the soil prior to incubation and experimental conditions (such as temperature, soil moisture content) during the incubation. Drying and rewetting, grinding and sieving the soil, better air circulation, and optimal availability of nutrients can increase potentially mineralizable N in the soil. The temperature dependence of N mineralization rate constant has usually been described by Q_{10} (temperature coefficient) and Arrhenius function, $k = Ae^{-B/T}$ in which k is the mineralization rate constant, T is the absolute temperature, and A and B are regression coefficients. There are considerable differences in the coefficients reported for Arrhenius function by various researchers. To make a comparison, the Arrhenius functions reported in different studies were normalized (Stanford, Frere, and Schwaninger, 1973; Campbell, Myers, and Weier, 1981; Addiscott, 1983; Nordmeyer and Richter, 1985) to 1.0 at 5°C (see Figure 14.3). As can be seen, there are large differences in the temperature adjustment factor used in different studies. In the kinetic models, the effect of temperature is generally incorporated by adjusting the rate coefficient and assuming the pool size to be constant at all temperatures, which was estimated solely from the N mineralization pattern at the highest temperature. Ellert and Bettany (1992) have proposed simultaneous fitting of the kinetic model and temperature response function to the N mineralization data. Recently, MacDonald, Zak, and Pregitzer (1995) have shown that it is the mineralizable N pool (N_0) rather than the rate constant (k) which is temperature dependent. MacDonald, Zak, and Pregitzer (1995) have observed an increase in mineralizable N pool with increase in temperature from 5 to 25°C. They argue that this may be due to a shift in the microbial community, changes in biochemical composition of the fraction mineralized, or changes in transport processes such as diffusion with temperature. In view of these results, the temperature dependence of N mineralization kinetics needs to be further investigated.

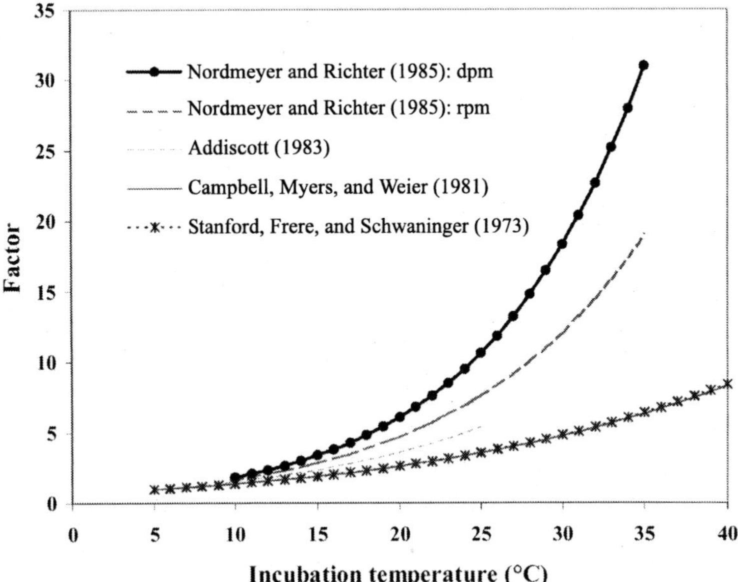

FIGURE 14.3. Correction factors for incorporating the effect of temperature on N mineralization parameters as used in different studies. Correction factors are normalized to 1.0 at 5°C. The lines for the data of Stanford, Frere, and Schwaninger (1973) and Campbell, Myers, and Weier (1981) are overlapping.

Mechanistic Models. Although simple functional models are designed to predict net N mineralization, mechanistic models attempt to simulate gross mineralization and associated immobilization (viz. mineralization-immobilization turnover). These are process-based models that try to include the best possible description of the processes involved. Based on current understanding of organic matter decomposition in soil, hypotheses and assumptions about various processes are made that are used to develop mathematical formulations of the decay process. In fact, there are almost limitless possibilities for making models of mineralization more mechanistic. These models generally characterize SOM in terms of its C and N content and, therefore, simulate the C and N cycle. Although models of this category are discussed in detail in Chapter 13, only a brief description is presented in this chapter.

Generally, the decomposition of organic matter is simulated by defining different fractions of organic matter as functional pools on the basis of their presumed chemical characteristics, fixed C/N ratio, and location in the soil. The number of pools in different models varies greatly in number, N and C content, and turnover rates. Small pools with high turnover rates and larger pools with slower turnover rates are distinguished (Henin, Monnier, and Turc, 1959; Jenkinson and Rayner, 1977; Van Veen and Frissel, 1981; Bradbury et al., 1993; Molina et al., 1983; Jenkinson et al., 1987; Parton et al., 1987; Hansen et al., 1991; Rijtema and Kroes, 1991). In each of these pools the turnover rates are simulated by first-order kinetics and the rate coefficients are modified by superimposing the effects of abiotic factors such as temperature, soil moisture, and clay content using empirical relations.

Some models consider microbial biomass as one of the defined pools (Hansen et al., 1991; Whitmore et al., 1991) while others do not include a biomass pool and model it as part of the stable organic matter/humus pool (Rijtema and Kroes, 1991). The biomass pool may further be divided into labile and resistant, dynamic and stable, or zymogenous and autochthonous pools. Models in this category do not really simulate growth and maintenance of microbial biomass explicitly; it only occurs in parameter values of the C and N reservoirs.

Models developed by Parnas (1975), Smith (1979), McGill et al. (1981), Knapp, Elliott, and Campbell (1983), and Grant, Juma, and McGill (1993) explicitly consider biomass kinetics to simulate C and N cycling during organic matter decomposition. Availability of organic C mainly controls the growth of soil microbial biomass, which in turn controls N immobilization and mineralization. These models (Smith, 1979; McGill et al., 1981; Van Veen and Frissel, 1981; Simkins, Mukherjee, and Alexander, 1986) use Monod (1942) kinetics to predict the decomposition rate (ξ_i) of the organic pool *i:*

$$\xi_i = \frac{\mu_{\max} \cdot S_i}{K_s + S_i} B \tag{14.42}$$

where, S_i is the substrate concentration in pool *i*, *B* is the size of the microbial biomass, $\mu_{\max}$ is the maximum specific growth rate, and K_s is the half-saturation constant for growth. Although several identical features are included in this kind of model, they differ considerably in the treatment of many processes. For example, while most of the models treat biomass turnover or death explicitly in relation to substrate/energy availability (Parnas, 1975; McGill et al., 1981; Van Veen and Frissel, 1981; Kersebaum and Richter, 1994), the PHOENIX model (McGill et al., 1981) additionally includes a function to describe microbial death through predation and density-dependent death of microorganisms and a floating C/N ratio of microbes.

Although the models incorporating biomass are explicitly valuable for understanding the processes involved in organic matter turnover in soil, usually, too large a number of input parameters is required to apply the models.

They require data on the rates of uptake of organic compounds by soil organisms, the efficiency of substrate use for biosynthesis and energy supply, and the nature and rates of release of organic products from the soil microbial biomass. Some parameters are difficult to obtain under field conditions. Use of information extrapolated from laboratory studies may not be very precise as the biomass behavior differs considerably in the soil and solution cultures. Independent tests of the biomass-based models have not been encountered. Mostly, the underlying concepts have been verified in a qualitative or semiquantitative manner.

Another class of models has been developed in which C and N fluxes in the soil are related to the abundance and activity of soil organisms, constituting the soil food web (Hunt et al., 1987; De Ruiter et al., 1993). In these models, organisms are classified as functional groups according to food choice and life-history parameters (Moore, Walter, and Hunt, 1988). Consumption rates among the groups of organisms are calculated based on biomass and turnover rates. Nitrogen mineralization rates are computed from consumption rates using information on energy conversion efficiencies and C:N ratios of the organisms. De Ruiter and Van Faassen (1994) compared a multicompartment model (modified form of Jenkinson and Rayner, 1977) with a food-web model to simulate N mineralization dynamics in an arable system. Both models simulated annual N mineralization rates close to the observed ones, but failed to predict N immobilization.

The food-web models require information on physiological parameters such as assimilation efficiency, production efficiency, death rate, etc., for different functional groups of soil biota. As these values are not constant throughout the year/season, such models can calculate N mineralization kinetics only during a period under observation and are thus of descriptive value. Furthermore, these models do not include dynamics of soil organic matter that need to be considered along with the dynamics of soil biota.

Nitrification

Models for nitrification range from simple empirical relationships (Seifert, 1980; Hadas et al.,1986) to a set of mathematical equations based on enzyme kinetics (Paul and Domsch, 1972). Mechanistic models use Michaelis-Menten kinetics and describe the rate of nitrification as a function of substrate (NH_4^+) concentration and several biological parameters such as number of bacteria and their specific growth rate (Van Veen and Frissel, 1981). Hadas et al. (1986) describe nitrate production and ammonia disappearance by using a modified logistic equation (Equation 14.43)

$$NO_3^- = \frac{a}{1+(a/NO_3^-{}_0 - 1)\exp[-ak(t-t_0)]} \tag{14.43}$$

where a and $NO_3^-{}_0$ are the initial and final nitrate concentrations, t_0 is the initial time, and k is a constant. Seifert (1980) has proposed the use of the Gompertz function for soil incubation studies

$$NO_3^- = NO_3^-{}_{max}\exp[-m\exp(-kt)] \tag{14.44}$$

where m and k are constants and $NO_3^-{}_{max}$ is the final nitrate concentration. Such empirical models may fit experimental data better than the kinetic models, but the coefficients in these models have no biological meaning. Description of nitrification process in the large-scale N regime models is usually based on zero- or first-order kinetics.

Fixation and Release of NH_4^+

A part of the inorganic N as NH_4^+ in soil may undergo fixation reactions, which results in entrapment of NH_4^+ ions in interlayer spaces of phyllosilicates, in sites similar to K^+ in micas (SSSA, 1984). The fixation may occur spontaneously in aqueous suspensions, or as a result of heating to remove interlayer water. The fixed ammonium, also called nonexchangeable or interlayer NH_4^+, cannot be extracted with neutral normal potassium salt solutions (SSSA, 1984). Ammonium ions in collapsed interlayer spaces are exchangeable only after expansion of the interlayer.

Soils vary in their capacity to fix NH_4^+ ranging from a few kilograms to several thousand kilograms per hectare in the plow layer. The content of native-fixed NH_4^+ and the differential NH_4^+-N fixation capacity of soils from different regions may be linked to parent material (Table 14.2), texture, clay content, clay mineral composition, potassium status of the soil and K saturation of the interlayers of 2:1 clay minerals, and moisture conditions. The nonexchangeable NH_4^+ can either be indigenous/native-fixed during the genesis of silicate minerals or recently fixed as a result of organic and inorganic fertilizer application and organic N mineralization. The native-fixed ammonium may be released on geological weathering and essentially unavailable for plant uptake (Smith, Power, and Kemper, 1994; Liu, Laird, and Barak, 1997). The recently fixed NH_4^+ may be released during crop growth period and thus be available for plant uptake (Nieder and Benbi, 1996). Early investigations concluded that only a very small amount of nonexchangeable NH_4^+-N was available to microorganisms and plants. Investigations in the 1980s and 1990s suggest that fixed NH_4^+ may be slowly released and partially used by crops. There are conflicting reports on

TABLE 14.2. Concentration of Fixed NH_4^+ in Soils of Different Texture and Parent Material

Soil texture	Clay content (percent)	Fixed NH_4^+ (mg kg^{-1} soil)
Coarse-textured soils		
(Diluvial sand, red sandstone, granite)	<5-19	10-80
Medium-textured soils		
(Loess, marsh, alluvial sediment, basalt)	10-40	60-270
Fine-textured soils		
(Limestone, boulder clay, clay stone)	30-46	90-460

Source: Compiled from a number of published studies.

the role of fixed NH_4^+ in N dynamics in the soil-plant system. The results differ because of differences in methodology, soil type, mineralogical composition, and agroclimatic conditions.

Availability of Nonexchangeable Ammonium to Plants and Microbes

Fixation and release of clay-fixed NH_4^+ is dependent upon chemical equilibria between the amounts of clay-fixed NH_4^+, the exchangeable NH_4^+, and the NH_4^+ in soil solution. Release of fixed NH_4^+ follows the Elovich model suggesting that it is a heterogeneous diffusion process (Steffens and Sparks, 1997). With the lowering of NH_4^+ concentration in soil solution, NH_4^+ ions diffuse out of interlayers. Therefore, factors such as fertilizer N application, plant cover, soil organic matter, microbial biomass, clay content, and clay mineral composition that affect concentration of NH_4^+ in soil solution may promote either release or fixation of NH_4^+.

Plant Availability. Most investigations on clay-fixed NH_4^+ have focused on the availability of this fraction to crops with conflicting results. Legg and Allison (1959) and Black and Waring (1972) found that the nonexchangeable NH_4^+ played a minor role in plant nutrition. Kowalenko and Cameron (1976) found that a substantial amount of fertilizer NH_4^+ bound in a nonexchangeable form was available to crops. Amounts ranging from 35 to 72 percent of fertilizer-derived fixed NH_4^+-N were released for uptake by ryegrass (Norman and Gilmour, 1987). As percentage of total nonexchangeable NH_4^+-N, the release of this fraction ranged from 4 to 25 percent in different soils (Osborne, 1976; Smith, Power, and Kemper, 1994; Steffens and Sparks, 1997).

Variable results on plant availability of nonexchangeable NH_4^+-N may be ascribed to the fact that some investigators distinguish between native- and recently fixed NH_4^+ while others do not. Recently fixed NH_4^+ is mainly fertilizer derived (Chen, Turner, and Dixon, 1989; Smith, Power, and Kemper, 1994), but it may also originate from mineralized soil organic matter. Apparently, recently fixed NH_4^+ is more available to plants than native, nonexchangeable NH_4^+ which is more tightly held (Black and Waring, 1972; Mengel and Scherer, 1981). This is probably because native-fixed NH_4^+ may be trapped more in the center of the interlayers, while recently fixed NH_4^+ may be largely retained in the peripheral zone of the interlayers (Scheffer and Schachtschabel, 1984). Differences in clay mineralogy and K saturation of the minerals also influence the release of clay-fixed NH_4^+. Mengel, Horn, and Tributh (1990) have found that only soils containing vermiculite and low in exchangeable K^+ released significant amounts of nonexchangeable ammonium as compared to soils with no vermiculite and high percentage of K saturation of the clay.

The mechanisms regulating the release and subsequent plant uptake are not completely understood. According to one argument (Mengel, Horn, and Tributh, 1990), plant roots deplete NH_4^+ concentration in soil solution to a very low level and, therefore, promote the release of nonexchangeable NH_4^+. According to another argument, it is the relative amount of ammonium and potassium in soil solution that governs the rate of release of these ions from interlayer positions (Hanway and Scott, 1956; Welch and Scott, 1960). When large amounts of these ions are present, they tend to exclude other cations from the interlayer positions because their unhydrated dimensions allow them to fit better, and they are not as strongly hydrated as sodium and calcium ions. Under field conditions, continued uptake of ammonium and potassium ions by growing plant roots may reduce concentrations of these ions so that the blocking effect is diminished and release of fixed ammonium from the soil can proceed at significant rates. However, due to similarity in the sizes of K^+ and NH_4^+ ions, the two ions are believed to compete with each other for surface-exchange sites.

Availability to Soil Microbes. It is not clear whether plant uptake relies primarily on direct assimilation of released NH_4^+ by plant roots or on processes mediated by soil microorganisms (Breitenbeck and Paramasivam, 1995). Early work concerning the influence of microbial biomass on clay-fixed NH_4^+ suggests that soil microbes play only a minor role because fixation greatly reduces the availability of NH_4^+ to both the nitrifiers and the heterotrophs. More recent studies suggest that heterotrophic microorganisms can rapidly assimilate NH_4^+ from the clay-fixed fraction (Drury and Beauchamp, 1991). These findings and results from field experiments during the cereal-growing season (Nieder et al., 1995, 1996) suggest that microbial biomass may exert a great influence on the dynamics of fixed NH4+. In the later studies, the minima of mineral N and, correspondingly, of the contents of fixed NH_4^+ occurred during phases of maximum amounts of microbial N and vice versa. Obviously, heterotrophic microbes bind soil N by promoting the release of fixed NH_4^+, primarily under a C-supplying plant cover.

The phenomenon of temporary fixation and release of added fertilizer NH_4^+ may help in reducing N losses from the soil-plant system. The extent of nitrate leaching has been found to partly depend on the NH_4^+ fixation capacity of the soil. For example, the losses of added fertilizer N ($^{15}NH_4$) were more than double in a Histosol with an extremely low NH_4^+-fixing capacity as compared to that from a gley soil with a high fixing capacity (Fischer et al., 1981). Similarly, NH_4-N fixation by clay minerals may also help in reducing NH_3 volatilization losses (Dou and Steffens, 1995). Therefore, clay fixation of NH_4^+ can provide a significant sink for fertilizer N that subsequently acts as a source of N for plant uptake.

CROP N UPTAKE

Plants take up nitrogen in its ionic forms of NH_4^+ and NO_3^-. The process of uptake involves the transport of water-soluble species to the roots followed by their absorption across the root surface. The uptake of a nutrient per unit root length *(U)* is given by (Equation 14.45) the product of the nutrient ion concentration in the soil solution at the root surface (C_l) and the root absorbing power (α) as:

$$U = 2\pi\, r\, \alpha\, C_l \qquad (14.45)$$

where r is the root radius (Nye and Tinker, 1977). The ion concentration at the root surface (C_l) depends on the root absorbing power and the rate of ion movement from the bulk soil toward the root. The root-absorbing power (α) for a nutrient specie and plant specie is rarely constant over a range of nutrient concentration. It depends upon the metabolic conditions and, therefore, changes during the crop growth period. Warncke and Barber (1973) developed empirical equations relating α to NH_4^+ and NO_3^- concentration for corn.

Mass flow and diffusion are the two major processes by which NO_3^- and NH_4^+ are transported to the root. Mass flow has been assumed to be the main mechanism for moving NO_3^- toward the roots (Renger and Strebel, 1976; Benbi, Prihar, and Cheema, 1991). This is based on the assumption that the amount of NO_3^- extracted from the soil depends only on the amount of water taken up by the crop. However, Liao and Bartholomew (1974) have shown that N uptake by corn can be less than, equal to, or greater than that predicted by mass flow alone, depending on a number of soil and plant factors. In situations where plant N content is high, plant processes can discriminate against N uptake so that the actual uptake is less than the amount transported to the root surface by the transpiration stream. In sandy soils, mass flow could account for only 40 to 60 percent of the N uptake by corn (Watts and Hanks, 1978). Strebel et al. (1980) report that during early development of spring wheat 50 percent of the ^{15}N-labeled fertilizer NO_3^- was transported by mass flow, but the percentage dropped with crop age so that in later stages more NO_3^- was transported by diffusion than by mass flow. In a study on corn, Okajima and Taniyama (1980) report that diffusion contributed 8.8 percent during 11 to 25 days after sowing and 59.7 percent during 25 to 35 days after sowing toward NO_3-N supply to the plant roots.

Convective flow of water toward roots in response to transpiration results in the transport of NH_4 and NO_3 to the roots along with the water. The concentration of these ions at the root surface decreases when the rate of root uptake exceeds the rate of supply of these ions by mass flow. Diffusion of NH_4 and NO_3 toward the roots then occurs due to concentration gradient, which can be described by Fick's law. Liao and Bartholomew (1974) have provided evidence that diffusion can play a major role and under certain conditions can be the dominant mechanism in N transport to the root system. Models have been developed to compute the relative contribution of mass flow and diffusion toward crop nutrient uptake (Davidson, Rao, and Selim, 1977; Phillips et al., 1976; Watts and Hanks, 1978; Stockle and Campbell, 1989).

Modeling Crop N Uptake

Like most of the other N-cycle processes, crop N uptake can be modeled at several levels of complexity. The approaches range from the simplest such as mass flow in the transpiration stream to the most complex such as considering the root geometry and flux of mineral N to plant roots. Benbi, Prihar, and Cheema (1991) have used the mass flow approach and observed that model-predicted N uptake for wheat grown in Punjab (India) matched closely with the measured N uptake (Figure 14.4). The predictions were closer in the relatively wet treatment suggesting that under dry moisture regime, in addition to mass flow, other processes of N uptake such as diffusion and root interception may have a relatively greater contribution. Watts and Hanks (1978) incorporate both mass flow and diffusion in their model. Schenk and Barber (1979) use the approach presented by Claassen and Bar-

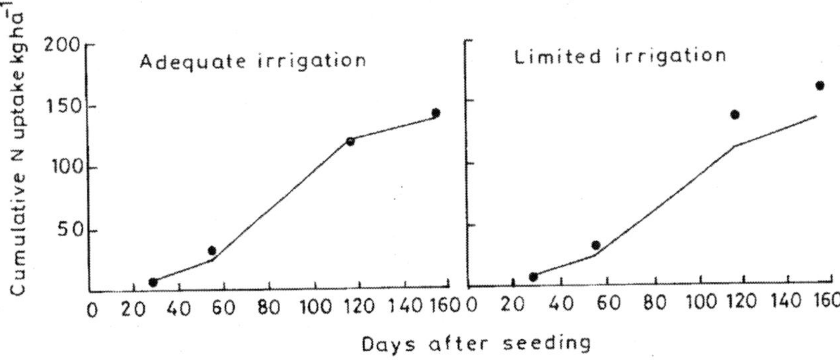

FIGURE 14.4. Modeled (line) versus measured (points) N uptake by wheat grown in Punjab, India. (*Source:* Adapted from Benbi, Prihar, and Cheema, 1991, p. 82.)

ber (1977) to predict the flux of nutrients to plant roots and their uptake by plants. In this approach, supply of N from the soil to the root is assumed to be from a combination of mass flow and diffusion. Crop uptake is assumed to follow a Michaelis-Menten relationship between concentration and influx. Details of the approach have been presented by Barber (1984).

There are also more fundamental approaches to modeling N uptake. Some (Hutson and Wagenet, 1992) assume that uptake of N depends on the length of root, and some (Kersebaum and Richter, 1991) explicitly simulate the flux of solute from the bulk soil to the root. Such models need to compute the length of root per unit volume and they assume (Hansen et al., 1991) that uptake occurs from a cylinder of soil around the root. The interaction between these cylinders of uptake when they intersect, as must happen, usually seems to be ignored. Tillotson and Wagenet (1982) and Hutson and Wagenet (1992) have modeled plant N uptake based on the theory of Nye and Tinker (1977; Equation 14.45). It involves the use of two parameters, root radius and root absorption coefficient (α), which are difficult to obtain under field conditions. Warncke and Barber (1973) showed that the value of α has to be established for different soils, crops, and N species under a range of concentrations.

Several models assume N uptake to be driven by demand from the plant. They calculate N demand for growth rates and the concentration of N required in various plant organs by accounting for gross photosynthesis, respiration, and the development of the plant (Spitters, van Keulen, and van Kraalongen, 1989; Hansen et al., 1991; Groot and De Willigen, 1991). Whitmore and Addiscott (1987) have computed demand using a simple variant of the logistic function to compute the pattern of N uptake *(Y)*, with thermal time *(x)*:

$$Y = \left(A^{-1/n} + e^{-kx}\right)^{-n} \tag{14.46}$$

where A is the maximum of Y, n is a shape factor, and k is a rate constant.

Although widely varying approaches have been used to model plant N uptake, there seems to be little evidence that more complex models give better simulations than simpler ones (De Willigen, 1991).

MODELING NITROGEN DYNAMICS IN THE SOIL-PLANT SYSTEM

A number of simulation models exist that describe some or all of the N-cycle processes occurring in soil-plant system. The level of resolution and the processes considered in different models vary depending on the goals and the intended use of the models. Tanji (1982) has reviewed system-level models and provided an in-depth description and appraisals of selected N models. A detailed examination of current models has been published in Groot, de Willigen, and Verberne (1991). Although some models are concerned with specific aspects of the N cycle such as ammonia volatilization (Rachhpal-Singh and Nye, 1986a), leaching (Burns, 1974), and denitrification (Smith, 1990), others simulate most of the major processes related to N dynamics in the soil-plant system. Several of these models aim to examine the overall effects of management practices on C and N flow through the system (Thornley and Verberne, 1989), or the long-term changes in soil N dynamics (Wolf, de Wit, and van Keulen, 1989). Others are designed to provide fertilizer recommendations (Neeteson, Greenwood, and Draycott, 1987; Richter, Kersebaum, and Uttermann, 1988). Models have also been developed with the specific purpose of examining aspects of pollution from organic wastes and from excessive fertilization (Donigian and Crawford, 1976; Rao, Davidson, and Jessup, 1981; Selim and Iskandar, 1980).

Although the concepts adopted in the modeling of various N-cycle processes have been discussed in the preceding sections, a brief description of some models simulating total N regime in the soil-plant system is presented as follows. The description of most of these models is essentially condensed from that presented on the Internet at Register of Ecological Models.

ANIMO (Agricultural NItrogen MOdel)

ANIMO (Rijtema and Kroes, 1991) simulates the C, N, and P cycles in unsaturated and saturated soil systems. The model was developed to analyze the leaching of N from the soil surface to groundwater and surface waters. Main processes modeled are: mineralization and immobilization (related to the decomposition of C); crop uptake (based on transpiration flux and/or diffusion); denitrification related to (partial and temporal) anaerobiosis and decomposing organic materials; oxygen and temperature distribution in the soil; nitrification (as a first-order process); desorption and adsorption of NH_4^+ and P to the soil complex; runoff; discharge to different surface water systems; and leaching to groundwater. The ANIMO model has a pseudo two-dimensional transport module to calculate solute transport to different drainage systems. The input to the model includes an extended soil water balance, and soil physical and chemical properties.

Benbi's Model

The model presented by Benbi, Prihar, and Cheema (1991) is made up of two submodels: a water balance submodel for simulating components of field water balance and a nitrogen submodel for simulating N dynamics in the soil-plant system. The N processes modeled are: urea hydrolysis, mineralization, nitrification, NH_3 volatilization, NO_3^- leaching, and plant N uptake. The input to the model includes initial soil NH_4-N and NO_3-N profiles, atmospheric evaporative demand, leaf-area development, and root-growth characteristics of the crop. Leaf-area development of the crop could be generated by using the crop-growth submodel (Benbi, 1994).

Bradbury's Model

The model presented by Bradbury et al. (1993) has a compartmental structure and runs on a weekly time step. Nitrogen input processes considered are atmospheric deposition and application of fertilizers or organic manures. Nitrogen loss pathways considered are denitrification (from 0-25 cm layer only and assumed to be proportional to the quantity of CO_2 produced by a given soil layer and its NO_3-N content), leaching (based mainly on piston flow and through bypass flow if the rainfall exceeds a specified threshold value), and volatilization. Organic nitrogen is contained within three of the model compartments—crop residues, soil microbial biomass, and humus. Inorganic nitrogen is held in two pools as NH_4^+ or NO_3^-. Nitrogen flows in and out of these inorganic pools as a result of mineralization, immobilization, nitrification (by a first-order process), leaching, denitrification, and plant uptake. The model is designed to be used in a "carry forward" mode—one year's run providing the input for the next, and so on. The model also allows the addition of ^{15}N as labeled fertilizer, and follows its progress through crop and soil. The model requires a description of the soil and the meteorological records for the site—mean weekly air temperature, weekly rainfall, and weekly evapotranspiration. The model has many ideas in common with SOILN (Bergström, Johnsson, and Torstensson, 1991), DAISY (Hansen et al., 1991), and NCSOIL (Molina et al., 1983).

CANDY (Carbon and Nitrogen Dynamics in Soils)

CANDY (Franko, Oelschlägel, and Schenk, 1995) simulates dynamics of soil N, temperature, and water to provide information about N uptake by crops, leaching, and water quality. CANDY uses a semicohort system to track litter decay, and calculates a biologically active time to allow comparison among sites. Decomposition of soil organic matter is described by multiple pools as: active organic matter (defined by decomposition constant, reaction constant, C/N ratio obtained by fitting), stable organic matter (defined by reaction constant, C/N ratio), and inert organic matter (obtained by fitting). Factors assumed to affect organic matter decomposition include soil moisture, soil temperature, clay content, and nitrogen content.

CENTURY (Grassland and Agroecosystem Dynamics Model)

CENTURY (Parton et al., 1987; Parton, Stewart, and Cole, 1988) is a process-based model that is used to describe C and nutrient dynamics for grasslands, agricultural lands, forests, and savannas. The model uses monthly time steps for simulations of up to several thousand years to examine the flows of C, N, and P. The input data required are monthly mean maximum and minimum temperatures, precipitation, soil texture and soil depth, plant N, P, K, and lignin content, atmospheric and soil N inputs, and initial soil C and N. The model has been used for global climate change research. It can be used to assess the impacts of regional climate change on a variety of important grassland ecosystems.

CREAMS (Chemicals, Runoff, and Erosion from Agricultural Management Systems)

CREAMS (Knisel, 1980) is a field scale model for predicting runoff, erosion, and chemical transport from agricultural management systems. It is applicable to field-sized areas. Water movement through the soil profile is modeled using a simple capacity approach, with flow occurring when a layer exceeds field capacity. The plant nutrient submodel of CREAMS has an N component that considers mineralization, nitrification, and denitrification processes. Plant uptake is estimated, and nitrate leached by percolation out of the root zone is calculated. The model estimates N and P transported with sediment by using enrichment ratios.

DAISY (Danish Simulation Model for Transformation and Transport of Energy and Matter in the Soil Plant Atmosphere System)

DAISY (Hansen et al., 1991) simulates crop production and dynamics of soil water and N under diverse agricultural management systems. DAISY includes a hydrological model with a soil water flow submodel. The N cycling processes considered are turnover of organic matter (based on C pools, microbial biomass, and first-order kinetics), mineralization/immobilization (a consequence of the carbon turnover), nitrification, denitrification, N uptake by plants (based on single root uptake and root density), and N transport (convection-dispersion equation) and leaching. The soil organic matter is divided into four fractions dependent upon history and origin. A fraction is further divided into two pools with fast and slow turnover rates. Each of the four main organic matter fractions is specified by a minimum of seven constants. Some of the required parameters may be difficult to access experimentally. The C and N submodel requires initialization of the chemical and biological processes and several parameters describing the transformation characteristics of the soil. These include rate constants for nitrification and denitrification, impedance factors, longitudinal dispersivity, NH_4-adsorption -isotherm, and average wet and dry deposition rates/concentrations. The model allows for the simulation of different management strategies and crop rotations.

Frissel and Van Veen Model

This model (Frissel and Van Veen, 1981) simulates N transformations in soil and is based on the assumption that SOM can be represented by several carbon and nitrogen pools. Nitrogen mineralization and immobilization is controlled by C/N ratio. Consideration is made for differences in decomposition rates of organic compounds in plant residues for amino acids, cellulose, lignin fractions, and microbial biomass. The biomass growth rate is controlled by the carbon availability from the added soil pool, and it is assumed that there is no change in the microbial population if no carbon is added to the soil. Nitrogen immobilization is proportional to biomass growth and it is assumed that mineralization occurs simultaneously and independently of immobilization.

LEACHM (Leaching Estimation and Chemistry Model)

LEACHM (Hutson and Wagenet, 1992) refers to a suite of four simulation models indented to describe water (LEACHW), pesticides (LEACHP), N and P regimes (LEACHN), and salinity in calcareous soils (LEACHC). The models simulate chemical fate and transport in transient-flow field situations as well as in laboratory columns subject to steady state or interrupted flow. Water flow and solute transport is described by the Richards and convective-dispersion equations, or by a modified mobile/immobile capacity (tipping bucket) concept. The model predicts amount of chemical leaching below the root zone and amount taken up by the plants. The model uses a daily time step. The input to the model includes soil physical properties, bulk density, particle size distribution, water retention characteristic, amount of N and P, daily maximum and minimum temperature, precipitation and evapotranspiration estimates.

MANAGE_N (Decision Support System for Nitrogen Management in Rice)

MANAGE_N (Riethoven, ten Berge, and Drenth, 1995) is a decision-support tool for nitrogen management in rice. When supplied with the proper parameter values, MANAGE_N provides information on the highest attainable yield versus fertilizer level, the optimal time for N application at each N input level, yield response to specific split N application schemes, and patterns of N uptake and biomass accumulation. The core of the MANAGE_N package is the explanatory (process-based) simulation model for biomass accumulation in rice as limited by nitrogen ORYZA_O (Drenth, ten Berge, and Riethoven, 1994).

MINERVA (Model for Simulation of Nitrogen Dynamics in Arable Soils)

This model (Kersebaum and Richter, 1991; Beblik, 1996) was designed for advisory purposes and uses relatively simple approaches so that it can be used under restricted data availability. The main processes considered are N mineralization, denitrification, transport of water (based on capacity approach) and N (using convection-dispersion equation), crop growth, and N uptake by crops. The submodel for N transformations follows the concept of net mineralization from two pools of potentially mineralizable N according to first-order kinetics. Pool sizes are derived as a fixed percentage of soil organic matter (13 percent) and the composition of different crop residues and manure. N mineralization is limited to the top 30 cm of the soil. Denitrification is simulated for the topsoil using Michaelis-Menten kinetics modified by reduction functions dependent on water-filled pore space and temperature. The maximum denitrification loss per time is related to the total carbon content of the plowing layer. The model operates on a daily weather data basis, soil information and management data including organic matter content, and C/N ratio of the plow layer, texture, etc. The extended version of the model can be used in combination with geographic information systems for site-specific simulations in medium-scaled regions. The process formulation in the model is similar to HERMES (nitrogen balance in plants simulation) (Kersebaum, 1995).

MOSOM (MOdelling Soil Organic Matter)

MOSOM (Verberne et al., 1990) describes C and N cycling in different soil types and predicts daily net mineralization. Soil organic matter is divided into four fractions (including microbial biomass pool) and three fractions of residues. Their chemical composition and spatial distribution in the soil determine the availability of organic materials as a substrate for microorganisms. The microbial biomass and mineralization rates are assumed to be affected by the extent of physical protection within the soil. Transformation rates are described by first-order kinetics. The inputs to the model include moisture and oxygen status, carbon dioxide pressure, fraction of decomposable, structural, and resistant material, and microbial biomass. The model is a descendent of the Frissel and Van Veen model (1981) and has concepts adopted from Van Veen and Paul (1981), Van Veen, Ladd, and Frissel (1984), and Van Veen, Ladd, and Amato (1985).

NCSOIL (Nitrogen and Carbon Transformation in Soil)

NCSOIL (Molina et al., 1983) describes the changes in organic forms of C and N, exchangeable plus soluble ammonium, and nitrate concentrations that result from the processes of residue decomposition, mineralization, immobilization, nitrification, denitrification, and nonsymbiotic N_2 fixation.

These transformation rates are soil water and temperature dependent. A mechanism is included to account for the increase in mineralization following a physical disturbance (tillage) of the soil. The model is built on the concept of catenary sequence of heterogeneous substrates. The soil organic C and N substrates are defined by their kinetic rate constants and their position in the model's structure. Residues are defined in terms of their chemical, or morphological, characteristics. Eight C and N pools are considered: the organic residues, pool I and pool II of the soil active organic phase, pool III of the stable organic phase, exchangeable plus soluble ammonium, nitrate, dinitrogen, and carbon dioxide. NCSOIL is a submodel of the NCSWAP model.

NCSWAP (Simulation Model of the Soil-Crop-Water Biosystem)

NCSWAP (Clay et al., 1989) simulates water, C and N dynamics in soil through the concept of water, N, and temperature stress, expressed as reduction factor functions. It compares crop yield from a reference yield which is modified to account for the influence of water, nitrogen, and temperature. The model can be used for prediction of nitrification, denitrification, NO_3-N leaching, N mineralization and immobilization, plant available NO_3-N and NH_4-N, and plant N uptake. The input to the model includes meteorological data, soil physical properties, and initial NO_3-N and NH_4-N profiles.

N_DICEA (Nitrogen Dynamics in Crop Rotations in Ecological Agriculture)

This nitrogen dynamics model is designed to calculate the availability of nitrogen in a crop rotation. It includes a simple water balance to estimate soil moisture content and drainage (Habets and Oomen, 1993). Mineralization of soil organic matter and added crop residues and organic manure are calculated using Janssen's one parameter formula for the decomposition of organic matter with a correction for soil moisture and temperature. Crop uptake is based on actual yield data and derived amounts of crop residues and roots. Fixation by legume crops is estimated. A nitrogen balance is made including mineralization, deposition, fixation, uptake, and (leaching) loss.

NLEAP (Nitrate Leaching and Economic Analysis Package)

Nitrate leaching and economic analysis package (NLEAP) is a process-oriented simulation model (Shaffer, Halverson, and Pierce, 1991) for predicting potential NO_3 leaching below the root zone and to groundwater supplies associated with agricultural practices. The model includes mineralization of soil organic matter and added organic matter. The model contains a fast and a slow pool for soil organic N, and mineral N pools for NO_3-N and NH_4-N. It uses basic information concerning on-farm management practices, soils, and climate to project N budgets and NO_3 leaching indices. The model requires soil properties and daily weather and management data for input. The model can be executed over a range of time scales, from a single event to annual. The spatial scale of the model is field to regional. The model has been calibrated/validated against lysimeter, field plot, and farm field data from eight states.

NTRM (Nitrogen-Tillage-Residue Management)

The NTRM model is a simulation model examining soil-plant interactions. The model is used to determine soil management strategies and the effect of climate change on crop yields and NO_3 leaching (Shaffer and Larson, 1982; Shaffer, 1985). The model combines basic process equations for water flow, solute transport, plant and root growth, tillage effects, climate interactions, nutrient transformations, soil chemical processes, and soil temperature. This is an application-oriented model and provides a means of evaluating management options, which include supplemental N, conservation tillage, and irrigation on the eroded profiles. The model requires soil properties and daily weather data and management options as input. The model uses daily time steps and examines field/plot to regional spatial scales using geographic information systems (GIS).

PHOENIX

The PHOENIX model (McGill et al., 1981) makes fundamental improvements in the coordinated treatment of C and N transfers, decomposition, mineralization, and soil organic matter dynamics, but omits fauna and includes a relatively simple plant model. Unique features of PHOENIX include: concurrent mineralization and immobilization of N; population density effects on decomposition; density-dependent death of microorganisms; fluctuating C/N ratio of microorganisms; the manner in which litter is partitioned which implies faster internal recycling of N than of C from fresh litter; a cascading system of SOM cycling; inclusion of two populations of microorganisms; and microbial death through predation. The model is intended to examine the interrelationship of C and N and their recycling within soil.

Smith

Smith (1979) has developed a general model to simulate the carbon, nitrogen, phosphorous, and potassium dynamics in soils. The processes modeled include: rainfall and runoff, soil hydrologic balance, soil moisture, soil temperature, and dynamics of C, N, P, and K. Most of the transformations are moderated by microbes, and the dynamics of microorganisms are explicitly represented. In addition to the simulation of biological aspects of decomposition, the model treats the physicochemical processes of precipitation, fertilizer and native mineral inputs, leaching loss, sorption of organic and inorganic ions on soil colloids, condensation between organic and aromatic compounds, and exchange reactions.

SOILN (Simulation Model for Nitrogen Conditions in Soils)

The SOILN model (Eckersten and Jansson, 1991) is designed to simulate transport and transformations of nitrogen in soils and its uptake by plants. It uses some of the output data from the SOIL model as input. SOILN includes the following processes: mineralization of humus, mineralization/immobilization of C and N fraction in crop residue and the manure, nitrification, denitrification, NO_3 leaching, and plant uptake. Nitrogen treatment by plants depends on the plants. Two plant submodels are included for describing plant N uptake either for crops or trees. The soil mineral N pools receive N by mineralization of litter and humus, nitrification, fertilization and deposition and loose nitrogen by immobilization to litter, nitrification, leaching, denitrification, and plant uptake.

TEM (Terrestrial Ecosystem Model)

TEM is a process-based ecosystem model (McGuire et al., 1993) that describes C and N dynamics of plants and soils for nonwetland ecosystems. It uses spatially referenced information on climate, elevation, soils, vegetation, and water availability, as well as soil- and vegetation-specific parameters to make monthly estimates of important C and N fluxes and pool sizes. Hydrological inputs for TEM are determined by a water balance model that uses the same climate data and soil-specific parameters as used in TEM. The TEM operates on a monthly time step and at a 0.5 degrees latitude/longitude spatial resolution. N uptake in the model is controlled by stochiometric C:N ratio of biomass production. In TEM, decomposition is a function of single soil organic C compartment, temperature, and soil moisture. The C and N pool sizes of vegetation and soil are affected by dynamic C and N fluxes (net primary productivity, litter fall C, decomposition, litter fall N, net N mineralization, N uptake, etc.).

WELL_N (Model for Sustainable Use of Nitrogen Fertilizer)

WELL_N is designed to predict crop N requirements for 24 arable and horticultural crops, together with the final N content of the crop and the potential for NO_3-N leaching to groundwater (Rahn, Greenwood, and Draycott, 1996). The model takes account of all the major factors affecting N status of field and crop including expected yield, geographical position of the field, local weather, and soil texture, as well as the contribution of N from previous crop residues.

WHNSIM (Water, Heat, and Nitrogen Simulation)

This model simulates water, heat, and N budget of agricultural fields. Water transport is modeled with the Richards equation, heat transport is simulated with the heat transport equation (advection and diffusion), and NO_3-N transport is described by the advection-dispersion equation (Huwe, 1987).

From the foregoing, it is apparent that there are a large number of models at different levels of complexity with a varying number of processes. Obviously, there exist a number of other models as well. However, there are very few reports on the verification and validation of these models, particularly by researchers other than the authors of the models. It seems that it is much easier to develop a new model than to verify or validate existing computer codes. Application of N dynamics models is often complicated by limited information on parameter values and appropriate boundary conditions. Generally, models are parameterized by calibration against other measurements or by back calculations. However, many different parameter combinations can produce similar good simulation results with respect to available calibration data (termed as equifinality). Schulz, Bevan, and Huwe (1999) have shown that for a simple N budget model applied to agricultural catchment, the equifinality of different parameter sets results in large uncertainty in the predictions. They also suggest that introducing more complexity into the process description is very unlikely to allow this uncertainty to be constrained.

EVALUATION, COMPARISON, AND APPLICATION OF MODELS

The evaluation and comparison of different models presented in this section is essentially condensed from the review by De Willigen (1991), Vereecken et al. (1991), and Richter and Benbi (1996). Vereecken et al. (1991) evaluated five complex models (ANIMO, DAISY, EPIC, RENLEM, and SWATNIT) with regard to their ability to simulate components of water and N regimes (N-mineralization, denitrification, nitrate leaching, and crop N uptake) on data sets from differentially cropped sandy soils from Denmark and the Netherlands. Five statistical criteria were used to evaluate simulations (Table 14.3).

According to Vereecken et al. (1991), the modeling efficiency for plant N uptake was higher as compared to NO_3^- leaching (Table 14.4). Considering the substantial variation in measured N uptake, the performance of the relevant submodels appeared to be good. From a comparison of 14 models De Willigen (1991) also concluded that aboveground variables were simulated more accurately than belowground variables. However, the problem with fertilizer recommendations is more related to belowground processes such as leaching coupled with water flow. For simulating water storage and movement, the models showed systematic differences between the capacity type (EPIC) and the "mechanistic" models (ANIMO, DAISY, and SWATNIT) for a Danish experimental site (Vereecken et al., 1991). Despite the different approaches for N mineralization and water movement, a comparison with measured nitrate leaching revealed that ANIMO, DAISY, and EPIC were similar in quality (Table 14.4). DAISY appeared to be a bit superior in behavior, because it explained 89 percent (EF) of the measured variability with a RMSE of only 27 percent, but it did so on a reduced sample of only 29 datasets out of 43. RENLEM showed a tendency to overestimate leaching, whereas SWATNIT did the opposite. EPIC overestimated leaching by about 40 kg ha^{-1}. Essentially there was no difference in simulation quality due to different approaches adopted in the models (Vereecken et al., 1991). Following this conclusion, the question may arise as to whether the quality of estimates by simulation of nitrate leaching has really improved in comparison, e.g., to the empirical graphical estimates of Kolenbrander (1981). The application of Kolenbrander's approach to data gathered by Walther (1989) shows that it is still a valuable approach (Richter and Benbi, 1996).

Most of the models for N mineralization used arbitrary approaches rather than empirical approaches and generally behaved expectedly poor. None of the models could account for the apparent mineral N loss observed shortly after application of fertilizers in late spring or early summer (De Willigen, 1991). The model discrepancies were related to lack of microbial pool. In

TABLE 14.3. Statistical Criteria for Evaluation of the Simulated Results According to Loague and Green (1991)

Criteria	Symbol	Calculation formula	Range	Optimum
Maximum Error	ME	$Max\lvert P_i - O_i \rvert_{i=1}^{n}$	≥0	0
Root Mean Square Error	RMSE	$\sqrt{\sum_{i-1}^{n}(P_i - O_i)^2 / n} * 100 / \overline{O}$	≥0	0
Coefficient of Determination	CD	$\sum_{i=1}^{n}\left(O_i - \overline{O}\right)^2 \Big/ \sum_{i=1}^{n}\left(P_i - \overline{O}\right)^2$	≥0	1
Modeling Efficiency	ME	$\left(\sum_{i=1}^{n}\left(O_i - \overline{O}\right)^2 - \sum_{i=1}^{n}(P_i - O_i)^2\right) \Big/ \sum_{i=1}^{n}\left(O_i - \overline{O}\right)^2$	≤1	1
Coefficient of Residual Mass	CRM	$\left(\sum_{i=1}^{n} O_i - \sum_{i=1}^{n} P_i\right) \Big/ \sum_{i=1}^{n} O_i$	≤1	0

P_i = predicted value i, O_i = observed value i, $\overline{O}$ = mean of observed values, n = number of data pairs

TABLE 14.4. Statistical Evaluation of the Performance of Different Models on Simulation of Crop N Uptake and Nitrate Leaching (kg ha^{-1} y^{-1}) on Common Data Sets from Europe

Model	n	ME	RMSE	CD	EF	CRM
Plant N uptake						
ANIMO	44	172.2	19.4	1.17	0.94	0.04
DAISY	29	60.0	15.2	0.92	0.97	-0.08
EPIC	44	191.0	22.9	0.92	0.90	0.00
SWATNIT	39	120.9	17.0	1.06	0.95	-0.05
Nitrate leaching						
ANIMO	43	61.8	36.2	0.96	0.78	0.00
DAISY	29	39.7	27.2	1.09	0.89	0.05
EPIC	43	53.7	37.1	1.17	0.77	0.01
RENLEM	43	55.4	53.4	0.83	0.52	0.27
SWATNIT	38	85.7	52.6	0.63	0.53	-0.27

Source: Adapted from Vereecken et al., 1991, pp. 327, 330.

n = number of observations; for an explanation of the abbreviations see Table 14.3.

addition, possible explanations of seasonal variation of soil mineral N on the basis of changing the relative extent of ammonification and nitrification, together with a change of NH_4^+ sorption or fixation, and seasonal variation of microbial biomass and/or N seems to be of major importance. To attain a higher degree of agreement between simulation and measurements, future developments in modeling should concern essential processes such as denitrification, ammonia volatilization, and ammonium exchange and fixation-defixation of NH_4^+ (Richter and Benbi, 1996).

SUMMARY

Nitrogen is an essential nutrient and has significance in environmental and health problems. The nitrogen forms such as NH_3, NH_4^+, NO_3^-, N_2, and organic N compounds undergo several transformations in the environment as part of what is called the N cycle. The N-cycle processes include the input processes (atmospheric wet and dry deposition, biological nitrogen fixation, and fertilization), the loss processes (erosion and runoff, NO_3^- leaching, NH_3 volatilization, and denitrification), the transformation processes (mineralization and immobilization, nitrification, fixation, and release of NH_4^+), and plant uptake. In this chapter, the theories behind various N-cycle processes, approaches to model these processes, and the extent of their occurrence in different soil-plant systems have been described.

Widely varying approaches have been adopted for modeling various N-cycle processes. For most of the processes, the modeling approaches range from simple empirical to highly complex, process-based mechanistic approaches. But there is no evidence that increasing the model complexity increases simulation quality. Soil variability appears to be the major problem with modeling and measurement of nitrate leaching particularly because of the small-scale heterogeneity that gives rise to mobile and immobile categories of water. Mineralization is one of the most difficult processes to model as the N source (soil and added organic matter) and the microbes that mineralize N are poorly characterized. Simple function models (ranging from first-order, single compartment to multicompartment) predict net N mineralization without any consideration for the separate processes of ammonification and nitrification. Kinetic models with two pools of mineralizable N appear to describe N mineralization fairly well. Although such models are a gross simplification of the organic matter turnover process, they are easy to parameterize and are in accordance with the principle of parsimony. Process-based mechanistic models attempt to simulate gross mineralization and associated immobilization. These models generally characterize SOM in terms of its C and N content, and therefore simulate the C and N cycle. These simulate organic matter decomposition by defining different fractions of organic matter as functional pools on the basis of their presumed chemical characteristics, fixed C/N ratio, and location in the soil. Although such models are valuable for understanding the processes, usually they require too large a number of input parameters, some of which are difficult to obtain under field conditions. Therefore, it is not surprising that there are very few reports on the verification and validation of these models. Few data sets are available for testing a range of models and a few models have been tested on a range of soils and climatic conditions.

REFERENCES

Addiscott, T.M. (1977). A simple computer model for leaching in structured soils. *Journal of Soil Science 28:* 554-563.

Addiscott, T.M. (1983). Kinetics and temperature relationships of mineralization and nitrification in Rothamsted soils with differing histories. *Journal of Soil Science 34:* 343-353.

Addiscott, T.M. (1996). Measuring and modeling nitrogen leaching: Parallel problems. *Plant and Soil 181:* 1-6.

Addiscott, T.M. and D. Cox (1976). Winter leaching of nitrate from autumn-applied calcium nitrate, ammonium sulphate, urea and sulphur-coated urea in bare soil. *Journal of Agricultural Science, Cambridge 87:* 381-389.

Addiscott, T.M. and R.J. Wagenet (1985). Concepts of solute leaching in soils: A review of modelling approaches. *Journal of Soil Science 36:* 411-424.

Amoozegar-Fard, A., D.R. Nielsen, and A.W. Warrick (1982). Soil solute concentration distributions for spatially varying pore water velocities and apparent diffusion coefficients. *Soil Science Society of America Journal 46:* 3-9.

Amundson, R.G. and E.A. Davidson (1990). Carbon dioxide and nitrogenous gases in the soil atmosphere. *Journal of Geochemical Exploration 38:* 13-41.

Andrén, O and K. Paustian (1987). Barley straw decomposition in the field: A comparison of models. *Ecology 68:* 1190-1200.

Arah, J.R.M. (1990). Diffusion-reaction models of denitrification in soil microsites. In *Denitrification in Soil and Sediment,* eds. N.P. Revsbech and J. Sorensen, New York: Plenum Press,pp. 245-258.

Arah, J.R.M. and K.A. Smith (1989). Steady state denitrification in aggregated soils: A mathematical model. *Journal of Soil Science 40:* 139-149.

Arah, J.R.M., K.A. Smith, I.J. Crichton, and H.S. Li (1991). Nitrous oxide production and denitrification in Scottish arable soils. *Journal of Soil Science 42:* 351-367.

Asman, W.A.H., M. A. Sutton, and J.K. Schjørring (1998). Ammonia: Emission, atmospheric transport and deposition. *New Phytologist 139:* 27-48.

Barber, S.A. (1984). *Soil Nutrient Bioavailability.* New York: John Wiley and Sons.

Bates, R.G. and G.D. Pinching (1950). Dissociation constant of aqueous ammonia at 0° to 50° from e.m.f. studies of the ammonium salt of a weak acid. *Journal American Chemical Society 72:* 1393-1396.

Bear, J. (1972). *Dynamics of Fluid in Porous Media.* New York: Elsevier.

Beauchamp, E.G., C. Gale, and J.C. Yeomans (1980). Organic matter availability for denitrification in soils of different textures and drainage classes. *Communications in Soil Science and Plant Analysis 11:* 1221-1233.

Bèauchamp, E.G., W.D. Reynolds, D. Brasche-Villeneuve, and K. Kirkby (1986). Nitrogen mineralization kinetics with different soil pretreatments and cropping histories. *Soil Science Society of America Journal 50:* 1478-1483.

Beblik, A.J. (1996). Description of the N-dynamics simulation model MINERVA. In *Nitrate Leaching from Arable Soils into the Groundwater of Differently Con-*

taminated Catchments Typical for Lower Saxony, Germany. Final Report, BEO No. 00339121 C. Bonn, Germany: Ministry of Science and Technology, 235 pp.

Benbi, D.K. (1990). Efficiency of nitrogen use by dryland wheat in a subhumid region in relation to optimizing the amount of available water. *Journal of Agricultural Science, Cambridge 115:* 7-10.

Benbi, D.K. (1994). Prediction of leaf area indices and yields of wheat. *Journal of Agricultural Sciences, Cambridge 122:* 13-20.

Benbi, D.K., C.R. Biswas, and J.S. Kalkat (1991). Nitrate distribution and accumulation in an Ustochrept soil profile in a long-term fertilizer experiment. *Fertilizer Research 28:* 173-178.

Benbi, D.K. and S.P.S. Brar (1994). Influence of soil organic carbon on the interpretation of soil test P for wheat grown on alkaline soils. *Fertilizer Research 37:* 35-41.

Benbi, D.K., S.S. Prihar, and H.S. Cheema (1991). A model to predict changes in soil moisture, NO_3-N content and N uptake by wheat. *Fertilizer Research 28:* 73-84.

Benbi, D.K. and J. Richter (1996). Nitrogen mineralization kinetics in sewage water irrigated and heavy metal treated sandy soils. In *Progress in Nitrogen Cycling Studies,* eds. O. Van Cleemput, G. Hofmann, and A. Vermoesen, Dordrecht, the Netherlands: Kluwer Academic Publishers, pp. 17-22.

Benbi, D.K. and J. Richter (2001). Nitrogen mineralization kinetics in soils: An assessment of some modelling approaches. In *Plant Nutrition: Food Security and Sustainability of Agro-ecosystems,* eds. W.J. Horst, M.K. Schenk, A. Bürkert, N. Claassen, H. Flessa, W.B. Frommer, H. Goldbach, H.W. Olfs, V. Römheld, B. Sattelmacher, et al., Dordrecht, the Netherlands: Kluwer Academic Publishers, pp. 946-947.

Benbi, D.K., R. Singh, G. Singh, K.S. Sandhu, R. Singh, and S. Saggar (1993). Response of dryland wheat to fertilizer nitrogen in relation to stored water rainfall and residual farm yard manure. *Fertilizer Research 36:* 63-70.

Bergstrom, D.W. and E.G. Beauchamp (1993). An empirical model of denitrification. *Canadian Journal of Soil Science 73:* 421-431.

Bergström, L., H. Johnsson, and G. Torstensson (1991). Simulation of soil nitrogen dynamics using SOILN model. *Fertilizer Research 27:* 181-188.

Beutier, D. and H. Renon (1978). Representation of NH_3-H_2S-H_2O, NH_3-CO_2-H_2O, and NH_3-SO_2-H_2O vapor-liquid equilibria. *Industrial Engineering and Chemical Processes: Design and Development 17:* 220-230.

Black, A.S. and S.A. Waring (1972). Ammonium fixation and availability in some cereal producing soils of Queensland. *Australian Journal of Soil Research 10:* 197-207.

Bobbink, R., D. Boxman, E. Fremstand, G. Heil, A. Houdijk, and J. Roelofs (1992). Critical loads for nitrogen eutrophication of terrestrial and wetland ecosystems based upon exchange in vegetation and fauna. In *Critical Loads for Nitrogen,* eds. P. Grennfelt and E. Thörnelöf, Report Nord 1992. Copenhagen, Denmark: Nordic Council of Ministeres, 41 pp.

Bonde, T.A. and T. Lindberg (1988). Nitrogen mineralization kinetics in soil during long-term aerobic laboratory incubations: A case study. *Journal of Environmental Quality 17:* 414-417.

Bonde, T.A. and T. Rosswall (1987). Seasonal variation of potentially mineralizable nitrogen in four cropping systems. *Soil Science Society of America Journal 15:* 1508-1514.

Bonde, T.A., J. Schnürer, and T. Rosswall (1988). Microbial biomass as a fraction of potentially mineralizable nitrogen in soils from long-term field experiments. *Soil Biology and Biochemistry 20:* 447-452.

Bouwman, A.F., D.S. Lee, W.A.H. Asman, F.J. Dentener, K.W. van der Hoek, and J.G.J. Olivier (1997). A global high-resolution emission inventory for ammonia. *Global Biogeochemical Cycles 11:* 561-587.

Bouwmeester, R.J.B. and P.L.G. Vlek (1981). Rate control of ammonia volatilization from rice paddies. *Atmospheric Environment 15:* 131-140.

Bradbury, N.J., A.P. Whitmore, P.B.S. Hart, and D.S. Jenkinson (1993). Modelling the fate of nitrogen in crop and soil in the years following application of ^{15}N labelled fertilizer to winter wheat. *Journal of Agricultural Sciences 121:* 363-379.

Breitenbeck, G.A. and S. Paramasivam (1995). Availability of ^{15}N-labeled nonexchangeable ammonium to soil microorganisms. *Soil Science 159:* 301-310.

Bremner, J.M. and D.W. Nelson (1968). Chemical decomposition of nitrite in soils. *Transactions 9th International Congress of Soil Science 2:* 495-503.

Broadbent, F.E. (1986). Empirical modeling of soil nitrogen mineralization. *Soil Science 141:* 208-210.

Brunner, W. and D.D. Focht (1984). Deterministic three half order kinetic model for microbial degradation of added C substrates in soil. *Applied and Environmental Microbiology 47:* 167-172.

Burns, I.G. (1974). A model for predicting the redistribution of salts applied to fallow soils after excess rainfall or evaporation. *Journal of Soil Science 25:* 165-178.

Burns, R.C. and R.W.F. Hardy (1975). *Nitrogen Fixation in Bacteria and Higher Plants.* Berlin: Springer Verlag.

Cabrera, M.L. (1993). Modeling the flush of nitrogen mineralization caused by drying and rewetting soils. *Soil Science Society of America Journal 57:* 63-66.

Cabrera, M.L. and D.E. Kissel (1988). Potentially mineralizable nitrogen in disturbed and undisturbed soil samples. *Soil Science Society of America Journal 52:* 1010-1015.

Campbell, C.A., R.J.K. Myers, and K.L. Weier (1981). Potentially mineralizable nitrogen, decomposition rates and their relationship to temperature for five Queensland soils. *Australian Journal of Soil Research 19:* 323-332.

Chae, Y.M. and M.A. Tabatabai (1986). Mineralization of nitrogen in soils amended with organic wastes. *Journal of Environmental Quality 15:* 193-198.

Chen, C.C., F.T. Turner, and J.B. Dixon (1989). Ammonium fixation by high-charge smectite in selected Texas Gulf Coast soils. *Soil Science Society of America Journal 53:* 1035-1040.

Chichester, F.W. and C.W. Richardson (1992). Sediment and nutrient loss from clay soils as affected by tillage. *Journal of Environmental Quality 21:* 587-590.

Childs, S.W. and R.J. Hanks (1975). Model for soil salinity effects on crop growth. *Soil Science Society of America Proceedings 39:* 617-622.

Claassen, N. and S.A. Barber (1977). Potassium influx characteristics of corn roots and interaction with N, P, Ca and Mg influx. *Agronomy Journal 69:* 860-864.

Clay, D.E., C.E. Clapp, J.A.E. Molina, and D.R. Linden (1989). Nitrogen-tillage-residue management: Simulating soil and plant behavior by the model NCSWAP. *Soil Science 147:* 319-325.

Dagan, G. and E. Bresler (1979). Solute dispersion in unsaturated heterogeneous soil at field scale. I. Theory. *Soil Science Society of America Journal 43:* 461-466.

Danckwerts, P.V. (1970). *Gas-Liquid Reactions.* London: McGraw-Hill.

Davidson, J.M., P.S.C. Rao, and H.M. Selim (1977). Simulation of nitrogen movement, transformations, and plant uptake in the root zone. In *Proceedings of the National Conference of Irrigation Return Flow Quality Management,* eds. J.P. Law and G.V. Skogerboe. Fort Collins, CO: Colorado State University Press, pp. 9-18.

De Ruiter, P.C., J.C. Moore, K.W. Zwart, L.A. Bouwman, J. Hassink, J. Bloem, J.A. de Vos, J.C.Y. Marinissen, W.A.M. Didden, G. Lebbink, et al. (1993). Simulation of nitrogen mineralization in the belowground food webs of two winter-wheat fields. *Journal of Applied Ecology 30:* 95-106.

De Ruiter, P.C. and Van Faassen (1994). A comparison between an organic matter dynamics model and a food web model simulating nitrogen mineralization in agro-ecosystems. In *Nitrogen Mineralization in Agricultural Soils,* eds. J.J. Neeteson and J. Hassink. Haren, the Netherlands: AB-DLO, pp. 207-219.

De Willigen, P. (1991). Nitrogen turnover in the soil-crop system: Comparison of fourteen simulation models. *Fertilizer Research 27:* 141-149.

Deans, J.R., J.A.E. Molina, and C.E. Clapp (1986). Models for predicting potentially mineralizable nitrogen and decomposition rate constants. *Soil Science Society of America Journal 50:* 323-326.

Del Grosso, S.J., W.J. Parton, A.R. Mosier, D.S. Ojima, A.E. Kulmala, and S. Phongpan (2000). General model for N_2O and N_2 gas emissions from soils due to denitrification. *Global Biogeochemical Cycles 14:* 1045-1060.

Demeyer, P., G. Hofman, and O.van Cleemput (1995). Fitting ammonia volatilization dynamics with a logistic equation. *Soil Science Society of America Journal. 59:* 261-265.

Denmead, O.T., J.R. Freney, and J.R. Simpson (1982). Dynamics of ammonia volatilization during furrow irrigation of maize. *Soil Science Society of America Journal 46:* 149-155.

Diaz-Fierros, F., M.C. Villar, F. Gil, M. Carballas, M.C. Leiros, T. Carballas, and A. Cabaneiro (1988). Effect of cattle slurry fractions on nitrogen mineralization in soil. *Journal of Agricultural Science, Cambridge 110:* 491-497.

Donigian, A.S. and N.H. Crawford (1976). *Modeling Pesticides and Nutrients on Agricultural Lands.* EPA-600/2-76-043, US Environmental Protection Agency, Athens, Georgia: Office of Research and Development, Environmental Research Laboratory.

Dou, H. and D. Steffens (1995). Recovery of ^{15}N labelled urea as affected by fixation of ammonium by clay minerals. *Zeitschrift für Pflanzenernaehrung und Bodenkunde 158:* 351-354.

Drenth, H., H.F.M. ten Berge, and J.J.M. Riethoven (Eds.) (1994). ORYZA simulation modules for potential and nitrogen limited rice production. *Simulation and System Analysis for Rice Production (SARP) Research Proceedings,* Wageningen, the Netherlands: DLO-Research Institute for Agrobiology and Soil Fertility, WAU- Department of Theoretical Production Ecology and Los Banos, International Rice Research Institute, 223 pp.

Drury, C.F. and E.G. Beauchamp (1991). Ammonium fixation, release, nitrification and immobilization in high- and low-fixing soils. *Soil Science Society of America Journal 55:* 980-985.

Drury, C.F., R.P. Voroney, and E.G. Beauchamp (1991). Availability of NH4+-N to microorganisms and the soil internal N cycle. *Soil Biology and Biochemistry 23:* 165-169.

Duxbury, J.M. (1990). Agriculture, nitrous oxide, and our environment. *New York's Food and Life Sciences Quarterly 20:* 28-31.

ECETOC (European Centre for Ecotoxicology and Toxicology of Chemicals) (1994). *Ammonia Emissions to Air in Western Europe.* Technical Report No. 62, Brussels, Belgium: Author.

Eckersten, H. and P.-E. Jansson (1991). Modelling water flow, nitrogen uptake and production for wheat. *Fertilizer Research 27:* 313-329.

Ellert, B.H. and J.R. Bettany (1992). Temperature dependence of net nitrogen and sulfur mineralization. *Soil Science Society of America Journal 56:* 1133-1141.

Emerson, K., R.C. Russo, R.E. Lund, and R.V.Thurston (1975). Aqueous ammonia equilibrium calculations: Effect of pH and temperature. *Journal of Fisheries Research Board of Canada 32:* 2379-2383.

Engler, R.M. and W.H. Patrick Jr. (1974). Nitrate removal from floodwater overlying flooded soils and sediments. *Journal of Environmental Quality 3:* 409-412.

Farquhar, G.D., R. Wetselaar, and B. Weir (1983). Gaseous nitrogen losses from plants. In *Gaseous Loss of Nitrogen from Plant-Soil Systems,* eds. J.R. Freney and J.R. Simpson. The Hague: Martinus Nijhoff, pp. 159-180.

Ferm, M. (1998). Atmospheric ammonia and ammonium transport in Europe and critical loads: A review. *Nutrient Cycling in Agroecosystems 51:* 5-17.

Firestone, M.K. (1982). Biological denitrification. In *Nitrogen in Agricultural Soils,* eds. F.J. Stevenson, J.M. Bremmer, R.D. Hauck, and D.R. Keeney. Madison, WI: American Society of Agronomy, Agronomy *22:* 289-236.

Firestone, M.K. and E.A. Davidson (1989). Microbiological basis of NO and N_2O production and consumption in soil. In *Exchange of Trace Gases Between Terrestrial Ecosystem and the Atmosphere,* Dahlem Workshop Report 47, eds. M.O. Andreae and D.S. Schimel. New York: John Wiley and Sons, pp. 7-21.

Fischer, W.R., T. Pfanneberg, E.A. Niederbudde, and R. Medina (1981). Transformation of ^{15}N-labelled ammonium in two soils differing in NH_4^+-fixing capacity. *Journal of Soil Science 32:* 409-418.

Focht, D.D. (1974). The effect of temperature, pH and aeration on the production of nitrous oxide and gaseous nitrogen—A zero-order kinetic model. *Soil Science 118:* 173-179.

Focht, D.D. and W. Verstraete (1977). Biochemical ecology of nitrification and denitrification. In *Advances in Microbial Ecology,* Volume 1, ed. M. Alexander. New York: Plenum Press, pp. 135-214.

Franko, U., B. Oelschlägel, and S. Schenk (1995). Simulation of temperature, water, and nitrogen dynamics using the model CANDY. *Ecological Modelling 81:* 213-222.

Freney, J.R. and O.T. Denmead (1992). Factors controlling ammonia and nitrous oxide emissions from flooded rice fields. *Ecological Bulletin 42:* 188-194.

Frissel, M.J. and J.A Van Veen (1981). Simulation model for nitrogen immobilization and mineralization. In *Modelling Wastewater Renovation by Land Disposal,* ed. I.K. Iskandar, New York: John Wiley and Sons, pp. 359-381.

Génermont, S. and P. Cellier (1997). A mechanistic model for estimating ammonia volatilization from slurry applied to bare soil. *Agricultural and Forest Meteorology 88:* 145-167.

Goulding, K.W.T. (1990). Nitrogen deposition to land from the atmosphere. *Soil Use and Management 6:* 61-63.

Goulding, K.W.T., N.J. Bailey, N.J. Bradbury, P. Hargreaves, M. Howe, D.V. Murphy, P.R. Poulton, and T.W. Willison (1998). Nitrogen deposition and its contribution to nitrogen cycling and associated soil processes. *New Phytologist 139:* 49-58.

Granli, T. and O.C Bøckman (1994). Nitrous oxide from agriculture. *Norwegian Journal of Agricultural Sciences. Supplement 12:* 7-128.

Grant, R.F., N.G. Juma, and B. McGill (1993). Simulation of carbon and nitrogen transformations in soil. *Soil Biology and Biochemistry 25:* 1317-1338.

Griffin, G.F. and A.F. Laine (1983). Nitrogen mineralization in soils previously amended with organic wastes. *Agronomy Journal 75:* 124-129.

Groot, J.J.R. and P. De Willigen (1991). Simulation of the nitrogen balance in the soils and a winter wheat crop. *Fertilizer Research 27:* 261-272.

Groot, J.J.R., P. De Willigen, and E.L.J. Verberne (Eds.) (1991). *Nitrogen Turnover in the Soil/Crop System.* Dordrecht, the Netherlands: Kluwer Academic Publishers.

Habets, A.S.J. and G.J.M. Oomen (1993). *Modellering van de sticstofdynamiek binnen gewasrotaties in de biologische landbouw.* Wageningen: Department of Ecological Agriculture, Wageningen Agricultural University.

Hadas, A., S. Feigenbaum, A. Feigin, and R. Portoy (1986). Nitrification rates in profiles of differently managed soil types. *Soil Science Society of America Journal 50:* 633-639.

Hagin, J., A. Amberger, G. Kruh, and E. Segall (1976). Outlines of a computer simulation model on residual and added nitrogen changes and transport in soils. *Zeitschrift für Pflanzenernaehrung und Bodenkunde. 4:* 443-445.

Hales, J.M. and D.R. Drewes (1979). Solubility of ammonia in water at low concentrations. *Atmospheric Environment 13:* 1133-1147.

Hansen, S., H.E. Jensen, N.E. Nielsen, and H. Svendsen (1991). Simulation of nitrogen dynamics and biomass production in winter wheat using the Danish simulation model DAISY. *Fertilizer Research 27:* 245-259.

Hanway, J.J., and A.D. Scott (1956). Ammonium fixation and release in certain Iowa soils. *Soil Science 82:* 379-386.

Harmel, R.D., R.E. Zartman, C. Mouron, D.B. Wester, and R.E. Sosebee (1997). Modeling ammonia volatilization from biosolids applied to semiarid rangelands. *Soil Science Society of America Journal 61:* 1794-1798.

Hauck, R.D. and R.W. Weaver (Eds.) (1986). *Field measurement of dinitrogen fixation and denitrification.* Soil Science Society of America Special Publication No. 18, Madison, WI: Soil Science Society of America, 115 pp.

Hengnirun, S.S., S. Barrington, S.O. Prasher, and D. Lyew (1999). Development and verification of a model simulating ammonia volatilization from soil and manure. *Journal of Environmental Quality 28:* 108-114.

Henin, S., G. Monnier, and L. Turc (1959). Un aspect de la dynamique des mattiéres organiques du sol. *Comptes rendus de l'Académie des Sciences-France 248:* 138-141.

Herridge, D.F. and F.J. Bergersen (1988). Symbiotic nitrogen fixation. In *Advances in Nitrogen Cycling,* ed. J.R. Wilson. Wallingford, Oxon, UK: CAB International, pp. 46-65.

Holtan-Hartwig, L. and O. C. Bøckman (1994). Ammonia exchange between crops and air. *Norwegian Journal of Agricultural Sciences Supplement No. 14:* 5-40.

Huijsmans, J.F.M. and R.M. de Mol (1999). A model for ammonia volatilization after surface application and subsequent incorporation of manure on arable land. *Journal of Agricultural Engineering Research 74:* 73-82.

Hunt, H.W. (1977). A simulation model for decomposition in grasslands. *Ecology 58:* 469-484.

Hunt, H.W., D.C. Coleman, E.R. Ingham, R.E. Ingham, E.T. Elliott, J.C. Moore, S.L. Rose, C.P.P. Reid, and C.R. Morley (1987). The detrital food web in a shortgrass prairie. *Biology and Fertility of Soils 3:* 57-68.

Hutson, J.L. and R.J. Wagenet (1992). LEACHM: Leaching estimation and chemistry model: A process-based model of water and solute movement, transformations, plant uptake and chemical reactions in the unsaturated zone. Version 3.0. Research Series No. 93-3, Ithaca, NY: Department of Soil, Crop and Atmospheric Sciences, Cornell University.

Huwe, B. (1987). *Deterministische und stochastische Ansaetze zur Modellierung des Stickstoffhaushaltes landwirtschaftlich genutzter Flächen auf unterschiedlichen Skalenniveaus.* Mitteilungen/Institut für Wasserbau 77, Universität Stuttgart. Stuttgart: Institut fü Wasserbau der Universität Stuttgart.

Inubushi, K. and H. Wada (1987). Easily decomposable organic matter in paddy soils: VII. Effect of various pretreatments on N mineralization in submerged soils. *Soil Science and Plant Nutrition 33:* 567-576.

Itier, B. and A. Perrier (1976). Présentation d'une étude analytique de l'advection. I Advection liée aux variations horizontales de concentration et de température. *Annals of Agronomy 27:* 111-140.

Jabro, J.D., J.D. Toth, Z. Dou, R.H. Fox, and D.D. Fritton. (1995). Evaluation of nitrogen version of LEACHM for predicting nitrate leaching. *Soil Science 160:* 209-217.

Jansson, S.L. and J. Persson (1982). Mineralization and immobilization of soil nitrogen. In *Nitrogen in Agricultural Soils,* ed. F.J. Stevenson. Madison, WI: American Society of Agronomy, pp. 229-252.

Jayaweera, G.R. and D.S. Mikkelsen (1990). Assessment of ammonia volatilization from flooded soil systems: A computer model. I. Theoretical aspects. *Soil Science Society of America Journal 54:*1447-1455.

Jayaweera, G.R., D.S. Mikkelsen, and K.T. Paw U (1990). Ammonia volatilization from flooded soil systems: A computer model. III. Validation of the model. *Soil Science Society of America Journal 54:* 1462-1468.

Jenkinson, D.S. (1990). An introduction to the global nitrogen cycle. *Soil Use and Management 6:* 56-61.

Jenkinson, D.S., P.B.S. Hart, J.H. Rayner, and L.C. Parry (1987). Modelling the turnover of organic matter in long-term experiments at Rothamsted. *INTECOL Bulletin 15:* 1-8.

Jenkinson, D.S. and J.H. Rayner (1977). The turnover of soil organic matter in some of the Rothamsted classical experiments. *Soil Science 123:* 298-305.

Juma, N.G. and E.A. Paul (1981). Use of tracer and computer simulation techniques to assess mineralization and immobilization of soil nitrogen. In *Simulation of Nitrogen Behavior in Soil Plant Systems,* eds. M.J. Frissel and J.A. Van Veen. Wageningen: Pudoc, pp. 145-154.

Jury, W.A. and D.R. Nielsen (1989). Nitrate transport and leaching mechanisms. In *Nitrogen Management and Ground Water Protection,* ed. R.F. Follett. Amsterdam: Elsevier, pp. 139-157.

Justine, J.K. and R.L. Smith (1962). Nitrification of ammonium sulphate in a calcareous soil as influenced by combination of moisture, temperature and levels of added N. *Soil Science Society of America Proceedings 26:* 246-250.

Kersebaum, K.C. (1995). Application of a simple management model to simulate water and nitrogen dynamics. *Ecological Modelling 81:* 145-156.

Kersebaum, K.C. and J. Richter (1991). Modeling nitrogen dynamics in a plant-soil system with a simple model for advisory purposes. *Fertilizer Research 27:* 272-281.

Kersebaum, K.C. and O. Richter (1994). A model approach to simulate C and N transformations through microbial biomass. In *Nitrogen Mineralization in Agricultural Soils,* eds. J.J. Neeteson and J. Hassink. Haren, the Netherlands: AB-DLO, pp. 221-230.

Kirk, G.J.D. and P.H. Nye (1991a). A model of ammonia volatilization from applied urea. V. The effect of steady-state drainage and evaporation. *Journal of Soil Science 42:* 103-113.

Kirk, G.J.D. and P.H. Nye (1991b). A model of ammonia volatilization from applied urea. VI. The effects of transient-state water evaporation. *Journal of Soil Science 42:* 115-125.

Knapp, E.B., L.F. Elliott, and G.S. Campbell (1983). Carbon, nitrogen and microbial biomass interrelationships during the decomposition of wheat straw: A mechanistic simulation model. *Soil Biology and Biochemistry 15:* 455-461.

Knisel, W.G. (Ed.) (1980). *CREAMS, a Field Scale Model for Chemicals, Runoff, and Erosion from Agricultural Management Systems.* Conservation Research

Report 26, United States Department of Agriculture (USDA) Agriculture Research Service (ARS). Washington, DC: USDA, Science and Education Administration, 653 pp.

Kolenbrander, G.J. (1969). Nitrate content and nitrogen loss in drainwater. *Netherlands Journal of Agricultural Science 17:* 246-255.

Kolenbrander, G.J. (1981). Leaching of nitrogen in agriculture. In *Nitrogen Losses and Surface Runoff,* ed. J.C. Brogan. Brussels: ECSC, EEC, EAEC, pp. 199-216.

Kowalenko, C.G., and D.R. Cameron (1976). Nitrogen transformations in an incubated soil as affected by combinations of moisture and temperature and adsorption-fixation of ammonium. *Canadian Journal of Soil Science 56:* 63-70.

Langford, A.O. and F.C. Fehsenfeld (1992). Natural vegetation as a source or sink for atmospheric ammonia: A case study. *Science 255:* 581-583.

LaRue, T.A. and T.G. Patterson (1981). How much nitrogen do legumes fix? *Advances in Agronomy 34:* 15-38.

Lee, D.S., I. Köhler, E. Grobler, F. Rohrer, R. Sausen, L. Gallardo-Kienner, J.G.J. Olivier, F.J. Dentener, and A.F. Bouwman (1997). Estimations of global NO_x emissions and their uncertainties. *Atmospheric Environment 31:* 1735-1749.

Legg, J.O. and F.E. Allison (1959). Recovery of ^{15}N-tagged nitrogen from ammonium-fixing soil. *Soil Science Society of America Proceedings 23:* 131-134.

Li, C., S. Frolking, and T.A. Frolking (1992). A model of nitrous oxide evolution from soil driven by rainfall events. I. Model structure and sensitivity. *Journal of Geophysical Research 97:* 9777-9796.

Liao, C.F.H. and W.W. Bartholomew (1974). Relationship between nitrate absorption and water transpiration by corn. *Soil Science Society of America Proceedings 38:* 472-477.

Lindemann, W.C. and M. Cardenas (1984). Nitrogen mineralization potential and nitrogen transformations of sludge-amended soils. *Soil Science Society of America Journal 50:* 323-326.

Lindemann, W.C., G. Connell, and N.S. Urquhart (1988). Previous sludge addition effects on nitrogen mineralization in freshly amended soil. *Soil Science Society of America Journal 52:* 109-112.

Liu, B., G. Li, and X. Zhao (2000). Effect of erosion on environmental quality of soil on southern loess plateau in China. In *Soil Erosion and Dryland Farming,* eds. J.M. Laflen, J. Titan, and C.H. Huang. Boca Raton, FL: CRC Press, pp.145-152.

Liu, Y.J., D.A. Laird, and P. Barak (1997). Release and fixation of ammonium and potassium under long-term fertility management. *Soil Science Society of America Journal 61:* 310-314.

Loague, K. and R.E. Green (1991). Statistical and graphical methods for evaluating solute transport models: Overview and application. *Journal of Contaminant Hydrology 7:* 51-73.

MacDonald, N.W., D.R. Zak, and K.S. Pregitzer (1995). Temperature effects on kinetics of microbial respiration and net nitrogen and sulfur mineralization. *Soil Science Society of America Journal 59:* 233-240.

Mahon, J.D. (1979). Environmental and genotypic effects on the respiration associated with symbiotic nitrogen fixation in peas. *Plant Physiology 63:* 892-897.

Mahon, J.D. and J.J. Child (1979). Growth response of inoculated peas *(Pisum sativum)* to combined nitrogen. *Canadian Journal of Botany 57:* 1687-1693.

Marion, G.M. and C.H. Black (1987). The effect of time and temperature on nitrogen mineralization in arctic tundra soils. *Soil Science Society of America Journal 51:* 1501-1508.

Marion, G.M., J. Kummerow, and P.C. Miller (1981). Predicting nitrogen mineralization in chaparral soils. *Soil Science Society of America Journal 45:* 956-961.

McConnaughey, P.K. and D.R. Bouldin (1985). Transient microsite models of denitrification. I. Model development. *Soil Science Society of America Journal 51:* 1501-1508.

McGill, W.B., H.W. Hunt, R.G. Woodmansee, and J.O. Reuss (1981). PHOENIX: A model of carbon and nitrogen dynamics in grassland soils. In *Terrestrial Nitrogen Cycles,* eds. F.E. Clark and T. Rosswall. Stockholm: Swedish Natural Science Research Council, pp. 49-115.

McGuire, A.D., L.A. Joyce, D.W. Kicklighter, J.M. Melillo, G. Esser, and C.J. Vorosmarty (1993). Productivity response of climax temperate forests to elevated temperature and carbon dioxide: A North American comparison between two global models. *Climate Change 24:* 287-310.

Mengel, K., D. Horn, and H. Tributh (1990). Availability of interlayer ammonium as related to root vicinity and mineral type. *Soil Science 149:* 131-137.

Mengel, K. and H.W. Scherer (1981). Release of nonexchangeable (fixed) soil ammonium under field conditions during the growing season. *Soil Science 131:* 226-232.

Moal, J.F., J. Martinez, F. Guiziou, and C.M. Coste (1995). Ammonia volatilization following surface-applied pig and cattle slurry in France. *Journal of Agricultural Science 125:* 245-252.

Moeller, M.B. and P.L.G. Vlek (1982). The chemical dynamics of ammonia volatilization from aqueous solutions. *Atmospheric Environment 16:* 709-717.

Molina, J.A.E., C.E. Clapp, and W.E. Larson (1980). Potentially mineralizable nitrogen in soil: The simple exponential model does not apply for the first 12 weeks of incubation. *Soil Science Society of America Journal 44:* 442-444.

Molina, J.A.E., C.E. Clapp, M.J. Shaffer, F.W. Chichester, and W.E. Larson (1983). NCSOIL, a model of nitrogen and carbon transformations in soil: Description, calibration, and behavior. *Soil Science Society of America Journal 47:* 85-91.

Monod, J. (1942). *Recherches sur la Croissanse des Cultures Bacteriennes.* Paris: Hermann.

Moore, J.C., D.E. Walter, and H.W. Hunt (1988). Arthropod regulation of micro- and mesobiota in belowground food webs. *Annual Review of Entomology 33:* 419-439.

Morgan, J.A. and W.J. Parton (1989). Characteristics of ammonia volatilization from spring wheat. *Crop Science 29:* 726-731.

Myrold, D.D. (1988). Denitrification in ryegrass and winter wheat cropping systems of western Oregon. *Soil Science Society of America Journal 52:* 412-416.

Neeteson, J.J., D.J. Greenwood, and A. Draycott (1988). A dynamic model to predict yield and optimum nitrogen fertilizer application rate for potatoes. In *Nitro-*

gen Efficiency in Agricultural Soils, eds. D.S. Jenkinson and K.A. Smith. Barking, United Kingdom: Elsevier, pp. 384-393.

Nieder, R. and D.K. Benbi (1996). Fixation of ammonium in arable soils: An overview. In *Transactions of the 9th Nitrogen Workshop,* Braunschweig, Germany. pp 47-50.

Nieder, R., H.P. Dauck, and D.K. Benbi (2001). Mineralization of newly accumulated nitrogen. In *Plant Nutrition: Food Security and Sustainability of Agroecosystems,* eds. W.J. Horst, M.K. Schenk, A. Bürkert, N. Claassen, H. Flessa, W.B. Frommer, H. Goldbach, H.W. Olfs, V. Römheld, B. Sattelmacher, et al. Dordrecht, the Netherlands: Kluwer Academic Publishers, pp. 940-941.

Nieder, R., E. Neugebauer, A. Willenbockel, K.C. Kersebaum, and J. Richter (1996). Nitrogen transformation in arable soils of Northwest Germany during the cereal growing season. *Biology and Fertility of Soils 22:* 179-183.

Nieder, R., E. Neugebauer, A. Willenbockel, and J. Richter (1995). Die Rolle der mikrobiellen Biomasse und des mineralisch fixierten Ammoniums bei den Stickstoff-Transformationen in niedersächsischen Löß-Ackerböden unter Winter-Weizen. I. Poolgrößenveränderungen. *Zeitschrift für Pflanzenernährung und Bodenkunde 158:* 469-475.

Nieder, R., G. Schollmayer, and J. Richter (1989). Denitrification in the rooting zone of cropped soils with regard to methodology and climate: A review. *Biology and Fertility of Soils 8:* 219-226.

Nielsen, D.R. and J.W. Biggar (1962). Miscible displacement. III. Theoretical considerations. *Soil Science Society of America Proceedings 26:* 216-221.

Nordmeyer, H. and J. Richter (1985). Incubation experiments on nitrogen mineralization in loess and sandy soils. *Plant and Soil 83:* 433-445.

Norman, R.J. and J.T. Gilmour (1987). Utilization of anhydrous ammonia fixed by clay minerals and soil organic matter. *Soil Science Society of America Journal 51:* 959-962.

Nuske, A. and J. Richter (1981). N mineralization in loess-parabrownearths: Incubation experiments. *Plant and Soil 59:* 237-247.

Nye, P.H. and Tinker, P.B. (1977). *Solute Movement in the Soil-Root System.* Oxford: Blackwells.

Okajima, H. and I. Taniyama (1980). Significance of mass flow in nitrate-nitrogen supply to plant roots. *Soil Science and Plant Nutrition 26:* 363-374.

Osborne, G.J. (1976). The significance of intercalary ammonium in representative surface and subsoils from southern New South Wales. *Australian Journal of Soil Research 14:* 381-388.

Parnas, H. (1975). Model for decomposition of organic material by microorganisms. *Soil Biology and Biochemistry 7:* 161-169.

Parton, W.J., A.R. Mosier, D.S. Ojima, D.W. Valentine, D.S. Schimel, K. Weier, and A.E. Kulmala (1996). Generalized model for N_2 and N_2O production from nitrification and denitrification. *Global Biogeochemical Cycles 10:* 401-412.

Parton, W.J., A.R. Mosier, D.S. Schimel (1988). Rates and pathways of nitrous oxide production in a shortgrass steppe. *Biogeochemistry 6:* 45-48.

Parton, W.J., D.S. Schimel, C.V. Cole, and D.S. Ojima (1987). Analysis of factors controlling soil organic matter levels in Great Plains grasslands. *Soil Science Society of America Journal 51:* 1173-1179.

Parton, W.J., J.W.B. Stewart, and C.V. Cole (1988). Dynamics of C, N, P and S in grassland soils: A model. *Biogeochemistry 5:* 109-131.

Paul, W. and K.H. Domsch (1972). Ein mathematisches modell für den nitrifikatioprozeß in den boden. *Archives Microbiology 87:* 77-92.

Peng, L., J. Wang, and C. Yu (1996). Soil nitrogen uptake by crops, nitrogen leaching and loss from eroded dryland. *Journal of Soil Erosion and Soil and Water Conservation 2:* 10-16.

Phillips, R.E., T. NaNagara, R.E. Zartman, and J.E. Legget (1976). Diffusion and mass flow of nitrate-nitrogen to plant roots. *Agronomy Journal 68:* 63-66.

Potter, C.S., P.A. Matson, P.M. Vitousek, and E.A. Davidson (1996). Process modeling of controls on nitrogen trace gas emissions from soils worldwide. *Journal of Geophysical Research 101:* 1361-1377.

Powlson, D.S., P.R. Poulton, T.M. Addiscott, and D.S. McCann (1989). Leaching of nitrate from soils receiving organic or inorganic fertilizers continuously for 135 years. In *Nitrogen in Organic Wastes Applied to Soils,* eds. J.A. Hansen and K. Henriksen. London: Academic Press, pp. 334-345.

Prihar, S.S., P.R. Gajri, D.K. Benbi, and V.K. Arora (2000). *Intensive Cropping: Efficient Use of Water, Nutrients, and Tillage.* Binghamton, NY: The Haworth Press.

Quemada, M. and M.L. Cabrera (1997). Temperature and moisture effects on C and N mineralization from surface applied clover residue. *Plant and Soil 189:* 127-137.

Quinn, J.A. and N.C. Otto (1971). Carbon dioxide exchange at the air-sea interface: Flux augmentation by chemical reaction. *Journal of Geophysical Research 76:* 1539-1549.

Rachhpal-Singh and P.H. Nye (1986a). A model of ammonia volatilization from applied urea. I. Development of the model. *Journal of Soil Science 37:* 9-20.

Rachhpal-Singh and P.H. Nye (1986b). A model of ammonia volatilization from applied urea. III. Sensitivity analysis, mechanisms, and applications. *Journal of Soil Science 37:* 31-40.

Rahn, C.R., D.J. Greenwood, and A. Draycott (1996). Prediction of nitrogen fertilizer requirement with the HRI WELL-N computer model. In *Progress in Nitrogen Cycling,* eds. O. van Cleemput, G. Hofman, A. Vermoesen. Dordrecht, the Netherlands: Kluwer Academic Publishers, pp. 255-258.

Rao, P.S.C., J.M. Davidson, and R.E. Jessup (1981). Simulation of nitrogen behaviour in the root zone of cropped land areas receiving organic wastes. In *Simulation of Nitrogen Behaviour of Soil-Plant System,* eds. M.J. Frissel and J.A. Van Veen. Wageningen, the Netherlands: Pudoc, pp. 81-95.

Rao, P.S.C., R.E. Jessup, and K.R. Reddy (1984). Simulation of nitrogen dynamics in flooded soils. *Soil Science 138:* 54-62.

Reddy, K.R. and W.H. Patrick Jr. (1980). Losses of applied $^{15}NH_4$-N, urea-^{15}N and organic ^{15}N in flooded soil. *Soil Science 130:* 326-333.

Renault, P. and P. Stengel (1994). Modeling oxygen diffusion in aggregated soils: I. Anareobiosis inside the aggregates. *Soil Science Society of America Journal 58:* 1017-1023.

Renger, M. and O. Strebel (1976). Nitratantlieferung an die Pflanzenwurzel als Funktion der Tiefe und der Zeit. *Landwirtshaft Forschung SH 33 (II):* 13-19.

Richter, G.M., A. Hoffmann, R. Nieder, and J. Richter (1989). Nitrogen mineralization in loamy arable soils after increasing the ploughing depth and ploughing grasslands. *Soil Use and Management 5:* 169-173.

Richter, J. and D.K. Benbi (1996). Modeling of nitrogen transformations and translocations. *Plant and Soil 181:* 109-121.

Richter, J., K.C. Kersebaum, and J. Uttermann (1988). Modelling of the nitrogen regime in arable field soils for advisory purposes. In *Nitrogen Efficiency in Agricultural Soils,* eds. D.S. Jenkinson and K.A. Smith. London: Elsevier, pp. 371-383.

Richter, J., A. Nuske, W. Habenicht, and J. Bauer (1982). Optimized N-mineralization parameters of loess soils from incubation experiments. *Plant and Soil 68:* 379-388.

Riethoven, J.J.M., H.F.M. ten Berge, and H. Drenth (Eds.) (1995). Software developments in the SARP project: a guide to applications and tools. *Simulation and System Analysis for Rice Production (SARP) Research Proceedings,* Wageningen, the Netherlands: DLO-Research Institute for Agrobiology and Soil Fertility, WAU- Department of Theoretical Production Ecology and Los Banos, International Rice Research Institute, pp. 301.

Rijtema, P.E. and J.G. Kroes (1991). Some results of nitrogen simulations with model ANIMO. *Fertilizer Research 27:* 189-198.

Riley, I.T. and M.J. Dilworth (1985). Cobalt status and its effects on soil populations of *Rhizobium lupini,* rhizosphere colonization and nodule initiation. *Soil Biology and Biochemistry 17:* 81-85.

Riley, W.J. and P.A. Matson (2000). NLOSS: A mechanistic model of denitrified N_2O and N_2 evolution from soil. *Soil Science 165:* 237-249.

Robertson, L.A. and J.G. Kuenen (1991). Physiology of nitrifying and denitrifying bacteria. In *Microbial Production and Consumption of Greenhouse Gases: Methane, Nitrogen Oxides and Halomethanes,* eds. J.E. Rogers and W.B. Whitman. Washington, DC: American Society for Microbiology, pp. 189-199.

Roelcke, M., Y. Han, S.X. Li, and J. Richter (1996). Laboratory measurements and simulations of ammonia volatilization from urea applied to calcareous Chinese loess soils. *Plant and Soil 181:* 123-129.

Rolston, D.E., P.S.C. Rao, J.M. Davidson, and R.E. Jessup. (1984). Simulation of denitrification losses of nitrate fertilizer applied to uncropped, cropped, and manure-amended field plots. *Soil Science 137:* 270-279.

Rose, C.W. and R.C. Dalal (1988). Erosion and runoff of nitrogen. In *Advances in Nitrogen Cycling in Agricultural Systems,* ed. J.R. Wilson. Wallingford, Oxon, UK: CAB International, pp. 212-235.

Ryle, G.J.A, C.E. Powell, and A.J. Gordon (1979). The respiratory costs of nitrogen fixation in soybean, cowpea, and white clover. I. Nitrogen fixation and the respiration of the nodulated root. *Journal of Experimental Botany 30:* 135-144.

Sadeghi, A.M., K.J. McInnes, D.E. Kissel, M.L. Cabrera, J.M. Koelliker, and E.T. Kanemasu (1988). Mechanistic model for predicting ammonia volatilization from urea. In *Ammonia Volatilization from Urea Fertilizers,* Bulletin Y-206, eds. B.R. Bock and D.E. Kissel. Muscle Shoals, AL: National Fertilizer Development Center, Tennessee Valley Authority, pp. 67-92.

Scheffer, F. and P. Schachtschabel (1984). *Lehrbuch der Bodenkunde.* Stuttgart, Germany: Ferdinand Enke Verlag.

Schenk, M.K. and S.A. Barber (1979). Phosphate uptake by corn as affected by soil characteristics and root morphology. *Soil Science Society of America Journal 43:* 880-883.

Schlesinger, W.H. and A.E. Hartley (1992). A global budget for atmospheric NH_3. *Biogeochemistry 15:* 191-211.

Schulz, K., K. Bevan, and B. Huwe (1999). Equifinality and the problem of robust calibration in nitrogen budget simulations. *Soil Science Society of America Journal 63:* 1934-1941.

Seifert, J. (1980). The effect of temperature on nitrification intensity in soil. *Folia Microbiology 25:* 144.

Seliga, H. (1993). The role of copper in nitrogen fixation in *Lupinus luteus* L. *Plant and Soil 155/156:* 349-352.

Selim, H.M. and I.K. Iskandar (1980). Simplified model for prediction of nitrogen behavior in land treatment of wastewater. *USA Cold Regions Research and Engineering Laboratory (CRREL) Report 80-12.*

Seyfried, M.S. and P.S.C. Rao (1988). Kinetics of nitrogen mineralization in Costa Rican soils: Model evaluation and pretreatment effects. *Plant and Soil 106:* 159-169.

Shaffer, M.J. (1985). Simulation model for soil erosion-productivity relationships. *Journal of Environmental Quality 14:* 144-150.

Shaffer, M.J., A.D. Halverson, and F.J. Pierce (1991). Nitrate leaching and economic analysis package (NLEAP): Model description and application. In *Managing Nitrogen for Groundwater Quality and Farm Profitability,* eds. L.F. Follett, D.R. Kenney, and R.M. Cruise. Madison, WI: Soil Science Society of America, pp. 285-322.

Shaffer, M.J. and W.E. Larson (Eds.) (1982). *Nitrogen-Tillage-Residue Management (NTRM) Model: Technical Documentation.* Research Report. St Paul, MN: United States Department of Agriculture (USDA)-Agriculture Research Service (ARS).

Sharpley, A.N., S.J. Smith, J.R. Williams, O.R. Jones, and G.A. Coleman (1991). Water quality impacts associated with sorghum culture in the southern Plains. *Journal of Environmental Quality 20:* 239-244.

Shoun, H., D.H. Kim, H. Uchiyama, and J. Sugiyama (1992). Denitrification by fungi. *FEMS Microbiological Letters 94:* 277-282.

Simkins, S., R. Mukherjee, and M. Alexander (1986). Two approaches to modelling kinetics of biodegradation by growing cells and application of a two-compartment model for mineralization kinetics in sewage. *Applied and Environmental Microbiology 51:* 1153-1160.

Singh, H. and D.K. Benbi (2000). Modelling nitrogen mineralization kinetics in semi-arid soils of Punjab. In *Proceedings International Conference on Managing Natural Resources for Sustainable Agricultural Production in the 21st Century,* Volume 2. New Delhi, India: Indian Society of Soil Science, pp. 232-233.

Smith, K.A. (1990). Anaerobic zones and denitrification in soil: Modelling and measurement. In *Denitrification in Soil and Sediment,* eds. N.P. Revsbech and J. Sorensen. New York: Plenum Press, pp. 229-244.

Smith, O.L. (1979). An analytical model of the decomposition of soil organic matter. *Soil Biology and Biochemistry 11:* 585-606.

Smith, S.J., J.F. Power, and W.D. Kemper (1994). Fixed ammonium and nitrogen availability indexes. *Soil Science 158:* 132-140.

Søderlund, R. and T. Rosswall (1982). The nitrogen cycle. In *The Handbook of Environmental Chemistry, Volume 1B. The Natural Environment and the Biogeochemical Cycles,* ed. O. Hutzinger. Heidelberg, Germany: Springer Verlag, pp. 60-81.

Sommer, S.G. and A.K. Ersboll (1996). Effect of air flow rate, lime amendments and chemical soil properties on the volatilization of ammonia from fertilizers applied to sandy soils. *Biology and Fertility of Soils 21:* 53-60.

Sommer, S.G. and J.E. Olesen (2000). Modelling ammonia volatilization from animal slurry applied with trail-hoses to cereals. *Atmospheric Environment 34:* 2361-2372.

Spitters, C.J.T., H. van Keulen, and D.W.G. van Kraalongen (1989). A simple and universal crop growth simulator: SUCROS 87. In *Simulation and System Management in Crop Production,* Simulation Monographs, eds. R. Rabbinge, S.A. Ward, and H. van Laar. Wageningen: Pudoc, pp. 147-181.

SSSA (Soil Science Society of America) (1984). *Glossary of Soil Science Terms.* Madison, WI: Soil Science Society of America.

Stanford, G., M.H. Frere, and D.H. Schwaninger (1973). Temperature coefficient of soil nitrogen mineralization. *Soil Science 115:* 321-323.

Stanford, G. and S.J. Smith (1972). Nitrogen mineralization potentials of soils. *Soil Science Society of America Proceedings 36:* 465-472.

Stanford, G., R.A. Vander Pol, and S. Dzienia (1975). Denitrification rates in relation to total and extractable soil carbon. *Soil Science Society of America Proceedings 39:* 284-289.

Steffens, D. and D.L. Sparks (1997). Kinetics of nonexchangeable ammonium release from soils. *Soil Science Society of America Journal 61:* 455-462.

Stevens, R.J., R.J. Laughlin, and D.J. Kilpatrick (1989). Soil properties related to the dynamics of ammonia volatilization from urea applied to the surface of acidic soils. *Fertilizer Research 20:* 1-9.

Stevenson, F.J. (Ed.) (1982). *Nitrogen in Agricultural Soils.* Madison, WI: American Society of Agronomy, Crop Science Society of America, Soil Science Society of America.

Stevenson, F.J. (1986). *Cycles of Soil: Carbon, Nitrogen, Phosphorus, Sulfur, Micronutrients.* New York: John Wiley and Sons.

Stockle, C.O. and G.S. Campbell (1989). Simulation of crop response to water and nitrogen: An example using spring wheat. *Transactions of the American Society of Agricultural Engineering 32:* 66-74.

Strebel, O., H. Grimme, M. Renger, and H. Fleige (1980). A field study with nitrogen-15 of soil and fertilizer nitrate uptake and of water withdrawal by spring wheat. *Soil Science 130:* 205-210.

Sutton, M.A., C.E.R. Pitcaim, and D. Fowler (1993). The exchange of ammonia between the atmosphere and plant communities. *Advances in Ecological Research 24:* 301-390.

Sutton, M.A., J.K. Schjørring, and G.P. Wyers (1995). Plant-atmosphere exchange of ammonia. *Philosophical Transactions of Royal Society of London A 351:* 261-278.

Tabatabai, M.A. and A.A. Al-Khafaji (1980). Comparison of nitrogen and sulfur mineralization in soils. *Soil Science Society of America Journal 44:* 1000-1006.

Tanji, K.K. (1982). Modeling of the soil nitrogen cycle. In *Nitrogen in Agricultural Soils,* ed. F.J. Stevenson. Madison, WI: American Society of Agronomy, pp. 721-772.

Tanji, K.K., L.D. Doneen, G.V. Ferry, and R.S. Ayers (1972). Computer simulation analysis on reclamation of salt-affected soils in San Joaquin Valley, California. *Soil Science Society of America Proceedings 36:* 127-133.

Thomas, G.W. and R.E. Phillips (1979). Consequences of water movement in macropores. *Journal of Environmental Quality 8:* 149-152.

Thornley, J.H.M. and E.L.J. Verberne (1989). A model of nitrogen flows in grassland. *Plant, Cell and Environment 12:* 863-886.

Tillotson, W.R. and R.J. Wagenet (1982). Simulation of fertilizer nitrogen under cropped conditions. *Soil Science 133:* 133-143.

Umarov, M.M. (1990). Biotic sources of nitrous oxide in the context of the global budget of nitrous oxide. In *Soils and the Greenhouse Effect,* ed. A.F. Bouwman. Chichester, UK: John Wiley and Sons, pp. 263-268.

van Cleemput, O., L. Uytterhaegen, and L. Baert (1987). Chemo-denitrification of groundwater. In *Proceedings 5th International CIEC Symposium: Protection of Water Quality from Harmful Emissions with Special Regard to Nitrate and Heavy Metals,* eds. E. Welte and I. Szabolcs. Balatonfüred, Hungary: Balatonfüred, pp. 301-311.

van der Molen, J., H.G. van Fassen, M.Y. Leclerc, R. Vriesema, and W.J. Chardon (1990). Ammonia volatilization from arable land after application of cattle slurry. 2. Derivation of a transfer model. *Netherlands Journal of Agricultural Science 38:* 239-254.

van Genuchten, M.Th. and P.J. Wierenga (1976). Mass transfer studies in sorbing porous media. I. Analytical solutions. *Soil Science Society of America Proceedings 40:* 473-480.

Van Veen, J.A. and M.J. Frissel (1981). Simulation model of the behavior of N in soil. In *Simulation of Nitrogen Behavior of Soil Plant Systems,* eds. M.J. Frissel and J.A. Van Veen. Wageningen: Pudoc, pp. 126-144.

Van Veen, J.A., J.N. Ladd, and M. Amato (1985). Turnover of carbon and nitrogen through the microbial biomass in a sandy loam and a clay soil incubated with

[^{14}C(4)] glucose and [^{15}N](NH_4)SO_4 under different moisture regimes. *Soil Biology and Biochemistry 17:* 747-756.

Van Veen, J.A., J.N. Ladd, and M.J. Frissel (1984). Modelling C and N turnover through the microbial biomass in soil. *Plant and Soil 76:* 257-274.

Van Veen, J.A. and E.A. Paul (1981). Organic carbon dynamics in grassland soils. 1. Background information and computer simulation. *Canadian Journal of Soil Science 61:* 185-201.

Verberne, E.L.J., J. Hassink, P. de Willigen, J.J.R. Groot, and J.A. Van Veen (1990). Modelling organic matter dynamics in different soils. *Netherlands Journal of Agricultural Science 38:* 221-238.

Vereecken, H., E.J. Jansen, M.J.D. Hackten Brooke, M. Swerts, R. Engelke, S. Fabrewitz, and S. Hansen (1991). Comparison of simulation results of five nitrogen models using different datasets. In *Soil and Groundwater Research. Report II: Nitrate in Soils.* Luxembourg: Commission of the European Communities, Luxembourg, pp. 321-338.

Vlek, P.L.G. and C.T. Craswell (1979). Effect of nitrogen source and management on ammonia volatilization losses from flooded rice soil systems. *Soil Science Society of America Journal 43:* 352-358.

Vlek, P.L.G. and J.W. Stumpe (1978). Effects of solution chemistry and environmental conditions on ammonia volatilization losses from aqueous systems. *Soil Science Society of America Journal 42:* 416-421.

Walther, W. (1989). The nitrate leaching out of soils and their significance for groundwater, results of long-term tests. In *Nitrogen in Organic Wastes Applied to Soils.* eds. J.A. Hansen and K. Henriksen. London: Academic Press, pp. 346-356.

Warncke, D.D. and S.A. Barber (1973). Ammonium and nitrate uptake by corn (*Zea mays* L.) as influenced by NH_4/NO_3 ratio. *Agronomy Journal 65:* 950-953.

Watson, C.J. (1990). The influence of soil properties on the effectiveness of PPD in reducing ammonia volatilization from surface applied urea. *Fertilizer Research 24:* 1-10.

Watts, D.G. and Hanks, R.J. (1978). A soil-water-nitrogen model for irrigated corn on sandy soils. *Soil Science Society of America Journal 42:* 492-499.

Weier, K.L., J.W. Doran, J.F. Power, and D.T. Walters (1993). Denitrification and the dinitrogen/nitrous oxide ratio as affected by soil water, available carbon, and nitrate. *Soil Science Society of America Journal 57:* 66-72.

Welch, L.F. and A.D. Scott (1960). Nitrification of fixed ammonium in clay minerals as affected by added potassium. *Soil Science 90:* 79-85.

White, R.E. (1985). A model for nitrate leaching in undisturbed structured clay soil during unsteady flow. *Journal of Hydrology 79:* 37-51.

White, R.E. (1987). A transfer function model for the prediction of nitrate leaching under field conditions. *Journal of Hydrology 92:* 207-222.

Whitehead, D.C. and N. Raistrick (1990). Ammonia volatilization from the five nitrogen compounds used as fertilizers following surface application to soils. *Journal of Soil Science 41:* 387-394.

Whiteman, W.G. (1923). A preliminary experimental confirmation of the two film theory of gas absorption. *Chemical and Metallurgical Engineering 29:* 146-148.

Whitmore, A.P. and T.M. Addiscott (1987). A function for describing nitrogen uptake, dry matter production and rooting in wheat crops. *Plant and Soil 101:* 51-60.

Whitmore, A.P., K.W. Coleman, N.J. Bradly, and T.M. Addiscott (1991). Simulation of nitrogen in soil and winter wheat crops: Modelling nitrogen turnover through organic matter. *Fertilizer Research 27:* 283-291.

Wijler, J. and C.C. Delwiche (1954). Investigations on denitrification process in soil. *Plant and Soil 5:* 155-169.

Wilson, J.R. (Ed.) (1988). *Advances in Nitrogen Cycling in Agricultural Ecosystems.* Wallingford, UK: CAB International.

Wolf, J., C.T. de Wit, and H. van Keulen (1989). Modeling long-term response to fertilizer and soil nitrogen. *Plant and Soil 120:* 11-22.

Xing, G.X. and Z.L. Zhu (1997). Preliminary studies on N_2O emission fluxes from upland soils and paddy soils in China. *Nutrient Cycling in Agroecosystems 49:* 17-22.

Chapter 15

Phosphorus Dynamics

Bnayahu Bar-Yosef

Phosphorus (P) dynamics in soils are the result of multiple simultaneous reactions that together determine the temporal P distribution in the soil-plant-microbial mass system. The individual processes that govern P dynamics are well known. They include: (1) partitioning of inorganic (Pi) and organic (Po) species between solid and liquid phases in soil; (2) mineralization of Po and immobilization of Pi; (3) transport of Pi and Po species in soil and on soil surface (runoff); and (4) uptake of Pi in relation to plant root and microbial activities. These processes and their effects on surface water eutrophication have been discussed and reviewed in the literature during the past three decades, but little attempt has been made to integrate them and evaluate their relative roles in determining the dynamics and fate of P in soil.

The objective of this chapter is to present a conceptual comprehensive P-dynamics model and to discuss its possible implementation in terms of both available and incomplete knowledge. The model is perceived as another step toward developing a universal platform that would describe P chemistry, transport, transformations, and uptake in soil, and would be a means of improving our understanding and control of P resources and its impact on the environment.

THE CONCEPTUAL MODEL

The heart of the model (Figure 15.1) is the soil solution pool, in which ions and soluble macromolecules can move and be taken up by plant roots and microorganisms. The pertinent ions are Pi and Po species, H^+/OH^-, NO_3^-, NH_4^+ and a predefined soil solution electrolyte, but only P species are presented in the figure. A sparingly soluble mineral, e.g., fluorite (CaF_2), controls the Ca^{2+} activity. The concentrations of the various Pi and Po species are determined by adsorption (*A*, mol P l^{-1} soil), precipitation/dissolution *(CR)*, occlusion (*OC*, Pi only), mineralization *(M)*, and uptake (*U*, Pi only), as described in Equations (15.1 and 15.2). Pi is added to the system

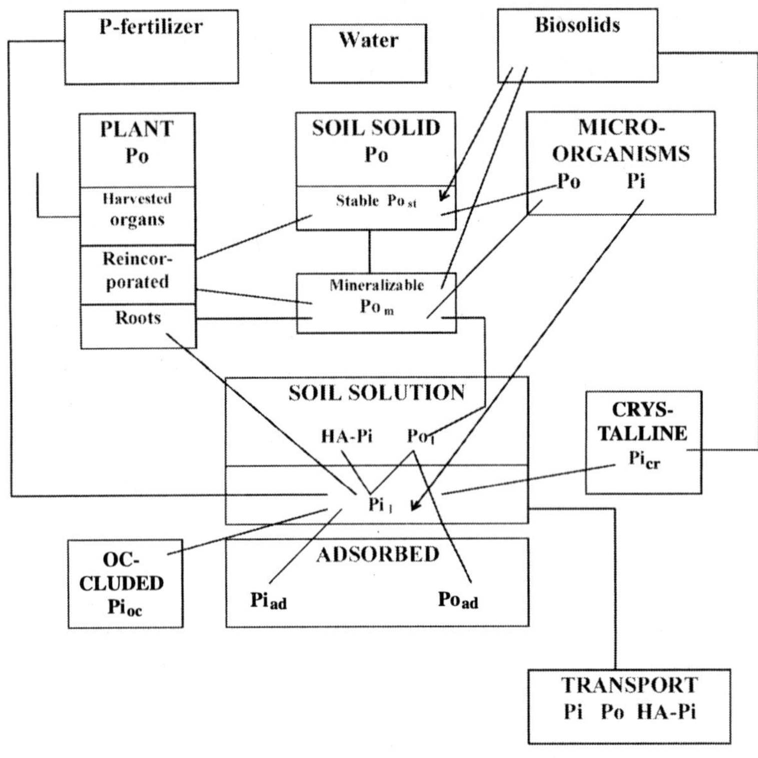

FIGURE 15.1. Flow diagram of main processes that determine P dynamics in soil. The organic-P (Po) is found in plant and microorganism tissue, in stable and mineralizable solid forms (Po_{st} and Po_m, respectively) in soil, and in mobile forms (Po_l- soil solution, and Po_{ad}- adsorbed sites on soil). The mobile forms are composed of inositol hexaphosphate (IHP), RNA, and glycerophosphate (see text). The inorganic-P (Pi), which is absorbed by plants and microorganisms, instantaneously transforms into Po; it is also partitioned between soil solution and solid phases (Pi_l and Pi_{ad}, respectively) and undergoes crystallization, occlusion, and complexation reactions (Pi_{cr}, Pi_{oc}, and HA-Pi, respectively).

by fertilizers, and Pi and Po are added by biosolids and plant and microbial residues. The Po is divided between the stable (or slowly mineralizable) pool, and the readily mineralizable pool. All species in the solution pool that are shown in Figure 15.1 move in the soil profile by diffusion and convection (Equation 15.3), and both of these processes depend strongly on soil water content (θ, $l\,l^{-1}$):

$$\frac{\partial Q}{\partial t} = -\frac{\sum_{i=1}^{n} \partial F_i}{\partial z} - (CR + OC) + M - U \qquad (15.1)$$

Here Q is the sum of the n mobile orthophosphate (o-P) species concentrations in the soil (discussed later) (mol l^{-1} soil, Equation 15.2), F_i is the flux of the ith species (mmol $cm^{-2}s^{-1}$, Equation 15.3), t is time and z is depth (cm). The other rate terms in Equation (15.1) were defined previously; their units are mol l^{-1}soil sec^{-1}. The individual processes in Equations (15.1), (15.2), and (15.3) will be discussed next.

$$Q = \sum_{1}^{n} (Ap_i + \theta \cdot C_i) \qquad (15.2)$$

$$F_i = -Dp_i \frac{\partial C_i}{\partial z} + q \cdot C_i \qquad (15.3)$$

In Equation (15.3) Dp is the ion diffusion coefficient in soil solution (discussed later, cm^2s^{-1}), Ci is the concentration of the ions of the ith species in soil solution (mol l^{-1}), $q = -K(h)(\frac{\partial h}{\partial z} + 1)$ h is the water pressure head (cm) that is related to θ via the soil water retention function, and $K(h)$ (cm s^{-1}) is the soil hydraulic conductivity function. The solution velocity v (cm s^{-1}) is defined as $v = q/\theta$. Since association/dissociation reactions are instantaneous, all the o-P species can be expressed in terms of $H_2PO_4^-$. Equations (15.1, 15.2, and 15.3) must also be solved for soluble Po species (see as follows), only in this case U, CR and $OC = 0$, and M should be specifically defined. Equations (15.1 and 15.3) are also used to compute the transport of the organo-P complex (HA-P, Figure 15.1). The rate terms in this case are replaced by a source term describing generation of this molecule in soil solution.

To solve Equations (15.1 and 15.2) the spatial and temporal q, θ, and pH must be known. The latter is a prerequisite for P speciation and pH-dependent adsorption calculations. Consequently, at each time step the solutions of Equation (15.1) must be preceded by solution of the Richards Equation (15.4) so that $\theta(z,t)$, $h(z,t)$ and $q(z,t)$ are known, and by solution for pH(z,t) which will be discussed as follows.

$$\frac{\partial \theta}{\partial t} = -\frac{\partial J}{\partial z} - W_u \qquad (15.4)$$

The W_u term describes the spatial water uptake rate by plant roots ($l\ l^{-1}soil\ s^{-1}$), which depends on root length and h distributions in the soil (see as follows). Discussion and applications of the Richards equation are outside the scope of this chapter, and interested readers are referred to, e.g., Hillel (1980) for a review of the effects of soil factors on the $K(h)$ and $h(\theta)$ functions.

The conceptual model presented herein differs from previously published models in four major respects:

1. It allows o-P sequestration by a high-molecular-weight, water-soluble organic carbon compound (HA-P in Figure 15.1). According to Jones, Salonen, and de Haan (1988) and Stevenson (1994), this compound might be humate. This mechanism may explain the enhanced P leaching in biosolids-amended soils in comparison with soils not amended with organic matter (Vetter and Steffens, 1981; Latterel et al., 1982; Sims, Simard, and Joern, 1998; Breeuwsma, Reijerink, and Schoumans, 1995).
2. Organic-P species (specified as follows) compete with Pi for common pH-dependent adsorption sites and move in the soil. This may be an additional mechanism to explain the enhanced P movement mentioned previously.
3. Microorganisms absorb P from soil solution only. This mechanism was suggested by Cole, Innis, and Stewart (1977) but was not included in later models.
4. It accounts for H^+/OH^- dynamics in soil and their excretion by roots in response to differential ammonium-nitrate uptake. This option is very important where soil-N and P models are run simultaneously.

Discussion of the P forms and pools presented in Figure 15.1 and material flow among them, mathematical formulation of most of the individual processes, and evaluation of flux rates in the continuity equation are the subjects of following sections in this chapter.

PHOSPHORUS FORMS IN PLANT AND SOIL AND EUTROPHICATION

Phosphorus is an essential element for all living organisms. Its total concentration in common plants and plant organs varies between ~2 and 20 g kg^{-1} dry matter, of which 29-88 percent is Pi, ~5-6 percent is phospholipid-P, 50-5 percent is nucleic acid-P, and 12-1 percent is inositol phosphate-P. The high nucleic acid and inositol P concentrations are found in seeds. Intermediate metabolites such as sugar and adenosine phosphates and phosphoproteins comprise only trace concentrations (Mengel and Kirkby, 1987).

In soils, Po makes up 20-80 percent of total P (Stevenson and Cole, 1999) and in soil solution, 20-70 percent may exist in organically bound forms (Iyamuremye and Dick, 1996). In accordance with P composition of plants, the soil Po is composed of three main groups:

1. The monoester inositol hexaphosphate [IHP, $C_6H_6(H_2PO_4)_6$] which comprises ~60 percent of total Po in soil.
2. Diesters of orthophosphate, mainly nucleotides (DNA and RNA), which constitute ~1-2 percent of total Po (later called NAP).
3. Phospholipids and phosphoglycerides (PLP), which together constitute ~5 percent of total Po (Stevenson and Cole, 1999).

The other components of Po (30-40 percent) can be found only as solid Po. The rate of mineralization of these compounds is yet unknown.

Inorganic P ions and ion pairs prevailing in soil differ in valence and in hydrated radius, which affect their adsorption, mobility, and uptake by plant roots. In calcareous soils, pH range of 5 to 9, the Pi species that should be accounted for in Equation (15.1) are $H_2PO_4^-$, HPO_4^{2-}, $CaH_2PO_4^+$, $CaHPO^o$, and $CaPO_4^-$. The pK values of the dissociation and stability constants of these compounds ($nH^+ + PO_4^{3-} = H_nPO_4^{(3-n)-}$ and $nH^+ + Ca^{2+} + PO_4^{3-} = CaH_nPO_4^{(n-1)+}$), are 19.573, 12.375, 20.923, 15.035, and 6.46, respectively (Schecher, 2001). In the absence of Ca-complexing agents the mole fraction of the sum of the Ca-P ion pairs in solution at pH 5.5, 7.0, 8.0, and 9.0 is 0.25, 0.65, 0.83, and 0.93, respectively (total P concentration = 1 mM; total Ca [Tca] = 10 mM). In the presence of citrate, a moderate Ca-complexing agent, at a concentration that equals TCa, the corresponding mole fractions are 0.06, 0.15, 0.27, and 0.49, showing that the concentration of the uncomplexed Pi ions increases considerably. Carbonate has a marginal effect on Ca-P ion pair concentrations (Bar-Yosef, 1996).

In acid soils, P forms ion pairs with Fe and Al. Citrate and oxalate form stable complexes with Fe^{3+} and Al^{3+} (Bar-Yosef, 1996), thereby increasing the mole fraction of free Pi ions in solution. Gardner et al. (1983) suggest that citrate forms a soluble ferric-hydroxyphosphate polymer ($Fe|O|OH|H_2PO_4|Cit$), which moves P from the bulk soil solution to root surfaces.

Phosphorus and other nutrients in the soil-root volume have a predominant effect on the overenrichment of surface and underground water with mineral nutrients (eutrophication). The results are excessive production of autotrophs, especially algae and cyanobacteria. This high productivity leads to high bacterial populations and enhanced respiration rates, which lead to hypoxia or anoxia in poorly mixed bottom waters, and at night in surface waters during calm, warm conditions (Correll, 1999). Low dissolved oxygen causes the loss of aquatic animals and the release of toxic materials. The toxins and the cyanobacterial bloom drastically reduce the water quality and in some cases even cause water bird casualties due to outbreaks of botulism (James and Havens, 1996). Although C and N are essential to the growth of aquatic biota, most of the attention has focused on P inputs, because of the difficulty in controlling the exchange of N and C between the atmosphere and water, and fixation of N by some blue-green algae. Thus, P is often the limiting element, and controlling its fluxes in runoff water and in deep field drainage is of prime importance in reducing the accelerated eutrophication of freshwater. Orthophosphate is the only form of P that autotrophs can assimilate, but inputs of Po must also be considered due to hydrolysis by extracellular enzymes. In saline oceans N generally becomes the element controlling aquatic productivity (Daniel, Sharpley, and Lemunyon, 1998).

Much progress has been made in understanding eutrophication processes and in constructing modeling frameworks for projecting the effectiveness of nutrient reduction strategies (Thomann and Linker, 1998). Simple steady state models (Vollenweider, 1976; Moustafa, 1998) successfully predict algal biomass as a function of total P input rate and outflow per unit of lake surface area. Dynamic models simulate algae growth as a function of orthophosphate uptake, algae decay and mineralization, Pi sorption by suspended particles and lake sediments, and short-distance P diffusion (Clement, Somlyoly, and Kenesos, 1998). More detailed models take into account organic-P and organic-N transformations as well (DUFLOW, 1992). Model and empirical results indicate that soluble-P concentration between 0.01 and 0.02 mg P l^{-1} in surface water are critical values above which eutrophication is accelerated (Daniel, Sharpley, and Lemunyon, 1998).

PHOSPHORUS TRANSPORT CONSIDERATIONS

Transport in Soil

The transport in soil is determined by the diffusion and convection flows (Equation 15.3), and by the partitioning of the ions between instantaneous adsorption sites in the soil and the solution phase (Equation 15.2). The coefficient *Dp* (Equation 15.3) is the sum of the diffusion and dispersion components (Tinker and Nye, 2000)

$$Dp = Dp^* + \lambda v; \; Dp^* = Dof\,\theta \qquad (15.5)$$

where Do is the diffusion coefficient in water (for H_2PO_4 5.OE-6 $cm^2\ s^{-1}$, Olsen and Kemper, 1968), f is the dimensionless soil impedance factor (<1), λ is the dispersivity (cm), and v and θ were defined previously. According to Olsen and Kemper (1968) the effect of θ on Dp^* is $Dp^* = Do\ a\ e^{b\theta}$, where b = 10 and a = 0.005 to 0.001 depending on the soil, while Mualem and Friedman (1991) found that $Dp^* = Do\ (\theta_s - \theta_r)^{1.5}\ S_e^{2.5}$, where θ_s and θ_r are saturated and residual θ and S_e is the effective saturation degree [$=(\theta - \theta_r)/(\theta_s - \theta_r)$]. The soil impedance decreases (higher soil tortuosity) as the clay content of the soil increases, and as θ decreases. More detailed information on soil factors that affect Dp^* can be found in a review by Jungk (1996).

To obtain $\partial Q/\partial t$ in Equation (15.1), Equation (15.2) is differentiated with respect to time. Under fixed pH and θ and by application of the chain rule

$$\frac{\partial Q}{\partial t} = \frac{dA}{dC}\frac{\partial C}{\partial t} + \theta\frac{\partial C}{\partial t} \tag{15.6}$$

The derivative dA/dC [$= b(C, \text{pH})$] is the buffering capacity of the system, and will be discussed in the next section; it is used in Equation (15.7) which describes P transport toward a single root in soil. For q values generated by water uptake by roots, λ is negligible and $Dp = Dp^*$

$$\left[b(C, pH) + \theta\right]\frac{\partial C}{\partial t} = \frac{\partial}{\partial r}(Dp\frac{\partial C}{\partial r}) + \frac{1}{r}(Dp\frac{\partial C}{\partial r}) + \frac{1}{r}\frac{\partial}{\partial r}(q_o r_o C) \tag{15.7}$$

Here q_o and r_o denote solution flux at the root surface, and the root radius, respectively, and the equation is written for a single P species for convenience. The most general boundary condition at the root surface is

$$-Dp\frac{\partial C}{\partial r} + q_o C = \alpha C \tag{15.8}$$

where α is a general ion permeability function. The outer boundary condition, found either at the edge of the root-affected soil volume or midway between adjacent roots, whichever is smaller, *(R)* is

$$\partial C_R / \partial r = 0 \tag{15.9}$$

Application of Equation (15.7) is discussed as follows.

Transport in Agricultural Runoff

The problem of contamination of agricultural runoff by chemicals in general and phosphorus in particular (Ahuja, Lehman, and Sharpley, 1983;

Sharpley and Smith, 1989; Sharpley et al., 1992; Grant et al., 1996; Pote et al., 1999) has stimulated considerable work, both experimental and theoretical, to understand and quantify this phenomenon (Knisel, 1980). Several models simulating runoff volume and solute concentration induced by continuous rainfall on a sloping soil have been published (Wallach, van Genuchten, and Spencer, 1989; Wallach and van Genuchten, 1990; Wallach and Shabtai, 1992). These models account for transient water infiltration, convective-dispersive solute transport in the soil, and rainfall rate. They also consider rate-limited mass transfer through a laminar boundary layer at the soil surface/runoff water interface. Models describing basin scale transport usually account for similar processes (Hopstaken and Ruijgh, 1994; Zhang, Tisdale, and Wagner, 1996; Cassel, 1998). Solutes in these models are assumed to be subject to linear equilibrium sorption isotherms or to first-order kinetics. The models do not account for biological processes in the field, or chemical processes that may affect soil hydraulic conductivity due to sealing and clay compression (Shainberg and Levy, 1992; Shainberg et al., 1997). The complexity of the physical runoff models and their unsatisfactory treatment of chemical processes may explain the fact that prediction of P transport in agricultural runoff has resorted to empirical relationships. For example, an empirical model by Sharpley and Smith (1989) assumes that the soluble phosphorus (SP) concentration of runoff can be predicted from the rate of P desorption and the effective depth of soil layer in which surface soil and runoff interact

$$P_r = K_{pr} \cdot P_s \cdot E \cdot \rho \cdot t^{\alpha} \cdot W_p^{\beta} / V_p \tag{15.10}$$

Here P_r is the average SP concentration in runoff for an individual event (mg l^{-1}), P_s is the extractable P content (in this case Bray, mg kg^{-1}) of surface soil (0-50 mm) before each runoff event, E is the effective soil depth (mm), ρ is the bulk density of soil (kg l^{-1}), t is the runoff event duration (min), W_p is the runoff water/soil (suspended sediment) ratio, V_p is the total runoff during the event (mm), and K_{pr}, α, and β are constants for a given soil. The ranges of values of these parameters in the cited study were 0.05-0.54, 0.12-0.17, and 0.3-0.62, respectively. To separate between SP and particulate P (PP), the runoff was filtered (0.45 μm) and SP determined in the filtrate. The PP concentration, which is also regarded as bioavailable, is determined by NaOH extraction (Sharpley et al., 1992). To get the best simulation of P_r, E should be a function of rainfall-runoff amount, intensity, and energy. As this function is usually unavailable without deterministic simulation, it was replaced by measured soil loss (kg ha^{-1}) (Sharpley and Smith, 1989):

$$\ln(E) = i + 0.576 \ln(\text{soil loss}) \quad (15.11)$$

where i is related to soil texture and ranges between –1.2 and –2.5. When soil loss is unknown, a constant E value of 0.3 mm is used.

Recently, mechanistic models were incorporated in models assessing ecological risks (Hession et al., 1996) and economic implications of P loading in runoff water (Govindasamy, Cochran, and Buchberger, 1994).

ADSORPTION-DESORPTION

Inorganic-P

Sorption of inorganic-P has been shown to be rapid during the first 24 hours *(h)* and to slow down considerably as equilibration progresses (Bar-Yosef et al., 1988). The fast reaction is attributed to adsorption (*A*, Equation 15.2), and the slower reaction to precipitation (*CR* in Figure 15.1; Enfield et al., 1981; van der Zee and van Riemsdijk, 1991; Overman, 1999). Another slow reaction—the diffusion of P into the soil mineral lattice (*OC* in Figure 15.1)—usually involves only a small fraction of the mobile P (Barrow, 1983) and therefore will not be further discussed in this chapter.

In most P transport models adsorption is assumed to be instantaneous. This is justified if the adsorption rate is much faster than changes in concentration caused by P transport in the soil. To evaluate this assumption, consider a soil with the following characteristics: (1) $\theta = 0.25$ l l^{-1} and $f = 0.5$. Inserting Do of $H_2PO_4^-$ (5.0E-6 cm^2s^{-1}), $\lambda = 0$, f and θ in Equation (15.5) yields $Dp = 0.625 \cdot 10^{-6}$ cm^2s^{-1}. (2) P adsorption *(A)* increases from 0 to 2.5 μmol P cm^{-3} soil in 24 *h*. (3) Consider three imaginary horizontal planes (a1, a2, a3) separated vertically by Δz = a1-a2 = a2-a3 = 0.5 cm, and such that Cp is 0.4, 0.2, and 0.1 μmol P ml^{-1} at planes a1, a2, and a3, respectively. We may now compare the adsorption rate ($\Delta A/\Delta t$) with $\Delta F/\Delta z$. The value of ΔF is equal to F_{a1a2} [$= 6.25 \cdot 10^{-7} \cdot (0.4 - 0.2)/0.5$] minus F_{a2a3} [$= 6.25 \cdot 10^{-7} \cdot (0.2 - 0.1)/0.5$], which yields $1.25 \cdot 10^{-7}$ μmol P cm^{-2}soil s^{-1}. Dividing by Δz gives $2.5 \cdot 10^{-7}$ μmol P cm^{-3} soil s^{-1}. The 24 *h* mean $\Delta A/\Delta t$ is $2.9 \cdot 10^{-5}$ μmol P cm^{-3}soil s^{-1}, which is indeed much greater than $\Delta F/\Delta z$.

Another problem associated with adsorption is its reversibility. Published data indicate that desorption, in most cases, is smaller than the original adsorption (Kafkafi, Posner, and Quirk, 1967; Barrow and Shaw, 1975; and a review by Sanyal and De Datta, 1991). Goldberg and Sposito (1985) suggest that adsorbed o-P becomes "kinetically irreversible" because of conversion of reversibly adsorbed monodentate P to the "irreversible" bidentate form. However, Goldberg and Sposito (1985) offer no satisfactory theory to explain those transformations, and it is possible that the experimental methods used to examine the desorption process (e.g., excessive dilution) have contributed to the observed irreversibility by modifying the surface characteristics. The lack of a convincing body of evidence and the possibility of artifacts explain the fact that the irreversibility phenomenon is still ignored in most P transport models.

An acceptable instantaneous adsorption model must account for the following factors: (1) the effect of concentration of various P species in solution on P adsorption *(Ap)*; (2) the effect of pH; and (3) the effect of the ionic strength *(I)* on *Ap* with a fixed quantity of P in the system. To account for the three factors, the model should include at least two adsorption layers: one for the potential-determining ions (H^+ and OH^-) and specifically adsorbed ions, and the other for the electrolyte ions (the diffuse layer). The three-factors prerequisite is important, as those are the variables that are most strongly affected by plant rhizospheres. Among published multilayer models (for a review see Dzombak and Morel, 1990; Goldberg, 1995) the simplest one (the two-layer model) is presented in Figure 15.2 and Equations 15.12.1 to 15.12.13. This model has been tested for P adsorption by iron oxides (Dzombak and Morel, 1990), but not by soils.

Equations 15.12.1-15.12.13:

$$XOH_2^+ = XOH^o + H^+ \qquad K_{a1}^{app}(XOH_2^+) = \gamma_1 (K_{a1}^{app})^{-1}(XOH^o)(H^+)$$
$$XOH^o = XO^- + H^+ \qquad K_{a2}^{app} \quad (XO^-) = \gamma_1^{-1} K_{a2}^{app}(XOH^o)/(H^+)$$
$$XOH^o + A^{3-} + 2H^+ = XHA^- + H_2O \qquad K_{2A}^{app}(XHA^-) = \gamma_1^2 \gamma_3 K_{2A}^{app}(XOH^o)(A^{3-})(H^+)^2$$
$$XOH^o + A^{3-} + H^+ = XA^{2-} + H_2O \qquad K_{1A}^{app\ (XA2-}) = \gamma_1 \gamma_3 K_{1A}^{app}(XOH^o)(A^{3-})(H^+)$$
$$H^+ + OH^- = H_2O \qquad K_w \quad (OH^-) = \gamma_1^{-2} K_w/(H^+)$$
$$H^+ + A^{3-} = HA^{2-} \qquad K_1 \quad (HA^{2-}) = \gamma_1 \gamma_3 K_1 (H^+)(A^{3-})$$
$$2H^+ + A^{3-} = H_2A^- \qquad K_2 \quad (H_2A^-) = \gamma_1^2 \gamma_3 K_2 (H^+)^2 (A^{3-})$$
$$K_i^{app} = K_i^{int} \exp[-\Delta Z\, F\, \psi/(R\, T)] \qquad \text{(Coulombic correction)}$$
$$TOTH = (H^+) - (OH^-) + (XOH_2^+) - (XO^-) + (XA^{2-}) + 2*(XHA^-) \quad \text{(Mole balance equations)}$$
$$TOTA = (HA^{2-}) + (H_2A^-) + (XHA^-) + (XA^{2-})$$
$$TOTXOH = (XA^{2-}) + (XHA^-) + (XOH^o) + (XO^-) + (XOH_2^+)$$
$$\sigma = [F/(A * S)] * [(XOH_2^+) - (XO^-) - (XHA^-) - 2*(XA^{2-})] \quad \text{(Surface charge)}$$
$$\sigma_d = (8\, R\, T\, \varepsilon\, \varepsilon_o\, c\, 10^3)^{1/2} \sinh(Z\, \psi\, F/2\, R\, T)$$ (diffuse layer charge=potential relationship for symmetrical electrolyte. At low Ψ and for T = 298K $\sigma_d = 0.1174 * c^{1/2} * \sinh(Z\, \Psi * 19.46)$.

XOH represents the surface functional group. The model parameters are defined as follows: A = specific surface area, m^2g^{-1}; c = electrolyte concentration, mol l^{-1}; A^{3-} = concentration of PO_4^{3-}, mol l^{-1}; F = Faraday ($=9.6485*10^4$ C mol^{-1}); F/(R T) = 38.92 V^{-1}; R = gas constant [=8.3144 V*C/(mol K)]; S = solid concentration, g l^{-1}; T = absolute temperature, K; Z = valence of ions in symmetrical electrolyte; Δ = change in the charge num-

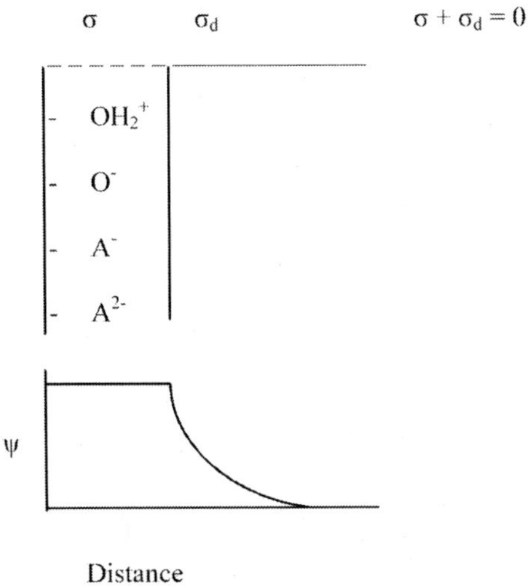

FIGURE 15.2. Adsorption planes of the two-layer model: A^-, A^{2-} specifically adsorbed ions located in the plane of constant Ψ. The pH-dependent charge density in this layer (σ) and the diffuse layer charge density (σ_d) must counterbalance each other.

ber of the surface species involved in a surface complexation reaction; σ = net surface charge density, C m^{-2} (or μ C μ^{-2}); Ψ = surface potential, V; ε and ε_0 are permitivity in the medium and in vacuum, respectively; γ_1, γ_2, and γ_3 are activity coefficients of mono, di, and trivalent ions, respectively.

In another group of models, P adsorption is calculated as a function of ion concentration and pH (but not of I, except for its effect on ionic strength). The widely used single-layer (constant capacitance) model (Goldberg and Sposito, 1984) that belongs to this group has been shown to provide a good simulation of P adsorption by soils and clay minerals (for a review see van Remsdijk and van der Zee, 1991). It has been utilized (with certain modifications) by Suarez and Simunek (1997) to describe B, As, and Se reactions and transport in soil, and by Grant and Heaney (1997) to simulate P movement in the soil profile. The formulation of the constant capacitance model is similar to the two-layers model, only the diffuse layer charge-potential relationship is replaced by the capacitance equation $\sigma = C\ S\ A\ \psi / F$, where C is the capacitance density (F m^{-2}) (Goldberg, 1995).

This also includes a phenomenological, modified Langmuir competitive adsorption model (Bar-Yosef et al., 1988). This model (Equation 15.13) entails only one equation and is differentiable with respect to both P and H^+, thus providing the buffering capacity in Equation (15.6)

$$A_i = \frac{T \cdot \sum_{i}^{n} K_i \cdot C_i}{1 + \sum_{j}^{m} K_j \cdot C_j} \tag{15.13}$$

where, $T = T_o \cdot e^{[(RPH/pH)-1] \cdot G}$ (mol l^{-1} soil), RPH is a soil constant that was found experimentally to be related to the pH at the inflection point of the surface titration curve, T_o and G are soil constants (mole l^{-1} soil and dimensionless, respectively), K is the Langmuir affinity constant (mol $l^{-1})^{-1}$, C is concentration in soil solution (mol l^{-1}), $i = 1,..n$ are the adsorbing P species (mentioned previously), and $j = 1,..n$ are the P species plus OH^-. Note that when hydroxyl adsorption is calculated, $i = 1$. Although empirical adsorption models can be criticized for lack of chemical consistency, it should be recalled that chemical models, also, are not free of some empiricism. For example, the capacitance term in the one-layer model cannot be independently determined (Goldberg, 1995); the modes of monodentate and bidentate P surface complexes have not yet been definitely identified by spectroscopic methods, therefore, the selection of surface species by the models is arbitrary (Persson, Nilsson, and Sjoberg, 1996; Suarez, Goldberg, and Su, 1998) and the solubility behavior of the solid phase is neglected (Schulthess and Sparks, 1988). Another problem, which is pertinent to P transport models, is the prediction of pH in soil as a function of time and space; as long as adequate tools for such prediction are not available, there is no justification for incorporating refined adsorption models that account for this effect in transport models.

Organic-P

In the conceptual model, three organic-P compounds (IHP, DNA+RNA, and PLP) can be partitioned between the soil solution (Po_l) and the soil solid phase, and can move in the soil (Figure 15.1). Meager information is available on the adsorption of these compounds by soils (for IHP see Anderson, Williams, and Moir, 1974). Nevertheless, adsorption isotherms of IHP and RNA by clay minerals could be recalculated from the data of Goring and Bartholomew (1952 and 1950, respectively) (Figures 15.3 and 15.4). The isotherms indicate that at similar concentration in solution and pH, adsorption of IHP slightly exceeds that of Pi. This indication is supported by data showing competition between IHP and Pi on common adsorption sites in Fe-oxide strips (Sharpley, Robinson, and Smith, 1995) and in soils (Evans, 1985; Helal and Dressler, 1989). Recent studies of IHP and Pi adsorption by goethite (Ognalaga, Frossard, and Thomas, 1994; Celi et al., 2001), goethite,

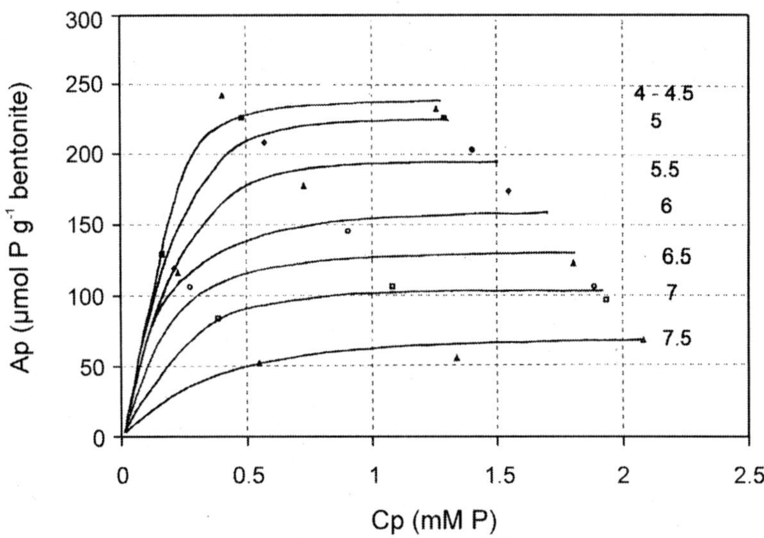

FIGURE 15.3. Adsorption of inositol-hexaphosphate (IHP) on bentonite (Ap) as a function of concentration in equilibrium suspension solution (Cp) and pH. Symbols are interpolated values of original data. Lines were hand fitted through interpolated points. (*Source:* Adapted from Goring and Bartholomew, 1950, p. 190.)

illite, and kaolinite (Celi et al., 1999, initial pH ~4.5), and calcite (Celi et al., 2000, pH 8.3) showed a Langmuir adsorption within a certain concentration range in solution (C_l), and a change in phase when C_l exceeded that range (Table 15.1). The data indicate that at pH < 7, Pi adsorption by goethite, illite, and kaolinite surpassed the adsorption of IHP; in calcite (higher pH) the adsorption of IHP was greater. The adsorption of Pi and IHP had similar effects on the release of OH^- from these clay minerals, and on their zeta potentials (data not presented). These findings support earlier indications that Pi and IHP share the same surface adsorption sites.

The adsorption rates of IHP and Pi were studied in a short-range, ordered precipitate of Al at initial pH of 6.8. Both rates were very fast during the first hour, and slowed down considerably between 1 and 24 h (Shang, Huang, and Stewart, 1990). A first-order kinetics model could describe the adsorption during both stages. As expected from molecule size considerations, the rate constant of Pi exceeded that of IHP. The activation energies of Pi and IHP adsorption were 48 and 89 KJ mol^{-1} of P adsorbed, respectively. The adsorption kinetics of Pi and IHP on short-range-ordered iron precipitate (Shang, Stewart, and Huang, 1992) was similar to that reported for the short-range Al precipitate.

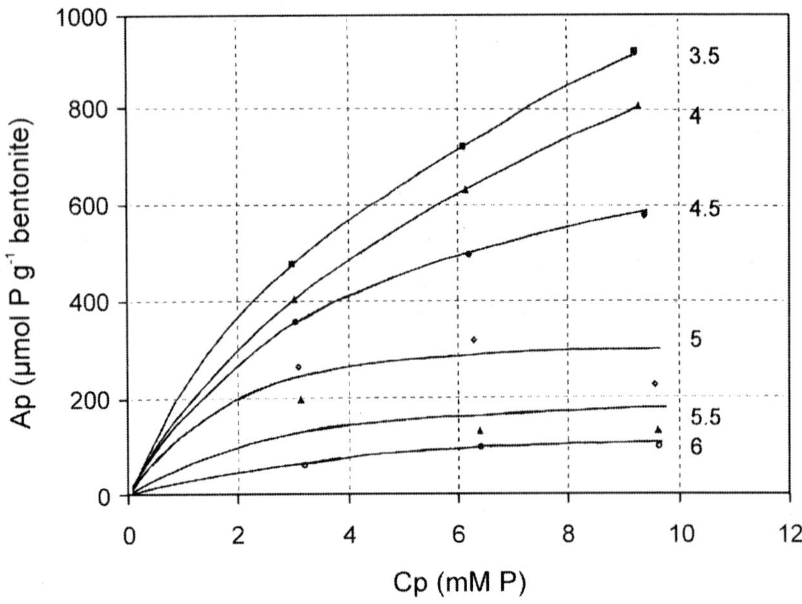

FIGURE 15.4. Adsorption of RNA on bentonite (Ap) as a function of concentration in equilibrium suspension solution (Cp) and pH. Symbols are interpolated values of original data. Lines were hand fitted through interpolated points. (*Source:* Adapted from Goring and Bartholomew, 1952, p. 153.)

Adsorption of PLP and glycerophosphates has been shown to be much smaller than that of IHP and NAP (Evans, 1985; Rauschkolb et al., 1976), but their concentrations in soil solution can, therefore, be higher, rendering them greater mobility and facilitating their deeper transport in the soil profile. Incorporating the three mobile Po species in one of the adsorption models concomitantly with the Pi species should enable the model to simulate the competition effect.

The water-soluble HA-Pi complex (Figure 15.1) has not yet been investigated with respect to stability or dependency of its complexation on pH. Jones, Salonen, and de Haan (1988) showed that humic materials having a nominal molecular weight of 10,000-20,000 sequestered Pi quite rapidly (in ~1 h) and maintained the complex in solution for three weeks or more. The complexation was pH dependent, and the retention was enhanced in the presence of ferric iron. It is expected that the adsorption of such a macromolecule by soil would be small because of its low charge density and large size. Sibanda and Young (1986) also reported reduced o-P adsorption by

TABLE 15.1. Langmuir Coefficients[a] of Adsorption Isotherms of Pi and IHP on Several Adsorbents (Ad) Characterized by Specific Surface Area (s) and Suspension Density (d).

Ad	P	b	K	pH	s	d	Source
		μmol Pm^{-2}	L mol^{-1}		m^{2}g^{-1}	gL^{-1}	
Goethite	IHP	0.64	8.0E3	4.5[b]	42	4.3	Celi et al., 1999
	Pi	2.4	5.6E3	4.5[b]	42	4.3	Celi et al., 1999
	IHP	0.45[c]	—	6.5	38	20	Ognalaga, Frossard, and Thomas, 1994
	Pi	1.7[c]	—	6.5	38	20	Ognalaga, Frossard, and Thomas, 1994
Fe-precipitate	IHP	60[cd]	—	6	—	—	Shang, Stewart, and Huang, 1992
	Pi	560[cd]	—	6	—	—	Shang, Stewart, and Huang, 1992
Al-precipitate	IHP	5.6	—	6.8	147	—	Shang, Huang, and Stewart, 1990
	Pi	5.8	—	4.7	147	—	Shang, Huang, and Stewart, 1990
Illite	IHP	0.38	1.0E5	4.5[b]	75	25	Celi et al., 1999
	Pi	1.0	7.5E3	4.5[b]	75	25	Celi et al., 1999
Kaolinite	IHP	0.27	1.0E5	4.5[b]	18	30	Celi et al., 1999
	Pi	0.79	1.6E3	4.5[b]	18	30	Celi et al., 1999
Calcite	IHP	17.8	1.5E4	8.3	2.8	15	Celi et al., 2000
	Pi	1.4	—	8.3	2.8	15	Celi et al., 2000

Source: Data compiled from cited literature.

[a] $A = K b C / (1 + K C)$, where K is the affinity constant (L mol^{-1}), b is maximum adsorption (μmol P m^{-2} unless otherwise stated), C is concentration in solution (mol P L^{-1}), and A is adsorption (units as b).
[b] Initial pH (increased by up to 1 unit with increasing adsorption).
[c] Hand fitted according to plateau of reported isotherm.
[d] b is given in μmol g^{-1}.

goethite, gibbsite, and soil in the presence of humic acid (at an initially high concentration of 16 g l^{-1}), but they attributed it to competition on o-P adsorption sites. The effect diminished with increasing pH and ceased at pH of ~7.6, which is consistent with the P complexation mechanism.

PRECIPITATION-DISSOLUTION KINETICS

The theory of nucleation, crystal growth, and dissolution has been reviewed recently by Stumm and Morgan (1996) and by Lasaga (1998). The process of crystal growth (or dissolution) is constrained by three main reactions: (a) transport of ions away from, or to, the dissolving crystal, (b) detachment (or attachment) of ions from (or to) the surface of the crystal, and (c) combination of (a) and (b) (Liu and Narasimhan, 1989). In reaction (a) the ions are detached quickly enough to build up to form a saturated solution adjacent to the surface; further dissolution will be inhibited unless the concentration is reduced below saturation by advective-diffusive transport to the surrounding solution ("transport-controlled dissolution"). The rate of dissolution (R, mol l^{-1} soil h^{-1}) can therefore be functionally related to the concentration, C, of the surrounding solution, and the saturation concentration C_S (mol l^{-1}) (Lasaga, 1998):

$$R = -K \cdot (C - C_s)^n \tag{15.14}$$

if $C > C_S$ crystal growth will take place. In Equation (15.14) K (l solution l soil^{-1} h^{-1}) is a temperature-dependent rate constant and n is the order of the reaction. In reaction (b), ion detachment is sufficiently slow so that ion buildup at the crystal surface cannot keep up with transport and the resulting concentration adjacent to the surface is practically the same as that of the surrounding solution ("surface-controlled dissolution"). Liu and Narasimhan (1989) report that, in many cases, ill-prepared sample surfaces result in diffusion-controlled dissolution kinetics, while careful analysis by scanning electron microscopy has revealed that, in fact, the rate-limiting step is mineral dissolution. Rate equations subject to "surface controlled" dissolution can be found in Lasaga (1998, p. 647).

A different approach to the description of mineral dissolution and precipitation is based on the standard collision theory relationship between chemical reaction stoichiometry and the rate expression (Smith and Jaffe, 1998). For a given solid S^* composed of N_s species the net rate of precipitation, R^{net} is expressed as

$$R_S^{net} = K_S \left[\prod_i^{N_S} [FIA_i]^{a_{i,s}} - K_{SP}^S \right] \tag{15.15}$$

where K_S is the rate coefficient specified for solid S*, FIA_i is the dissolved free ion activity of component *i*, $\alpha_{i,s}$ is the stoichiometric coefficient of the *i*th component of solid S*, and K^{s}_{SP} is the solubility product of solid S*. The net rate for the *i*th dissolved component of the solid, R^i_S, is $R^i_S = -\alpha_{i,s} R_s^{net}$. Inskeep and Bloom (1985) have used this equation to describe calcite precipitation rate, *R*, as: $R = \gamma^2 K_f S \{[Ca^{2+}][CO_3^{2-}] - K_{SP}\gamma^{-2}\}$, where $K_f = 118$ $l^2\ mol^{-1}\ m^{-2}\ s^{-1}$, S = surface area ($m^2\ l^{-1}$), and γ is the activity coefficient. A similar approach was applied by Musvoto et al. (2000a) in their kinetic-based model predicting salt precipitation from the weak acid/base system.

Most studies on P dissolution, sorption, and transport in soil have used a diffusion-controlled kinetic expression similar to Equation (15.14) with an o-P species replacing *C* (Ferguson, Jenkins, and Eastman, 1973; Shaviv and Shachar, 1989; Enfield et al., 1981). The choice of C_s (Equation 15.14) and K_{SP} (Equation 15.15) is arbitrary, because in highly supersaturated solutions containing Ca and P, dicalcium phosphate dihydrate (DCPD, $CaHPO_4 \cdot 2H_2O$, $pK_{sp} = 6.6$) and amorphous calcium phosphate [ACP, $Ca_3(PO_4)_2 \cdot xH_2O$, $pK_{sp} = 24 - 25.5$] are phases that precipitate first, with DCPD precipitating at pH < 7 and ACP at higher pHs. This behavior complies with the empirical "rule of stages" (Musvoto et al., 2000a) which states that the least stable solid will precipitate, with the initially formed metastable precursor species converting with time to the more thermodynamically stable tricalcium phosphate [TCP, $Ca_3(PO_4)_2$ $pK_{sp} = 32.6$], octacalcium phosphate [OCP, $Ca_4H(PO_4)_3 \cdot 2.5H_2O$, $pK_{sp} = \sim 46$], and hydroxyapatite [HAP, $Ca_{10}(PO_4)_6(OH)_2$, $pK_{sp} = 49 - 57$]. The crystal growth of the stable species is very slow such that the conversion takes a long time, usually between a minimum of 1 month to a number of years (Musvoto et al., 2000b). Because of this behavior Musvoto et al. (2000a) have used ACP to simulate Ca-P precipitation in wastewater, while Cho (1991) assumes that the governing Ca-P mineral in Ca saturated systems is DCPD. The rate of DCPD precipitation *(R)* has been described by Cho (1991) as a second-order kinetics:

$$R = r_p\,(HPO_4^{2-})[(Ca^{2+}) - (K_{sp}/HPO_4^{2-})] \tag{15.16}$$

where r_p is a system rate constant. Experimental data by Grossl and Inskeep (1991) reveal that *R* of DCPD in an initially supersaturated solution (11.5 mM P, 12.5 mM Ca, and constant pH and temperature of 5.7 and 25°C, respectively) varied between 17 $\mu mol\ l^{-1}\ s^{-1}$ (0-5 minutes) and 0.22 $\mu mol\ l^{-1}\ s^{-1}$ (5-120 minutes). These rates can be compared with the mean rate of hydroxyapatite (HAP) precipitation (initial P and Ca 0.86 mM) which Inskeep and Silvertooth (1988) have reported to be 0.007 $\mu mol\ l^{-1}\ s^{-1}$ (0-700 minutes). In these two previous references *R* was described as $R = \gamma^2 K_f S [Ca^{2+}][C_p]$ where C_p represents HPO_4^{2-} and PO_4^{3-} concentrations, respectively. The units were defined previously. Assuming that R in the laboratory experiment and in the soil solution are equal to each other allows estimation of the crystallization flux *(CR)* in Equation (15.1). For è = 0.25 l l^{-1}, a precipitation rate of 0.2 ìmol $l^{-1}s^{-1}$ corresponds to 5.0E–5μmol P cm^{-3}soil s^{-1}, which is similar to the estimated 0-24 h mean adsorption rate (R_A). As the degree of supersaturation diminishes, R decreases and becomes appreciably smaller than R_A. It appears, however, that organic acids prevailing in soil significantly reduce K_f (Inskeep and Silvertooth, 1988; Grossl and Inskeep, 1991), therefore, R under field conditions is expected to be smaller than in uncontaminated solutions. In the presence of high NH_4^+ concentration, which may occur in soils loaded with biosolids, particularly under conditions that inhibit nitrification, o-P may precipitate as struvite ($MgNH_4PO_4 \cdot 6H_2O$, pK_{sp} between 12.6 and 13.15, Lowenthal, Kornmuller, and van Heerden, 1994; Buchanan, Mote, and Robinson, 1994; Web and Ho, 1992).

In models in which adsorption and precipitation are not separated, P sorption/desorption is often described by continuous empirical time functions. A notable example is a model by Barrow (1983) that stipulates that adsorption precedes a slower solid-state diffusion of P into the bulk of reactive minerals. The model has been used to predict a variety of time-dependent P sorption data (Bolan, Barrow, and Posner, 1985), but the good agreement may be a result of the eight fitting parameters involved, and is not a proof of the validity of the suggested mechanism (van der Zee and van Riemsdijk, 1991). Other rate expressions that have been used to describe the overall sorption kinetics of P in soil and clay mineral suspensions include first- and second-order reactions with respect to o-P solution concentration, a parabolic-diffusion model, a power function, and the Elovich equation. For a review of these empirical models in soils the reader is referred to Sanyal and De Datta (1991). A general review of the kinetics of soil chemical reactions can be found in Aharoni and Sparks (1991).

A nonmechanistic approach to the approximation of P kinetics in soil, based on three inorganic-P (Pi) pools, has been suggested by Jones et al. (1984), who assume that the rate of P transition (R_{la}) from a "labile" pool P_{il} (to which fertilizer P is added and from which plants absorb P) to an "active" pool P_{ia} can be described by:

$$R_{la} = K_p{*}(P_{il} - P_{ileq}){*}F_{im}{*}F_{it} \tag{15.17}$$

where $P_{ileq} = P_{ia}{*}F_l/(1 - F_l)$, F_l is the "soil P availability index," defined as $F_l = \Delta P_{il}/P_f$, and ΔP_{il} is the change in P_{il} caused by the application of fertilizer P to the soil at a rate P_f. The value of P_{il} is determined by soil extraction with an anion exchange resin. The rate constant K_P was assigned a value of 0.1 day^{-1} for optimum values of T and θ, and was corrected for other T and θ values according to empirical bell-shaped functions F_{im} and F_{ip}, respectively, that varied between 0 and 1. The transition from P_{il} to P_{ia} reflects the rapid reaction of P with the soil following fertilization ($P_{il} > P_{ileq}$), while replenishment of labile P takes place when $P_{il} < P_{ileq}$. The slow sorption is envisioned as a transfer of inorganic-P from P_{ia} to the third pool P_{is} (stable P_i). The rate is

$$R_{as} = K_{as}\ (4{*}P_{ia} - P_{is}) \tag{15.18}$$

where K_{as} = exp (-177*F_l – 7.05). No partitioning between solution and solid phases is considered in this model.

ORGANIC-P TRANSFORMATIONS

Assessments of Po mineralization range from rough estimations to complex simulations of the process. Barber (1995) has assumed the rate of Pi release to be equal to the rate of organic matter (OM) decomposition multiplied by the P concentration in the OM: in a 0-40-cm soil layer with soil bulk density of 1.25 kg l^{-1}, 2 percent OM content, 2 percent OM decomposition per year, and 0.5 percent P in the OM, the rate of Pi release would be 10 kg Pi ha^{-1} per year (y^{-1}). According to Dalal (1979) the $t_{1/2}$ of clover P mineralization in a soil of pH 6 was shown to be 10.8 weeks for roots and 12.4 weeks for the canopy. Grierson, Comerford, and Jokela (1998) have compared several kinetic models and have concluded that a zero-order model best describes the mineralization of Po in undried sandy soils.

Traditionally, simulation of Po mineralization involves a set of simultaneous rate equations accounting for several organic-P pools in the soil and mass transfer among them (Blair and Boland, 1978; Jones et al., 1984; Williams, 1987; Wolf et al., 1987; Parton et al., 1989; Fardeau, 1995). In all the cited models forward and backward first-order reactions with respect to Po concentration in the source pool were used. The models differ in the number of pools and in the parametric values assigned to the rate constants. As an example, the P submodel of the EPIC simulator (Jones et al., 1984) is presented. This model first calculates the rate of decomposition of crop residue in soil *(R_{or})*. It employs a predefined rate constant *(K_{or})*, the size of the residue pool *(O_r)*, and four empirical reduction functions which account for the effects of soil temperature *(F_{ot})*, θ (F_θ), C/N ratio *($F_{c/n}$)*, and C/P ration *($F_{c/p}$)* effects:

$$R_{or} = K_{or} * O_r * (F_{ot} * F_\theta)^{1/2} * \{\min (F_\theta, F_{c/p})\} \tag{15.19}$$
$$F_{c/p} = \exp\{-0.693{*}[C_r/(P_{or} + P_{il}) - 200]/200\}$$

$$F_{c/n} = \exp\{-0.693*[C_r/(N_i + N_{or}) - 25]/25\}$$
$$F_\theta = \theta/\theta_{.03Mpa}, \text{ and } C_r = 0.4*O_r$$

Here N_i and P_{il}, and N_{or} and P_{or}, are the N and P contents (kg kg^{-1} soil) of the inorganic and residue pools, respectively. The value of K_{or} decreases from 0.8 d^{-1} when the ratio $O_r/O_{r\,\text{initial}} > 0.8$ to 0.05 day^{-1} when $0.1 < O_r/O_{r\,\text{initial}} < 0.8$, and to 0.001 day^{-1} at lower ratios.

The second step is to calculate the P uptake rate by the microbial mass (R_{upr}). It is assumed that a certain fraction of the carbon in R_{or} (= 0.4) is converted into microbial mass. This mass consumes P (uptake) according to the microbial P concentration (P_m/O_m):

$$R_{upr} = 0.4*0.4*R_{or}*P_m/O_m \tag{15.20}$$

The P_m/O_m varies between 0.01 and 0.02 according to P_{il}. When $P_{il} > 10$ kg P ha^{-1}, $P_m/O_m = 0.02$; when $P_{il} < 10$, $P_m/O_m = 0.01 + 0.001*P_{il}$. The next step is to calculate Po mineralization. This involves both the Po in the decaying crop residue (R_{pr}, from which R_{or} was calculated) and the Po in the stable pools (R_{pos}),

$$R_{pr} = R_{or}*P_{or}/O_r \tag{15.21}$$

$$R_{pos} = K_{os}*P_{os}*\{\min(F_\theta, F_{c/p})\} \tag{15.22}$$

The rate constant K_{os} was assigned a value of 0.0003 day^{-1} in moist soils; P_{or} and P_{os} are the sizes of the residue and the stable P pools. The net mineralization or immobilization of P from all sources (R_p) is given by Equation (15.21). The factor 0.8 accounts for the 20 percent of R_{pr} incorporated in P_{os}.

$$R_p = 0.8*R_{pr} + R_{pos} - R_{upr} \tag{15.23}$$

In a soil owing a residual pool (O_r) of 20 g OM kg^{-1}, and a rate constant $K_{or} = 0.1$ day^{-1}, the decomposition rate of crop residue (R_{or}) under conditions of optimal T, θ and C/N ratio is 1 g OM kg^{-1} d^{-1} (Equation 15.19). Assuming $P_{or}/O_r = 1/500$, P_{os} and K_{os} of 0.012 g P kg^{-1} soil and $3{\cdot}10^{-4}$ d^{-1}, and P_m/O_m and R_{or} of 0.015 and 1 g OM kg^{-1} d^{-1}, respectively, and inserting these parameters in Equations (15.21), (15.22), and (15.20), respectively, yields R_{pr}, R_{pos}, and R_{upr} of $2{\cdot}10^{-3}$, $3.6{\cdot}10^{-6}$, and $2.4{\cdot}10^{-3}$ g P kg^{-1} d^{-1}. Under such conditions the net mineralization (R_p, Equation 15.23) is negative, namely, immobilization (R_{upr}) > $R_{pr} + R_{pos}$. The flux R_{pr} is unrealistically large (~1800 kg P ha^{-1} day^{-1}), which underscores the role of the reduction functions in the simulation. Despite some shortcomings of this approach, no better model is available in the literature. The following is a short discussion of the shortcomings of the EPIC model and suggested improvements that have already been incorporated in the conceptual model (Figure 15.1).

First, the EPIC model does not account for adsorption of organic-P species and their transport in the soil. In the conceptual model Po is divided into IHP, NAP, and PAP that are selectively partitioned between the solid and liquid phases of the soil, and a fourth fraction (PNS) that comprises the rest of the Po, and cannot be found in the soil solution. To be executable, the four compounds must constitute known, time-invariant fractions of Po of each of the organic pools in Figure 15.1, whereas Po is estimated from its known concentration in the OM. It is still unknown whether the whole of the IHP, NAP, and PAP is available for adsorption, or whether part of them is found in OM that is not in direct contact with the soil matrix.

Second, in the EPIC model, P uptake by microorganisms is defined as consumption, namely, incremental increase in biomass multiplied by an assumed P concentration in this mass. In the conceptual model (Figure 15.1) it is proposed to use an M&M flux-concentration relationship to determine P uptake by microorganisms, on the assumption that absorption is from the soil solution only (Scow and Hutson, 1992). This approach is consistent with the significant correlation Chauhan, Stewart, and Paul (1981) found between microbial P uptake and solution P concentration. Since P uptake by the microbial mass has been estimated to be three to five times as great as that by plants (Cole, Innis, and Stewart, 1977; Sanyal and DeDatta, 1991), such an approach would more realistically describe temporal P concentrations in soil solution, leaching, and uptake by plants (Bosatta and Agren, 1991; Linquist, Singleton, and Cassman, 1997). Unfortunately, M&M data for P uptake by microorganisms is scant or entirely unavailable.

Third, in EPIC, 40 percent of the C from decomposed residue is allocated to the microbial mass. This arbitrary allocation can be replaced by the Monod kinetic model, which simulates the microbial growth rate (dB/dt) as a function of substrate concentration S (mol l^{-1}) and population density (g biomass l^{-1}) (Simkins and Alexander, 1984; Scow and Hutson, 1992; Rifai and Bedient, 1994):

$$dB/dt = B*U_{max}*S/(K_s + S) - R_{sl} \tag{15.24}$$

Here U_{max} is the maximum specific growth rate, K_s the half-saturation constant for growth (mol l^{-1}), and R_{sl} is the rate of sloughing ($R_{sl} = s*B$, where s is a fraction). The C and P released by the sloughing should be allocated to appropriate pools in the soil. The substrate that limits growth is most likely the C concentration in the soil solution (DOC), which is a function of R_{or}. The functional relationship between DOC and R_{or} is still unknown. Use of the Monod model emphasizes importance of DOC transport in soil in determining development of microorganisms.

The concentration of P in the microbial mass (P_m/O_m) is determined by both the P uptake and the microbial mass. Deviations of P_m/O_m from a predetermined optimal concentration can be used in principle to adjust microbial growth by introducing a bell-shaped correction function with value of 1 at the optimal concentration. When the microbial and root masses are known, the rate of phosphatase release can be estimated and used to estimate the rate of Po mineralization from the OM pools. However, at present, only meager information is available (Tarafdar and Jungk, 1987; Beck, Fusseder, and Kraus, 1989) on factors affecting acid and alkaline phosphatase release by different crops and microflora, and rates of release per unit mass.

P UPTAKE AND ACTIVITY AND GROWTH OF ROOTS

Plant roots have a major influence on P dynamics in the soil: (1) they provide a sink for Pi and a source of Po; (2) they take up water, thus generating mass flow toward the roots and variations in soil θ, and the latter affects the apparent P diffusion coefficient (Equation 15.5) as well as the Po mineralization rate in the soil (Equation 15.19); (3) root excretions (particularly protons, and carboxylic and aminocarboxylic anions) affect P partitioning between solution and solid phases in the soil; and (4) the root distribution in the soil, which can be modified by horticultural practices (Bar-Yosef, 1999), affects water and P distributions, and, in turn, soluble P transport toward underground water. These factors and the interrelationships among them are discussed in the following section, with reference to P dynamics and transport modeling.

P Uptake

Numerous studies have shown that the M&M model (Equation 15.25) satisfactorily describes the P uptake flux (Fp, mol cm^{-1}root h^{-1}) as a function of P concentration in a well-stirred solution (Cp_b, mol l^{-1}) (Barber, 1995). Under such conditions Cp_b may adequately represent the true concentration at the root surface (Cp_a). The coefficients Fp_{max} and K_M denote the maximum Fp, and the Cp_a at which $Fp = 0.5*F_{max}$, respectively.

$$Fp = Fp_{max} * Cp_a / (K_M + Cp_a) \tag{15.25}$$

When $Cp_a < K_M$, equation (15.25) becomes $Fp = \alpha\ Cpa$, which is easier to incorporate in transport models. The effect of v (Equation 15.5) on α (root absorbing power, l cm^{-1} root h^{-1}) can be evaluated from Dalton, Raats, and Gardner (1975), who developed a model relating solute and water uptake by roots. Nye (1977) showed that over the range of nutrient concentrations usually occurring in soil the effect of v is small, and á is determined by the active uptake component of the solute flux.

A compilation of reported M&M coefficients for P in some field crops can be found in Bar-Yosef (1999). In some cases (Barber, 1995) Cp_a in Equation (15.25) is replaced by $Cp_a - C_{min}$ to allow for P efflux (Jungk et al., 1990). The disadvantage is that C_{min} (the concentration for which the net influx is zero) is an additional parameter and its inclusion does not necessarily improve the model. It is noted that the solution of Equation (15.1) is P concentration in the bulk soil solution (Cp_b), while to solve Equation (15.25) Cp_a must be used. To evaluate Cp_a from the computed Cp_b, Bar-Yosef (1999) applied the steady-state solution of Equation (15.7) subject to Equations (15.8 and 15.9) (Equation 15.26, see Olsen and Kemper, 1968). Note that α in Equation (15.8) was replaced by $\alpha = Fp_{max} /(K_M + Cp_a)$ so that the M&M equation is used to describe the boundary condition at the root surface:

$$Cp_a = \frac{B}{(K_M / Cp_a + 1)} + \left[Cp_b - \frac{B}{(K_M / Cp_a + 1)} \right] A \tag{15.26}$$

Here $A = (b / a)^{W/2\pi Dp}$, b and a, respectively, are the midway distance between the roots and the root radius, $B = Fp_{max}/W$, and Dp was defined in Equation (15.5). W is related to the solution velocity toward the root, v_{root} (cm h^{-1}) by $W = v_{root}\, 2\pi a\theta$. At the end of each time step in the solution of Equation (15.1), P in each soil subvolume is "remixed" to give a new and uniform C_b, and a new quasi-steady-state concentration profile toward the roots is assumed, from which Cp_a (Equation 15.26) and Fp (Equation 15.25) are calculated during the next Δt. When $K_M > Cp_a$

$$Cp_b / Cp_a = 1 / A - B(1 / A - 1) / K_M \tag{15.27}$$

The M&M equation cannot account for the influence on Fp of the P status in the plant. If the status can be expressed in terms of P concentration inside the root (Cp_l, mol l^{-1}) (Aoalsteinsson and Jensen, 1990), then the flux can be written

$$Fp = K_T * (Cp_a - Cp_l) \tag{15.28}$$

where K_T (cm h^{-1}) is the root ion transfer coefficient. A similar approach was used to simulate W, only in this case the concentration difference has been replaced by root water pressure head (ψ_a, cm) and water pressure head at the root surface (ψ_l), and K_T by a coefficient K_{TW} (h^{-1}) that accounts for the soil-root interface hydraulic conductance and geometry (Gardner, 1960):

$$W = K_{TW} * (\psi_a - \psi_l) \tag{15.29}$$

Hillel (2000) used the same expression, but employed total water potential (matric+gravitational+osmotic) instead of ψ. The water uptake rate by plant (W_w) equals to $W_w = RL * W$.

It is interesting to estimate the rate of P uptake per unit soil volume per unit time (U, Equation 15.1) in relation to other fluxes in that equation. According to Bar-Yosef (1999), root concentration of tomato plants in soil is ~50 mg dry root kg^{-1} soil. Considering 5 percent DM content in roots, fresh root bulk density of 1 kg l^{-1} and root radius of 0.02 cm yields ~ 800 cm root kg^{-1} soil. The M&M constants of a tomato plant are: $F_{max} = 6.6 \cdot 10^{-13}$ mol P cm^{-1} s^{-1} and $K_M = 436 \cdot 10^{-6}$ M (Bar-Yosef, 1999). For $Cp_a = Cp_b = 10^{-4}$ M and the tomato F_{max} and K_M, Equation (15.25) yields $F = 1.23 \cdot 10^{-13}$ mol P cm^{-1} s^{-1}. The P uptake rate Q under these conditions is $Q = RL \cdot F = 9.8 \cdot 10^{-11}$ mol kg^{-1} s^{-1}, or (for d=1.25 kg soil l^{-1}) $U = 1.2 \cdot 10^{-7}$ μmol P cm^{-3} s^{-1}. This result is similar to the $\Delta F/\Delta z$ value, and much smaller than the rate of transfer of adsorbed P to the solution phase ($\Delta A/\Delta t$) evaluated there.

Root Growth and Distribution in Soil

Phosphorus uptake rate by plants (Up, mol pl^{-1} s^{-1}) is $Up = \int_{z=0}^{Z} (Fp \cdot RL)dz$,

where z is the soil depth (cm). Therefore, RL affects not only Up, but also its distribution in the soil. Root growth [$RL(z)$] depends on the plant genetic code (Zobel, 1975), on physical and chemical conditions in the soil (for review see Hoogenboom, 1999), and on the allocation of carbohydrates from the canopy to the roots (Hoogenboom and Huck, 1986). Various attempts have been made to simulate these effects. Timlin, Acock, and van Genuchten (1996) computed carbon partitioning by assuming that the allocation to roots was a function of leaf water potential, whereas Fishman et al. (1984) considered it to be determined by the relative sink power of plant organs, defined as the carbohydrate content of the organ divided by the total content in the plant.

Root growth models (for review see Hoogenboom, 1999) distinguish between potential and actual root growth rates. The potential root growth rate in a given soil subvolume is described as a first-order reaction with respect to root length, *RL* (cm):

$$dRL / dt = \beta RL \tag{15.30}$$

where *t* is time, and is a temperature-dependent rate constant (h^{-1}). The potential growth rate for the entire plant is $\int_{z=0}^{z} (\beta \cdot RL)dz$. If the potential growth rate exceeds the actual rate of carbon allocation by the canopy, translocation to the individual soil subvolumes is proportional to their relative root growth rate. Values of â that were used to simulate corn and cotton growth in soil and their temperature dependency can be found in Bar-Yosef, Lambert, and Baker (1982). Once the root length in a given soil cell exceeds a critical length the roots can extend into neighboring cells (their number depends on whether a one-, two- or three-dimensional system is simulated). The fraction extended into each cell is determined by its relative θ (relative to other neighboring cells), but dependency on the concentrations of O_2 and NO_3^- (Bar-Yosef and Lambert, 1979) and P (Drew, Saker, and Ashley, 1973; Aoalsteinsson and Jensen, 1990) may also be included. Dynamic root growth models must account for the transitions of a unit length of root from an active state to suberized state (reduced uptake efficiency), and finally to a sloughed state (Bar-Yosef, Lambert, and Baker, 1982). The C and Po of the sloughed root are allocated to the readily mineralizable pool (Figure 15.1). The modeling of root morphology and architecture (Clausnitzer and Hopmans, 1994), which may group roots according to their radius and extension angle, is too complex to be applied in P dynamics modeling. The same problem applies to simulation of root geometry by using fractal branching models (van Noordwijk, Spek, and Dewilligen, 1994; Lynch et al., 1997). Using logistic equations to describe root growth (e.g., by integrating Equation 15.30; Williams, 1948) without considering environmental factor effects is unsatisfactory, since it does not account for the feedback relations between root extension and physical and chemical conditions in the wetted soil volume.

Root Activity

Excess anion over cation uptake is accompanied by H^+ influx into the root to preserve cell electroneutrality. Under field conditions this excess usually occurs in the presence of nitrates, and results in a pH increase in the rhizosphere. The absorbed nitrates must be reduced prior to their incorporation in the plant: $NO_3^- + 8H^+ + 8e^- = NH_3 + OH^- + H_2O$ (Marschner, 1995). The free OH^- is immobile in the phloem, therefore, it has to be neutralized in leaves by carboxylation and formation of phloem-mobile carboxylic anions (Marschner, 1995). Some of the carboxylic anions are excreted to the root-surrounding solution, thus avoiding an equivalent influx of protons. When cation uptake exceeds anion uptake (usually when ample NH_4^+ or K^+ fertilizers are applied), electroneutrality is maintained by H^+ efflux, which decreases the outer solution pH. A consequence of this mechanism is that different NH_4/NO_3 ratios in soil solution result in different solution pH and carboxylate concentrations around roots. As discussed, such variations should be taken into account when simulating P speciation, adsorption, and precipitation in the soil, H^+/OH^- propagation, and dissolution of Ca-P and Al-P minerals through complexation of Ca and Al and reduction of the free ion activity in solution. The subject of root excretions and their effects on the availability of phosphorus as outlined has been reviewed by Bar-Yosef (1996). Specific influences of pH and carboxylates on P reactions have been described in the pertinent sections.

Quantitative relationships between rates of carboxylate and proton release per unit length of root and NO_3 and NH_4 uptake rates are scarce. Imas et al. (1997b) report that below a certain threshold, NO_3 uptake rate by tomato plants (55-95 μmol per plant per 6h, depending on plant age) the citrate exudation rate is negligible (<0.05 μmol plant^{-1} per 6h). Above this threshold value an incremental citrate release of ~0.4 μmol plant^{-1} per 6h in response to an incremental NO_3 uptake rate of 10 μmol plant^{-1} per 6h was observed for 37-day-old plants. The shoot and root fresh weights at this age were 28 and 17 g plant^{-1}, respectively. The relationship between the NH_4 uptake and the H^+ efflux showed that for NH_4 uptake rates below 300 μmol plant^{-1} per 6h the efflux was negligible. Above this threshold the H^+ efflux rate was 15 μmol plant^{-1} per 6h for an NH_4 uptake rate increment of ~50 μmol plant^{-1} per 6h (for plants of the same age and weight).

The problem of incorporating root excretions in soil-P models has hardly been addressed in the literature. One approach is to calculate a quasi-steady state (Gardner et al., 1983) or steady efflux (Nye, 1986) concentration profile of a given exudate as a function of distance from the root, and to incorporate it in a model that computes P uptake (Gillespie and Pope, 1990). This will be further elaborated for the case of proton efflux in the next section. However, when the radial flow toward roots is included in a larger model that computes vertical (and possibly lateral) ion transport, the obtained exudate profile overlaps with the main model concentration profile (Equation 15.1), and as discussed in the case of P, mixing based on mass balance in each soil cell must be calculated prior to advancing the solution by another

Δt. Another approach is to mix the entire quantity of exudate released at Δt in the entire soil cell volume; this would reduce the number of parameters compared with the previous approach [diffusion coefficients in soil solution (Darrah, 1991) and adsorption isotherm slopes], but would decrease the effects of exudates on uptake by diluting their effective concentration in the soil.

Bar-Yosef (1996) concluded that in the range of exudate concentration of 0.1-2 mM that is expected in soil solution only citrate and, possibly, oxalate might be effective in modifying P mobility in soil. The high efficiency of citrate in releasing adsorbed Pi from montmorillonite and kaolinite (Kafkafi et al., 1988; Traina et al., 1987), and from Fe-oxides (Nagarajah, Posner, and Quirk, 1968) has been experimentally proven, and simulated (Bowden et al., 1980). Citrate was also shown to release adsorbed RNA and IHP from kaolinite and bentonite (Goring and Bartholomew, 1952).

Another aspect of root activity is phosphatase release by roots and its influence on Po mineralization (Bar-Yosef, 1996). Confounding effects by microbial and mycorrhizal phosphatases and conflicting experimental results on the contribution of root phosphatases to Po mineralization (Joner et al., 1995) prohibit a quantitative analysis of this important mechanism.

pH Propagation in Soil

Because of the crucial effect of pH on P partitioning in the soil, the proton/hydroxyl dynamics must be taken into consideration as well. This is particularly important in the presence of plants and H^+/OH^- release by the roots. The continuity equation for H^+ is:

$$\theta \frac{\partial(H)}{\partial t} = -\frac{\partial F_H}{\partial z} - \frac{\partial A_H}{\partial t} - R_{base} + n \cdot U_{NH4} \tag{15.31}$$

Where: R_{base}, $\frac{\partial(HPO_4 + HCO_3 + OH)}{\partial t}$, $F_H = -D_{PH}\frac{d(H)}{dz} + v(H)$, U_{NH4} = NH_4^+ uptake rate (mols l^{-1} soil h^{-1}), n = mol H^+ excreted per mole NH_4^+ consumed, parentheses denote concentration in soil solution (mol l^{-1}), A_H is its adsorption (mol l^{-1} soil, described in Equation 15.12 or Equation 15.13), and F is flux (mol cm^{-2} h^{-1}). Applying the chain rule, and differentiating R_{base} after expressing the three ions in terms of (H^+), Pco_2 and $H_2PO_4^-$ yields:

$$\frac{\partial A_H}{\partial t} = \frac{dA_H}{d(H^*)} \frac{\partial(H^+)}{\partial t} \tag{15.32}$$

$$\frac{\partial(HPO_4 + HCO_3 + OH)}{\partial t} = \frac{\partial(H)/\partial(t)}{(H)^2} \cdot [K_w + k_{p1} \cdot H_2PO_4 + k_1 k_2 \cdot Pco_2] \tag{15.33}$$

Here $pK_{pl} = 7.2$, $pk_d = 6.38$, and $pk_s = 1.41$. The use of partial CO_2 pressure (Pco_2) implies the assumption that the system is found in equilibrium with atmospheric CO_2. Inserting these constants, as well as $H_2PO_4^- = 10^{-4}$ M, $Pco_2 = 3.5 \cdot 10^{-4}$ bar, q = 0.3 and pH 7 in Equation (15.33) yields $\theta \cdot R_{base} = [\partial H / \partial t]*360$ mol l^{-1}soil h^{-1}. Choosing the H/OH adsorption model described in Equation (15.13) and differentiating with respect to (H^+) gives

$$\frac{dA_H}{d(H^+)} = A_H \cdot \left\{ \frac{1}{H^+} \cdot \left[1 + \frac{RPH \cdot G}{2.303 \cdot (pH)^2} \right] - \frac{K_H}{1 + \sum K_i C_{Pi} + K_H \cdot (H^+)} \right\} \tag{15.34}$$

Defining the terms in {} (Equation 15.34) as Y_H and inserting Equations (15.34), (15.33), and (15.32) in Equation (15.31) yields:

$$\left\{ A_H \cdot Y_H + \frac{(K_w + K_{p1} \cdot H_2PO_4 + k_1 \cdot k_2 \cdot Pco_2)}{(H^+)^2} + \theta \right\} \frac{\partial(\mathrm{H}^+)}{\partial z'} = D_{PH} \frac{\partial^2 F_H}{\partial z^2} - q \frac{\partial(H)}{\partial z} + n \cdot U_{NH4} \tag{15.35}$$

Note that the activity coefficients in Equation (15.35) are assumed to be 1. Considering again pH = 7 and $H_2PO_4^- = 10^{-4}$ M, and inserting them together with $A_H = 200$ µmol l^{-1} soil, RPH = 6.5, G = 2, $K_P = 10^3$ L/mol, and $K_H = 10^4$ l mol^{-1} in Equation (15.34) yields = $2.23 \cdot 10^3$ (the HPO_4^- species was neglected in this calculation). Altering the pH from 7 to 5 and keeping the other factors unchanged yields $\partial A_H / \partial(\mathrm{H}^+) = 260$ (mol H^+ l^{-1} soil)/(mol H^+ l^{-1} solution). The last value is somewhat smaller than the multiplier of $\partial H / \partial t$ in the R_{base} term mentioned previously.

To account for the NH_4 uptake rate, its dynamics in soil must also be simulated. This subject is not addressed in this review; interested readers are referred to Wagenet and Hutson (1987) for such simulation. However, to evaluate the possible role of NH_4^+ in rhizosphere acidification consider an example according to which NH_4^+ uptake rate is tenfold greater than the estimated rate of P uptake by tomato plant (1.2×10^{-7}µmol P cm^{-3}soil s^{-1}). Converting this value to units used herein yields 4.3 ìmol NH_4 l^{-1}soil h^{-1}. In order to estimate *n* (Equation 15.31) one can employ the ratio of 15 ìmol H^+ released per 50 μmol NH_4^+ absorbed cited previously. This gives (4.3·15/50 =) 1.3 μmol H^+ released per h per l soil. This rate is small relative to the cation exchange capacity (as a measure of the H^+ buffering capacity) of soils even if multiplied by ten days of continuous NH_4^+ uptake and proton release. A different estimate is obtained if the released protons are bound within the "root affected" soil volume only. Considering a distance of 0.2 cm from the root surface as representing this volume (Jungk, 1996), a root length and radius of 1000 cm l^{-1} soil and 0.02 cm, respectively, a soil volume of 120-cm^3 soil is obtained. This corresponds to (1.3·1000/120 =) 10.8 μmol H^+ l^{-1}affected soil h^{-1}. The example indicates that under larger ammonium uptake fluxes and proton release/ammonium uptake ratios, and smaller root affected soil volumes this mechanism may significantly affect the rhizosphere pH.

Two factors that have been omitted from the equations are: (1) proton consumption during the mineralization of organic N following application of biosolids to the soil (Hadas and Bar-Yosef, 1984) and their release during nitrification; (2) proton release by roots of certain crops under severe P deficiency (Imas et al., 1997a). The first factor should be evaluated in an N dynamics model. The second factor seems to be less important and can be neglected.

MODELS DESCRIBING P-DYNAMICS IN SOIL

Published models describing P dynamics in the soil account for only some of the processes and reactions described in Figure 15.1. The available models can be grouped as: (1) simple models aimed to predict long-term crop response to fertilizer P (Wolf et al., 1987; Sinclair and Johnstone, 1995). These models employ empirical first-order rate equations that describe available P concentration in the soil (P_A) as a function of P fertilizer application, coupled with a logistic function of dry matter yield versus P_A. The models have been used to simulate crop response to rate and time of P application. (2) Models that simulate P transport in sterilized soil, which implies disregard of organic-P transformations in the soil. Mokady and Zaslavsky (1966) simulated movement and retention of o-P following its addition as P-fertilizer by accounting for diffusion and convection, fast retention sinks, and a time-dependent sink. Enfield et al. (1981) and Gerritse (1989) developed similar models using an equation similar to (15.12) to account for o-P precipitation. Later o-P transport models (Mansell et al., 1985; Notodarmojo et al., 1991) incorporated more complex, time-dependent P retention functions, developed by Barrow and Shaw (1975a, 1979). Despite concomitant variations in P and pH during P transport, adsorption and precipitation/dissolution, the only model that accounted for the simultaneous P and pH propagation was the one by Kirk and Nye (1985), which simulated P dissipation from P fertilizer granules. The simulation of temporal and spatial pH variations was done by means of an equation similar to (15.29), without plants and assuming that o-P adsorption was independent of pH. These models showed that P movement from added P fertilizer does not exceed 10-20 cm in most soils. More recent empirical data (Jensen et al., 1998; Stamm et al., 1998; Simard, Beauchemin, and Haygarth, 2000) indicate, however, that under certain conditions, P transport can extend appreciably deeper than predicted in these models. This was attributed to preferential flow of soil solution in soil cracks or macropores created by soil worms. (3) Included in the third category are models that simulate P uptake from the soil by single roots (see Equations 15.7, 15.25, 15.26). In their review, Olsen and Kemper (1968) solved Equation (15.7) analytically subject to different boundary conditions at the root surface and midway between two roots. A numerical solution that assumed nonlinear instantaneous Langmuir adsorption and M&M uptake kinetics was published by Bar-Yosef, Bresler, and Kafkafi (1972), and later, with slight modifications, by Nye, Brewster, and Bhat (1975) and Barber and Cushman (1981). Shaviv, Shnek, and Ravina (1992) modeled P uptake by a single root by including P sorption kinetics. A review on models in this group and how they can be used to improve P fertilization efficiency was presented by Barber (1995). The effects of proton and organic acid excretions on the uptake of P and other ions by a single root have been meagerly treated in the literature. Bar-Yosef, Fishman, and Talpaz (1980) simulated this effect for zinc ions, and verified the model experimentally in a sand-clay mixture with porous tubes simulating roots (Bar-Tal, Bar-Yosef, and Chen, 1991). (4) Soil/crop-uptake models. A comprehensive model that accounts for P transport, specific adsorption via surface complexation (the constant capacitance model), precipitation/dissolution, formation of ion pairs, and P uptake by plants was developed by Grant and Heaney (1997). It accounts for all P species normally found in the soil solution in the presence of Fe^{3+}, Ca^{2+}, and Mg^{2+}; it calculates ionic activity coefficients, and it considers organic-P mineralization and inorganic-P mobilization. It does not account for feedback relations between root growth and θ and nutrient concentrations in soil, root excretions and their effects on P reactions, or dynamics of pH in soil. This model is part of the larger ecosystem model, ECOSYS (Grant, 1993). Another model in this category is part of an erosion-productivity impact (EPIC) model, developed by Jones et al. (1984). This model is based on concepts of the organic and inorganic soil P cycles described by Cole, Innis, and Stewart (1977). The inorganic-P chemistry and plant uptake subroutines are based on simplified assumptions. Tests should be made of whether incorporation of currently available knowledge would improve the performance of this and Grant and Heaney's (1997) models.

SUMMARY

The individual processes that determine short- and long-term phosphorus behavior in soil have been reviewed and their contributions to fluxes in the continuity Equation (15.1) have been evaluated. The processes can be divided into three categories: (1) processes that have been satisfactorily investigated and consequently incorporated in P-soil and P-soil-crop models; (2) processes that have been individually studied, are known to be important in simulating P dynamics in soil, but have not yet been incorporated in P models; (3) processes that are potentially important in simulating P dynamics in soil, but have been so far insufficiently investigated and are premature for inclusion in P-soil and P-soil-crop models. Processes in category (1) include Pi transport, adsorption, precipitation, mineralization/immobilization, and uptake. Processes in category (2) include pH-dependent adsorption, transport, and sorption of IHP, RNA, and PLP, dynamics of root growth and distribution in soil in relation to physical and chemical conditions in the soil-root volume and carbohydrate status and partitioning in plant, depletion of o-P concentration at the root surface under given o-P concentration in the bulk soil, and H^+/OH^- excretion by roots and propagation in the bulk soil. Processes in category (3) include release of organic C into the soil solution and its transport in soil and utilization by microorganisms, Pi retention by mobile macro-organic molecules, and Pi uptake by microorganisms from the soil solution via an M&M type mechanism. Future research should focus on improving current P-dynamics models by incorporating in them category 2 processes, and extending the understanding of category 3 processes and reactions. Once sufficient progress on this front is made, a modular platform for simulating all the discussed P processes can be developed. Such a platform can be used to predict the fate of P in soil and its impact on the environment, and improve P management decision making under wide range of growth, soil, and environmental conditions.

REFERENCES

Aharoni, C. and D.L. Sparks (1991). Kinetics of soil chemical reactions—A theoretical treatment. In *Rates of Soil Chemical Processes,* eds. D.L. Sparks and D.L.

Suarez. Madison, WI: American Society of Soil Science, Special Publication 27, pp. 1-18.

Ahuja, L.R., O.R. Lehman, and A.N. Sharpley (1983). Bromide and phosphate in runoff water from shaped and clodly soil surfaces. *Soil Science Society of America Journal 47:*746-748.

Anderson, G., E.G. Williams, and J.O. Moir (1974). A comparison of the sorption of inorganic orthophosphate and inositol hexaphosphate by six acid soils. *Journal of Soil Science 25:*51-62.

Aoalsteinsson, S. and P. Jensen (1990). Influence of temperature on root development and phosphate influx in winter wheat grown at different P levels. *Physiologia Plantarum 80:*69-74.

Barber, S.A. (1995). *Soil Nutrient Bioavailability* (Second Edition). New York: John Wiley & Sons.

Barber, S.A. and J.H. Cushman (1981). Nitrogen uptake model for agronomic crops. In *Modeling Wastewater Renovation: Land Treatment,* ed. I.K. Iskandar. New York: Wiley-Interscience, pp. 382-409.

Barrow, N.J. (1983). A mechanistic model for describing the sorption of phosphate by soil. *Journal of Soil Science 34:*733-750.

Barrow, N.J. and T.C. Shaw (1975a). The slow reactions between soil and anions: 2. Effect of time and temperature on the decrease in phosphate concentration in the soil solution. *Soil Science 119:*167-177.

Barrow, N.J. and T.C. Shaw (1975b). The slow reactions between soil and anions: 5. Effects of period of prior contact on the desorption of phosphate from soils. *Soil Science 119:*311-320.

Barrow, N.J. and T.C. Shaw (1979). Effects of solution and vigour of shaking on the rate of phosphate adsorption by soils. *Journal of Soil Science 30:*67-76.

Bar-Tal, A., B. Bar-Yosef, and Y. Chen (1991). Validation of a model of the transport of zinc to an artificial root. *Journal of Soil Science 42:*399-411.

Bar-Yosef, B. (1996). Root excretions and their environmental effects – Influence on the availability of phosphorus. In *Plant Roots: The Hidden Half* (Second Edition), eds. Y. Waisel, A. Eshel, and U. Kafkafi. New York: Dekker, pp. 581-604.

Bar-Yosef, B. (1999). Advances in fertigation. *Advances in Agronomy 65:*1-77.

Bar-Yosef, B., E. Bresler, and U. Kafkafi (1972). Uptake of phosphorus by plants growing under field conditions: I. Theoretical model and experimental determination of its parameters. *Soil Science Society of America Journal 36:*780-794.

Bar-Yosef, B., S. Fishman, and H. Talpaz (1980). A model of zinc movement to single roots in soil. *Soil Science Society of America Journal 44:*1271-1279.

Bar-Yosef, B., U. Kafkafi, R. Rosenberg, and G. Sposito (1988). Phosphorus adsorption by kaolinite and montmorillonite: I. Effect of time, ionic strength and pH. *Soil Science Society of America Journal 52:*1580-1585.

Bar-Yosef, B. and J.R. Lambert (1979). Corn and cotton root growth in response to osmotic potential and oxygen and nitrate concentration in nutrient solutions. In *The Soil Root Interface,* eds. J.C. Harley and R.S. Russell. London: Academic Press, pp. 287-299.

Bar-Yosef, B., J.R. Lambert, and D.N. Baker (1982). Rhizos: A simulation of root growth and soil processes. Sensitivity analysis and validation for cotton. *Transaction American Society of Agricultural Engineering 25:*1268-1273.

Beck, E., A. Fusseder, and M. Kraus (1989). The maize root system in situ: Evaluation of structure and capability of utilization of phytate and inorganic soil phosphates. *Zeitschrift für Pflanzenernahrung und Bodenkunde 152:*159-167.

Blair, G.J. and O.W. Boland (1978). The release of P from plant material added to soil. *Australian Journal of Soil Research 16:*101-111.

Bolan, N.S., N.J. Barrow, and A.M. Posner (1985). Describing the effect of time on sorption of phosphate by iron and aluminum hydroxides. *Journal of Soil Science 36:*187-197.

Bosatta, E. and G.I. Agren (1991).Theoretical analysis of carbon and nutrient interactions in soil under energy-limited conditions. *Soil Science Society of America Journal 55:*728-733.

Bowden, J.W., S. Nagarajah, N.J. Barrow, and A.M. Posner (1980). Describing the adsorption of phosphate, citrate and selenite on variable-charge mineral surface. *Australian Journal of Soil Research 18:*49-60.

Breeuwsma, A., J.G.A. Reijerink, and O.F. Schoumans (1995). Impact of manure accumulation and leaching of phosphate in areas of intensive livestock farming. In *Animal Waste and the Land-Water Interface,* ed. K. Steele. New York: Lewis Publisher, pp. 239-249.

Buchanan, J.R., C.R. Mote, and R.B. Robinson (1994). Struvite control by chemical treatment. *Transactions of the American Society of Agricultural Engineers 37:*1301-1308.

Cassel, E.A. (1998). Modeling phosphorus dynamics in ecosystems: Mass balance and dynamic simulation approaches. *Journal of Environmental Quality 27:*293-298.

Celi, L., S. Lamacchia, F.A. Marsan, and E. Barberis (1999). Interaction of inositol hexaphosphate on clays: Adsorption and charging phenomena. *Soil Science 164:*574-585.

Celi, L., S. Lamacchia, F.A. Marsan, and E. Barberis (2000). Interaction of inositol phosphate with calcite. *Nutrient Cycling in Agroecosystems 57:*271-277.

Celi, L., M. Presta, F. Ajmore-Marsan, and E. Barberis (2001). Effects of pH and electrolytes on inositol hexaphosphate interaction with geothite. *Soil Science Society of America Journal 65:*753-760.

Chauhan, B.S., J.W.B. Stewart, and E.A. Paul (1981). Effect of labile inorganic P status and organic carbon addition on the microbial uptake of P in soil. *Canadian Journal of Soil Science 61:*373-385.

Cho, C.M. (1991). Phosphate transport in Ca-saturated systems: I. Theory. *Soil Science Society of America Journal 55:*1275-1281.

Clausnitzer, V. and J.W. Hopmans (1994). Simultaneous modeling of transient three-dimensional root growth and soil water flow. *Plant and Soil 164:*299-314.

Clement, A.L., L. Somlyoly, and L. Koncsos (1998). Modeling the P retention of the Kis-Balaton upper reservoir. *Water Science and Technology 37:*113-120.

Cole, C.V., G.S. Innis, and J.W.B. Stewart (1977). Simulation of phosphorus cycling in semiarid grasslands. *Ecology 58:*1-15.

Correll, D.L. (1999). Phosphorus: A rate limiting nutrient in surface waters. *Poultry Science 78:*674-682.

Dalal, R.C. (1979). Mineralization of carbon and phosphorus from C^{14} and P^{32} labeled plant material added to soil. *Soil Science Society of America Journal 43:*913-916.

Dalton, F.N., P.A.C. Raats, and W.R. Gardner (1975). Simultaneous uptake of water and solutes by plants roots. *Agronomy Journal 67:*334-339.

Daniel, T.C., A.N. Sharpley, and J.L. Lemunyon (1998). Agricultural phosphorus and eutrophication: A symposium overview. *Journal of Environmental Quality 27:*251-257.

Darrah, P.R. (1991). Measuring the diffusion coefficient of rhizosphere exudates in soil. II. The diffusion of sorbing compounds. *Journal of Soil Science 42:*421-434.

Drew, M.C., L.R. Saker, and T.W. Ashley (1973). Nutrient supply and the growth of seminal root system in barley. *Journal of Experimental Botany 24:*1189-1202.

DUFLOW (1992). A micro-computer package for the simulation of unsteady flow and water quantity processes in one dimensional channel systems. The Netherlands: Public Works Department (Rijkswaterstaat).

Dzombak, D.A. and F.M.M. Morel (1990). *Surface Complexation Modeling. Hydrous Ferric Oxide.* New York: John Wiley & Sons.

Enfield, C.G., T. Phan, D.M. Walters, and R. Ellis Jr. (1981). Kinetic model for phosphate transport and transformation in calcareous soils. I. Kinetics of transformation. *Soil Science Society of America Journal 45:*1059-1064.

Evans, A. Jr. (1985). The adsorption of inorganic-P by a sandy soil as influenced by dissolved organic compounds. *Soil Science 140:*251-255.

Fardeau, J.C. (1995). Dynamics of phosphate in soils. An isotopic outlook. *Fertilizer Research 45:*91-100.

Ferguson, J.F., D. Jenkins, and J. Eastman (1973).Calcium phosphate precipitation at slightly alkaline pH values. *Water Pollution Control Federation 45:*620-631.

Fishman, S., B.H. Talpaz, M. Dinar, Y. Arazi, Y. Rozman, and S. Varshavski (1984). Phenomenological model of dry matter partitioning among plant organs for simulation of potato growth. *Agricultural Systems 14:*159-169.

Gardner, W.K., D.G. Parbery, D.A. Barber, and L. Swinden (1983). The acquisition of phosphorus by *Lupinus albus* L. v. the diffusion of exudates away from roots: A computer simulation. *Plant and Soil 72:*13-29.

Gardner, W.R. (1960). Dynamic aspects of water availability to plants. *Soil Science 89:*63-73.

Gerritse, R.G. (1989). Simulation of phosphate leaching in acid soils. *Australian Journal Soil Research 27:*55-56.

Gillespie, A.R. and P.E. Pope (1990). Rhizosphere acidification increases phosphorus recovery of black locust. II. Model prediction and measured recovery. *Soil Science Society of America Journal 54:*538-541.

Goldberg, S. (1995). Adsorption models incorporated into chemical equilibrium models. In *Chemical Equilibrium and Reaction Models,* eds. A. Loeppert, R.H. Loeppert, A.P. Schwab, S. Goldberg, D.E. Kissel, and G.H. Heichel. Madison, WI: American Society of Soil Science, Special Publication 42, pp. 75-95.

Goldberg, S. and G. Sposito (1984). A chemical model of phosphate adsorption by soils. I. Reference oxide minerals. *Soil Science Society of America Journal 48:*772-778.

Goldberg, S. and G. Sposito (1985). On the mechanism of specific phosphate adsorption by hydroxylated mineral surfaces: A review. *Communications in Soil Science and Plant Analysis 16:*801-821.

Goring, C.A.I. and W.V. Bartholomew (1950). Microbial products and soil organic matter. III. Adsorption of carbohydrate phosphates by clays. *Soil Science Society of America Proceedings 15:*189-194.

Goring, C.A.I. and W.V. Bartholomew (1952). Adsorption of mononucleotides, nucleic acids, and nucleoproteins by clays. *Soil Science 74:*149-164.

Govindasamy, R., M.J. Cochran, and E. Buchberger (1994). Economic implications of phosphorus loading policies for pasture land application of poultry litter. *Water Resources Bulletin 30:*901-910.

Grant, R.F. (1993). ECOSYS. In *Global Changes in Terrestrial Ecosystems Focus 3 Wheat Network. Model and Experimental Data.* Oxford University, Oxford: International Geosphere-Biosphere Program, pp. 13-23.

Grant, R.F. and D.J. Heaney (1997). Inorganic phosphorus transformation and transport in soils: Mathematical modeling in ECOSYS. *Soil Science Society of American Journal 61:*752-764.

Grant, R., A. Laubel, B. Kronvang, H.E. Andersen, L.M. Svendsen, and A. Fuglsang (1996). Loss of dissolved and particulate phosphorus from arable catchments by subsurface drainage. *Water Research 30:*2633-2642.

Grierson, P.F., N.B. Comerford, and E.J. Jokela (1998). Phosphorus mineralization kinetics and response of microbial phosphorus to drying and rewetting in a Florida Spodosol. *Soil Biology and Biochemistry 30:*1323-1331.

Grossl, P.R. and W.P. Inskeep (1991). Precipitation of dicalcium phosphate dihydrate in the presence of organic acids. *Soil Science Society of America Journal 55:*670-675.

Hadas, A. and B. Bar-Yosef (1984). Effect of pelleting, temperature and soil type on mineral N release from poultry and dairy manures. *Soil Science Society of America Journal 47:*1129-1133.

Helal, H.H. and A. Dressler (1989). Mobilization and turnover of soil phosphorus in the rhizosphere. *Zeitschrift für Planzenernahrung und Bodenkunde 152:*175-180.

Hession, W.C., D.E. Storm, C.T. Haan, S.L. Burks, and M.D. Maltock (1996). A watershed-level ecological risk assessment methodology. *Water Resources Bulletin 32:*1039-1054.

Hillel, D. (1980). *Fundamentals of Soil Physics.* New York: Academic Press.

Hillel, D. (2000). *Environmental Soil Physics.* San Diego: Academic Press.

Hoogenboom, G. (1999). Modeling root growth and impact on plant development *Acta Horticulturae 507:*241-251.

Hoogenboom, G. and M.G. Huck (1986). ROOTSIMU v4.0. A dynamic simulation of root growth, water uptake, and biomass partitioning in a soil-plant-atmosphere continuum, update and documentation. Auburn University, Auburn, AL: Agronomy and Soils Department Series No. 109.

Hopstaken, C.F. and E.F.W. Ruijgh (1994). Modeling N and P loads on surface and groundwater due to land use. In *Contamination of Groundwater,* eds. D.C. Adriano, A.K. Iskandar, and I.P. Murarka. Northwood, United Kingdom: Scientific Review, pp. 161-188.

Imas, P., B. Bar-Yosef, U. Kafkafi, and R. Ganmore-Neumann (1997a). Phosphate induced carboxylate and proton release by tomato roots. *Plant and Soil 191:*35-39.

Imas, P., B. Bar-Yosef, U. Kafkafi, and R. Ganmore-Neumann (1997b). Release of carboxylic anions and protons by tomato roots in response to ammonium/nitrate ratio and pH in nutrient solution. *Plant and Soil 191:*27-34.

Inskeep, W.P. and P.R. Bloom (1985). An evaluation of rate equations for calcite precipitation kinetics at Pco_2 less than 0.01 atm and pH greater than 8. *Geochimica Cosmochimica Acta 49:*2165-2180.

Inskeep, W.P. and J.C. Silvertooth (1988). Inhibition of hydroxyapatite precipitation in the presence of fulvic, humic and tannic acids. *Soil Science Society of America Journal 52:*941-946.

Iyamuremye, F. and R.P. Dick (1996). Organic amendments and phosphorus sorption by soils. *Advances in Agronomy 56:*139-185.

James, R.T. and K.E. Havens (1996). Algal bloom probability in a large subtropical lake. *Water Resources Bulletin 32:*995-1006.

Jensen, M.B., P.R. Joergensen, H.C.B. Hansen, and N.E. Nielsen (1998). Biopore mediated subsurface transport of dissolved orthophosphate. *Journal of Environmental Quality 27:*1130-1137.

Joner, E.J., J. Magid, T.S. Gahoonia, and I. Jakobson (1995). P depletion and activity of phosphatase in the rhizosphere of mycorrhizal and nonmycorrhizal cucumber (*Cucumis sativus* L.). *Soil Biology and Biochemistry 27:*1145-1151.

Jones, C.A., C.V. Cole, A.N. Sharpley, and J.R. Williams (1984). A simplified soil and plant phosphorus model: I. Documentation. *Soil Science Society of America Journal 48:*800-804.

Jones, R.I., K. Salonen, and H. de Haan (1988). Phosphorus transformations in the epilimnion of humic lakes: Abiotic interactions between dissolved humic materials and phosphate. *Freshwater Biology 19:*357-369.

Jungk, A.O. (1996). Dynamics of nutrient movement at the soil-root interface. In *Plant Roots: The Hidden Half* (Second Edition), eds. Y. Waisel, A. Eshel, and U. Kafkafi. New York: Dekker, pp. 581-604.

Jungk, A., C.J. Asher, D.G. Edwards, and D. Meyer (1990). Influence of phosphate status on phosphorus uptake kinetics of maize *(Zea mays)* and soybean *(Glycine max).* In *Plant Nutrition—Physiology and Applications,* ed. M.L. van Beusichem. New York: Kluwer Academic Publishers, pp. 135-142.

Kafkafi, U., B. Bar-Yosef, R. Rosenberg, and G. Sposito (1988). Phosphorus adsorption by kaolinite and montmorillonite. II. Organic anion competition. *Soil Science Society of America Journal 52:*1585-1589.

Kafkafi, U., A.M. Posner, and J.P. Quirk (1967). Desorption of phosphate from kaolinite. *Soil Science Society of America Proceedings 32:*559-562.

Kirk, G.J.D. and P.H. Nye (1985). The dissolution and dispersion of dicalcium phosphate dihydrate in soil. I. A predictive model for a planar source. *Journal of Soil Science 36:*445-459.

Knisel, W.G. (Ed.) (1980). *CREAMS: A Field Scale Model for Chemicals, Runoff, and Erosion from Agricultural Management Systems.* USDA-ARS Conservation Research Report 26. Washington, DC: US Government Print Office.

Lasaga, A.C. (1998). *Kinetic Theory in the Earth Sciences.* Princeton, NJ: Princeton University Press.

Latterel, M.B., R.H. Dowdy, E.C. Clapp, W.E. Larson, and D.R. Linden (1982). Distribution of phosphorus in soils irrigated with municipal wastewater effluent: Five year study. *Journal of Environmental Quality 11:*124-128.

Linquist, B.A., P.W. Singleton, and K.G. Cassman (1997). Inorganic and organic phosphorus dynamics during a build up and decline of available phosphorus in an ultisol. *Soil Science 162:*254-264.

Liu, C.W. and T.N. Narasimhan (1989). Redox-controlled multiple-species reactive chemical transport. I. Model development. *Water Resources Research 25:*869-882.

Lowenthal, R.E., U.R.C. Kornmuller, and E.P. van Heerden (1994). Modeling struvite precipitation in anaerobic treatment systems. *Water Science and Technology 30*(12):107-116.

Lynch, J.P., K.L. Nielsen, R.D. Davis, and A.D. Jablokow (1997). SimRoot: Modeling and visualization of root systems. *Plant and Soil 188:*139-151.

Mansell, R.S., P.J. McKenna, E. Flaig, and M. Hall (1985). Phosphate movement in columns of sandy soil from a wastewater-irrigated site. *Soil Science 140:*59-68.

Marschner, H. (1995). *Mineral Nutrition of Higher Plants* (Second Edition). London: Academic Press.

Mengel, K. and E.A. Kirkby (1987). *Principles of Plant Nutrition* (Third Edition). Bern: International Potash Institute.

Mokady, R.S. and D. Zaslavsky (1966). Movement and fixation of phosphate applied to soil. *Transaction Commonwealth II & IV,* ed. ISSS. Aberdeen: International Society of Soil Science, pp. 329-334.

Moustafa, M.Z. (1998). Long-term equilibrium phosphorus concentration in the Everglades as predicted by a Vollenweider-type model. *Water Resources Association of America Journal 34:*135-147.

Mualem, Y. and S.P. Friedman (1991). Theoretical prediction of electrical conductivity in saturated and unsaturated soil. *Water Resources Research 27:*2771-2777.

Musvoto, E.V., M.C. Wentzel, R.E. Loewenthal, and G.A. Ekama (2000a). Integrated chemical-physical processes modeling. I. Development of a kinetic-based model for mixed weak acid/base systems. *Water Research 34:*1857-1867.

Musvoto, E.V., M.C. Wentzel, R.E. Loewenthal, and G.A. Ekama (2000b). Integrated chemical-physical processes modeling. II. Simulating aeration treatment of anaerobic treatment of anaerobic digester supernatants. *Water Research 34:*1868-1880.

Nagarajah, S., A.M. Posner, and J.P. Quirk (1968). Desorption of phosphate from kaolinite by citrate and bicarbonate. *Soil Science Society of America Proceedings 32:*507-510.

Notodarmojo, S., G.E. Ho, W.D. Scott, and G.B. Davis (1991). Modelling phosphorus transport in soil and groundwater with two consecutive reactions. *Water Research 25:*1205-1216.

Nye, P.H. (1977). The rate limiting step in plant nutrient absorption from soil. *Soil Science 123:*292-297.

Nye, P.H. (1986). Acid-base changes in the rhizosphere. *Advances in Plant Nutrition 2:*129-153.

Nye, P.H., J.L. Brewster, and K.K.S. Bhat (1975). The possibility of predicting solute uptake and plant growth response from independently measured soil and plant characteristics. III. The growth and uptake of onion in a soil fertilized to different initial levels of phosphate and a comparison of the results with model predictions. *Plant Soil 42:*197-226.

Ognalaga, M., E. Frossard, and F. Thomas (1994). Glucose-1-phosphate and myo-inositol hexaphosphate adsorption mechanisms on goethite. *Soil Science Society of America Journal 58:*332-337.

Olsen, S.R. and W.D. Kemper (1968). Movement of nutrients to plant roots. *Advances in Agronomy 20:*91-151.

Overman, A.R. (1999). Langmuir-Hinshelwood model of soil phosphate kinetics. *Communications in Soil Science and Plant Analysis 30:*109-119.

Parton, W.J., R.L. Sanford, P.A. Sanchez, and J.W.B Stewart (1989). Modeling soil organic matter dynamics in tropical soils. In *Dynamics of Soil Organic Matter in Tropical Ecosystems,* eds. D.C. Coleman, J.M. Oades, and G. Uehara. Honolulu, HI: University of Hawaii, pp. 153-171.

Persson, P., N. Nilsson, and S. Sjoberg (1996). Structure and bonding of orthophosphate ions at the iron oxide-aqueous interface. *Journal Colloid Interface Science 177:*263-275.

Pote, D.H., T.C. Daniel, D.J. Nichols, A.N. Sharpley, P.A. Moore, D.M. Miller, and D.R. Edwards (1999). Relationship between P level in three ultisols and P concentration in runoff. *Journal of Environmental Quality 28:*170-175.

Rauschkolb, R.S., D.E. Rolston, R.J. Miller, A.B. Carlton, and R.G. Burau (1976). Phosphorus fertilization with drip irrigation. *Soil Science Society of America Proceedings 40:*68-72.

Rifai, H.S. and P.B. Bedient (1994). Modeling contaminant transport and biodegredation in groundwater. In *Contamination of Groundwater*, eds. D.C. Adriano, A.K. Iskandar, and I.P. Murarka. Northwood: Scientific Review, pp. 221-278.

Sanyal, S.K. and K. De Datta (1991). Chemistry of phosphorus transformation in soil. *Advances in Soil Science 16:*2-120.

Schecher, W. (2001). *Thermochemical Data Used in Mineql+ Version 4.5.* Hallowell, ME: Environmental Research Software.

Schulthess, C.P. and D.L. Sparks (1988). A critical assessment of the surface adsorption models. *Soil Science Society of America Journal 52:*92-97.

Scow, K.M. and J. Hutson (1992). Effect of diffusion and sorption on the kinetics of biodegredation: Theoretical considerations. *Soil Science Society of America Journal 56:*119-127.

Shainberg, I. and G.J. Levy (1992). Physico-chemical effects of salts upon infiltration and water movement in soils. *Advances in Soil Science 20:*37-93.

Shainberg, I., G.J. Levy, J. Levin, and D. Goldstein (1997). Aggregate size and seal properties. *Soil Science 162:*470-478.

Shang, C., P.M. Huang, and J.W.B. Stewart (1990). Kinetics of adsorption of organic and inorganic phosphates by short-range ordered precipitates of aluminum. *Canadian Journal of Soil Science 70:*461-470.

Shang, C., J.W.B. Stewart, and P.M. Huang (1992). pH effect on kinetics of adsorption of organic and inorganic phosphates by short-range ordered aluminum and iron precipitates. *Geoderma 53:*1-14.

Sharpley, A.N., J.S. Robinson, and S.J. Smith (1995). Bioavailable phosphorus dynamics in agricultural soils and effects on water quality. *Geoderma 67:*1-15.

Sharpley, A.N. and S.J. Smith (1989). Prediction of soluble P transport in agricultural runoff. *Journal of Environmental Quality 18:*313-316.

Sharpley, A.N., S.J. Smith, O.R. Jones, W.A. Beng, and G.A. Coleman (1992). The transport of bioavailable P in agricultural runoff. *Journal of Environmental Quality 12:*30-35.

Shaviv, A. and N. Shachar (1989). A kinetic-mechanistic model of phosphorus sorption in calcareous soils. *Soil Science 148:*172-178.

Shaviv, A., M. Shnek, and I. Ravina (1992). Modeling nutrient uptake considering sorption kinetics: Phosphorus uptake from a one dimensional rhizosphere. *Journal of Plant Nutrition 15:*1099-1114.

Sibanda, H.M. and S.D. Young (1986). Competitive adsorption of humus acids and phosphate on goethite, gibbsite and two tropical soils. *Journal of Soil Science 37:*197-204.

Simard, R.R., S. Beauchemin, and P.M. Haygarth (2000). Potential for preferential pathways of P transport. *Journal of Environmental Quality 29:*97-105.

Simkins, S. and M. Alexander (1984). Models for mineralization kinetics with the variables of substrate concentration and population density. *Applied Environmental Microbiology 47:*1299-1306.

Sims, J.T., R.R. Simard, and B.C. Joern (1998). Phosphorus loss in agricultural drainage. Historical perspectives and current research. *Journal of Environmental Quality 27:*277-293.

Sinclair, A.G. and P.D. Johnstone (1995). Ability of a dynamic model of phosphorus to account for observed patterns of response in a series of phosphate fertilizer trials on pasture. *Fertilizer Research 40:*21-29.

Smith, S.L. and P.R. Jaffe (1998). Modeling the transport and reaction of trace metals in water saturated soils and sediments. *Water Resources Research 34:*3135-3147.

Stamm, C.H., H. Flühler, R.G. Gachter, J. Leuenberger, and H. Wunderli (1998). Preferential transport of phosphorus in drained grassland soils. *Journal of Environmental Quality 27:*515-522.

Stevenson, F.J. (1994). *Humus Chemistry: Genesis, Composition, Reactions* (Second Edition). New York: John Wiley & Sons.

Stevenson, F.J. and M.A. Cole (1999). *Cycles of Soil: Carbon, Nitrogen, Phosphorus, Sulfur and Microelements.* (Second Edition). New York: John Wiley & Sons.

Stumm, W. and J.J. Morgan (1996). *Aquatic Chemistry: Chemical Equilibria and Rates in Natural Waters.* New York: Wiley.

Suarez, D.L., S. Goldberg, and C. Su (1998). Evaluation of oxyanion adsorption mechanisms on oxides using FTIR spectrometry and electrophoretic mobility. In *Mineral-Water Interfacial Reactions. Kinetics and Mechanisms,* eds. D.L. Sparks and T.J. Grundl, ACS Symposium Series 715. Washington, DC: ACS, pp. 136-178.

Suarez, D.L. and J. Simunek (1997). UNSATCHEM: Unsaturated water and solute transport model with equilibrium and kinetic chemistry. *Soil Science Society of America Journal 61:*1633-1646.

Tarafdar, J.C. and A. Jungk (1987). Phosphatase activity in the rhizosphere and its relation to the depletion of soil organic phosphorus. *Biology and Fertility of Soils 3:*199-204.

Thomann, R.V. and L.C. Linker (1998). Contemporary issues in watershed and water quality modeling for eutrophication. *Water Science Technology 37*(3): 93-102.

Timlin, D., B. Acock, and R. van Genuchten (Eds.). (1996). *2DSOIL: A Modular Simulation of Soil and Root Processes.* Release 03. Remote sensing and modeling laboratory. Beltsville, MD: USDA.

Tinker, P.B. and P.H. Nye (2000). *Solute Movement in the Rhizosphere* (Second Edition). New York: Oxford University Press.

Traina, S.J., G. Sposito, G.R. Bradford, and U. Kafkafi (1987). Kinetic study of citrate effects on orthophosphate solubility in an acidic, montmorillonitic soil. *Soil Science Society of America Journal 51:*1483-1487.

van der Zee, S.E.T.M. and W.H. van Riemsdijk (1991). Model for the reaction kinetics of phosphate with oxides and soils. In *Interactions at the Soil Colloid-Soil Solution Interface,* eds. G.H. Bolt, M.F.L. De Boodt, M.H.B. Hayes, M.B. McBride, and E.B.A. De Stooper. Amsterdam: Kluwer Academic Publishers, pp. 205-239.

van Noordwijk, M., L.Y. Spek, and P. Dewilligen (1994). Proximal root diameter as predictor of total root size for fractal branching models. I. Theory. *Plant and Soil 164:*107-117.

van Remsdijk, W.H. and S.E.T.M. van der Zee (1991). Comparison of models for adsorption, solid solution and surface precipitation. In *Interactions at the Soil Colloid-Soil Solution Interface,* eds. G.H. Bolt, M.F.L. De Boodt, M.H.B. Hayes, M.B. McBride, and E.B.A. De Stooper. Amsterdam: Kluwer Academic Publishers, pp. 241-256.

Vetter, H. and G. Steffens (1981). Phosphate accumulation in soil profile and P losses after application of animal manure. In *Phosphorus in Sewage Sludge and Animal Waste Slurries,* eds. T.W.G. Hucker and G. Catroux. Boston: Reidel, pp. 309-331.

Vollenweider, R.A. (1976). Advances in defining critical loading levels of phosphorus in lake eutrophication. *Memorie dell'Instituto Italiano di Idrobiologia 33:*53-83.

Wagenet, R.J. and J.L. Hutson (1987). LEACHM: Leaching Estimation and Chemistry Model, a process-based model of water and solute movement, transformations, plant uptake and chemical reactions in the unsaturated zone, Continuum 2. Department of Agronomy. Ithaca, NY: Cornell University.

Wallach, R. and R. Shabtai (1992). Surface runoff contamination by soil chemicals: Simulations for equilibrium and first-order kinetics. *Water Resources Research 28:*167-173.

Wallach, R. and M.T. van Genuchten (1990). A physically based model for predicting solute transfer from soil solution to rainfall-induced runoff water. *Water Resources Research 26:*2119-2126.

Wallach, R., M.T. van Genuchten, and W.F. Spencer (1989). Modeling solute transfer from soil to surface runoff: The concept and effective depth of transfer. *Journal of Hydrology 109:*307-317.

Web, K.M. and G.E. Ho (1992). Struvite ($MgNH_4PO_4{\cdot}6H_2O$) solubility and its application to a piggery effluent problem. *Water Science and Technology 26:*2229-2232.

Williams, G.R. (1987). The coupling of biogeochemical cycles of nutrients. *Fertilizer Research 4:*61-75.

Williams, R.F. (1948). The effect of phosphorus supply on the rate of intake of phosphorus and nitrogen and upon certain aspects of phosphorus metabolism in gramineous plants. *Australian Journal of Scientific Research 1:*333-361.

Wolf, J., C.T. De Wit, B.H. Janssen, and D.J. Lathwell (1987). Modeling long-term crop response to fertilizer phosphorus. I. The model. *Agronomy Journal 79:*445-451.

Zhang, J., T.S. Tisdale, and R.A. Wagner (1996). A basin scale phosphorus transport model for south Florida. *Applied Engineering Agriculture 12:*321-327.

Zobel, R.W. (1975). The genetics of root development. In *The Development and Functions of Roots,* eds. J.G. Bolt and D.T. Clarkson. London: Academic Press, pp. 261-275.

Chapter 16

Potassium Dynamics

Tatiana V. Karpinets
Duncan J. Greenwood

Most crops contain large quantities of potassium (K). It plays a dominant role in the maintenance of plant turgor, stomatal movement, cell expansion, cation-anion balance, pH, phloem transport, protein synthesis, and the activation of many enzymes. Good crops can remove more than 300 kg K ha^{-1}. A typical arable soil in Europe contains about 40,000 kg K ha^{-1} but only about 40 kg K ha^{-1} is immediately available for any crop; the remainder is locked up in minerals and only becomes slowly available. The difference between crop demand for K and the ability of soil to meet that demand can therefore be considerable. It is usually met in richer countries by the application of manures and fertilizers and often, excessive amounts are applied and there is much waste. Most people in the world, however, live in countries that cannot afford to import potassium fertilizer and export more potassium in crops than they import so that their soils become progressively depleted of potassium (Cooke, 1986). Potassium requirements vary considerably depending on the crop, soil, and weather and it is important that K-fertilizer practices are efficient.

To achieve this objective it is necessary to take account of the many K fluxes in the soil-plant system. They are very complex and include chemical, physical, microbiological, biological, and geological processes, which interact with one another. Experimental studies cannot embrace all these interactions simultaneously. To study complicated processes in the soil-plant system, whether by experiment or simulation, they must be divided into subsystems. There exists a wealth of scientific information about the various potassium processes in different soils and plants. Some of these processes have been described mathematically in considerable detail, which facilitates interpretation of experimental results and improves understanding of the processes. Few attempts have been made to incorporate this knowledge into more inclusive models (Baldwin, Nye, and Tinker, 1973; Claassen and Barber, 1976; Barber, 1984; Claassen, Syring, and Jungk, 1986).

In spite of the significance of such models, there are some drawbacks. First, there are many soil and plant parameters that are difficult to estimate in the field. They make such models too complex for practical use. Another drawback is that they place too much emphasis on a limited number of processes and not enough on others. The net result is that simulation of the whole regime is unsatisfactory. It is often better to study the processes in less detail and to take better account of the synergism between the more important processes.

In this chapter the different modeling concepts used for the description of K dynamics are presented, in particular an approach based on the self-organization concept to predict changes over a period of years and an approach based on detailed analyses of the processes to predict K dynamics over the period of a single arable crop will be presented.

LONG-TERM K DYNAMICS IN SOIL-PLANT SYSTEMS

Self-Organization in the Soil-Plant System

Self-organization is an important feature of open nonlinear systems that are far from thermodynamic equilibrium and have energetic maintenance from the environment. Soil is also an open and nonlinear system. It has a number of positive and negative feedback effects and the functioning of the processes is constantly supported by solar energy. Therefore, according to the laws of nonequilibrium thermodynamics, all the soil regimes including potassium must be self-regulated. In natural soils, as a result of this self-regulation, steady stationary states of the soil regimes are established that are characterized by steady stationary values of the system variables, for example exchangeable soil K. Stationary means that the values of the system variables do not change with time. Their steadiness or stability means that the values of the system variables are insensitive to changes in external influences. Stability of these variables is a result of self-organization of the soil system. It means that if a natural soil is subjected to external influences then the processes and fluxes are directed toward the system variables returning to their steady state values. In fact, this self-regulation ability defines long-term dynamics of K in soil-plant systems, provided they are not subjected to strong external influences.

Main Components of the K Regime and Their Characteristics

To understand self-organization mechanisms of the K regime in soil, they must be presented as a system of the most important components (system variables) and the main processes that link these system variables. Structure and function of every component and their time hierarchy and interactions with the environment must also be taken into account. Modern scientific literature indicates an appropriate presentation; there is much experimental data and several important reviews (Arnold, 1970; Munson, 1985; Sparks and Huang, 1985; Evangelou, Wang, and Phillips, 1994; Syers, 1998). The main components of the K regime in soil-plant systems and crucial transformations between them are given in the scheme in Figure 16.1.

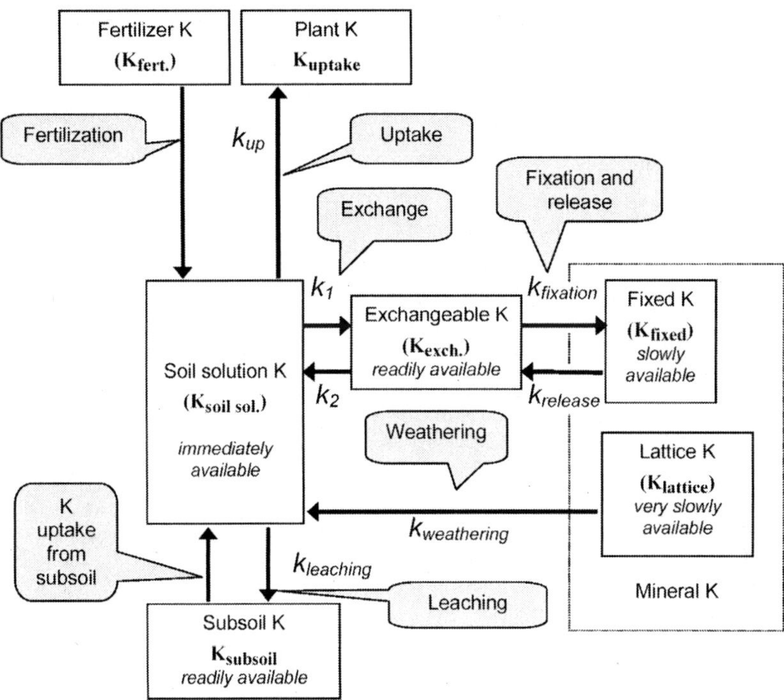

FIGURE 16.1. The main components of the K regime in soil-plant systems and the crucial transformations between them. $K_{soil\ sol.\ (solution)}$, $K_{exch.\ (exchangeable)}$, K_{fixed}, $K_{lattice}$, and $K_{subsoil}$ are potassium in the soil solution, exchangeable, fixed, lattice, and in the subsoil; K_{uptake} is the potassium uptake by plants; $K_{fert.(fertilizer)}$ is the potassium that is incorporated in soil as fertilizer; k_1, k_2 are rate constants for K adsorption and desorption; $k_{fixation}$, $k_{release}$ are rate constants for K fixation and release; $k_{weathering}$, $k_{leaching}$ are rate constants for weathering and leaching; $k_{up.\ (uptake)}$ is a rate constant for K uptake.

Soil K Forms

Potassium exists in soil in different forms that differ in their reserves, their amounts, their mobility, and their availability to plants; they have different rate constants for transformation and may be determined, approximately, by various chemical procedures.

The main soil K forms and their characteristics are defined as follows:

1. Soil solution K ($K_{soil\ sol.}$) is very mobile and immediately available for plants.
2. Exchangeable K ($K_{exch.}$) is held on cation exchange positions on the surface of clay minerals and organic substances. Sites differ in the readiness with which they release K in exchange for other cations. Exchangeable K is less mobile than solution K but generally readily available to plants because of its rapid equilibration with K in soil solution.
3. Fixed K (K_{fixed}) is K held on adsorption sites with high specificity for K ions that are mainly in inter-lattice positions of expanded illites, montmorillonites, and vermiculites (Rich, 1972). Fixed K is less mobile than exchangeable K and becomes available to plants slowly. According to mineralogical studies, however, there is a continuum in the affinities of absorption sites for K from those where absorption is very strong and to those where it is weak (Robert, 1992). In consequence, the distinction between exchangeable and fixed K is somewhat arbitrary and depends on the way they are determined.
4. Lattice or matrix K ($K_{lattice}$) is bonded in the crystal structures of K clay minerals and is released to plants only by weathering of the minerals. Release of matrix K is, however, irreversible (Kirkman et al., 1994), which contrasts with the reversible release of fixed potassium.

The relative mobility and availability to plant roots of the different forms of K are:

$$K_{soil\ sol.} > K_{exch.} > K_{fixed} > K_{lattice}$$

The quantities of each of these forms are important in the self-organization of the K regime. Exchangeable-K in soil is usually measured by extraction with solutions of neutral salts such as 1 **M** NH_4NO_3. The estimation of fixed and lattice K forms is less precise. These forms have been determined by extraction with acid, ion exchange resins, or salt solutions. They have also been determined by electroultrafiltration, pot experiments, and by determination of the soil clay fraction (Gorbunov, 1978). Despite uncertainties in the methods of estimation, there is no doubt that the overwhelming majority of soils have far smaller reserves of $K_{soil\ sol.}$ and $K_{exch.}$ than of K_{fixed} and $K_{lattice}$, which is much less available to plants.

Potassium Concentrations in Soils and Minerals

The contribution of the different minerals to the structural, fixed, and exchangeable soil-K is given in Table 16.1. Igneous rocks and their derivatives, and metamorphic and sedimentary rocks are the primary source of K in soils. The K content of igneous rocks varies between 1 and 6 percent depending on how they were formed (Lawton, 1955). The main K-bearing minerals in these rocks and the sand and silts that are derived from them are feldspars and micas.

The micas weather less slowly than the feldspars. Sands and silts have relatively small surface areas but weather to form secondary or clay minerals. As clays have a large surface area they can have a profound impact on the K-economy in soils.

According to Schroeder and Gething (1984) micas weather according to the scheme presented in Figure 16.2. All these minerals are of 2:1 type with increasing capacity to accommodate K^+ between the layers, increasing surface areas, and cation exchange capacities, but decreasing K contents as the degradation proceeds toward montmorillonite and vermiculite.

The breakdown of feldspars produces silica, hydrous oxides of Al and Fe, K^+, and other cations. With intense leaching and with the pH ranging from 5 to 9, both the cations and silica are leached from the profile so that aluminium and iron oxides accumulate. If leaching is less severe and considerable quantities of silica remain but few alkaline or alkaline earth ions

TABLE 16.1. The Role of Minerals in the Dynamics of Soil K

Forms of soil K		
Mineral		
Structural (lattice)	**Fixed**	**Exchangeable**
K-feldspars	Vermiculitic minerals	Clay minerals
Micas	Certain short-range ordered aluminosilicates	Sesquioxides (silica minerals)
Other K-bearing minerals	Zeolites	
	K-taranakite (a phosphate mineral)	

Source: Adapted from Huang, 1989.

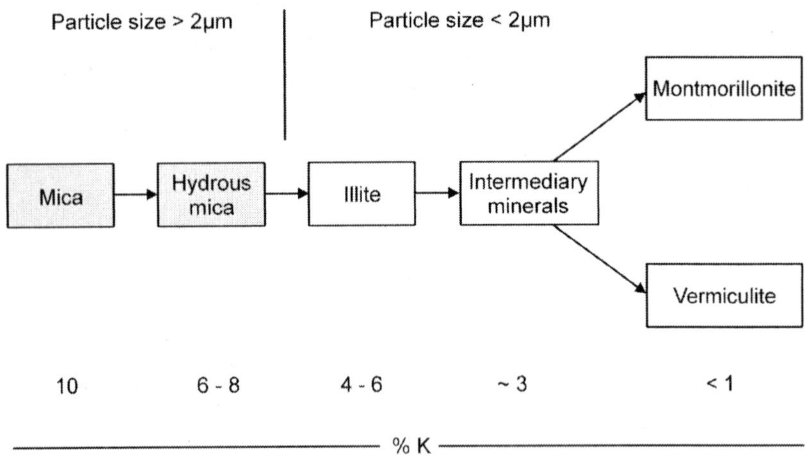

FIGURE 16.2. Formation of 2:1 clay minerals from micas. (*Source:* Adapted from Schroeder and Gething, 1984, p. 21.)

are left in solution, then kaolinite is formed. This clay mineral has adjacent layers that are joined tightly and so closely together that no K can be held between them. Potassium is held on the few negative charges on the outside of the crystal structure but then only in small amounts. Soils containing large quantities of aluminium and iron oxides and kaolinite occur in the wet, humid tropics and include the laterites. Breakdown of feldspars in the presence of ample quantities of silicia, K^+ ions, and other alkaline cations leads to the formation of illites and other 2:1 minerals which can degrade as described previously (Greenland and Hayes, 1978; Schroeder and Gething, 1984).

Soils with different climatic regimes, mineralogy, and texture may contain very different amounts of total K. Temperate, neutral soils may contain up to about 3.6 percent or about 140 t of K ha^{-1} in the top 30 cm of soil (Lawton, 1955), whereas in wet, humid tropics levels can be as low as 0.05 percent or about 0.6 t ha^{-1} (Phetchawee et al., 1985). Generally only a very small proportion of the total amount of K in soils is exchangeable or in solution and is thus readily available to crops (Table 16.2). Even so, levels of exchangeable soil-K (1**M** NH_4NO_3 extractable) vary enormously as a result of plant growth and nutrient uptake, and fertilizer and manurial practices; levels vary from 20 to 1,500 mg of K $kg^{-1}$of soil with a typical value in a West European arable soil of about 150 mg of K kg^{-1}. In such a soil, the fixed-K can be at least 2,000 mg kg^{-1} (Welte and Niederbudde, 1965) or about 8 t of

TABLE 16.2. Relative Quantities of Different Forms of K in Various Soils

Representative soils	K of soil solution, % from total	Exchange-able K, % from total	Mineral K: Fixed, % from total	Mineral K: Lattice, % from total	Total K, $g \cdot kg^{-1}$ soil
East European arable soils (chernozem)	0.8[1]		36.2[3]	63.0[7]	16.6
U.S. coastal plain soils					
sandy loam	4.1[2]		5.3[4]	90.6[7]	1.8
loamy sand	3.3[2]		6.0[4]	90.7[7]	1.5
U.S. continental soils					
kaolinitic	0.060[5]	1.3[6]	90.4[7]		3.3
mixed	0.056[5]	2.5[6]	89.6[7]		8.9
smectitic	0.025[5]	1.2[6]	91.8[7]		15.8

Source: Calculated for chernozems from Gorbunov,1978; for U.S. coastal plain soils from Martin and Sparks, 1983; for U.S. continental soils from Sharpley, 1990.

[1]1.0 **M** NH_4OAc(pH 7.0)
[2]$CaCl_2$ extractable
[3] calculated by multiplication of K in fraction, which is less than 0.001 mm, by the amount of this K fraction in soil
[4]1 **M** HNO_3 extractable
[5]water-soluble
[6]values exclude water-soluble K
[7]calculated as difference between total K and other forms

K ha^{-1}. Soils containing substantial quantities of vermiculites, montmorillonites or smectites have much fixed-K and have a considerable K-buffer capacity.

The amount of K held on the excess negative charge on the surface of particles, the cation exchange capacity, and the K held between the layers of 2:1 minerals is especially important for crop growth. The larger the cation capacity the greater is the potential ability of the soil to hold exchangeable K. Typical cation exchange capacities of the various clay minerals are montmorillonite 800-1,500 me kg^{-1}, illite 100-400 me kg^{-1}, and kaolinite 30-150 me kg^{-1} of mineral (Toth, 1955). Soil organic matter can have exchange capacities that reach 3,000 meq kg^{-1} of dry organic matter (Russell, 1973) which emphasizes the importance of soil organic matter in the supply of K to crops.

Processes Influencing Soil K Dynamics

The main processes that determine the dynamics of K forms in soils include: (1) weathering, (2) interchange between exchangeable and fixed K, (3) interchange between soil solution K and exchangeable K, and (4) plant uptake from the subsoil and leaching. The role of these processes in self-regulation of K in soil is discussed as follows.

Weathering

Weathering deforms or breaks down crystal lattices in clay minerals and releases K into the soil solution (Rich, 1972; Tributh, 1987). Some of this solution K is absorbed by plants, some is retained as exchangeable K, which is generally readily available to plants, and some is fixed in vermiculite-type minerals, if present, and then is only slowly available to plants (Sparks and Huang, 1985).

Soil K depletion by plant roots promotes the transformation of illites to smectites, particularly in the topsoil (Tributh et al., 1987). Plant-induced transformations of minerals may be especially significant in the rhizosphere. For example, according to Hinsinger and Jaillard (1993) roots act by depleting the soil solution K, thus shifting the exchange equilibrium so as to release interlayer K and expand the interlayer space of the phyllosilicate leading to the transformation of the phlogopite into vermiculite. These processes may therefore stabilize soil solution K and exchangeable K in the rhizosphere. It is, however, notable that no change in mica concentrations as a result of continuous cropping and K-depletion was detected in samples of soil taken between 1856 and 1987 from plots in the U.K. Broadwalk experiment (Singh and Goulding, 1997). Discrepancies in this experiment between K balances and measured changes in total K in the arable layer were attributed to plant uptake of K from the subsoil and to K leaching. Thus, during long-term cropping without K fertilizer, both weathering and changes in subsoil K tend to drive the exchangeable K in the plow layer toward a level characteristic of the soil. From the results of long-term trials in which K fertilizer and other K inputs were withheld, while nitrogen and phosphorous treatments were imposed, the average K removal by a single crop on a sandy clay loam soil was up to 47 kg K ha^{-1} y^{-1} (Johnston, 1986).

If weathering is considered as a surface-controlled reaction, then the general rate equation for this process may be derived from transition state theory (Sverdrup, 1990). The PROFILE model, based on this approach, gives a good correlation between the calculated weathering rates and those measured in the field (Sverdrup and Warfringe, 1993). Even so, there is a difficulty in reconciling the relative rates of weathering of the different minerals with those obtained by direct measurement in the field, possibly because of difficulties in accounting for diffusion-controlled processes of transformation (Wilson, 1992).

It appears that the detailed mathematical description of the various processes is not yet possible. Nevertheless, as a first approximation, the process may be regarded as a first-order irreversible reaction defined in terms of chemical kinetics.

Exchange

Exchange is the main process that provides K to plants and maintains stationary and steady K concentrations in the soil solution. In calcareous soils, K ions in solution are exchanged with divalent cations, especially calcium and to a lesser extent magnesium, adsorbed on the solid phase.

$$K_{\text{soil sol.}} + Ca_{\text{exch.}} \Leftrightarrow K_{\text{exch.}} + Ca_{\text{soil sol.}} \quad (16.1)$$

where $Ca_{\text{exch.}}$ is exchangeable Ca that is adsorbed on soil particles and $Ca_{\text{soil sol.}}$ is Ca in soil solution.

Quantity intensity relationships (Q/I relationships) have a strong theoretical basis (Evangelou, Wang, and Phillips, 1994) and have long been used to characterize potassium in soil (Beckett, 1964). Quantity is the exchangeable K, $K_{\text{exch.}}$, and intensity (I) is the activity ratio which is approximately equal to $[K]/([Ca] + [Mg])^{0.5}$ in many nonsaline soils. This expression for activity ratio applies to soils in which Ca and Mg are the main exchangeable cations, but after modification has been satisfactorily used with acid soils (Tinker, 1964). An important advantage of defining intensity in this way, rather than as concentration, is that it is virtually unaffected by the ratio of soil to solution and by considerable change in ionic concentration. To determine the Q/I curve of a soil, the exchangeable K is measured and samples of soil are shaken with dilute $CaCl_2$ solution, typically 0.005**M**, containing different amounts of potassium for a short period after which the K, Ca, and Mg concentrations in solution are measured. The change in exchangeable K is calculated from the change in concentration of the soil solution and I is determined from the equilibrium concentrations of K, Ca, and Mg.

Parameters of the Q/I curve are very dependent on clay mineralogy and their measurement may help to elucidate clay properties; they are also altered after fertilization and after cropping if these processes are accompanied by changes in fixed-K (Niederbudde, 1986). By shaking soils for a long period a different Q/I curve is obtained from that obtained after a short period of shaking. The difference between the two curves provides a measure of the release or fixation of potassium (Beckett, 1971; Matthews and Beckett, 1962). Studies using this procedure support the view that fixation or release is determined by the extent to which the soil contains more or less K than its equilibrium percentage K saturation (Matthews and Beckett, 1962).

Equations of exchange and of adsorption are used for formal descriptions of the ion-exchange process in soils. Several models have been proposed that are mainly based on the thermodynamic principles but make different approximations to describe the cation-exchange isotherm in terms of a small number of measurable constants (Gapon, 1933; Beckett, 1964; Dufey and Delvaux, 1989; Bond, 1995). Exchange phenomena in soils are, however, complicated because they depend on the multi-cation nature of exchange equilibria, complex ion behavior in soil solution, and energetic dispersion of soil particle surfaces (Bolt, Sumner, and Kamphort, 1963; Jardine and Sparks, 1984; Talibudeen et al., 1978). For these reasons and also because of the absence of well-defined quantitative approaches to the description of these processes, it can be argued that mechanistic models of K exchange equilibrium are primarily aimed at providing a better understanding of the mechanisms of adsorption. It is necessary to simplify them and insert some empirical parameters before such equations can be used in dynamic models of K response and applied to predict optimal fertilizer practices.

K Fixation and Release

K fixation and release are the main processes that support steady stationary levels of exchangeable K when K fertilization or K uptake takes place. Potassium fixation and release may significantly affect K uptake by plants, especially on soils with high reserves of fixed K and when significant deviations of exchangeable K from steady stationary levels take place after fertilization or exhaustive cropping (Eagle, 1967; Medvedeva, 1983). Studies of K behavior in the rhizosphere show that fixed K may be released in sufficient amounts to double the amount of exchangeable K within a radius of 1 mm from the root surface (Jungk, Claassen, and Kuchenbuch, 1982). The contribution of fixed K to crop uptake increases with the demand of the crop and with the length of time for which the crops have been grown without K fertilizer (Goulding, 1984).

Drying and rewetting a soil alters the exchangeable K content in a direction determined by the percentage K saturation of the cation exchange capacity compared with that when equilibrium is established and the soil neither fixes nor releases K into the exchangeable fraction (Matthews and Sherrell, 1960; Addiscott and Johnston, 1975). This means that in soils of low K status exchangeable K will be fixed and in soils of high K status fixed K will be released after wetting and drying, i.e., the alteration of these processes reestablishes the equilibrium between exchangeable and fixed K forms. Irreversible K fixation may take place in soils containing a high proportion of montmorillonite clays after a large number of wetting-drying cycles as a result of specific structural reorganization (Gaultier and Mamy, 1979).

Alternate drying and wetting of the soil can be used to provide a simple means of measuring the soil's ability to fix and release K. It enables the rate constants of K fixation ($k_{fixation}$) and release ($k_{release}$) to be determined (Karpinets, 2000). This approach is based on considering that K fixation and release are first-order reversible reactions:

$$K_{exch} \underset{k_{release}}{\overset{k_{fixation}}{\Leftrightarrow}} K_{fixed} \tag{16.2}$$

The kinetics of this reaction can be calculated in terms of two parameters, B and $T_{1/2}$, defined as follows.

Parameter B characterizes the relation between the rate constants of K fixation and release:

$$k_{release} /(k_{fixation} + k_{release}) = B \tag{16.3}$$

It defines the exchangeable K as a proportion of the sum of the exchangeable and fixed K in soil. Its value can be obtained by a procedure based on alternate wetting-drying of soil as described by Karpinets (1992) or by investigating the distribution of the added K between the water-soluble, exchangeable, and fixed K forms as described by Sharpley (1990) in soils of different mineralogical content.

Parameter $T_{1/2}$ defines the time it takes for 50 percent of the reequilibration between exchangeable and fixed K to take place after incorporating fertilizer K:

$$k_{fixation} + k_{release} = (1/ T_{1/2}) \cdot \ln 2 \tag{16.4}$$

Rough estimates of this parameter can be obtained from measurements in experiments in which changes of exchangeable K content have been investigated after adding K fertilizer or in the experiments in which the kinetics of K desorption have been studied. Measurements of apparent K desorption rate coefficients in different soils including soils with different texture have given values ranging from 0.0114 min^{-1} to 0.0018 $hour^{-1}$, which correspond to $T_{1/2}$ between 1 hour and 16 days (Martin and Sparks, 1983; Ioannou, Dimirkou, and Paschalidis, 1994). As the rates of K fixation and release depend mainly on the type of the clay minerals and their relative amounts, $T_{1/2}$ is generally constant for soils of the same type and texture.

From equations (16.3) and (16.4) the rate constants for K fixation and release are calculated as:

$$k_{release} = (\ln 2/T_{1/2}) \cdot B \qquad (16.5)$$

$$k_{fixation} = (\ln 2/T_{1/2}) \cdot (1-B) \qquad (16.6)$$

Leaching and Plant K Uptake from Subsoil Layers

Leaching mainly depends on K concentration in soil solution, soil texture, and water regime. Field experiments have indicated that K movement from the arable layer into the subsoil increases on sandy soils, with a decrease in the soil organic matter content and farmyard manure incorporation (Johnston, Goulding, and Mercer, 1993; Sparks, 1980; Lipkina, 1986). K movement to the subsoil also increases with an increase of K fertilization (Pieri and Oliver, 1986) and the volume of drainage (Johnston and Goulding, 1993). A potassium leaching model (KLEACH) has been developed from laboratory studies of K adsorption in soils of different texture and has been modified for the nonequilibrium conditions often found in field soils (Johnston, Goulding, and Mercer, 1993). This model predicts the movements of K down the profile and the risk of K losses by leaching.

If the K status of the arable soil layer significantly increases as a result of application of K fertilizer, an increase of the K content of the subsoil also takes place (Kochl, 1984; Fotyma and Gosek, 1991). On the other hand, if the level of K in the arable layer falls as a result of cropping while fertilizer K is withheld then plants will absorb more K from the subsoil (Singh and Goulding, 1997). Such increased absorption of K from the subsoil also occurs if the soil water content in the surface layers falls and thus reduces transport of K to the roots (Kobsarenko, 1983). Therefore, the subsoil appears, like other soil K reserves, to be a buffer that supports the self-organization of the K regime in the soil-plant system.

Conceptual Scheme for K Transformations in the Soil-Plant System

If one takes into account the properties of the main soil K forms, their reserves, mobility, and availability for plants and the time hierarchy of their transformation, a conceptual scheme of K regime may be suggested in the form of a simple physical model (Figure 16.3).

Three main soil K pools in this conceptual model are represented by the communicating containers with the sizes that correspond to the reserve of K (immediate, intermediate, and potential) in these pools at the steady state. The flexibility of the wall of each container corresponds to the mobility of the pool. The first container with a more flexible wall contains the immediate K reserve with the most mobile K that is absorbed by plants and leached into the subsoil. Containers and pools are joined according to the scheme of K transformations in a soil-plant system presented in Figure 16.1. The connections provide K flows in the system.

At the steady state, the K contents in the containers and pools are at levels that characterize the soil when K fertilization and K uptake are absent. If the amount of K in the first container increases (by fertilization), it stimulates K

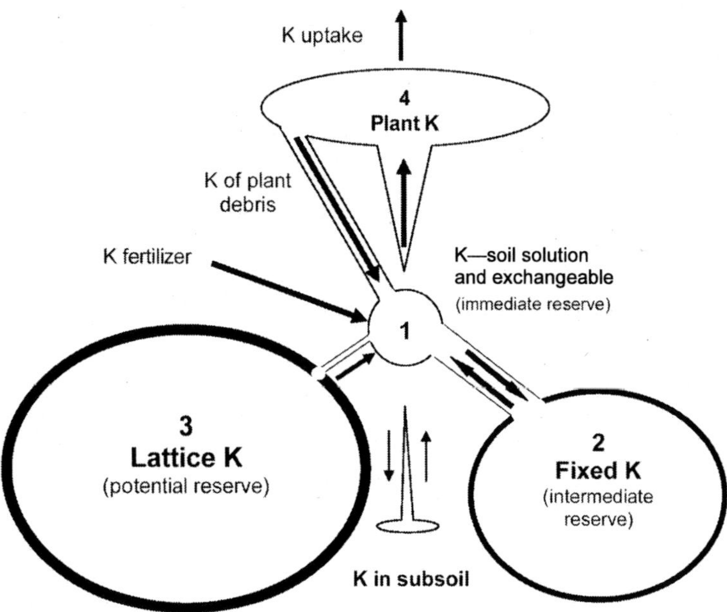

FIGURE 16.3. Scheme of the conceptual physical model of the K regime in the soil-plant system.

flow from this container into the second one and it also increases K uptake by plants and the extent of leaching. If K uptake decreases the amount of K in the first container, K flow from the second and third container into it is stimulated and leaching depressed. As the reserve of exchangeable K is more labile than that of fixed K it is restored more rapidly to its steady value.

Fertilization and exhaustive cropping mostly affect the amount of K in the second container (fixed K) but do not affect the sizes of the containers. Therefore, neither of these influences affects the steady stationary state of the system, but only compel it to function in an unsteady state. High, unsteady K contents in the immediate reserve may become stationary if K fertilizers are frequently added to compensate for additional K flows into plants, into fixed K forms, and into the subsoil. K fertilization also increases the K concentration in the plant, sometimes without any effect on plant yield, which leads to waste of potassium (luxury consumption). In consequence, high plant yields on heavily K fertilized soils and the effect of "luxury consumption" may be considered to be the result of self-regulation of the soil-plant system as it moves toward steady stationary values of the soil K forms. Exhaustive cropping can also lead to a stationary condition. As most soils have large second and third containers with intermediate and potential K reserves and also a subsoil K pool, they can provide considerable K flows into the first container. This will provide K for the plant and at the same time maintain a low unsteady stationary value of the immediate K reserve in the first container.

Much experimental data confirms the validity of this simple physical model of K of the soil-plant system and thus supports the view that the entirety of the various processes fit within the self-organizational concept. It is well known that low K soils have very stable contents of exchangeable K and, vice versa, high K soils have very unstable levels of exchangeable soil K. Exhaustive cropping and removal of the plant K results in the exchangeable K falling to a minimum value below which it is very difficult to reduce (Sharpley and Buol, 1987; Fotyma and Gosek, 1991). By contrast, high exchangeable potassium levels in soils are very unstable and decrease if potassium fertilizer is not applied (Lipkina, 1986; Chuyan and Karpinets, 1986; Prokoshev and Sokolova, 1990). In long-term experiments with K fertilizers high rates of fertilizer K are needed to maintain high soil-K levels, otherwise the annual decrease is much greater than on low K soils (Kochl, 1984; Chuyan and Karpinets, 1986; Fotyma and Gosek, 1991). According to the study of Singh and Goulding (1997) the more negative the K balance during long-term cropping the more "extra" K appears in the plow layer from the subsoil for reasons that are not clear. By the same principal, the more positive the K balance, the greater the K leaching. These results demonstrate the self-regulatory ability of different soils in maintaining the steady stationary state of K regime characterized by steady stationary values of soil K reserves.

Synergetic Model of K Dynamics in the Soil-Plant System and Its Qualitative Analysis

According to the formal schemes of K transformation in a soil-plant system (Figures 16.1 and 16.3) and the mathematical description of these processes, the entire set of processes may be presented as a combination of consecutive and parallel reactions. The description of the reactions may be obtained from chemical kinetics. As a result, the main features of the dynamic changes of the K forms will be presented in the form of differential equations, which make up the mathematical basis of the synergetic model of K regime in soil-plant systems (Karpinets and Lipkina, 1992):

$$dK_{soil\ sol.}/dt = k_{weathering} \cdot K_{lattice} - k_{leaching} \cdot K_{soil\ sol.} - k_1 \cdot Ca_{exch.} \cdot K_{soil\ sol.} + k_2 \cdot Ca_{soil\ sol.} \cdot K_{exch.} \quad (16.7)$$

$$dK_{exch.}/dt = k_1 \cdot Ca_{exch.} \cdot K_{soil\ sol.} - k_2 \cdot Ca_{soil\ sol.} \cdot K_{exch.} - k_{fixation} \cdot K_{exch.} + k_{release} \cdot K_{fixed} \quad (16.8)$$

$$dK_{fixed.}/dt = k_{fixation} \cdot K_{exch.} - k_{release} \cdot K_{fixed} \quad (16.9)$$

Notations in the equations correspond to Figure 16.1.

This mathematical model is too complicated to permit an accurate analytical solution. Moreover, such a solution would not highlight the most important dynamic properties and mechanisms of self-organization of the K regime. Qualitative analyses of the model may, however, be used to estimate stationary values of main K forms, their stability and dependence on basic soil properties.

According to this analysis, applied to the equations 16.7-16.9, the stationary states for exchangeable ($K_{exch.(st.)}$), fixed K ($K_{fixed(st.)}$), and K of soil solution ($K_{soil\ sol.(st.)}$) may be given as:

$$K_{soil\ sol.(st.)} = (k_{weathering} \cdot K_{lattice} + k_2 \cdot Ca_{soil\ sol.} \cdot K_{exch.}) / (k_{leaching} + k_1 \cdot Ca_{exch.}) \quad (16.10)$$

$$K_{exch.(st.)} = (k_{weathering} \cdot k_1 \cdot Ca_{exch.} / (k_{leaching} \cdot k_2 \cdot Ca_{soil\ sol.})) \cdot K_{lattice} \quad (16.11)$$

$$K_{fixed.} = (k_{weathering} \cdot k_1 \cdot k_{fixation} \cdot Ca_{exch.} / (k_{leaching} \cdot k_2 \cdot k_{release} \cdot Ca_{soil\ sol.})) \cdot K_{lattice} \quad (16.12)$$

By analyzing these equations the steady state potassium regimes of various soils can be estimated. The steady stationary contents of these soil K forms, which characterize this steady regime, will be greater in soils with (1) a high weathering rate, which is the case in slightly and moderately weathered soils whose parent rocks are rich in potassium containing primary minerals; (2) a high value for the equilibrium constant between soil solution and exchangeable K (k_1/k_2) and likewise for the ratio $Ca_{exch.}$ and $Ca_{soil\ sol.}$, which is the case in soils with high base saturation, high organic matter content, and cation exchange capacity of mineral particles; (3) a high value for the equilibrium constant between fixed and exchangeable K ($k_{fixation} / k_{release}$), which depends on the amount of trioctahedral minerals in soils; and (4) a low value for the rate constant for leaching ($k_{leaching}$), which mainly depends on soil texture.

The stability of these stationary states may be analyzed by the analytical method of Lyapunov (1892) by considering small deviations of the analyzed potassium forms from the stationary values in the equations 16.7-16.9. According to this analysis, if steady stationary values of the exchangeable and fixed K forms are subjected to disturbing influences (such as uptake by plant and fertilizer application), the dynamic changes of these K forms will have the character demonstrated in Figure 16.4. Any perturbation will intensify those K processes in the soil-plant system that promotes the renewal of steady stationary values of the various soil K forms.

Quantitative Estimation of Steady Stationary Soil K Forms

The dynamic changes of the artificially created (by application of K fertilizer) levels of soil K forms during exhaustive cropping can be used to estimate steady stationary levels of the various K forms (Karpinets, 1995). According to the analysis of the synergetic model, changes of exchangeable K ($dK_{exch.}/dt$) are proportional to the initial deviation of the artificially created level of exchangeable K ($K_{exch.art.}$) from the steady value of this K form:

$$dK_{exch.}/dt = - A \cdot (K_{exch.art.} - K_{exch.(st.)}), \qquad (16.13)$$

where $K_{exch.(st.)}$ is the steady stationary value of exchangeable K, and A is a coefficient that defines the intensity of the soil processes directed toward return of steady stationary content of exchangeable K.

This relationship is illustrated by the results of a microfield experiment on chernozems in which soil samples with artificially created levels of exchangeable K were subjected to the exhaustive cropping. After that, increments of the exchangeable K (extracted by acetic acid) were analyzed. The results of the experiment were regressed to get the linear relationship between the changes of this K form and its initial artificial levels (Figure 16.5).

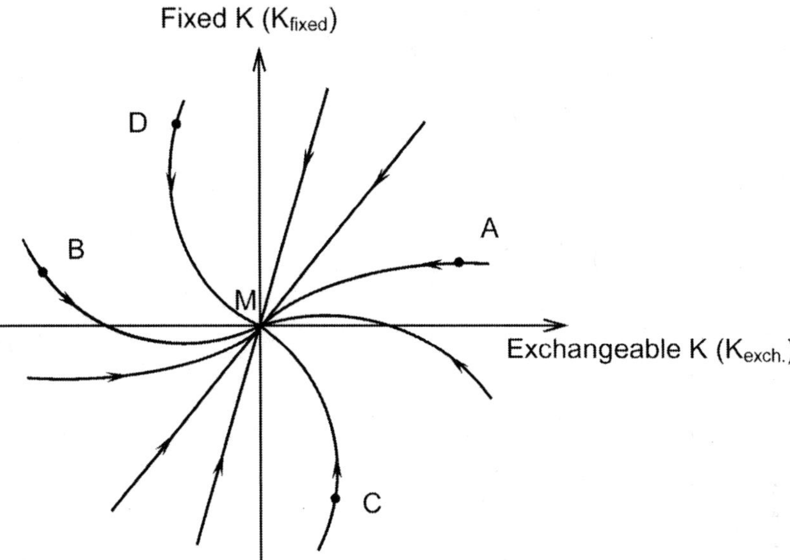

FIGURE 16.4. The dynamics of exchangeable and fixed K forms after deviations from their steady stationary values $K_{exch.(st.)}$ and $K_{fixed(st.)}$ (point M) caused by external influences (plant K uptake and K fertilization). Point A corresponds to soil with fixed K content near $K_{fixed(st.)}$ and a high level of exchangeable K (for example, as a result of one-time application of high rate of K fertilizer). Point B corresponds to a soil with fixed K content near $K_{fixed(st.)}$ and low level of exchangeable K (for example as a result of one-time K exhaustive cropping with only nitrogen and phosphorous fertilizers). Point C corresponds to an exhausted soil with a low level of fixed K and an exchangeable K near $K_{exch.(st.)}$, which is restored by K fertilization. Point D corresponds to a heavily fertilized soil with a high level of fixed K and an exchangeable K near $K_{exch.(st.)}$, which is restored by exhaustive cropping. According to the model, the decrease of exchangeable (point A) and fixed K (point D) after K application takes place because of plant K uptake and leaching. The increase of exchangeable (point B) and fixed K (point C) after exhaustive cropping takes place because of plant K uptake from the subsoil and because of weathering. (*Source:* Derived from Karpinets and Lipkina, 1992, p. 14.)

The steady stationary level corresponds to the exchangeable K content when no changes in this form take place.

The tendency of soil systems to restore their steady state of K regime reveals a possible practical application of equation 16.13 for predicting long-term dynamics in soils with high reserves of fixed K. Experimental support

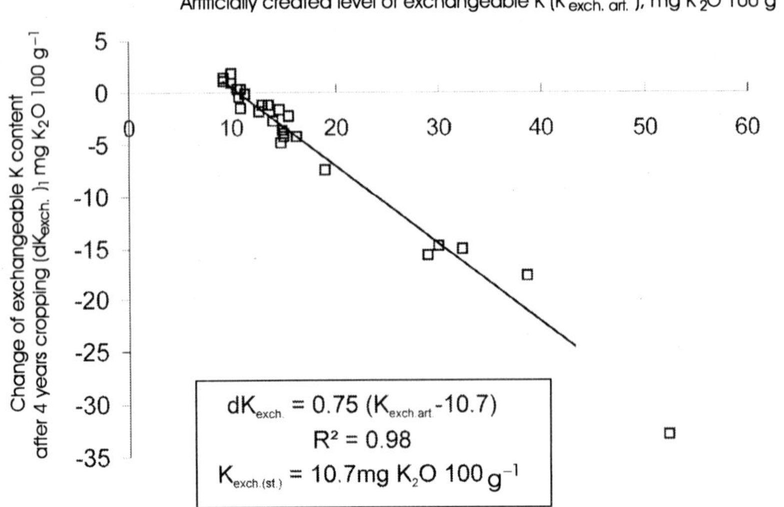

FIGURE 16.5. Experimental data (points) and regression line for changes of exchangeable K forms (0.5 **M** CH_3COOH extractable) after four years of exhaustive cropping against artificially created levels on typical and leached chernozems in the forest-steppe zone of Russia. (*Source:* Karpinets, 1995, p. 89.)

for this view is given in Figure 16.6, where the results of eight long-term field trials on clay loam and silt loam chernozems (Grib, Gorobets, and Krasnyuk, 1978; Shagaev and Michailina, 1978; Sevastyanova et al., 1978; Leontiev, 1981; Ponomarev, 1982; Berdnikov, Chilenko, and Berdnikova, 1983; Goncharenko et al.,1983; Nosko, Kuchir, and Rozdaibeda, 1988) are summarized. It gives the long-term changes of exchangeable K forms (0.5 **M** CH_3COOH extractable) plotted against the initial deviations from steady stationary state levels of exchangeable K in experiments where the soils have different K statuses and there are negative or slightly positive K balances during the crop rotations (Figure 16.6). The regression line obtained in the microfield experiments, with different artificial levels of the exchangeable soil K, gave a good prediction of the long-term changes of exchangeable soil K in field experiments. No relationship was found between the change in exchangeable K and the K balance in the same field trials (Figure 16.7). This plot ignored the effects of the differences between the actual and the stationary levels of exchangeable K. The comparison between Figures 16.6 and 16.7 emphasizes the dominance of the self-regulating soil processes, compared with the effects of K balance, on the processes

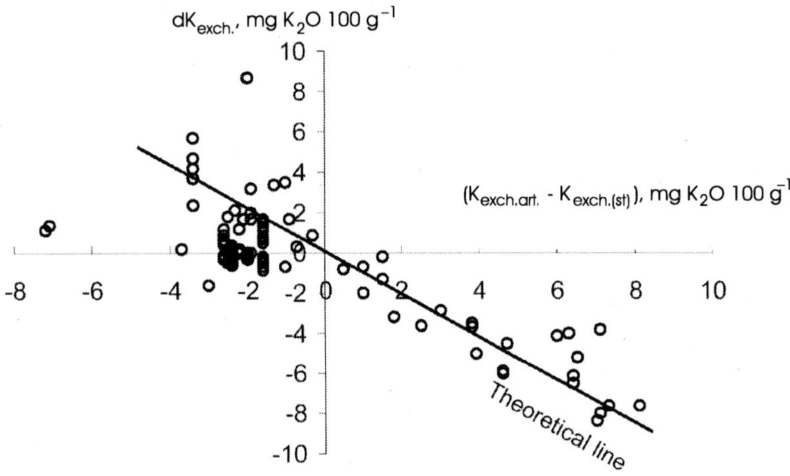

FIGURE 16.6. The relationship between the changes of the artificial levels of exchangeable K forms (0.5 **M** CH_3COOH extractable) after cropping ($dK_{exch.}$) with their initial deviation from the steady stationary state ($K_{exch.art.} - K_{exch.(st)}$). Points correspond to the experimental data of long-term field trials on clay loam and silt loam chernozems. The theoretical line is defined by the equation $dK_{exch.} = K_{exch.art.} - 10.7$, where 10.7 is steady stationary level of this K form obtained from the microfield experiments for these soils (see Figure 16.4). (*Source:* Derived from: Grib, Gorobets, and Krasnyuk, 1978, pp.78-141; Shagaev and Michailina, 1978, pp.175-190; Sevastyanova et al., 1978, pp.4-19; Leontiev, 1981, pp.108-120; Ponomarev, 1982, pp. 70-93; Berdnikov, Chilenko, and Berdnikova, 1983, pp.85-100; Goncharenko et al., 1983, pp.126-143; Nosko, Kuchir, and Rozdaibeda, 1988, pp. 3-14.)

in chernozems that force the exchangeable K toward steady stationary levels in the long term.

K DYNAMICS IN SOILS AND PLANTS DURING CROP GROWTH

The synergetic model given previously may predict the total increase in K uptake by the crop but it does not permit calculation of the time course of uptake during the growing season and does not consider factors that may influence yield. For accurate mathematical descriptions of the soil and plant K processes, attention needs to be given to plant demand (defined by potential growth and plant K concentration), root distribution in soil, K transport to

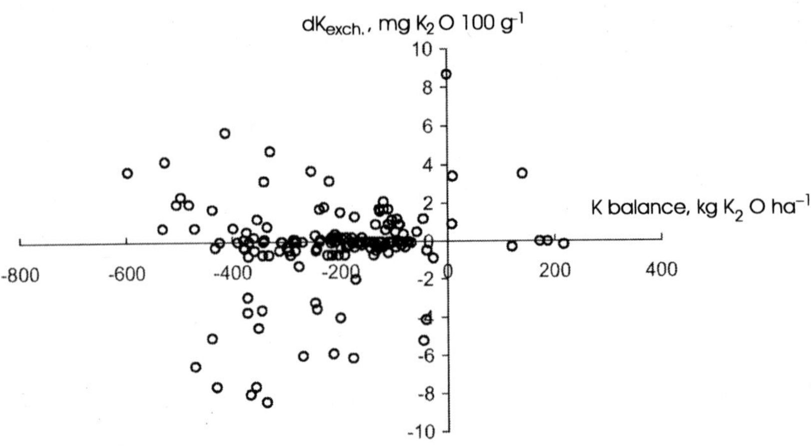

FIGURE 16.7. The relationship between the changes of the artificial levels of exchangeable K forms (0.5 **M** CH_3COOH extractable) after cropping ($dK_{exch.}$) with the existing K balance. Points correspond to the experimental data of long-term field trials on clay loam and silt loam chernozems. (*Source:* Shagaev and Michailina, 1978, pp. 175-190; Sevastyanova et al., 1978, pp. 4-19; Leontiev, 1981, pp. 108-120; Ponomarev, 1982, pp. 70-93; Berdnikov, Chilenko, and Berdnikova, 1983, pp. 85-100; Goncharenko et al., 1983, pp. 126-143; Nosko, Kuchir, and Rozdaibeda, 1988, pp. 3-14.)

roots, exchange between liquid and solid soil phases, and fixation and release of soil K.

Comprehensive models accounting for these variables are starting to be developed for predicting K dynamics in soils and crops during the growing season and for predicting crop response under different weather conditions (Greenwood and Karpinets, 1997a). They could become as important as some of those concerned with nitrogen and other agronomic practices (Tinker and Nye, 2000; Hartkamp, Jeffrey, and Hoogenboom, 1999). A common scheme of interactions between the main factors that define K dynamics of the soil-plant system is presented in Figure 16.8. Existing multiparameter mechanistic models of K dynamics in the soil-plant system describe each of the various processes in varying degrees of detail. Analysis of these processes and their mathematical description is discussed in the following sections.

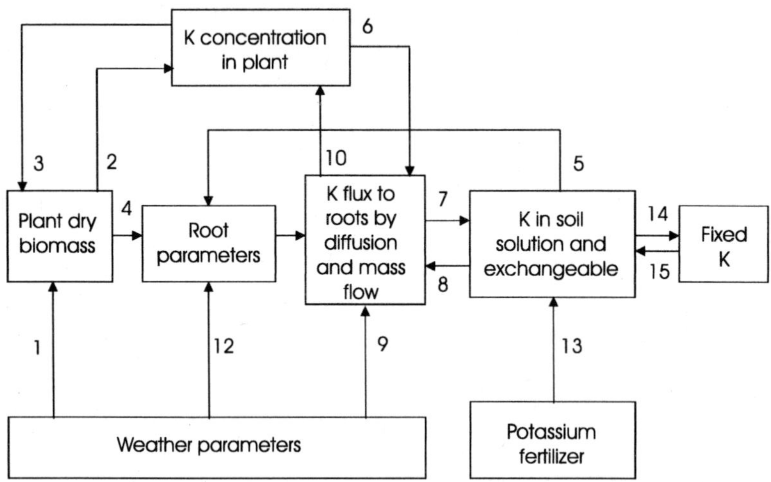

FIGURE 16.8. Common scheme of the interactions between the main parameters that define K dynamics of the soil-plant system. The numbers refer to those in the body of the text where the processes are described.

Plant Growth As a Function of Plant K Concentration

Potassium deficiency in soil affects K concentration and thus growth rate (Barnes, Greenwood, and Cleaver, 1976). One of the possible approaches to describe this process is to calculate an empirical coefficient R(T) for the effect of suboptimal K concentration (percent) in plant dry matter on the increase in plant weight. This coefficient is a function of actual plant K (P_k) and critical K (P_{crtk}) percentages in the plant dry matter:

$$R(T) = (1 + K_{hlf})/(1 + K_{hlf}/R_{tpy}(T)) \tag{16.14}$$

in which

$$R_{tpy}(T) = \min[P_k(T)/P_{crtk}(W), 1] \tag{16.15}$$

where K_{hlf} is set equal to 0.6, $P_k(T)$ is the percent K of the plant dry matter at time T, W is its weight in t ha^{-1} of the entire plants excluding fibrous roots.

The critical percent K, P_{crtk}, (the minimum percent K needed for maximum growth rate) declines as the crop grows. The photosynthetic parts of plants contain high concentrations of nutrients such as N and K and the structural and storage components contain low concentrations. As plants grow, the weight of photosynthetic tissue as a proportion of total plant weight declines and thus the concentrations of K and N also decline. The decline in the critical percentage N in the plant dry matter, P_{ctrn}, of a number of crops has been found to be related to increase in plant dry weight (W) when expressed in t ha^{-1} by

$$P_{crtn}(W) = 1.35(1+3e^{-0.26W}) \tag{16.16}$$

It has also been found that critical to percent K, P_{crtk} (W) is proportional to $P_{crtn}(W)$ so that

$$P_{crtk}(W) = G_{opt} \cdot P_{crtn}(W) \tag{16.17}$$

where G_{opt} is the coefficient that is characteristic of a crop (Greenwood and Stone, 1998).

Root Growth As a Function of Plant Mass and Soil K

Root growth has a decisive influence on plant K uptake from soil. Since root growth depends on plant growth, soil and climatic conditions, and interactions among them, a mechanistic approach to mathematical description is rather difficult. However, in most published K-uptake models, the required root growth relationships are empirically derived. The parameters in such models include root radius, rate of root growth, and the half distance between roots (Barber, 1984). Also included are empirical relationships for root length as a function of plant mass that have been derived from experiments (Gerwitz and Page, 1973; Greenwood et al., 1982; Greenwood and Karpinets, 1997a). Little attention has been given to the fact that not only the plant mass (process 4, Figure 16.8) but also the stage of plant development (Barraclough, 1989), the distribution of major nutrients in soil (process 5) (Drew, 1975), and weather parameters (process 12) can affect root growth.

The ability of roots to absorb K is not uniform throughout the root system. Thus Claassen and Jungk (1984) have shown that the most active parts of some root systems are root hairs. In addition, some believe that the ability of roots to absorb K ceases sometime after they have been formed which affects modeling (Greenwood and Karpinets, 1997a).

In all the models, roots are uniformly distributed in the arable layer. In the Barber-Cushman model (Barber and Cushman, 1981) the soil is visualized as consisting of cylinders of soil around each of the daily increments of root growth. Potassium uptake is calculated from K within the cylinder using K concentration at the outer surface of the cylinder as one of the boundary conditions. An increase in root density is accommodated by decreasing the radius of the soil cylinder. Another way of taking account of root competition (Greenwood and Karpinets, 1997a) is to consider that fresh roots always grow in the undepleted soil volume and take up K from a zone around them, which is considered to be a depletion zone. The calculation of the radius and thus the volume of this zone is based on the fact that the root mean-square displacement of a molecule diffusing from a point source in two dimensions is given by $2 \cdot (D_s \cdot T_{df})^{0.5}$, where D_s is the diffusion coefficient of K through soil and T_{df} is time from the formation of the root increment (Tinker and Nye, 2000). Changes in the fixed, exchangeable, and solution forms of K in the soil are calculated for each of these zones. These depletion zones are updated each day to take account of their expansion into soil that has not been depleted of K. If the total volume of the depletion zones exceeds the volume of the rooting layer, then any new roots are assumed to enter soil which has an exchangeable K concentration equal to average of that in the entire soil volume.

K Transport to Roots

Potassium transport to roots is important in modeling K uptake by the plant. It depends on plant demand (process 6, Figure 16.8), soil parameters (processes 7 and 8), root distribution (process 11), and weather parameters (process 9). In turn, K transport affects plant K concentration (process 10) and plant growth (process 3). Generally, mathematical simulation of K transport to roots calculates K fluxes by diffusion and mass flow. Comparison of these two fluxes (Barber, 1984; Barraclough, 1990) indicates that under temperate conditions diffusion is the main mechanism of K transport to roots. When evapotranspiraton is considerable, mass flow becomes significant. It can be estimated as the product of the evaporation rate and K concentration in soil solution.

The K flux (F, $\mu g \cdot cm^{-2} \cdot day^{-1}$) to the root surface by diffusion is described in the models according to Fick's law and together with mass flow may be written by a differential equation (Tinker and Nye, 2000):

$$F = -D_S \cdot \frac{\partial K_{exch.}}{\partial r} + V_o \cdot K_{soil\ sol.} \qquad (16.18)$$

where $K_{exch.}$ is the exchangeable K in soil in $\mu g \cdot cm^{-3}$, $K_{soil\ sol.}$ is the concentration in soil solution in $\mu g\ cm^{-3}$, V_o is the rate of water uptake in $cm^3\ cm^{-2} day^{-1}$, r is the radial distance from the root axis in cm, and D_S is the effective diffusion coefficient in $cm^2 day^{-1}$.

Greenwood and Karpinets (1997a,b) believe that diffusion of K to a root in soil can be thought of as diffusion to the surfaces of a cylinder of radius a (cm), that are maintained at zero concentration and are placed in an infinite amount of soil. As a close approximation (Crank, 1957) the flux (F) after different periods of time can be calculated from:

$$F = D_s \cdot K_{exch.} \cdot \{(\pi\, G_{tpy})^{-0.5} + 0.5\text{-}0.25 \cdot (G_{tpy}/\pi)^{0.5} + 0.125 \cdot G_{tpy}\} /a \;|G_{tpy} \leq 3 \quad (16.19)$$

$$F = 2 \cdot D_s \cdot K_{exch.} \cdot \{[\ln(4\, G_{tpy}) - 2\alpha]^{-1} \text{-} \alpha/[\ln(4\, G_{tpy}) - 2\alpha]^2\}/a \;|G_{tpy} > 3 \quad (16.20)$$

where G_{tpy}, a temporary dimensionless coefficient, defined by the expression $G_{tpy}=D_s.T_{df}/a^2$ in which T_{df} is time in days from the formation of the increment in root length, α=0.5772 is Euler's constant. In practice, the soil water content and thus D_s often vary from day to day, therefore in most simulations the daily cumulative value of D_s expresses the average value of D_s multiplied by T_{df}. This coefficient may be calculated from:

$$D_s = D_{wtsr} \cdot G_{tort} \cdot \Theta \cdot A_{grad} \cdot F_c\, ([Ca] + [Mg])^{0.5}, \quad (16.21)$$

where D_{wtsr} is the diffusion coefficient of K through water, Θ is the volumetric soil water content, ([Ca] + [Mg]) is sum of Ca and Mg cations (**M**) in soil solution, G_{tort} is a tortuosity factor, which as an approximation to the data in Tinker and Nye (2000) is set equal to Θ, and A_{grad} is the gradient of activity ratio of soil solution ($\mathbf{M}^{0.5}$) versus exchangeable K (μg· cm^{-3} of soil), F_c is a factor to convert from molarity to μg cm^{-3} and is equal to $39 \cdot 10^3$ μg of K $cm^{-3} \cdot M^{-1}$. To determine this gradient the activity ratio at different levels of exchangeable K may be determined as described by Matthews and Beckett (1962) and then an empirical relationship fitted to the results. The reason for using activity ratios is that they are insensitive to changes in soil water content and to substantial variations in ionic strength (Beckett, 1971). The use of activity ratios also enables account to be taken of the dependence of K concentration in soil solution on the concentrations of other cations, which are usually mainly calcium and magnesium. Their concentrations are approximately equal to the concentration of anions so that when nitrate is the dominant anion the relationship enables account to be taken of the effect on nitrate on K transport to the roots. Thus equation (16.21) enables the diffusion coefficient to be modified for a range of variables.

Influence of Plant Demand and K Fixation and Release on K Transport to Roots

The effect of plant demand on K transport is estimated in most K models through the changes of K concentration at the root surface, i.e., K uptake from soil solution by roots is described by the Michaelis-Menten relationship between uptake and the K concentration at the root surface. This equation is considered as one of the boundary conditions for the differential equation 16.18, which describes the transport of K to the root by diffusion and mass flow (Barber, 1984; Claassen, Syring, and Jungk, 1986).

This approach has some drawbacks. Barraclough (1990) argues that if severe depletion occurs near the root (Jungk, Claassen, and Kuchenbuch, 1982), insufficient K may be transported to the roots to meet observed K uptake because of failure of the forgoing models to include interchange between fixed and exchangeable K. Another weakness is that use of the Michaelis-Menten kinetics does not incorporate the effect of changes in plant-K status during growth on the ability of the roots to absorb K, yet the concentration of a nutrient in the aboveground parts of the plant has been repeatedly shown to have a large effect on the uptake of that nutrient by plant roots (Marschner, 1995).

This feedback effect of plant K on uptake may be estimated by a coefficient, G_f, which characterizes the depressive effects of plant K concentration (P_k) on uptake (Greenwood and Karpinets, 1997a). The relation between G_f and the ratio of P_k to the maximum possible K concentration in a plant of the same weight (P_{maxk}) has been obtained from root uptake/solution studies (Glass, 1975; Siddiqi and Glass, 1982). It was deduced from the data that G_f is 1 when the ratio is less than a given value (G_{zero}) but declines in a negative exponential manner with increase in the ratio according to:

$$G_f = \min(1, e^{-G_k((P_k/P_{maxk}) - G_{zero})}) \tag{16.22}$$

with $G_k = 2.8$ and $G_{zero} = 0.3$ for all crops.

The maximum possible percentage K in plant dry matter (P_{maxk}) may be corrected for the effect of plant dry weight (W) in a similar way to the critical K concentration in equations (16.16) and (16.17). Thus if G_{max} is a proportionality constant, characteristic of the crop, and P_{critn} is the critical nitrogen concentration then:

$$P_{maxk}(W) = G_{max} \cdot P_{critn}(W) \tag{16.23}$$

Both critical and maximum possible K concentrations are proportional to critical nitrogen concentration throughout growth. Values of the proportionality constants for 16 different vegetable species were obtained from the multilevel K-fertilizer experiments (Greenwood and Stone, 1998).

The interchange between exchangeable and fixed K forms (processes 14 and 15, Figure 16.8) considerably affects K uptake by plants, especially in soils that are rich in fixed K, but have low levels of exchangeable K and low K concentrations in soil solution. Not only the soil K but also the applied K fertilizer (process 13) is involved in the slow interchange between the exchangeable and fixed K forms. In the models, any fertilizer K incorporated in soil is instantly converted to exchangeable K. One of the approaches to describe the effect of K fixation and release on K uptake is to consider the interchange between the exchangeable and fixed K forms as a first-order reversible reaction (see equation16.2). In this case the dynamic changes of the exchangeable K in soil are calculated as:

$$dK_{exch.}/dt = - k_{fixation} \cdot K_{exch.} + k_{release} \cdot K_{fixed} \quad (16.24)$$

where K_{fixed} is calculated as described by Karpinets (1992):

$$K_{fixed} = K_{exch.} \cdot (1 - B)/B \quad (16.25)$$

Parameter B and the rate constants $k_{fixation}$ and $k_{release}$ may be estimated as previously described (equations 16.5 and 16.6).

VALIDATION AND APPLICATION OF MULTIPARAMETER MECHANISTIC K MODELS

Mechanistic models include plant, soil, and weather parameters. Their detailed measurement is necessary to calibrate, validate, and apply any such model. The ease of numerical estimation of model parameters is important in the choice of mathematical procedures for describing the various processes because it determines the potential usefulness of the model. A major problem with some existing models is that they have parameters that are difficult to determine; these parameters include root growth and the rate constants for K fixation and release. In consequence, the practical usefulness of the models is restricted. Poor estimation of the parameters' values, their averaging and indirect definition may also cause serious discrepancies between simulation and experimental results.

Experimental tests of the validity of the Barber-Cushman model (Barber and Cushman, 1981) have been reported. Some tests have shown agreement between measured and simulated K accumulation in plants grown on low K soils (Barber, 1984; Chen and Barber, 1990; Teo, Beyrouty, and Gbur, 1992). The model has also successfully predicted potassium concentration gradients near roots of plants grown under artificial conditions (Claassen, Syring, and Jungk, 1986). Other reports have indicated that the model is unsatisfactory for some crops and soils (Silva, Magalhaes, and Barber, 1991; Seiffert et al., 1995) especially high-K soils (Seward, 1991) and soils which fix K (Brouder and Cassman, 1994).

The validity of the model described by Greenwood and Karpinets (1997a,b) was tested against the results of single year, multilevel K-fertilizer experiments on wheat, summer cabbage, and turnips. Reasonably good agreement was obtained between simulated values and measurements of plant mass, K concentration in the plants and K activity ratio of soil at harvest and at intervals during the growing season. Good agreement was generally obtained for plant weight and plant percent K at harvest in less detailed experiments on 10 other crops. The model also satisfactorily predicted the small changes in exchangeable K (0.5 **M** CH_3COOH extractable) that occurred during the growth of sugar beets in field experiments on a typical chernozem in Russia (Karpinets, 2000).

Crop and soil response models developed by Greenwood and Karpinets (1997a) have more readily available inputs than other models and may predict such output parameters as dry crop biomass and K, total K uptake, and the exchangeable K in soil. This is exactly the information that is needed by farmers or farmers' advisers to improve fertilizer practice. It also simulates the effect of weather conditions on plant and soil processes. Simulations with the model indicate that, in central England, no response of 10 crops to K fertilizer would be likely on soils containing more than 170 mg of 1 **M** ammonium nitrate extractable-K/kg of soil with clay contents of between 15 and 45 percent (Greenwood and Karpinets, 1997b). The results of the calculation experiments on silt loam chernozems indicate that under optimal weather conditions and nitrogen and phosphorous regimes, it is possible to obtain sugar beet yields of about 47-50 t/ha without incorporation of K fertilizer if the soil has 0.5 **M** CH_3COOH extractable K content of more than 10 mg K_20/100g (Karpinets, 2000). Special efforts have been made to produce a version of this K-response model that is easy to use and readily available. It runs interactively and quickly on the Internet at <http://www.qpais.co.uk/moda-djg/potass.htm> and has been run thousands of times by users throughout the world.

The usefulness of crop and soil K response models is, however, restricted by the problem of calibrating them for differences in conditions. A promising way of increasing the application of mechanistic models would be to include a module in the modeling packages that would enable users to calibrate the model for their conditions. The usefulness of these models could be further improved by introduction of more rapid and accurate methods of chemical analysis such as those based on new ion selective techniques that appear to permit the monitoring of cations in the field (Evangelou, Wang, and Phillips, 1994).

Despite these developments, much uncertainty still exists about how to adjust K fertilizer advice for differences in crops and conditions. Most advice is based on the results of fertilizer trials. Little use has been made of the vast amount of fundamental knowledge about key soil and plant processes. The models that are now being developed will enable this fundamental knowledge to be "packaged" so that it can, for the first time, be applied in practice. Such models should enable more reliable predictions to be made with fewer fertilizer trials and thus at much less cost. They could be useful as an aid to the development of national policies designed to make the most effective use of available fertilizer. Sensitivity analyses of the models should enable the relative importance of different crop characteristics in potassium nutrition to be determined and thus help the improvement of cultivars by breeding and genetic manipulation. There are a number of separate ways in which potassium research and modeling could be of benefit worldwide. Internet technology could greatly help in making these models of immediate practical help.

SUMMARY

The K processes in soil are self-regulated and can be characterized by the stationary state with steady contents of various forms of soil K. Soil systems support the steady stationary K regimes through increasing or decreasing the rates of the different soil processes: K fixation, release, leaching, weathering, and uptake by plants. The existence of steady stationary levels of the different soil K forms has been confirmed by microfield experiments and by long-term field experiments with increasing rates of K fertilizer. A synergetic model of the K regime in the soil-plant system, based on the self-organization theory, permits the estimation of the steady stationary contents of the various forms of soil K from the results of such experiments. These steady stationary contents form the basis for the prediction of the dynamic changes of these K forms during long-term cropping. This has been demonstrated by the results of eight long-term field experiments on clay loam and silt loam chernozems.

K dynamics in soils and plants during vegetation may be predicted in multiparameter mechanistic models. Such models include mathematical descriptions of plant growth as a function of the climate and K concentration in the plant, and root growth as a function of plant mass. Potassium transport to roots is influenced by plant demand and K fixation and release in soil. Experimental validation of multiparameter mechanistic models of the K dynamics and examples of their application are considered as the most reliable and cost-effective tools for integrating information on soil and plant potassium and for forecasting the influence of the different processes in the ecosystem.

REFERENCES

Addiscott, T.M. and A.E. Johnston (1975). Potassium in soils under different cropping systems. 3. Non-exchangeable potassium in soils from long-term experiments at Rothamsted and Woburn. *Journal of Agricultural Science, Cambridge 84:* 513-524.

Arnold, P.W. (1970). *The Behaviour of Potassium in Soils.* Proceedings of the Fertilizer Society, Number 115, York, UK, 30 pp.

Baldwin, J.P., P.H. Nye, and P.B. Tinker (1973). Uptake of solutes by multiple root systems from soil. III. A model for calculating the solutes uptake by a randomly dispersed root system developing in a finite volume of soil. *Plant and Soil 38:* 621-635.

Barber, S.A. (1984). *Soil Nutrient Bioavailability: A Mechanistic Approach.* New York: John Wiley and Sons.

Barber, S.A. and J.H. Cushman (1981). Nitrogen uptake model for agronomic crops. In *Modeling Waste Water Renovation Land Treatment,* ed. J.K. Iskander. New York: Wiley-Interscience, pp. 382-409.

Barnes, A., D.J. Greenwood, and T.J. Cleaver (1976). A dynamic model for the effects of potassium and nitrogen fertilizers on the growth and nutrient uptake of crops. *Journal of Agricultural Science, Cambridge, 86:* 225-244.

Barraclough, P.B. (1989). Root growth, and nutrient uptake by field crops under temperate conditions. *Aspects of Applied Biology 22:* 227-233.

Barraclough, P.B. (1990). Modeling K uptake by plants from soil. In *Proceedings of the 22nd Colloquium of the International Potash Institute.* Worblaufen-Bern: International Potash Institute, pp. 217-230.

Beckett, P.H.T. (1964). Studies on soil potassium. II. The "immediate" Q/I relations of labile potassium in the soil. *Journal of Soil Science 15:* 9-23.

Beckett, P.H.T. (1971). Potassium potential—A review. *Potash Review 30:* 1-41.

Berdnikov, A.M., A.G. Chilenko, and K.G. Berdnikova (1983). The effect of different systems of fertilization of agricultural crops in the forest-steppe zone of the Ukraine Republic. In *Results of Studies of Long-Term Trials with Fertilizer in South and West Regions.* Moscow: All-Union Scientific Research Institute of Fertilizer and Agropedology, pp. 85-100 (in Russian).

Bolt, J.A., M.E. Sumner, and A. Kamphort (1963). A study of the equilibria between three categories of potassium in an illitic soil. *Soil Science Society of America Proceedings 27:* 294-299.

Bond, W.J. (1995). On the Rothmund-Kornfeld description of cation exchange. *Soil Science Society of America Journal 59:* 436-443.

Brouder, S.M. and K.G. Cassman (1994). Evaluation of a mechanistic model for potassium uptake by cotton in vermiculitic soil. *Soil Science Society of America Journal 54:* 1174-1183.

Chen, J.H. and S.A. Barber (1990). Soil pH and phosphorus and potassium uptake by maize evaluated with an uptake model. *Soil Science Society of America Journal 54:* 1032-1036.

Chuyan, G.A. and T.V. Karpinets (1986). Some features of potassium accumulation in eroded soils. In *Scientific and Technical Bulletin of All-Union Scientific Research Institute of Agriculture and Soil Erosion 2/49.* Kursk: All-Union Scientific Research Institute of Agriculture and Soil Erosion Control, pp. 48-52 (in Russian).

Claassen, N. and S.A. Barber (1976). Simulation model for nutrient uptake from soil by a growing plant root system. *Agronomy Journal 69:* 860-864.

Claassen, N. and A. Jungk (1984). Effect of K uptake rate, root growth and root hairs on potassium uptake efficiency of several plant species. *Zeitschrift für Pflanzenernährung und Bodenkunde 147:* 276-189.

Claassen, N., K.M. Syring, and A. Jungk (1986). Verification of a mathematical model by simulating potassium uptake from soil. *Plant and Soil 95:* 209-220.

Cooke, G.W. (1986). The intercontinental transfer of plant nutrients. In *Proceedings of the 13th IPI Conference.* Worblaufen-Bern: International Potash Institute, pp. 267-287.

Crank, J. (1957). *The Mathematics of Diffusion.* Oxford: Clarendon Press.

Drew, M.C. (1975). Root and shoot growth in barley. *New Physiologist 75:* 479-490.

Dufey, J.I. and B. Delvaux (1989). Modeling potassium-calcium exchange isotherms in soils. *Soil Science Society of America Journal 53:* 1297-1299.

Eagle, D.L. (1967). Release of non-exchangeable potassium from certain soils. In *Soil Potassium and Magnesium.* London: Ministry of Agriculture, Fisheries and Food, HMSO, pp. 49-54.

Evangelou, V.P., J. Wang, and R.E. Phillips (1994). New developments and perspectives on soil potassium quantity/intensity relationships. *Advances in Agronomy 52:* 173-227.

Fotyma, S. and S. Gosek (1991). Long term potassium fertilization in Poland. *Potash Review 2:* 1-3.

Gapon, E.N. (1933). Theory of exchange adsorption in soils. *Journal of General Chemistry of the USSR 3:* 144-152.

Gaultier, J.P. and J. Mamy (1979). Evolution of exchange properties and crystallographic characteristics of biionic K-Ca montmorillonite submitted to alternate wetting and drying. In *Proceedings of the VI International Clay Conference.* Amsterdam: Elsevier, pp. 167-175.

Gerwitz, A. and E.R. Page (1973). An empirical mathematical model to describe plant root systems. *Journal of Applied Ecology 11:* 773-781.

Glass, A. (1975). The regulation of potassium absorption in barley roots. *Plant Physiology 56:* 377-380.

Goncharenko, B.Yu., L.P. Shodeeva, L.A. Tkarch, and R.P. Gladkich (1983). Study of the effect of mineral fertilizer and manure on the productivity of a vegetable rotation and on soil characteristics. In *Results of Studies in Long-Term Trials with Fertilizer in South and West Regions.* Moscow: All-Union Scientific Research Institute of Fertilizer and Agropedology, pp. 126-143 (in Russian).

Gorbunov, N.I. (1978). *Soil Mineralogy and Physical Chemistry.* Moscow: Kolos.

Goulding, K.W.T. (1984). The availability of potassium in soils to crops as measured by its release to a calcium-saturated cation exchange resin. *Journal of Agricultural Science, Cambridge 97:* 261-296.

Greenland, D.J. and M.H.B. Hayes. (1978). Soils and soil chemistry. In *The Chemistry of the Soil Constituents,* eds. D.J. Greenland and M.H.B. Hayes. Chichester, UK: John Wiley and Sons, pp. 1-27.

Greenwood, D.J., A. Gerwitz, D.A. Stone, and A. Barnes (1982). Root development of vegetable crops. *Plant and Soil 68:* 75-96.

Greenwood, D.J. and T.V. Karpinets (1997a). Dynamic model for the effect of K-fertilizer on crop growth, K-uptake and soil-K in arable cropping. 1. Description of the model. *Soil Use and Management 13:* 178-183.

Greenwood, D.J. and T.V. Karpinets (1997b). Dynamic model for the effect of K-fertilizer on crop growth, K-uptake and soil-K in arable cropping. 2. Field test of the model. *Soil Use and Management 13:* 184-189.

Greenwood, D.J. and D.A. Stone (1998). Prediction and measurement of the decline in the critical-K, the maximum-K and total cation plant concentration during the growth of field vegetable crops. *Annals of Botany 82:* 871-991.

Grib, N.I., N.F. Gorobets, and I.M. Krasnyuk (1978). The effect of rates and relationships of main types of fertilizers on crops productivity in rotation, yield quality and soil fertility. In *Results of Studies in Long-Term Trials with Fertilizer.* Moscow: All-Union Scientific Research Institute of Fertilizer and Agropedology, pp. 78-141 (in Russian).

Hartkamp, A.D., W.W. Jeffrey, and G. Hoogenboom (1999). Interfacing geographic information systems with agronomic modeling: A review. *Agronomy Journal 91:* 761-772.

Hinsinger, P. and B. Jaillard (1993.). Root-induced release of interlayer potassium and vermiculitization of phlogopite as related to potassium depletion in the rhizosphere of ryegrass. *Journal of Soil Science 44:* 525-534.

Huang, P.M. (1989). Feldspars, olivines, pyroxenes, and amphiboles. In *Minerals in Soil Environments,* eds. J.B. Dixon and S.B. Weed. Madison, WI: Soil Science Society of America, pp. 975-1050.

Ioannou, A., A. Dimirkou, and Ch. Paschalidis (1994). Kinetics of potassium desorption by alfisols of Greece. *Communications in Soil Science and Plant Analysis 25:* 1355-1372.

Jardine, P.M. and D.L. Sparks (1984). Potassium-calcium exchange in multireactive soil system: I. Kinetics. *Soil Science Society of America Journal 48:* 39-45.

Johnston, A.E. (1986). Potassium fertilization to maintain K balance under various farming systems. In *Nutrient Balances and the Need for Potassium.* Proceedings of the 13th IPI Congress, Reims, France. Warblaufen-Bern: International Potash Institute, pp. 199-226.

Johnston, A.E. and K.W.T. Goulding (1993). Potassium concentrations in surface and groundwaters and the loss of potassium in relation to land use. In *Potassium in Ecosystems.* Proceedings of 23rd Colloquium of the International Potash Institute, Prague, Czechoslovakia. Basel: International Potash Institute, pp. 135-158.

Johnston, A.E., K.W.T. Goulding, and E. Mercer (1993). Potassium leaching from a sandy loam soil. *Potash Review 4:* 1-16.

Jungk, A., N.A. Claassen, and R. Kuchenbuch (1982). Potassium depletion of the soil-root interface in relation to soil parameters and root properties. In *Proceedings IX International Plant Nutrition Colloquium,* ed. A. Scaife. Warwick University, Coventry, UK: Commonwealth Agricultural Bureau, pp. 250-255.

Karpinets, T.V. (1992). Estimation of K fixation and release in soil by two consecutive methods. In *Potassium in Ecosystems.* 23rd Colloquium of the International Potash Institute, Prague, Czechoslovakia. Worblaufen-Bern: International Potash Institute, pp. 391-395.

Karpinets, T.V. (1995). Determination of stable stationary contents of forms of potassium in soils. *Eurasian Soil Science 27/9:* 88-97.

Karpinets, T.V. (2000). Modeling of potassium regime in soil-plant system. (Doctoral thesis, Kursk: All-Russian Institute of Agriculture and Soil Erosion Control) (in Russian).

Karpinets, T.V. and G.S. Lipkina (1992). Stable steady states of the potassium regime in soil. *Eurasian Soil Science 24/7:* 9-17.

Kirkman, J.H., A. Basker, A. Surapaneni, and A.N. MacGregor (1994). Potassium in the soils of New Zealand—A review. *New Zealand Journal of Agricultural Research 37:* 207-227.

Kobsarenko, V.I. (1983). Importance of subsoil layers in phosphorous and potassium plant supply. In *Soil Fertility and Ways of Its Increase.* Moscow: Kolos, pp. 84-91 (in Russian).

Kochl, A. (1984). Potassium balances in a series of field experiments. In *Nutrient Balances and Fertilizer Needs in Temperate Agriculture.* Proceedings of 18th Colloquium of the International Potash Institute, Gardone-Riviera, Italy. Worblaufen-Bern: International Potash Institute, pp. 177-185.

Lawton, K. (1955). Chemical composition of soils. In *Chemistry of the Soil,* ed. F.E. Bear. New York: Reinhold Publishing, pp. 53-84.

Leontiev, A.K. (1981). Combinations and relationships of mineral fertilizer and manure in grain-cultivated crop rotation. In *Results of Studies in Long-Term Trials With Fertilizer.* Moscow: All-Union Scientific Research Institute of Fertilizer and Agropedology, pp. 108-120 (in Russian).

Lipkina, G.S. (1986). Concentration of mobile potassium compounds in heavily fertilised loam Sod-Podzolic soils. *Pochvovedeniye 12:* 69-75 (in Russian).

Lyapunov, A.A. (1892). *Common task about the stability of movement.* Kharkov, Ukraine: Kharkov State University (in Russian).

Marschner, H. (1995). *Mineral Nutrition of Higher Plants.* Second Edition, London: Academic Press.

Martin, H.W. and D.L. Sparks (1983). Kinetics of nonexchangeable potassium release from two coastal plain soils. *Soil Science Society of America Journal 47:* 883-887.

Matthews, B.C. and P.H.T. Beckett (1962). A new procedure for studying the release and fixation of potassium ions in soil. *Journal of Agricultural Science, Cambridge 58:* 59-64.

Matthews, B.C. and C.G. Sherrell (1960). Effect of drying on exchangeable potassium of Ontario soils and the relation of exchangeable potassium to crop yield. *Canadian Journal of Soil Science 40:* 35-41.

Medvedeva, O.P. (1983). Non-exchangeable fixed potassium from fertilizer as an indicator of plant supply with available potassium. *Agrochemistry 11:* 25-31 (in Russian).

Munson, R.D. (1985). *Potassium in Agriculture.* Madison, WI: American Society of Agronomy.

Niederbudde, E.A. (1986). Factors affecting potassium release and fixation in soils. In *Transactions of XIII Congress of International Society of Soil Science,* ed. ISSS. Hamburg: International Society of Soil Science, vol. VI, pp. 1155-1167.

Nosko, B.S., H.A. Kuchir, and V.G. Rozdaibeda (1988). Modeling of soil agrochemical characteristics. In *Modeling of Soil Fertility and Optimisation of Fertilizer in Different Natural Agricultural Zones.* Moscow: All-Union Scientific Research Institute of Fertilizer and Agropedology 90, pp. 3-14 (in Russian).

Phetchawee, S., C. Kanareugsa, C. Sittibusaya, and H. Khunathai (1985). Potassium availability in the soils of Thailand. In *Proceedings of the 19th Colloquium of the International Potash Institute.* Worblaufen-Bern: International Potash Institute, pp. 167-196.

Pieri, C. and R. Oliver (1986). Assessment of K losses in tropical cropping systems of francophone Africa and Madagascar. In *Nutrient Balances and Needs for Potassium.* Bern: International Potash Institute, pp. 73-92.

Ponomarev, A.I. (1982). The effect of fertilizers in typical chain of crop rotation on podzolic leashed chernozems of forest-steppe chernozemic zone of Russia. In *Results of Studies in Long-Term Trials with Fertilizer.* Moscow: All-Union Scientific Research Institute of Fertilizer and Agropedology, pp. 70-93 (in Russian).

Prokoshev, V.V. and T.A. Sokolova (1990). Soil properties and potassium behaviour. In *Proceedings of the 22nd Colloquium of the International Potash Institute.* Worblaufen-Bern: International Potash Institute, pp. 99-115.

Rich, C.I. (1972). Potassium in soil minerals. In *Potassium in Soil.* Proceedings of the 9th Colloquium of the International Potassium Institute, Federal Republic of Germany. Worblaufen-Bern: International Potash Institute, pp. 3-19.

Robert, M. (1992). K-fluxes in soils in relation to parent material and pedogenesis in tropical, temperate and arid climates. In *Potassium in Ecosystems.* Proceedings of 23rd Colloquium of the International Potash Institute, Prague, Czechoslovakia. Worblaufen-Bern: International Potash Institute, pp. 25-44.

Russell, E.W. (1973). *Soil Conditions and Plant Growth.* London: Longman.

Schroeder, D. and P.A. Gething (1984). *Soils—Facts and Concepts.* Bern: International Potash Institute.

Seiffert, S., J. Kaselowsky, A. Jungk, and N. Claassen (1995). Observed and calculated potassium uptake by maize as affected by soil water content and bulk density. *Agronomy Journal 87:* 1070-1077.

Sevastyanova, V.V., B.Yu. Goncharenko, L.P. Shodeeva, and L.A. Tkarch (1978). Study of the effect of mineral fertilizer and manure on productivity of vegetable rotation and soil characteristics. In *Results of Studies in Long-Term Trials with Fertilizer.* Moscow: All-Union Scientific Research Institute of Fertilizer and Agropedology, pp. 4-19 (in Russian).

Seward, P.D. (1991). The development of a mechanistic mathematical model to predict the uptake of potassium by wheat plants from soil. (PhD thesis, University of Reading).

Shagaev, V.Ya. and N.V. Michailina (1978). The effect of long-term fertilization on crop yield, quality of production and soil characteristics. In *Results of Studies in Long-Term Trials With Fertilizer.* Moscow: All-Union Scientific Research Institute of Fertilizer and Agropedology, pp. 175-190 (in Russian).

Sharpley, A.N. (1990). Reaction of fertilizer potassium in soils of different mineralogy. *Soil Science 149:* 44-51.

Sharpley, A.N. and S.W. Buol (1987). Relationship between minimum exchangeable potassium and soil taxonomy. *Communications in Soil Science and Plant Analysis 18:* 601-614.

Siddiqi, M.Y. and A.D.M. Glass (1982). Simultaneous consideration of tissue and substrate potassium concentrations in K+ uptake kinetics: A model. *Plant Physiology. 69:* 283-285.

Silva, F.L.I.M., J.R. Magalhaes, and S.A. Barber (1991). Potassium uptake by sweet corn predicted by simulation. *Pesquisa-Agropecuaria-Brasileira 26:* 431-438.

Singh, B. and K.W.T. Goulding (1997). Change with time in the potassium content and phyllosilicates in the soil of the Broadbalk continuous wheat experiment at Rothamsted. *European Journal of Soil Science 48:* 651-659.

Sparks, D.L. (1980). Chemistry of soil potassium in Atlantic coastal plain soils: A review. *Communications in Soil Science and Plant Analysis 11:* 435-449.

Sparks, D.L. and P.M. Huang (1985). Physical chemistry of soil potassium. In *Potassium in Agriculture,* ed. R.D. Munson. Madison, WI: America Society of Agronomy, pp. 201-276.

Sverdrup, H.U. (1990). *The Kinetics of Base Cation Release due to Chemical Weathering.* Lund: Lund University Press.

Sverdrup, H. and P. Warfringe (1993). Calculating field weathering rates using a mechanistic geochemical model PROFILE. *Applied Geochemistry 8:* 273-283.

Syers, J.K. (1998). *Soil and Plant Potassium in Agriculture.* York, UK: The Fertilizer Society.

Talibudeen, O., J.D. Beasley, P. Lane, and N. Rajendran (1978). Assessment of soil potassium reserves available to plant roots. *Journal of Soil Science 29:* 207-218.

Teo, Y.H., C.A. Beyrouty, and E.E. Gbur (1992). Evaluating a model for predicting nutrient uptake by rice during vegetative growth. *Agronomy Journal 84:* 1064-1070.

Tinker, P.B. (1964). Studies on soil potassium 3. Cation activity ratios in acid Nigerian soils. *Journal of Soil Science 15:* 24-34.

Tinker, P.B. and P.H. Nye (2000). *Solute Movement in the Rhizosphere.* Oxford: Oxford University Press.

Toth, S.J. (1955). Colloid chemistry of soils. In *Chemistry of the Soil,* ed. F.E. Bear. New York: Reinhold Publishing, pp. 85-106.

Tributh, H. (1987). Development of K containing minerals during weathering and suitable methods for their determination. In *Methodology in Soil-K Research.* Proceedings of 20th Colloquium of the International Potash Institute held in Baden bei Wien, Austria. Worblaufen-Bern: International Potash Institute, pp. 65-83.

Tributh, H., E. Boguslawski, A. Lieres, and K. Mengel (1987). Effect of potassium removal by crops on transformation of illitic clay minerals. *Soil Science 6:* 404-409.

Welte, E. and E.A. Niederbudde (1965). Fixation and availability of potassium in loess-derived and alluvium soils. *Journal of Soil Science 16:* 116-120.

Wilson, M.J. (1992). K-bearing minerals and their K-release rates in different climates. In *Potassium in Ecosystems.* Proceedings of 23rd Colloquium of the International Potash Institute, Prague, Czechoslovakia. Worblaufen-Bern: International Potash Institute, pp. 45-57.

Chapter 17

Secondary Nutrients: Sulphur, Calcium, and Magnesium

Surinder Saggar
Nanthi S. Bolan

The persistence of life on earth depends on the cycling of essential elements, including sulphur (S), calcium (Ca), and magnesium (Mg), which are considered secondary to plant nutrient requirements. Losses of these nutrients need to be quantified and replaced to sustain agricultural production systems so that they continue to be productive over a long period. New management practices are required that make more efficient use of nutrient inputs and reduce nonproduct nutrient losses. However, an improvement in the knowledge of the processes controlling the cycling and bioavailability of these nutrients is essential to better understand the nature of soil resources and the development of models that may help both to describe the current status of these nutrients in the soil ecosystem and to predict future changes. This chapter reviews and develops an understanding of the nature of the secondary nutrients (S, Ca, and Mg), and the factors influencing their cycling through soils, plants, and animals.

Sulphur, the thirteenth-most abundant element, comprises about 0.1 percent of the earth's crust. In nature, this element, with six valence electrons, can exist in a wide variety of organic and inorganic combinations, in various states of oxidation ranging from -2 (sulphide) to +6 (sulphate), and in solid, liquid, or gaseous forms. Research into soil S dynamics blossomed in the 1970s because of the realization of its importance in plant nutrition and the lack of understanding of soil S as a source for plants. Since then prodigious advances have been made toward understanding the S dynamics in terrestrial ecosystems (e.g., see reviews by Freney and Williams, 1983; Tabatabai, 1984; Freney, 1986; Howarth, Stewart, and Ivanov, 1992; Mitchell, David, and Harrison, 1992; Nguyen and Goh, 1994; Eriksen, Murphy, and Schnug, 1998; Janzen and Ellert, 1998; Saggar, Hedley, and Phimsarn, 1998) and plant S metabolism (Thompson, Smith, and Madison, 1986; Hell and Rennenberg, 1998).

Calcium, the fifth-most abundant element in the earth's crust (approximately 3.64 percent), is an important component of cell walls. Magnesium, the eighth-most abundant element of earth's crust, is the central component of chlorophyll. In many soil components, Ca and Mg have similar properties. Although both are regarded as macronutrients, they are not always considered in fertilizer management. One of the reasons for this is that Ca and Mg are added to soil as accessory elements in many fertilizers and liming materials. Increasing use of Ca-free ammonium phosphate fertilizers and reduced use of liming materials have resulted in increasing incidences of Ca deficiency in soils. Increasing incidences of Mg deficiency are being observed because of decreased use and reduced concentration of Mg in potassium (K) fertilizer, kainite. Furthermore, accelerated soil acidification caused by modern agricultural practices has reduced exchangeable basic nutrient cations resulting in Ca and Mg deficiency in soils.

In this chapter, only a brief description of the inputs and reaction in soils, and plant and animal requirements of S, Ca, and Mg in the agroecosystems as a background to subsequent discussion on modeling of these nutrients is presented. The reader is referred to appropriate reviews that provide in-depth coverage of respective topics.

INPUT TO SOILS

Sulphur

In the parent materials of soils, S may occur in some primary minerals as sulphide (S^-) or sulphate (SO_4^{2-}). Over 2,000 S-bearing minerals with S contents ranging from 7 to 53 percent are known to occur in soils. Igneous rocks generally have greater total S contents (0.05-0.30 percent) than the sedimentary rocks with S^- dominance. Silicate minerals generally contain <100 mg S kg^{-1}. Soil S includes both organic and inorganic compounds, the proportions and forms of which vary depending on soil type, depth in profile, climate, and cultural conditions. Inorganic soil S represents a relatively small proportion (<25%) of total S, comprising chiefly soluble, adsorbed, insoluble, and coprecipitated SO_4^{2-} and S^-. Elemental S (S°) and other compounds of lower oxidation state than SO_4^{2-}, mostly present as transitory reaction intermediates, are less prevalent in the soil environment. The organic S mainly originates as plant and animal residues that are subsequently decomposed and remetabolized by soil organisms. The main forms of organic S deposited in soils include S-containing amino acids and sulfonates, in which S is directly bonded to carbon (C-S), and also the true organic esters of sulphuric acid (C-O-S), in which S is bonded to oxygen in the form of

C-O-SO_3^- linkages. Sulphamates may also be found, in which S occurs in the form of N-O-SO_3^- and N-SO_3^- groups. Although most soil S may be present in the organic form, on an annual basis only a small percentage of this fraction (as little as 2 percent) may enter the active S cycling pool, which supplies plant uptake. Recognition of the importance of soil microbial populations in cycling and transforming S in the plant root zone has led to efforts to measure microbial pool size and activity as approaches to solving the enigma of S availability in soils (Saggar, Bettany, and Stewart, 1981a,b; Wu et al., 1994; Benerjee and Chapman, 1996). The significance of microbial S in soil is seen in its position as a boundary between inorganic and organic S, the catalyst of the interconversion between the two forms (Saggar, Hedley, and Phimsarn, 1998).

The primary S input to soils is in S-containing fertilizers. Historically, high amounts of S have been applied as incidental constituents of N, P, and K fertilizers. A wide range of S fertilizers have now been developed to cover a wide range of environmental and farming conditions (see Boswell and Gregg, 1998). Rates of application to pastures and crops are generally between 10-30 kg ha^{-1} yr^{-1}. Sulphur also enters the soils through rainfall and dry deposition (Syers, Skinner, and Curtin, 1987), and these inputs can vary (1-30 kg ha^{-1} yr^{-1}) with distance from coasts and proximity to industrialized areas. Eriksen, Murphy, and Schnug (1998) have described the beneficial and harmful effects of atmospheric S accretions on soil-plant-water environments.

Calcium

Calcium in soils is found mainly in minerals such as feldspar, calcite, dolomite, and apatite. Calcium sulphate (gypsum), which occurs in arid soils, and calcium carbonate (calcite), which occurs in calcareous soils, are the two important calcium minerals controlling Ca concentration in these soils. Soils developed from calcite and dolomite are alkaline in reaction, and a high pH and the presence of Ca favor the formation of Ca humate complexes and account for the dark color of soils. The Ca content of soils depends on the type of parent materials and the extent of weathering. Although most soils contain 0.1-5.0 percent Ca, some of the calcareous soils contain over 20 percent Ca (Bruce, 1999).

Calcium deficiency in soils can be overcome by adding Ca-containing compounds. Traditionally, superphosphate has been used as the major source of phosphorus, which also supplies Ca to soils. The Ca in single superphosphate is gypsum ($CaSO_4.2H_2O$), which is readily soluble in soil solution. In addition to superphosphate, the other two most commonly used

Ca compounds are lime and gypsum. Lime is added mainly to overcome the problems associated with soil acidification; gypsum is used both as an S source and as an amendment to improve the physical conditions of soils. Mixing lime with ammonium-based fertilizers causes ammonia volatilization, and hence this practice should not be followed. Gypsum can readily be mixed with ammonium-based fertilizers.

Magnesium

Magnesium is a normal component of both igneous and sedimentary rocks and of the soils developed from such rocks. Soils developed from basic rocks (diabase, basalts, and limestone) generally contain high levels of Mg (0.27-2.86 percent) and those developed on coastal sand and granite and sandstones contain low levels of Mg (0.01-0.34 percent) (Aitken and Scott, 1999). In most soils, Mg is present in minerals and in organic matter as exchangeable cation, and also in soil solution (Mokwunye and Melsted, 1972). However, the majority of soil Mg is present in forms that are not readily available to the plant. Magnesium exists in primary minerals such as biotite, serpentine, olivine, and hornblende, and in the secondary silicate clay minerals chlorite, vermicullite, illite, and montmorillonite. The chemical formulae of some of the important silicate mineral species carrying Mg are shown in Table 17.1. In nature, however, the composition varies over a range, depending on the extent of substitution in the crystalline structure of these mineral species. For example, montmorillonite in its ideal state contains no Mg but Mg replaces isomorphically a part of Al ions in the lattice of montmorillonite (Smeck, Saif, and Bingham, 1994). Magnesium is also present in various other mineral forms in soils (see Table 17.2).

Magnesium deficiency in soils can be overcome by adding Mg fertilizers such as serpentine superphosphate, epsom salt, dolomite, and calcined magnesite (Table 17.2) (Augustin, Mindrup, and Meiwes, 1997). Soluble in water, epsom salt is used as a fast-release Mg source, and is the most satisfactory Mg source used for foliar application to plants. The other fertilizers are insoluble in water, and are used as a slow-release source.

Dolomite, which contains both Ca and Mg, is more effective in acid soils because the Mg is brought into solution by the acid soil. It is the most widely used source of Mg for application to soils, both as an ingredient of mixed fertilizers and as a separate amendment for liming. It has the advantage of neutralizing free acid found in certain phosphate materials and of counteracting the potential acidity of the fertilizer N. There will rarely be any need for additional Ca and Mg for any crop in which natural acid-soil and clima-

TABLE 17.1. Magnesium Minerals in Soils

Magnesium minerals	Chemical formula	Magnesium content ($g\ kg^{-1}$)
Fosterite	Mg_2SiO_4	320-350
Pyrope	$3MgO.Al_2O_3.3SiO_2$	60-130
Iolite	$H_2(Mg,Fe)_4Al_8Si_{10}O_{37}$	50-80
Diopside	$CaMg(SiO_3)_2$	20-140
Augite	$CaMg(SiO_3)_2$	45-100
Enstatite	$MgSiO_3$	180-220
Actinolite	$Ca(Mg,Fe)_3Si_4O_{12}$	100-160
Hornblende	CaMg metasilicate	10-90
Serpentine	$H_4Mg_3Si_2O_9$	19-26
Talc	$H_2Mg3Si_4O_{12}$	160-200
Phlogopite	$H_3Mg_3Al(SiO_4)_3$	130-180
Biotite	$(H,K)_2(Mg,Fe)_2Al_2Si_3O_{12}$	10-160
Clinochlore	$H_8(Mg,Fe)_5Al_2Si_3O_{18}$	100-120

tic conditions require a more or less regular liming program and where dolomite is applied at regular intervals.

All the slightly soluble Mg fertilizers listed in Table 17.2 are acid-neutralizing materials and would thus reduce the acid-forming potential of the fertilizer to which they might be added. Magnesium silicates, including magnesite, are not effective sources of Mg for plants. Selectively, calcined dolomite in which the Mg component is oxidized to MgO is more reactive than dolomite. Calcined magnesite, currently available as a granular product (Granmag), can be added to soils or dusted onto pastures as a source of Mg. Both these materials are not suitable for mixing with ammonium containing fertilizers.

REACTIONS IN SOILS

Sulphur

Both abiotic and biotic soil processes control S dynamics. Abiotic processes include physiochemical reactions such as SO_4^{2-} sorption-desorption and precipitation-dissolution with soil mineral surfaces; biotic processes in-

TABLE 17.2. Magnesium Fertilizers

Magnesium minerals	Chemical formula	Mg solubility (mmol L^{-1})	Solubility product (pK_{SP})	Magnesium content (g kg^{-1})
Dolomite	$MgCO_3.CaCO_3$	0.038	17.09	100
Calcined dolomite	$MgO.CaCO_3$			160
Hydrated dolomite	$MgO.Ca(OH)_2$			170
Magnesite	$MgCO_3$	0.076	8.24	260
Brucite	$Mg(OH)_2$	0.091	11.41	360
Magnesia	MgO	0.150		560
Kieserite	$MgSO_4.H_2O$	4943		160
Epsom salt	$MgSO_4.7H_2O$	127.3	90	90
Kainite	$MgSO_4.KCl.3H_2O$			70
Langbeinite	$2MgSO_4.K_2SO_4$			110
Fosterite	Mg_2SiO_4	$0.067.\ 10^{-3}$	28.11	320-350

clude the plant and microbial assimilation of SO_4^{2-} into organic S and the mineralization of plant and animal residues by soil microflora and microfauna. A more complete discussion of the mechanisms and metabolic pathways of various S transformations in different agroecosystems is available elsewhere (Fitzgerald, 1976, 1978; Freney and Williams, 1983; Freney, 1986; Mitchell, David, and Harrison, 1992; Eriksen, Murphy, and Schnug, 1998; Janzen and Ellert, 1998; Saggar, Hedley, and Phimsarn, 1998). The rate of S mineralization is controlled by factors influencing the growth of microorganisms and their release or production of extracellular hydrolytic enzymes. These factors include moisture, temperature, pH, food (energy) supply as influenced by soil depth, the presence and absence of plants (plant roots), organic matter addition, and the amount of SO_4^{2-} present. Similar factors have been shown to influence the level of sulphatase enzymes in soil.

Calcium

Calcium is present in three major forms in soils: in soil minerals, in exchangeable sites, and in soil solution (Bruce, 1999). Calcium ions released from fertilizers undergo cation exchange reactions similar to K ions. The adsorption sites of the inorganic colloids are not very selective for Ca. The adsorption of Ca to organic colloids and especially to the humic acids is very specific. Thus in soils containing large amounts of Ca (calcareous soils), the humic acids are mainly present in the form of Ca humate. Ca adsorbed onto both inorganic and organic colloids tends to equilibrate with Ca in solution. Most mineral soils contain enough levels of Ca in solution, and their exchangeable sites are well saturated with Ca to meet crop demand adequately. In acid peat soils, the natural Ca content can be so low that plants suffer from Ca deficiency and require Ca-containing fertilizers. Under acid conditions, the presence of calcite (lime) would only be transient, and it would readily dissolve to give Ca ions that would be held on exchange sites and in the soil solution. In acid soils, most of the Ca would exist in soluble forms as Ca ions.

The availability of Ca in soils for plant uptake is as follows: solution > exchangeable > mineral. There is equilibrium between the various forms of Ca that allows the release of Ca from less available forms to more available forms (Bruce, 1999). As with other cations, the requirements for Ca application to plants are based on the exchangeable soil Ca test. Since most soils contain adequate amounts of Ca, and traditionally, superphosphate, which contains significant amounts of Ca, Ca deficiency has never been a problem in most soils. Field calibrations of Ca soil tests are not available. A simple guide that may be useful to overcome Ca deficiency in soils is to maintain an adequate Ca saturation of the exchange sites.

Problems with Ca solubility and availability are seldom encountered in most soils, although very acid soils and particularly some subsoils may be deficient in Ca. The amounts of Ca and other basic cations decrease as the soil becomes more acid. Leaching of nitrate and sulphate ions induces the leaching of basic cations such as Ca and Mg. Deficiency of basic cations, in particular Ca and Mg, is likely to occur in peat soils, pumice soils, highly leached, acid, sandy soils, and sandy soils that have received high rates of K fertilizers. When an excessive amount of K fertilizer is added to soil it is possible that some of the Ca ions retained onto the cation exchange sites will be replaced by K ions, inducing leaching of Ca.

Magnesium

Soil Mg is also subdivided into soluble, rapidly exchangeable, slowly exchangeable, and structural (mineral) forms (Metson and Brookes, 1975). This arbitrary subdivision accounts for differences in bioavailability (Hailes, Aitken, and Menzies, 1997), and the bioavailability of Mg in soils which is as follows: solution > exchangeable > mineral (Rice and Kamprath, 1968). Only a small fraction of the total Mg is present in soil solution and, depending on the soil, the majority of the Mg is present in other forms. Plants absorb Mg from soil solution, which is buffered by the readily exchangeable forms that, in turn, are slowly replenished by soil reserves that include slowly exchangeable and structural forms (Simard, deKimpe, and Zizka, 1992). There is equilibrium between the various forms of Mg that allows the release of Mg from less available forms to more available forms.

Most soils hold important reserves of Mg in primary and secondary minerals (Mayland and Wilkinson, 1989). These reserves are reported to contribute a large proportion of the Mg needs of annual and perennial crops (Christenson and Doll, 1978). The rate of release of Mg for plant uptake depends on the rate of weathering of these minerals (Saif, Smeck, and Bingham, 1997). Kinetics of Mg release from soils and various soil fractions have often being investigated using cation exchange resins, dilute salts, and organic acids (Simard, deKimpe, and Zizka, 1992). These studies have shown that as weathering proceeds, the slowly exchangeable Mg originates from progressively coarser particle size fractions. The release of slowly exchangeable Mg increases as the particle size is decreased. Furthermore, surface coatings on mineral surfaces act as semipermeable barriers, limiting contact between the soil solution and fresh mineral surfaces, thus reducing the rate of mineral weathering and, consequently, the rate of Mg release to soil solution (Courchesne, Turmel, and Beauchemin, 1996).

The amounts of Mg and other basic cations decrease as the soil becomes more acid. Aluminium-saturated soils, however, will readily react with Mg in solution to form Mg-Al double hydroxides (Smeck, Saif, and Bingham, 1994; Saif, Smeck, and Bingham, 1997) resulting in the retention of Mg. Deficiency of basic cations, in particular Mg, is likely to occur in peat soils, pumice soils, highly leached, acid, sandy soils, and sandy soils that have received high rates of K fertilizers.

Magnesium deficiency is common in soils that are low in exchangeable Mg, light textured, highly leached, acidic, and low in cation exchange capacity. In addition, induced deficiency may occur on some soils as a result of nutrient imbalances. Soils that are heavily fertilized, particularly with materials lacking in Mg or high in Ca, K, and NH_4 can also induce Mg deficiencies. By growing crops that require high levels of Mg throughout the growing season (e.g., tobacco, citrus, potatoes, cotton, and soybeans), Mg deficiency in soils may be exacerbated or induced (Sinclair, 1980).

PLANT UPTAKE

Sulphur

Sulphur is usually available to plants as soluble SO_4^{2-} from soil, but is mainly required by cellular processes in its reduced form of S^{2-} (Hell and Rennenberg, 1998). Besides soil SO_4^{2-}, plants can take up S from the atmosphere (SO_2), and from foliar spray (SO_4^{2-} and S°) through the stomata of their leaves (Noggle, Meagher, and Jones, 1986). Soil SO_4^{2-} enters the plant in the root hair region and is translocated from roots to shoots as inorganic SO_4^{2-}. The rate of SO_4^{2-} uptake decreases with increased SO_4^{2-} supply. Unlike N and other nutrients, S is often taken up in excess of plant requirements and stored in various ester sulphates or as free sulphate (Rennenberg, 1984; Mayland and Robbins, 1994), probably because the plant root cells are not equipped to prevent excess uptake of SO_4^{2-}. When plants are subjected to S stress, mature leaves retain SO_4^{2-}, even when young, newly emerging leaves appear chlorotic (Clarkson, Hawkesford, and Davidian, 1993), because SO_4^{2-} mobilization in leaves is too slow to maintain the growth of younger shoots under deficiency conditions. The most important factors governing the plant uptake of S are soil moisture content (plant available water), SO_4^{2-} concentration in soil solution, soil SO_4^{2-} buffer capacity, plant root distribution, and the actual evapotranspiration from the crop.

The primary metabolic fate of assimilated S is the synthesis of amino acids (cysteine, cystine, and methionine) and proteins. These S-containing amino acids and the S-containing peptides glutathione and γ-glutamylcysteine, account for the bulk of S in most plants (~90 percent). Sulphur also occurs in various compounds of uncertain function, including a wide range of sulphate esters, polysaccharide SO_4^{2-}, glucosinolates, and lipid SO_4^{2-}.

Calcium

Plants take up calcium as Ca^{2+} cation. It is an important constituent of plant cell walls and is involved in maintaining the turgidity of plants (Christiansen and Foy, 1979). Calcium is required for cell elongation and cell division; it activates enzymes, particularly those that are membrane bound, and is important in membrane permeability and the maintenance of cell integrity. Low content of Ca in storage organs induces high membrane permeability and allows solute diffusion in these tissues. This is obviously of importance in fruits and storage organs that accumulate large amount of sugars from phloem (Terblanche and Wooldridge, 1979). Calcium moves very slowly in plants and is not transported from older leaves to younger leaves. As Ca deficiency symptoms are often noticed first in the growing tips, a continuous supply of Ca is required for these tips (Hanger, 1979; Ferguson, 1979).

Magnesium

Plants take up magnesium as Mg^{2+} cation. It is the central unit and the only metal ion of chlorophyll in plant leaves and it cannot be substituted by other metal ions. An insufficient supply of Mg reduces chlorophyll formation, which is likely to affect the photosynthetic ability of the plants. Because Mg is a structural element of chlorophyll, it is assigned a dominant role in the life of the plant. But it is not only on this account that Mg is indispensable to plants; its capacity for forming complexes with water can be of even greater significance, since in this way Mg has a controlling action on the swelling of plasma. Mg is involved in the production of starch during photosynthesis, and plays a major role in the functions of many enzymes in plants, as discussed in detail by Krauss and Gauch (1959).

Mg deficiency symptoms generally show up rather clearly in the leaves, which is understandable, since the leaf is the plant organ in which assimilation of photosynthate occurs. As a rule, Mg deficiency leads to chlorosis of the leaves, a formation of pale yellow spots and streaks that results from local failure in the formation of green leaf pigment. The occurrence of chlorosis is not the only indication of Mg deficiency; often variegated coloration of the leaves as well as their decay are observed. Unlike Ca, Mg is readily mobile in plants; it moves from older to younger leaves under Mg deficiency, and the deficiency symptoms therefore show first on the older leaves of the plant as Mg is withdrawn from them.

PLANT REQUIREMENTS

Sulphur

Plants contain about the same amounts of S as P, the usual range being 0.2-0.5 percent on a dry weight basis (Stevenson and Cole, 1999). However, plants grown in soils well supplied with S often contain 1 percent or more, part of which is present as SO_4^{2-}. Legumes are particularly sensitive, but other crops, including cereals, brassica, and tea, are affected by S deficiency. Sulphur is of the same importance to trees as it is to most other plants (Lambert and Turner, 1998). The foliage SO_4^{2-} concentration can vary between 0.027 and 0.085 percent. However, S deficiency is known to occur at foliage SO_4^{2-} levels of 0.008 percent. When S availability is increased in soils, the relative amount of SO_4^{2-} as a proportion of total S increases in the plants. Because much of the organic S in plants is in the amino acids in proteins, and the protein N: organic S ratio in leaves is quite constant at about 15:1, protein synthesis is restricted in S-deficient plants.

The plant S pool in a given agroecosystem (from a few kilograms to almost 100 kg ha^{-1} $year^{-1}$) is determined by crop type, nutritional S status, and crop yield (Janzen and Ellert, 1998). The spatial variability of S uptake in a grazed pasture can be short range, depending on the influence of slope, aspect, soil moisture content, and the influence of stock grazing and camping behavior on soil fertility (Saggar, Hedley, et al., 1990; Saggar, Mackay, et al., 1990).

Calcium

To a large extent, Ca content of plants is genetically controlled, and little affected by the Ca supply in the root medium, provided Ca availability is adequate for normal plant growth (Loneragan and Snowball, 1969). Ca deficiency is characterized by a reduction in growth of meristematic tissues, in which the affected tissue becomes soft due to the dissolution of the cell walls. Calcium deficiency is rarely seen in field crops but is often observed in fruit crops such as apples. Of all the mineral elements, calcium has the greatest impact on the post-harvest quality of pipfruit (Terblanche and

Wooldridge, 1979). Ca deficiency resulting from an undersupply of Ca to fruit and storage tissues is often observed. In apples, the disease is called bitter pit, and in tomatoes it is known as blossom-end rot.

Magnesium

Plants differ markedly in their response to a Mg deficiency in soil. In general, buckwheat is sensitive, corn intermediate, and small grains, grasses, and clover are only slightly responsive to Mg fertilization. The deficiency is initially characterized by an interveinal chlorosis, although, in acute stages, the leaf may be generally deficient in both green and yellow pigments, and there may be necrosis in the areas of the leaf first affected by the deficiency. A deficiency of Mg also induces the formation of anthocyanins in some plant species such as cotton. One of the most important Mg deficiency diseases was identified in tobacco. Commonly referred to as "sand drown," it occurs when tobacco leaves contain less than 0.25 percent Mg on a dry weight basis. Similarly, when pasture concentration is less than 0.20 percent, Mg deficiency leads to "grass tetany" in grazing animals (McNaught and Dorofaeff, 1965; Bolan et al., 2000).

INTERACTIONS WITH OTHER NUTRIENTS

Sulphur

Sulphur, C, N, and P cycles are interdependent because of the stoichiometric relationship in organic matter, and the interactions of S with these elements have been extensively discussed elsewhere (see Ivanov and Freney, 1983; Bolin and Cook, 1983; Parton, Stewart, and Cole, 1988; Howarth and Stewart, 1992). The S cycle is also intimately related to the cycles of two essential trace elements, selenium (Se) and molybdenum (Mo) (Howarth and Stewart, 1992). The chemical forms of Se such as selenate, selenite, and organic and inorganic selenide are similar to sulphate, sulphite, and sulphides; in the selenide form, Se substitutes for S proteins. The presence of SO_4^{2-} in the soil generally reduces plant Se. This may either result from direct antagonism or may simply reflect a dilution of plant Se due to increased plant growth (Mikkelsen, Page, and Bingham, 1978). Similarly, SO_4^{2-} can inhibit the assimilation of selenate by some bacteria, and sulphate-reducing bacteria are capable of reducing selenate. Molybdate is very similar in stereochemistry to SO_4^{2-}, which can inhibit molybdate assimilation (Howarth and Cole, 1985).

Calcium

In acid soils, Ca deficiency may be induced as a result of high levels of soluble Al, which reduces Ca uptake. Liming to neutralize exchangeable Al usually supplies adequate Ca for growth. Organic complexes are potential repositories for Ca when acid soils are limed, with the native Al being replaced by Ca. The uptake of Ca can be competitively depressed by the presence of other cations such as K and NH_4, which are rapidly taken up by roots. Imbalances with Mg, NH_4, and K can create Ca availability problems under certain circumstances and Ca availability is then a function of the ratio of Ca to other cations in the soil solution (Simpson, Corey, and Summer, 1979). Continuous use of diammonium phosphate fertilizer (DAP), along with muriate of potash (KCl) in dairy pastures, causes Ca deficiency in pastures. Ca and Mg deficiency in pastures is likely to reduce the cattle's blood Ca and Mg concentration.

Magnesium

Magnesium concentration in plants is depressed by high concentrations of other cations, such as Ca, K, NH_4, and Al (Ellis, 1979). Field trials have indicated that continuous use of ammonium-based fertilizer such as diammonium phosphate (DAP) along with muriate of potash (KCl) in dairy pastures results in Mg deficiency in pastures (Bolan et al., 2000). Assuming a daily intake of 15 kg dry matter by dairy cattle, increasing levels of both DAP and KCl are likely to decrease the supply of Mg for the animal uptake. The addition of KCl increases the concentration of exchangeable K but decreases exchangeable Ca and Mg, resulting from the displacement of these cations by K ions added through the fertilizer. Magnesium ions released from fertilizers undergo cation exchange reactions similar to K ions. When excessive amounts of K fertilizer are added to soil there is a possibility that some of the Mg ions retained on the cation exchange sites will be replaced by K ions, inducing leaching of these cations.

Lime-induced reductions in tissue Mg level and Mg uptake by plants have been observed (Hossner and Doll, 1970; Juo and Uzu, 1977; Christenson and Doll, 1978). This results from an increase in Mg adsorption due to an increase in pH-dependent adsorption sites in soils containing variable charge components (Grove and Sumner, 1985). Liming not only creates new amorphous Al polymers, but also changes the charge character of their surfaces. More negative charge sites are formed on such variable surface charge materials, and cation adsorption would therefore be favored. Magnesium would be favored over other cations for continued adsorption by such materials because of the presence of Mg sink (Grove, Sumner, and Syers, 1981). Liming has been shown to increase the leaching potential of Mg mainly because the Ca added through lime exchanges with Mg on the soil surfaces, leading to the leaching of Mg in the soil solution. Similarly, the addition of Ca compounds such as gypsum and lime has been shown to decrease the concentration of Mg in soil solution (Edmeades and Judd, 1980).

ANIMAL REQUIREMENTS

Sulphur

The S requirements of sheep and cattle, expressed on a dietary basis, are 1.3 to 1.7 g S kg^{-1} dry matter. Like plants, most S in animals is present as S amino acids in the body proteins, including wool, hair, and milk. Other important S-containing compounds, sulphated polysaccharides, are present in cartilage, tendons, and heparin. Sulphur is metabolized in the grazing ruminant by a variety of biochemical pathways. In the rumen, both dietary inorganic and some organic S are reduced to sulphide S, which is then utilized by the rumen microflora to synthesize S amino acids for microbial protein (Bray and Till, 1975). Sulphur as microbial protein enters the small intestine, is broken down by proteolytic enzymes, and absorbed as S amino acids to be utilized by the animal tissue. The unabsorbed S is excreted in the feces as organic S, and the products of S metabolism within the animal (free or esterified SO_4^{2-}, taurine, thiosulphate, sulphocyanates, etc.) are excreted in the urine. Sulphur in urine is generally present as SO_4^{2-}-S (80-85 percent), ester-SO_4^{2-} (5-8 percent) and carbon-bonded (amino acids) (10-14 percent) (Georgievskii, 1981).

Calcium

Calcium, as one of the constituents of hydroxyapatite minerals, forms the matrix of bones and teeth. It is involved in nerve function, contraction of muscles, and blood clotting. Calcium-binding proteins such as calmodulin play a central role in cellular regulation in animals. A decrease in blood Ca levels in recently calved dairy cattle can cause a disorder known as milk fever, which is characterized by restlessness, muscle tremors, and sometimes coma. Animals with milk fever are treated with Ca borogluconate administrated subcutaneously or intravenously. The ratio of the different basic cations in herbage influences animals relative absorption of these cations. For example, the increased intake of K in New Zealand has been identified as the major cause of a decrease in the absorption of Ca in animals resulting in milk fever (Mason and Young, 1999).

The supply of Ca in animals is monitored from the dietary cation-anion difference (DCAD) values in the pasture:

$$DCAD = ([Na^+] + [K^+]) / ([Cl^-] + [SO_4^{2-}]) \quad (17.1)$$

where [ ] = milliequivalents kg^{-1} dry matter.

Dietary cation-anion balance is important because animals attempt to maintain systemic acid-base balance and osmotic pressure in order both to protect the integrity of cells and membranes and to optimize biochemical and physiological processes (Wilson, 1999). When herbage with excess cations over anions (positive DCAD) is fed to animals, the concentration of alkali ions such as bicarbonate increases in body fluids resulting in alkalosis. Conversely, when feed with a surplus of anions (negative DCAD) is ingested, the concentration of acidic hydrogen ions increases and metabolic acidosis occurs. There have been conflicting reports on the optimum levels of DCAD values required for dairy cattle (Roche, 1997). It has, however, been shown that diets with high DCAD values tend to increase the incidence of milk fever, and the supplementation of pre-calving rations with anionic salts (low DCAD values) reduces the incidence of milk fever (Beede, 1992; Wilson, 1996). In New Zealand, the DCAD values for most pastures range from 200 to 800 meq kg^{-1} DM, indicating a cation surplus for anions. This is normally the outcome of the luxury K uptake by pasture, resulting in a high concentration of K in the herbage (Wilson, 1996; Bolan et al., 2000).

Magnesium

Magnesium plays an important role in the enzymatic metabolism of carbohydrates, lipids, proteins, and nucleic acids. Generally, Mg is an activator for the numerous enzymes such as phosphatases and the enzyme-catalyzing reactions involving ATP (adenosine triphosphate), which split and transfer phosphate groups. It is also involved in nerve conduction and muscular contraction. Deficiency of Mg in blood plasma can cause a disorder known as hypomagnesemia (grass tetany or staggers), which usually occurs in dairy cows in the early part of lactation. Magnesium deficiency in animals can be overcome by regular use of Mg salts as a drench or water-trough treatment, as a lick, or in pasture after foliar application. In New Zealand, increased intake of K has been identified as the major cause for decreased absorption of Mg in animals, which results in tetany and convulsion. Irrigation of pasture with dairy shed effluents rich in K has been shown to increase the incidence of Mg deficiency, which leads to grass staggers.

In pasture and fodder crops, the grass staggers index (GSI) (Equation 17.2) is used to predict the chances for the occurrence of Mg deficiency.

$$GSI = [K^+]/[Ca^{2+}]+[Mg^{2+}] \quad (17.2)$$

where, [ ] = milliequivalents kg^{-1} dry matter.

A GSI value of >2.2 has been suggested to enhance the risk of grass staggers (Mason and Young, 1999), a condition generally linked with animal serum Mg levels less than 1.0-1.5 mg100 ml^{-1}, compared with normal levels of 1.7-3.0 mg100 ml^{-1}. Bolan et al. (2000) have obtained GSI values from 1.4 to 3.5, and the values increased with increasing levels of K through dairy shed effluent irrigation. Application of epsom salt decreased GSI values, mainly due to an increase in the concentration of Mg.

MODELING CONCEPTS AND APPLICATIONS

Sulphur

Agriculture in Australia and New Zealand is dominated by a legume-based pastoral agroecosystem, and most of the recent developments in modeling S research in pastoral systems that originated in these two countries have been discussed extensively elsewhere (McCaskill and Blair, 1988; Blair et al., 1993; Nyugen and Goh, 1994; Boswell and Gregg, 1998; Saggar, Hedley, and Phimsarn, 1998; Watkinson and Bolan, 1998). The focus of these models was to provide the quantitative information required for the design of fertilizer forms and management practices that will reduce S loss and increase S recycling.

A computerized model (Ministry of Agriculture and Fisheries [MAF], Soil Fertility Service [SFS] model) that calculates S mass balance for various pasture systems has been used to make maintenance S recommendation for pastures in New Zealand (Sinclair and Saunders, 1984). This model enables the formulation of fertilizer recommendations (amounts, forms, and timing of application) to replace net losses.

Increasing amounts of S° have been used as slow-release S fertilizers for pastures (Boswell and Gregg, 1998). As discussed earlier, the S° has to be oxidized to SO_4^{-}-S before being taken up by the plants. The rate of oxidation depends upon the size of S° particles. To improve selection of sizes of S° particles suitable for application, the mechanisms of S° oxidation and factors affecting S° oxidation have been modeled using empirical (Blair et al., 1993) and mechanistic models (Janzen and Bettany, 1987; Watkinson and Bolan, 1998). Janzen and Bettany (1987) have noted from the work of Fox et al. (1964) that the rate of S° oxidation was linearly related to the initial surface area of the particles. This suggests that as the surface decreases with oxidation, the rate of oxidation per unit surface area remains constant, i.e., the rate of oxidation is proportional to the surface area, as represented in the following equation:

$$dm/dt = -kS_a \quad (17.3)$$

where, m is the mass of S° that has dissolved from the surface after time t, k is the dissolution rate constant, and S_a is the total area of the spheres.

This is analogous to the approach used by Swartzenruber and Barber (1965) in modeling the dissolution of limestone in soils. This equation can be solved for conditions required by model of S° particles of increasing complexity of shape and size. For a simple system of particles of spherical size with equal size, the equation can be simplified to:

$$M/M_o = (1\text{-}kt/rD_o)^3 \quad (17.4)$$

where, M_o is the initial mass, M is the mass at time t, r is the density, and D_o is the initial radius of the spherical particle.

This equation has been very effective in predicting the rate of oxidation of S° under controlled conditions (Watkinson, 1989). Under field conditions, however, it may not be possible to maintain a uniform size and shape of S° particles in fertilizer materials. Watkinson and Lee (1994) have developed equations to calculate the oxidation rate of S° with a range of particle sizes and shapes (refer to the review by Watkinson and Bolan, 1998).

McCaskill and Blair (1988) have developed a pseudo-mechanistic computer simulation model for predicting perennial pasture growth and S uptake. The model calculates potential pasture growth achievable under optimum conditions, with temperature, moisture, light, and S as reduction factors. The transformations of S in the model are based on the dynamics of soil organic matter, but include plant uptake, fertilizer release, and leaching. Although the accuracy of the model has not been verified, it is the first attempt at a comprehensive model. Using an alternative approach to McCaskill and Blair (1988), Heng (1991) and Phimsarn (1991) have developed simple models of pasture S uptake using the relationships between actual daily evapotranspiration (AET), average root density in the soil profile, soil solution SO_4^{2-} levels predicted from extractable soil SO_4^{2-} levels, and soil SO_4^{2-} buffer capacities. The soil solution concentration for a given amount of extractable sulphate is given by:

$$C_i = (E_s/a_i)^{(1/bi)} \quad (17.5)$$

where C_i is sulphate concentration (kg S mm^{-1} depth ha^{-1}), E_s is the amount of extractable sulphate (kg S ha^{-1}), and a_i and b_i are coefficients of the Freundlich isotherm (empirical constants).

Using a more complicated water balance model, Heng (1991) produced a convection-dispersion model to predict S leaching from a permanent pasture. Both models (Heng, 1991; Phimsarn, 1991) were able to demonstrate the effect of different types of S fertilizers on SO_4^{2-} leaching losses. The models were, however, unable to predict the SO_4^{2-} leaching during periods of increased mineralization or after superphosphate application.

Saggar, Hedley, et al. (1990) and Saggar, Mackay, et al. (1990) used the mass balance approach to account for S inputs and outputs in grazed pastures. They applied a simple regression analysis to construct a model that described the influences of the slope and aspect of hill-country farms on the returns of plant S and P to the soil. Using this model, they were able to predict large losses of S from superphosphate-fertilized pasture on soils of low S retention and relatively smaller losses on soils of high S retention. They also found that the recovery of applied S was dependent on pasture growth rate and S uptake, transfer losses of excretal S to stock camps, S removal in animal products, S immobilization in organic matter, and the extent of S leaching. These studies suggest that S applied as superphosphate has not been utilized efficiently in New Zealand pastoral systems, especially where high rates of fertilizer have been applied. Using this mass-balance model approach for irrigated sheep-grazed pastures, Nguyen and Goh (1992) have estimated that 50 percent of S fertilizer applied at rates well above the pasture maintenance S requirements is lost to leaching. Sakadevan, Hedley, and Mackay (1993) have shown that a major proportion of annual S leaching loss in hill pastures is derived from the mineralization of soil organic matter. These studies demonstrate that excretal S transfers to stock camps and leaching losses beyond the plant rooting depth are the major S outputs from these grazed pastures.

McGill and Cole (1981) developed a conceptual model of S transformations in soil in which microbial C, N, and S transformations were linked, and ester-SO_4^{2-} and C-bonded S fractions were treated separately. They proposed two distinct mechanisms of organic S mineralization: the biological mineralization of C-bonded S resulting from C mineralization for energy demand, and the biochemical mineralization of ester-SO_4^{2-} regulated by microbial demand. Although this concept was useful in constructing computer simulations of S mineralization in soil, it did not account for the complexity of S mineralization processes (Janzen and Ellert, 1998; Saggar, Hedley, and Phimsarn, 1998).

How S processes have been formulated in a wide range of S models used for both grassland and forest systems has been reviewed by Mitchell and

Fuller (1988). They observed that no attention has been paid to organic S dynamics in models that have been applied to forest systems. Parton, Stewart, and Cole (1988) have developed the CENTURY model to simulate the dynamics of C, N, P, and S in cultivated and grassland soils. With further refinements, a complete ecosystem model has now been developed to simulate the long-term dynamics of C, N, P, and S for grassland, cropping, forest, and savanna systems (Metherell et al., 1993). In this model, different aboveground plant production submodels are linked to a common soil organic matter submodel that simulates the flow of C, N, P, and S through plant litter and inorganic and organic soil pools. However, the S submodel has not been as well tested as the N and P submodels. For example, even the P submodel in CENTURY is not able to describe P dynamics in tropical soils (Gijsman et al., 1996). CENTURY is designed more for calculating N, P, and S flux rates than for predicting their response.

Recent reductions in gaseous emissions from industrial sources in the northern hemisphere have resulted in more widespread incidences of S deficiency, but there appears to be little development and application of crop models that include S as a growth-limiting nutrient element. In many parts of the world, soil and plant S analyses are still used to make fertilizer S recommendations for a wide range of crops.

Calcium

Lime is used to overcome Ca deficiency and the problems associated with soil acidification. Lime produces alkaline hydroxyl ions (OH^-) that neutralize the acid H^+ ions and thereby decrease the activity of Al^{3+} and Mn^{2+}, and increase Ca^{2+}. There are a range of materials containing lime, which vary in their ability to neutralize acidity, including agricultural lime ($CaCO_3$), burnt lime (CaO), slaked lime ($Ca[OH]_2$), dolomite ($CaMg[CO_3]_2$), and slag ($CaSiO_3$). The acid neutralizing value of liming materials is expressed in terms of calcium carbonate equivalent (CCE). The amount of liming material required to correct soil acidity depends on a number of factors that include the neutralizing value of the liming material, the pH buffering capacity of the soil, the form and amount of fertilizer used, and the carrying capacity of the farm.

The neutralizing value of limestone particles is affected by both the rate of dissolution (which is highly sensitive to pH), and the rate of diffusion of Ca away from the particle surface. The rate at which the soil pH is affected by a lime application depends on the distance the diffusing Ca and OH ions must move before the zones of neutralization around individual limestone particles overlap. The rate of diffusion of Ca and OH ions in soil solution is very slow; hence, finely ground limestone uniformly mixed with soils provides more particles per unit volume of soil, less distance between particles, and therefore more rapid neutralization of the soil.

Swartzenruber and Barber (1965) presented a mathematical model to predict the rate of lime particle dissolution based on the concept of equal diameter reduction. The equal diameter reduction hypothesis assumes that (1) the initial mass of limestone is present as spheres of uniform size, density, and composition, and (2) the rate of loss of mass is directly proportional to the instantaneous surface area of the spheres. The mathematical formulation of the second assumption is similar to the one presented for S° (Equation 17.4). From Equation 17.4 they showed that:

$$M/M_0 = 1-(1-ct)^3 \qquad (17.6)$$

where M_0 is the initial mass, M is the dissolved mass, c is equal to $2k/rD_0$ where r is the density, and D_0 is the initial diameter of the spherical particle.

Commonly known as the >cube root equation=, this equation is widely used to predict the dissolution of insoluble fertilizer materials such as S° (Watkinson, 1989) and Mg (Mitchell et al., 1999) fertilizers.

Warfvinge and Svendrup (1989) developed a conceptual model for the dissolution of lime in soils. It incorporated the following four chemical reaction systems in soils:

1. Dissolution of lime in stagnant aqueous systems
2. Cation exchange reactions with Ca and exchangeable acidity
3. Leaching and accumulation of dissolved components
4. The carbonate equilibrium system

Based on this conceptual model, they also observed that the rate of dissolution depends on the particle size.

In most agricultural systems Ca is added through liming materials used to correct soil acidity. Because crops differ in their sensitivity to soil acidity, lime requirements also vary with crop. Lime requirement is generally calculated from the pH buffering capacity of the soil that is defined as the amount of base or acid required to change the pH by one unit per unit weight of soil. The pH buffering capacity is measured by allowing the soil to react with a buffer solution and, after a period of reaction, the equilibrium pH is measured. For example, the widely used Shoemaker-McLean-Pratt (SMP) method uses a mixed acetate, triethanolamine, p-nitrophenol, and chromate buffer that provides for a liner buffer response to acidity in the pH range from 5 to 7.5 (McLean, Terwiler, and Eckert, 1977). Possible equilibrium pH values versus the quantity of lime required to raise the soil pH to a desired value have been tabulated (Sims, 1996).

Since the pH buffering capacity of a soil is controlled by soil components, empirical relationships have been obtained between lime requirements and soil components. One such equation that is very commonly used in New Zealand to calculate lime requirement (LR) in pasture soils is given as follows (Cornforth and Sinclair, 1994):

$$LR\ (Mg/ha) = 28.7 - 4.8\ (\text{initial pH}) + 0.17(\%C) \quad (17.7)$$

Recently, the New Zealand Pastoral Agricultural Research Institute Limited developed decision support software programs (AgResearch PKS Lime Program) to calculate the lime response in terms of yield increase and economic return.

Magnesium

A number of soil test methods are used to predict Mg availability to plants. These include 1N ammonium acetate exchangeable Mg, percent CEC saturated with Mg, and indices of relative availability (pMg in soil solution, pMg / pCa in solution, 2 pMg / p K in solution) (Draycott and Durrant, 1970; Doll and Lucas, 1973; Ennde and Evers, 1997). The ability of these indices to predict the availability of Mg to plants varies depending on the relative concentration of Mg, Ca, and K in soil solution (Rahmatullah and Baker, 1981). Similar to K and Ca, the requirements for Mg application to pasture are based on the exchangeable soil Mg test. Field calibrations for Mg soil tests are available that enable the soil testing service to predict the Mg status of the soils and to make necessary Mg fertilizer recommendations.

Recently, Mitchell et al. (1999) examined the validity of the cube root equation (Equation 17.5) to predict the dissolution of a range of Mg fertilizers. They observed that within each size class, the changing surface area of the particles controlled the rate of dissolution of dolomite over time. The specific dissolution rates increased with decreases in particle size. Swartzenruber and Barber (1965) suggested that the variation of the specific rate of dissolution with particle size may be due to the geometric mean of the particle sizes not being a representative measure of the range of diameters of the limestone particles within each size fraction.

Deficiencies of nutrient elements such as Ca and Mg are less widespread and sufficiently infrequent than N, P, K, and S. Therefore, it would appear that these elements are not of wide enough interest to justify their inclusion in crop/forest models (Probert and Keating, 2000). Most of the modeling work on Ca and Mg in agricultural ecosystems is concentrated on their use as liming materials rather than essential plant nutrients.

Research in the past 10 to 20 years has led to increased understanding of S processes in agricultural and terrestrial ecosystems. However, there is limited knowledge of the processes controlling Ca and Mg dynamics. Challenges that confront the modelers of S, Ca, and Mg dynamics include insufficient field data to account for the complexity of processes.

SUMMARY

Modeling secondary nutrient processes in soil-plant systems is essential to improving the understanding of the key factors controlling their dynamics and to provide the quantitative information required for the design of fertilizer and management practices that will increase their agronomic effectiveness through reduced losses and increased recycling. Plant and soil S models developed over the last two decades have provided a greater insight into the fundamental processes. Sulphur applied as superphosphate is not being utilized efficiently in the grazed hill-country pastoral systems, especially where high rates of fertilizer are applied, and a major proportion of annual S leaching loss is derived from the mineralization of soil organic matter. From these grazed pastures, excretal S transfers to stock camps and leaching losses beyond the plant rooting depth are the major S outputs. There are few data available for a wide range of crops that are suitable for developing an understanding of soil organic S fractions responsible for short- and long-term S supply. Despite more widespread incidences of S deficiency due to decreased atmospheric inputs in the northern hemisphere, there appears to be little development and application of crop models that include S as a growth-limiting nutrient element. In many parts of the world, soil and plant S analyses are still used to make fertilizer S recommendations for a wide range of crops.

Ca and Mg are applied to agricultural systems mainly as components of liming materials such as agricultural lime and dolomite which are used to overcome soil acidity. Most modeling work involves the measurement of rates of dissolution of liming materials and the agronomic response obtained by overcoming the problems associated with soil acidification. A limited amount of modeling work has been done in relation to the value of Ca and Mg as plant nutrients mainly because the deficiencies of these nutrients are not widespread and these nutrients are not often included in fertilizer programs. As deficiencies of these two elements increase, more examination should be done on their dynamics in soil/plant/animal systems so that modeling work can be carried out to predict their requirements in agricultural systems.

REFERENCES

Aitken, R.L. and Scott, B.J. (1999). Magnesium. In *Soil Analysis: An Interpretation Manual,* eds. K.I. Peverill, L.A. Sparrow, and D.I. Reuter. Collingwood, Australia: CSIRO Publishing, pp. 255-262.

Augustin, S., M. Mindrup, and K.J. Meiwes (1997). Soil chemistry. In *Magnesium Deficiency in Forest Ecosystem,* eds. R.F. Huttl and W. Schaaf. London: Kluwer Academic Publishers, pp. 255-273.

Beede, E. (1992). The DCAD concept: Transition rations for dry pregnant cows. *Feedstuffs 64:* 12.

Benerjee, M.R. and S.J. Chapman (1996). The significance of microbial biomass sulphur in soil. *Biology and Fertility of Soils 22:* 116-125.

Blair, G.J., R.D.B. Lefroy, M. Dana, and G.C. Anderson (1993). Modelling of sulfur oxidation from elemental sulfur. *Plant and Soil 155/156:* 379-382.

Bolan, N. S., D.J. Horne, G.F. Wilson, D. Dawood, and N. Selvarajah (2000). Composition and nutritive value of pasture irrigated with farm effluents. In *Soil Research: A Knowledge Industry for Land-Based Exporters,* eds. L.D. Currie and P. Loganathan. Proceedings of workshop, Palmerston North, New Zealand: Massey University, pp. 157-164.

Bolin, B. and R.B. Cook (1983). *The Major Biogeochemical Cycles and their Interactions.* Chichester: Wiley.

Boswell, C.C. and P.E.H. Gregg (1998). Sulphur fertilizers for grazed pasture systems. In *Sulphur in the Environment,* ed. D.G. Maynard. New York: Marcel Dekker Inc., pp. 95-135.

Bray, A.C. and A.R. Till (1975). Metabolism of sulphur in the gastrointestinal tract. In *Digestion and Metabolism in Ruminant,* eds. I.W. McDonald and A.C.I. Warner. Armidale: The University of New England Publishing Unit, p. 243.

Bruce, R.C. (1999). Calcium. In *Soil Analysis: An Interpretation Manual,* eds. K.I. Peverill, L.A. Sparrow, and D.J. Reuter. Collingwood, Australia: CSIRO Publishing, pp. 247-254.

Christenson, D.R. and E.C. Doll (1978). Magnesium uptake from exchangeable and nonexchangeable sources in soils as measured by intensive cropping. *Soil Science 126:* 166-168.

Christiansen, M.N. and C.D. Foy (1979). Fate and function of calcium in tissue. *Communications in Soil Science and Plant Analysis 10:* 427-442.

Clarkson, D.T., M.J. Hawkesford, and J.-C. Davidian (1993). Membrane and long-distance transport of sulphate. In *Sulphur Nutrition and Assimilation in Higher Plants: Regulatory, Agricultural and Environmental Aspects,* eds. L.J. De Kok, I. Stulen, H. Rennenberg, C. Brunold, and W.E. Rauser. The Hague: SPB Academic Publishers, pp. 3-19.

Cornforth, I.S. and A.G. Sinclair (1994). *Fertilizer Recommendations for Pasture and Crops in New Zealand.* Wellington: Ministry of Agriculture and Fisheries.

Courchesne, F., M. Turmel, and P. Beauchemin (1996). Magnesium and potassium release by weathering in Spodosols: Grain surface coating effects. *Soil Science Society of America Journal 60:* 1188-1196.

Doll, E.C. and R.E. Lucas (1973). Testing soil for potassium, calcium and magnesium. In *Soil Testing and Plant Analysis,* eds. L.M. Walsh and J.D. Beaton. Madison, WI: Soil Science Society of America, pp. 133-151.

Draycott, A.P. and M.J. Durrant (1970). The relationship between exchangeable soil magnesium and response by sugar beet to magnesium sulphate. *Journal of Agriculture Science (Cambridge) 75:* 137-143.

Edmeades, D.C. and M.J. Judd (1980). The effects of lime on the magnesium status and equilibria in some New Zealand topsoils. *Soil Science 129:* 156-161.

Ellis, R. Jr. (1979). *Influence of Soil, Liming, Magnesium, Potassium and Nitrogen on Magnesium Composition of Plants,* Madison, WI: American Society of Agronomy, Special Publication.

Ennde, H.P. and F.H. Evers (1997). Visual magnesium symptoms and threshold values. In *Magnesium Deficiency in Forest Ecosystem,* eds. R.F. Huttl and W. Schaaf. London: Kluwer Academic Publishers, pp. 3-22.

Eriksen, J., M.D. Murphy, and E. Schnug (1998). The soil sulphur cycle. In *Sulphur in Agroecosystems,* ed. E. Schnug. Netherlands: Kluwer Academic Press, pp. 39-73.

Ferguson, I.B. (1979). The movement of calcium in non-vascular tissue of plants. *Communications in Soil Science and Plant Analysis 10:* 217-224.

Fitzgerald, J.W. (1976). Sulfate ester formation and hydrolysis: A potentially important yet often ignored aspect of the sulphur cycle of aerobic soils. *Bacteriological Reviews 40:* 698-721.

Fitzgerald, J.W. (1978). Naturally occurring organo sulphur compounds in soils. In *Sulphur in the Environment, Part II Ecological Impacts,* ed. J. O. Nriagu. New York: Wiley, pp. 391-443.

Fox, R.L., H.M. Atesalp, D.H. Kampbell, and H.F. Rhodes (1964). Factors influencing the availability of sulfur fertilisers to alfalfa and corn. *Soil Science Society of America Proceedings 28:* 406-408.

Freney, J.R. (1986). Form and reactions of organic sulphur compounds in soils. In *Sulphur in Agriculture,* ed. M.A. Tabatabai. Madison, WI: Agronomy Monograph 27, American Society of Agronomy, pp. 207-232.

Freney, J.R. and C.H. Williams. (1983). The sulphur cycle in soil. In *Global Biogeochemical Sulphur Cycle,* eds. M.V. Ivanov and J.R. Freney. New York: Wiley, pp. 129-201.

Georgievskii, V.I. (1981). The physiological role of macronutrients. In M*ineral Nutrition of Animals,* eds. V.I. Geogievskii, B.N. Annenkov, and V.T. Samokhin. London: Butterworth and Co., pp. 91-170.

Gijsman, A.J., A. Oberson, H. Tiessen, and D.K. Friesen (1996). Limited applicability of the CENTURY model to highly weathered tropical soils. *Agronomy Journal 88:* 894-903.

Grove, J.H. and M.E. Sumner (1985). Lime induced magnesium stress in corn: Impact of magnesium and phosphorus availability. *Soil Science Society of America Journal 49:* 1192-1196.

Grove, J.H., M.E. Sumner, and J.K. Syers (1981). Effect of lime on exchangeable magnesium in variable surface charge soils. *Soil Science Society of America Journal 45:* 497-500.

Hailes, K.J., R.L. Aitken, and N.W. Menzies (1997). Magnesium in tropical and subtropical soils from northeastern Australia. I. Magnesium fractions and interrelationships with soil properties. *Australian Journal of Soil Research 35:* 615-627.

Hanger, B.C. (1979). The movement of calcium in plants. *Communications in Soil Science and Plant Analysis 10:* 171-192.

Hell, R. and H. Rennenberg (1998). The plant sulphur cycle. In *Sulphur in Agroecosystems,* ed. E. Schnug. Netherlands: Kluwer Academic Press, pp. 135-173.

Heng, L.K. (1991). Measurement and modelling of chloride and sulphate leaching from a mole-drained soil. (PhD Thesis, Massey University, Palmerston North, New Zealand.)

Hossner, L.R. and E.C. Doll (1970). Magnesium fertilisation of potatoes as related to liming and potassium. *Soil Science Society of America Proceedings 34:* 772-774.

Howarth, R.W. and J.J. Cole (1985). Molybdenum availability, nitrogen limitation, and phytoplankton growth in natural waters. *Science 229:* 653-655.

Howarth, R.W. and J.W.B. Stewart (1992). Interactions of sulphur with other elements cycles in ecosystems. In *Sulphur Cycling on the Continents,* eds. R.W. Howarth, J.W.B. Stewart, and M.V. Ivanov. New York: Wiley, pp. 67-84.

Howarth, R.W., J.W.B. Stewart, and M.V. Ivanov (1992). *Sulphur Cycling on the Continents.* New York: Wiley.

Ivanov, M.V. and J.R. Freney (1983). *The Global Biogeochemical Sulphur Cycle.* Chichester: Wiley.

Janzen, H.H. and J.R. Bettany (1987). Measurement of sulfur oxidation in soils. *Soil Science 143:* 444-452.

Janzen, H.H. and B.H. Ellert (1998). Sulphur dynamics in cultivated, temperate agroecosystems. In *Sulphur in the Environment,* ed. D.G. Maynard. New York: Marcel Dekker Inc., pp. 11-44.

Juo, A.S.R. and F.O. Uzu (1977). Liming and nutrient interactions in two Ultisols from southern Nigeria. *Plant and Soil 47:* 419-430.

Krauss, R.W. and H.G. Gauch (1959). Roles of magnesium in plants. In *Magnesium and Agriculture,* eds. G.C. Anderson, E.M. Jencks, and D.J. Horvath. Proceedings of the Symposium, Morgantown, WV: West Virginia University, pp. 39-61.

Lambert, M.J. and J. Turner (1998). Sulphur nutrition and cycling in southern hemisphere temperate and subtropical forest ecosystems. In *Sulphur in the Environment,* ed. D.G. Maynard. New York: Marcel Dekker Inc., pp. 263-294.

Loneragan, J.F. and K. Snowball (1969). Calcium requirements of plants. *Australian Journal of Agriculture Research 20:* 465-478.

Mason, I. and B.S. Young (1999). The influence of potassium on crop quality and animal health in land treatment systems. In *Proceedings of 46th Conference of New Zealand Water and Waste Association.* Auckland, New Zealand: New Zealand Water and Waste Association, pp. 114-119.

Mayland, H.F. and C.W. Robbins (1994). Sulfate uptake by salinity-tolerant plant species. *Communications in Soil Science and Plant Analysis 25:* 2523-2541.

Mayland, H.F. and S.R. Wilkinson (1989). Soil factors affecting magnesium availability in plant-animal systems: A review. *Journal of Animal Science 67:* 3437-3444.

McCaskill, M.R. and G.J. Blair (1988). Development of a simulation model of sulphur cycling in grazed pastures. *Biogeochemistry 5:* 165-181.

McGill, W.B. and C.V. Cole (1981). Comparative aspects of cycling of organic C, N, S and P through soil organic matter. *Geoderma 26:* 267-286.

McLean, E.O., J.F. Terwiler, and D.J. Eckert (1977). Improved SMP buffer method for determination of lime requirement of acid soils. *Communications in Soil Science and Plant Analyses 8:* 667-675.

McNaught, K.J. and F.D. Dorofaeff (1965). Magnesium deficiency in pastures. *New Zealand Journal of Agriculture Research 8:* 555-572.

Metherell, A.K., L.A. Harding, C.V. Cole, and W.J. Parton (1993). CENTURY Soil Organic Model Environment. Technical Documentation, Agroecosystem Version 4.0, USDA-ARS, Colorado State University, Colorado.

Metson, A.J. and J.M. Brookes (1975). Magnesium in New Zealand soils. II. Distribution of exchangeable "reserve" magnesium in the main soil groups. *New Zealand Journal of Experimental Agriculture 18:* 317-335.

Mikkelsen, R.L., A.L. Page, and F.T. Bingham (1978). Factors affecting selenium accumulation in agricultural crops. In *Selenium in Agriculture and the Environment,* ed. L.W. Jacobs . Madison, WI: Soil Science Society of America, Special Publication Number 23, pp. 65-94.

Mitchell, A.D., P. Loganathan, T.W. Payne, and R.W. Tillman (1999). Magnesium fertiliser dissolution rates in pumice soils under *Pinus radiata. Australian Journal of Soil Research 38:* 753-767.

Mitchell, M.J., M.B. David, and R.B. Harrison (1992). Sulphur dynamics of forest ecosystems. In *Sulphur Cycling on the Continents,* eds. R.W. Howarth, J.W.B Stewart, and M.V. Ivanov. New York: Wiley, pp. 215-260.

Mitchell, M.J. and R.D. Fuller (1988). Models of sulfur dynamics in forest and grassland ecosystems with emphasis on soil processes. *Biogeochemistry 5:* 133-163.

Mokwunye, A. and S.W. Melsted (1972). Magnesium forms in selected temperate and tropical soils. *Soil Science Society of America Proceedings 36:* 762-764.

Nguyen, M.L. and K.M. Goh (1992). Nutrient cycling and losses based on a mass-balance model in grazed pastures receiving long-term superphosphate application in New Zealand. 2. Sulphur. *Journal of Agricultural Science, Cambridge 119:* 107-122.

Nguyen, M.L. and K.M. Goh (1994). Sulphur cycling and its implications on sulphur fertiliser requirements of grazed grassland ecosystems. *Agriculture Ecosystems and Environment 49:* 173-206.

Noggle, J.C., J.F. Meagher, and U.C. Jones (1986). Sulphur in the atmosphere and its effects on plant growth. In *Sulfur in Agriculture,* ed. M.A. Tabatabai. Madison, WI: Agronomy Monograph 27, American Society of Agronomy, pp. 251-278.

Parton, W.J., J.W.B. Stewart, and C.V. Cole (1988). Dynamics of C, N, P, and S in grassland soils: A model. *Biogeochemistry 5:* 109-132.

Phimsarn, S. (1991). A comparison of the fate of elemental sulphur and sulphate sulphur based fertilizers in pasture soils. (PhD Thesis, Massey University, Palmerston North, New Zealand.)

Probert, M.E. and B.A. Keating (2000). What should be included in crop and forest models? *Agriculture, Ecosystems and Environment 82:* 273-281.

Rahmatullah, H. and D.E. Baker (1981). Magnesium accumulation by corn (*Zea mays* L.) as a function of potassium-magnesium exchange in soils. *Soil Science Society of America Journal 45:* 899-903.

Rennenberg, H. (1984). The fate of excess sulphur in higher plants. *Annual Review of Plant Physiology 35:* 121-153.

Rice, H.B. and E.J. Kamprath (1968). Availability of exchangeable and non-exchangeable Mg in sandy coastal plain soils. *Soil Science Society of America Proceedings 32:* 386-388.

Roche, J. (1997). Dairy Research Institute Ellinbank, Australia Report, p. 65.

Saggar, S., J.R. Bettany, and J.W.B. Stewart (1981a). Measurement of microbial sulphur in soil. *Soil Biology and Biochemistry 13:* 493-498.

Saggar, S., J.R. Bettany, and J.W.B. Stewart (1981b). Sulphur transformation in relation to carbon and nitrogen in incubated soils. *Soil Biology and Biochemistry 13:* 499-511.

Saggar, S., M.J. Hedley, A.G. Gillingham, J.S. Rowarth, S. Richardson, N.S. Bolan, and P.E.H. Gregg (1990). Predicting the fate of fertilizer sulphur in grazed hill country pasture by modelling the transfer and accumulation of soil phosphorus. *New Zealand Journal of Agriculture Research 33:* 129-138.

Saggar, S., M.J. Hedley, and S. Phimsarn (1998). Dynamics of sulphur transformations in grazed pastures. In *Sulphur in the Environment,* ed. D.G. Maynard. New York: Marcel Dekker Inc., pp. 45-94.

Saggar, S., A.D. Mackay, M.J. Hedley, M.G. Lambert, and D.A. Clark (1990). A nutrient transfer model to explain the fate of phosphorus and sulphur in grazed hill-country pasture. *Agriculture Ecosystems and Environment 30:* 295-315.

Saif, H.T., N.E. Smeck, and J.M. Bingham (1997). Pedogenic influence on base saturation and calcium/magnesium ratios in soils of southeastern Ohio. *Soil Science Society of America Journal 61:* 509-515.

Sakadevan, K., M.J. Hedley, and A.D. Mackay (1993). Mineralisation and fate of soil sulphur and nitrogen in hill pastures. *New Zealand Journal of Agricultural Research 36:* 271-281.

Simard, R.R., C.R. deKimpe, and J. Zizka (1992). Release of potassium and magnesium from soil fractions and its kinetics. *Soil Science Society of America Journal 56:* 1421-1428.

Simpson, C.R., R.B. Corey, and M.E. Sumner (1979). Effect of varying Ca:Mg ratios on yield and composition of corn and alfalfa. *Communications in Soil Science and Plant Analysis 10:* 153-162.

Sims, J.T. (1996). Lime requirement. In *Methods of Soil Analysis. Part II. Chemical Methods,* ed. D.L. Sparks, Madison, WI: Soil Science Society of America, pp. 491-515.

Sinclair, A.G. and W.H.M. Saunders (1984). Sulfur. In *Fertilizer and Lime Recommendations for Pastures and Crops in New Zealand,* Second Edition, eds. I.S.

Cornforth and A.G. Sinclair. Wellington, New Zealand: Ministry of Agriculture and Fisheries, pp. 15-17.

Sinclair, A.H. (1980). Availability of magnesium to rye grass from soils during intensive cropping in the glasshouse. *Journal of Agriculture Science 96:* 635-642.

Smeck, N.E., H.T. Saif, and M. Bigham (1994). Formation of a transient magnesium aluminium double hydroxide in soils of southeastern Ohio. *Soil Science Society of America Journal 58:* 470-476.

Stevenson, F.J. and M.A. Cole (1999). *Cycles of Soil Carbon, Nitrogen, Phosphorus, Sulphur, Micronutrients.* New York: Wiley.

Swartzenruber, D. and S.A. Barber (1965). Dissolution of limestone particles in soil. *Soil Science 100:* 287-291.

Syers, J.K., R.J. Skinner, and D. Curtin (1987). Soil and fertilizer sulphur in UK agriculture. In *Proceedings of the Fertilizer Society London 264:* 43 pp.

Tabatabai, M.A. (1984). Importance of sulphur in crop production. *Biogeochemistry 1:* 45-62.

Terblanche, J.H. and L.G. Wooldridge (1979). The redistribution and immobilisation of calcium in apple trees with special reference to bitter pit. *Communications in Soil Science and Plant Analysis 10:* 195-215.

Thompson, J.F., I.K. Smith, and J.T. Madison (1986). Sulphur metabolism in plants. In *Sulphur in Agriculture,* ed. M. A. Tabatabai. Madison, WI: Agronomy Monograph 27, American Society of Agronomy, pp. 59-122.

Warfvinge, P. and H. Svendrup (1989). Modeling limestone dissolution in soils. *Soil Science Society of America Journal 53:* 44-51.

Watkinson, J.H. (1989). Measurement of the oxidation rate of elemental sulfur in soils. *Australian Journal of Soil Research 27:* 365-375.

Watkinson, J.H. and N.S. Bolan (1998). Modeling the rate of elemental sulphur oxidation in soils. In *Sulphur in the Environment,* ed. D.G. Maynard. New York: Marcel Dekker Inc., pp. 135-172.

Watkinson, J.H. and A. Lee (1994). Kinetics of field oxidation of elemental sulphur in New Zealand pastoral soils and the effects of soil temperature and moisture. *Fertilizer Research 37:* 59-68.

Wilson, G.F. (1996). Mineral imbalances in the diet of dairy cows. Effect of fertilisers and novel solutions to calcium and magnesium problems in dairy cows. *Proceedings Massey Dairy Farmers Conference 48,* 140 pp.

Wilson, G.F. (1999). The DCAD concept and its relevance to grazing sheep. In *Proceedings of Society of Sheep and Beef Cattle Veterinarians.* Palmerston North, New Zealand: Massey University, pp. 149-155.

Wu, J., A.G. O'Donnell, Z.L. He, and J.K. Syres (1994). Fumigation-extraction method for the measurement of microbial biomass-S. *Soil Biology and Biochemistry 26:* 117-125.

Chapter 18

Micronutrients and Toxic Elements

Ronald G. McLaren

In addition to the major nutrients, several other elements required in relatively small amounts, the micronutrients (Table 18.1), are also essential for the healthy growth and reproduction of plants and animals. Conversely, some of these micronutrients, if present in excessive concentrations, may actually be toxic to living organisms, e.g., copper (Cu) and zinc (Zn). In addition, other nonessential elements, such as lead (Pb), cadmium (Cd), mercury (Hg), and arsenic (As) (Table 18.1) can also have toxic effects on plants, livestock, and human health. Because of the predominantly metallic nature of many of these toxic elements they are commonly referred to as heavy metals, although the name is also frequently used to encompass nonmetals such as As and selenium (Se). As a result of the serious effects that these elements have on food production, and on human and environmental health, research into micronutrients and toxic elements in soils has been intense. A particular focus has been on attempting to predict and understand the occurrence of elemental deficiencies and toxicities. Modeling of the processes involving micronutrients and toxic elements in soils is increasingly being refined as a means of improving such predictions and for assessing the transport and impact of contaminant elements in the environment. In the past, both micronutrients and toxic elements have often been referred to as "trace elements," and to simplify the presentation of this review the term has been used where appropriate throughout this chapter.

SOURCES OF TRACE ELEMENTS IN SOILS

Natural Sources

For most soils under natural conditions, trace elements are inherited predominantly from the parent rocks from which they are developed. There is considerable information available on the variation in elemental composition among rocks, particularly with regard to trace elements (e.g., Mitchell,

TABLE 18.1. The Micronutrients and Toxic Elements

Essential micronutrients	Toxic elements
Essential micronutrients required by both plants and animals	*Toxic elements of greatest environmental significance*
Chlorine (Cl)	Aluminium (Al)
Cobalt (Co)[A]	Arsenic (As)
Copper (Cu)	Cadmium (Cd)
Iron (Fe)	Chromium (Cr)
Manganese(Mn)	Copper (Cu)
Molybdenum (Mo)	Fluorine (F)
Nickel (Ni)[AB]	Lead (Pb)
Silicon (Si)[AB]	Mercury (Hg)
Zinc (Zn)	Molybdenum (Mo)
Essential micronutrients required by plants only	Nickel (Ni)
	Selenium (Se)
Boron (B)	Zinc (Zn)
Essential micronutrients required by animals only	*Toxic elements of relatively low environmental significance* [C]
Chromium (Cr)	Antimony (Sb)
Iodine (I)	Beryllium (Be)
Selenium (Se)	Gold (Au)
Arsenic (As)[B]	Rubidium (Rb)
Fluorine (F)[B]	Silver (Ag)
Tin (Sn)[B]	Thallium (Tl)
Vanadium (V)[B]	Tin (Sn)
	Uranium (U)
	Vanadium (V)

Source: Mengel and Kirkby, 1978; Underwood, 1981; Kabata-Pendias and Pendias, 1992.

[A]Not shown to be essential for all plants.
[B]Essentiality observed only under highly specialized conditions. No practical significance in livestock nutrition.
[C]May be of importance in specific situations.

1964; Kraskopf, 1972). Some of these data are summarized in Table 18.2 which shows considerable variation in trace element concentrations among different types of both igneous and sedimentary rocks. In the case of young or immature soils in particular, it might be expected that the trace element content would strongly reflect those of the parent rocks. This is indeed the case, but in older soils the degree of weathering undergone during soil development may make the correlation less obvious (Mitchell, 1964).

TABLE 18.2. Mean Micronutrient and Toxic Metal Contents of Major Rock Types (mg kg^{-1})

	Igneous rocks			Sedimentary rocks		
	Ultra-mafic	Mafic	Felsic	Limestone	Sandstone	Shales
Ag	0.06	0.1	0.04	0.12	0.25	0.07
As	1	1.5	1.5	1	1	13 (1-900)
Au	0.003	0.003	0.002	0.002	0.003	0.0025
Cd	0.12	0.13	0.09	0.028	0.05	0.22 (<240)
Co	110	35	1	0.1	0.3	19
Cr	2980	200	4	11	35	90
Cu	42	90	13	5.5	30	39
Hg	0.004	0.01	0.08	0.16	0.29	0.18
Mn	1040	1500	400	620	460	850
Mo	0.3	1	2	0.16	0.2	2.6
Ni	2000	150	0.5	7	9	68
Pb	14	3	24	5.7	10	23
Sb	0.1	0.2	0.2	0.3	0.05	1.5
Se	0.13	0.05	0.05	0.03	0.01	0.5
Sn	0.5	1.5	3.5	0.5	0.5	6
Tl	0.0005	0.08	1.1	0.14	0.36	1.2
U	0.03	0.43	4.4	2.2	0.45	3.7
V	40	0.36	72	45	20	130
Zn	58	100	52	20	30	120

Source: Adapted from Alloway (1990, p. 31).

During the weathering process, rocks and constituent minerals are decomposed as a result of hydrolysis and redox reactions, etc., and trace elements may be released as soluble ions, some of which may be leached from the developing soil altogether. However, many ions will be reprecipitated as, or incorporated into, new secondary minerals such as phyllosilicate clays and Fe and Mn hydrous oxides and oxides. Other ions will be sorbed onto the surfaces of such materials, or be bound by accumulating organic materials. The net effect is that the concentration and forms of trace elements in mature soils may well differ substantially from those in the soil parent material.

Anthropogenic Sources

Atmospheric Deposition

Trace elements in the atmosphere are derived from both natural and anthropogenic sources (Table 18.3), including volcanic activity, wind-blown dusts, coal and refuse burning, metal smelting, and automobile emissions (Fergusson, 1990). Deposition from the atmosphere may take place close to the original source of contamination, e.g., around a smelter or, in other cases, the contaminant elements may travel vast distances before deposition. Even soils in rural areas are likely to be subjected to a continuous low-level input of anthropogenic trace elements from the atmosphere.

Agricultural Fertilizers, Manures, and Pesticides

Many fertilizers contain a range of elements in trace concentrations, including both essential micronutrients and potentially toxic elements (Swaine, 1962). In recent years, the accumulation of Cd in soils as a result of the long-term application of phosphate fertilizers has raised particular concerns (McLaughlin et al., 1996). Limestones are also known to contain trace concentrations of micronutrients and toxic elements.

Animal manures applied to the land can also be a source of toxic element contamination in some situations. Cu, Zn, and As have all been used in livestock production systems to promote animal growth, particularly with pigs and poultry, resulting in high concentrations of these elements in the animal manure.

The use of metal-containing pesticides in agriculture and horticulture has also resulted in the accumulation in the soil of elements such as Cu and Hg (in fungicides), and Pb and As (in insecticides).

TABLE 18.3. Natural and Anthropogenic Emissions of Trace Metals in the Atmosphere

	Emissions (10^8 g/yr)	
Element	**Natural[A]**	**Anthropogenic**
Ag	0.6	50
As	28.0 (210)[B]	780
Cd	2.9	55
Co	70	50
Cr	580	940
Cu	190	2600
Hg	0.4 (250)[B]	110
Mn	6100	3200
Mo	11	510
Ni	280	980
Pb	40	4000
Sb	9.8	380
Se	4.1 (30)[B]	140
Sn	52	430
V	650	2100
Zn	360	8400

Source: Adapted from Salomons and Förstner (1984, p. 101).

[A]Soil dust, volcanic dust, and volcanic emanation fluxes.
[B]Vapor emissions of volatile species from land and sea.

Urban and Industrial Wastes

Land treatment of waste materials is becoming widespread as a result of an increasing number of restrictions on waste disposal into rivers and the marine environment, and because of problems associated with the landfilling of wastes (McLaren and Smith, 1996). Although many of these wastes contain valuable nutrients and organic matter, some also contain a range of contaminants including toxic elements. Probably the waste of most concern is sewage sludge produced during the treatment of sewage from large urban and industrial areas. This waste is known to contain a whole range of micronutrients and toxic elements (McLaren and Smith, 1996).

However, other effluents and sludges produced by individual factories can also be sources of toxic metals, e.g., tannery wastes (Cr) and wastes from pulp and paper mills (Cd, Cr, Cu, Pb, Ni, Hg, and Zn).

Mining and Smelting

In addition to atmospheric contamination by smelters referred to previously, the metalliferous mining industry can also result in soil contamination with toxic elements in other ways. For example, toxic elements can be widely dispersed in dusts, effluents, and seepage water. Unfortunately, there are also many situations where tailings discharged into rivers have polluted alluvial soils downstream from mines during flooding. Probably the best-known example of this is the Cd contamination of soils in the Ichi and Maruyama river basins in Japan (Asami, 1981).

FORMS OF TRACE ELEMENTS IN SOILS

Irrespective of their original source, trace elements in soils occur in a variety of forms and associations (Table 18.4). The distribution of elements among these forms is very important for the processes influencing their bioavailability and mobility.

Soil Solution Forms

Concentrations of trace elements in soil solutions are generally extremely low, often just a few μg l^{-1} (below 10^{-5} to 10^{-6} M), even in contaminated soils (e.g., Kabata-Pendias and Pendias, 1992). Such concentrations represent a very small proportion of the total element concentrations in soil.

TABLE 18.4. The Forms of Micronutrients and Toxic Elements in Soil

1. Soil solution forms
2. Exchangeable ions
3. Ions specifically sorbed by inorganic colloids
4. Ions complexed or chelated by organic colloids
5. Elements present in decomposing organic materials and the soil biomass
6. Ions occluded by, or structural components of, secondary minerals and other inorganic compounds
7. Elements present in the structures of primary minerals

However, the soil solution is the most mobile pool of elements in the soil and the immediate source for uptake by plants and soil organisms. In the soil solution, elements occur either as simple ions (e.g., Cu^{2+}, Cd^{2+}, MoO_4^{2-}) or in the form of complexes with a range of inorganic and organic ligands. The dominant inorganic ligands involved in complexing trace elements are OH^-, HCO_3^-, CO_3^{2-}, Cl^-, and SO_4^{2-}, forming such complexes as $CrOH_2^+$, $PbHCO_3^+$, $CuCO_3^0$, $CdCl^+$, and $NiSO_4^0$ (Sposito, 1983). Potential organic ligands range from simple organic acids such as citric and oxalic, to complex humic and fulvic acid molecules. For some elements, Cu in particular, a high proportion of the total solution concentration is likely to be present in organically complexed forms (e.g., McBride and Blasiak, 1979).

Exchangeable Ions

Trace elements that occur as cationic species might be expected to take part in normal cation exchange mechanisms in the soil (i.e., nonspecific, electrostatic attraction of cations to negatively charged soil colloids). Similarly, elements that occur in anionic forms may interact with anion exchange sites. However, it seems unlikely that ion exchange is a particularly important mechanism for the trace elements (Swift and McLaren, 1991; Barrow, 1999). The concentrations of major nutrient ions (e.g., Ca^{2+}, Mg^{2+}, K^+, NO_3^-, SO_4^{2-}) in the soil solution are several orders of magnitude greater than those of the trace elements. Thus, mass action (competition) effects should prevent the occurrence of significant concentrations of trace elements on exchange sites. This is confirmed by the observation that only extremely small concentrations of exchangeable trace elements are determined in soils using standard techniques of extraction, i.e., extraction with salt solutions (McLaren and Crawford, 1973a; Shuman, 1979).

Ions Specifically Sorbed by Inorganic Soil Colloids

The sorption of trace elements by inorganic soil colloids such as aluminosilicate clays and iron (Fe), aluminium (Al), and manganese (Mn) oxides and hydroxides is considered to occur mainly by specific sorption mechanisms (Swift and McLaren, 1991). This term implies that the soil colloid shows a preference for the sorbed substance, and thus cations such as Cu^{2+} or Zn^{2+} may be sorbed despite the presence of much higher concentrations of Ca^{2+} or Mg^{2+} (Barrow, 1999). Unlike the mechanism of ion exchange that involves electrostatic bonding only, specific sorption involves the formation of a stable chemical bond between a trace element ion and certain functional groups at the surface of the soil colloid. Some examples proposed for the types of mechanisms involved are shown in Figure 18.1.

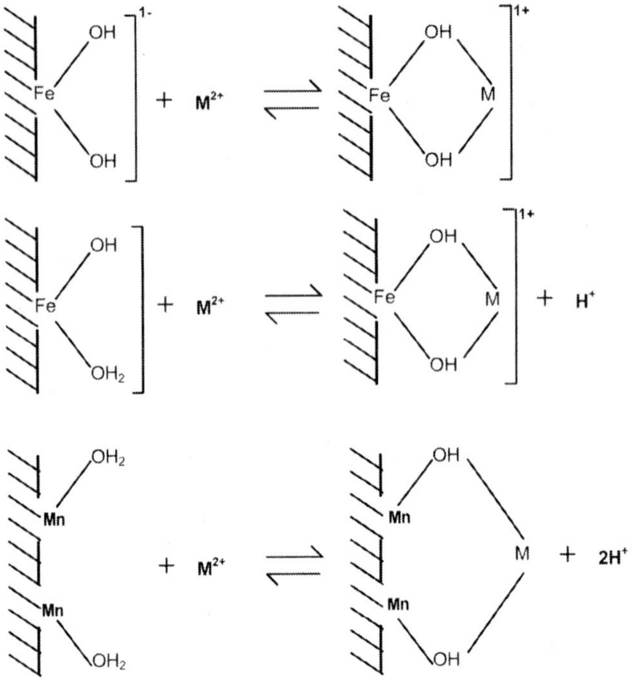

FIGURE 18.1. Some proposed mechanisms for the specific adsorption of trace element cations (M^{2+}) by inorganic soil colloids. (*Source:* McLaren and Cameron, 1996, p. 239.)

Ions Complexed or Chelated by Organic Colloids

As with inorganic colloids, soil organic colloids are able to bind substantial concentrations of trace elements, and with some elements (e.g., Cu), the bonds formed are very strong. It is generally assumed that bonding is the result of complex formation involving the various functional groups (carboxyl, hydroxyl, etc.) found in soil organic matter in general, and humic substances in particular (Stevenson, 1982). Figure 18.2 shows an example of the complexing of Cu involving carboxyl and phenolic hydroxyl groups. When ring structures are formed by this type of reaction, the element is said to be chelated. The complexation of trace elements by organic colloids can also be regarded as an example of a specific sorption mechanism.

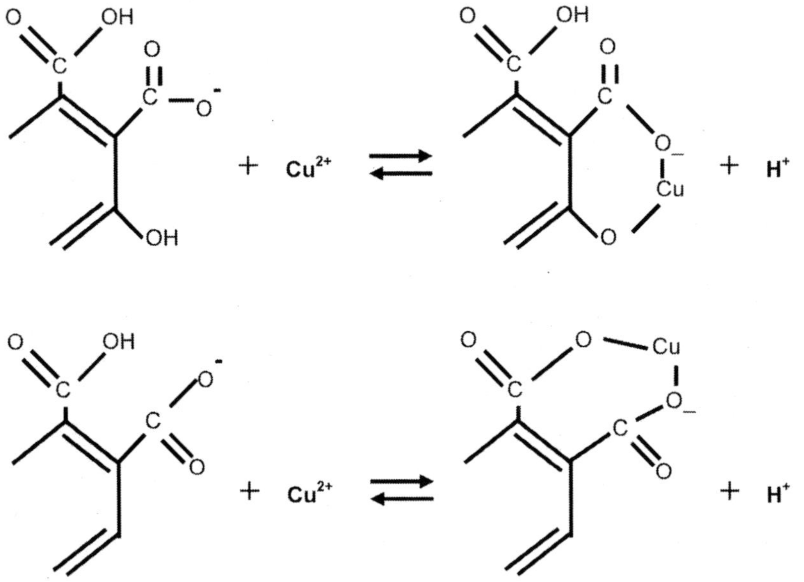

FIGURE 18.2. Complexing of copper by soil organic matter. (*Source:* McLaren and Cameron, 1996, p. 240.)

Elements Present in Decomposing Organic Materials and the Soil Biomass

All plant and animal material added to the soil contains a range of trace elements that will be released into the soil as the organic materials are decomposed. Since very few trace elements have significant structural functions in plants and animals (Bowen, 1966), it seems likely that they will be released into the soil fairly early in the decomposition process. Thus, apart from those elements sorbed by colloidal organic particles as described previously, the proportions of the total trace elements in the soil that remain associated with decomposing organic materials are likely to be relatively small.

Similarly, the concentrations of trace elements contained within living soil organisms would also seem likely to account for extremely small proportions of the total soil concentrations. For example, Sparling and Berrow (1985), using a fumigation-extraction technique, concluded that concentrations of trace elements in the soil microbial biomass were very small, even compared to those present as exchangeable ions. Even for earthworms, which can accumulate appreciable concentrations of trace elements, particularly in contaminated soils, the total amounts involved are very small compared with the total soil concentrations (Lee, 1985).

Ions Occluded by, or Structural Components of, Secondary Minerals and Other Inorganic Compounds

Trace elements sorbed onto the surfaces of Fe, Al, and Mn oxides may eventually be occluded by further growth of the oxides. Alternatively, trace elements may be substituted in the structures of oxide minerals during their formation in soils. For example, there is good evidence for the substitution of trace metals in the structures of a range of Fe oxides and oxyhydroxides (e.g., Gilkes and McKenzie, 1988; Cornell and Schwertmann, 1996). In some cases, it may also be possible that trace elements are precipitated in the form of simple inorganic compounds such as carbonates, phosphates, sulphides, or hydroxides. However, this is most likely to happen only in heavily contaminated soils, under waterlogged or high pH conditions, or in calcareous soils. For most soils there is little evidence for the existence of such compounds (Swift and McLaren, 1991).

Elements Present in the Structures of Primary Minerals

As noted, in the absence of anthropogenic inputs, trace elements in soils are derived from the parent rocks from which they are developed. Many soils, particularly young, poorly weathered ones, will still contain a proportion of the primary minerals inherited from these rocks.

FACTORS AFFECTING THE SOLUBILITY AND BIOAVAILABILITY OF TRACE ELEMENTS

To be available for uptake by plants (or soil microorganisms), trace elements must be present in the soil solution. The ability of the soil to control trace element concentrations in the soil solution is a major factor affecting their availability to plants. Figure 18.3 schematically shows the relationships and equilibria between trace elements in the soil solution and the other forms in the soil.

Solution concentrations are essentially controlled in two main ways: (1) by the solubility of solid phase compounds containing the element of interest, and (2) by sorption/desorption reactions at the surfaces of soil colloids. Solubility of their respective oxides is probably the main factor controlling solution concentrations of Al, Fe, and Mn, but for all other trace

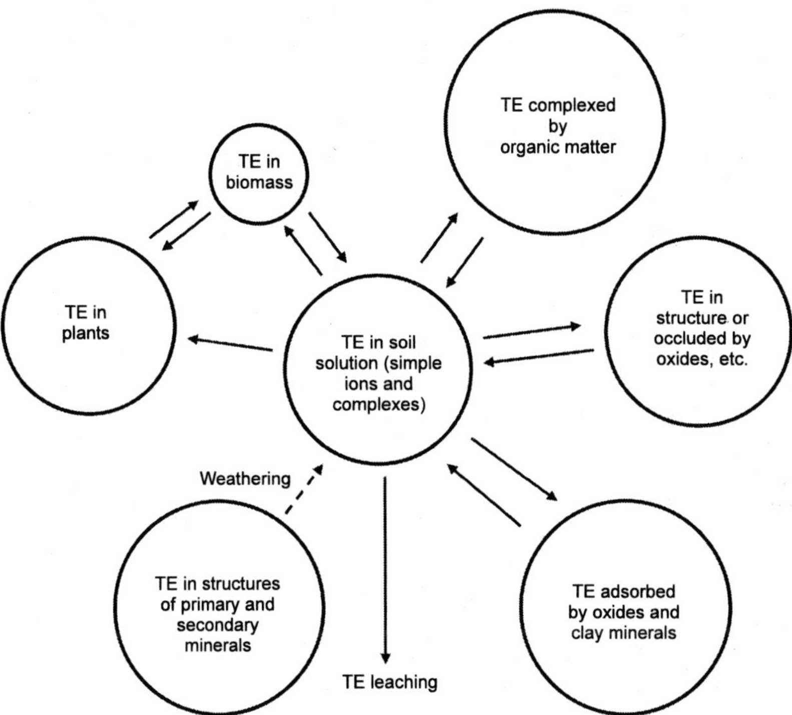

FIGURE 18.3. Relationships between forms of trace elements (TE) in soils. (*Source:* Adapted from McLaren and Cameron, 1996, p. 241.)

elements, sorption/desorption reactions are likely to be the major types of mechanisms involved (Swift and McLaren, 1991). Those trace elements present in the crystal structures of primary and secondary silicate minerals, or occluded by oxides (unless soil conditions favor their dissolution), will clearly be unavailable for plant uptake.

The role of the soil biomass in trace element availability is by no means clear. As discussed previously, only small amounts of trace elements will be present in the biomass, but the mineralization of micronutrients from decaying organic materials, and the continual cycling of trace elements through the soil microbial population could have a significant effect on maintaining solution concentrations. In addition, the conditions within the root rhizosphere may be important and are discussed later.

Chemical Speciation of Trace Elements in the Soil Solution

There is considerable evidence that the chemical speciation of trace elements in solution affects their availability and/or toxicity to plants (Parker, Chaney, and Norvell, 1995). For example, in many studies, plant uptake of the trace metals has shown a high level of correlation with the activity of free, uncomplexed metal ions in solution, e.g., Cu^{2+} (Graham, 1981) and Cd^{2+} (Cabrera, Young, and Rowell, 1988), rather than with total element concentrations. Such studies have been interpreted as demonstrating that complexed forms of trace elements in solution are unavailable for plant uptake. However, other studies have suggested that this is not necessarily the case. For example, over 40 years ago, DeKock and Mitchell (1957) demonstrated increased uptake of Al and other trivalent metals by chelators such as ethylenediaminetetraacetic acid (EDTA) and diethylenetriaminepentaacetic acid (DTPA). More recently, Smolders and McLaughlin (1996) have demonstrated that Cl-complexed Cd may also be available for plant uptake (in addition to Cd^{2+}).

Total Trace Element Content

Clearly, soluble trace element concentrations in soils are likely to be influenced to some extent by the total concentrations of trace elements present in the soil. Thus in uncontaminated soils, trace element bioavailability is likely to be related to the nature of the soil parent material and the degree of soil weathering. This is the case with micronutrient deficiencies common in soils derived from felsic igneous rocks, sandstones, and highly leached podzolic soils. In the case of contaminated soils, solution trace element concentrations are also likely to increase with the total contaminant loading. However, the source of the contaminant trace elements may also be important. For example, trace elements added to the soil as simple metal salts may initially be much more soluble than those applied in materials such as sewage sludge.

Soil Composition

Sorption/desorption reactions at the surfaces of soil colloids can play a substantial role in controlling the solution concentrations of many trace elements. The presence in the soil of those colloids able to sorb trace elements will therefore have a major influence in controlling trace element availability to plants. Soil organic matter in particular has a large capacity to sorb or complex trace elements, and in many studies this has been observed to be a dominant soil constituent affecting sorption, e.g., for Cu (McLaren and Crawford, 1973b), for Cd (Gray et al., 1998), and for Hg (Yin et al., 1996).

Trace elements may also be sorbed by clay minerals and the oxides of Fe, Al, and Mn. However, the results of many studies suggest that, in the short-term at least, these constituents play a relatively minor role in maintaining solution trace element concentrations compared to the overriding dominance of soil organic matter. The lack of significant correlations between trace element sorption and soil clay and oxide (Fe, Al, Mn) contents could be due to the low amounts of these constituents, and therefore their small contribution to the colloidal fraction of the soil compared to organic matter. Alternatively, the effects of these materials may actually be masked by coatings of organic matter.

There are undoubtedly some exceptions to this generalization. For instance, in some soils, Mn oxides appear to have a strong influence on Co solution concentrations, and Co availability to plants (Tiller, Honeysett, and Hallsworth, 1968). In other soils, Zn sorption is strongly influenced by clay content (Elrashidi and O'Connor, 1982). Certainly, in the long-term, soil oxide fractions do accumulate trace elements and therefore may act as substantial reservoirs for trace elements, without necessarily having a direct influence on solution concentrations.

Soil pH

Soil pH is recognized as having a major influence on the availability of trace elements. For those trace elements that occur predominantly as cations (e.g., Cu^{2+}, Co^{2+}, Pb^{2+}, etc.) availability to plants is highest in acid soils (possibly causing toxicities), and decreases as the soil pH increases. Conversely, trace elements such as Mo and Se that occur as anions are most available in soils of high pH and least available in acid soils.

The influence of soil pH on trace element availability is due mainly to its effect on the reactions controlling trace element concentrations in the soil solution. Under acid conditions, sorption of trace element cations by soil colloids is at a minimum, and thus solution concentrations are relatively high. In addition, the solubilities of Fe and Mn oxides are high under low pH conditions. As soil pH rises, sorption of trace element cations increases and the solubility of oxides decreases. The sorption of those elements that occur in anionic forms decreases with increasing soil pH, and hence solution concentrations and availability increase.

A complicating factor with certain trace element cations (e.g., Cu and Pb) is that, as soil pH increases, metal solubility reaches a minimum between pH 6 and 7 and then rises again at even higher pH values (e.g.,

McBride and Blasiak, 1979; Bruemmer, Gerth, and Herms, 1986). This is mainly due to increased solubility of organic matter at high pH, causing the retention of trace elements in solution in the form of soluble organic complexes.

Soil Redox Conditions

The trace elements Fe and Mn can occur in more than one oxidation state. In freely drained aerobic soils they occur predominantly in their higher oxidation states as various Fe and Mn oxide minerals. Such minerals are relatively insoluble, and hence the availability of Fe and Mn to plants is at a minimum. As soils become anaerobic due to waterlogging, the redox potential decreases and transformation to the more soluble reduced forms of these elements (Fe^{2+} and Mn^{2+}) takes place. This effectively results in the dissolution of oxide materials, and substantial increases in solution concentrations and bioavailability of Fe and Mn. In addition, since substantial concentrations of many other trace elements are often sorbed or occluded by Fe and Mn oxides in soils (e.g., Le Riche and Weir, 1963; Taylor and McKenzie, 1966), solubilization of the oxides also releases these other elements into solution.

Root Rhizosphere and Mycorrhizal Effects

Plant roots can drastically alter the physical and chemical properties in their immediate vicinity, i.e., the rhizosphere. The processes taking place in the rhizosphere can have a major influence on the dynamics of trace elements in soil surrounding plant roots, and therefore on their availability to the plant (Hinsinger, 2001). In particular, root-induced acidification or alkalinization of the rhizosphere can affect solid/solution phase trace element equilibria and chemical speciation, including the dissolution of soil minerals.

For those trace elements that can occur in various oxidation states (e.g., Fe, Mn, Cr, As), changes in redox potential in the rhizosphere may also play a role in their mobilization or immobilization. Complexation or chelation of trace elements by root exudates is yet another way in which rhizosphere processes can affect trace element solubility, speciation, and thus plant availability.

Processes involving trace elements in the rhizosphere are further modified by the presence of mycorrhizal fungi. Indeed it may be difficult to distinguish between the effects of plant roots and those due to mycorrhizae since both are able to alter rhizosphere pH, release exudates, and solubilize trace element-containing minerals.

Desorption

The immediate bioavailability of trace elements depends primarily on their concentration and speciation in the soil solution, i.e., the intensity factor of supply. However, continuing bioavailability depends on the soil's ability to release trace elements from the solid phase to replenish those removed by plant uptake, i.e., the quantity or capacity factor of supply. It is now generally accepted that, in the medium- to long-term, solution concentrations of trace elements are most likely to be controlled by sorption-desorption reactions at the surfaces of soil colloids (Swift and McLaren, 1991).

Desorption of trace elements previously sorbed onto intact soils and individual soil constituents (organic matter, oxides, and clays) shows a variety of trends, depending on the natures of both the surface and the trace element being studied. These trends range from (a) complete desorption as observed for example, in the case of Se (as selenite) desorption from gibbsite (Hingston, Posner, and Quirk, 1974), and B desorption from the clay minerals illite and montmorillonite (Hingston, 1964), to (b) significant desorption but with a proportion of trace element retained as observed for Co sorbed by humic acid (McLaren, Lawson, and Swift, 1986), and (c) minimal desorption, with a high proportion retained by the surface, for example, Cu and Co sorbed by soil oxide materials (McLaren, Williams, and Swift, 1983; McLaren, Lawson, and Swift, 1986). Many trace elements fall into category (c), appearing to be strongly retained by soils and showing limited reversibility of sorption (desorption).

As with sorption, desorption of trace elements from soils is affected markedly by soil pH. In particular, desorption of both native and applied trace metals have been observed to decrease with increasing soil pH (e.g., Gray et al., 1998). Figure 18.4 shows the effect of pH on the desorption of native soil Cd caused by repeated equilibration of soil in Cd-free weak electrolyte (0.01 M $Ca[NO_3]_2$).

In the case of trace elements added to soils as fertilizers or contaminants, there is also good evidence that with increasing contact time (aging) between soil and trace element, the ability of the element to desorb from the soil may decrease (Barrow, 1986b; Gray et al., 1998). This phenomenon has been linked to observations that, following the rapid sorption of trace elements by soil oxide materials, continuing slow reactions between the trace elements and the oxides take place (Bruemmer, Gerth, and Tiller, 1988;

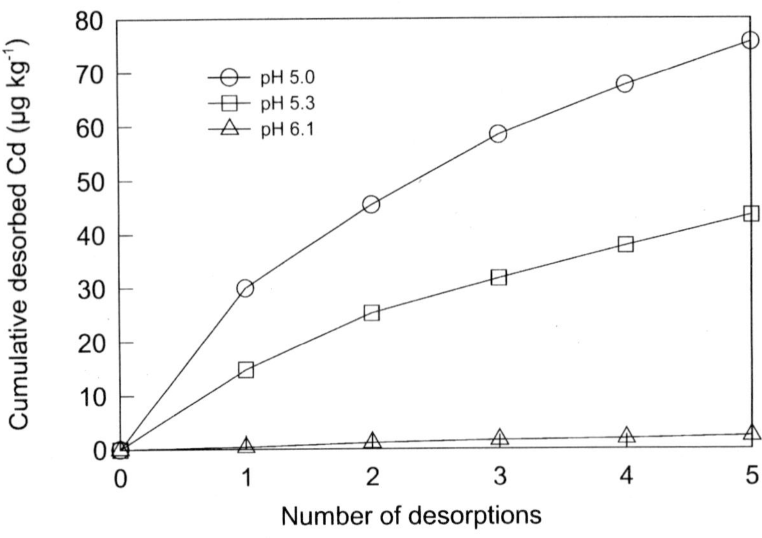

FIGURE 18.4. Effect of pH on desorption of native Cd from Te Kowhai silt loam. (*Source:* Adapted from Gray et al., 1998, p. 203.)

Backes et al., 1995). Similar aging effects have been observed for the desorption of Cd and Co from soil clay fractions (McLaren, Backes, et al., 1998), and for the desorption of Cu from humic acids (Rate, McLaren, and Swift, 1993). Whether such processes involve solid-state diffusion of trace elements into the lattice structures of oxides or clays, which would be expected to be extremely slow, or diffusion of ions into very small pores and interparticle spaces, remains to be determined (McBride, 1991). In the case of trace element (Cu) binding by humic acids, it has been suggested that the decrease in desorption on aging is due to slow complexation-induced conformational changes of the humic molecules (Rate, McLaren, and Swift, 1993).

Whatever the mechanisms, there is evidence to link such slow reactions with decreased availability of trace elements to plants. For example, Brennan, Gartrell, and Robson (1980) have reported reduced Cu availability to plants with increasing contact time between added Cu and the soil. Similar observations have been made by Brennan (1990) in relation to the availability of Zn to plants. More recently, Hamon et al. (1998) have produced evidence that Cd added to soils in superphosphate fertilizer becomes less bioavailable with time.

MOBILITY OF TRACE ELEMENTS IN SOILS

With some exceptions, trace elements tend to be sorbed relatively strongly by soils, and only small concentrations are present in the soil solution. They are therefore generally considered to be rather immobile in soils and their leaching downward through the soil profile is minimal. However, evidence is accumulating that the leaching of both trace element cations and anions does occur, and in some circumstances may result in significant movement of trace elements down through the soil profile and into the groundwater.

Evidence for Leaching

The suggestion by many researchers that no substantial long-term leaching losses of trace elements occur from surface soils contaminated by inorganic or organic (e.g., sludges) sources of trace elements, is often based on the observation that little trace element accumulation is observed in the subsoil. For example, studies by Chang et al. (1984) and Williams et al. (1985) both showed that, despite large applications of sewage sludge to soils, most trace metals did not appear to move below the depth of sludge incorporation. However, this assumption ignores the possibilities that (1) drainage leachate may move through preferential flow channels in the subsoil, thus bypassing most of the soil mass, and (2) sorption of some trace elements in the subsoil may be minimal due to the dominance of organically complexed or colloid-associated forms. The possibility that trace elements might be transported to lower depths in the soil through cracks and macropores has been suggested by Dowdy et al. (1991). Such a process could be conducive to significant leaching of trace elements without markedly increasing concentrations in the subsoil (Camobreco et al., 1996).

Apart from data from field studies, knowledge of metal movement through soils is based primarily on studies using homogenized, repacked soil columns (Camobreco et al., 1996). Such studies, even when using coarse-textured soils, have tended to support the contention that cationic trace elements are basically immobile. In recent years, however, studies using undisturbed soil columns (or lysimeters) have demonstrated the potential importance of preferential flow for trace element leaching. For example, Camobreco et al. (1996) compared the movement of Cd, Zn, Cu, and Pb through undisturbed and homogenized soil columns. The homogenized columns retained all added metals, whereas substantial amounts of all four metals were present in the effluent from the undisturbed columns. More recently, McLaren et al. (1999) have demonstrated increased metal leaching from undisturbed soil lysimeters treated with metal-spiked sewage sludge.

In this study there were clear examples of preferential flow of metals taking place, with increased metal concentrations in leachates occurring well before a pore volume of water had passed through the lysimeters. For example, Figure 18.5 shows the cumulative leaching of Ni from control and sludge-treated lysimeters of a sandy forest soil.

For some trace elements, principally those occurring in anionic forms, leaching may take place much more rapidly than for the trace element cations. In the case of As, there are certainly several good examples of substantial leaching through soils. For example, Peryea and Creger (1994) have observed considerable movement of As to depths greater than 120 cm in soils contaminated with lead arsenate pesticide residues. Similarly McLaren, Naidu, et al. (1998) have reported substantial downward movement of As in soils at cattle dip sites in Australia. In studies with undisturbed soil columns contaminated with timber preservative containing Cu, Cr, and As, it has also been shown that substantial amounts of Cr (as dichromate) and As (as arsenate) can be leached through the soil (Carey et al., 1996).

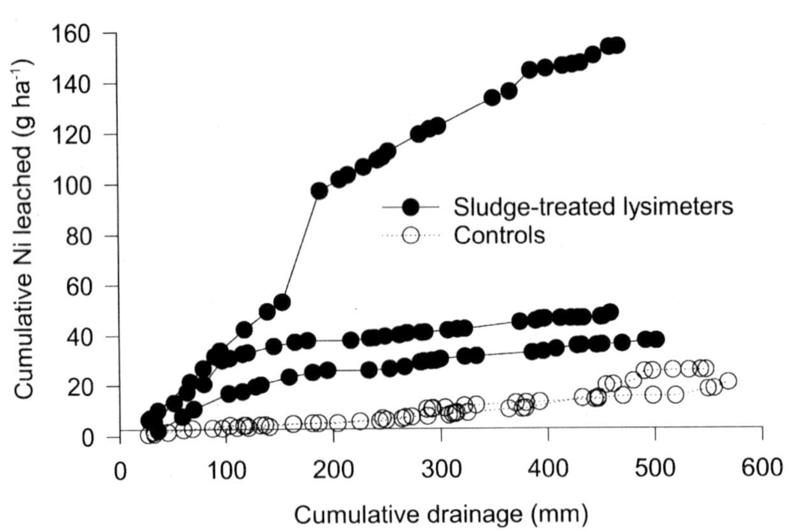

FIGURE 18.5. Cumulative Ni leached from control and sludge-treated lysimeters of Waikuku soil. (*Source:* Adapted from McLaren et al., 1999, p. 256.)

MODELING OF TRACE ELEMENT SPECIATION IN SOIL SOLUTION

Although physical methods are available for the direct determination of some individual trace element species (e.g., specific ion electrodes, ion chromatography, spectrophotometry, polarography), many have poor precision, lack the required sensitivity, and are extremely time consuming to undertake. For many trace element species there are indeed no reliable methods for direct determination. Against this background, a number of computer-based chemical equilibrium models have been developed to predict the chemical speciation of trace elements (and major elements) in soil solutions (e.g., Sposito and Mattigod, 1980).

The input data required to use these models consists of the total trace element and ligand concentrations in soil solution, and other relevant determinations such as pH, pE, or the partial pressure of CO_2. The models commonly use the mass action principle that relates the activities of free trace element ions, free ligands, and the various trace element-ligand complexes by means of known equilibrium constants. The various mass action relationships are linked with mass balance equations, resulting in a set of linear equations. The equilibrium speciation of each component is then obtained by solving the set of linear equations (Mattigod, 1995). Since soil solutions are in contact with soil particles, some models also include simulation of processes such as solubility equilibria, including precipitation and dissolution reactions, and sorption by colloid surfaces. An excellent description of the basic chemical principles and various equations used in predicting chemical speciation in solution is given by Lumsdon and Evans (1995).

Probably the best-known and most widely used computer speciation models are GEOCHEM (Sposito and Mattigod, 1980) and its descendent GEOCHEM-PC (Parker, Norvell, and Chaney, 1995), and MINTEQA2 (Allison and Brown, 1995). Although each of these programs contains a large database of equilibrium constants for the formation of all potential complexes between up to 50 cations and 99 different ligands, users select only those components relevant to a specific problem. Indeed, output from the models suggests that for most trace elements, speciation in soil solution is dominated by relatively few principal species (Table 18.5). However, it would be fair to say that evaluation of speciation models has not been extensive, mainly because of the lack of suitable methods for the determination of many chemical species. Certainly some limitations to computer speciation modeling are recognized. Some of the major ones include the lack of reliable thermodynamic data for some of the species suspected of being present in soil solutions, problems with the reliable determination of carbonate

TABLE 18.5. The Main Chemical Species of Some of the Common Trace Elements in Soil Solutions (Oxidizing Conditions)

Trace element	Species
Cd	Cd^{2+}, $CdSO_4^0$, $CdCl^+$, $CdHCO_3^+$
Cr	$CrOH_2^+$, CrO_4^{2+}, CrO_4^{2-}, $Cr(OH)_4^-$
Cu	Cu-Org, Cu^{2+}, $CuCO_3$
Fe	Fe^{2+}, $FeSO_4^0$, $FeH_2PO_4^+$, $FeOH^{2+}$, $Fe(OH)_2$, $FeCO_3^0$, $FeHCO_3$, $Fe(OH)_3^0$, Fe-Org
Mn	Mn^{2+}, $MnSO_4^0$, Mn-Org, $MnCO_3^0$, $MnHCO_3^+$
Mo	$H_2MoO_4^0$, $HMoO_4^-$, MoO_4^{2-}
Ni	Ni^{2+}, $NiSO_4^0$, $NiHCO_3^+$, Ni-Org, $NiCO_3$
Pb	Pb^{2+}, Pb-Org, $PbSO_4^0$, $PbHCO_3^+$, $PbCO_3^0$, $Pb(CO3)_2^{2-}$, $PbOH^-$
Zn	Zn^{2+}, $ZnSO_4^0$, $ZnHCO_3^+$, $ZnCO_3^0$

Source: Adapted from Alloway (1990, p. 19).

Org = organic complexes

equilibria (Suarez, 1995), and difficulties in characterizing the organic ligands present in soil solutions. This latter problem is of particular importance for many trace elements, since they are known to form strong complexes with a variety of organic ligands. Most users of speciation models attempt to overcome this problem by assuming that the dissolved organic carbon in their samples is present either (1) as a defined mixture of simple organic acids whose reactions with trace elements are well characterized, or (2) as fulvic acids for which stability constants of complexes with trace elements have been previously determined (e.g., Sposito et al., 1982).

MODELING THE DISTRIBUTION OF TRACE ELEMENTS BETWEEN SOIL SOLUTION AND SOLID PHASES

In this chapter, the accumulation of trace elements by the soil solid phase has been referred to by the general term "sorption." Sposito (1984) has suggested the use of this term to avoid the implication that accumulation was due to either adsorption or precipitation processes. However, the nature of the association of trace elements with the solid phase assumes particular importance when attempting to model the interrelationship with trace elements in soil solution. The possible occurrence of trace elements as discrete precipitates, or as ions adsorbed on the surfaces of soil colloids, certainly influences the type of model required to explain solution-solid phase trace element relationships.

The Solubility Product Model

This approach considers that soil solution trace element concentrations are controlled by the solubility of solid-phase compounds (precipitates, minerals) containing the trace element of interest. This hypothesis, as discussed by Lindsay (1972), states that whenever the concentration of a trace element in the soil solution falls below the equilibrium solubility of a solid phase compound of which the trace element is a constituent, that phase dissolves. Conversely, if the concentration of a micronutrient exceeds the equilibrium concentration of a compound, that compound begins to precipitate. In this way, solid phase compounds buffer trace element concentrations in the soil solution. Unfortunately, apart from Fe, Al, and Mn, the existence of definite mineral forms of trace elements in the soil has not been demonstrated convincingly, although some exceptions may occur in grossly contaminated soils (e.g., lead carbonates in soils contaminated with lead shot) (Storgaard Jorgensen and Willems, 1987).

In general, soil solution concentrations of trace elements in many soils tend to be much lower than those predicted from the known solubilities of compounds that might be expected to occur in soils (Brümmer et al., 1983; Brümmer, Gerth, and Herms, 1986). Figure 18.6 illustrates this point for Cu and Zn. Below pH 6-7, almost all of the observed soil solution concentrations are below those predicted by solubility data for the respective hydroxides, phosphates, or carbonates of these elements.

Adsorption Isotherms

In contrast to the solubility product model, soil solution/solid phasc trace element relationships do appear to be well characterized by various adsorption models. These range from empirical models such as the Langmuir and Freundlich adsorption isotherms, that have little theoretical basis, to chemical (surface complexation) models that treat ion adsorption as complexation reactions analogous to complex formation in solution.

Langmuir and Freundlich Isotherms

Numerous examples exist in the literature in which the sorption of trace elements by soils can be fitted by empirical equations such as the Langmuir,

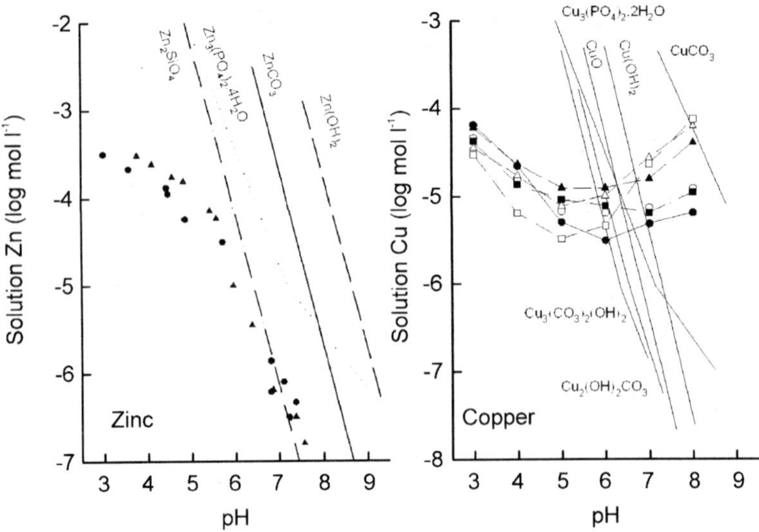

FIGURE 18.6. Actual soil solution Zn and Cu concentrations (symbols) compared to the solubility of individual Zn and Cu compounds. (*Source:* Adapted from Brümmer, Gerth, and Herms, 1986, pp. 392, 394.)

Fruendlich, or other equations (e.g., McLaren and Crawford, 1973b; Kurdi and Doner, 1983). The trace element solution concentrations predicted by these equations at the low end of the range are generally consistent with the concentrations actually observed in soils. However, as noted by Harter and Smith (1981), although these models are often excellent at describing adsorption, they are simply numerical relationships used to curve-fit data. The parameters obtained using these types of models are valid only for the conditions under which the experiment was carried out. Models such as the Langmuir or Freundlich cannot be used to predict adsorption under changing conditions of pH, ionic strength, and temperature. Neither can they deal with the problems of aging and limited reversibility of adsorption (desorption) discussed previously.

Surface Complexation Models

Unlike empirical models, surface complexation models treat ion adsorption as complexation reactions analogous to complex formation in solution (Goldberg, 1995). For example, general equations for surface complexation of trace metal ions can be written as (Hohl, Sigg, and Stumm, 1980):

$$SOH + M^{m+} \Leftrightarrow SOM^{(m-1)} + \quad H^+ \tag{18.1}$$

$$2SOH + M^{m+} \Leftrightarrow (SO)_2M^{(m-2)} + 2H^+ \tag{18.2}$$

where SOH represents the surface functional group, and M is a metal ion of charge m^+. Surface complexation models define surface species, chemical reactions, equilibrium constant expressions, surface activity coefficients, mass and charge balance, and consider the charge of both the adsorbate and adsorbent. They are clearly much more complex than the empirical adsorption isotherm models and are able to model pH and ionic strength effects. There are several different types of surface complexation models, each differing in their structural representation of the solid-solution interface, that is, the location and hydration status of the adsorbed ions (Goldberg, 1995).

The brevity of this chapter prevents a detailed discussion of the various surface complexation models, but several excellent reviews are already available on the subject (e.g., Barrow, 1985; Goldberg, 1995). The models discussed in these reviews include the constant capacitance model, the diffuse and generalized two-layer models, the triple-layer model, and the one-pK model. All have been used reasonably successfully to model trace metal ion adsorption in reasonably simple systems involving pure clay or oxide materials. However, application of surface complexation models to model reactions in whole soils has been somewhat limited.

The Mechanistic Model of Barrow

The model developed by Barrow and co-workers at the University of Western Australia (Barrow, 1987) is essentially a mixture of the constant capacitance model and the triple-layer model, and has also been called the four-layer model. However, it differs from the surface complexation models described previously in that it lacks chemical reactions and equilibrium constant expressions. In addition, unlike the surface complexation models, the total number of surface sites and the maximum adsorption are adjustable parameters.

The original adsorption model was used successfully to describe adsorption of the trace metals Cu, Pb, and Zn on the iron oxide, goethite (Barrow et al., 1981). However, it was recognized that, despite the complexity of the four-layer model, it nevertheless represented a gross oversimplification of the real situation in soils where there is a marked heterogeneity in the nature and properties of colloid surfaces. To accommodate this, a distribution function was introduced to account for differences in binding strength or surface potential at different sites (Barrow, 1983). The mechanistic model was then further extended to describe rates of adsorption and the effect of temperature (Barrow, 1986a), including modeling of the diffusive penetration of ions into the interior of colloids following the initial adsorption. The mechanistic model has not been tested extensively with regard to trace elements, although it has been successful in describing the reactions of Zn with a limited range of soils, including modeling the effects of time, temperature, and pH on Zn sorption (Barrow, 1986b; Barrow, 1986c).

Unfortunately, with the mechanistic model in particular, many of the parameters required are not amenable to measurement in "real" systems. The model is fitted to data by iterative procedures, i.e., by initially guessing at values for unknown parameters, followed by repeated adjustment of values to "best fit" the observed data. When values for a substantial number of parameters are obtained by this procedure, there can be many combinations, which produce good fits with the observed data. Therefore, such models are of limited value for extrapolation and prediction in new situations.

Semiempirical Model for Predicting Trace Element Concentrations or Activities in Soil Solution

The solubility and adsorption models described previously have undoubtedly helped to increase the understanding of the reactions of trace elements in soils in relation to solution-solid phase equilibria. However, it can be argued that they have not been developed nor are they suitable for providing simple predictions or estimates of trace element concentrations in "real" soil solutions. In an attempt to provide such information, based on the determination of a few key soil properties, McBride, Sauvé, and Hendershot (1997) have recently developed a semiempirical equation (Equation 18.3) that relates trace metal activity of soil solutions to the soil's pH, organic matter content (OM), and total trace metal content (M_T):

$$pM = \text{constant} - \log(M_T OM^{-1}) + y\text{pH} \qquad (18.3)$$

where pM is the negative logarithm to base 10 of the solution activity of trace metal M.

Based on Equation (18.3), McBride, Sauvé, and Hendershot (1997) have also provided the following general equation for total solution trace element concentrations, M_s:

$$\log[M]_S = a + b\text{pH} + c\log M_T + d\log OM \qquad (18.4)$$

where the coefficients (a, b, c, and d) can have a positive or negative sign.

In spite of the several assumptions made in their derivation, equations of type (18.3) and (18.4) have been shown to be reasonably good predictors of Cu activity and solubility, Pb activity and solubility, and Zn and Cd solubility in soils (McBride, Sauvé, and Hendershot, 1997; Sauvé, McBride, and Hendershot, 1998; Gray et al., 1999). For example, Figure 18.7 shows the comparison of predicted and measured soluble Cd concentrations from combined data sets of McBride, Sauvé, and Hendershot (1997) and Gray et al. (1999). It should be pointed out, however, that such relationships have been established for soils from relatively long-contaminated sites and thus it is reasonable to conclude that they represent a near steady-state condition (McBride, Sauvé, and Hendershot, 1997).

KINETICS OF TRACE ELEMENT REACTIONS IN SOILS

The models described in the preceding sections are all essentially equilibrium-based models. However, since soils are seldom, if ever at equilibrium with respect to trace element ion transformations, the appropriateness of equilibrium studies to simulate field conditions is questionable. To properly understand the fate of trace elements in soils, and particularly to comprehend

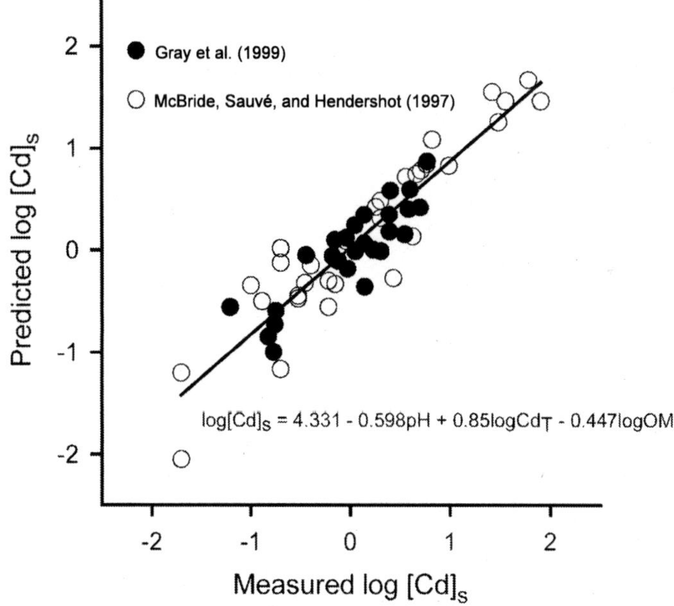

FIGURE 18.7. Relation between measured and predicted soluble Cd values (μg L^{-1}) for the combined data sets of McBride, Sauvé, and Hendershot (1997) and Gray et al. (1999). (*Source:* Adapted from Gray et al., 1999, p. 132.)

their mobility and bioavailability with time, kinetic investigations are necessary (Sparks, 1995). Compared to the kinetic studies undertaken with major nutrients, there have been relatively few studies with trace elements, and these have been concerned mainly with the kinetics of sorption or desorption processes. These studies have been undertaken using both batch and flow techniques and have involved experiments with both whole soils (Amacher et al., 1986; Lin and Xue, 1994) and individual soil components, e.g., oxides (Backes et al., 1995; Grossl et al., 1997), clays (McLaren, Backes, et al., 1998; Eick, Naprstek, and Brady, 2001), and organic matter (Bunzl, Schmidt, and Sansoni, 1976; Randle and Hartmann, 1987).

Trace element sorption and desorption kinetics data from these studies have been described by a number of different models, including zero-, first-, and second-order, Elovich, power function, and parabolic diffusion. Excellent descriptions and discussions of these various models can be found in a number of publications (Amacher et al., 1986; Sparks, 1995). Studies have shown that in many cases several of these models can equally well describe trace element sorption and desorption kinetics data. Unfortunately, it is not usually possible to use the fit of experimental data to these models alone as evidence for the mechanisms of trace element sorption and desorption processes (Sparks, 1995). Time-dependent phenomena in soils are predominantly a combination of true chemical kinetics and transport-controlled kinetics involving processes such as film, particle, surface, and subsurface diffusion.

Kinetic data are clearly important to the understanding of the fate of trace elements in soils. However, we are far from the stage where experimentally derived kinetic parameters can be used, with confidence, in models to predict such phenomena as trace element mobility and bioavailability.

MODELING OF TRACE ELEMENT LEACHING FROM SOIL

The realization that trace elements do leach from soils under some circumstances has led to a demand for predictive models of trace element leaching, particularly in relation to toxic elements such as As, Cr(VI), and Cd. A limited number of attempts have indeed been made to model trace element transport in soils, ranging from simple to extremely complex models. However, it would be fair to say that most, if not all, attempts to model trace element leaching have not progressed past experimental studies carried out with soil columns in the laboratory. It is not the intention here to go into the details of all of the proposed models, but rather to highlight some of the issues involved in the modeling of trace element transport. For those readers who require more detailed information on the subject, the recent text by Selim and Amacher (1996) is highly recommended.

Leaching Predictions Based on Soil Solution Trace Element Concentrations

Probably the most simple model for estimating trace element leaching from soils is by Holm et al. (1998) to estimate leaching outflow of Cd and Zn from contaminated soils. In this study, leaching outflow (g of metal per hectare per year) is estimated as:

$$\text{Outflow } (\text{g ha}^{-1}\ \text{y}^{-1}) = 0.01\text{P} \times \text{C} \qquad (18.5)$$

where P (mm y^{-1}) is the net water percolation through the soil, and C is the metal concentration (μg L^{-1}) in the soil solution. Clearly, this model assumes that the metal concentration in the soil solution remains constant throughout the year, and that drainage leachate has an identical composition to the soil solution. Unfortunately, under conditions of high-intensity rainfall and infiltration, the maintenance of a constant metal concentration in the soil solution is unlikely to occur. In particular, water passing rapidly through macropores is unlikely to attain equilibrium with metals associated with the soil solid phase.

Thus, realistically, Equation (18.5) is likely only to provide an estimate of the maximum potential leaching of trace metals from the soil. However, this may be adequate for many risk assessment purposes.

Transport Models Based on the Convective-Dispersion Equation

The relatively small number of mechanistic models that have been developed and applied specifically to the transport of trace elements through soils are based on various formulations of the convective-dispersion equation (Sidle, Kardos, and van Genuchten, 1997; Amoozegar-Fard, Fuller, and Warrick, 1984). For example, Sidle, Kardos, and van Genuchten (1997) have used the following formulation as the starting point for modeling trace metal transport in field soils:

$$(\rho/\theta)(\delta S/\delta t) + (\delta C/\delta t) = D(\delta^2 C/\delta x^2) - v_p(\delta C/\delta x) \qquad (18.6)$$

where C = concentration of trace element in solution, S = amount of the trace element adsorbed, D = dispersion coefficient, v_p = average pore water velocity, ρ = soil bulk density, θ = volumetric water content, x = linear distance in direction of flow, and t = time.

The main differences between published trace metal transport models generally relate to the ways in which metal retention-release mechanisms are dealt with, i.e., the solution-solid phase equilibrium. In some cases, adsorption equations such as Freundlich's and Langmuir's are used to model the solution-solid phase equilibrium and assume complete reversibility (Sidle, Kardos, and van Genuchten, 1997). In other cases, a linear equilibrium adsorption mechanism is incorporated into the convection-dispersion equation. In yet others, kinetic equations are incorporated to allow for time-dependent reactions (Amoozegar-Fard, Fuller, and Warrick, 1984). The resulting models tend to be relatively complex, and although good fits are often obtained with data from soil column leaching experiments, parameters such as dispersion and rate coefficients are not amenable to independent determination. Thus values for the model parameters obtained by data fitting are generally not transferable to other soils and situations.

Some attempts have been made to relate leaching model parameter values and calculated trace metal leaching velocities to individual soil properties such as organic matter content, soil pH, and clay or iron oxide contents (Amoozegar-Fard, Fuller, and Warrick, 1984). However, serious validation of this approach for predictive purposes in the field does not appear to have been undertaken.

APPLICATIONS OF TRACE ELEMENT MODELING IN SOILS

Prediction of Trace Element Bioavailability and Toxicity

The ability to predict accurately the bioavailability, toxicity, and plant uptake of trace elements is of major importance for both food production and the risk assessment of contaminated sites. The realization that chemical speciation of trace elements in solution affects their bioavailability and toxicity has resulted in considerable use of computer speciation models such as GEOCHEM or GEOCHEM-PC to predict trace element bioavailability to plants. In addition, the semiempirical model of McBride, Sauvé, and Hendershot (1997) for estimating trace element activities in solution could also prove to be valuable for the prediction of trace element bioavailability and toxicity. For example, Sauvé et al. (1998) have demonstrated good relationships between free metal activities of Pb and Cu, calculated using the semiempirical model of McBride, Sauvé, and Hendershot (1997), and toxic effects on plants, soil organisms, and soil microbial processes. In addition to their use for predictive purposes, computerized speciation models have also become widely used in plant mineral nutrition research, e.g., Parker, Chaney, and Norvell (1995).

Development of Soil Quality Criteria

In recent years, considerable attention has been directed to determining "safe levels" for burdening the soil environment with toxic elements. This has been done essentially to aid the development of strategies for the control of toxic element additions to the soil and to assess the risk associated with existing levels of contamination. However, contrasting methods have been used in setting toxic element limits in soils, resulting in widely different numerical limits being set for the same element (McGrath et al., 1994). The one commonality among all current sets of soil toxic element limits, however, is that they are quantified solely on the basis of total metal concentrations in the soil. This, in spite of the fact that it is generally accepted that it is the solubility of elements that controls their bioavailability and mobility in the soil.

McBride, Sauvé, and Hendershot (1997) have suggested that the semi-empirical equations described previously (Equations 18.3 and 18.4) may be useful in arriving at guidelines for metal loading limits for soils. Metal solubilities or activities in soil solutions that correspond to toxicity thresholds for soil organisms or plants can be entered into Equations 18.6 or 18.7 and the equations can be solved for different soil pH and organic matter values to estimate total metal concentrations (M_T) that would cause toxicity. An example of this approach is provided by Sauvé et al. (1998) in relation to Cu and Pb toxicity in soils. These researchers developed total soil metal quality criteria derived for 50 and 25 percent inhibition endpoints derived from toxicological data for plants, soil organisms, and soil microbial processes (Table 18.6). Such an approach to developing soil quality criteria undoubtedly deserves further investigation.

TABLE 18.6. Total Soil Metal Quality Criteria Derived from 50 and 25 Percent Inhibition Endpoints Obtained from Toxicological Data for Plants, Soil Organisms, and Soil Microbial Processes

	Total soil metal quality criteria (mg kg^{-1})			
pH	5.5	6	6.5	7
$pPb^{2+}_{50\%} = 8.3$	177	415	972	2276
$pb^{2+}_{25\%} = 9.5$	7	16	36	84
$pCu^{2+}_{50\%} = 7.7$	103	265	684	1766
$pCu^{2+}_{25\%} = 9.6$	8	20	52	135

Source: Adapted from Sauvé et al., 1998, p. 1487.

Trace Element Balance Models

Compared to the extremely complex nutrient cycling and balance models developed for some elements, for example, nitrogen, there have been few attempts to produce comprehensive balance models for the trace elements. However, alongside the development of soil quality criteria for toxic metals, there is considerable interest in being able to predict trends with time in total metal concentrations in soils. As described by Moolenaar, Van Der Zee, and Lexmond (1997), this can be achieved by means of a relatively simple dynamic balance model:

$$dM_T/dt = A - L - U \qquad (18.7)$$

where M_T is the total content of metal in the soil, A is the input rate of metal at the soil surface, L is the leaching rate at the lower boundary of the system, and U is the removal by harvesting plants. Unfortunately, in many situations, reliable data on trace metal inputs to and losses from soils are not available. In such cases, estimates may be possible using some of the semiempirical modeling approaches described previously (e.g., for estimating leaching, Equation 18.5). Verification of such models is also not without problems, since this requires good long-term data sets of trace metal trends in soils. Nevertheless, some interesting examples of trace element balances have been undertaken, particularly in relation to Cd (e.g., Tjell and Christensen, 1992; Jeng and Singh, 1995).

SUMMARY

As discussed in this chapter, many of the models developed to describe trace element reactions and processes in soil are essentially mathematical relationships used to curve-fit experimental data. Such models may, in some cases, help to provide evidence for the mechanisms by which trace elements interact with soils, but at this point in time, most appear to have little practical application. Parameters required to run the models can often only be determined by iterative procedures and are unlikely to be readily transferable. Of all the models developed to describe trace element reactions in soil, it is probably the most simple, empirical, or semiempirical models that show the greatest possibilities for practical application.

REFERENCES

Allison, J. D. and D. S. Brown. (1995). MINTEQA2/PRODEFA2—A geochemical speciation model and interactive preprocessor. In *Chemical Equilibrium and Reaction Models,* eds. R. H. Loeppert, A. P. Schwab, and S. Goldberg. Madison, WI: Soil Science Society of America, Inc., pp. 241-252.

Alloway, B. J. (1990). *Heavy Metals in Soils*. Glasgow: Blackie Academic and Professional.

Amacher, M. C., J. Kotuby-Amacher, H.M. Seklim, and I. K. Iskandar. (1986). Retention and release of metals by soils—Evaluation of several models. *Geoderma 38*: 131-154.

Amoozegar-Fard, A., W. H. Fuller, and A. W. Warrick. (1984). An approach to predicting the movement of selecting polluting metals in soils. *Journal of Environmental Quality 13*: 290-297.

Asami, T. (1981). The Ichi and Maruyama River basins: Soil pollution by cadmium, zinc, lead and copper discharged from Ikuno Mine. In *Heavy Metal Pollution in Soils of Japan,* eds. K. Kitagishi and I. Yamane. Tokyo: Japan Scientific Societies Press, pp. 125-136.

Backes, C. A., R. G. McLaren, A. W. Rate, and R. S. Swift. (1995). Kinetics of cadmium and cobalt desorption from iron and manganese oxides. *Soil Science Society of America Journal 59*: 778-785.

Barrow, N. J. (1983). A mechanistic model for describing the sorption and desorption of phosphate by soil. *Journal of Soil Science 34*: 733-750.

Barrow, N. J. (1985). Reactions of anions and cations with variable-charge soils. *Advances in Agronomy 34*: 183-230.

Barrow, N. J. (1986a). Testing a mechanistic model. I. The effects of time and temperature on the reaction of fluoride and molybdate with a soil. *Journal of Soil Science 37*: 267-275.

Barrow, N. J. (1986b). Testing a mechanistic model. II. The effects of time and temperature on the reaction of zinc with a soil. *Journal of Soil Science 37*: 277-286.

Barrow, N. J. (1986c). Testing a mechanistic model. IV. Describing the effects of pH on zinc retention by soils. *Journal of Soil Science 37*: 295-302.

Barrow, N. J. (1987). *Reactions with Variable-Charge Soils*. Dordrecht: Martinus Nijhoff Publishers.

Barrow, N. J. (1999). The four laws of chemistry: The Leeper lecture 1998. *Australian Journal of Soil Research 37*: 787-829.

Barrow, N. J., J. W. Bowden, A. M. Posner, and J. P. Quirk. (1981). Describing the adsorption of copper, zinc and lead on a variable charge mineral surface. *Australian Journal of Soil Research 19*: 309-404.

Bowen, H. J. M. (1966). *Trace Elements in Biochemistry*. London: Academic Press.

Brennan, R. F. (1990). Reaction of zinc with soil affecting its availability to subterranean clover. II. Effect of soil properties on the relative effectiveness of applied zinc. *Australian Journal of Soil Research 28*: 303-310.

Brennan, R. F., J. W. Gartrell, and A. D. Robson. (1980). Reactions of copper with soil affecting its availability to plants. I. Effect of soil type and time. *Australian Journal of Soil Research 18*: 447-459.

Brümmer, G. W., J. Gerth, and U. Herms. (1986). Heavy metal species, mobility and availability in soils. *Zeitschrift fur Pflanzenernahrung Bodenkunde 149*: 382-398.

Brümmer, G. W., J. Gerth, and K. G. Tiller. (1988). Reaction kinetics of the adsorption and desorption of nickel, zinc and cadmium by goethite. I. Adsorption and diffusion of metals. *Journal of Soil Science 39*: 37-52.

Brümmer, G., K. G. Tiller, U. Herms, and P. M. Clayton. (1983). Adsorption-desorption and/or precipitation-dissolution processes of zinc in soils. *Geoderma 31*: 337-354.

Bunzl, K., W. Schmidt, and B. Sansoni. (1976). Kinetics of ion exchange in soil organic matter. IV. Adsorption and desorption of Pb^{2+}, Cu^{2+}, Cd^{2+}, Zn^{2+} and Ca^{2+} by peat. *Journal of Soil Science 27*: 32-41.

Cabrera, D., S. D. Young, and D. L. Rowell. (1988). The toxicity of cadmium to barley plants as affected by complex formation with humic acid. *Plant and Soil 105*: 195-204.

Camobreco, V. J., B. K. Richards, T. S. Steenhuis, J. H. Peverly, and M. B. McBride. (1996). Movement of heavy metals through undisturbed and homogenized soil columns. *Soil Science 161*: 740-750.

Carey, P. L., R. G. McLaren, K. C. Cameron, and J. R. Sedcole. (1996). Leaching of copper, chromium, and arsenic through some free-draining New Zealand soils. *Australian Journal of Soil Research 34*: 583-597.

Chang, A. C., J. E. Warneke, A. L. Page, and L. J. Lund. (1984). Accumulation of heavy metals in sewage sludge-treated soils. *Journal of Environmental Quality 13*: 87-91.

Cornell, R. M. and U. Schwertmann. (1996). *The Iron Oxides*. Weinheim: VCH Publishers.

DeKock, P. C. and R. L. Mitchell. (1957). Uptake of chelated metals by plants. *Plant and Soil 84*: 55-62.

Dowdy, R. H., J. J. Latterell, T. D. Hinesly, R. B. Grossman, and D. L. Sullivan. (1991). Trace metal movement in an Aeric Ochraqualf following 14 years of annual sludge applications. *Journal of Environmental Quality 20*: 119-123.

Eick, M. J., B. R. Naprstek, and P. V. Brady. (2001). Kinetics of Ni(II) sorption and desorption on kaolinite: Residence time effects. *Soil Science 166*: 11-17.

Elrashidi, M. A. and G. A. O'Connor. (1982). Influence of solution composition on sorption of zinc by soils. *Soil Science Society of America Journal 46*: 1153-1158.

Fergusson, J. E. (1990). *The Heavy Elements: Chemistry, Environmental Impact and Health Effects*. Oxford, England: Pergamon Press.

Gilkes, R. J. and R. M. McKenzie. (1988). Geochemistry of manganese in soil. In *Manganese in Soils and Plants,* eds. R. D. Graham, R. J. Hannam, and N. C. Uren. Dordrecht: Kluwer Academic Publishers, pp. 23-35.

Goldberg, S. (1995). Adsorption models incorporated into chemical equilibrium models. In *Chemical Equilibrium and Reaction Models,* eds. R. H. Loeppert,

A. P. Schwab, and S. Goldberg. Madison, WI: Soil Science Society of America, Inc., pp. 75-95.

Graham, R. D. (1981). Absorption of copper by plants. In *Copper in Soils and Plants,* eds. J. F. Loneragan, A. D. Robson, and R. D. Graham. Sydney: Academic Press, pp. 141-163.

Gray, C. W., R. G. McLaren, A. H. C. Roberts, and L. M. Condron. (1998). Sorption and desorption of cadmium from some New Zealand soils: Effect of pH and contact time. *Australian Journal of Soil Research 36*: 199-216.

Gray, C. W., R. G. McLaren, A. H. C. Roberts, and L. M. Condron. (1999). Solubility, sorption and desorption of native and added cadmium in relation to properties of soils in New Zealand. *European Journal of Soil Science 50*: 127-137.

Grossl, P. R., M. Eick, D. L. Sparks, S. Goldberg, and C. C. Ainsworth. (1997). Arsenate and chromate retention mechanisms on goethite. 2. Kinetic evaluation using a pressure-jump relaxation technique. *Environmental Science and Technology 31*: 321-326.

Hamon, R. E., M. J. McLaughlin, R. Naidu, and R. Correll. (1998). Long-term changes in cadmium bioavailability in soil. *Environmental Science and Technology 32*: 3699-3703.

Harter, R. D. and G. Smith. (1981). Langmuir equation and alternate methods of studying "adsorption" reactions in soils. In *Chemistry of the Soil Environment,* ed. R. H. Dowdy. Madison, WI: Soil Science Society of America, Inc., pp. 167-182.

Hingston, F. J. (1964). Reactions between boron and clays. *Australian Journal of Soil Research 2*: 83-95.

Hingston, F. J., A. M. Posner, and J. P. Quirk. (1974). Anion adsorption by goethite and gibbsite. II. Desorption of anions from hydrous oxide surfaces. *Journal of Soil Science 25*: 16-26.

Hinsinger, P. (2001). Bioavailability of trace elements as related to root-induced chemical changes in the rhizosphere. In *Trace Elements in the Rhizosphere,* eds. G. R. Gobran, W. W. Wenzel, and E. Lombi. Boca Raton, FL: CRC Press, pp. 25-41.

Hohl, H., L. Sigg, and W. Stumm. (1980). Characterisation of surface chemical properties of oxides in natural waters. *Advances in Chemistry Series 189*: 1-31.

Holm, P. E., T. H. Christensen, S. E. Lorenz, R. E. Hamon, H. C. Domingues, E. M. Sequeira, and S. P. McGrath. (1998). Measured soil water concentrations of cadmium and zinc in plant pots and estimated leaching outflows from contaminated soils. *Water, Air and Soil Pollution 102*: 105-115.

Jeng, A. S. and B. R. Singh. (1995). Cadmium status of soils and plants from a long-term fertility experiment in southeast Norway. *Plant and Soil 175*: 67-74.

Kabata-Pendias, A. and H. Pendias. (1992). *Trace Elements in Soils and Plants,* Second Edition. Boca Raton, FL: CRC Press, Inc.

Krauskopf, K. B. (1972). Geochemistry of micronutrients. In *Micronutrients in Agriculture,* eds. J. J. Mortvedt, P. M. Giordano, and W. L. Lindsay. Madison, WI: Soil Science Society of America, Inc., pp. 7-40.

Kurdi, F. and H. E. Doner. (1983). Zinc and copper sorption and interaction in soils. *Soil Science Society of America Journal 47*: 873-876.

Le Riche, H. H. and A. H. Weir. (1963). A method of studying trace elements in soil fractions. *Journal of Soil Science 14*: 225-235.

Lee, K. E. (1985). *Earthworms: Their Ecology and Relationships with Soils and Land Use*. Sydney: Academic Press.

Lin, Y.-S. and J.-H. Xue. (1994). pH effect on kinetics of heavy metal sorption in soils. *Pedosphere 4*: 225-231.

Lindsay, W. L. (1972). Inorganic phase equilibria of micronutrients in soils. In *Micronutrients in Agriculture,* eds. J. J. Mortvedt, P. M. Giordano, and W. L. Lindsay. Madison, WI: Soil Science Society of America, Inc., pp. 41-57.

Lumsdon, D. G. and L. J. Evans. (1995). Predicting chemical speciation and computer simulation. In *Chemical Speciation in the Environment,* eds. A. M. Ure and C. M. Davidson. Glasgow: Blackie Academic and Professional, pp. 86-134.

Mattigod, S. V. (1995). Chemical equilibrium and reaction models: Applications and future trends. In *Chemical Equilibrium and Reaction Models,* eds. R. H. Loeppert, P. A. Schwab, and S. Goldberg. Madison, WI: Soil Science Society of America, Inc., pp. 1-5.

McBride, M. B. (1991). Processes of heavy and transition metal sorption by soil minerals. In *Interactions at the Soil Colloid-Soil Solution Interface,* eds. G. H. Bolt, M. F. De Boodt, M. H. B. Hates, and M. B. McBride. Dordrecht: Kluwer Academic Publishers, pp. 149-175.

McBride, M. B. and J. J. Blasiak. (1979). Zinc and copper solubility as a function of pH in an acid soil. *Soil Science Society of America Journal 43*: 866-870.

McBride, M. B., S. Sauvé, and W. Hendershot. (1997). Solubility control of Cu, Zn, Cd and Pb in contaminated soils. *European Journal of Soil Science 48*: 337-346.

McGrath, S. P., A. C. Chang, A. L. Page, and E. Witter. (1994). Land application of sewage sludge: Scientific perspectives of heavy metal loading limits in Europe and the United States. *Environmental Reviews 2*: 108-118.

McLaren, R. G., C. A. Backes, A. W. Rate, and R. S. Swift. (1998). Cadmium and cobalt desorption kinetics from soil clays: Effect of sorption period. *Soil Science Society of America Journal 62*: 332-337.

McLaren, R. G. and K. C. Cameron. (1996). *Soil Science: Sustainable Production and Environmental Protection,* Second Edition. Auckland: Oxford University Press.

McLaren, R. G. and D. V. Crawford. (1973a). Studies on soil copper. I. The fractionation of copper in soils. *Journal of Soil Science 24*: 172-191.

McLaren, R. G. and D. V. Crawford. (1973b). Studies on soil copper. II. The specific adsorption of copper by soils. *Journal of Soil Science 24*: 443-452.

McLaren, R. G., D. M. Lawson, and R. S. Swift. (1986). Sorption and desorption of cobalt by soils and soil components. *Journal of Soil Science 37*: 413-426.

McLaren, R. G., R. Naidu, J. Smith, and K. G. Tiller. (1998). Fractionation and distribution of arsenic in soils contaminated by cattle dip. *Journal of Environmental Quality 27*: 348-354.

McLaren, R. G. and C. J. Smith. (1996). Issues in the disposal of industrial and urban wastes. In *Contaminants and the Soil Environment in the Australasia-Pacific Region,* eds. R. Naidu, R. S. Kookana, D. P. Oliver, S. Rogers, and M. J. McLaughlin. Dordrecht: Kluwer Academic Publishers, pp. 183-212.

McLaren, R. G., M. D. Taylor, T. Hendry, and L. Clucas. (1999). Leaching of metals and nutrients from soils treated with metal-amended sewage sludge. In *Best Management Practices for Production,* eds. L. D. Currie, M. J. Hedley, D. J. Horne, and P. Loganathan. Massey University, Palmerston North: Fertilizer and Lime Research Centre, pp. 251-260.

McLaren, R. G., J. G. Williams, and R. S. Swift. (1983). Some observations on the desorption and distribution behaviour of copper with soil components. *Journal of Soil Science 34*: 325-331.

McLaughlin, M. J., K. G. Tiller, R. Naidu, and D. P. Stevens. (1996). Review: The behaviour and environmental impact of contaminants in fertilizers. *Australian Journal of Soil Research 34*: 1-54.

Mengel, K. and E. A. Kirkby. (1978). *Principles of Plant Nutrition.* Worblaufen-Bern, Switzerland: International Potash Institute.

Mitchell, R. L. (1964). Trace elements in soils. In *Chemistry of the Soil,* Second Edition, ed. F. E. Bear. New York: Reinhold, pp. 320-368.

Moolenaar, S. W., S. E. A. T. M. Van Der Zee, and T. M. Lexmond. (1997). Indicators of the sustainability of heavy-metal management in agro-ecosystems. *The Science of the Total Environment 201*: 155-169.

Parker, D. R., R. L. Chaney, and W. A. Norvell. (1995). Chemical equilibrium models: Applications to plant nutrition research. In *Chemical Equilibrium and Reaction Models,* eds. R. H. Loeppert, A. P. Schwab, and S. Goldberg. Madison, WI: Soil Science Society of America, Inc., pp. 163-200.

Parker, D. R., W. A. Norvell, and R. L. Chaney. (1995). GEOCHEM-PC—A chemical speciation program for IBM and compatible personal computers. In *Chemical Equilibrium and Reaction Models,* eds. R. H. Leoppert, A. P. Schwab, and S. Goldberg. Madison, WI: Soil Science Society of America, Inc., pp. 253-269.

Peryea, F. J. and T. L. Creger. (1994). Vertical distribution of lead and arsenic in soils contaminated with lead arsenate pesticide residues. *Water, Air and Soil Pollution 78*: 297-306.

Randle, K. and E. H. Hartmann. (1987). Application of the continuous flow stirred cell (CFSC) technique to adsorption of zinc, cadmium and mercury on humic acids. *Geoderma 40*: 281-296.

Rate, A. W., R. G. McLaren, and R. S. Swift. (1993). Response of copper(II)-humic acid dissociation kinetics to factors influencing complex stability and macromolecular conformation. *Environmental Science and Technology 27*: 1408-1414.

Salomons, W. and U. Förstner. (1984). Metals in the atmosphere. In *Metals in the Hydrocycle.* Berlin: Springer-Verlag, pp. 99-137.

Sauvé, S., A. Dumestre, M. B. McBride, and W. H. Hendershot. (1998). Derivation of soil quality criteria using predicted speciation of Pb^{2+} and Cu^{2+}. *Environmental Chemistry and Toxicology 8*: 1481-1489.

Sauvé, S., M. McBride, and W. Hendershot. (1998). Soil solution speciation of lead(II): Effects of organic matter and pH. *Soil Science Society of America Journal 62*: 618-621.

Selim, H. M. and M. C. Amacher. (1996). *Reactivity and Transport of Heavy Metals in Soils.* Boca Raton, FL: Lewis Publishers.

Shuman, L. M. (1979). Zinc, manganese and copper in soil fractions. *Soil Science 127*: 10-17.

Sidle, R. C., L. T. Kardos, and M. T. van Genuchten. (1977). Heavy metal transport model in a sludge-treated soil. *Journal of Environmental Quality 6*: 438-443.

Smolders, E. and M. J. McLaughlin. (1996). Chloride increases cadmium uptake in Swiss chard in a resin-buffered nutrient solution. *Soil Science Society of America Journal 60*: 1443-1447.

Sparks, D. L. (1995). Kinetics of metal sorption reactions. In *Metal Speciation and Contamination of Soil,* eds. H. E. Allen, C. P. Huang, and G. W. Bailey. Boca Raton, FL: Lewis Publishers, pp. 35-58.

Sparling, G. P. and M. L. Berrow. (1985). Effect of air drying, g-irradiation and chloroform fumigation of soil on extractability of trace elements. *Journal of Agricultural Science, Cambridge 104*: 223-226.

Sposito, G. (1983). The chemical forms of trace metals in soils. In *Applied Environmental Geochemistry,* ed. I. Thornton. London: Academic Press, pp. 123-170.

Sposito, G. (1984). *The Surface Chemistry of Soils*. Oxford: Oxford University Press.

Sposito, G., F. T. Bingham, S. S. Yadav, and C. A. Inouye. (1982). Trace metal complexation by fulvic acid extracted from sewage sludge: II. Development of chemical models. *Soil Science Society of America Journal 46*: 51-56.

Sposito, G. and S. V. Mattigod. (1980). *A Computer Program for the Calculation of Chemical Equilibria in Soil Solutions and Other Natural Water Systems*. Riverside, CA: Kearney Foundation of Soil Science, University of California.

Stevenson, F. J. (1982). *Humus Chemistry, Genesis, Composition, Reactions*. New York: Wiley.

Storgaard Jorgensen, S. and M. Willems. (1987). The fate of lead in soils: The transformation of lead pellets in shooting-range soils. *Ambio 16*: 11-15.

Suarez, D. L. (1995). Carbonate chemistry in computer programs and application to soil chemistry. In *Chemical Equilibrium and Reaction Models,* eds. R. H. Loeppert, A. P. Schwab, and S. Goldberg. Madison, WI: Soil Science Society of America, Inc., pp. 53-73.

Swaine, D. J. (1962). *The Trace Element Content of Fertilizers*. Technical Communication No. 52. Farnham Royal, England: Commonwealth Agricultural Bureau.

Swift, R. S. and R. G. McLaren. (1991). Micronutrient adsorption by soils and soil colloids. In *Interactions at the Soil Colloid-Soil Solution Interface,* eds. G. H. Bolt, M. F. De Boodt, M. B. H. Hayes, and M. B. McBride. Dordrecht: Kluwer Academic Publishers, pp. 257-292.

Taylor, R. M. and R. M. McKenzie. (1966). The association of trace elements with manganese minerals in Australian soils. *Australian Journal of Soil Research 4*: 29-39.

Tiller, K, G., J. L. Honeysett, and E. G. Hallsworth. (1968). The isotopically exchangeable form of native and applied cobalt in soils. *Australian Journal of Soil Research 7*: 43-56.

Tjell, J. C. and T. H. Christensen. (1992). Sustainable management of cadmium in Danish agriculture. In *Impact of Heavy Metals on the Environment,* ed. J.-P. Vernet. Amsterdam: Elsevier, pp. 273-286.

Underwood, E. J. (1981). *The Mineral Nutrition of Livestock,* Second Edition. Slough, England: Commonwealth Agricultural Bureaux.
Williams, D. E., J. Vlamis, A. H. Pukite, and J. E. Corey. (1985). Metal movement in sludge-treated soils after six years of sludge addition. 2. Nickel, cobalt, iron, manganese, chromium and mercury. *Soil Science 140*: 120-125.
Yin, Y., H. E. Allen, Y. Li, C. P. Huang, and P. F. Sanders. (1996). Adsorption of mercury (II) by soil: Effects of pH, chloride, and organic matter. *Journal of Environmental Quality 25*: 837-844.

Chapter 19

Agrochemicals

Thilo Streck

The term "agrochemicals" stands for a large variety of synthetic compounds used to increase the production of agricultural crops. It subsumes pesticides used to protect plants against weeds and various diseases as well as growth regulators used to enhance the standing of plants. Pesticides are also commonly classified according to their target organisms (e.g., herbicides, insecticides, fungicides, nematicides), their chemical structure (e.g., organophosphates, triazines, carbamates), or their mode of action (e.g., systemic or nonsystemic).

According to the Committee on the Future Role of Pesticides in U.S. Agriculture (2000), in 1997 world pesticide sales amounted to $31 billion. Although consumption grows at a yearly rate of about 3 percent, pesticides are used in only about one-third of the world's cropped area (Yudelman, Ratta, and Nygaard, 1998). More than half the pesticides (in monetary value) are applied in North America and Western Europe. In these areas, herbicides make up the largest fraction, while in the developing countries insecticides account for the largest share. For example, in the United States herbicides have a market share of 57 percent, while in India insecticides account for 67 percent of the agrochemical market (Committee on the Future Role of Pesticides in U.S. Agriculture, 2000; Anonymous, 2001).

Recently, some (e.g., Müller, 2000) have broadened the term agrochemicals to include fertilizers, arguing that in agriculture fertilizers are as important as crop protection chemicals. Regarding the relevant processes in soils, however, the distinction should be made between organic and inorganic chemicals. Since the fate of the latter is treated in other chapters of this book this chapter adheres to the more traditional nomenclature and is exclusively devoted to the fate of organic agrochemicals.

The key processes affecting the fate of agrochemicals in the soil-plant system are sorption and degradation. Sorption (as well as the reverse process, desorption) determines the partitioning between the dissolved or gaseous phase and the soil matrix and thus between mobile and immobile phases.

Degradation withdraws the agrochemicals from the system, but the substances formed may be hazardous as well. Degradation pathways may be biotic or abiotic. Biotic pathways are considered more important (Battersby, 1990). Therefore, the travel time of agrochemicals through the topsoil, which is a function of the water flux on the one hand and of sorption on the other, plays a major role insofar as the agrochemicals can leave the biologically most active soil horizon before degrading and reach the groundwater. Table 19.1 lists soil-water partitioning coefficients and half-lives for a number of agrochemicals, which express the extent of sorption and the viability to degradation, respectively. As demonstrated by Gustafson (1989) a combination of these properties may be used to explain the appearance of agrochemicals in groundwater.

TABLE 19.1. Properties of Selected Agrochemicals

Agrochemical	**k_{oc} ($dm^3\ kg^{-1}$)**	**Half-life in soil (days)**
Leachers		
Aldicarb	17	7
Atrazine	107	74
Metolachlor	99	44
Picloram	26	206
Transition		
Alachlor	161	14
Carbofuran	55	37
Dieldrin	12,100	934
Fonofos	5,105	25
Nonleachers		
DDT	213,600	38,200
Dicamba	511	25
Endosulfan	2,040	120
Trifluralin	7,950	83

Source: Adapted from Gustafson (1989).

k_{oc} represents carbon normalized distribution coefficient
Classification regarding leachability was done by the California Department of Food and Agriculture as stated in Gustafson (1989).

Evidence exists of losses by leaching particularly with water flowing along preferential flow paths. Less is known about the extent to which the more strongly sorbing agrochemicals may be leached by cotransport with organic particles. Loss of agrochemicals with eroding soil will particularly affect the quality of surface waters, but will not be treated in this chapter. Agrochemicals may also leave the soil by volatilization, either during and shortly after application or after diffusion through the soil air. Loss by plant uptake is usually neglected. Agrochemicals may cause severe health problems, especially if exposure is direct or dietary, e.g., during application or through consumption of contaminated food. Since the late 1970s when pesticides were first detected in groundwater, there has been a growing concern about contamination of ground and surface water. In industrialized countries, a considerable fraction of samples taken from shallow groundwater have shown one or more agrochemicals or metabolites at detectable concentrations. For instance, agrochemicals or metabolites were detected in 54 percent of 1,034 wells and springs sampled within a countrywide water-quality assessment program in the United States (Kolpin, Barbash, and Gilliom, 1998). In 4 percent of the samples, concentrations were above 1 $\mu g\ l^{-1}$. In Germany, agrochemicals or metabolites were detected in 19 percent of 12,886 samples, while 1 percent was above 1 $\mu g\ l^{-1}$ (Länderarbeitsgemeinschaft Wasser (LAWA), 1997). In both countries, atrazine and its metabolite deethylatrazine were the agrochemicals most frequently found.

SORPTION AND DEGRADATION IN SOIL

Sorption

Equilibrium Sorption

The concentration of an agrochemical sorbed to the soil matrix, S (mg kg^{-1}), is a function of its concentration in the soil solution, C (mg l^{-1}). Solute partitioning can be conveniently determined in laboratory batch experiments. Sorption isotherms are measured by addition of different amounts of a target compound to soil suspensions, analysis of the solutions after equilibration and centrifugation, and calculation of the sorbed phase concentrations from mass balance. Although the term sorption isotherm was originally restricted to equilibrium situations, it is now used in a more general sense, not least because in practice equilibration is difficult to verify.
In the linear case the sorption isotherm reads:

$$S = k_D C \tag{19.1}$$

where k_D ($dm^3\ kg^{-1}$) denotes the distribution or partitioning coefficient. For nonionic organic chemicals, measured k_D values can be referenced to the mass fraction of organic carbon in soil, f_{oc}, yielding the carbon normalized distribution coefficient (k_{oc}), which is fairly specific for each substance. Conversely, if the k_{oc}-value of a substance is known its k_D value can be assessed from:

$$k_D = f_{oc}\, k_{oc} \tag{19.2}$$

For a number of organic compounds, k_{oc} values have been estimated from measured sorption data and tabulated (Rao and Davidson, 1980). For specific compounds, regression equations are available that allow estimates of k_{oc} values from octanol-water partition coefficients (see Green and Karickhoff, 1990, for a useful scheme on the estimation of distribution coefficients).

Nonlinear sorption of agrochemicals is often described by the Freundlich equation

$$S = kC^m \tag{19.3}$$

where k ($mg^{1-m}\,L^m\,kg^{-1}$) and m (unitless) denote Freundlich coefficient and exponent, respectively. The latter is in general smaller than unity indicating a relative decrease of sorption with increasing solution phase concentration.

Nonlinear sorption is sometimes modeled with the Langmuir isotherm. This isotherm is almost linear at small concentrations while at higher concentrations sorbed phase concentrations reach saturation. There are many modifications of the Langmuir isotherm, e.g., to account for sigmoidality or competition by other compounds. Most of these modifications have only been used with laboratory data and are difficult to apply in field situations. As an example, accounting for competition effects between several agrochemicals applied requires the solution of a set of coupled transport equations. However, as concentrations of agrochemicals in field soils are generally small, competition effects are usually neglected.

Kinetic Sorption

Laboratory studies have shown that sorption of many organic chemicals occurs in two steps. A fast initial sorption is followed by a slow approach to equilibrium (Brusseau, Jessup, and Rao, 1989; Ball and Roberts, 1991). The first step is completed within minutes or, at the maximum, hours, while the latter step may take weeks or even months, if not longer (Kleineidam et al., 1999). Such observations contradict the widespread belief that equilibration of agrochemicals added to soil suspensions is fast and completed within one or two days. As a practical consequence, sorption experiments that allow only 24 to 48 hours for equilibration cannot be expected to yield correct (i.e., equilibrium) sorption isotherms.

In sorption-desorption experiments it has frequently been observed that adsorption and desorption curves do not coincide (Figure 19.1). This phenomenon, known as sorption nonsingularity or sorption hysteresis, has been attributed to a number of factors, ranging from centrifugation artifacts to steric hindrance. However, a simple and often satisfactory explanation is that equilibrium is not attained in the different experimental steps. Using a kinetic sorption model Altfelder, Streck, and Richter (2000) evaluated sorption and desorption data measured by different procedures. In addition, they revisited published data that had often been referenced as evidence for true

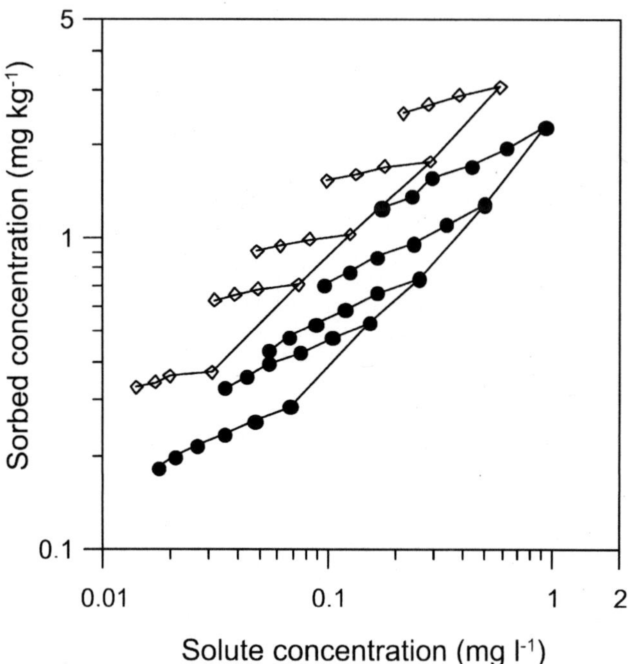

FIGURE 19.1. Sorption-desorption isotherms of the herbicide chlortoluron for a sandy soil. At each sorption or desorption step, the time period allowed for equilibration is one and thirty days, respectively. The investigated soil contained 1 percent organic carbon and was collected from the plowing horizon of a cambisol in Northern Germany. (*Source:* Altfelder and Streck, unpublished data.)

sorption hysteresis. In all cases, nonsingularity could be traced back to kinetic causes. Figure 19.1 clearly demonstrates that the results of sorption-desorption experiments depend on the time periods allowed for equilibration.

Evidence has shown that slow sorption is due to diffusion into soil organic matter that can be viewed as a polymeric network (Brusseau, Jessup, and Rao, 1991). According to more recent investigations (Pignatello and Xing, 1996; Pignatello, 1998) organic sorbents are made up of rubbery regions, where sorption is due to partitioning and linear in nature, and of glassy regions, where sorption occurs by hole-filling mechanisms and is rather nonlinear. Since the glassy regions are believed to be situated in the interior of the sorbent, sorption nonlinearity should increase with time, an effect which has indeed been observed in laboratory experiments. An alternative hypothesis states that the diffusion process occurs in pores of mineral particles covered by organic material. This would mean that diffusion may be better described by a retarded diffusion approach. This hypothesis has been put forward mainly by scientists working with aquifer material, the organic carbon content of which is relatively low. In surface soils with generally high organic matter contents, it seems highly likely that intraorganic matter diffusion is more relevant than intraparticle diffusion.

Regarding the relevance of sorption kinetics, laboratory evidence has been supported by several common field observations. Kinetic sorption offers a simple explanation for the aging effect (Scribner et al., 1992). This term is used to describe the phenomenon that the behavior of agrochemicals in the field depends on contact time. A related finding is that the distribution coefficients measured in the field appear to increase with time. In fact, some modelers routinely account for the increase of the k_D value over time, using a formula proposed by Walker (1987). How far the formation of bound residues can also be explained by intrasorbent diffusion still remains a question (Gevao, Semple, and Jones, 2000). The importance of diffusive processes for the formation of bound residues is indicated by the fact that the amount of residues extracted depends on the extraction time allowed in the experimental protocol (Alexander, 1995).

The diffusion process can be modeled in a simplified manner by taking a mass transfer approach. Since the velocity of sorption appears to vary, the sorbent is divided into two compartments with different accessibility for the dissolved chemical. The total sorbed phase concentration is then the sum of the local sorbed phase concentrations, S_1 and S_2:

$$S = fS_1 + (1-f)S_2 \tag{19.4}$$

where f denotes the fraction of the sorption sites that are readily accessible. Sorption at these sites is assumed to be instantaneous (Equation 19.3):

$$S_1 = kC^m \tag{19.5}$$

A simple linear first-order approach to describe the solute transfer into the interior of the organic matter yields

$$(1-f)\frac{\partial S_2}{\partial t} = \alpha(S_1 - S_2) \tag{19.6}$$

where α (s^{-1}) is a rate coefficient, which can be related to the respective mass transfer coefficient (Streck et al., 1995). When combined with the appropriate equations for mass conservation and transport, the kinetic sorption Equations (19.4) through (19.6) can be applied to processes in both batch and field.

This model, referred to as the two-stage model, is similar to the two-site model introduced by Selim, Davidson, and Mansell (1976), which was later implemented in the popular computer program CXTFIT. Since Selim, Davidson, and Mansell's model was originally derived on chemical grounds, sorption sites were distinguished by assuming different sorption strengths. Many years later, after having put forward the hypothesis of intraorganic matter diffusion, Brusseau and Rao (1989) and Brusseau, Jessup, and Rao (1991) showed that the two-site model can be interpreted as a mass transfer approach to intraorganic matter diffusion. With some redefinition of variables and parameters, two-site and two-stage models are equivalent provided sorption at the fast sites is instantaneous (Streck et al., 1995).

Although the two-site model is almost routinely fitted to breakthrough curves determined in laboratory column experiments, kinetic models have only rarely been tested against field data. In the few published studies (Boesten, van der Pas, and Smelt, 1989; Streck et al., 1995; Streck and Richter, 1999) the agreement between measured and predicted concentration profiles in soil was good to excellent. In these studies, kinetic sorption parameters were independently estimated from laboratory batch experiments. Hence, field and laboratory processes appear to be fully compatible. Since the batch method is simple and cost-effective, this finding is very encouraging regarding practical application.

These field studies have revealed, however, that rate parameters determined in laboratory experiments are useful for modeling solute transport in the field only if the time scales of the laboratory experiment and the field transport process are about equal. The reason is that rate parameters of mass transfer models are not readily applicable to different time scales if the sorption process is diffusive in nature. Boesten, van der Pas, and Smelt (1989) monitored the fate of two herbicides under field conditions. To account for long-term sorption they added a third compartment to their two-site model. The respective kinetic parameters were estimated from laboratory data collected up to 260 days. Unfortunately, such long-term investigations are costly and, because of the simultaneous occurrence of sorption and degradation, not easily accomplished.

Using the simplifying assumption of free diffusion into the sorbent, Streck and Richter (1999) have shown that the rate parameter of the two-stage model should decrease with the square root of time. Although this assumption is only reasonable at far-equilibrium conditions, it yields an upper bound as to how the rate coefficient should decrease over time. The square-root-of-time dependence is in accordance with numerical results by Young and Ball (1995). These researchers investigated the time dependence by running numerical breakthrough experiments using a hypothetical medium made up of spheres of equal radii. Although at the beginning of the sorption process the rate parameter indeed decreases with the square root of time, it approaches a constant value near equilibrium, which corresponds to the so-called Glückauf approximation. It is noteworthy that this approximation is usually applied without paying attention to the restriction that it only holds close to equilibrium. Based on their results, Streck and Richter (1999) proposed a simple equation, which allows application of rate parameters estimated from short-time experimental data to larger time scales. Although the application to their field data was successful it should be kept in mind that the recalculated rate parameter cannot be more than a rough approximation.

In principle, this problem can be avoided if the diffusion process is modeled explicitly by writing a partial differential equation for intrasorbent diffusion. Yet, this approach unfortunately requires an assumption about the geometry of the sorbent. It is often assumed that the soil is made up of uniform spheres. This appears to be an acceptable approximation, at least for the fraction of organic matter present as particles. However, the uniformity assumption will hardly be met in natural soils. Accounting for nonuniformity will considerably increase the necessary computational efforts, but data on the size distribution of organic particles are generally limited. Even if the model input is restricted to statistical information (Culver et al., 1997) the data to assess the statistical distributions will often not be available.

From the mathematical perspective, sorbent heterogeneity can be most easily accounted for with the mass-transfer approach. However, each compartment added increases the number of parameters and hence the cost for the experiments to assess them. To keep the number of parameters small, Chen and Wagenet (1997) presented a multirate approach, which assumes that partitioning coefficients form a continuous distribution. Multirate approaches can be shown to be equivalent to the diffusion approach, provided that the distribution of the rate parameter is selected in a specific manner (Haggerty and Gorelick, 1995). Even if only a few parameters are needed, however, their proper estimation will require that all relevant time scales be sampled. Models with an arbitrary number of compartments have been used in theoretical studies for the sake of generality (Roth and Jury, 1993; Streck and Piehler, 1998).

Degradation

In the soil environment, agrochemicals may be transformed along biotic or abiotic pathways. In general, both degradation pathways occur simultaneously. Biotic and abiotic steps may alternate in a single pathway. In the deeper soil layers abiotic processes are more important. Degradation by photolysis is restricted to a thin soil layer where light can penetrate. Biotic degradation is dominant in the root zone where the biological activity is high. The distinction between biotic and abiotic processes is especially difficult in the field, and from a modeler's perspective it is not really necessary.

Complete decomposition of potentially hazardous agrochemicals is most effectively achieved by microbial degradation. The rate of biodegradation widely differs for the different agrochemicals. It not only depends on soil type, but also on environmental variables such as temperature, water content, and nutrient availability. It may also depend on the concentration of the agrochemical itself. Factors affecting biodegradation are discussed by Aislabie and Lloyd-Jones (1995).

A chemical's half-life is defined as the time after which half of it has been degraded. In the field, the half-lives of agrochemicals may vary between a few days to several years (Rao and Davidson, 1980). Some agrochemicals with very long half-lives, such as DDT (dichlorodiphenyltrichloroethane), have been observed to accumulate in food chains.

Microbial activity may also result in polymerization and conjugation reactions, leading to the incorporation of agrochemicals in soil organic matter. Although the concentration of an agrochemical in the soil solution is effectively reduced with degradation, there is concern about the environmental fate of accumulating bound residues. Refer to Gevao, Semple, and Jones (2000) for a recent review of the formation of bound residues in soils and the possible environmental impact of this process.

The simplest way to model degradation is to assume a first-order process

$$\frac{\partial C_t}{\partial t} = -\mu_t C_t \tag{19.7}$$

where C_t (mg kg^{-1}) is the total concentration of the chemical and μ_t (s^{-1}) is a first-order degradation coefficient. Equation (19.7) assumes that the rate of degradation is the same in all soil phases. Although this assumption is not likely to hold true, it is often made nonetheless, since degradation in the different phases is difficult to quantify (Scow, 1993). In general, sorption is believed to effectively reduce the susceptibility of an agrochemical to degradation. Consequently, degradation may be assumed to be restricted to the solution phase concentration:

$$\frac{\partial C_t}{\partial t} = -\mu_s \theta C \tag{19.8}$$

where θ (unitless) is the volumetric water content while, again, m_s (s^{-1}) is a degradation constant. Comparison of Equations (19.7) and (19.8) reveals that it differs from μ_t by the factor $C_t/(\theta C)$. Hence, the original equations should be carefully inspected when values for first-order coefficients are taken from literature. From a practical point of view, μ_t may be considered a lumped parameter, which therefore depends more strongly on the prevailing environmental conditions.

Gustafson and Holden (1990) showed that biodegradation is no longer first-order when the spatial variability of the degradation constant μ_t is taken into account. Assuming a gamma distribution for μ_t, they tested their new approach against a number of published data sets. Compared to the usual first-order approach, parameter fits could be improved in most cases. Other researchers have accounted for deviations from first-order behavior in a more empirical manner by assuming a second-order or even fractional order process. Experimental data often show that degradation depends on the concentration of the chemical investigated. This behavior can be modelled by the Michaelis-Menten equation (Richter, Diekkrüger, and Nörtersheuser, 1996):

$$\frac{\partial C}{\partial t} = -\frac{V_m C}{K_m + C} \tag{19.9}$$

where V_m (mg l^{-1} s^{-1}) denotes the maximum rate of the reaction and K_m (mg l^{-1}) is the Michaelis-Menten constant. Note that, if $C << K_m$, the reaction is approximately first-order. Equation (19.9) can be derived by considering enzyme kinetics in solution. Despite its theoretical basis, its application in soil science is rather empirical. Consequently, the two parameters have to be estimated from degradation experiments. Upon rewriting the Michaelis-Menten equation in total concentrations, it may be assumed to hold in heterogeneous systems as well. However, it seems more consistent to assume that biodegradation is restricted to the solution phase. In this case, Equation (19.9) should change to:

$$\frac{\partial C_t}{\partial t} = -\frac{\theta V_m C}{K_m + C} \tag{19.10}$$

Biodegradation is largely affected by the prevailing environmental conditions, the most important being soil temperature and soil moisture (Figure 19.2). Effects of temperature and moisture are incorporated into models by measuring degradation under controlled no-flow conditions in the laboratory. The experimental data are then used to estimate the parameters of response curves. Response curves can easily be incorporated into more complex models. The approach implies that the conditions of laboratory and field experiments are compatible, which, in practice, cannot always be guar-

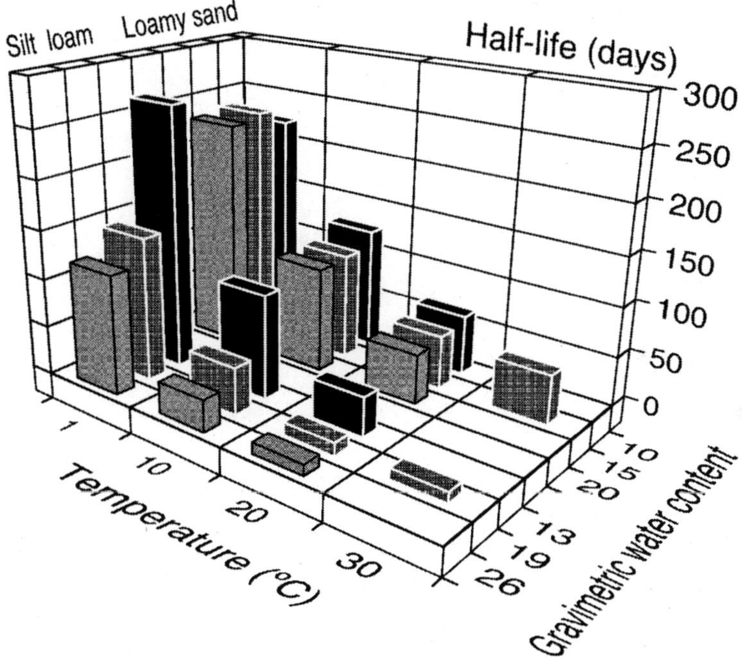

FIGURE 19.2. Half-life of the herbicide chlortoluron as a function of temperature in two soils at different gravimetric water contents. Half-lives were estimated by Heiermann (1998) from laboratory degradation curves. The soils contained 1.2 percent (silt loam) and 1.1 percent (loamy sand) organic carbon.

anteed. Also, the compatibility of the modeling approaches applied to the laboratory data and in the field should be carefully checked.

Temperature effects are most often accounted for by assuming an Arrhenius-type relationship (Zander et al., 1999):

$$\mu(T) = \mu_0 \exp\left(-\frac{a}{T}\right) \tag{19.11}$$

where $\mu(T)$ (s^{-1}) is a rate coefficient (e.g., μ_t or μ_s in Equation 19.7 or 19.8, respectively), T (K) denotes the temperature and μ_0 and a are adjustable parameters. Equation (19.11) works well at moderate temperatures (up to about 25-30°C). At higher temperatures, however, the Arrhenius approach may fail since it does not take into account optimum and maximum temperatures for biodegradation. These may be described by a suitable function, e.g., the empirical function proposed by O'Neill (Richter, Diekkrüger, and Nörtersheuser, 1996). This function contains four adjustable parameters. Compared to the Arrhenius function, which contains two parameters, more experimental data will thus be required for parameter estimation. Obviously, reliable estimates of the optimum or maximum temperature can only be obtained if degradation data are measured over a wide temperature range. If such data are not available, expert knowledge may help to assess the shape of the temperature response curve.

According to Walker (1987) the effect of moisture can be accounted for by:

$$\mu(\theta) = \beta\theta^{\gamma} \tag{19.12}$$

where β and γ are adjustable parameters. To account for the observation that degradation decreases at water contents close to saturation, Richter, Diekkrüger, and Nörtersheuser (1996) proposed the equation:

$$\mu(\theta) = \vartheta\left(\frac{\theta}{\theta_c}\right)^{x} \exp\left(1 - \left(\frac{\theta}{\theta_c}\right)^{x}\right) \tag{19.13}$$

where ϑ, χ, and θ_c are adjustable parameters.

In practice, soil moisture and temperature will act simultaneously. This can be taken into account by combining the selected functions into (Richter, Diekkrüger, and Nörtersheuser, 1996):

$$\mu(T, \theta) = \mu(T)\mu(\theta) \tag{19.14}$$

Parameter estimation requires that degradation data are available for different combinations of soil temperature and moisture.

TRANSPORT IN SOIL

Leaching

Convective-Dispersive Transport

Most agrochemicals are strongly sorbed to the soil matrix so that their displacement is very slow. In many instances, however, a small fraction of the agrochemical applied passes very quickly through the upper soil layers. Since this fraction is considerably less susceptible to degradation, it may cause groundwater pollution. This phenomenon, known as preferential transport or transport along preferential flow paths, will be discussed in the next section. First, we will concentrate on the bulk flow, which can in general be described by the convection-dispersion equation. The derivation of this equation starts from a statement of mass balance, for simplicity written here in one dimension

$$\frac{\partial C_t}{\partial t} = -\frac{\partial J_s}{\partial z} + \sum_i r_i \tag{19.15}$$

where C_t (mg dm^{-3}) and J_s (mg m^{-2} s^{-1}) denote total concentration and solute flux density, respectively, t and z stand for the time and space coordinates and r_i for scalar source or sink terms. The latter may account for processes such as microbial degradation or plant uptake.

The total concentration can be written as the sum of the different phase concentrations:

$$C_t = \theta C + \rho S + \varepsilon C_g \tag{19.16}$$

where ρ and ε stand for the soil bulk density (kg dm^{-3}), and the volume fraction of air-filled pores, respectively. If the partial pressure of the chemical is sufficiently low, concentration and flow in the gas phase can be neglected. Assuming that the transport process is convective-dispersive the solute flux density is given by:

$$J_s = -D_s^w \frac{\partial C}{\partial z} + J_w C \tag{19.17}$$

where D_s^w (m^2 s^{-1}) and J_w (m s^{-1}) denote the apparent dispersion coefficient and the volumetric water flux density, respectively. The transport of volatile agrochemicals will be treated in the following section.

Equations (19.15), (19.16), and (19.17) can now be combined, yielding the convection-dispersion equation:

$$\frac{\partial \theta C}{\partial t}+\frac{\partial \rho S}{\partial t}=\frac{\partial}{\partial z}\left(D_s^w \frac{\partial C}{\partial z}\right)-\frac{\partial J_w C}{\partial z}+\sum_i r_i \tag{19.18}$$

Solving Equation (19.18) requires the specification of (1) water contents, θ, and water flux densities, J_w, which are usually obtained from solving the appropriate equations for the water regime, (2) source and sink terms, r_i, and (3) the relationship between sorbed and solution phase concentrations, that is, either a sorption isotherm or a set of kinetic equations. Although they may partly cancel, errors in the input terms will in general severely affect the accuracy of the solution. It can be shown, however, that simulation results are rather insensitive to errors in water content if the solute considered undergoes strong sorption in soil (Streck and Piehler, 1998).

Preferential Flow

Bulk displacement of agrochemicals is generally very slow. As mentioned, however, a small fraction of mass applied is often transported with soil water which rapidly flows along cracks or macropores. The existence of such preferential flow paths has been demonstrated in various soils using dye tracer techniques. Cracks develop in heavier soils upon soil swelling and shrinking and may persist for years. Macropores are most often biopores, which are formed by earthworm activity or root growth. The number of earthworm burrows and other biopores can be as high as 600 per square meter (Poier and Richter, 1995). Single macropores may extend to considerable soil depth, earthworm pores, e.g., as deep as 2 m. Preferential transport can also be observed in sandy soils, particularly if soils are water repellent (Ritsema et al., 1993). In this case, preferential water flow results from instabilities leading to the formation of fingers with higher water content and, hence, higher conductivity.

The effect of preferential flow paths on the transport of agrochemicals is demonstrated in Figure 19.3. Michaelsen and Rexilius (1997) applied two herbicides, carbetamide and dimefuron, and also bromide as a nonsorbing tracer to a tile-drained field soil (sandy loam) and monitored the concentrations in the drain outlet. The distance between soil surface and drainage lines was 1 m. With the first rainfall, seven days after application, both herbicides appeared in the drainage water at high concentrations, together with bromide. Mass loss was about 1 percent of mass applied. With each of the following rainfalls, the herbicides showed new peak concentrations, decreasing with time.

Flury (1996) reviewed a large number of field experiments and concluded that up to five percent of mass applied is lost by preferential trans-

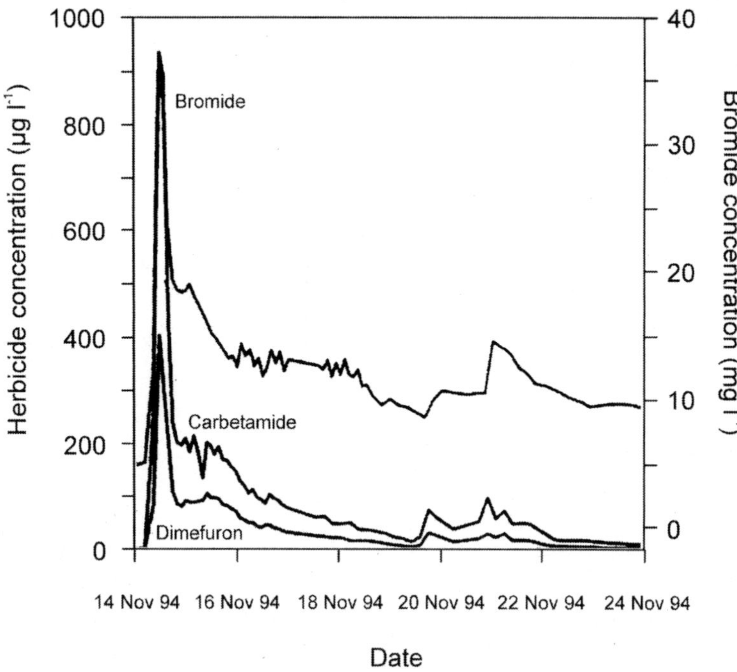

FIGURE 19.3. Concentrations of bromide, carbetamide, and dimefuron in the drainage water of a sandy loam soil in northern Germany. Day of application was November 7, 1994, and first rain fell on November 14, 1994. (*Source:* Michaelsen and Rexilius, 1997, p. 57, translated.)

port. Unless application is followed by heavy rainfall, mass loss can be expected to be smaller than one percent. Preferential transport depends on soil tillage (conservation tillage versus plowing, superficial application versus mechanical incorporation), initial water content, type of irrigation, and time between application and first rainfall.

According to capillary theory, water flow from the soil domain to and along a macrostructure (crack or pore) may only occur when the soil domain is close to water saturation. When flowing down the macrostructure, water and solute will reenter the soil domain at locations where the soil is dry or where the macrostructure ends, e.g., at the plow sole (Stange, Diekkrüger, and Nordmeyer, 1998). Proper modeling of the transfer between macrostructure and soil domain is complicated by the fact that hydraulic conductivity of the soil may vary with distance from the macrostructure. For example, walls of earthworm burrows have been shown to have a lower conductivity than the surrounding soil. Moreover, there is some debate regarding the flow mechanism in macropores (Germann, 1987).

The present ideas of the underlying physics are in qualitative accordance with observations, but in reality the problem of modeling the processes of preferential water flow and transport have not really been solved yet. Apart from a few exceptions (Thoma and Priesack, 1993, who solved the hydrodynamic equations for water flow in a three-dimensional pore) current approaches use a two-domain concept. In the soil domain water regime and solute transport are modeled with the Richards equation and the convection-dispersion equation. Gerke and van Genuchten (1993) used the same approach (with different parameters) in the macropore domain. More investigators, however, assume a simple power law to calculate the water flux in the macropore domain (Jarvis, Bergström, and Dik, 1991; Germann and Beven, 1985):

$$J_{w,m} = K_{s,m}\left(\frac{\theta_m}{\theta_{s,m}}\right)^a \tag{19.19}$$

where $J_{w,m}$ (m s^{-1}), $K_{s,m}$ (m s^{-1}), θ_m, and $\theta_{s,m}$ denote water flux density, hydraulic conductivity, actual and saturated water content in the macropore region, while a is an adjustable parameter. The exchange of water and solutes is modeled with a first-order approach where at a given depth water and solute transfer are proportional to the average matric potential in the soil domain and inversely proportional to the half-width between macropores. The model has already been applied under field conditions, but some of the parameters are usually estimated by calibration (Jarvis et al., 1991). In fact, at the present state of knowledge, the independent estimation of the parameters of the two-domain approach seems difficult, if not impossible.

Cotransport

Natural soil solutions may contain considerable amounts of organic matter. Measurable organic carbon is termed DOC (dissolved organic carbon), no matter if it is really dissolved or bound to particulates (for a further description of DOC see Chapter 13). As outlined by Totsche, Knabner, and Kögel-Knabner (1996) the impact of DOC on the transport of agrochemicals may be quite complex. Depending on the relative strength of sorption that occurs between agrochemicals, DOC and the stationary soil organic matter, the apparent solubility and thereby the transport velocity of agrochemicals may either be reduced or enhanced. Moreover, DOC in the soil solution may vary in space and time, with regard to both mass and composition.

The formation of complexes between agrochemicals and DOC is believed to be fast (Knabner, Totsche, and Kögel-Knabner, 1996). Even though transport with DOC is thus relatively easy to calculate by introducing an additional term in Equation (19.17), the necessary information will hardly be available in the field. A difficult problem arises if cotransport occurs along preferential flow paths. A modeling study of particle transport in macroporous soils was recently presented by Jarvis, Villholth, and Ulén (1999).

Presuming that DOC concentrations are in steady state or even constant with depth, DOC-affected transport can be modeled by means of effective isotherms (Knabner, Totsche, and Kögel-Knabner, 1996; Totsche, Knabner, and Kögel-Knabner, 1996). Since standard protocols make no effort to distinguish between free and carrier-bound chemicals in solution, such sorption isotherms implicitly account for DOC effects. However, DOC concentration and chemical composition are not necessarily the same in the batch and in the field, particularly if the formation of DOC depends on the soil to solution ratio. According to Chiou (1989), for agrochemicals with $k_{oc} < 5$ dm^3 kg^{-3} the effect of DOC can be neglected, presumed that DOC < 100 mg l^{-1}.

Soil Heterogeneity

The practical application of environmental fate models for agrochemicals is hindered by the heterogeneity of field soils (Jury et al., 1987). Although soil layering can generally be traced back to the different processes occurring during soil genesis, horizontal heterogeneity is often considered to be essentially stochastic. Even in apparently homogeneous fields, the coefficient of variation of agrochemical distribution coefficients may be 30-40 percent (Elabd, Jury, and Cliath, 1986; Bunte and Pestemer, 1991). Moreover, initial and boundary conditions will also vary. Due to lack of information, in general, the problem cannot be resolved by considering the variability of parameters and other input variables explicitly. In fact, it is most often assumed that solute transport at the field scale can be modeled with the one-dimensional convection-dispersion equation, using "effective" parameters. Yet, the convective-dispersive approach is usually taken solely for practical reasons. It is often overlooked that taking this approach implies an assumption on the lateral mixing process, which can be seen as the key to modeling solute transport at the field scale (Flühler, Durner, and Flury, 1996). Besides transport distance (or time), the degree of mixing also depends on the correlation length of the relevant model input variables. Close to the soil surface, lateral solute mixing is negligible. The degree of mixing increases with the transport distance, until complete mixing is achieved. In more theoretical terms, solute transport starts as a stochastic-convective process and asymptotically develops to a convective-dispersive process (Jury and Roth, 1990). Only the latter is represented by the convection-dispersion equation. Unfortunately, the proper selection of the transport model can, at present, only be judged from experiment. Regarding the relatively short transport distances in field problems, however, the convective-dispersive approach is probably often not justified if applied at the field scale.

As long as lateral solute mixing is negligible, the transport process can be thought of as occurring in an ensemble of isolated soil columns (Dagan, 1993). This approach has been termed as the stream-tube or parallel soil columns (PSC) approach (Jury and Roth, 1990). In each column, solute transport is assumed to be convective-dispersive. Since local dispersion is generally small, the process is approximately stochastic-convective. In many studies, local dispersion is neglected completely. Each column is characterized by a set of local parameters, which are characterized by their statistics (marginal or joint probability density function). The boundary conditions, e.g., flux of irrigation water or mass of chemical applied, may also be given as statistical distributions. Since the model equations are frequently solved by means of the Monte Carlo method, PSC simulations are often simply termed "Monte Carlo simulations." The solution gives ensembles of concentration profiles or breakthrough curves. Most applications only aim at the mean (field-scale) concentration profiles or breakthrough curves, but the variance may also be of interest in some cases. This type of Monte Carlo simulation is not to be confused with the application of the Monte Carlo method to assess the uncertainty of modeling predictions, as, e.g., by Soutter and Musy (1998) in their regional study on pesticide leaching to groundwater. In fact, both techniques may as well be combined to account for the uncertainty of spatially variable input variables ("double-looping") (Cullen and Frey, 1999).

The water regime at the field scale can also be modeled by taking a PSC approach. In each column, water flow and water content are then modeled with the Richards equation. As mentioned previously, however, the variability of water contents has only a slight impact on the transport of strongly sorbing solutes. Moreover, the impact of spatially variable water flow tends to be overridden by the spatial variability of sorption (van der Zee and van Riemsdijk, 1987). As a consequence, simulation of the water regime can often be considerably simplified.

Although Monte Carlo simulations have become a standard tool to model transport of agrochemicals in soils, results have rarely been tested against field data. The published studies are not fully comprehensible in that the spatial distributions of only some of the parameters have been measured. Moreover, regarding the agreement between modeled and measured data the results of the studies are not consistent (Gish et al., 1986; Indelman et al., 1998; Streck and Richter, 1999). Due to the homogenizing effect of soil tillage, the impact of soil heterogeneity on bulk transport of agrochemicals in plowed topsoils is likely to be negligible (Streck and Richter, 1999).

VOLATILIZATION

The preceding sections have dealt with only nonvolatile agrochemicals. However, depending on the vapor pressure and the environmental conditions, diffusion through the air-filled pores in soil may contribute to the total flux of the chemical considered. The other flux components, convection as well as dispersion, and diffusion in the solution phase, may even be negligible. If no artificial pressure gradient is induced (such as in remediation operations), convection with the flowing soil air can in general be neglected. Volatilization from soil and plant surfaces shortly after application as well as spray drift may also cause significant losses of agrochemicals to the atmosphere.

Solution and gas phase (soil air) concentrations of agrochemicals are related by Henry's law (Jury, Gardner, and Gardner, 1991):

$$C_g = k_H C \tag{19.20}$$

where C_g denotes the gas phase concentration and k_H is the dimensionless form of Henry's constant. In dilute solutions, k_H is related to the more common form of Henry's constant in pressure units (Pa), k_H, by (Jury, Gardner, and Gardner, 1991):

$$k_H = \frac{M_w}{\rho_w RT} k_H^* \tag{19.21}$$

where M_w and ρ_w are molecular mass and density of water, respectively, and R denotes the universal gas constant. Note that Equation (19.20) assumes equilibrium between solution and gas phases. To account for the flux in the gas phase Equation (19.17) can be rewritten as:

$$J_s = -D_s^* \frac{\partial C}{\partial z} + J_w C \tag{19.22}$$

where

$$D_s^* = D_s^w + k_H D_s^g \tag{19.23}$$

is an apparent liquid and vapor diffusion/dispersion coefficient (m^2 s^{-1}). The vapor diffusion coefficient can be calculated from the binary diffusion coefficient in air, D^g (m^2 s^{-1}), by

$$D_s^g = \xi_g D^g \tag{19.24}$$

where ξ_g is the tortuosity factor. Several formulae are available to assess the value of ξ_g from air and water contents of the soil (Jin and Jury, 1996). It should be stressed, however, that due to the very important impact of soil structure, e.g., the existence of macropores, these formulae do not work well in undisturbed soils. The presence of only a few air-filled macropores, for example, may considerably enhance vapor diffusion in wet soils. On the other hand, a plow pan or surface crust may seriously reduce vapor diffusion under otherwise favorable conditions.

Inserting Equation (19.22) into Equation (19.15) and using Equations (19.16) and (19.20) yields

$$\frac{\partial \theta C}{\partial t} + \frac{\partial \rho S}{\partial t} + k_H \frac{\partial \varepsilon C}{\partial t} = \frac{\partial}{\partial z}\left(D_s^* \frac{\partial C}{\partial z} \right) - \frac{\partial J_w C}{\partial z} + \sum_i r_i \tag{19.25}$$

which can be solved without difficulty provided an additional (kinetic or equilibrium) sorption equation is given and the source and sink terms, r_i, are specified. If storage and flux in the soil solution can be neglected, Equation (19.25) simplifies to

$$\frac{\partial \varepsilon C_g}{\partial t} + \frac{\partial \rho S}{\partial t} = \frac{\partial}{\partial z}\left(D_s^g \frac{\partial C_g}{\partial z} \right) + \sum_i r_i \tag{19.26}$$

Although the vapor pressure of even the most volatile agrochemicals is comparably low, volatilization from the soil surface may be considerable, since the concentration of agrochemicals in the air is generally close to zero. On the other hand, however, the agrochemicals have to cross the boundary layer between soil and atmosphere. Assuming steady state and zero background concentration in the atmosphere, the vapor flux density across the boundary layer is given by (Jury, Spencer, and Farmer, 1983):

$$J_{s,\,boundary} = -D^g \frac{C_g(0,t)}{d} \tag{19.27}$$

where $C_g(0,t)$ stands for the concentration of the chemical at the soil surface, and d (m) denotes the thickness of the boundary layer.

Jury, Spencer, and Farmer (1983, 1984) have used an analytical solution of Equation (19.25), subject to Equation (19.27), to assess the environmental fate of a number of common agrochemicals. They concluded that volatilization may greatly affect the persistence of many of the screened chemicals in soil.

PLANT UPTAKE

The accumulation of a chemical in a crop relative to that of the soil on which the crop grows is often expressed by an empirical index, the bioconcentration factor *(BCF).* The *BCF* is defined as the ratio of measured concentrations of the chemical in plant material and in soil. Both concentrations are calculated on a dry weight basis. Based on a literature review, Travis and Arms (1988) have listed *BCF* values for a number of organic chemicals. In this noncomprehensive list, the *BCF* values of the agrochemicals vary between 0.0008 (herbicide benfluralin) and 7 (herbicide aldicarb). *BCF* values have been found to be negatively correlated to octanol-water partition coefficients. Since the octanol-water partition coefficient reflects the water solubility of a chemical this correlation indicates that the latter plays an important role regarding the transfer of agrochemicals from soil to plants.

Agrochemicals are taken up by plants through different mechanisms (Trapp et al., 1990). In the mass balance Equation (19.15), plant uptake is considered through a scalar sink term. For agrochemicals with moderate to high water solubility the most important uptake path is by convection with the transpiration water:

$$r_T = Q_T C \tag{19.28}$$

where Q_T (s^{-1}) denotes the water uptake by plant roots and C represents the concentration of the agrochemical in the soil solution. The transpiration stream may freely flow through the apoplast, but, since it cannot pass the Casparian strip, it must cross a biological membrane before entering the symplast. As a consequence, composition may change. In the literature, this has been accounted for by introducing a reflection coefficient (Lindstrom, Boersma, and McFarlane, 1991) or the transpiration stream concentration factor (Briggs, Bromilow, and Evans, 1982). Upon insertion of the latter, τ, Equation (19.28) changes to

$$r_T = \tau Q_T C \tag{19.29}$$

For nonionic organic chemicals, τ can be calculated from the octanol-water partition coefficient by applying the empirical relationship (Briggs, Bromilow, and Evans, 1982):

$$\tau = 0.784 \exp\left(-\frac{(\log K_{OW} - 1.78)^2}{2.44} \right) \tag{19.30}$$

When plant uptake is active, a radial depletion (or accumulation) zone may develop around the plant root, giving rise to a radial diffusive flux to the root. In cylindrical coordinates, the radial flux density, J_r (mg m^{-2} s^{-1}), is given by:

$$J_r = -D_s^w \frac{\partial C}{\partial r} \tag{19.31}$$

where r denotes the radial coordinate and D_s^w is the apparent dispersion coefficient defined in the transport section. Most often, the diffusive flux to a single root of length l is approximated by the stationary flux, J_r^* (mg s^{-1}), across the depletion zone which extends from r_1 to r_2

$$J_r^* = 2\pi l \, D_s^w \frac{C_2 - C_1}{\ln r_2 - \ln r_1} \tag{19.32}$$

where r_1 and r_2 are the radii of root and depletion zone, respectively, while C_1 and C_2 stand for the dissolved concentrations at the root surface and in the bulk solution, respectively. For the sake of simplicity, it is usually assumed that all roots have the same diameter. By replacing D_s^w by D_s^*, as discussed in the preceding section, Equation (19.32) can be changed to represent the diffusive fluxes in both soil water and air. If the root length l is replaced by the root length density in soil (m m^{-3}) the right side of Equation (19.32) gives the diffusive flux as a sink term, r_D, which may directly be inserted into Equation (19.15).

To assess possible contamination of food, it is not sufficient to model the total uptake of an agrochemical by the crop. The aim should rather be to assess the concentrations in specific plant parts, in particular those to be processed for food production (e.g., wheat grain). The redistribution of agrochemicals among different plant parts has been modeled by several researchers (Boersma, Lindstrom, and McFarlane, 1988; Lindstrom, Boersma, and McFarlane, 1991; Trapp et al., 1990; Trapp, McFarlane, and Matthiess, 1994; Trapp, 2000). Despite their difference in complexity, the models have a similar structure. They use the linear compartment approach to account for fluxes and storage changes in the soil-plant system. The soil is represented as one, well-mixed compartment. This approach leads to a system of ordinary differential equations, the solution of which yields the concentrations of the agrochemical considered in the soil and plant compartments over time. Uptake models have been applied to data from laboratory and outdoor lysimeter experiments (by Boersma, Lindstrom, and McFarlane, 1991; Trapp et al., 1990; Gayler et al., 1995), but not yet thoroughly tested against field data.

SUMMARY

Disregarding volatilization during or shortly after application, the environmental fate of agrochemicals is mainly controlled by sorption and degradation in soil. Both processes are often nonlinear. If the focus is on field problems, however, the linearity assumption, frequently made by modelers, may be a sufficient approximation. Sorption and desorption of many agrochemicals are slow, with equilibration times of weeks or even months, which is probably due to diffusion within the soil organic matter. This process may explain, at least partly, the so-called aging effect as well as the formation of bound residues in soil. In view of the equilibration times allowed in the current protocols, slow sorption may prevent the correct determination of the equilibrium partitioning coefficient in the laboratory. Its impact on a field transport process depends on the time scales of sorption and transport processes. Both sorption and degradation are severely affected by the natural heterogeneity of field soils.

Although agrochemicals undisputedly contribute to the growing demand for food in the world, the possible contamination of groundwater is an important problem, both in developed and developing countries. Whereas the frequent detection of agrochemicals in shallow groundwater cannot be explained by matrix flow, there is evidence that transport essentially occurs along preferential flow paths, a process that at present cannot be modeled in a predictive manner. To what extent cotransport with particles, especially with dissolved organic matter, may contribute to leaching of agrochemicals to groundwater is not yet clear. For more volatile agrochemicals, diffusion in the vapor phase may also be an important pathway to the environment.

REFERENCES

Aislabie, J. and G. Lloyd-Jones (1995). A review of bacterial degradation of pesticides. *Australian Journal of Soil Research 33:* 925-942.

Alexander, M. (1995). How toxic are toxic chemicals in soils? *Environmental Science & Technology 29:* 2713-2717.

Altfelder, S., T. Streck, and J. Richter (2000). Nonsingular sorption of organic compounds in soil: The role of slow kinetics. *Journal of Environmental Quality 29:* 917-925.

Anonymous (2001). Asian market recovers. *Agrow World Crop Protection News 370.* Richmond, Surrey, UK: PJB Publications,

Ball, W.P. and P.V. Roberts (1991). Long-term sorption of halogenated organic chemicals by aquifer material. 1. Equilibrium. *Environmental Science & Technology 25:* 1223-1237.

Battersby, N.S. (1990). A review of biodegradation kinetics in the aquatic environment. *Chemosphere 21:* 1243-1284.

Boersma, L., F.T. Lindstrom, and C. McFarlane (1988). Uptake of organic chemicals by plants: A theoretical model. *Soil Science 146:* 403-417.

Boersma, L., F.T. Lindstrom, and C. McFarlane (1991). Mathematical model of plant uptake and translocations of organic chemicals: Applications to experiments. *Journal of Environmental Quality 20:* 137-146.

Boesten, J.J.T.I., L.J.T. van der Pas, and J.H. Smelt (1989). Field test of a mathematical model for non-equilibrium transport of pesticides in soil. *Pesticide Science 25:* 187-203.

Briggs, G.G., R.H. Bromilow, and A.A. Evans (1982). Relationship between lipophilicity and root uptake and translocation of non-ionised chemicals by barley. *Pesticide Science 13:* 495-504.

Brusseau, M.L., R.E. Jessup, and P.S.C. Rao (1989). Modeling the transport of solutes influenced by multiprocess nonequilibrium. *Water Resources Research 25:* 1971-1988.

Brusseau, M.L., R.E. Jessup, and P.S.C. Rao (1991). Nonequilibrium sorption of organic chemicals: Elucidation of rate-limiting processes. *Environmental Science & Technology 25:* 134-142.

Brusseau, M.L. and P.S.C. Rao (1989). The influence of sorbate-organic matter interactions on sorption nonequilibrium. *Chemosphere 18:* 1691-1706.

Bunte, D. and W. Pestemer (1991). Horizontale und vertikale Variabilität bodenkundlicher Kenndaten und deren Einfluß auf das Verhalten von Pflanzenschutzmitteln auf landwirtschaftlich genutzten Flächen. *Nachrichtenbatt des Deutschen Pflanzenschutzdienstes 43:* 238-244.

Chen, W. and R.J. Wagenet (1997). Description of atrazine transport in soil with heterogeneous nonequilibrium sorption. *Soil Science Society of America Journal 61:* 360-371.

Chiou, C.T. (1989). Theoretical considerations of the partition uptake of nonionic organic compounds to soil organic matter. In *Reactions and Movement of Organic Chemicals in Soils,* eds. B. L. Sawhney and K. Brown. Madison, WI: Soil Science Society of America, Special Publication 22, pp. 1-29.

Committee on the Future Role of Pesticides in U.S. Agriculture (2000). *The Future Role of Pesticides in US Agriculture.* Washington, DC: National Academy Press.

Cullen, A. and H. Frey (1999). *Probabilistic Techniques in Exposure Assessment: A Handbook for Dealing with Variability and Uncertainty in Models and Inputs.* New York: Plenum Press.

Culver, T.B., S.P. Hallisey, D. Sahoo, J.J. Deitsch, and J.A. Smith (1997). Modeling the desorption of organic contaminants from long-term contaminated soil using distributed mass transfer rates. *Environmental Science & Technology 31:* 1581-1588.

Dagan, G. (1993). The Bresler-Dagan model of flow and transport: Recent theoretical developments. In *Water Flow and Solute Transport in Soils,* eds. D. Russo and G. Dagan. Berlin, Germany: Springer-Verlag, pp. 13-32.

Elabd, H., W.A. Jury, and M.M. Cliath (1986). Applicability of the steady state flow assumption for solute advection in field soils. *Environmental Science & Technology 20:* 256-260.

Flühler, H., W. Durner, and M. Flury (1996). Lateral solute mixing processes: A key to understanding field-scale transport of water and solutes. *Geoderma 70:* 165-183.

Flury, M. (1996). Experimental evidence of transport of pesticides through field soils: A review. *Journal of Environmental Quality 25:* 25-45.

Gayler, S., S. Trapp, R. Schroll, and H. Behrendt (1995). Uptake of terbuthylazine and its medium polar metabolites into maize plants. *Environmental Science and Pollution Research 2:* 98-103.

Gerke, H.H. and M.T. van Genuchten (1993). A dual-porosity model for simulating the preferential movement of water and solutes in structured porous media. *Water Resources Research 29:* 305-319.

Germann, P.F. (1987). The three modes of water flow through a vertical pipe. *Soil Science 144:* 153-154.

Germann, P. and K. Beven (1985). Kinematic wave approximation to infiltration into soils with sorbing macropores. *Water Resources Research 21:* 990-996.

Gevao, B., K.T. Semple, and K.C. Jones (2000). Bound pesticide residues in soils: A review. *Environmental Pollution 108:* 3-14.

Gish, T., W. Zhuang, C.S. Helling, and P.C. Kearney (1986). Chemical transport under no-till field conditions. *Geoderma 38:* 251-259.

Green, R.E. and S.W. Karickhoff (1990). Sorption estimates for modeling. In *Pesticides in the Soil Environment: Processes, Impacts, and Modeling,* ed. H.H. Cheng. Madison, WI: Soil Science Society of America, pp. 351-399.

Gustafson, D.I. and L.R. Holden (1990). Nonlinear pesticide dissipation in soil: A new model based on spatial variability. *Environmental Science & Technology 24:* 1032-1038.

Gustafson, D.J. (1989). Groundwater ubiquity score: A simple method for assessing pesticide leachability. *Environmental Toxicology and Chemistry 8:* 339-357.

Haggerty, R. and S.M. Gorelick (1995). Multiple-rate mass transfer for modeling diffusion and surface reactions in media with pore-scale heterogeneity. *Water Resources Research 31:* 2383-2400.

Heiermann, M. (1998). Untersuchungen zum Verhalten von Herbiziden im Boden als Grundlage für Simulationsrechnungen im Herbst und Winter. (Doctoral thesis, Humboldt-University Berlin, Germany.)

Indelman, P., I. Touber-Yasur, B. Yaron, and G. Dagan (1998). Stochastic analysis of water flow and pesticides transport in a field experiment. *Journal of Contaminant Hydrology 32:* 77-97.

Jarvis, N.J., L. Bergström, and P.E. Dik (1991). Modelling water and solute transport in macroporous soil. II. Chloride breakthrough under non-steady flow. *Journal of Soil Science 42:* 71-81.

Jarvis, N.J., P. Jansson, P.E. Dik, and I. Messing (1991). Modelling water and solute transport in macroporous soil. II. Model description and sensitivity analysis. *Journal of Soil Science 42:* 59-70.

Jarvis, N.J., K.G. Villholth, and B. Ulén (1999). Modelling particle mobilization and leaching in macroporous soil. *Euopean Journal of Soil Science 50:* 621-632.

Jin, Y. and W.A. Jury (1996). Characterizing the dependence of gas diffusion coefficient on soil properties. *Soil Science Society of America Journal 60:* 66-71.

Jury, W.A., W.R. Gardner, and W.H. Gardner (1991). *Soil Physics.* Fifth edition. New York: John Wiley & Sons.

Jury, W.A. and K. Roth (1990). *Transfer Functions and Solute Movement Through Soil: Theory and Applications.* Basel, Switzerland: Birkhäuser.

Jury, W A., D. Russo, G. Sposito, and H. Elabd (1987). The spatial variability of water and solute transport properties in unsaturated soil. I. Analysis of property variation and spatial structure with statistical models. *Hilgardia 55:* 1-32.

Jury, W.A., W.F. Spencer, and W.J. Farmer (1983). Behavior assessment model for trace organics in soil, I. Model description. *Journal of Environmental Quality 12:* 558-563.

Jury, W.A., W.F. Spencer, and W.J. Farmer (1984). Behavior assessment model for trace organics in soil, III. Application of screening model. *Journal of Environmental Quality 13:* 573-579.

Kleineidam, S., H. Rügner, B. Ligouis, and P. Grathwohl (1999). Organic matter facies and equilibrium sorption of phenanthrene. *Environmental Science & Technology 33:* 1637-1644.

Knabner, P., K.U. Totsche, and I. Kögel-Knabner (1996). The modeling of reactive solute transport with sorption to mobile and immobile sorbents. 1. Experimental evidence and model development. *Water Resources Research 32:* 1611-1622.

Kolpin, D.W., J.E. Barbash, and R.J. Gilliom (1998). Occurrence of pesticides in shallow groundwater of the United States: Initial results from the national water-quality assessment program. *Environmental Science & Technology 32:* 558-566.

Länderarbeitsgemeinschaft Wasser (LAWA) (1997). *Bericht zur Grundwasserbeschaffenheit: Pflanzenschutzmittel.* Berlin: Kulturbuchverlag.

Lindstrom, F.T., L. Boersma, and C. McFarlane (1991). Mathematical model of plant uptake of organic chemicals: Development of the model. *Journal of Environmental Quality 20:* 129-136.

Michaelsen, J. and Rexilius, L. (1997). *Untersuchungen zum Migrationsverhalten von Pflanzenschutzmitteln nach Herbstapplikation auf einer gedränten Ackerfläche-Ergebnisse aus 1994/95.* Mitteilungen aus der Biologischen Bundesanstalt für Land-und Forstwirtschaft, Berlin: Parey Buchverlag. Volume 328, pp. 49-61.

Müller, F. (2000). *Agrochemicals: Composition, Production, Toxicology, Applications.* Weinheim, Germany: VCH Publishing.

Pignatello, J.J. (1998). Soil organic matter as a nanoporous sorbent of organic pollutants. *Advances in Colloid and Interface Science 76-77:* 445-467.

Pignatello, J.J. and B. Xing (1996). Mechanisms of slow sorption of organic chemicals to natural particles. *Environmental Science & Technology 30:* 1-11.

Poier, K.R. and J. Richter (1995). Earthworm burrows as a structural element of intensively used arable Loess soils. In *Soil Structure: Its Development and Function,* eds. K.H. Hartge and B.A. Stewart. Boca Raton, FL: CRC Lewis Publishers, pp. 175-195.

Rao, P.S.C. and J.M. Davidson (1980). Estimation of pesticide retention and transformation parameters required in non-point source pollution models. In *Environmental Impact of Nonpoint Source Pollution,* eds. M.R. Overcash and J.M. Davidson. Ann Arbor, MI: Ann Arbor Science Publishers, pp. 23-67.

Richter, O., B. Diekkrüger, and P. Nörtersheuser (1996). *Environmental Fate Modelling of Pesticides.* Weinheim, Germany: VCH Verlagsgesellschaft.

Ritsema, C.J., L.W. Dekker, J.M.H. Hendrickx, and W. Hamminga (1993). Preferential flow mechanism in a water repellent sandy soil. *Water Resources Research 29:* 2183-2193.

Roth, K. and W.A. Jury (1993). Linear transport models for adsorbing solutes. *Water Resources Research 29:* 1195-1203.

Scow, K.M. (1993). Effect of sorption-desorption and diffusion processes on the kinetics of biodegradation of organic chemicals in soil. In *Sorption and Degradation of Pesticides and Organic Chemicals in Soil,* ed. D.M. Linn. Madison, WI: Soil Science Society of America and American Society of Agronomy, pp. 73-114.

Scribner, S.L., T.R. Benzing, S. Sun, and S.A. Boyd (1992). Desorption and bioavailability of aged simazine residues in soil from a continuous corn field. *Journal of Environmental Quality 21:* 115-120.

Selim, H.M., J.M. Davidson, and R.S. Mansell (1976). "Evaluation of a two-site adsorption-desorption model for describing solute transport in soils." (Paper presented at Summer Computer Simulation Conference. Washington, DC: Simulation Councils Inc.).

Soutter, M. and A. Musy (1998). Coupling 1d Monte-Carlo simulations and geostatistics to assess groundwater vulnerability to pesticide contamination on a regional scale. *Journal of Contaminant Hydrology 32:* 25-39.

Stange, C.F., B. Diekkrüger, and H. Nordmeyer (1998). Measurement and simulation of herbicide transport in macroporous soils. *Pesticide Science 52:* 241-250.

Streck, T. and H. Piehler (1998). On field scale dispersion of strongly sorbing solutes in soils. *Water Resources Research 34:* 2769-2773.

Streck, T., N.N. Poletika, W.A. Jury, and W.J. Farmer (1995). Description of simazine transport with rate-limited, two-stage, linear and nonlinear sorption. *Water Resources Research 31:* 811-822.

Streck, T. and J. Richter (1999). Field-scale study of chlortoluron movement in a sandy soil over winter. II. Modeling. *Journal of Environmental Quality 28:* 1824-1831.

Thoma, M. and E. Priesack (1993). Coupled porous and free flow in structured media. *Zeitschrift für angewandte Mathematik und Mechanik:* T566-T569.

Totsche, K.U., P. Knabner, and I. Kögel-Knabner (1996). The modeling of reactive solute transport with sorption to mobile and immobile sorbents 2. Model discussion and numerical simulation. *Water Resources Research 32:* 1623-1634.

Trapp, S. (2000). Modeling uptake into roots and subsequent translocation of neutral and ionisable organic compounds. *Pest Management Science 56:* 767-778.

Trapp, S., M. Matthies, I. Scheunert, and E.M. Topp (1990). Modeling the bioconcentration of organic chemicals in plants. *Environmental Science & Technology 24:* 1246-1252.

Trapp, S., C. McFarlane, and M. Matthiess (1994). Model for uptake of xenobiotics into plants: Validation with bromacil experiments. *Environmental Toxicology and Chemistry 13:* 413-422.

Travis, C.C. and A.D. Arms (1988). Bioconcentration of organics in beef, milk, and vegetation. *Environmental Science & Technology 22:* 271-274.

van der Zee, S.E.A.T.M. and W.H. van Riemsdijk (1987). Transport of reactive solute in spatially variable soil systems. *Water Resources Research 23:* 2059-2069.

Walker, A. (1987). Evaluation of a simulation model for prediction of herbicide movement and persistence in soil. *Weed Research 27:* 143-152.

Young, D.F. and W.P. Ball (1995). Effects of column conditions on the first-order rate modeling of nonequilibrium solute breakthrough. *Water Resources Research 31:* 2181-2192.

Yudelman, M., A. Ratta, and D. Nygaard (1998). *Pest Management and Food Production: Looking to the Future.* Food, Agriculture, and the Environment, Discussion Paper 25. Washington, DC: International Food Policy Research Institute.

Zander, C., T. Streck, T. Kumke, S. Altfelder, and J. Richter (1999). Field-scale study of chlortoluron movement in a sandy soil over winter. I. Experiments. *Journal of Environmental Quality 28:* 1817-1823.

Chapter 20

Crop Growth and Development

Gerrit Hoogenboom

The main objective of modeling crop growth and development within the soil-plant system is to predict crop yield. Crop yield can consist of different products, depending on the crop that is being grown and the product that will be harvested. For cereals, in most cases, the grain is harvested, although maize and sorghum can also be grown for silage. For grain legumes such as soybeans, the seeds are harvested while for peanuts the pods are harvested. For potatoes and similar crops the tubers or roots are harvested. For vegetables normally the fresh product is harvested in the form of fruits, such as tomatoes, cucumbers, peppers, and eggplant, or the entire crop, such as lettuce and endive. Green bean is grown as a vegetable when the immature pods are harvested, but it is also grown as an agronomic crop for seed in the form of dry beans or common beans. Other crops that are important include fruit crops and flowers. For many crops, especially vegetables and flowers, the quality of the crop with respect to composition and other characteristics is as important as total production. For some agronomic crops, including wheat, quality and composition has also become important. Modeling crop yield is therefore not as easy as it initially seems, as it is not only a function of the type of crop, but also the product that is being harvested. Despite this complexity, crop modelers have attempted to develop simple or complex simulation models for most of these crops. They have used a systems analysis approach to clearly define the individual components of the soil-plant-atmosphere system that are important and the interactions between the plant, the soil, and the atmosphere. These interactions include physical, chemical, as well as biological processes.

When modeling crop growth and development, researchers have modeled at different scales, starting at the cellular level (Milthorpe and Newton, 1963; Molz and Ferrier, 1982) to complex processes, such as photosynthesis (Charles-Edwards, 1981). Others have started at the process level and moved to a complete plant or crop (Hoogenboom, Jones, and Boote, 1992). Recently, White and Hoogenboom (1996) and Hoogenboom et al. (1997)

tried to bridge the gap between biotechnology and crop physiology through the development of a gene-based simulation model. Based on crop-level models, researchers have modeled crop yield, production, and resource use for individual farmers' fields, farms, watersheds, regions, and even countries. In many of these cases, crop models have been linked to some type of spatial analysis tool, such as geographic information systems (GIS) (Engel et al., 1997; Lal et al., 1993). A detailed review can be found in Hartkamp, White, and Hoogenboom (1999). With the recent interest in global warming, the potential impact of climate change has been modeled at fairly large scales. Crop models have been used to asses changes in crop production at a district or state level to countries and the entire world (Adams et al., 1990; Rosenzweig et al., 1995; Alexandrov and Hoogenboom, 2000a,b; Iglesias, Rosenzweig, and Pereira, 2000; Rosenzweig and Parry, 1994). The latest impact assessment report for climate change in the United States included several sections that were based on crop models to not only simulate crop yield and production, but also future resource needs such as water (National Assessment Synthesis Team, 2000).

The question can then be asked, what is a crop model, or how do you model crop growth and development? Depending on whom you ask, this question can result in many different answers. Most scientists agree that the ultimate objective is to model yield and crop production. However, most scientists cannot come to an agreement on how to develop or implement a model for a particular crop or application. In one of the earliest modeling papers, Baker and Curry (1976, p. 201) stated that, "Model building is as much an art as science." In 1994, the American Society of Agronomy (ASA) organized a symposium titled "Use and Abuse of Crop Simulation Models," to extend on he ideas of "Sense and Nonsense in Modeling," published by Passioura (1973) during the early years when models were introduced into agriculture. In this ASA symposium, four different perspectives were presented on modeling (Baker, 1996). Passioura (1996) differentiated between models that can be used to improve our understanding of physiology and the interactions with the environment versus those that can be used to help improve crop management. The former ones are associated with science to discover how things work, while the latter ones are associated with engineering to solve problems. Monteith (1996) stated that modelers have a tendency to overemphasize the validity of their models, rather than identifying the circumstances in which a model fails or including error limits for yield and other modeled variables. He encouraged simplification of models by removing components that do not contribute to the numerical precision of the models (Stockle, 1992). Sinclair and Seligman (1996) described crop modeling as being heuristic tools similar to ecological models; there are limitations with respect to the capabilities of these models to simulate the real world system. Nevertheless, Sinclair and Seligman (1996) concluded that models could be important tools for teaching and research as well as management and administrative applications. Boote, Jones, and Pickering (1996) presented the most positive overview of the potential for model applications, as well as the limitations associated with the models. They included several examples of modeling application for research, management, and policy decisions.

HISTORY

Traditionally, modeling of crop growth and development has not been a well-recognized area of research. One of the main causes for this has been the rather slow adaptation of computers and associated software in basic agronomic research. In many cases, agricultural scientists have also shown a disinterest in mathematics. However, this does not apply to all disciplines within agronomy. One can find soil physics, agrometeorology, and others in which mathematics and the implementation of models have been a key component of the research. Examples of these modeling applications can be found in Chapter 10 and Chapter 12 of this book. As a result, crop modelers have had a rather difficult time finding acceptance for modeling specific aspects of crop growth and development within the agronomic discipline.

The first crop models in the United States were developed by agricultural engineers who were interested in applying an overall systems approach to describe a farming system. Stapleton (1970) and Stapleton and Meyers (1971) developed the first model for cotton. Some of these models were based on differential equations to simulate photosynthesis, respiration, and a simple carbon balance, but these models did not account for either vegetative and reproductive development or interaction with the soil (Duncan et al., 1967; Curry, 1971; Curry and Chen, 1971; Splinter, 1974).

In Europe, the former Department of Theoretical Production Ecology of the Wageningen Agricultural University has been one of the leaders in the development and application of crop simulation models. Brouwer and de Wit (1968) and de Wit, Brouwer, and Penning de Vries (1970) developed some of the early crop growth models. Through the "School of de Wit" an extensive group of crop modelers has been trained that has had a large impact on the modeling community (de Wit and Goudriaan, 1974; van Keulen, 1975; Goudriaan, 1977; de Wit et al., 1978; Penning de Vries et al., 1989; Bouman et al., 1996). The underlying approach of this modeling group is based on systems ecology in which the state of the system can be expressed at any point in time and changes of the system can be expressed through mathematical terms. The use of continuous and dynamic simulation models is equivalent to the models that have been developed previously to describe industrial systems (Forrester, 1961).

In the United Kingdom, the initial modeling efforts were conducted at the Glasshouse Crops Research Institute in Littlehampton, West Sussex, for greenhouse crops. Thornley and Hurd (1974) developed one of the first models for tomatoes, based on a detailed growth analysis of plants grown in water culture (Hurd and Thornley, 1974). Most of the models developed by this group were very physiologically based; one of the main objectives was to understand the underlying processes of growth, such as photosynthesis and respiration (Thornley, 1977; Thornley, Hand, and Wilson, 1992), partitioning (Thornley, 1995), or development (Thornley, 1981; Thornley, Hurd, and Pooley, 1981). Many of these processes were modeled at a fairly detailed level in which mathematics and plant physiology were integrated (Charles-Edwards, 1981; Rose and Charles-Edwards, 1981).

DESCRIPTION OF THE APPROPRIATE PROCESSES

A crop simulation model in its simplest form is a mathematical implementation of a real-world system. Detailed mathematical models and associated equations can be found in Charles-Edwards (1982) and Charles-Edwards, Doley, and Rimmington (1986); these equations are still valid today. One can define crop yield as a function of biomass and harvest index. As stated in a previous section, most modelers associate harvest index with harvestable seed yield, but it can include other crop components as well. The following equations provide a qualitative set of descriptors that relate final crop yield to environmental variables as well as genetics.

$$\text{Yield} = \text{f (biomass, harvest index)} \tag{20.1}$$

where:

$$\text{Biomass} = \text{f (growth, development)} \tag{20.2}$$

$$\text{Harvest Index} = \text{f (stress, genetics)} \tag{20.3}$$

In this simplified approach, the main determinants for total biomass are growth and development. The harvest index assumes that a certain fraction of total biomass can be harvested as an economic product. This fraction could be as high as 1 for crops that are harvested for silage, such as corn, 0.2 to 0.3 for grain legumes, such as soybean, or 0 for crops that are grown as a ground cover. Critical in this overall process, however, is to differentiate between the two distinct growth and development components. Growth in general defines the total amount of biomass produced through photosynthesis, accounting for maintenance and growth respiration, and partitioning to the different components. Development relates to the duration of the various growth phases and defines when a crop changes growth stages, such as from vegetative to reproductive through the start of flowering, or from the start of flowering to first pod and first seed occurrence, as well as maturity.

$$\text{Growth} = f\text{ (environment, stress, genetics)} \quad (20.4)$$

Environmental factors that control growth mainly affect photosynthesis and include the total amount of incoming solar radiation, the CO_2 concentration of the air, as well as temperature, especially when crops are grown under nonoptimum conditions. Temperature can also affect respiration and partitioning. The underlying interaction with the environment, however, is defined by genetics. The main stress factors that limit growth are drought, nitrogen, and other nutrient deficits, as well as interactions with biotic factors, such as pests and diseases.

$$\text{Development} = f\text{ (environment, stress, genetics)} \quad (20.5)$$

Genetics are the key factor that defines the interaction with the various environmental and stress factors. Genetics define the differences between crops as well as the differences between individual cultivars or varieties. The main environmental factors that control development are temperature and photoperiod, especially for those crops that are sensitive to changes in day length. Drought stress, as well as a phosphorus deficit can also impact development. Recent reports have shown that CO_2 can also affect the rate of development for certain crops.

$$\text{Environment} = f\text{ (temperature, solar radiation, } CO_2\text{, photoperiod)} \quad (20.6)$$

The main environmental factors that control growth (Equation 20.4) and development (Equation 20.5) were defined previously. Temperature has a major impact on both development as well as growth, while solar radiation's main impact is on growth. Photoperiod affects development and sometimes also has an impact on biomass partitioning between the various plant components. The impact of CO_2 is on growth, especially through photosynthesis. The main environmental factor that is missing from this list is rainfall. In most cases, rainfall does not directly affect either growth or development, but only indirectly affects growth and development through stress factors, especially drought stress and flooding stress. Both stress factors are caused by an interaction of different components, such as soil water-holding capacity, rainfall, potential evapotranspiration, and plant water uptake.

$$\text{Stress} = \text{f (abiotic factors, biotic factors, genetics)} \quad (20.7)$$

The main abiotic stress factors are drought stress and nutrient stress, especially nitrogen and phosphorus deficits. Each stress factor has a unique impact on the many different physiological processes, therefore making it very difficult to model. There is also a very strong interaction with the genetic composition of a particular crop and cultivar. An example is photoperiod sensitivity in soybean, in which some cultivars, grown at northern latitudes are insensitive to photoperiod, while soybeans grown in the tropical environments are extremely sensitive to photoperiod. Some crops, such as common bean, have shown an interaction between temperature and the level of photoperiod sensitivity. Without detailed experimental data, however, it is very difficult to develop models that include these effects. Biotic factors include interactions with insects, diseases, nematodes, and weeds. Although specific models exist for pests and diseases, the actual modeling of the impact of biotic factors has not been developed. One of the main reasons is that each crop is affected by multiple pests and diseases, resulting in a very complex modeling problem. Also, pests and diseases are affected by distribution dynamics, such as the boundary layer meteorology and aerodynamics of the atmosphere. The interaction of pests and diseases on crops includes many uncertainties that cannot easily be simulated.

$$\text{Genetics} = \text{f (cultivar, species)} \quad (20.8)$$

Although breeders might disagree with the separation of genetics between cultivars and species, from a modeling perspective they are two different categories. In many cases, species characteristics are defined through parameters and look up functions that define the interactions of the crop with the various environmental factors. For instance, the CO_2 compensation points as described by the equations of Charles-Edwards, Doley, and Rimmington (1986) show differences between various C3 plant species and can be considered a species-specific parameter. Other examples include the effect of temperature on growth, development, and various other processes. Cultivar characteristics are more specific and allow a modeler to define variables and functions that are unique for a particular cultivar, especially for management applications. Examples are the number of photothermal days to flowering or maturity, or the sensitivity to photoperiod. These are parameters that are quite commonly used in modeling applications.

CONCEPTS IN MODELING THE PROCESSES

Modeling is a very individualistic discipline. Each modeler of a crop or cropping system seems to have his or her own ideas on how to develop a model and how to implement it. This has resulted in hundreds of models, of which only a few are currently being used. Even with respect to concepts, scientists do not seem to agree. Passioura (1996) differentiates between the models that are scientifically based and are used to help understand the physiological processes versus the engineering type models that are mainly used to help solve problems (Passioura, 1973). In some cases, these are also referred to as mechanistic and functional models (Addiscott and Wagenet, 1985). However, when evaluating current applications of crop simulation models as well as the models that are being used for these applications, these models seem to be mixed. Both the mechanistic and functional type models are being used to address scientific issues and help improve physiological understanding. Some of the same models are also being used for more practical crop management applications and on-farm decision support systems.

Hodges (1991b) defines the functional models as empirical models, in which the simplest empirical model is a linear regression equation. Crop modelers who try to model at the detailed physiological process level have not been very receptive to this type of modeling approach. However, it is commonly used by agroclimatologists. Ritchie (1994) identifies three groups of models: statistical models, mechanistic models, and functional models. Statistical models have mainly been applied for large area yield predictions in which yield is correlated to one or more weather variables, including monthly temperatures and precipitation totals (Thompson, 1969a,b; Alexandrov and Hoogenboom, 2001). Ritchie (1994) identifies mechanistic models to mainly be associated with defining the instantaneous rate of processes that may rapidly change on short time scales, such as photosynthesis and transpiration. Functional models are defined as models that incorporate simplified approaches to describe the more complex processes. One of the main differences between functional models and mechanistic models is that functional models operate on a daily or larger time step, while mechanistic models operate at a much smaller time step. Other modelers might disagree with this statement.

Besides evaluating the level of complexity at which a crop process is modeled, one can also evaluate the level of detail of modeling the crop processes in general. This approach has been used extensively by the School of de Wit (Bouman et al., 1996). Phase 1 type models include potential production, in which water is nonlimiting. At this phase only solar radiation and temperature are model inputs and all other factors are assumed to be optimal. Phase 2 type models include water-limited production; precipitation is defined as input for the model and the soil and plant water balances are simulated. Phase 3 type models include nitrogen-limited production; fertilizer is defined as an input and the soil and plant nitrogen balance are simulated, in addition to the other processes simulated for Phase 1 and 2. Phase 4 is pest- and nutrient-limited production. This is a very complex phase, as it includes multiple levels of nutrients and multiple levels of pests and diseases.

Development

A detailed overview of the various approaches to model phenology is presented in Hodges (1991a). It includes examples of empirical models as well as mechanistic models. These models simulate both vegetative (Kiniry et al., 1991) and reproductive development (Major and Kiniry, 1991). Hodges (1991a) also presents several implementations of models for specific crops, such as maize (Kiniry and Bonhomme, 1991), wheat (Hodges and Ritchie, 1991; Rickman and Klepper, 1991), and cotton (Baker and Landivar, 1991; Jackson, 1991). A similar overview is presented in Hanks and Ritchie (1991), in which simulation models for wheat, maize, and soybean phasic development are described (Jones, Boote, et al., 1991; Kiniry, 1991; Ritchie, 1991; Ritchie and Nesmith, 1991).

Most models use a degree-day approach, in which the number of degree-days above a certain threshold value is accumulated. These models can also include some type of maximum threshold for temperature that prevents a further increase in the accumulation of degree-days when the temperature rises above this maximum threshold value. These thresholds are also referred to as the cardinal temperatures. In some models, the degree-day concept is converted to relative values to make it easier to integrate the effect of photoperiod on development (Jones, Boote, et al., 1991). Models that include both the temperature and photoperiod effect on development are normally referred to as photothermal models. For each phenological stage, a certain number of degree-days or photothermal days is required. Most models assume that a change in phenological stage is the equivalent of 50 percent of the plants in the field having reached that stage. Critical stages in plant development include germination and emergence, first leaf occurrence, start of flower initiation, start of flowering, start of first pod or ear, start of first seed or grain, and physiological maturity. The period between each stage is normally referred to as a phase. The duration of a particular phase is controlled by the number of degree-days or photothermal days. This duration is unique for each crop and cultivar or variety.

Growth

For modeling growth, there are in general two different approaches. The first one is mechanistic and models photosynthesis as a function of incoming solar radiation, CO_2 concentration, air or leaf temperature, leaf area index, and one or more stress factors. In most cases these models operate at a time step of less than one day. The pool of carbohydrates available for growth is first reduced by a small fraction that is a function of total biomass to account for maintenance respiration. The second approach is more functional and converts incoming solar radiation directly into available carbohydrates, based on light use efficiency. Many of these functional models operate at a daily or larger time step.

Both the mechanistic and functional methods to calculate photosynthesis result in a pool of carbohydrates or biomass that has to be distributed to the different plant components for new growth to occur. Again, in general there are two different approaches to partition carbohydrates to the various plant components, using either a sink-driven or source-driven approach. In the source-driven approach, all carbohydrates are partitioned based on the total amount of carbohydrates in the pool and partitioning coefficients. Each plant component, including leaves, stems, petioles, roots, and reproductive structures, such as pods, seeds, ears, and grains, has its own coefficient. These partitioning coefficients can be static or vary as a function of crop age, stress, or other plant and environmental variables. In the sink-driven approach, the plant components determine their individual needs based on certain growth functions, which are determined by size, age of the plant component, environmental conditions, stress, and other factors. The amount required for new growth is then combined with the carbohydrate demands of all components. Based on the size of the pool, carbohydrates for new growth are partitioned accordingly. Most models also use a conversion efficiency to account for the cost of growing new tissue, sometimes called growth respiration. If one knows the composition of new growth tissue, growth respiration can be calculated quite accurately based on the biochemical conversion processes that are involved. For instance, the grain legume model CROPGRO defines composition in the form of percentage protein, lipids, carbohydrates, lignins, and other components for leaves, stems, roots, seeds, and pods (Boote, Jones, and Hoogenboom, 1998).

APPLICATIONS OF CROP MODELS

To present a comprehensive review of the different crop models as well as their applications is rather difficult and is beyond the scope of this chapter. In fact, many of these reviews have already been published. Instead, a few crops have been selected from both horticulture as well as agronomy and some examples are presented.

Horticultural Crops

Despite the diversity of horticultural crops, which consist of fruits, vegetables, and flowers, a fair number of models have been developed for them. Gary, Jones, and Tchamitchian (1998) recently presented a detailed review of horticultural models. They state that one of the main accomplishments has been an exchange of information between modelers, which is a unique outcome given the wide range of philosophies that modelers have. One means of information exchange has been regular workshops on modeling of plant growth and control for greenhouse environments, sponsored by the International Society for Horticultural Science (ISHS) (Challa, 1985; Bar-Yosef and Seginer, 1999).

Fruits

One of the key issues for modeling fruit crops is to simulate fresh fruit weight. Genard and Huguet (1996) and Fishman and Genard (1998) developed a model that simulates both dry and fresh mass for individual fruits. Several models have been developed for peaches, both in Europe as well as in the United States. Genard, Baret, and Simon (2000) developed a detailed three-dimensional model to simulate photosynthesis and light interception for individual peach trees as a function of tree density. The same research group also developed a model for peach trees at the fruit-bearing level (Genard et al., 1998; Lescourret, BenMimoun, and Genard, 1998) and for simulating peach phenology (Lescourret, Inizan, Genard, 2000). Grossman and DeJong (1994) developed a simulation model called "Peach" which has been used in several applications to study the relationship between reproductive development, dry-matter partitioning, and yield of peaches (DeJong, 1999; Berman et al., 1998).

Although grapes are economically a very important crop, especially due to the wine industry, only a few models that simulate grape growth, development, and yield seem to exist (Schultz, 1995; Bindi et al., 1997). A model for kiwi fruit was presented by Lescourret et al. (1998, 1999). In addition to predicting fresh fruit yield, vernalization is another issue that is important for fruits. Vernalization is normally simulated through the accumulation of chilling hours, where different crops and varieties have different threshold values. Nesmith and Bridges (1992) presented a chilling hour model for blueberries, while Linsleynoakes and Allan (1994) conducted a comparison of two commonly used chilling hour models for peaches. Most of the other models for fruit crops relate to predicting phenology as an input for integrated pest management decision support systems (Vanderzwet et al., 1994; Degrandihoffman et al., 1995; Timmer and Zitko, 1996; Cross and Burgess, 1998) and other management decisions, such as harvest scheduling (Malezieux et al., 1994).

Vegetables

A significant number of models have been developed for vegetable crops, especially for those crops that are raised in greenhouses, such as tomatoes, cucumbers, and bell pepper. Gary, Jones, and Tchamitchian (1998), Lentz (1998), and Marcelis, Heuvelink, and Goudriaan (1998) have recently presented detailed reviews on models for horticultural crops. As stated previously, Thornley and Hurd (1974) developed the first simulation model for tomatoes. Hoogenboom (1980) developed one of the first tomato models under the auspices of the School of de Wit family of models (Bouman et al., 1996). The tomato model TOMSIM is based on the model SUCROS (Spitters, van Keulen, and van Kraalingen, 1989) and was also developed by the Wageningen Agricultural University, the Netherlands (Heuvelink, 1996, 1997, 1999). The TOMGRO model is based on the SOYGRO family of models (Hoogenboom, Jones, and Boote, 1992); it was developed as a collaborative effort between the University of Florida, and the Agricultural Research Organization in Israel (Jones, Dayan, et al., 1991; Dayan et al., 1993a,b). The main application of both the TOMSIM and TOMGRO crop models is an optimization of greenhouse climate control in order to improve tomato production (Seginer and Ioslovich, 1998). Bertin and Heuvelink (1993) conducted a comparison of the performance for these two models.

Most of the models for greenhouse crops do not include a soil water and nitrogen balance. These models mainly concentrate on simulating the interaction between the aboveground canopy and the conditions inside the greenhouse. Recent studies have included the simulation of tomatoes for outdoor conditions in which the simulation of soil water and nutrients are critical (McNeal et al., 1995; Irmak and Jones, 2000). Other vegetable crops for which models have been developed include kohlrabi (Wiebe, 1995), radish (Engeleneigles and Erwin, 1997; Heissner, 2000), and watermelon (Soltani, Anderson, and Hamson, 1995).

Ornamentals

Most of the models for ornamentals deal with simulating development, shoot elongation (Karlsson and Heins, 1994), and timing of flowering. The most common crops for which models have been developed include chrysanthemums and roses. Production of chrysanthemums for greenhouse production has been studied extensively (Hiden and Larsen, 1994; Larsen and Hiden, 1995; Larsen and Persson, 1999) as well as for roses (Hopper, Hammer, and Wilson, 1994; Pasian and Lieth, 1994, 1996). Dayan, Fuchs, Dayan, et al. (2001) and Dayan, Fuchs, Plaut, et al. (2001) developed a model for roses primarily for application in the optimization of climate control to improve growth and development, both qualitatively as well as quantitatively.

Fisher, Lieth, and Heins (1996) developed a model to predict flower elongation and time to start of flowering for Easter lily. Fisher and Lieth (2000) recently extended this work into the development of a decision support system called LilyDate, which can be used to optimize greenhouse temperature to target particular harvest dates. Other bulb and tuber flowering crops for which models have been developed include narcissus (Hanks, 1996) and Jerusalem artichoke (Denoroy, 1996). For harvest and marketing scheduling, it is especially important to understand the impact of temperature and other environmental factors on both vegetative and reproductive development of potted plants. Examples of simulation models for potted plants include poinsettia (Fisher, Heins, and Lieth, 1996) and African violet (Faust and Heins, 1994). Caprio (1993) developed a model for lilac, a flowering shrub. A very interesting application is the orchid model, developed by Kindlmann and Balounova (1999). The main objective of this model was to develop a relationship between the aboveground plant components, especially leaf area, and the belowground tubers. The model can be used to estimate tuber size based on leaf size without having to destructively harvest the plants.

Agronomic Crops

The number of agronomic crops is relatively small compared to the number of horticultural crops. As a result, many different models have been developed for agricultural crops. For instance, for wheat, at least 20 crop simulation models have been presented in the literature. An extensive review of agronomic models was published by Whisler et al. (1986). Similarly to the previous sections, in this review we will mainly concentrate on recent developments related to agronomic crop models and concentrate on a few crops.

Root and Tuber Crops

Potato. One of the most important belowground crops is potato in which yield is expressed through tuber yield. Several simple, degree-based models have been developed to simulate potato development (Lee, Sterrett, and Henninger, 1992; Arazi, Wolf, and Marani, 1993). Nath and Moore (1992) developed a developmental aging model, based on the Richards equation. The Richards equation is a mathematical approach that is commonly used for the development of simulation models. Nath and Moore (1992) used potato tubers as an example to predict the kinetics of growth and development. Prihar et al. (1995) compared several simple radiation-use, efficiency-based models for potato. Trooien and Heermann (1992a) developed a model to specifically simulate potato leaf area. They also presented detailed steps for both model development and evaluation (Trooien and Heermann, 1992b,c). Another issue that is important for potatoes is biomass partitioning, because it is different from the priority schemes normally applied to partitioning between vegetative and reproductive structures in grain crops (Kooman and Rabbinge, 1996).

Several comprehensive models for potatoes have been developed that include a simulation of growth and development as well as soil and plant water and nitrogen balances. MacKerron (1985, 1987) and MacKerron and Waister (1985) developed a "simple" model to simulate potato growth and development. Jefferies and Heilbronn (1991) and Jefferies, Heilbronn, and MacKerron (1991) expanded this approach by developing a model that simulates the impact of drought stress on growth. This model was then applied as a tool to simulate drought responses of different potato genotypes and to assess drought tolerance (Jefferies, 1993a,b; Jefferies and MacKerron, 1993). A more comprehensive model is the SUBSTOR model developed by Griffin, Johnson, and Ritchie (1993). SUBSTOR is a component of the Decision Support System for Agrotechnology Transfer (DSSAT) suite of crop simulation models (Tsuji, Uehara, and Balas, 1994; Hoogenboom, Wilkens, and Tsuji, 1999). Another comprehensive model is SIMPOTATO, which is very similar to the SUBSTOR model (Hodges, Johnson, and Johnson, 1992; Hodges, 1998). The SUBSTOR model simulates both growth and development and predicts tuber yield; it has been applied in several countries (Mahdian and Gallichand, 1995; Travasso, Caldiz, and Saluzzo, 1996). Fleisher et al. (2000) modified SUBSTOR for controlled-environment studies as part of a study to evaluate the potential for growing potatoes in space. A detailed comparison of various potato models can be found in Kabat et al. (1995).

Because of the importance of water for potato production, several applications have used potato models for irrigation management (Singh, Brown, and Barr, 1993; Singh et al., 1993; Hamer, Carr, and Wright, 1994; Shaykewich et al., 1998). Ejieji and Gowing (2000) developed a dynamic model for irrigation scheduling. This decision support system includes options for real-time management and accounts for short-term weather forecasts (Gowing and Ejieji, 2001). Other applications that have been developed with potato models include cultivar evaluation (Gordon, Brown, and Dixon, 1996), development of breeding strategies (Spitters and Schapendonk, 1990), and forecasting for seed potato production (Nemecek et al., 1996). With the recent developments in precision farming and decision-support systems, potato models have also been linked to GIS for spatial applications (Han et al., 1995) and remote sensing data (Finke, 1992) to address the spatial variability of potato yield (Verhagen, 1997).

Cassava. An important root crop especially for the tropics is cassava. Gutierrez, Wermelinger, et al. (1988) developed a cassava model that simulates the carbon, water, and nitrogen dynamics for cassava-based cropping systems in Africa. This model has mainly been applied for integrated pest management (Gutierrez, Yaninek, et al., 1988; Gutierrez, Neuenschwander, et al., 1988). El-Sharkawy and Cock (1990) and El-Sharkawy et al. (1990) conducted detailed physiological studies for cassava. They developed various relationships between photosynthesis, growth, and biomass partitioning to the roots. Matthews and Hunt (1994) developed the model GUMCAS, a comprehensive crop simulation model for cassava. GUMCAS is part of the DSSAT modeling system mentioned previously (Tsuji, Uehara, and Balas, 1994; Hoogenboom, Wilkens, and Tsuji, 1999).

Grain Crops

Wheat. Wheat is one of the most important agronomic crops in the world. A large number of comprehensive models have therefore been developed that simulate growth, development, and yield for wheat. Under the auspices of the Global Change and Terrestrial Ecosystems (GCTE) group, several crop networks have been established for evaluation and improvement of crop simulation models. The wheat network has been the most active network; it has organized annual workshops where wheat modelers and experimentalists meet to exchange data and compare individual models. The network also has developed standard data sets from many locations across the world for model evaluation and comparison. In Global Change and Terrestrial Ecosystems (1994), nineteen different crop simulation models for wheat are presented with a short summary of all the processes simulated by each model. As part of the GCTE initiative, Porter, Jamieson, and Wilson (1993) and Jamieson, Porter, et al. (1998), compared the most commonly used wheat models, including AFRCWHEAT2 (Ewert, Porter, and Honermeier, 1996), CERES-Wheat (Ritchie et al., 1998), Sirius (Jamieson, Semenov, et al., 1998), SUCROS2, and SWHEAT. Moulin and Beckie (1993) compared CERES-Wheat and EPIC, a generic simulation model, for predicting spring wheat yield in Canada. A rather controversial study was conducted by Landau et al. (1998) who compared yield performance of various models against observed regional yield levels in the United Kingdom.

Many different approaches have been discussed in the literature to model wheat development, although one would expect that modeling of wheat development is relatively simple (Amir and Sinclair, 1991a; McMaster et al., 1991; Jamieson et al., 1995; Jamieson, Brooking, et al., 1998; Rickman, Waldman, and Klepper, 1996; Jame, Cutforth, and Ritchie, 1998; Yan and Hunt, 1999; Yan and Wallace, 2000). An accurate simulation of growth and especially the interaction with the soil water and nitrogen balances are other issues that are important for wheat production (Amir and Sinclair, 1991b; Teittinen, Karvonen, and Peltonen, 1994; Schultz and Mirschel, 1995; O'Leary and Connor, 1996a,b; Arora, Singh, and Singh, 1997; Meinke, Hammer, van Keulen, Rabbinge, and Keating, 1997; Arora and Gajri, 1999). Applications of wheat models have also been extended to study the entire cropping system, including the impact on water pollution and long-term sustainability (Pfeil et al., 1992a,b; Amir and Sinclair, 1996; Pannkuk, Stockle, and Papendick, 1998; Asseng et al., 1998) and decision management (Woodruff, 1992). Most simulation models only predict seed yield in the form of final seed quantity. However, for wheat, as well as for many other crops, the quality of the product also has become very important. Teittinen, Karvonen, and Peltonen (1994) developed a wheat model that predicts both the final seed quantity as well as the quality of the seed through seed composition.

Several applications have extended the single point predictions to spatial applications for regional wheat yield forecasting (Supit, 1997; Supit and van der Goot, 1999). In some cases yield forecasting has been associated with climate variability and the southern oscillation index (Rimmington and Nicholls, 1993). Linkage of wheat models with remote sensing can potentially help improve these yield forecasts (Maas, 1993). Because of its agronomic importance, wheat models have also been used extensively to study the potential impact of climate change on wheat production (Mearns, Rosenzweig, and Goldberg, 1996; Mearns et al., 2001; Rosenzweig and Tubiello, 1996; Semenov et al., 1996; Ewert, van Oijen, and Porter, 1999).

Rice. Rice is the most important food crop in the world. However, the number of rice simulation models that have been developed is rather limited. One of the main reasons is that rice is mainly grown in tropical regions, compared to wheat and other crops. As was discussed previously in the section on history of crop models, most of the modeling efforts were started in Europe and the United States. Examples of phenology models for rice can be found in Yin, Kropff, Nakagawa, et al. (1997), Yin, Kropff, Horie, et al. (1990), Dingkuhn, Wopereis, and Miezan (1998), Wu and Wilson (1998), and Fukai (1999). For rice, simulation of the water balance is also very critical, especially for lowland and flooded conditions (Wopereis et al., 1993, 1994, 1996; Bouman et al., 1994; ten Berge et al., 1995). The more comprehensive crop simulation models for rice include ORYZA (Casanova, Goudriaan, and Bosch, 2000) and CERES-Rice (Ritchie et al., 1998). Several rice models have been used to study the interactions with weeds (Graf, Rakotobe, et al., 1990; Graf, Gutierrez, et al., 1990; Graf et al., 1991; Caton, Foin, and Hill, 1999a,b) or to design cultivars that are more competitive or higher yielding than the traditional cultivars (Bastiaans et al., 1997; Sheehy, Dionora, and Mitchell, 2001). The rice models have also been used in climate change studies (Bachelet and Gay, 1993; Bachelet, Herstrom, and Brown, 1993; Singh and Ritchie, 1993).

Sorghum and Millet. Growth, development, and yield for sorghum has been modeled quite frequently, both for grain sorghum (Rosenthal and Gerik, 1990; Heiniger, Vanderlip, and Welch, 1997; Sinclair, Muchow, and Monteith, 1997) as well as for forage sorghum (Baez-Gonzalez and Jones, 1995a,b; Fritz et al., 1997). Because sorghum grows well in tropical environments, sorghum models have been used extensively for a wide range of applications in Australia (Robertson et al., 1993; Hammer, Carberry, and Muchow, 1993; Robertson and Fukai, 1994). The link between climate risks and sorghum productivity has been studied extensively (Hammer and Muchow, 1994; Muchow, Hammer, and Vanderlip, 1994). Only a few models exist for millet, a crop normally grown in arid environments. Daamen (1997) and Boegh et al. (1999) presented detailed micrometeorological models for measuring water and vapor fluxes. The CERES-Millet (Ritchie et al., 1998) is part of the DSSAT system (Tsuji, Uehara, and Balas, 1994; Hoogenboom, Wilkens, and Tsuji, 1999). This model has been applied on a limited basis for several studies in West Africa (Fechter et al., 1991), including as a yield forecasting tool for a famine early warning system (Thornton et al., 1997).

Barley and Rye. These are minor crops for which only a few models exist (Grant, 1993; Murphy et al., 1994; Wagner-Riddle et al., 1997). The CERES-Barley model (Ritchie et al., 1998) has been used for a few applications related to evaluation of yield potential as it relates to local conditions (Travasso and Magrin, 1998; Zalud, Trnka, and Dubrovsky, 2000).

As stated previously, this chapter does not provide scope for a detailed review of all crop simulation models. We have concentrated on presenting examples of models and applications for major crops such as wheat, rice, and potato, and minor crops, including cassava, sorghum, millet, barley, and rice. For most crops, simulation models have been developed, including soybean, peanut, dry bean, pea, chickpea and various other grain legumes, as well as maize, cotton, sugarcane, sunflower, canola, and other crops. Examples of several of these models and their applications can be found in Hoogenboom (2000) and Tsuji, Hoogenboom, and Thornton (1998).

GENERIC CROP SIMULATION MODELS

Many of the applications of crop simulation models are conducted with only a few crop simulation models, which in many cases can handle multiple species. Most of these generic models provide a detailed simulation of plant growth and development, as well as a soil and plant water and nitrogen balance. One of the most comprehensive plant models is the software package Plantgro, developed by Hackett and Vanclay (1998). The Plantgro Web site states that it can handle more than 100 different species, including eight grain crops, seven root crops, and nine miscellaneous crops. However, examples of successful applications of this model seem to be limited, especially in the literature.

The Erosion-Productivity Impact Calculator (EPIC) model was originally developed to determine the impact of erosion on crop productivity (Williams et al., 1989; Jones, Dyke, et al., 1991). The inputs for this model are relatively simple, especially for nonmodelers. It has, therefore, been used for a wide range of applications, including erosion, pollution, sustainability, climate change, and others (Edwards et al., 1994; Easterling et al., 1996; Brown and Rosenberg, 1997, 1999; Bernardos et al., 2001; Mearns et al., 2001). The model can simulate yield for most of the important agronomic crops; it lists 66 different species for the most recent version. A modified version of EPIC, called EPIC-Phase, was developed for weather and soil conditions in France (Cabelguenne et al., 1990, 1997; Cabelguenne, Debaeke, and Bouniols, 1999). Brisson et al. (1998) developed the model STICS (Simulateur mulTIdisciplinaire pour les Cultures Standard), based on many components of other models. STICS is also a generic model that can handle different crops, such as wheat and maize, and can simulate productivity for different cropping systems.

In Australia, the University of Queensland's Agricultural Production Systems Research Unit (APSRU) developed the agricultural production system simulator (APSIM) (McCown et al., 1996). Models that have been developed as part of APSIM are those that are important in Queensland, including wheat, sorghum, sugarcane, and peanut. Several successful applications have been developed by the APRSU group, including on-farm applications for farmers and climate risks associated with agriculture (Probert et al., 1995, 1998; Meinke, Hammer, van Keulen, and Rabbinge, 1997; Meinke, Rabbinge, et al., 1997; Keating and Meinke, 1998; Keating et al., 1999). The development of APSIM has received strong support from the local industry. Unfortunately, due to intellectual property rights of the Australian Government, availability of the system and especially the source code is limited.

One of the most widely used modeling systems across the world is DSSAT. This system was initially developed under the auspices of the International Benchmark Sites Network for Agrotechnology Transfer (IBSNAT) Project, which was funded by the United States Agency for International Development (US AID) from 1982 through 1993. Since the termination of the IBSNAT Project, development of the DSSAT system has continued through collaboration among various universities and research institutes across the world. The DSSAT software system version 3.5 (Tsuji, Uehara, and Balas, 1994; Hoogenboom, Wilkens, and Tsuji, 1999) includes crop simulation models for 17 different crops. The main models are the generic cereal model CERES for wheat, maize, barley, sorghum, millet, and rice (Ritchie et al., 1998) and the grain legume model CROPGRO for soybean, peanut, dry bean, and chickpea (Boote, Jones, and Hoogenboom, 1998). Other models include the SUBSTOR model for potato, the CROPSIM model for cassava, OILCROP for sunflower, and CANEGRO for sugarcane. All crop simulation models that are part of DSSAT use the same soil water balance (Ritchie, 1998) and soil nitrogen balance (Godwin and Singh, 1998). This makes it especially easy for applications by users, such as strategic applications for risk and sustainability analysis (Thornton and Hoogenboom, 1994; Thornton et al., 1995). A detailed overview of various applications of the DSSAT models as it relates to irrigation and fertilizer management, breeding, climate change, food security, land use, and related applications can be found in Tsuji, Hoogenboom, and Thornton (1998).

SUMMARY

Computer models that simulate growth, development, and yield have been developed for more than 30 years. These models vary in complexity as well as in detail, ranging from simple heat unit and chilling hour equations to calculate development, to comprehensive models that predict crop yield through the simulation of growth and development and interactions with the soil and plant water and nitrogen balances. A generic approach has been presented that integrates growth, development, and yield with crop and cultivar genetics and the interactions with weather, soil, and other environmental variables. Models have been developed for most horticultural and agronomic crops. Examples of models have been presented for several crops, including tomatoes, chrysanthemums, roses, potatoes, wheat, and rice. These models have been used for a wide range of applications, including greenhouse climate control, irrigation and fertilizer management, precision management, pesticide applications, cultivar selection, yield forecasting, regional land use, and climate change. Although a large number of crop simulation models exist, most of the applications are conducted with generic models, such as EPIC, APSIM, and the DSSAT cropping system models. It is expected that the agronomic community will become more receptive to the use and application of crop simulation models in the near future, especially with the rapid integration of computer hardware and software in the various agricultural disciplines.

REFERENCES

Adams, R.M., C. Rosenzweig, R.M. Peart, J.T. Ritchie, B.A. McCarl, J.D. Glyer, R.B. Curry, J.W. Jones, K.J. Boote, and L.H. Allen Jr. (1990). Global climate change and US agriculture. *Nature 345:*219-224.

Addiscott, T.M. and R.J. Wagenet (1985). Concepts of solute leaching in soils: A review of modeling approaches. *Journal of Soil Science 36:*411-424.

Alexandrov, V.A. and G. Hoogenboom (2000a). The impact of climate variability and change on major crops in Bulgaria. *Agricultural and Forest Meteorology 104:*315-327.

Alexandrov, V.A. and G. Hoogenboom (2000b). Vulnerability and adaptation assessments of agricultural crops under climate change in the southeastern USA. *Theoretical and Applied Climatology 67:*45-63.

Alexandrov, V.A. and G. Hoogenboom (2001). Climate variation and crop production in Georgia, USA, during the 20th century. *Climate Research 17:*33-43.

Amir, J. and T.R. Sinclair (1991a). A model of the temperature and solar-radiation effects on spring wheat growth and yield. *Field Crops Research 28:*47-58.

Amir, J. and T.R. Sinclair (1991b). A model of water limitation on spring wheat growth and yield. *Field Crops Research 28:*59-69.

Amir, J. and T.R. Sinclair (1996). A straw mulch system to allow continuous wheat production in an arid climate. *Field Crops Research 47:*21-31.

Arazi, Y., S. Wolf, and A. Marani (1993). A prediction of developmental stages in potato plants based on the accumulation of heat units. *Agricultural Systems 43:*35-50.

Arora, V.K. and P.R. Gajri (1999). Evaluation of a crop growth-water balance model for analysing wheat responses to climate- and water-limited environments. *Field Crops Research 59:*213-224.

Arora, V.K., C. Singh, and K. Singh (1997). Comparative assessment of soil water balance under wheat in a subtropical environment with simplified models. *Journal of Agricultural Science Cambridge 128:*461-468.

Asseng, S., I.R. Fillery, G.C. Anderson, P.J. Dolling, F.X. Dunin, and B.A. Keating (1998). Use of the APSIM wheat model to predict yield, drainage, and NO_3 leaching for a deep sand. *Australian Journal of Agricultural Research 49:*363-377.

Bachelet, D. and C.A. Gay (1993). The impacts of climate change on rice yield: A comparison of four model performances. *Ecological Modelling 65:*71-93.

Bachelet, D., A. Herstrom, and D. Brown (1993). Rice production and climate change: Design and development of a GIS database to complement simulation models. *Landscape Ecology 8:*77-91.

Baez-Gonzalez, A.D. and J.G.W. Jones (1995a). Models of sorghum and pearl millet to predict forage dry matter production in semi-arid Mexico. 1. Simulation models. *Agricultural Systems 47:*133-145.

Baez-Gonzalez, A.D. and J.G.W. Jones (1995b). Models of sorghum and pearl millet to predict forage dry matter production in semi-arid Mexico. 2. Regression models. *Agricultural Systems 47:*147-159.

Baker, C.H. and R.B. Curry (1976). Structure of agricultural simulators: A philosophical review. *Agricultural Systems 1:*201-218.

Baker, D.N. and J.A. Landivar (1991). The simulation of plant development in GOSSYM. In *Predicting Crop Phenology,* ed. T. Hodges. Boca Raton, FL: CRC Press, pp. 153-170.

Baker, J.M. (1996). Use and abuse of crop simulation models. *Agronomy Journal 88:*689.

Bar-Yosef, B. and I. Seginer (1999). Models for plant growth and control of the shoot and root environments in greenhouses. *Acta Horticulturae 507:*1-317.

Bastiaans, L., M.J. Kropff, N. Kempuchetty, A. Rajan, and T.R. Migo (1997). Can simulation models help design rice cultivars that are more competitive against weeds? *Field Crops Research 51:*101-111.

Berman, M.E., A. Rosati, L. Pace, Y.L. Grossman, and T.M. DeJong (1998). Using simulation modeling to estimate the relationship between date of fruit maturity and yield potential in peach. *Fruit Varieties Journal 52:* 229-235.

Bernardos, J.N., E.F. Viglizzo, V. Jouvet, F.A. Lertora, A.J. Pordomingo, and F.D. Cid (2001). The use of EPIC model to study the agroecological change during 93 years of farming transformation in the Argentine pampas. *Agricultural Systems 69:*215-234.

Bertin, N. and E. Heuvelink (1993). Dry-matter production in a tomato crop: Comparison of two simulation models. *Journal of Horticultural Science 68:*995-1011.

Bindi M., F. Miglietta, B. Gozzini, S. Orlandini, and L. Seghi (1997). A simple-model for simulation of growth and development in grapevine (*Vitis-vinifera* l). 1. Model description. *Vitis 36:*67-71.

Boegh, E., H. Soegaard, T. Friborg, and P.E. Levy (1999). Models of CO_2 and water vapour fluxes from a sparse millet crop in the Sahel. *Agricultural and Forest Meteorology 93:*7-26.

Boote, K.J., J.W. Jones, and G. Hoogenboom (1998). Simulation of crop growth: CROPGRO model. In *Agricultural Systems Modeling,* eds. R.M. Peart and R.B. Curry. New York: Marcel Dekker, pp. 651-692.

Boote, K.J., J.W. Jones, and N.B. Pickering (1996). Potential uses and limitation of crop models. *Agronomy Journal 88:*704-716.

Bouman, B.A.M., H. van Keulen, H.H. van Laar, and R. Rabbinge (1996). The "School of de Wit" crop growth simulation models: A pedigree and historical overview. *Agricultural Systems 52:*171-198.

Bouman, B.A.M., M.C.S. Wopereis, M.J. Kropff, H.F.M. ten Berge, and T.P. Tuong (1994). Water use efficiency of flooded rice fields. II. Percolation and seepage losses. *Agricultural Water Management 26:*291-304.

Brisson, N., B. Mary, D. Ripoche, M.H. Jeuffroy, F. Ruget, B. Nicoullaud, P. Gate, F. Devienne-Barret, R. Antonioletti, C. Durr, et al. (1998). STICS: A generic model for the simulation of crops and their water and nitrogen balances. I. Theory and parameterization applied to wheat and corn. *Agronomie 18:*311-346.

Brouwer, R. and C.T. de Wit (1968). A simulation model of plant growth with special attention to root growth and its consequences. In *Proceedings of the 15th Easter School of Agricultural Sciences.* Nottingham, United Kingdom: University of Nottingham, pp. 224-242.

Brown, R.B. and N.J. Rosenberg (1997). Sensitivity of crop yield and water use to change in a range of climatic factors and CO_2 concentrations: A simulation study applying EPIC to the central USA. *Agricultural and Forest Meteorology 83:* 171-203.

Brown, R.B. and N.J. Rosenberg (1999). Climatic change impacts on the potential productivity of corn and winter wheat in their primary United States growing regions. *Climatic Change 41:*73-107.

Cabelguenne, M., P. Debaeke, and A. Bouniols (1999). EPIC-PHASE, a version of the EPIC model simulating the effects of water and nitrogen stress on biomass and yield, taking account of developmental stages: Validation on maize, sunflower, sorghum, soybean and winter wheat. *Agricultural Systems 60:*175-196.

Cabelguenne, M., P. Debaeke, J. Puech, and N. Bosc (1997). Rcal timc irrigation management using the EPIC-PHASE model and weather forecasts. *Agricultural Water Management 32:*227-238.

Cabelguenne, M., C.A. Jones, J.R. Marty, P.T. Dyke, and J.R. Williams (1990). Calibration and validation of EPIC for crop rotations in Southern France. *Agricultural Systems 33:*153-171.

Caprio, J.M. (1993). Flowering dates, potential evapotranspiration and water-use efficiency of *Syringa vulgaris* L. at different elevations in the western United States of America. *Agricultural and Forest Meteorology 63:*55-71.

Casanova, D., J. Goudriaan, and A.D. Bosch (2000). Testing the performance of ORYZA1, an explanatory model for rice growth simulation, for the Mediterranean conditions. *European Journal of Agronomy 12:*175-189.

Caton, B.P., T.C. Foin, and J.E. Hill (1999a). A plant growth model for integrated weed management in direct-seeded rice. I. Development and sensitivity analyses of monoculture growth. *Field Crops Research 62:*129-143.

Caton, B.P., T.C. Foin, and J.E. Hill (1999b). A plant growth model for integrated weed management in direct-seeded rice. II. Validation testing of water-depth effects and monoculture growth. *Field Crops Research 62:*145-155.

Challa, H. (1985). Report of the working party on crop growth models. *Acta Horticulturae 174:*169-175.

Charles-Edwards, D.A. (1981). *The Mathematics of Photosynthesis and Productivity.* London: Academic Press.

Charles-Edwards, D.A. (1982). *Physiological Determinants of Crop Growth.* Sydney, Australia: Academic Press.

Charles-Edwards, D.A., D. Doley, and G.M. Rimmington (1986). *Modelling Plant Growth and Development.* Sydney, Australia: Academic Press.

Cross, J.V. and C.M. Burgess (1998). Strawberry fruit yield and quality responses to flower bud removal: A simulation of damage by strawberry blossom weevil *(Anthonomus rubi). Journal of Horticultural Science and Biotechnology 73:* 676-680.

Curry, R.B. (1971). Dynamic simulation of plant growth. I. Development of a model. *Transactions of the American Society of Agricultural Engineers 14:*946-959.

Curry, R.B. and L.H. Chen (1971). Dynamic simulation of plant growth. II. Incorporation of actual daily weather and partitioning of net photosynthesis. *Transactions of the American Society of Agricultural Engineers 14:*1170-1175.

Daamen, C.C. (1997). Two source model of surface fluxes for millet fields in Niger. *Agricultural and Forest Meteorology 83:*205-230.

Dayan, E., M. Fuchs, E. Dayan, and E. Matan (2001). Synchrony of rose development. In *Fourth International Symposium on Mathematical Modelling and Simulation in Agricultural and Bio-Industries.* Haifa, Israel: Technion, pp. 1-15.

Dayan, E., M. Fuchs, Z. Plaut, E. Presnov, E. Matan, M. Dinar, A. Solphoy, U. Mugira, N. Pines, and J. Ben-Asher (2001). A model to describe rosebush growth based on the plants' morphological parameters. In *Fourth International Symposium on Mathematical Modelling and Simulation in Agricultural and Bio-Industries.* Haifa, Israel: Technion, pp. 1-21.

Dayan, E., H. van Keulen, J. W. Jones, I. Zipori, D. Shmuel, and H. Challa (1993a). Development, calibration and validation of a greenhouse tomato growth model: I. Description of the model. *Agricultural Systems 43:*145-163.

Dayan, E., H. van Keulen, J. W. Jones, I. Zipori, D. Shmuel, and H. Challa (1993b). Development, calibration and validation of a greenhouse tomato growth model: II. Field calibration and validation. *Agricultural Systems 43:*165-183.

de Wit, C.T., R. Brouwer, and F.W.T. Penning de Vries (1970). The simulation of the photosynthetic systems. In *Prediction and Measurement of Photosynthetic productivity, Proceedings of the IBP/PP Technical Meeting, Trebon, September 1969,* ed. I. Setlik. Wageningen, the Netherlands: Center for Agricultural Publishers and Documentation (Pudoc), pp. 47-70.

de Wit, C.T. and J. Goudriaan (1974). *Simulation of Ecological Processes.* Simulation monographs. Wageningen, the Netherlands: Centre for Agricultural Publishing and Documentation (Pudoc).

de Wit, C.T., et al. (1978). *Simulation of Assimilation, Respiration and Transpiration of Crops.* Simulation monographs. Wageningen, the Netherlands: Centre for Agricultural Publishing and Documentation (Pudoc).

Degrandihoffman, G., D. Mayer, I. Terry, and D.H. Li (1995).Validation of PC-REDAPOL—Fruit-set prediction model for apples. *Journal of Economic Entomology 88:*965-972.

DeJong, T.M. (1999). Developmental and environmental control of dry-matter partitioning in peach. *Hortscience 34:*1037-1040.

Denoroy, P. (1996). The crop physiology of *Helianthus tuberosus* L.—A model orientated view. *Biomass and Bioenergy 11:*11-32.

Dingkuhn, M., M.C.S. Wopereis, and K.M. Miezan (1998). Rice crop duration and leaf appearance rate in a variable thermal environment. I. Development of an empirically based model. *Field Crops Research 57:*1-13.

Duncan, W.G., R.S. Loomis, W.A. Williams, and R. Hanau (1967). A model for simulating photosynthesis in plant communities. *Hilgardia 38:*181-205.

Easterling, W.E., X. Chen, C. Hays, J.R. Brandle, and H. Zhang (1996). Improving the validation of model-simulated crop yield response to climatic change: An application to the EPIC model. *Climate Research 6:*263-273.

Edwards, D.R., V.W. Benson, J.R. Williams, T.C. Daniel, and R.G. Gilbert (1994). Use of the EPIC model to predict runoff transport of surface-applied inorganic fertilizer and poultry manure constituents. *Transactions of the American Society of Agricultural Engineers 37:*403-409.

Ejieji, C.J. and J.W. Gowing (2000). A dynamic model for responsive scheduling of potato irrigation based on simulated water-use and yield. *Journal of Agricultural Science Cambridge 135:*161-171.

El-Sharkawy, M.A. and J.H. Cock (1990). Photosynthesis of Cassava *(Manihot esculenta). Experimental Agriculture 26:*325-340.

El-Sharkawy, M.A., J.H. Cock, J.K. Lynam, A.D.P. Hernàndez, and F.L. Cadavid (1990). Relationships between biomass, root-yield and single-leaf photosynthesis in field-grown cassava. *Field Crops Research 25:*183-201.

Engel, T., G. Hoogenboom, J. W. Jones, and P. W. Wilkens (1997). AEGIS/WIN—A computer program for the application of crop simulation models across geographic areas. *Agronomy Journal 89:*919-928.

Engeleneigles, G. and J.W. Erwin (1997). A model-plant for vernalization studies. *Scientia Horticulturae 70:*197-202.

Ewert, F., J. Porter, and B. Honermeier (1996). Use of AFRCWHEAT2 to predict the development of main stem and tillers in winter triticale and winter wheat in northeast Germany. *European Journal of Agronomy 5:*89-103.

Ewert, F., M. van Oijen, and J.R. Porter (1999). Simulation of growth and development processes of spring wheat in response to CO_2 and ozone for different sites and years in Europe using mechanistic crop simulation models. *European Journal of Agronomy 10:*231-247.

Faust, J.E. and R.D. Heins (1994). Modeling inflorescence development of the African violet (*Saintpaulia ionantha* Wendl). *Journal of the American Society for Horticultural Science 119:*727-734.

Fechter, J., B.E. Allison, M.V.K. Sivakumar, R.R. van der Ploeg, and J. Bley (1991). An evaluation of the SWATRER and CERES-Millet models for southwest Niger. In *Soil Water Balance in the Sudano-Sahhelian zone.* Eds. M.K. Sivakumar, J.S. Wallace, C. Renard, and C. Giroux. Wallingford, United Kingdom: International Association of Hydrological Sciences, pp. 505-513.

Finke, P.A. (1992). Integration of remote sensing data in the simulation of spatially variable yield of potatoes. *Soil Technology 5:*257-270.

Fisher, P.R., R.D. Heins, and J.H. Lieth (1996). Quantifying the relationship between phases of stem elongation and flower initiation in poinsettia. *Journal of the American Society for Horticultural Science 121:*686-693.

Fisher, P.R. and J.H. Lieth (2000). Variability in flower development of Easter lily (*Lilium longiflorum* Thunb.): Model and decision-support system. *Computers and Electronics in Agriculture 26:*53-64.

Fisher, P.R., J.H. Lieth, and R.D. Heins (1996). Modeling flower bud elongation in Easter lily (*Lilium longiflorum* Thunb.) in response to temperature. *HortScience 31:*349-352.

Fishman, S. and M. Genard (1998). A biophysical model of fruit growth: Simulation of seasonal and diurnal dynamics of mass. *Plant Cell and Environment 21:*739-752.

Fleisher, D.H., J. Cavazzoni, G.A. Giacomelli, and K.C. Ting (2000). Adaption of SUBTOR for hydroponic, controlled environment white potato production. American Society of Agricultural Engineers Paper 004089, St. Joseph, Michigan: American Society of Agricultural Engineers.

Forrester, J.W. (1961). *Industrial Dynamics.* Boston, MA: Massachusetts Institute of Technology-Press.

Fritz, J.O., R.L. Vanderlip, R.W. Heiniger, and A.Z. Abelhalim (1997). Simulating forage sorghum yields with SORKAM. *Agronomy Journal 89:*64-68.

Fukai, S (1999). Phenology in rainfed lowland rice. *Field Crops Research 64:*51-60.

Gary, C., J.W. Jones, and M. Tchamitchian (1998). Crop modelling in horticulture: State of the art. *Scientia Horticulturae 74:*3-20.

Genard, M., F. Baret, and D. Simon (2000). A 3D peach canopy model used to evaluate the effect of tree architecture and density on photosynthesis at a range of scales. *Ecological Modelling 128:*197-209.

Genard, M. and J.G. Huguet (1996). Modeling the response of peach fruit-growth to water-stress. *Tree Physiology 16:*407-415.

Genard, M., F. Lescourret, M. BenMimoun, J. Besset, and C. Bussi (1998). A simulation model of growth at the shoot-bearing fruit level. II. Test and effect of source and sink factors in the case of peach. *European Journal of Agronomy 9:*189-202.

Global Change and Terrestrial Ecosystems (GCTE) (1994). *Global Change and Terrestrial Ecosystems Focus 3 Wheat Network: 1993 Model and Experimental Meta data.* Report No. 2. Canberra, Australia, Global Change and Terrestrial Ecosystems.

Godwin, D.C. and U. Singh (1998). Nitrogen balance and crop response to nitrogen in upland and lowland cropping systems. In *Understanding Options for Agricultural Production,* eds. G.Y. Tsuji, G. Hoogenboom, and P.K. Thornton. Dordrecht, the Netherlands: Kluwer Academic Publishers, pp. 55-77.

Gordon, R., D.M. Brown, and M.A. Dixon (1996). Evaluation of a cultivar-sensitive soil water model for the potato crop. *Canadian Journal of Soil Science 76:*275-283.

Goudriaan, J. (1977). *Crop Micrometeorology: A Simulation Study.* Simulation monographs. Wageningen, the Netherlands: Centre for Agricultural Publishing and Documentation (Pudoc).

Gowing, J.W. and C.J. Ejieji (2001). Real-time scheduling of supplemental irrigation for potatoes using a decision model and short-term weather forecasts. *Agricultural Water Management 47:*137-153.

Graf, B., M. Dingkuhn, F. Schnier, and V. Coronel (1991). A simulation model for the dynamics of rice growth and development: Part III—Validation of the model with high-yielding varieties. *Agricultural Systems 36:*329-349.

Graf, B., A.P. Gutierrez, O. Rakotobe, P. Zahner, and V. Delucchi (1990). A simulation model for the dynamics of rice growth and development: Part II—The competition with weeds for nitrogen and light. *Agricultural Systems 32:*367-392.

Graf, B., O. Rakotobe, P. Zahner, V. Delucchi, and A.P. Gutierrez (1990). A simulation model for the dynamics of rice growth and development: Part I—The carbon balance. *Agricultural Systems 32:*341-365.

Grant, R.F. (1993). Simulation of competition between barley and wild oats under different managements and climates. *Ecological Modelling 71:*269-287.

Griffin, T.S., B.S. Johnson, and J.T. Ritchie (1993). *A Simulation Model for Potato Growth and Development: SUBSTOR-Potato Version 2.0.* IBSNAT Research Report Series 2. Honolulu, Hawaii: University of Hawaii.

Grossman, Y.L. and T.M. DeJong (1994). PEACH—A simulation-model of reproductive and vegetative growth in peach-trees. *Tree Physiology 14:*329-345.

Gutierrez, A.P., P. Neuenschwander, F. Schulthess, H.R. Herren, J.U. Baumgaertner, B. Wermelinger, B. Löhr, and C.K. Ellis (1988). Analysis of biological control of cassava pests in Africa. II. Cassava mealybug *Phenacoccus manihoti. Journal of Applied Ecology 25:*921-940.

Gutierrez, A.P., B. Wermelinger, F. Schulthess, J.U. Baumgaertner, H.R. Herren, C.K. Ellis, and J.S. Yaninek (1988). Analysis of biological control of cassava pests in Africa. I. Simulation of carbon, nitrogen and water dynamics in cassava. *Journal of Applied Ecology 25:*901-920.

Gutierrez, A.P., J.S. Yaninek, B. Wermelinger, H.R. Herren, and C.K. Ellis (1988). Analysis of biological control of cassava pests in Africa. III. Cassava green mite *Mononychellus tanajoa. Journal of Applied Ecology 25:*941-950.

Hackett, C. and J.K. Vanclay (1998). Mobilizing expert knowledge of tree growth with the PLANTGRO and INFER systems. *Ecological Modelling 106:*233-246.

Hamer, P.J.C., M.K.V. Carr, and E. Wright (1994). Crop production and water-use. III. The development and validation of a water-use model for potatoes. *Journal of Agricultural Science Cambridge 123:*299-311.

Hammer, G.L., P.S. Carberry, and R.C. Muchow (1993). Modelling genotypic and environmental control of leaf area dynamics in grain sorghum. I. Whole plant level. *Field Crops Research 33:*293-310.

Hammer, G.L. and R.C. Muchow (1994). Assessing climatic risk to sorghum production in water-limited subtropical environments. I. Development and testing of a simulation model. *Field Crops Research 36:*221-234.

Han, S., R.G. Evans, T. Hodges, and S.L. Rawlins (1995). Linking a geographic information system with a potato simulation model for site-specific crop management. *Journal of Environmental Quality 24:*772-777.

Hanks, G.R. (1996). Variation in the growth and development of narcissus in relation to meteorological and related factors. *Journal of Horticultural Science 71:*517-532.

Hanks, J. and J.T. Ritchie, eds. (1991). *Modeling Plant and Soil Systems.* Agronomy 31. Madison, WI: American Society of Agronomy-Crop Science Society of Agronomy-Soil Science Society of Agronomy Publishers.

Hartkamp, A.D., J.W. White, and G. Hoogenboom (1999). Interfacing geographic information systems with agronomic modeling: A review. *Agronomy Journal 91:*762-772.

Heiniger, R.W., R.L. Vanderlip, and S.M. Welch (1997). Developing guidelines for replanting grain sorghum: I. Validation and sensitivity analysis of the SORKAM sorghum growth model. *Agronomy Journal 89:*75-83.

Heissner, A. (2000). A tuber and leaf growth model of radish (*Raphanus sativus* L. var. *sativus*): Parameter estimation. *Gartenbauwissenschaft 65:*218-226.

Heuvelink, E. (1996). Dry-matter partitioning in tomato—Validation of a dynamic simulation-model. *Annals of Botany 77:*71-80.

Heuvelink, E. (1997). Effect of fruit load on dry matter partitioning in tomato. *Scientia Horticulturae 69:*51-59.

Heuvelink, E. (1999). Evaluation of a dynamic simulation model for tomato crop growth and development. *Annals of Botany 83:*413-422.

Hiden, C. and R.U. Larsen (1994). Predicting flower development in greenhouse-grown chrysanthemum. *Scientia Horticulturae 58:*123-138.

Hodges, T., ed. (1991a). *Predicting Crop Phenology.* Boca Raton, FL: CRC Press.

Hodges, T. (1991b). Modeling and programming philosophies. In *Predicting Crop Phenology,* ed. T. Hodges. Boca Raton, FL: CRC Press, pp. 101-105.

Hodges, T. (1998). Water and nitogen applications for potato: Commercial and experimental rates compared to a simulation model. *Journal of Sustainable Agriculture 13:*79-90.

Hodges, T., S. L. Johnson, and B.S. Johnson (1992). A modular structure for crop simulation models: Implemented in the SIMPOTATO model. *Agronomy Journal 84:*911-915.

Hodges, T. and J.T. Ritchie (1991). The CERES-Wheat phenology model. In *Predicting Crop Phenology,* ed. T. Hodges. Boca Raton, FL: CRC Press, pp. 133-141.

Hoogenboom, G. (1980). Tomato simulator. Simulation of the Growth of Tomatoes in a Greenhouse. (MSc Thesis. Wageningen, the Netherlands: Department of Theoretical Production Ecology. Agricultural University.)

Hoogenboom, G. (2000). Contribution of agrometeorology to the simulation of crop production and its applications. *Agricultural and Forest Meteorology 103:*137-157.

Hoogenboom, G., J.W. Jones, and K.J. Boote (1992). Modeling growth, development and yield of grain legumes using SOYGRO, PNUTGRO, and BEANGRO: A Review. *Transactions of the American Society of Agricultural Engineers 35:*2043-2056.

Hoogenboom, G., J.W. White, J. Acosta-Gallegos, R.G. Gaudiel, J.R. Myers, and M.J. Silbernagel (1997). Evaluation of a crop simulation model that incorporates gene action. *Agronomy Journal 89:*613-620.

Hoogenboom, G., P.W. Wilkens, and G.Y. Tsuji, eds. (1999). *Decision Support System for Agrotechnology Transfer (DSSAT) Version 3 Volume 4.* Honolulu, HI: University of Hawaii.

Hopper, D.A., P.A. Hammer, and J.R.A. Wilson (1994). A simulation model of Rosa hybrida growth response to constant irradiance and day and night temperatures. *Journal of the American Society for Horticultural Science 119:*903-914.

Hurd, R.G. and J.H.M. Thornley (1974). An analysis of the growth of young tomato plants in water culture at different light integrals and CO_2 concentrations. I. Physiological aspects. *Annals of Botany 38:*375-388.

Iglesias, A., C. Rosenzweig, and D. Pereira (2000). Agricultural impacts of climate change in Spain: Developing tools for a spatial analysis. *Global Environmental Change 10:*69-80.

Irmak, A. and J.W. Jones (2000). Use of crop simulation to evaluate antitranspirant effects on tomato growth and yield. *Transactions of the the American Society of Agricultural Engineers 43:*1281-1289.

Jackson, B.S. (1991). Simulating yield development using the cotton model COTTAM. In *Predicting Crop Phenology,* ed. T. Hodges. Boca Raton, FL: CRC Press, pp. 171-180.

Jame, Y.W., H.W. Cutforth, and J.T. Ritchie (1998). Interaction of temperature and daylength on leaf appearance rate in wheat and barley. *Agricultural and Forest Meteorology 92:*241-249.

Jamieson, P.D., I.R. Brooking, J.R. Porter, and D.R. Wilson (1995). Prediction of leaf appearance in wheat: A question of temperature. *Field Crops Research 41:*35-44.

Jamieson, P.D., I.R. Brooking, M.A. Semenov, and J.R. Porter (1998). Making sense of wheat development: A critique of methodology. *Field Crops Research 55:*117-127.

Jamieson, P.D., J.R. Porter, J. Goudriaan, J.T. Ritchie, H. van Keulen, and W. Stol (1998). A comparison of the models AFRCWHEAT2, CERES-Wheat, Sirius, SUCROS2, and SWHEAT with measurements from wheat grown under drought. *Field Crops Research 55:*23-44.

Jamieson, P.D., M.A. Semenov, I.R. Brooking, and C.A. Francis (1998). Sirius: A mechanistic model of wheat response to environmental variation. *European Journal of Agronomy 8:*161-179.

Jefferies, R.A. (1993a). Responses of potato genotypes to drought. I. Expansion of individual leaves and osmotic adjustment. *Annals of Applied Biology 122:*93-104.

Jefferies, R.A. (1993b). Use of a simulation model to assess possible strategies of drought tolerance in potato (*Solanum-tuberosum* L.). *Agricultural Systems 41(1):*93-104.

Jefferies, R.A. and T.D. Heilbronn (1991). Water stress as a constraint on growth in the potato crop. 1. Model development. *Agricultural and Forest Meteorology 53:*185-196.

Jefferies, R.A., T.D. Heilbronn, and D.K.L. MacKerron (1991). Water stress as a constraint on growth in the potato crop. 2. Validation of the model. *Agricultural and Forest Meteorology 53:*197-205.

Jefferies, R.A. and D.K.L. MacKerron (1993). Responses of potato genotypes to drought. II. Leaf area index, growth and yield. *Annals of Applied Biology 122:*105-112.

Jones, C.A., P.T. Dyke, J.R. Williams, J.R. Kiniry, V.W. Benson, and R.H. Griggs (1991). EPIC: An operational model for evaluation of agricultural sustainability. *Agricultural Systems 37:*341-350.

Jones, J.W., K.J. Boote, S.S. Jagtap, and J.W. Mishoe (1991). Soybean development. In *Modeling Plant and Soil Systems,* eds. J. Hanks and J.T. Ritchie. Agronomy 31. Madison, WI: American Society of Agronomy-Crop Science Society of Agronomy-Soil Science Society of Agronomy Publishers, pp. 71-90.

Jones, J.W., E. Dayan, L.H. Allen, H. van Keulen, and H. Challa (1991). A dynamic tomato growth and yield model (TOMGRO). *Transactions of the American Society of Agricultural Engineers 34:*663-672.

Kabat, P., B. Marshall, B.J. van den Broek, J. Vos, and H. van Keulen, eds. (1995). *Modelling and Parameterization of the Soil-Plant-Atmosphere System. A Comparison of Potato Growth Models.* Wageningen, the Netherlands: Wageningen.

Karlsson, M.G. and R.D. Heins (1994). A model of chrysanthemum stem elongation. *Journal of the American Society for Horticultural Science 119:*403-407.

Keating, B.A. and H. Meinke (1998). Assessing exceptional drought with cropping systems simulator: A case study for grain production on northwest Australia. *Agricultural Systems 57:*315-332.

Keating, B.A., M.J. Robertson, R.C. Muchow, and N.I. Huth (1999). Modelling sugarcane production systems. I. Development and performance of the sugarcane module. *Field Crops Research 61:*253-271.

Kindlmann, P. and Z. Balounova (1999). Energy partitioning in terrestrial orchids—A model for assessing their performance. *Ecological Modelling 119:* 167-176.

Kiniry, J.R. (1991). Maize phasic development. In *Modeling Plant and Soil Systems,* eds. J. Hanks and J.T. Ritchie. Agronomy 31. Madison, WI: American Society of Agronomy-Crop Science Society of Agronomy-Soil Science Society of Agronomy Publishers, pp. 55-70.

Kiniry, J.R. and R. Bonhomme (1991). Predicting maize phenology. In *Predicting Crop Phenology,* ed. T. Hodges. Boca Raton, FL: CRC Press, pp. 115-131.

Kiniry, J.R., W.D. Rosenthal, B.S. Jackson, and G. Hoogenboom (1991). In *Predicting Crop Phenology,* ed. T. Hodges. Boca Raton, FL: CRC Press, pp. 29-42.

Kooman, P.L. and R.R. Rabbinge (1996). An analysis of the relation between dry matter allocation to the tuber and earliness of potato crop. *Annals of Botany 77:*235-242.

Lal, H., G. Hoogenboom, J.-P. Calixte, J.W. Jones, and F.H. Beinroth (1993). Using crop simulation models and GIS for regional productivity analysis. *Transactions of the American Society of Agricultural Engineers 36:*175-184.

Landau, S., R.A.C. Mitchell, V. Barnett, J.J. Colls, J. Craigon, K.L. Moore, and R.W. Payne (1998). Testing winter wheat simulation models' predictions against observed UK grain yields. *Agricultural and Forest Meteorology 89:*85-99.

Larsen, R.U. and C. Hiden (1995). Predicting leaf unfolding in flower induced shoots of greenhouse-grown chrysanthemum. *Scientia Horticulturae 63:*225-239.

Larsen, R.U. and L. Persson (1999). Modelling flower development in greenhouse chrysanthemum cultivars in relation to temperature and response group. *Scientia Horticulturae 80:*73-89.

Lee, G.S., S.B. Sterrett, and M.R. Henninger (1992). A heat-sum model to determine yield and onset of internal heat necrosis for Atlantic potato. *American Potato Journal 69:*353-362.

Lentz, W. (1998). Model applications in horticulture: A review. *Scientia Horticulturae 74:*154-174.

Lescourret, F., M. BenMimoun, and M. Genard (1998). A simulation model of growth at the shoot-bearing fruit level. I. Description and parameterization for peach. *European Journal of Agronomy 9:*173-188.

Lescourret, F., N. Blecher, R. Habib, J. Chadoeuf, D. Agostini, O. Pailly, B. Vaissiere, and I. Poggi (1999). Development of a simulation model for studying kiwi fruit orchard management. *Agricultural Systems 59:*215-239.

Lescourret, F., M. Genard, R. Habib, and O. Pailly (1998). Pollination and fruit-growth models for studying the management of kiwifruit orchards. 2. Models behavior. *Agricultural Systems 56:*91-123.

Lescourret, F., O. Inizan, and M. Genard (2000). Flower anthesis variability and influence on peach abscission and early growth variabilities. *Canadian Journal of Plant Science 80:*129-136.

Linsleynoakes, G.C. and P. Allan (1994). Comparison of 2 models for the prediction of rest completion in peaches. *Scientia Horticulturae 59:*107-113.

Maas, S.J. (1993). Within-season calibration of modeled wheat growth using remote sensing and field sampling. *Agronomy Journal 85:*669-672.

MacKerron, D.K.L. (1985). A simple model of potato growth and yields. Part II. Validation and external sensitivity. *Agricultural and Forest Meteorology 34:* 285-300.

MacKerron, D.K.L. (1987). A weather-driven model of potental yield in potato and its comparison with achieved yields. *Acta Horticulturae 214:*85-94.

MacKerron, D.K.L. and P.D. Waister (1985). A simple model of potato growth and yield. Part I. Model development and sensitivity analysis. *Agricultural and Forest Meteorology 34:*241-252.

Mahdian, M.H. and J. Gallichand (1995). Validation of the SUBSTOR model for simulating soil water content. *Transactions of the American Society of Agricultural Engineers 38:*513-520.

Major, D.J. and J.R. Kiniry (1991). Predicting daylength effects on phenological processes. In *Predicting Crop Phenology,* ed. T. Hodges. Boca Raton, FL: CRC Press, pp. 15-28.

Malezieux, E., J.B. Zhang, E.R. Sinclair, and D.P. Bartholomew (1994). Predicting pineapple harvest date in different environments, using a computer-simulation model. *Agronomy Journal 86:*609-617.

Marcelis, L.F.M., E. Heuvelink, and J. Goudriaan (1998). Modelling biomass production and yield of horticultural crops: A review. *Scientia Horticulturae 74:*83-111.

Matthews, R.B. and L.A. Hunt (1994). GUMCAS: A model describing the growth of cassava (*Manihot esculenta* L. Crantz). *Field Crops Research 36:*69-84.

McCown, R.L., G.L. Hammer, J.N.G. Hargreaves, D.P. Holzworth, and D.M. Freebairn (1996). APSIM: A novel software system for model development, model testing and simulation in agricultural systems research. *Agricultural Systems 50:*255-271.

McMaster, G.S., B. Klepper, R.W. Rickman, W.W. Wilhelm, and W.O. Willis (1991). Simulation of shoot vegetative development and growth of unstressed winter wheat. *Ecological Modelling 53:*189-204.

McNeal, B.L., J.M.S. Scholberg, J.W. Jones, C.D. Stanley, A.A. Csizinszky, and T.A. Obreza (1995). Application of a greenhouse tomato-growth model (TOMGRO) to field-grown tomato. *Florida Soil and Crop Science Society Proceedings 54:*86-93.

Mearns, L.O., T. Mavromatis, E. Tsvetsinskaya, C. Hays, and W. Easterling (2001). Comparative responses of EPIC and CERES crop models to high and low spatial resolution climate change scenarios. *Journal of Geophysical Research 104:* 6623-6646.

Mearns, L.O., C. Rosenzweig, and R. Goldberg (1996). The effect of changes in daily and interannual climatic variability of CERES-Wheat: A sensitivity study. *Climatic Change 32:*257-292.

Meinke, H., G.L. Hammer, H. van Keulen, and R. Rabbinge (1997). Improving wheat simulation capabilities in Australia from a cropping systems perspective. III. The integrated wheat model (I_WHEAT). *European Journal of Agronomy 8:*101-116.

Meinke, H., G.L. Hammer, H. van Keulen, R. Rabbinge, and B.A. Keating (1997). Improving wheat simulation capabilities in Australia from a cropping systems perspective: Water and nitrogen effects on spring wheat in a semiarid environment. *European Journal of Agronomy 7:*75-88.

Meinke, H., R. Rabbinge, G.L. Hammer, H. van Keulen, and P.D. Jamieson (1997). Improving wheat simulation capabilities in Australia from a cropping systems perspective. II. Testing simulation capabilities of wheat growth. *European Journal of Agronomy 8:*83-99.

Milthorpe, F.L. and P. Newton (1963). Studies on the expansion of the leaf surface. III. The influence of radiation on cell division and leaf expansion. *Journal of Experimental Botany 14:*483-495.

Molz, F.J. and J.M. Ferrier (1982). Mathematical treatment of water movement in plant cells and tissue: A review. *Plant Cell and Environment 5:*191-206.

Monteith, J.L. (1996). The quest for balance in crop modeling. *Agronomy Journal 88:*695-697.

Moulin, A.P. and H.J. Beckie (1993). Evaluation of the CERES and EPIC models for predicting spring wheat grain yield over time. *Canadian Journal of Plant Science 73:*713-719.

Muchow, R.C., G.L. Hammer, and R.L. Vanderlip (1994). Assessing climatic risk to sorghum production in water-limited subtropical environments: Effects of planting date, soil water at planting, and cultivar phenology. *Field Crops Research 36:*235-246.

Murphy, D.P.L., M.H. Dahab, G.C.L. Wyseure, and J.R. O'Callaghan (1994). The simulation of the effects of shading on the growth and yield of winter barley *(Hordeum distichum). Computers and Electronics in Agriculture 11:*347-350.

Nath, S.R. and F.D. Moore (1992). Growth analysis by the 1st, 2nd, and 3rd derivatives of the Richards function. *Growth Development and Aging 56:*237-247.

National Assessment Synthesis Team (2000). *Climate Change Impacts on the United States. The Potential Consequences of Climate Variability and Change.* Report for the U.S. Global Change Research Program. Cambridge, United Kingdom: Cambridge University Press.

Nemecek, T., J.O. Derron, O. Roth, and A. Fischlin (1996). Adaptation of a crop-growth model and its extension by a tuber size function for use in a seed potato forecasting system. *Agricultural Systems 52:*419-437.

Nesmith, D.S. and D.C. Bridges (1992). Modeling chilling influence on cumulative flowering—A case-study using tifblue rabbiteye blueberry. *Journal of the American Society for Horticultural Science 117:*698-702.

O'Leary, G.J. and D.J. Connor (1996a). A simulation model of the wheat crop in response to water and nitrogen supply. I. Model construction. *Agricultural Systems 52:*1-29.

O'Leary, G.J. and D.J. Connor (1996b). A simulation model of the wheat crop in response to water and nitrogen supply. II. Model validation. *Agricultural Systems 52:*31-55.

Pannkuk, C.D., C.O. Stockle, and R.I. Papendick (1998). Evaluating CropSyst simulation of wheat management in a wheat-fallow region of the US Pacific Northwest. *Agricultural Systems 57:*121-134.

Pasian, C.C. and J.H. Lieth (1994). Nondestructive dry-matter estimation of rose shoot leaves, stems, and flower buds using regression models. *HortScience 29:*162-164.

Pasian, C.C. and J.H. Lieth (1996). Prediction of rose shoot development—Model validation for the cultivar 'cara mia' and extension to the cultivars 'royalty' and 'sonia.' *Scientia Horticulturae 66:*117-124.

Passioura, J.B. (1973). Sense and nonsense in crop simulation. *Journal of the Australian Institute of Agricultural Science 39:*181-183.

Passioura, J.B. (1996). Simulation models: Science, snake oil, education or engineering? *Agronomy Journal 88:*690-694.

Penning de Vries, F.W.T., D.M. Jansen, H.F.M. ten Berge, and A. Bakema (1989). *Simulation of Ecophysiological Processes of Growth in Several Annual Crops.* Simulation Monographs. Wageningen, the Netherlands: Centre for Agricultural Publishing and Documentation (Pudoc).

Pfeil, E.V., W. Hundertmark, F.D. Thies, and P. Widmoser (1992a). Calibration of the simulation model "Ceres Wheat" under conditions of soils with shallow watertable and temperate climate. Part 1: Limitations in the applicability of the original model and necessary modifications. *Zeitschrift fur Pflanzenernährung und Bodenkunde 155:*323-326.

Pfeil, E.V., W. Hundertmark, F.D. Thies, and P. Widmoser (1992b). Calibration of the simulation model "Ceres Wheat" under conditions of soils with shallow watertable and temperate climate. Part 2: Verification of the modified model "CERES-Wheat." *Zeitschrift fur Pflanzenernährung und Bodenkunde 155:*327-331.

Porter, J.R., P.D. Jamieson, and D.R. Wilson (1993). Comparison of the wheat simulation models AFRCWHEAT2, CERES-Wheat and SWHEAT for non-limiting conditions of crop growth. *Field Crops Research 33:*131-157.

Prihar, S.S., V.K. Arora, G. Singh, and R. Singh (1995). Estimating potato-tuber yield in a subtropical environment with simple radiation-based models. *Experimental Agriculture 31:*65-73.

Probert, M.E., P.S. Carberry, R.L. McCown, and J.E. Turpin (1998). Simulation of legume-cereal systems using APSIM. *Australian Journal of Agricultural Research 49:*317-327.

Probert, M.E., B.A. Keating, J.P. Thompson, and W.J. Parton (1995). Modelling water, nitrogen, and crop yield for a long-term fallow management experiment. *Australian Journal of Experimental Agriculture 35:*941-950.

Rickman, R.W. and E.L. Klepper (1991). Tillering in wheat. In *Predicting Crop Phenology,* ed. T. Hodges. Boca Raton, FL: CRC Press, pp. 73-83.

Rickman, R.W., S.E. Waldman, and B. Klepper (1996). MODWht3: A development-driven wheat growth simulation. *Agronomy Journal 88:*176-185.

Rimmington, G.M. and N. Nicholls (1993). Forecasting wheat yields in Australia with the southern oscillation index. *Australian Journal of Agricultural Research 44:*625-632.

Ritchie, J.T. (1991). Wheat phasic development. In *Modeling Plant and Soil Systems,* eds. J. Hanks and J.T. Ritchie. Agronomy 31. Madison, WI: American Society of Agronomy-Crop Science Society of Agronomy-Soil Science Society of Agronomy Publishers, pp. 31-54.

Ritchie, J.T. (1994). Classification of crop simulation models. In *Crop Modeling and Related Environmental Data: A Focus on Applications for Arid and Semi-arid Regions in Developing Countries,* eds. P.F. Uhlir and G.C. Carter. Paris, France: CODATA, pp. 3-25.

Ritchie, J.T. (1998). Soil water balance and plant stress. In *Understanding Options for Agricultural Production,* eds. G.Y. Tsuji, G. Hoogenboom, and P.K. Thornton. Dordrecht, the Netherlands: Kluwer Academic Publishers, pp. 41-54.

Ritchie, J.T. and D.S. Nesmith (1991). Temperature and crop development. In *Modeling Plant and Soil Systems,* eds. J. Hanks and J.T. Ritchie. Agronomy 31. Madison, WI: American Society of Agronomy-Crop Science Society of Agronomy-Soil Science Society of Agronomy Publishers, pp. 5-29.

Ritchie, J.T., U. Singh, D.C. Godwin, and W.T. Bowen (1998). Cereal growth, development and yield. In *Understanding Options for Agricultural Production,* eds. G.Y. Tsuji, G. Hoogenboom, and P.K. Thornton. Dordrecht, the Netherlands: Kluwer Academic Publishers, pp. 79-98.

Robertson, M.J. and S. Fukai (1994). Comparison of water extraction models for grain sorghum under continuous soil drying. *Field Crops Research 36:*145-160.

Robertson, M.J., S. Fukai, G.L. Hammer, and M.M. Ludlow (1993). Modelling root growth of grain sorghum using the CERES approach. *Field Crops Research 33:*113-130.

Rose, D.A. and D.A. Charles-Edwards (1981). *Mathematics and Plant Physiology.* London: Academic Press.

Rosenthal, W.D. and T.J. Gerik (1990). Application of a crop model to evaluate cultural practices and optimize dryland grain sorghum yield. *Journal of Production Agriculture 3:*124-131.

Rosenzweig, C., L.H. Allen Jr., L.A. Harper, S.E. Hollinger, and J.W. Jones, eds. (1995). *Climate Change and Agriculture: Analysis of Potential International Impacts.* American Society of Agronomy Special publication no. 59. Madison, WI: American Society of Agronomy-Crop Science Society of Agronomy-Soil Science Society of Agronomy Publishers, 382 pp.

Rosenzweig, C. and M.L. Parry (1994). Potential impact of climate change on world food supply. *Nature 367:*133-138.

Rosenzweig, C. and F.N. Tubiello (1996). Effects of changes in minimum and maximum temperature on wheat yields in the central US: A simulation study. *Agricultural and Forest Meteorology 80:*215-230.

Schultz, A. and W. Mirschel (1995). Simulating soil water balance, nitrogen behavior and biomass components, using the agroecosystem model AGROSIM-WinterWheat and data from the north German Krummbach catchment. *Ecological Modelling 81:*133-144.

Schultz, H.R. (1995). Grape canopy structure, light microclimate and photosynthesis. 1. A 2-dimensional model of the spatial-distribution of surface-area densities and leaf ages in 2 canopy systems. *Vitis 34:*211-215.

Shaykewich, C.F., G.H.B. Ash, R.L. Raddatz, and D.J. Tomasiewicz (1998). Field evaluation of water use model for potatoes. *Canadian Journal of Soil Science 78:*441-448.

Seginer, I. and I. Ioslovich (1998). Seasonal optimization of the greenhouse environment for a simple two-stage crop growth model. *Journal of Agricultural Engineering Research 70:*145-155.

Semenov, M.A., J. Wolf, L.G. Evans, H. Eckersten, and A. Iglesias (1996). Comparison of wheat simulation models under climate change. II. Application of climate change scenarios. *Climate Research 7:*271-281.

Sheehy, J.E., M.J.A. Dionora, and P.L. Mitchell (2001). Spikelet numbers, sink size and potential yield in rice. *Field Crops Research 71:*77-85.

Sinclair, T.R., R.C. Muchow, and J.L. Monteith (1997). Model analysis of sorghum response to nitrogen in subtropical and tropical environments. *Agronomy Journal 89:*201-207.

Sinclair, T. and N.G. Seligman (1996). Crop modeling: From infancy to maturity. *Agronomy Journal 88:*698-704.

Singh, G., D.M. Brown, and A.G. Barr (1993). Modelling soil water status for irrigation scheduling in potatoes I. Description and sensitivity analysis. *Agricultural Water Management 23:*329-341.

Singh, G., D.M. Brown, A.G. Barr, and R. Jung (1993). Modelling soil water status for irrigation scheduling in potatoes. II. Validation. *Agricultural Water Management 23:*343-358.

Singh, U. and J.T. Ritchie (1993). Simulating the impact of climate change on crop growth and nutrient dynamics using the CERES-Rice model. *Journal of Agricultural Meteorology 48:*819-822.

Soltani, N., J.L. Anderson, and A.R. Hamson (1995). Growth analysis of watermelon plants grown with mulches and rowcovers. *Journal of the American Society for Horticultural Science 120:*1001-1009.

Spitters, C.J.T. and A.H.C.M. Schapendonk (1990). Evaluation of breeding strategies for drought tolerance in potato by means of crop growth simulation. *Plant and Soil 123:*193-203.

Spitters, C.J.T., H. van Keulen, and D.W.G van Kraalingen (1989). A simple and universal crop growth simulator: SUCROS 87. In *Simulation and System Management in Crop Protection,* eds. R. Rabbinge, S.A. Ward, and H.H. van Laar. Wageningen, the Netherlands: Centre for Agricultural Publishing and Documentation (Pudoc), pp. 147-181.

Splinter, W.E. (1974). Modeling of plant growth for yield prediction. *Agricultural Meteorology 14:*243-253.

Stapleton, H.N. (1970). Crop production system simulation. *Transactions of the American Society of Agricultural Engineers 13:*110-113.

Stapleton, H.N. and R.P. Meyers (1971). Modeling subsystems for cotton—The cotton plant simulation. *Transactions of the American Society of Agricultural Engineers 14:*950-953.

Stockle, C.O. (1992). Canopy photosynthesis and transpiration estimates using radiation interception models with different levels of detail. *Ecological Modelling 60:*31-44.

Supit, I. (1997). Predicting national wheat yields using a crop simulation and trend models. *Agricultural and Forest Meteorology 88:*199-214.

Supit, I. and E. van der Goot (1999). National wheat yield prediction of France as affected by the prediction level. *Ecological Modelling 116:*203-225.

Teittinen, M., T. Karvonen, and J. Peltonen (1994). A dynamic model for water and nitrogen limited growth in spring wheat to predict yield and quality. *Journal of Agronomy and Crop Science 172:*90-103.

ten Berge, H.F.M., K. Medtselaar, M.J.W. Jansen, E.M. de San Agustin, and T. Woodhead (1995). The SAWAH riceland hydrology model. *Water Resources Research 31:*2721-2732.

Thompson, L.M. (1969a). Weather and technology in the production of wheat in the production of wheat in the United States. *Journal of Soil Water Conservation 24:*219-224.

Thompson, L.M. (1969b). Weather and technology in the production of corn in the US corn belt. *Agronomy Journal 61:*453-456.

Thornley, J.H.M. (1977). Growth, maintenance and respiration: A re-interpretation. *Annals of Botany 41:*1191-1203.

Thornley, J.H.M. (1981). Organogenesis. In *Mathematics and Plant Physiology,* eds. D.A. Rose and D.A. Charles-Edwards. London: Academic Press, pp. 49-65.

Thornley, J.H.M. (1995). Shoot: Root allocation with respect to C, N and P: An investigation and comparison of resistance and teleonomic models. *Annals of Botany 75:*391-405.

Thornley, J.H.M., D.W. Hand, and J.W. Wilson (1992). Modelling light absorption and canopy net photosynthesis of glasshouse row crops and application to cucumber. *Journal of Experimental Botany 43:*383-391.

Thornley, J.H.M. and R.G. Hurd (1974). An analysis of the growth of young tomato plants in water culture at different light integrals and CO_2 concentrations. II. A mathematical model. *Annals of Botany 38:*389-400.

Thornley, J.H.M., R.G. Hurd, and A. Pooley (1981). A model of growth of the fifth leaf of tomato. *Annals of Botany 48:*327-340.

Thornton, P.K., W.T. Bowen, A.C. Ravelo, P.W. Wilkens, G. Farmer, J. Brock, and J.E. Brink (1997). Estimating millet production for famine early warning: An application of crop simulation modelling using satellite and ground-based data in Burkina Faso. *Agricultural and Forest Meteorology 83:*95-112.

Thornton, P.K. and G. Hoogenboom (1994). A computer program to analyze single-season crop model outputs. *Agronomy Journal 86:*860-868.

Thornton, P.K., G. Hoogenboom, P.W. Wilkens, and W. T. Bowen (1995). A computer program to analyze multiple-season crop model outputs. *Agronomy Journal 87:*131-136.

Timmer, L.W. and S.E. Zitko (1996). Evaluation of a model for prediction of postbloom fruit drop of citrus. *Plant Disease 80:*380-383.

Travasso, M.I., D.O. Caldiz, and J.A. Saluzzo (1996). Yield prediction using the substor-potato model under Argentinean conditions. *Potato Research 39:*305-312.

Travasso, M.I. and G.O. Magrin (1998). Utility of CERES-barley under Argentine conditions. *Field Crops Research 57:*329-333.

Trooien, T.P. and D.F. Heermann (1992a). Measurement and simulation of potato leaf area using image processing. I. Model development. *Transactions of the American Society of Agricultural Engineers 35:*1709-1712.

Trooien, T.P. and D.F. Heermann (1992b). Measurement and simulation of potato leaf area using image processing. II. Model results. *Transactions of the American Society of Agricultural Engineers 35:*1713-1718.

Trooien, T.P. and D.F. Heermann (1992c). Measurement and simulation of potato leaf area using image processing. III. Measurement. *Transactions of the American Society of Agricultural Engineers 35:*1719-1721.

Tsuji, G.Y., G. Hoogenboom, and P.K. Thornton, eds. (1998). *Understanding Options for Agricultural Production.* Systems Approaches for Sustainable Agricultural Development. Dordrecht, the Netherlands: Kluwer Academic Publishers.

Tsuji, G.Y., G. Uehara, and S. Balas, eds. (1994). *Decision Support System for Agrotechnology Transfer (DSSAT) Version 3.* Honolulu, HI: University of Hawaii.

van Keulen, H. (1975). *Simulation of Water Use and Herbage Growth in Arid Regions.* Simulation Monographs. Wageningen, the Netherlands: Centre for Agricultural Publishing and Documentation (Pudoc).

Vanderzwet, T., A.R. Biggs, R. Heflebower, and G.W. Lightner (1994). Evaluation of the maryblyt computer-model for predicting blossom blight on apple in West Virginia and Maryland. *Plant Disease 78:*225-230.

Verhagen, J. (1997). Site-specific fertiliser application for potato production and effects on N-leaching using dynamic simulation modelling. *Agriculture, Ecosystems and Environment 66:*165-175.

Wagner-Riddle, C., T.J. Gillespie, L.A. Hunt, and C.J. Swanton (1997). Modeling a rye cover crop and subsequent soybean yield. *Agronomy Journal 89:*208-218.

Whisler, F.D., B. Acock, D.N. Baker, R.E. Fye, H.F. Hodges, J.R. Lambert, H.E. Lemmon, J.M. McKinion, and V.R. Reddy (1986). Crop simulation models in agronomic systems. *Advances in Agronomy 40:*141-208.

White, J.W. and G. Hoogenboom (1996). Simulating effects of genes for physiological traits in a process-oriented crop model. *Agronomy Journal 88:*416-422.

Wiebe, H.J. (1995). Validation and application of a simulation-model for flower formation of kohlrabi depending on vernalization and devernalization. *Gartenbauwissenschaft 60:*97-101.

Williams, J.R., C.A. Jones, J.R. Kiniry, and D.A. Spanel (1989). The EPIC crop growth model. *Transactions of the American Society of Agricultural Engineers 32:*497-511.

Woodruff, D.R. (1992). 'WHEATMAN' a decision support system for wheat management in subtropical Australia. *Australian Journal of Agricultural Research 43:*1483-1499.

Wopereis, M.C.S., B.A.M. Bouman, M.J. Kropff, H.F.M. ten Berge, and A.R. Maligaya (1994). Water use efficiency of flooded rice fields. I. Validation of the soil-water balance model SAWAH. *Agricultural Water Management 26:*277-289.

Wopereis, M.C.S., B.A.M. Bouman, T.P. Tuong, H.F.M. Berge, and M.J. Kropff (1996). *ORYZA_W: Rice Growth Model for Irrigated and Rainfed Environments.* Wageningen, Netherlands: Research Institute for Agrobiology and Soil Fertility.

Wopereis, M.C.S., J.H.M. Wösten, H.F.M. ten Berge, T. Woodhead, and E.M. de San Agustin (1993). Comparing the performance of a soil-water balance model using measured and calibrated hydraulic conductivity data: A case study for dryland rice. *Soil Science 156:*133-140.

Wu, G.W. and L.T. Wilson (1998). Parameterization, verification, and validation, of a physiologically complex age-structured rice simulation model. *Agricultural Systems 56:*483-511.

Yan, W. and L.A. Hunt (1999). An equation for modelling the temperature response of plants using only the cardinal temperatures. *Annals of Botany 84:*607-614.

Yan, W. and D.H. Wallace (2000). Simulation and prediction of plant phenology for five crops based on photoperiod * temperature interaction. *Annals of Botany 81:*705-716.

Yin, X., M.J. Kropff, T. Horie, H. Nakagawa, H. Centeno, D. Zhu, and J. Goudriaan (1997). A model for photothermal responses of flowering in rice: Model description and parameterization. *Field Crops Research 51:*189-200.

Yin, X., M.J. Kropff, H. Nakagawa, T. Horie, and J. Goudriaan (1997). A model for photothermal responses of flowering in rice. *Field Crops Research 51:*201-211.

Zalud, Z., M. Trnka, and M. Dubrovsky (2000). Change of spring barley production potential using crop model CERES-Barley. *Rostlinna Vyroba 46:*423-428.

Chapter 21

Modeling Crop Responses to Plant Growth Regulators

Kambham Raja Reddy
Harry F. Hodges

Plant hormones are a group of naturally occurring, organic substances that are synthesized in one part of the plant and translocated to another part to influence a whole range of physiological and developmental processes at low concentrations. Changes in hormone concentration and sensitivity mediate an array of developmental processes in plants, many of which interact with environmental factors. There are five major classes of plant hormones grouped under two major categories based on their modes of action such as growth promoters (e.g., auxins, gibberellins, and cytokinins) and growth inhibitors (e.g., abscisic acid and ethylene). More recently, a few other compounds, namely, brassinosteroids, polyamines, jasmonic acid, and salicylic acid, have also attained the status of possible plant hormones in plant development (see Crozier et al., 2000; and references cited).

Since the 1928 discovery of auxins, humans have been intrigued by the possibility of manipulating plant growth and development. Numerous plant processes are known to be controlled by plant growth regulators, but the mechanisms of control have remained elusive. Extensive work has been expended to determine their occurrence, chemistry, modes of action, and their biological roles in governing plant processes (Salisbury and Ross, 1992). Although this work is far from complete, considerable progress has been made (Crozier et al., 2000). Plant growth regulators have been implicated as having a role in several cellular processes as well as influencing numerous processes at the organismal level in agronomic crops such as cotton (Cothern, 1994; Oosterhuis, 1994; and references cited).

THE USES OF PLANT GROWTH REGULATORS

Applications of different chemicals with hormonal effects or hormone-like effects have found their way into agriculture and have become important production practices, particularly to horticulturists. Chemicals are routinely applied in grape and blueberry production to induce seedless fruit. Other chemicals are used to induce root development of vegetative cuttings, to induce thinning in apples, or to manage plant height of several species of greenhouse-grown flowers and nursery-produced ornamental plants. Dormancy, juvenility, sex determination, floral induction, fruit set, growth and ripening, tuberization, rooting, and induction of desirable fruit color also may be enhanced/modified with phytohormones. Lodging may be reduced and other desirable effects may be produced in some agronomic crops by applying chemicals that cause a reduction in stem growth.

Chemicals used for hormonal control of some specific process or to induce a specific effect usually have fairly narrow concentration ranges that are suitable to attain the desired effect. The timing of the chemical application and the environmental conditions immediately before or following their application may also play an important role in determining the plant responses. The purpose of this chapter is to encourage the development of crop models that will aid producers who use plant growth regulators in their management more effectively. Invariably, plant responses to plant growth regulators interact with environmental conditions and growth stage. These complications are usually too complex or unknown to mentally assimilate and use for routine crop management decisions. A mechanistic crop growth model that routinely calculates soil conditions and growth processes facilitates estimating the crop response to a crop growth regulator. In this chapter, we present the concepts for developing a model for mepiquat chloride (MC) which is used to regulate cotton crop growth. Hopefully, these concepts will encourage others to develop models of plant responses to other growth regulators that will improve their crop management.

THE NEED FOR MODEL DEVELOPMENT

Mepiquat chloride (1, 1-dimethylpiperidium chloride) is a plant growth regulator known to affect stem elongation and several other processes in cotton (Cothern, 1994; Reddy et al., 1995; and references cited). It is the most widely used plant growth regulator in cotton production. Mepiquat chloride, sold commercially as PIX or under several trade names, behaves as an antigibberellin chemical. Approximately 2.3 million kilograms of MC are used annually on 48 percent of the cotton grown in the United States.

The application rate and timing are made more difficult by variable plant size, growth rate, and changing conditions caused by unpredictable weather. Application, followed by weather conditions that induce slow growth, may result in decreased yields. Because of variable crop responses to MC, the recommendations for its use have tended toward applications of small amounts with the expectation that additional MC would be applied, depending on subsequent weather and responses to prior applications. Such conservative recommendations are the result of both the inability to accurately predict the plant responses to the chemical and the uncertainty of future weather. Clearly, there is a need for more accurate prediction of the effects of MC and other plant growth regulators.

If one knew the nature and magnitude of the plant responses to the chemical, then one would be able to model crop behavior and consequently more accurately approximate timing and amounts of a plant growth regulator needed to attain the desired effects. Obviously, one should be able to model plant responses more accurately with more complete information regarding the plant responses to the chemical and its interactions with environmental variables. For example, it is assumed that the entire chemical intercepted by the crop is absorbed. Plants have different amounts and qualities of epicuticular wax, depending on the species and environmental conditions (Das et al., 1979; Rao and Reddy, 1980). Leaf wax may slow absorption by physically separating the applied chemical from the epidermis and by forming a chemical barrier to protect the epidermis; but these factors were assumed insignificant in this study. If, however, the appropriate wax/absorption information were available, the model could be made more mechanistic and absorption would be appropriately accounted for if a waxy cuticle became a barrier in certain environments.

The first thing one needs to consider in modeling a growth regulator is the primary and meaningful growth processes that the growth regulator may influence. The processes in which one is attempting to induce changes may include secondary or side effects of the chemical that may be positive or negative, insofar as crop production is concerned, but the effect should be known and quantified. Mechanistic crop models provide quantitative information about recent or current growth rates, the predicted end-of-the-season plant height with selected weather scenarios, and crop yields in each weather, soil, and management combination tested. Such valuable information may be used in a variety of ways to improve the decision-making process leading to improved crop yields, reduced risks of high crop losses, or other crop management goals.

Extensive use of MC has shown that its application may lead to lower cotton yields than would have been obtained without its use, yet producers continue to use it. They do so because it allows them to manage their crops for higher yield objectives by managing irrigation and nitrogen fertilizer practices more intensively. However, with higher yield goals and their associated management practices, certain risks become greater. Such practices induce more vegetative growth, taller plants, increased insect control problems, and a boll-rot disease syndrome. The judicious application of MC under these high yield-goal management practices lowers plant height, reduces leaf area, causes mature fruit to be produced in lower positions on the plant, and frequently allows high yields to be produced with less risk of yield losses to pests. In other words, MC applications allow more intensive crop management by causing canopy development that is less susceptible to boll-rot disease syndrome and is more accessible to insect-control measures. The primary effect of this plant growth regulator is not to directly increase crop yields but rather to lower the risks of crop losses to pests and diseases and to generally reduce losses due to biotic factors and factors that may be more severe in intensively managed environments.

To quantify cotton-plant responses to MC and develop a model that can be used to improve MC-application management in the context of other crop management factors, Reddy et al. (1995) developed a database and a MC model that can be used in conjunction with a cotton crop model, GOSSYM/COMAX. GOSSYM/COMAX is a cotton simulation model and an expert system that is used extensively in commercial cotton production (McKinion et al., 1989). It can also be used as a management-decision aid for nitrogen fertilization, irrigation scheduling, and several other applications (Ladewig and Thomas, 1982).

CROP GROWTH MODEL

The development, characteristics, and applications of GOSSYM/COMAX have been previously described (Baker, Lambert, and McKinion, 1983; McKinion et al., 1989; Boone, Porter, and McKinion, 1995; Reddy, Hodges, and McKinion, 1997; Hodges et al., 1998). Briefly, GOSSYM, an acronym coming from the word *Gossypium,* the genus of cotton, is a cotton crop simulation model that is linked to a rule-based expert system called COMAX, CrOp MAnagement eXpert.

GOSSYM is a materials balance model that accounts for carbon, nitrogen, and water in the plant and soil root zone. It simulates crop responses to environmental variables such as solar radiation, temperature, rain/irrigation, and wind as well as variations in soil and cultural practices. Growth and development are estimated and a record is kept of leaf and fruit age. GOSSYM estimates growth and development rates by calculating potential rates for daily temperatures assuming other conditions are not limiting, then it corrects the potential rates by intensity of environmental stresses (Baker, Lambert, and McKinion, 1983; Baker and Landivar, 1991; Reddy, Hodges, and McKinion, 1997; Hodges et al., 1998; Nobel, 2000). GOSSYM provides its users with the size and stage of the crop as well as its present growth rate and the intensity of the stress factors. So, a grower can assume certain future weather (days and weeks) conditions to determine yield estimates depending on the maturity of the crop. With the addition of a MC module and a harvest-aid chemical module, a producer can simulate the effect of these chemicals using different timing, amounts, weather scenarios, and other cultural practice variables to attain an optimum set of management practices for soil and environment.

The Expert System, COMAX

Briefly, COMAX is an expert system that was explicitly developed for working with the crop simulation model, GOSSYM, first developed by Lemmon (1986) and later modified by McKinion et al. (1989) and others (Hodges et al., 1998). COMAX is a forward-chaining, rule-based system that contains an inference engine, a file maintenance system for the simulation model requirements, a database system for the knowledge base, and "user friendly," menu-driven systems for user interaction. The inference engine applies rules to: (1) set up weather and cultural practice input data files, including plant growth regulator applications used by the GOSSYM program; (2) execute the GOSSYM program; and (3) interpret the model results and make recommendations on timing and amounts of irrigation, fertilizers, plant growth regulators, and harvest-aid chemicals. For more detailed information on COMAX, see Hodges et al. (1998). The GOSSYM/COMAX system is more properly called a model-based-reasoning system than an expert system. Such a model/expert captures the expertise of the modeling team's ability to develop the model, to interpret its results, and to suggest management practices, e.g., apply PIX. The availability of such a crop growth model is highly desirable for any intensively managed crop, and can be used either alone or with other emerging, newer technologies to provide useful plant growth and development information.

DATABASE AND MODEL DEVELOPMENT

Depending on the nature and mode of action of plant growth regulators, specific and appropriate experiments need to be designed to quantify the information for developing crop models. Reddy et al. (1995) conducted an experiment to generate a model database for MC-induced changes in cotton.

They quantified the effects of MC on primary growth processes such as stem elongation, leaf area expansion, and photosynthesis as functions of MC tissue concentration. From these relationships growth-reduction algorithms were developed to modify stem extension, leaf area expansion, and photosynthesis (Figure 21.1). When developing the MC model, Reddy et al. (1995) assumed that (1) MC is systemic, it moves freely throughout the plants, and it may influence any growing structure; (2) it is not metabolized; (3) it is lost from the plant only by abscission of plant parts or root death; (4) there are no interactions between MC and water deficits (Reddy, Trent, and Acock, 1992; Fernandez, Cothern, and McInnes, 1992); and (5) all cultivars respond similarly to MC (Schott and Schroeder, 1979; Reddy, Trent, and Acock, 1992).

Calculation of Mepiquat Chloride in the Plant Tissue

When MC is applied, GOSSYM calculates the intercepted portion of the chemical based on the method of application and converts the information

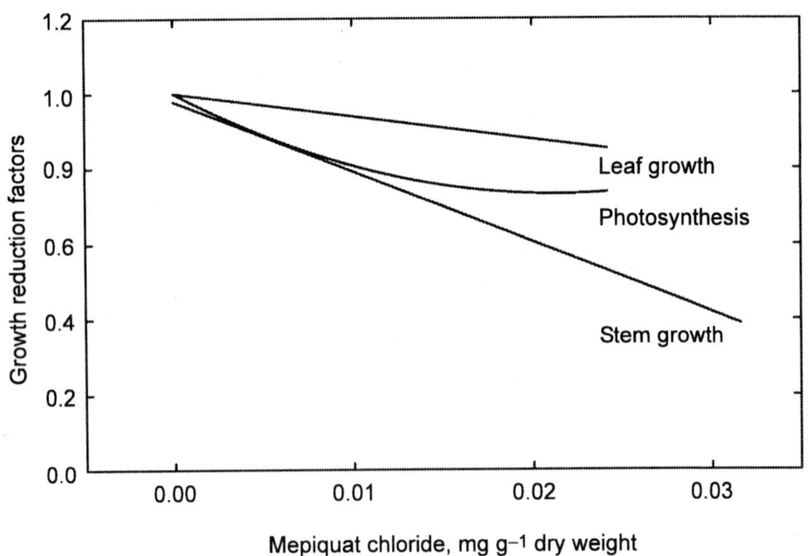

FIGURE 21.1. Relationship between plant tissue mepiquat chloride concentration and rates of stem elongation, leaf area expansion, and leaf photosynthesis expressed as fractions of the rates at zero MC concentration. (*Source:* Reduction functions were calculated from the data presented by Reddy et al., 1995, pp. 1102, 1103.)

into scientific units of an active ingredient. The MC intercepted by the crop canopy has been calculated by using the logic for light interception algorithms developed by Baker, Hesketh, and Weaver (1978). The MC received by the plant is estimated as if the chemical were broadcast; however, if MC is applied as a banded application on the top of the crop row, it is assumed that the intercepted portion is 90 percent of the applied chemical. The amount of MC applied is added to any existing concentration in the plant tissue, and the number of MC applications are updated and summed. The calculation of MC concentration on a tissue dry weight basis is based on the assumptions of Reddy (1993) and Landivar et al. (1992) with slight modifications. GOSSYM has been used to calculate plant dry weight and weight loss due to abscission of plant parts and root sloughing on a daily basis after MC application (for details see Reddy et al., 1995).

Calculation of Growth-Process Reduction Factors

The next step in the model is to calculate growth-process reduction factors caused by MC. Based on the results of Reddy et al. (1995), the primary effects of MC are on stem elongation, leaf area expansion, and photosynthesis (Figure 21.1). Mathematical relationships have been developed for plant height (stem elongation), leaf area expansion rates, and photosynthesis, assuming that these factors are equal to one at zero MC concentration (Figure 21.1). The reduction factors were used to decrease plant height, leaf area, and photosynthesis in various subroutines in GOSSYM.

MODEL VALIDATION AND APPLICATION

The cotton model, GOSSYM/COMAX, including the MC subroutine, is mechanistic in nature. The MC subroutine fits with other subroutines dealing with various processes and the tissue concentration and the effects are calculated daily. The predictive ability of GOSSYM/COMAX with the MC subroutine was tested for plant height, mainstem nodes, and yield data collected across the U.S. Cotton Belt. The data sets included a wide range of environmental conditions, a variety of cultural practices, and diverse genetic resources. They also comprised both single and multiple rates of MC applications on different dates during the growing season.

The model performance statistics are shown in Table 21.1. The data sets include a total of 50 cropping systems from 1987 to 1992. The data sets comprised irrigated and rain-fed conditions, three or more cultivars with several soils, and cultural practices where MC was applied as a tool to manipulate crop growth and development. Data on all traits were not collected

TABLE 21.1. Performance Statistics of the Simulation Model GOSSYM/COMAX with MC Subroutine for Plant Height, Mainstem Nodes, and Yields

Cultivar	N	n	Slope	R^2
Plant Height, cm				
DES 119	6	30	0.913*	0.96
DPL 20	6	41	0.966	0.96
DPL 50	14	91	1.018	0.98
Pooled data	26	162	0.980	0.97
Mainstem Nodes, no. plant $^{-1}$				
DES 119	6	30	0.993	0.98
DPL 20	6	41	0.985	0.99
DPL 50	14	91	1.023**	0.99
Pooled data	26	162	1.007	0.99
Lint Yield, Mg ha^{-1}				
DES 119	8	8	0.920	0.97
DPL 20	11	11	1.151*	0.97
DPL 50	14	14	1.073	0.97
GC510	4	4	0.822*	0.99
Pooled data	37	37	1.006	0.96

Source: Reddy et al., 1995, p. 1103.

*,** significantly different from one at 0.05, 0.01 probability levels.
N = Number of fields sampled for each cultivar.
n = the total number of observations in all the fields.

on every crop system, and therefore the number of observations for each trait is not additive. The R^2 values for plant height, mainstem nodes, and yields between observed and simulated ranged from 0.95 to 0.99. Leaf area was not compared in this model validation because the data were not available. The values of predicted and actual plant height and mainstem nodes using the MC model are also shown (Figure 21.2 and 21.3). A good linear relationship was obtained between observed and predicted values with points falling about 1:1 line. The performance of the model was also judged from the predictive capability of the model on yields across the U. S. Cotton Belt (Figure 21.4 and Table 21.1). The data include 37 management units grown over a range of weather conditions between 1987 and 1992, with diverse varieties, soil types, and management practices with several rates of

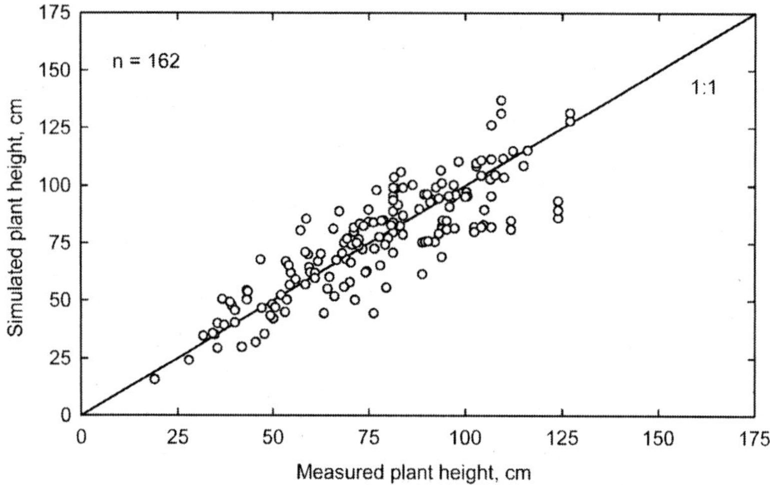

FIGURE 21.2. Comparison of measured and simulated plant height for three cultivars. The data collected from plants grown on various soil types, weather conditions, and management practices were used for model validation. They are from multiple observations in time from the same plots. (*Source:* Reddy et al., 1995, p. 1104.)

MC application. The cotton simulation model, GOSSYM/COMAX, with the MC model may be used on farms as a management tool for optimization of the plant growth regulator.

ON-FARM USE OF CROP MODELS AND TECHNOLOGY TRANSFER

The effects of a plant growth regulator may be modeled to improve crop management decisions. To do so, the important effects of the plant growth regulator need to be modeled on the crop as a function of tissue concentration. By knowing the growth rates of the crop via a crop model, one can more precisely tune the timing and rates of the growth regulator needed to obtain the desired crop responses.

Reddy (1995) also used a similar approach to develop a harvest-aid chemical model, PREP. PREP is a precursor to ethylene and has the effect of causing a rapid abscission of leaves and immature fruit and hastens boll opening. This facilitates more efficient harvesting and either this or other chemicals applied for that purpose are used on virtually all the commer-

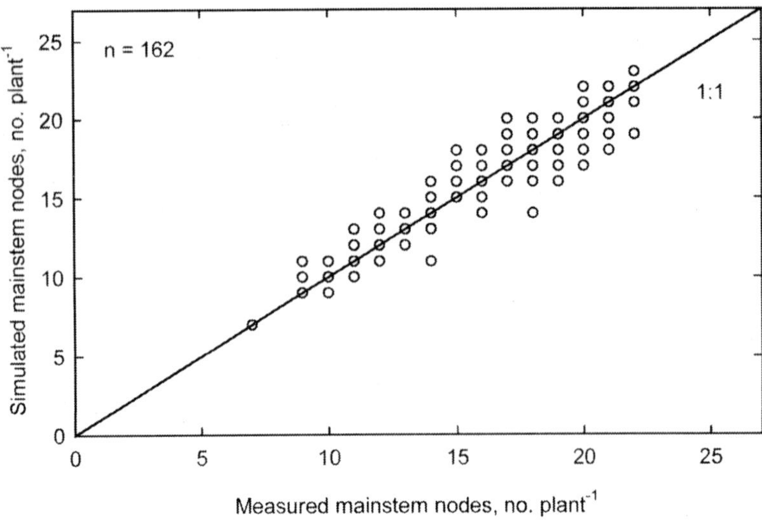

FIGURE 21.3. Comparison of measured and simulated mainstem nodes for three cultivars. The data collected from plants grown on various soil types, weather conditions, and management practices were used for model validation. They are from multiple observations in time from the same plots. (*Source:* Reddy et al., 1995, p. 1104.)

cially grown cotton where mechanical picking machines are used. Its effectiveness also depends on weather and crop condition, so a model of crop responses to timing and amount applied in the context to GOSSYM/COMAX provides growers with additional valuable information.

The concepts and principles involved in providing new technologies for improved agricultural production are continually being updated and tested (Reddy, Hodges, and McKinion, 1997). By agriculture's very nature, crops are produced in complex environments with varying weather, soils, cultural practices, and genetic resources. Introduction of any new technologies into such a complex system requires considerable time, experience, and understanding of the whole crop production system (Beinroth et al., 1980; Whisler et al., 1986). These requirements are always expensive and sometimes simply not available.

A partial solution to technology-transfer cost in time and testing may be found in crop production models (Reddy, Hodges, and Reddy, 1990). Crop models that are sufficiently mechanistic and validated will accurately predict crop performance in response to physical conditions, and may be used as vehicles to test new technology. In this chapter, we have shown how a

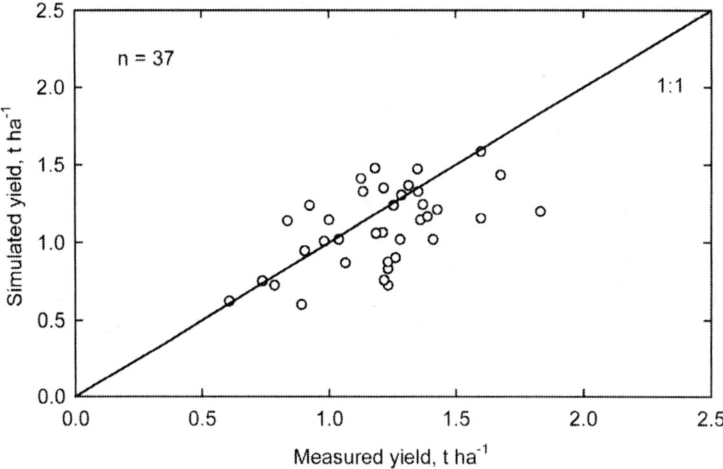

FIGURE 21.4. Comparison of measured and simulated yields for four cultivars. The data collected from plants grown on various soil types, weather conditions, and management practices across the U.S. Cotton Belt were used for model validation. (*Source:* Reddy et al., 1995, p. 1104.)

crop model/expert system, GOSSYM/COMAX, may be used to improve the usefulness of a plant growth regulator. The biological responses to the plant growth regulator were determined, the mathematical equations were developed and introduced into a more general crop model, and the modified crop model was evaluated in a number of field production environments.

This technique provides the grower, or the production manager, a decision aid that allows testing of the effects of the plant growth regulator at different rates, timing of application(s), and in combinations with other production inputs such as fertilizer or irrigation. The production manager can conduct many model runs in a short period, even before the crop is planted or the material is purchased. The model can also be tested in whatever management combinations are desired with weather records from several years. Managers can then develop a sense of the interactions of the plant growth regulator effects with weather and management factors before adopting the new practice. By using the crop model to develop this new information at the production-manager level, several years of field testing, demonstrations, and production-manager level education can be avoided. Consequently, the cost and time to deliver the new technology can be reduced with this technique. The average value of financial benefit of crop model use estimated by growers is approximately $50 per acre and more experienced, innovative, and computer-literate producers value the cotton model at much more (Ladewig and Thomas, 1992). The model and expert system are continuously being upgraded, their uses are being extended, and they are being made more user friendly (Reddy, Hodges, and McKinion, 1997; Hodges et al., 1998). The concepts employed for developing a growth-regulator model may be extended for other chemicals including foliar fertilizers. The concepts of crop model applications may also be extended to many other additions of newer technologies for precision agriculture such as geographical information systems (GIS), geographical position systems (GPS), intelligent implements, and remote sensing into crop production systems to precisely manage crops both spatially and temporally.

SUMMARY

Plant hormones, which are signal molecules present in trace amounts, mediate a whole array of growth and developmental processes in plants, many of which involve interactions with environmental factors. Applications of chemicals with hormonal effects or hormonelike effects have found their way into agriculture and have become important production practices. The plant growth status at the time of the chemical application and the environmental conditions immediately before or following their application usually play an important role in determining plant responses to the chemical and the economics of its use. Resource managers need information that can ease the decision-making process for efficient resource management. The purpose of this chapter was to provide frameworks for experimental protocol, data collection and analysis, and developing and validating a model for mepiquat chloride. Hopefully, these concepts will encourage others to develop models of plant responses to other growth regulators that will improve their crop management.

REFERENCES

Baker, D. N., J. D. Hesketh, and R. E. C. Weaver (1978). Crop architecture in relation to yield. In *Crop Physiology,* ed. U. S. Gupta, New Delhi, India: Oxford and IBH Publishing Company. pp. 110-136.

Baker, D. N., J. R. Lambert, and J. M. McKinion (1983). *GOSSYM: A simulator of cotton crop growth and yield.* South Carolina Agricultural Experimental Station Technical Bulletin 1089. pp. 135.

Baker, D. N. and J. A. Landivar (1991). The simulation of plant development in GOSSYM. In *Predicting Crop Phenology,* ed. T. Hodges, Boca Raton, FL: CRC Press. pp. 153-170.

Beinroth, F. H., G. Uehara, J. A. Silva, R. W. Arnold, and F. B. Caday (1980). Agrotechnology transfer in the topics based on soil taxonomy. *Advances in Agronomy 33:* 303-339.

Boone, M. Y. L., D. O. Porter, and J. M. McKinion (1995). *RHIZOS 1991: A simulator of row crop rhizospheres.* United States Department of Agriculture, Agriculture Research Service, Bulletin 113, Washington, DC: Govt. Printing Press Office. p. 175.

Cothern, J. T. (1994). Use of growth regulators in cotton production. In *Challenging the Future, Proceedings of the World Cotton Research Conference I,* eds. G. A. Constable and N. W. Forrester, Brisbane, Australia. pp. 6-24.

Crozier, A., Y. Kamiya, G. Bhishop, and T. Yokota (2000). Biosynthesis of hormones and elicitor molecules. In *Biochemistry and Molecular Biology of Plants,* eds. B. B. Buchanan, W. Gruissem, and R. L. Jones, Rockville, MD: American Society of Plant Physiologists. pp. 850-929.

Das, V. S. R., K. R. Reddy, Ch. M. Krishna, S. S. Murthy, and J. V. S. Rao (1979). Transpirational rates in relation to the quality of leaf epicuticular waxes. *Indian Journal of Experimental Biology 17:* 158-163.

Fernandez, C. J., J. T. Cothern, and K. J. McInnes (1992). Carbon and water economies of well-watered and water-deficit cotton plants treated with mepiquat chloride. *Crop Science 32:* 175-180.

Hodges, H. F., F. D. Whisler, S. M. Bridges, K. R. Reddy, and J. M. McKinion (1998). Simulation in crop management—GOSSYM/COMAX. In *Agricultural Systems Modeling and Simulation,* eds. R. M. Peart, and R. B. Curry, New York: Marcel Dekker, Inc. pp. 235-281.

Ladewig, H. and J. K. Thomas (1992). *A follow-up evaluation of GOSSYM-COMAX cotton program.* The Texas A & M University System, College Station, TX: Texas Agricultural Extension Service. pp. 47.

Landivar, J. A., S. Zypman, D. J. Lawlor, J. Vasek, and C. Crenshaw (1992). The use of an estimated plant PIX concentration for the determination of timing and rate of application. In *Proceedings-Beltwide Cotton Conference,* eds. D. J. Herber and D. A. Ritcher, Memphis, TN: National Cotton Council. pp. 1047-1049.

Lemmon, H. E. (1986). COMAX: An expert system for cotton crop management. *Science 223:* 29-33.

McKinion, J. M., D. N. Baker, F. D. Whisler, and J. R. Lambert (1989). Application of the GOSSYM/COMAX system to cotton crop management. *Agricultural Systems 31:* 55-65.

Nobel, P. S. 2000. Crop ecosystem responses to climatic change: Crassulacean acid metabolism crops. In *Climate Change and Global Crop Productivity,* eds. K. R. Reddy and H. F. Hodges, Wallingford, Oxon, UK: CABI Publishing, pp. 315-331.

Oosterhuis, D. M. (1994). Effects of PGR IV on the growth and yield of cotton: A review. In *Challenging the Future, Proceedings of the World Cotton Research Conference I,* eds. G. A. Constable and N. W. Forrester, Brisbane, Australia. pp. 29-39.

Rao, J. V. S., and K. R. Reddy (1980). Seasonal variation in leaf epicuticular wax of some semiarid shrubs. *Indian Journal of Experimental Biology 18:* 495-499.

Reddy, K. R., M. L. Boone, A. R. Reddy, H. F. Hodges, S. Turner, and J. M. McKinion (1995). Developing and validating a model for a plant growth regulator. *Agronomy Journal 87:* 1100-1105.

Reddy, K. R., H. F. Hodges, and J. M. McKinion (1997). Crop modeling and applications: A cotton example. *Advances in Agronomy 59:* 225-290.

Reddy, V. R. (1993). Modeling mepiquat chloride-temperature interactions in cotton: The model. *Computers Electronics in Agriculture 8:* 227-236.

Reddy, V. R. (1995). Modeling ethephon and temperature interaction in cotton. *Computers Electronics in Agriculture 13:* 27-35.

Reddy, V. R., D. N. Baker, and H. F. Hodges (1990). Temperature and mepiquat chloride effects on cotton canopy architecture. *Agronomy Journal 82:* 190-195.

Reddy, V. R., H. F. Hodges, and K. R. Reddy (1990). Streamlining agricultural information from the scientist to the economic users through simulation models. In *Proceedings of Asia-Pacific Society of Agricultural Engineering,* eds. V. M. Salokhe and S. G. Illangantileke, Bangkok, Thailand: Asian Institute of Technology, Volume 4. pp. 1213-1222.

Reddy, V. R., A. Trent, and B. Acock (1992). Mepiquat chloride and irrigation versus cotton growth and development. *Agronomy Journal 84:* 930-933.

Salisbury, F. B. and C. W. Ross (1992). *Plant Physiology,* Fourth Edition, Belmont, CA: Wadsworth Publishing Company.

Schott, P. E. and M. Schroeder (1979). Modification of the growth of *Gossypium* spp. (cotton) by the plant growth regulator mepiquat chloride. In *Proceedings of the Plant Growth Regulator Working Group,* Lake Alfred, FL: Plant Growth Regulator Society of America. pp. 250-265.

Whisler, F. D., B. Acock, D. N. Baker, R. E. Fye, H. F. Hodges, J. R. Lambert, H. E. Lemon, J. M. McKinion, and V. R. Reddy (1986). Crop simulation models in agronomic systems. *Advances in Agronomy 40:* 142-208.

Chapter 22

Modeling: Potential and Limitations

Thomas M. Addiscott

Thirty years ago, computer modeling was in its infancy in soil and plant science and its practitioners were viewed with mild suspicion by many other scientists. Today, computer modeling plays an increasingly important part in the decision making by those who determine environmental policy, and few project proposals are complete without some mention of modeling. It could be asked why a chapter on the *potential* of computer modeling in soil and plant science is needed at all. The main reason is that, as implied by the title, modeling has a rather limited *further* potential unless there is a clear understanding of the *limitations* attached to the use of models. The "millennium bug" proved to be a "paper tiger," but there are equally insidious problems which may arise if models are used too uncritically, particularly in conjunction with geographic information systems. This chapter will discuss the limitations on modeling with reference to the issues of scale and hierarchy, determinism and stochasticity, error, nonlinearity, and the validation and parameterization of models. It then goes on to consider, briefly, current applications and finally some new ideas that show potential for development in soil and plant modeling.

SCALE AND HIERARCHY

The problem of scale is generally recognized as one of the most important issues faced by soil modelers, if not the most important. Before discussing it, the meaning of the word *scale* itself needs to be clarified. A large-scale development project, for example, is likely to cover a substantial area of land. To a cartographer, however, a large-scale map summarizes a smaller land area on a given sheet of paper than a small-scale map. One way to prevent misunderstanding is to refer to "coarse" and "fine" rather than "large" and "small" scales. Alternatively, *scale* can be used as an umbrella term and speak of large and small areas of land. The concept of hierarchy is closely associated with that of scale because of the concern with a hierarchy of land units. Plots are parts of fields and fields parts of farms, and the hierarchy continues through catchments, regions, and countries.

Scale and the Applicability of Models

Policymakers often need models that can be used in pursuance of environmental objectives at regional or national scales, but many models are developed at field and plot scale. Either models that are applicable at all scales or some means of assessing the scales at which various models are applicable is needed. Models applicable at all scales provide the lesser of the two challenges, because any model whose parameters are entities that have the same meaning at all scales should be applicable at all scales (Dumanski, Pettapiece, and McGregor, 1998). These authors cite kilograms per hectare as such an entity, and the volumetric water content is another—it has exactly the same meaning and roughly the same value for 1 km^2 of soil as it does for 1 cm^2 of soil.

The problem of assessing the scales at which models are applicable can be tackled through the scale diagram (Hoosbeek and Bryant, 1992). This diagram uses the pedon as the *i*th or base level for the hierarchy and defines other levels with reference to it (Table 22.1). Each level comprises a plane within which model attributes are placed between *mechanistic* and *empiri-*

TABLE 22.1. The Hierarchical Levels of the Scale Diagram

Scale	Unit
i + 4	Region
i + 3	Interacting catchments
i + 2	Catena or catchment
i + 1	Field (Polypedon)
i	Pedon
i + 1	Profile horizon
i + 2	Peds, aggregates
i + 3	Mixtures
i + 4	Molecular
Contrasting model attributes at each i level	
Hoosbeek and Bryant (1992)	
Mechanistic vs. Empirical	

Source: Derived from Hoosbeek and Bryant, 1992, p. 186.

cal on one axis and *quantitative* and *qualitative* on the other. Models can also be classified by other criteria; Addiscott and Wagenet (1985) suggested a classification in which the first distinction was between deterministic and stochastic models and further distinction was made between mechanistic and functional, or nonmechanistic models.

By using the diagram we can ask whether models retain their positions in the classification regardless of the hierarchical level, or whether the terms deterministic and mechanistic, for example, become more or less appropriate as a model is moved from one hierarchical level to another. This should help to assess the scales at which various models are applicable. The diagram also brings the question of model validation into focus. Is a validation at one scale relevant at another, and can models be validated satisfactorily at the scale of the catchment or the region? Very similar questions apply to the parameterization of models and to the transferability of parameters across scales.

As well as asking what model is appropriate at a particular scale the problem of upscaling and downscaling a particular model needs to be addressed. This was tackled by Smith (1996) who considered issues such as changes in the scope of the model and the heterogeneity in the values of its inputs. Upscaling decreases the accuracy of inputs and increases their heterogeneity, whereas downscaling increases the accuracy needed. Smith was also concerned with the problem of validation at different scales and error propagation, which is discussed later in the chapter, and the interesting question of whether the number of processes that need to be included changes with the scale at which the model is used. This latter point relates to the issue of determinacy discussed as follows.

DETERMINACY AND RANDOMNESS

The classification system of Addiscott and Wagenet (1985) made the distinction between deterministic and stochastic models for transport processes, but this distinction has come under question. Sposito, Jury, and Gupta (1986) showed that the deterministic convection-dispersion equation for solute transport can be formulated as a stochastic transfer function, and Scotter, White, and Dyson (1993) suggested that a similar function lurks behind the apparently deterministic Burns (1974) leaching equation.

Decoherence

Whether processes are deterministic or stochastic is related to the issue of scale, and there is a link in the concept of decoherence (Stewart, 1995).

Classical physics was based mainly on observations made at large scales and was dominated on the theoretical side by Newton's laws and Laplacian determinism. Modern quantum physics is concerned with very small scales and is largely indeterminate. The planets obey Newton's law, but particles within the atoms that make up the planets obey the laws of chance. Decoherence is the term used to describe the loss of indeterminacy that seems to occur as small quantum systems are aggregated to make up larger ones. In short, it expresses changes in determinacy with scale and therefore has a bearing on the selection of appropriate models.

If decoherence occurs in soils and landscapes, large areas of land should behave in a more determinate and therefore predictable way than small volumes of soil. This may seem counter-intuitive, but denitrification provides an example. In small volumes of soil active microsites or "hot spots" control denitrification (Parkin, 1987) and these are unpredictable with respect to their location and behavior. The process remains difficult to predict at the scale of the field, but Groffman and Tiedje (1989) and Corre, Van Kessel, and Pennock (1996) have shown how predictive relationships for denitrification become easier to establish at the scale of the landscape. These depend on the fact that soil wetness, which strongly influences denitrification, can usually be assessed from the topography. A soil on a spur tends to be relatively dry, and one on a footslope is wetter. These topographical factors can potentially be combined with soil, climatic, and seasonal factors and with land use to provide models for denitrification. The concept of decoherence is applicable to temporal as well as spatial scale, as is shown in a later section.

Deterministic or Stochastic Models

Webster (2000) has recently added a new dimension to the discussion on determinacy and randomness by questioning whether soil variation is as random as it usually appears. He concluded that, because physical laws must apply, what appears random is chaotic, using the modern scientific sense of the word. Chaos is deterministic, although its outcome may be difficult to distinguish from that of random processes. This suggests that, although stochastic models may be useful for some practical purposes, our main focus should be on deterministic models and what determinism really means. We also need to be aware that the use of a deterministic model might in some circumstances lead to a chaotic outcome (Addiscott and Mirza, 1998).

Laplace's Concept of Determinism

A deterministic model is usually defined (Addiscott and Wagenet, 1985) as one which presumes that a system or a process operates such that a given sequence of events leads to a uniquely definable outcome. There is, however, more to determinism than this, and we need to go back to the statement by Laplace which underlies determinism and was quoted by Stewart (1995) as:

> An intellect which at any given moment knew all the forces which animate Nature and the mutual positions of all the beings that comprise it, if this intellect was vast enough to submit its data to analysis, could condense into a single formula the movement of the greatest bodies of the universe and that of the lightest atom: for such an intellect nothing could be uncertain, and the future just like the past would be present before its eyes. (p. 107)

Laying aside the theological implications of this statement and any doubts we may have about the vastness of our intellect, we are left with two problems which have to be faced by those wishing to use deterministic models:

- We do not know all the forces that operate within a single hierarchical level, let alone within the whole system.
- We are very rarely, if ever, fully able to define the initial state of the system.

The applicability in the present context of Laplace's concept and of the whole structure of determinism depends on our ability to resolve these problems. Indeed, our ability to predict using models at any hierarchical level may simply depend on the degree of uncertainty attached to our knowledge of the dominant forces acting at that level and the initial state of the system. Neither will be the same at all levels. The weather, which can be measured with reasonable certainty, may be the dominant force acting on nitrate leaching at catchment level whereas the dominant force at the level of the soil aggregate is probably a concentration gradient which can be measured only with the greatest difficulty and is therefore uncertain. We may well find too that a "broad brush" assessment of the initial state of the catchment based on readily available information gives a more satisfactory definition of the initial state than can be achieved at the level of the aggregate. It seems clear that anyone intending to develop a model or use an existing one needs to ask themselves two questions at the outset.

- How well do they understand the dominant forces acting in the system they wish to model?
- How reliable an assessment do they have of the initial state of the system?

It is unwise to proceed with the model without satisfactory answers to these questions.

ERROR AND NONLINEARITY

No model is perfect and no modeler infallible, so models tend to propagate error. This also happens because model inputs have various forms of error associated with them. This error propagation is potentially a trap for the unwary modeler, particularly in soil science where so many model inputs are subject to error in the statistical sense of *variation.* Soil scientists who develop models need to ensure that both they and the users of the models are aware of the sources of error and can test for it and take remedial action where needed. They also need to consider the propagation of error by soil models, particularly the interaction which occurs between error (variation) and nonlinearity, and the suppression and exaggeration of error. Error propagation takes on an extra dimension when models are used in conjunction with geographic information systems. This topic is discussed briefly later but readers wishing to explore this topic in full are referred to the excellent monograph of Heuvelink (1998a).

Forms and Sources of Error

The meaning of the word *error* depends greatly on the context in which it is used. Heuvelink (1998a, p. 9) suggested that in a modeling context error is the "difference between reality and our representation of reality," and the main forms of error discussed as follows are those he defined.

Model Error

The most basic form of error in modeling is some kind of fault in the model itself. This may arise simply from *error in concept,* meaning that the modeler's idea of the nature and functioning of the system being modeled is simply wrong. No specific test for this form of error can be defined, but the error may be exposed in a sensitivity analysis and should definitely be exposed if the model is evaluated (or validated) against measured data (Whitmore, 1991; Smith, Smith, and Addiscott, 1996). Both Heuvelink (1998b) and Jansen (1998) stress the importance of validation in assessing model error and distinguishing its effects from those of input error. The remedial action against model error is basically to change the model. Changes of scale may also necessitate a change in the model, usually of a simplifying nature, and Heuvelink (1998b) and Jansen (1998) discuss the implications of such scale-led changes.

Even if the concept is correct, there is still the possibility of *error in translation,* that is, in the process of converting the concept into a set of equations or computer code. This kind of error should be revealed in the verification of the model, in which the equations or computer code are examined to ensure that they properly represent the concept, and the remedial action should be obvious. The distinction between verification and validation can be seen from the Latin roots of the words. The former derives from *verus,* meaning true, and truth is absolute; something is either true or not true. Few of us will achieve a model that is absolutely true to reality in all particulars, and *verification* is usually used to describe the process of ensuring that the equations or computer code are true to the concept of the model. Validation comes from *validus,* meaning strong. This word carries no implications of the absolute, because strength is defined in relative terms or with respect to a defined purpose. There is no such thing as an absolute validation of a model, because statistical tests can show only the probability of its refutal, which may be small but is always finite. These issues are discussed subsequently in greater detail (see also Addiscott, Smith, and Bradbury, 1995).

Input Error

Both soil properties used as model parameters and weather information used as data are subject to error. This error may simply take the form of a *mistake* but, although we need to guard against error in this sense, few helpful generalizations can be made about mistakes. We are therefore more often concerned with the statistical usage of the word "error," that is, *variation.* With a parameter, we have the apparent anomaly that it is a constant in a particular case but also subject to variation. Many soil properties used as parameters do vary in space, but so long as the moments of the distribution of values do not change with time, there is no problem. A parameter which varies with time is not usable unless it varies in a clearly defined way.

With both forms of input, the error may spring from natural variation and error introduced during measurement or estimation. Jansen (1998) discussed how these and other sources of error can be quantified in the process of error or uncertainty analysis.

Output Error

Both model error and input error can lead to output error. Not much can be said usefully about the propagation of model error, or of input error where it implies mistakes in data. However, the relation between input and output error in terms of variation is the substance of *error propagation* and is not only interesting but central to the proper use of models. One reason why the topic is so important is that more is involved than the simple propagation of statistical error from input to output. If the model is nonlinear, then the error in the input contributes to the value of the *mean* of the output as well as the error. Thus, to understand error propagation fully, we need to consider the meaning and implications of nonlinearity.

Nonlinearity

Nonlinearity per se is a very simple concept. If plotting a relation gives a straight line, the relation is a linear one; if it does not, the relation is nonlinear. The equation $y = 2x$ gives a straight line, but $y = 2/x$ does not. The former is linear and the latter is nonlinear. The consequences of nonlinearity are less simple, but they cannot be ignored because of their influence on the propagation of error through models.

The Interaction Between Nonlinearity and Error

Addiscott and Tuck (2001) illustrated this interaction through the familiar Arrhenius relationship,

$$k = A\exp(-B/T), \tag{22.1}$$

in which k is a rate constant, T is the temperature (K), a variate, and A and B are the parameters of the equation. The rate constant k is linearly related to A but its relation to B is strongly nonlinear. Input error (variation) was introduced to the relation by assigning three values each to A and B. When A was varied, B was kept at its mean value and vice versa. The contribution of the input error to the mean of k was assessed by estimating this mean first by evaluating k from Equation (22.1) for each value of A or B and taking the mean (the "evaluate first" option), and then by taking the mean of A or B and evaluating the equation (the "average first" option). The output error attached to k was evaluated as the standard deviation (SD) when the first method was used.

When A was varied, the mean of k was the same for both options (see Table 22.2). However, when B was varied, the estimate of the mean of k ob-

TABLE 22.2. The Interaction Between Nonlinearity and Error in the Arrhenius Relationship

		μ_k		
Parameter varied	**Central value**	Evaluate first	Average first	Φ_k
A (mmole kg^{-1})	0.5×10^9	0.0211	0.0211	4.21×10^{-3}
B (K)	7000	0.841	0.0211	1.44

Source: Adapted from Addiscott and Tuck, 2001, p. 131.
The *A* and *B* coefficients were each given three values, a central value corresponding to that for mineralization, and values 20% greater or smaller. The mean, μ_k, of *k* was calculated from Equation (22.1) in two ways: by calculating it from the three individual values of *A* or *B* and taking the mean (evaluate first), or from the central (mean) value of *A* or *B* (average first). When *A* was varied, *B* was kept constant and vice versa. The standard deviation of *k*, F_k, was calculated with the evaluate first option. *T* was set at 293° K.

tained by the evaluate first option was forty times greater than that obtained from the average first option. This disparity resulted from the nonlinearity of Equation (22.1) with respect to *B*, and which also greatly increased the output error, as measured by the SD of *k*. The reason for the disparity can be seen from some equations presented by Rao, Rao, and Davidson (1977) to arise from the products of the second partial derivatives of *k* with respect to *A* and *B* and the variances of *A* and *B* (Addiscott, 1993). A linear relation has second partial derivatives of zero, which is why there was no difference between the two methods of evaluating the mean of *k* when *A* was varied. Rao, Rao, and Davidson (1977) also show the origin of the error in *k;* the variance of *k* is approximately the sum of the products of the squares of the first partial derivatives of *k* with respect to *A* and *B* and their variances; this relationship is also discussed by Heuvelink (1998b). The equations of Rao, Rao, and Davidson (1977) emphasize that correlation between parameters or inputs needs to be taken into account in such computations, a point also made by Jansen (1998).

The exercise brought out the point that a model can be linear with respect to one parameter or variable but nonlinear with respect to another, so any statement about nonlinearity needs to specify both the input and the output. There may obviously be differing degrees of nonlinearity.

Assessing Error Propagation

Virtually all the properties measured for use as parameters in soil models have an appreciable (statistical) error, and those parameters that are not measured but inferred by other means may have similar but unknown errors.

This, together with the interactions of the type discussed previously, shows that some means of assessing error propagation is needed, particularly if a model is used by people other than the developer. The equations of Rao, Rao, and Davidson (1977) are useful for showing how the interactions originate, but they are not universally applicable for assessing error propagation. Not all functions can be differentiated, and for those that can, the Taylor series expansion on which the method depends may not converge rapidly, as is the case for B in the Arrhenius relationship. There is also an obvious problem with computer models comprising several thousand lines of code.

A further complication is that there may be interactions in which a change in the error of one parameter may alter the propagation by the model of the error in another. There is clearly a need for some form of analysis for error propagation and the interactions between errors, and Addiscott and Tuck (2001) evolved the procedure outlined as follows for this purpose.

Error and Nonlinearity Interaction Analysis

The procedure assesses the interaction between nonlinearities in the model and the propagation of errors in its parameters through four tests, which are applied to *specific combinations* of parameter and output.

1. *First test for nonlinearity.* Does plotting the output from the model against the value of the parameter give a straight line or a curve?
2. *Second test for nonlinearity.* Is there a disparity between the "evaluate first" and "average first" procedures when the parameter has (statistical) error?
3. *First error propagation test.* Are the errors in the parameters suppressed or exaggerated as they are propagated through the model to the output?
4. *Second error propagation test.* Does the error in one parameter influence the propagation of the error in another?

Addiscott and Tuck (2001) used the solute leaching intermediate model (SLIM) solute leaching model (Addiscott and Whitmore, 1991) to show how such an analysis could be applied. The outcome showed that the evaluate first and average first options gave clearly differing results for some parameters and thereby provided a test for nonlinearity that is a useful supplement to the simple plotting test. The analysis also emphasized that we cannot make a general statement about nonlinearity in SLIM or any other model that has more than one parameter or output. We must refer to the effects on a *particular output* of nonlinearity with respect to a *specified pa-*

rameter. The analysis also showed that when the errors of the parameters were propagated through the model to the outputs some were exaggerated and others were suppressed. These effects differed between the various outputs of the model. The other key point it brought out was that the error of one parameter could clearly alter the effects of the error of another on both the mean and error of an output.

These interactions have wide-ranging implications, particularly in spatial averaging, where the evaluate first or interpolate first question often arises. Addiscott and Bailey (1990) found a considerable discrepancy between the two procedures, and this problem was also addressed by Stein et al. (1991) and Heuvelink and Pebesma (1999). This discrepancy raises an uncertainty in large-scale simulations in which nonlinear models are used, particularly where there is an implicit assumption that using spatially averaged parameters is equivalent to applying spatial averaging to the model output.

One interesting point to arise from the application of the error and nonlinearity interaction tests was that there were major differences between the patterns of behavior shown in the tests by the simulated mean concentrations for the whole season and those for four individual days. These suggested that the concentrations were more predictable at the longer than at the shorter temporal scale. Similar issues can arise with respect to *spatial* scale (Heuvelink and Pebesma, 1999), and both need to be discussed in the context of the phenomenon of decoherence described earlier.

The patterns shown by the concentrations for the four individual days suggested that for these outputs the model is strongly nonlinear with respect to the permeability parameter in the model, and this implies the possibility of chaotic behavior by the model. However, Addiscott and Mirza (1998) noted that this may be difficult to establish because chaotic behavior depends on the system being nonautonomous as well as nonlinear and occurs only within certain ranges in the inputs to the system. Chaos is discussed further as follows.

VALIDATION AND PARAMETERIZATION

The word *validation* is used widely to describe the process of testing and evaluating models. What the word means is essentially a philosophical question, but how the process should be effected is an operational and statistical matter. Both aspects are discussed in this section. The parameterization of models, that is, the selection of appropriate parameters for a particular context, is a closely related process which shares many of the problems of validation, making it relevant to discuss the two processes together.

The Philosophical Question

Validation

In discussing model validation, we need to ask first how science actually happens. The philosopher John Stuart Mill, building on the ideas of Francis Bacon in the seventeenth century, proposed that science progresses through the process of *induction,* in which declarations of fact arising from the evidence of the senses lead with certainty to the truth of general laws. Medawar (1967), however, argued that the two key processes in science were simply *having an idea* and *trying it out.* This approach had been formalized as the *hypothetico-deductive* system by the philosopher Karl Popper (Popper, 1959). Having an idea, framing a hypothesis, proposing a theory, or developing a model are fairly similar mental activities, and all carry the need for critical evaluation, which must include the possibility of refutal. Popper (1992) quoted Einstein (1920) as an example of the approach needed, when the latter wrote of his most famous theory, "If the red shift of the spectral lines due to the gravitational potential should not exist, then the general theory of relativity will be untenable" (Popper, 1992, p. 38).

Medawar (1967, p. 165) stated that, "The first strongly reasoned and fully argued exposition of a hypothetico-deductive system is unquestionably Karl Popper's"; he noted, however, that quite a large part of the system had been propounded "as a matter of learned discourse rather than critical analysis" by William Whewell. The system is also not without more recent rivals, but all recent theories of scientific development agree that the concept of induction is no longer tenable.

Because we have to allow for the possibility of refutal, there is a serious question as to whether we can ever *validate* a model, or an idea, theory, or hypothesis. Popper (1992, p. 149) cautioned that, "We can never justify a theory. But we can sometimes justify our preference for a theory . . . for a theory may stand up to criticism better than its competitors." This suggests that Popper would have told us that, although we may be able to discriminate between models, we can never validate a model in the sense of proving that it is entirely right. The apparently widespread impression that a model is justified if a regression of a simulated variate on a measured variate achieves $P < 0.05$ must therefore be dispelled. The fact that there is only a 1 in 20 chance that the model is refuted does not prove that it is right; it merely proves that there is a 1 in 20 chance that it is wrong.

Despite his cautionary note about justifying models or theories, Popper's comments on Einstein and his other writings suggest that he would have insisted on thorough-going critical evaluation for our models. St. Paul's great theme of "justification by faith" was not intended for modelers, and for modeling to retain its integrity we must use a robust, critical evaluatory procedure for any model that is to be used to provide advice. No procedure, however, can do more than indicate the probability of a model being refuted. It cannot prove that it works and, as Popper once observed, we have to live with a certain uncertainty.

Parameterization

Even the best of models is useless without reliable parameters. There are various ways of obtaining values for parameters, but they are not all of equal status (Addiscott, Smith, and Bradbury, 1995).

- The best option for getting the value of a parameter is direct measurement without any fitting. There is no risk that the value obtained will have been corrupted by a deficiency in the model or the fitting procedure, and the most likely problem, particularly for a soil parameter, may be (statistical) error leading to the complications discussed previously.
- Where the parameter cannot be measured directly, it is best obtained by fitting to data that is not of the type to be simulated. This may involve *direct fitting,* in which a parameter for an expression within the model is obtained by fitting to measurements that relate directly to that expression. The resulting parameter value will reflect any inadequacies in the expression or the fitting procedure but will not be affected by shortcomings in other parts of the model.
- Less satisfactory is *indirect fitting* which involves fitting the whole model to data of the type to be simulated (but not the actual set). If all the parameters are obtained simultaneously, it is fully possible that uncertainty will result because a variety of parameter sets will be found, all of which give more or less equally satisfactory fits (Beven and Binley, 1992). Even if only one parameter has to be obtained by fitting, the others being measured, its value will be influenced by any deficiency anywhere in the model.
- The least satisfactory practice of all is obtaining parameters by fitting to the data to be simulated. This shows that the model does work but not that it is of any practical use.

There is a decrease in the value of the information obtained from the model as you proceed from the first option to the fourth.

Operational and Statistical Aspects

Both the validation of a model and the derivation of a parameter value by fitting involve assessing the goodness of fit of simulations to measurements and are, from an operational viewpoint, rather similar procedures. Both depend considerably on the quality of the data, and Whitmore (1991) made the distinction between two cases:

1. When none or few of the measurements were replicated
2. When most or all of the measurements were replicated

Parameters will normally need to be obtained before the model can be validated, so parameterization will be discussed first. The procedures described as follows are those of Whitmore (1991), but Loague and Green (1991) have also suggested useful approaches to validation and a fuller review of the topic was provided by Smith, Smith, and Addiscott (1996).

Parameterization

1. When there is little or no replication, the best parameters can be selected by minimizing the sum of squares of the deviations between simulation and measurement.
2. When the replication is adequate, Whitmore's (1991) LOFIT (lack of fit) procedure should be used. This partitions the sum of squares of the deviations between simulation and measurement into those attributable to lack of fit and to pure error. Parameters can be selected by minimizing the lack of fit component and reducing it to zero if possible.

Validation

1. Where there was little or no replication, the best procedure is to assess both the degree of association and the degree of agreement between simulation and measurement, using the product moment correlation coefficient for the former and the mean difference for the latter. The correlation coefficient alone is not sufficient because it can be 0.99 even when the simulations overestimate the measurements by a factor of ten.
2. Where the replication is adequate, Whitmore's LOFIT procedure can again be used. Models can be evaluated or compared by comparing the mean square lack of fit to the mean square error, using the F-test to

assess the significance of the ratio. This tends to be a very stringent test of a model.

Some other points made by Whitmore (1991) are helpful. It helps greatly if a set of data has been obtained with the specific intent of validating a model or models, so that all the information needed is available. The experiment must be designed carefully with appropriate sampling, bearing in mind that a poor sampling arrangement might lead to the rejection of a good model or the acceptance of a poor one. Replication is advisable, and the model should be tested throughout its intended range of application.

Nonlinearity and Error

The interaction between nonlinearity and error discussed earlier in the chapter can cause problems in both the parameterization and validation of models. Addiscott, Smith, and Bradbury (1995) explained how this happens and showed some very simple examples in which ignoring the (statistical) error gave misleading results in both procedures. The problem can be alleviated in part by using Whitmore's (1991) LOFIT procedure, and the nonlinearity and error propagation tests recently proposed by Addiscott and Tuck (2001) should show in advance whether the model is sufficiently nonlinear to cause trouble of this kind.

APPLICATION OF MODELS

Decision Support

One of the main uses to which soil and plant models are being put is in advisory systems for farmers and other land users and for policymakers. The term "decision support" is now often used because such systems provide help in the process of making decisions. Converting a model into a decision-support system is not something to be undertaken lightly, as was found when the SUNDIAL nitrogen turnover model (Bradbury et al., 1993; Smith, Bradbury, and Addiscott, 1996) was recently converted into the SUNDIAL-FRS decision-support system for nitrogen fertilizer application.

The first part of the exercise was a consultation program in which farmers were contacted with the aid of the Association of Independent Crop Consultants to find out what they would require of such a system, what units they used, and what levels of computing capacity and skill they had available (Smith et al., 1997). All were, needless to say, somewhat variable and the farmers managed to think of more than 50 crops they wished to have included. The next stage was to try to incorporate the farmers' requirements in a "front end" attached to the model. This had to be formulated in a way that enabled them to enter details for all the individual fields they wished to include. These details included soil type and depth, crop to be grown, field drains, and estimated atmospheric nitrogen input. Details of previous cropping, fertilizer and manure application (including timing), and yield were also included. The front end allowed for the entry of weekly weather data for the farm, often collected by the farmer.

The initial version of the system was sent out to the farmers for trial and, where possible, the modifications they suggested were incorporated in the subsequent version. Further refinements had to be made before SUNDIAL-FRS was formally launched and allowance was made for backup to the system to allow further refinements to be made where useful. The overall development of the system took about five years.

One key point about this system is that it includes its own validation system so that the farmer or his or her advisor is able to check the function of the system against data from the farm. This emphasizes the point that it is essential for any decision-support system to be properly validated. Consideration also needs to be given to the other modeling pitfalls to which this chapter has drawn attention, that is, the problem associated with scale, determinacy, error propagation, and nonlinearity.

Use of Models with Geographic Information Systems

Models are now used increasingly with geographic information systems (GIS) in decision-support systems. Geographic information systems produce neat, attractively presented, color maps which are very convincing, even when based on seriously flawed information or models. It is a most effective way of propagating error from scientists to policymakers and planners, and this redoubles the need for proper evaluation of the models used.

It is not just the models that need evaluation. Heuvelink (1998a) has recently drawn attention to error propagation problems inherent in the use of GISs. He concluded that current GISs need some fundamental changes if they are to handle and manage uncertain spatial data in a professional manner. The most important of these changes is that these systems must be amenable to analysis of error propagation, methods for which he provided. His cautionary note on the use of GIS is very important, and his message must be conveyed to those who develop GIS-based decision-support systems and especially to those who use them without knowing what goes into them.

HOW MIGHT SOIL MODELING DEVELOP FURTHER?

We are in the age of postmodernism. One side of this phenomenon is that received wisdom is no longer valid, and the expert has no authority. Politicians defer to focus groups rather than deciding issues on principle. Environmental militants seem to get more credence than experienced scientists. We have the concept of the "stakeholder." What kinds of models are appropriate in the postmodernist age?

Stakeholder Involvement in Modeling

To begin with, the idea of the stakeholder is a useful one. In the section on decision support of key stakeholders in the SUNDIAL-FRS system, the farmers and their advisors had a major input into the way the system evolved. This was not only helpful (in the main) to the modelers, it also conferred on the stakeholders a sense of "ownership" which, we hope, encouraged them to test and eventually use the system. Much more is to be gained than lost in the involvement of stakeholders in the development of practically oriented models, particularly if we wish them to be used in real life. This does not necessarily abnegate the principles laid down by Popper (1959) and Medawar (1967). Someone still has to have the idea and someone has to test it, but part of the testing is done on a collective basis. The modelers themselves are key stakeholders. The involvement of stakeholders in modeling is almost certainly a positive approach.

Questioning Authority

The idea of questioning received wisdom and authority should not shock any decent scientist. It is the way science progresses, and the critical comparison of old and new theories or models is central to the philosophy of Popper (1992). One area where the laying aside of received wisdom proved useful is the transport of water and solutes in the soil. Models for solute leaching derived from classical soil water physics were greatly troubled by the heterogeneity of the soil. The much simpler models of Burns (1974, 1975), Addiscott (1977), and Corwin, Waggoner, and Rhoades (1991) proved very useful, mainly because they were much less affected by nonlinearity and statistical error in their parameters than the classical theory. As discussed in an earlier section, the nature of the Burns (1975) leaching equation has in turn come under question, and this questioning must go on for soil modeling to develop.

Environmental Militants versus Scientists

The credence given to environmental militants in preference to experienced scientists is another phenomenon to which Karl Popper has made a useful contribution. Magee (1985, p. 89) commented as follows on some of the ideas in Popper's book *The Open Society and Its Enemies,* and particularly his sociopsychological concept of "the strain of civilization."

> So from the beginning of critical thought, with the pre-Socratics, the developing tradition of civilization has had running parallel to it (or perhaps it would be more accurate to say running *within* it) a tradition of reaction against the strain of civilization, which produced accompanying philosophies of return to the womblike security of a precritical or tribal society, or of advance to a Utopia. Both (reactionary and utopian ideals) reject existing society and claim that a more perfect one is to be found at some other point in time. Hence both tend to be violent and yet romantic. (p. 89)

This statement seems strongly relevant to the current reaction against science, and the ideals that "tend to be both violent and yet romantic" (Magee, 1985, p. 89) call to mind the recent destruction of scientific trials designed to assess the risks associated with genetically modified crops. How should scientists respond?

It is clearly desirable to involve farmers as stakeholders in the development of decision-support systems developed for their benefit. However, although those who react against science could be argued to be stakeholders in its progress, it is far from clear that they can usefully be involved in the development of models unless they accept that decisions about environmental and related matters need to be decided on the basis of scientific, probability based risk assessment underpinned by Popper's philosophy. To see why this is necessary, we need look no further than the "nitrate problem." Vast sums of money are spent in the European Commission (EC) each year in removing from potable water something (nitrate) which is produced naturally in our bodies because it is the fuel for our body's defense system against gastroenteritis (Benjamin, 2000).

Top-Down and Bottom-Up Processes

What, if anything, can the modeler learn from politicians' recent enthusiasm for the focus group? The political process is a top-down process, with policy being passed down to the people from the government, usually via the civil service. The people get their say at election time but they have very limited influence at other times. The focus group inverts the process into a bottom-up one in which the people can feed policy ideas to the politicians. This raises an analogous question: Is our modeling a top-down or bottom-up endeavor? Modeling based on Popper's (1959) hypothetico-deductive principal begins as a top-down process, the having of the idea which is imposed in a speculative way on the data, but the data can refute the idea, in an essentially bottom-up process. The idea is then reformulated and retested, leading to an iterative, top-down-bottom-up cycle.

One problem in the use of focus groups is that "the people" rarely have a coherent idea of what they want. The soil-plant system, on the other hand, "knows" exactly how it works, and maybe it can tell us if approached in the right way. If so, the bottom-up approach probably has more potential in modeling soil-plant systems than in politics. This leads us to the final topics of the chapter, chaos and complexity.

Chaos and Complexity

Chaos

The phenomenon of chaos emerged briefly in the discussion about determinacy and randomness and may prove useful for interpreting some physical occurrences in the soil. Chaotic dynamics is a consequence of mathematics itself and can appear in various physical systems (Rasband, 1990). It is completely different from random behavior. A chaotic system evolves in a deterministic way, with the current state of the system always depending on the previous state. Chaos always arises from nonlinearity in the system, but this is not the only factor involved. The system also has to have more than one degree of freedom, that is, be nonautonomous. One of the commonest causes of chaotic behavior is a feedback loop in the system. Many processes in the soil are nonlinear and at least some of these must be non-autonomous or have feedback loops, suggesting that chaotic dynamics could result. Chaos manifests itself in various forms, of which fractals have probably had the most impact in soil science, although Phillips (1993) has found evidence that pedogenesis is a chaotic process.

The nonlinearity of soil processes is reflected by the nonlinearity of the models used to simulate them. If such models are nonautonomous or have feedback loops, they could behave in a chaotic manner, suggesting that it could be useful to be able to test for the possibility of chaotic behavior. One hallmark of a chaotic system is extreme sensitivity to small changes in initial conditions (Gleick, 1987), suggesting that examining the effects of small changes in the initial values of state variables should be a way of testing models for potential chaotic behavior. Some effort was devoted to such tests, particularly on the convection-dispersion equation, but without eliciting the kind of strong reaction to small changes that would have indicated chaotic behavior. The reason such behavior was not found is probably that chaos only occurs within certain ranges of a system. If water and solute flows do show chaotic behavior, it is possible to reformulate the theoretical approach to these flows through a "chaotic analysis" such as that made by Rodrigues-Iturbe et al. (1991) for continental-scale water flows. This has not yet been done for water flows in soil.

Complexity

Complex systems lie on the border between chaos and order, and one analogy suggested for them is the phase transition between two states of matter. Just as water molecules at the phase transition may alternate between being held together as liquid water and flying apart as individual molecules in steam, so a component of a complex system may move between the chaotic part of the system, where it is subject to change, and the ordered part, where it, and possibly the change, are stabilized. Very interesting forms of behavior can emerge from such systems, which are described as *emergent* systems (Waldrop, 1993). One important feature of an emergent system is that the whole is more than the sum of the parts.

There are two reasons for soil scientists to take an interest in complex systems: (1) the soil itself may well be a complex system in terms of the theory; and (2) soils, catchments, and landscapes tend to behave such that the whole is more than the sum of the parts, and may therefore be emergent systems.

Modeling emergent systems is somewhat different from mechanistic modeling. In mechanistic modeling, prior understanding of the system determines the nature of the model, but with emergent systems it is often the model, which may be quite simple, that reveals the nature of the system (Waldrop, 1993). Such modeling is very much a bottom-up rather than a top-down process.

The complex behavior characteristics of emergent systems can often be described by quite simple rules, and one of the most elegant examples of this is found in the "boids" (Reynolds, cited by Waldrop, 1993). Flocks of birds such as starlings show remarkable collective formation flying in which the birds seem to flow in a cohesive unit and to avoid obstacles without losing this cohesion. The "boids" computer program simulated this behavior quite remarkably by giving each "boid" three simple rules:

1. Maintain a minimum distance from other objects, including other "boids," in the environment.
2. Try to match velocities with "boids" in the neighborhood.
3. Try to move toward the perceived center of mass of "boids" in the neighborhood.

Each of these rules applied to individual "boids" (and none of them said, "form a flock"), so the formation of the flock and its flight pattern, like other emergent phenomena, occurred from the bottom up. Catchments, and even soil profiles, are essentially collections of interacting flow pathways, and it will be interesting to see if they show emergent behavior that can be simulated by rules that are as simple as those governing the "boids" and which do not predispose the system toward its known behavior.

SUMMARY

Computer models are now widely used as aids to decisions about environmental policy. Both those who develop models and those who use them need a clear understanding of the limitations on their use imposed by the attributes of the models, the parameters and data they use, and the systems to which they are applied. These limitations suggest that we also need to give careful thought to the future directions of modeling in soil-plant systems.

One of the most important issues is that of scale, the problem of treating consistently the hierarchy of land units to be found in any area of land. Many models are developed at plot or field scale whereas policymakers often need models that are applicable at regional or national scale. We therefore need either models that are applicable at all scales or some means of assessing the scales at which particular models are appropriate. The Hoosbeek and Bryant (1992) scale diagram is helpful for the latter purpose. Some modelers perceive soil processes as deterministic while others treat them as stochastic, but which assumption is appropriate may depend on the scale at which the model is used. This is suggested by the phenomenon of decoherence, which expresses the change in determinacy with scale. The apparent distinction between deterministic and stochastic models has been brought into question by the recent suggestion that soil processes which appear to be stochastic may in fact be deterministic-chaotic.

Models inevitably propagate error. This happens not only because they contain faults, but also because model inputs have various forms of error associated with them, notably error in the statistical sense of variation. Any nonlinearity in the model will interact with this statistical error, such that the propagation of this error influences the mean of the output and may either exaggerate or suppress the error of the input when it is transmitted to the output. The processes need to be understood if models are to be used reliably, and a form of error and nonlinearity interaction analysis has been suggested. Models are now frequently used with geographic information systems (GIS), and this has intensified the need for testing because there are error propagation problems inherent in the use of GIS as well as in models.

The process of evaluating or validating a model is essential if it is to be used for decision-making. There are well-established statistical methods for this purpose. These are also relevant to the parameterization of models when this involves fitting the model against a set of data. Both validation and parameterization can be affected by the interaction between error and nonlinearity.

The future progress of modeling in soil-plant science depends on both users and modelers. The involvement of a wider range of stakeholders seems likely, and there are interesting questions as to who these stakeholders should be. New developments in modeling may well involve greater inputs from chaos theory and complexity theory than have occurred so far.

REFERENCES

Addiscott, T.M. (1977). A simple computer model for leaching in structured soils. *Journal of Soil Science 28:* 554-563.

Addiscott, T.M. (1993). Simulation modeling and soil behaviour. *Geoderma 60:* 15-40.

Addiscott, T.M. and N.J. Bailey (1990). Relating the parameters of a leaching model to the percentages of clay and other components. In *Field-Scale Solute and Water Flux in Soils,* eds. K. Roth, H. Flühler, W.A. Jury, and J.C. Parker. Basel: Birkhaüser Verlag, pp. 209-221.

Addiscott, T.M. and N.A. Mirza (1998). New paradigms for modeling mass transfers in soils. *Soil and Tillage Research 47:* 105-109.

Addiscott, T., J. Smith, and N. Bradbury (1995). Critical evaluation of models and their parameters. *Journal of Environmental Quality 24:* 803-807.

Addiscott, T.M. and G. Tuck (2001). Non-linearity and error in modeling soil processes. *European Journal of Soil Science 52:* 129-138.

Addiscott, T.M. and R.J. Wagenet (1985). Concepts of solute leaching in soils: A review of modeling approaches. *Journal of Soil Science 36:* 411-424.

Addiscott, T.M. and A.P Whitmore (1991). Simulation of solute leaching in soils of differing permeabilities. *Soil Use and Management 7:* 94-102.

Benjamin, N. (2000). Nitrates in the human diet—Good or bad? *Annales de Zootechnologie 49:* 207-216.

Beven, K.J. and A. Binley (1992). The future of distributed models: Model calibration and uncertainty prediction. *Hydrological Processes 6:* 279-298.

Bradbury, N.J., A.P Whitmore, P.B.S Hart, and D.S Jenkinson (1993). Following the fate of nitrogen in crop and soil following application of ^{15}N-labelled fertilizer to winter wheat. *Journal of Agricultural Science 121:* 363-379.

Burns, I.G. (1974). A model for predicting the redistribution of salts applied to fallow soils after excess rainfall or evaporation. *Journal of Soil Science 25:* 165-178.

Burns, I.G. (1975). An equation to predict the leaching of surface-applied nitrate. *Journal of Agricultural Science, Cambridge 85:* 443-454.

Corre, M.D., C. Van Kessel, and D.J. Pennock (1996). Landscape and seasonal patterns of nitrous oxide emissions in a semiarid region. *Soil Science Society of America Journal 60:* 1806-1815.

Corwin, D.L., D.L. Waggoner, and J.L. Rhoades (1991). A functional model of solute transport that accounts for bypass. *Journal of Environmental Quality 20:* 647-658.

Dumanski, J., W.W Pettapiece, and R.J. McGregor (1998). Relevance of scale-dependent approaches for integrating biophysical and socio-economic information and development of agroecological indicators. *Nutrient Cycling in Agroecosystems 50:* 13-22.

Einstein, A. (1920). *Relativity: The Special and the General Theory: A Popular Exposition.* London: Methuen.

Gleick, J. (1987). *Chaos.* New York: Penguin.

Groffman, P.M. and J.M. Tiedje (1989). Denitrification in north temperate forest soils: Relationships between denitrification and environmental factors at the landscape scale. *Soil Biology and Biochemistry 21:* 621-626.

Heuvelink, G.B.M. (1998a). *Error Propagation in Environmental Modeling with GIS.* London: Taylor and Francis.

Heuvelink, G.B.M. (1998b). Uncertainty analysis in environmental modeling under a change of spatial scale. *Nutrient Cycling in Agroecosystems 50:* 255-264.

Heuvelink, G.B.M. and E.J Pebesma (1999). Spatial aggregation and soil process modeling. *Geoderma 89:* 47-65.

Hoosbeek, M.R. and R.B. Bryant (1992). Towards the quantitative modeling of pedogenesis—A review. *Geoderma 55:* 183-210.

Jansen, M.J.W. (1998). Prediction error through modeling concepts and uncertainty from basic data. *Nutrient Cycling in Agroecosystems 50:* 247-253.

Loague, K. and R.E. Green (1991). Statistical and graphical methods for evaluation solute transport models: Overview and application. *Journal of Contaminant Hydrology 7:* 51-73.

Magee, B. (1985). *Popper.* London: Fontana.

Medawar, P.B. (1967). *The Art of the Soluble.* London: Methuen.

Parkin, T.B. (1987). Soil microsites as a source of denitrification variability. *Soil Science Society of America Journal 51:* 1194-1199.

Phillips, J.D. (1993). Chaotic evolution of some coastal plain soils. *Physical Geography 14:* 566-580.

Popper, K.R. (1959). *The Logic of Scientific Discovery.* London: Routledge.

Popper, K.R. (1992). *Unended Quest: An Intellectual Autobiography.* London: Routledge.

Rao, P.S.C., P.V. Rao, and J.M. Davidson (1977). Estimation of the spatial variability of the soil-water flux. *Soil Science Society of America Journal 41:* 1208-1209.

Rasband, S.N. (1990). *Chaotic Dynamics of Nonlinear Systems*. New York: Wiley.

Rodrigues-Iturbe, I., D. Entekhabi, J.-S. Lee, and R.L. Bras (1991). Non-linear dynamics of soil moisture at climate scales. 2. Chaotic analysis. *Water Resources Research 27:* 1907-1915.

Scotter, D.R., R.E. White, and J.W. Dyson (1993). The Burns leaching equation. *Journal of Soil Science 44:* 25-33.

Smith, J.U. (1996). Models and scale: Up- and down-scaling. *Quantitative Approaches in Systems Analysis 6:* 25-42.

Smith, J.U., N.J. Bradbury, and T.M. Addiscott (1996). SUNDIAL: A PC-based system for simulating nitrogen dynamics in arable land. *Agronomy Journal 88:* 38-43.

Smith, J.U., A.G. Dailey, M.J. Glendining, N.J. Bradbury, T.M. Addiscott, P. Smith, A. Bide, D. Boothroyd, E. Brown, R. Cartwright, et al. (1997). Constructing a nitrogen fertilizer recommendation system: What do farmers want? *Soil Use and Management 13:* 225-228.

Smith, J.U., P. Smith, and T.M. Addiscott (1996). Quantitative methods to evaluate and compare soil organic matter (SOM) models. In *Evaluation of Soil Organic Matter Models,* eds. D.S. Powlson, P. Smith, and J.U. Smith. NATO ASI Series Volume I38, Heidelberg: Springer Verlag, pp. 181-199.

Sposito, G., W.A. Jury, and V.K. Gupta (1986). Fundamental problems in the stochastic convection-dispersion model of solute transport in aquifers and field soils. *Water Resources Research 22:* 77-88.

Stein, A., J. Staritsky, J. Bouma, A.C. van Eijnsbergen, and A.K. Bregt (1991). Simulation of moisture deficits and areal interpolation by universal cokriging. *Water Resources Research 27:* 1963-1973.

Stewart, I. (1995). *Nature's Numbers*. London: Wiedenfeld and Nicholson.

Waldrop, M.M. (1993). *Complexity: The Emerging Science at the Edge of Order and Chaos*. London: Viking.

Webster (2000). Is soil variation random? *Geoderma 97:* 149-163.

Whitmore, A.P. (1991). A method for assessing the goodness of computer simulation of soil processes. *Journal of Soil Science 42:* 289-299.

Index

Page numbers followed by the letter "f" indicate figures; those followed by the letter "t" indicate tables.

UW 0951197 0